1 MONTH OF
FREE
READING

at
www.ForgottenBooks.com

By purchasing this book you are eligible for one month membership to ForgottenBooks.com, giving you unlimited access to our entire collection of over 1,000,000 titles via our web site and mobile apps.

To claim your free month visit:
www.forgottenbooks.com/free1084052

ISBN 978-0-331-50624-2
PIBN 11084052

THE

PROCEEDINGS

OF THE

LINNEAN SOCIETY

OF

NEW SOUTH WALES.

FOR THE YEAR

1896.

Vol. XXI.

WITH SIXTY-ONE PLATES.

Sydney :
PRINTED AND PUBLISHED FOR THE SOCIETY
BY
F. CUNNINGHAME & CO., 146 PITT STREET,
AND
SOLD BY THE SOCIETY.
—
1896.

SYDNEY :
F. CUNNINGHAME AND CO., PRINTERS,
PITT STREET.

CONTENTS OF PROCEEDINGS, 1896,

PART I. (No. 81).

(Issued July 16th, 1896.)

* Issued separately as a Supplement to this Part.

CONTENTS.

PART II. (No. 82).

PART III. (No. 83).

CONTENTS.

PART III. *(continued).*

Note.—On pp. 378, 380, and 381, *for* Plate XXII. *read* Plate XXI ; and *for* Plate XXIII., *read* Plate XXII.

PART IV. (No. 84.)

(Issued May 31st, 1897)

PART IV. *(continued)*.

CORRIGENDA.

Page 50, after line 20 add—Pl. vi. figs. 4-7.

Page 71, line 32—for *schomburghii* and *hayi* read *schomburgkii* and *kayi*.

Page 85, line 16—for *C. albitarsis* read *E. albitarsis*.

Page 150, line 14—for *C. adelaidæ* read *C. tumidipes*.

Page 171, line 20—for *clypeus* read *clypeal*.

Page 173, line 20—for *Ceratoglossus* read *Ceratoglossa*.

Page 180—omit line 2.

Page 181, line 5—omit " South Australia," et seq.

Page 182, line 27—for *C. adelaidæ* read *C. tumidipes*.

Page 195, line 18—for *C. adelaidæ*, Blk., read *C. tumidipes*, Sl.

Page 253, line 7—for *C. adelaidæ* read *C. tumidipes*.

Page 253, line 27—for *on* read *in*.

Page 254, line 29—for *C. adelaidæ* read *C. tumidipes*.

Page 255, line 31—for *C. tenuipes* read *C. gracilipes*.

Page 314, line 24—for PUNCTULATUM read PUNCTULATUS.

Page 326, line 11—for Tome xlvii. read Tome xlii.

Page 345, line 30—for *Canthurus* read *Cantharus*.

Page 351, line 3—for *Canthurus* read *Cantharus*.

Page 378, line 5—for Plates xxii.-xxiii read Plates xxi.-xxii.

Page 378, line 7—for Plate xxii. read Plate xxi.

Page 380, line 3—for Plate xxiii read Plate xxii.

Page 381, line 10—for Plate xxii read Plate xxi.; for Plate xxiii. read Plate xxii.

Page 381, line 19—for Plate xxiii. read Plate xxii.

Page 430, line 8—for *phulicifolia* read *phylicifolia*.

Page 430, line 23—for *A. ixophylla* read *A. ixiophylla*.

Page 537, line 9—for *bruneicornis* read *brunneicornis*.

Page 567, line 13—for *Pipettelella* read *Pipettella*.

Page 758, line 25—for Naturliche read Naturliche.

LIST OF PLATES.

PROCEEDINGS 1896.

PROCEEDINGS

OF THE

LINNEAN SOCIETY

OF

NEW SOUTH WALES.

WEDNESDAY, 25TH MARCH, 1896.

The Ordinary Monthly Meeting of the Society was held in the Linnean Hall, Ithaca Road, Elizabeth Bay, on Wednesday evening, March 25th, 1896.

The President, Henry Deane, Esq , M.A., F.L.S., in the Chair.

The President gave notice that upon requisition he convened a Special General Meeting to be held on April 29th, to take precedence of the Monthly Meeting. *Business* · The Hon. Treasurer to move for the insertion in Rule xxiii. of an additional clause providing for the countersigning of all cheques drawn on behalf of the Society.

DONATIONS.

(Received since the Meeting in November, 1895.)

Manchester Museum, Owens College—Studies in Biology. Vol. iii. (1895): Catalogue of the Hadfield Collection of Shells from the Loyalty Islands. *From the Museum.*

Perak Government Gazette. Vol. viii. Nos. 27-31 (Oct.-Dec. 1895); Vol. ix. Nos. 1-3 (Jan. 1896). *From the Government Secretary.*

Royal Society of Victoria—Transactions. Vol. iv. (1895). *From the Society.*

Imperial University, Japan—Calendar, 1894-95. *From the President.*

College of Science, Imperial University, Japan—Journal. Vol. ix. Part 1 (1895). *From the Director.*

Société Royale Linnéenne de Bruxelles—Bulletin. 21me. Année. Nos. 1-3. (Nov. 1895-Jan. 1896). *From the Society.*

McAlpine's "Systematic Arrangement of Australian Fungi, together with Host-Index and List of Works on the Subject." (4to. 1895). *From the Trustees of the Free Public Library, Melbourne.*

Geological Society, London—Quarterly Journal Vol li. Part 4 (No. 204, Nov. 1895): Vol. lii. Part 1 (No. 205, Feb. 1896) : Geological Literature, &c., 1895. *From the Society.*

Zoologischer Anzeiger. xviii. Jahrg. Nos. 487-492 (Oct.-Dec. 1895); xix. Bd Nos. 493-495 (Jan -Feb. 1896). *From the Editor.*

Senckenbergische Naturforschende Gesellschaft in Frankfurt a/M.—Bericht, 1895. *From the Society.*

Société Géologique de Belgique — Annales. T. xx. 4° Liv. (1892-93); T. xxii. 2me. Liv. (Sept. 1895). *From the Society.*

American Naturalist. Vol. xxix. Nos. 347-348 (Nov.-Dec. 1895); Vol. xxx. Nos. 349-350 (Jan.-Feb. 1896). *From the Editors.*

Victorian Naturalist. Vol. xii. Nos. 8-11 (Nov. 1895-Feb. 1896). *From the Field Naturalists' Club of Victoria.*

American Geographical Society—Bulletin. Vol. xxvii. No. 3 (1895). *From the Society.*

Hamilton Association—Journal and Proceedings for 1894-95. *From the Association.*

Geological Survey of Canada—Palæozoic Fossils. Vol i. (1861-65); Vol. ii Part i. (1874); Vol. iii. Parts i.-ii. (1884 and 1895): Maps of the Principal Auriferous Creeks in the Cariboo Mining District, British Columbia, Nos. 364-372, 379-390 and 550-551 : Sheet No. 11, S. W. Nova Scotia : Eastern Townships Map—Quebec. N.-E. Quarter Sheet; Rainy River Sheet—Ontario. *From the Director.*

American Museum of Natural History—Bulletin. Vol. vii. (1895), Sig. 20-24, pp. 305-388 (Sept.-Dec. 1895). *From the Museum.*

Department of Mines, Perth, W. A.—" Mining Handbook to the Colony of Western Australia." 2nd Edition (1895). By H. P. Woodward, J.P., F.G S. *From the Secretary for Mines.*

Bureau of Agriculture, Perth, W A.—Journal. Vol. ii. Nos. 25-27 (Nov.-Dec. 1895); Vol. iii. Nos. 1-5 (Jan -Mar. 1896). *From the Secretary.*

Naturwissenschaftlicher Verein zu Osnabruck—Jahresbericht, 1893-94. *From the Society.*

Michigan Fish Commission—Bulletin. No. 5 (1895). *From the Commission.*

Harvard College, Cambridge, Mass.—Bulletin of the Museum of Comparative Zoology. Vol. xxvii. Nos. 4-6 (Aug.-Nov. 1895): Annual Report of the Curator, 1894-95. *From the Curator.*

Asiatic Society of Bengal—Journal. n.s. Vol. lxiv. (1895), Part i. No. 2 : Proceedings, 1895. Nos. vii.-viii. (July-Aug.). *From the Society.*

Société de Physique et d' Histoire Naturelle de Genève—Mémoires. T. xxxii. Première Partie (1894-95). *From the Society.*

Royal Microscopical Society—Journal. 1895. Parts 5 and 6 (Oct. and Dec.). *From the Society.*

Geological Survey of India—Records. Vol. xxviii. Part 4 (1895). *From the Director.*

Hooker's "Icones Plantarum." (Fourth Series). Vol. v. Parts ı.-ii. (Nov. 1895-Jan. 1896). *From the Bentham Trustees.*

K. K. Zoologisch-botanische Gesellschaft in Wien—Verhandlungen. Jahrgang, 1895. xlv. Band 8-10 Hefte. *From the Society.*

Australasian Journal of Pharmacy. Vol. x. No. 120 (Dec. 1895); Vol. xi. Nos. 121-123 (Jan.-Mar. 1896). *From the Editor.*

Pharmaceutical Journal of Australasia. Vol. viii. Nos. 11-12. (Nov.-Dec. 1895); Vol. ix. Nos. 1-2 (Jan.-Feb. 1896). *From the Editor.*

Pamphlet entitled "Stratigraphical Notes on the Georgina Basın," &c (1895). By R. L. Jack, F.G.S., F.R.G.S. *From the Author.*

Nederlandsche Entomologische Vereeniging—Tijdschrift voor Entomologie. Deel xxxviii. Afl. i. (1894-95). *From the Society.*

L'Académie Impériale des Sciences de St. Pétersbourg— Bulletın. vᵉ. Série. T. ii. No. 5 (May, 1895); T. iii. No. 1 (June, 1895). *From the Academy.*

Troisième Congrès International de Zoologie—Guide Zoologique: Communications Diverses sur les Pays-Bas (1895). *From the Netherlands Natural History Society, Helder.*

Société d' Horticulture du Doubs, Besançon—Bulletin. n.s. Nos. 59-60 (Nov.-Dec. 1895): Série Illustrée. No. 1 (Jan. 1896). *From the Society.*

Pamphlet entitled "On Mediterranean and New Zealand Reteporæ," &c. (1895). By A. W. Waters, F.L.S. *From the Author.*

Cambridge Philosophical Society — Proceedings. Vol. viii. Part 5; Vol. ix. Part ı. (1895). *From the Society.*

Zoological Society, London—Abstracts. 19th Nov., 3rd Dec., 17th Dec. 1895, 14th Jan. 1896, Feb. 4th : Proceedings, 1895. Part iii. : Transactions. Vol. xiii. Part 11 (Oct. 1895). *From the Society.*

Royal Society of South Australia—Transactions. Vol. xix. Part ii. (Dec. 1895). *From the Society.*

Société Royale de Géographie d'Anvers — Bulletin. T. xx. 2me-3me Fascs. (1895-96) : Mémoires. T. iv. *From the Society.*

Department of Agriculture, Sydney—Agricultural Gazette. Vol. vi. Parts 11-12 (Nov.-Dec. 1895); Vol. vii. Parts 1-2 (Jan.-Feb. 1896). *From the Hon. the Minister for Mines and Agriculture.*

Johns Hopkins University Circulars. Vol. xv. No. 121 (Oct. 1895). *From the University.*

Naturwissenschaftlicher Verein des Reg.-Bez., Frankfurt a/O. —Helios. xiii. Jahrg. 1895. Nos. 1-6 (Ap.-Sept.): Societatum Litteræ. ix. Jahrg. 1895. Nos. 4-9 (Ap.-Sept.). *From the Society.*

Scottish Microscopical Society—Proceedings, 1894-95. *From the Society.*

Kaiserliche Mineralogische Gesellschaft, St. Pétersbourg — Materialien zur Geologie Russlands. Bd. xvii (1895). *From the Society.*

U.S. Department of Agriculture—Division of Ornithology and Mammalogy—Bulletin. No 6 (1895): Division of Entomology—Bulletin. n.s. Nos. 1-2 (1895) *From the Secretary of Agriculture.*

Connecticut Academy of Arts and Sciences—Transactions. Vols. i-iii ; Vol. iv. Parts 1-2; Vols v.-vi. (1866-85). *From the Academy.*

Bombay Natural History Society—Journal. Vol ix. No. 5; Vol. x. No. 1 (Oct.-Nov., 1895). *From the Society.*

Société Entomologique de Belgique—Annales. T. xxxvi. (1892); xxxviii. (1894): Mémoires. i. (1892). *From the Society.*

Pamphlet entitled "Analyses of the Artesian Waters of New South Wales," &c. By J. C. H. Mingaye, F.C.S. No 2 (1895). *From the Author.*

Entomological Society of London—Proceedings, 1895. Parts iv.-v. *From the Society.*

Museo de La Plata—Revista. T. vi. Part ii. (1895). *From the Director.*

Australian Museum, Sydney—Records. Vol. ii. No. 7. (Jan., 1896). *From the Trustees.*

Académie Royale des Sciences et Lettres de Danemark, Copenhague—Bulletin. Année, 1895. No. 2 (April-May). *From the Academy.*

Kongliga Svenska Vetenskaps-Akademie – Handlingar. Bd. xxvi. (1894-95): Bihang. Bd. xx. Afd. i.-iv. *From the Academy.*

Journal of Conchology. Vol. viii. No. 5 (Jan., 1896). *From the Conchological Society of Great Britain and Ireland.*

Naturhistorischer Verein der Preussischen Rheinlande, Westfalens, and des Reg.-Bez., Osnabruck—Verhandlungen. lii. Bd. Erste Halfte, (1895): Sitzungsberichte der Niederrheinischen Gessellschaft fur Natur- und Heilkunde zu Bonn. 1895. Erste Halfte. *From the Society.*

Société Belge de Microscopie—Annales. T. xix. 2me. Fasc. (1895). *From the Society.*

Archiv for Mathematik og Naturvidenskab. Bd. xvii. Hefte 1-4 (Ap., 1894, Aug., 1895). *From the Royal University of Upsal.*

Upsala Universitets Mineralogisk-Geologiska Institution.— Meddelanden. Nos. 11 and 17-19. *From the Royal University of Norway.*

Société Royale Malacologique de Belgique—Annales. T. xxvii. (1892) : Procès-Verbaux. T. xxi. (1892), pp. 75-86. (Nov.-Dec.) ; T. xxii. (1893); T. xxiii. (1894); T. xxiv. (1895), pp. 1-83 (Jan.-May). *From the Society.*

Société Nationale des Sci. Nat. et Math. de Cherbourg— Mémoires. T. xxix. (1892-95). *From the Society.*

Geelong Naturalist. Vol. v. No. 2 (Jan., 1896). *From the Geelong Field Naturalists' Club.*

Royal Society, London—Proceedings. Vol. lviii Nos. 349-352 (Aug.-Nov., 1895); Vol. lix. No. 355 (Jan., 1896). *From the Society.*

Royal Irish Academy—Transactions. Vol. xxx. Parts 15-17 (Feb.-Dec., 1895) : Proceedings. Third Series. Vol. iii. No. 4 (Dec., 1895) : List of Members, 1895. *From the Academy.*

Entomologiska Foreningen i Stockholm—Entomologisk Tidskrift. Årg. 16, 1895. Haft 1-4. *From the Society.*

Queensland Geological Survey—Report on the Leichhardt Gold Field and other Mining Centres in the Cloncurry District. 1895 (No. 208). By W. H. Rands. *From the Director.*

Sydney Observatory—Results of Rain, River, and Evaporation Observations made in New South Wales during 1894 under the Direction of H. C. Russell, B.A , C.M.G., F.R.S , Govt. Astronomer. *From the Director.*

Bureau of American Ethnology—Bulletin W. (No. 23) [1894]. *From the Bureau.*

Public Library, Museums, and National Gallery, Melbourne— Report of the Trustees, 1894. *From the Trustees.*

University of Melbourne—Examination Papers . Matric. (Nov., 1895) ; Annual (Oct. and Dec., 1895). *From the University.*

Comité Geologique, St. Pétersbourg—Bulletin. Supplément au T. xiv., 1894 : Mémoires. Vol. x. No. 4 (1895). *From the Committee.*

Department of Agriculture, Victoria—Three Reports by Messrs. Sinclair and Irvine. Guides to Growers, Nos. 6-7, 18-20, and 22. *From C. French, Esq , F.L.S.*

Gordon Technical College, Geelong—The Wombat. Vol. i. No. 2 (1895). *From the College.*

L'Institut Colonial de Marseille—Annales. Vol. ii. (1895). *From the Institution.*

Revista de Sciencias Naturaes e Sociaes. Vol. iv. No. 14 (1896). *From the Directors.*

OBSERVATIONS ON THE RELATIONS OF THE ORGAN OF JACOBSON IN THE HORSE.

By R. Broom, M.D., B.Sc.

(Plate i.)

In Herzfield's recent paper "Ueber das Jacobson'sche Organ des Menschen und der Säugethiere "* he calls attention to the peculiarity in the Horse in that in it there is no naso-palatine canal opening into the mouth, and that the duct of Jacobson, instead of opening into the naso-palatine canal as in most higher mammals, opens into a deep depression in the nasal floor. This condition he found to exist in both the Horse and the Ass, and he states that according to Gratiolet† a similar condition is found in the Camel and Giraffe.

As I had from my studies on the organ of Jacobson in different Orders come to the conclusion that though the degree of development of the organ may vary greatly in different genera the type on which it is formed is remarkably uniform in each Order, I naturally became anxious to find the explanation of how it was that the organ in the Horse differed apparently so remarkably from the normal Ungulate type as found in the Sheep.

Being fortunate in having in my possession the head of a fœtal Horse I have made a study of the relations of the organ by means of a series of vertical sections. Though the examination of a younger specimen would doubtless have been even more

---

* Zoolog. Jahrbuch, Abtheil. für Anatomie und Ontogenie. Bd. iii. 1889.
† Recherches sur l'organe de Jacobson.   aris, 1845.

satisfactory, as the present series sufficiently elucidates the nature of the peculiarity, I think it well to publish the present results.

The Horse differs from most mammals in having the premaxillaries developed in such a way as to carry the palate forward in advance of the nares and forming a sort of rostrum—a condition seen in a much greater degree in the Tapir. As a result of this development a large portion of the anterior part of the nasal septum is clasped between the premaxillaries, and the lateral cartilages, which in most mammals become the "cartilages of the nasal floor," are here confined by the premaxillaries and prevented from developing laterally to any great degree, and seem to compensate for the want of lateral expansion by developing downwards.

Figure 1, Plate I., represents a section immediately behind the point where the premaxillary gives off its palatine process. A portion of the lateral cartilage (*l.c.*) is seen passing downwards from the nasal septum (*n.s*) between the premaxillary and the palatine process. A little below it may be observed an oval cartilage cut across—this is an anterior process from the lateral cartilage. It passes well forward, approaching nearer to the palate, and ending a little behind the rudimentary papilla. The most noteworthy peculiarity of this section is that there is no trace of the naso-palatine canal to be seen, nor is there in any anterior section. Even by the sides of the papilla, where the anterior opening of the canal would be expected, I have failed to find even a rudiment.

A little distance behind the plane of figure 1 the anterior process of the lateral cartilage is seen united with the main part, which though still attached to the nasal septum is becoming constricted off. In a slight concavity on the inferior end of the cartilage is found on this plane the anterior closed end of the imperfect naso-palatine canal (*n.p.c.*).

In figure 3 the naso-palatine canal is found to have a distinct lumen, and on its inner side it is supported by a small downward cartilaginous process.

In the next succeeding planes the relation of the duct to the cartilages is very similar, but the lateral cartilage is found becoming shorter and broader and detaching itself from the nasal septum (fig. 4).

On reaching the plane shown in fig. 5 the nasal cavity is found to be approaching the lateral cartilage, which here becomes for the first time a "nasal-floor cartilage" proper. At its outer angle it is seen sending up a process which further back is found to represent the rudimentary cartilage of the nasal wall. Here the naso-palatine canal is seen flattened out and about to give off Jacobson's duct. The inner part or Jacobson's duct is almost surrounded by cartilage.

In figure 6 the ducts are seen separated, and a cartilaginous partition passes between them.

In the following figure the outer part of the cartilage is seen detached, while the inner forms a complete investment for Jacobson's duct. Between the two portions of the divided lateral cartilage is found the naso-palatine canal about to open into the nasal cavity.

Behind this region the organ and its cartilages are found quite to follow the ordinary mammalian form.

It will be observed that the points in which the Horse differs from the normal type are these :—(1) occlusion or absence of the anterior part of the naso-palatine canal, leading to the secretion from Jacobson's organ passing backwards into the nasal cavity by the upper part of the naso-palatine canal; and '2) the anterior processes of cartilage usually given off from the nasal-floor or lateral cartilage and passing forward supporting Jacobson's duct and the naso-palatine canal, here for the greater part remain united with the lateral cartilage. In the absence of even a trace of the canal in its anterior part, it is doubtful whether the anterior cartilaginous process represents Jacobson's or Stenson's cartilages or a fusion of both—probably the latter.

In almost all other respects there is a close agreement between the condition of parts in the Horse and those in most other Ungulates.

Fig. 10 shows a section of part of the nose of a very small fœtal Calf. Here both Jacobson's and Stenson's cartilages are well developed and seem distinct from the broad nasal-floor cartilage. If this be compared with figures 4 or 5 the close resemblance will be seen; in fact the only marked difference is that in the Horse the cartilages of Jacobson and Stenson are united with the nasal-floor cartilage, in the Calf distinct. But all the corresponding parts can easily be observed.

Figure 11 represents a section of the fœtal Calf corresponding to figure 6 in the Horse. Here the duct cartilages are united with the nasal-floor cartilage as in the Horse. The resemblance is, however, somewhat marred by the enormous development of the cartilage of the nasal wall in the Calf. Such variations in cartilaginous development, however, occur in very nearly allied forms as the Cat and Dog.

The agreement of figure 12 with figure 8 is most striking.

The peculiarities in the Horse are probably due to the strong development of the premaxillary bones leading to the occlusion of the anterior part of the naso-palatine canal and to the vertical direction assumed by the lateral cartilage permitting the duct cartilages to remain united with the main body.

The similar condition in the Camel is probably accounted for by the fact that its very recent ancestors had remarkably well developed incisors, e.g , *Protolabes* from the Upper Miocene of Oregon.

In the Giraffe the explanation is not very manifest.

I have to acknowledge my indebtedness to Dr. John Mackie and Mr. A. Robb, F.R.C.V.S., of Glasgow, for the fœtal horse examined, and to Mr. Alf Swan, of Taralga, for the fœtal calf.

REFERENCES TO PLATE I.

a.l.c., anterior process of lateral cartilage; *J.c.*, Jacobson's cartilage; *J.d.*, Jacobson's duct; *J.o.*, Jacobson's organ; *l c.*, lateral cartilage; *Mx.*, maxillary; *n.f.c.*, nasal-floor cartilage; *n.p.c.*, naso-palatine canal, *n.w.c.*, nasal-wall cartilage, *n.s.*, nasal septum; *p.Pmx.*, palatine process of pre-maxillary; *Pmx.*, premaxillary.

Figs. 1- 9.—Transverse vertical sections through snout of fœtal Horse (head length about 7·5 c.m.) × 7.

Figs. 10-12.—Transverse vertical sections through snout of fœtal Calf (head length about 2 c.m.) × 30.

Dotted portion represents cartilage; parts shaded by lines represent the regions of ossification.

DESCRIPTIONS OF FURTHER HIGHLY ORNATE BOOMERANGS FROM NEW SOUTH WALES AND QUEENSLAND.

By R. Etheridge, Junr., Curator of the Australian Museum.

(Plates II.-V.)

The boomerangs described in the present communication may be regarded as supplementary to those of an ornate nature figured in these "Proceedings,"* and the "Macleay Memorial Volume."† They are from the collections of Dr. J. C. Cox, and Messrs. P. R. Pedley and N. Hardy, and my best thanks are due to these gentlemen for the loan of the weapons.

The first five boomerangs generally resemble one of those first referred to,‡ where the incised ornament consists of loops returned on themselves, either continuous along the whole length of the weapon or disconnected one from the other.

The most highly ornate of the five (Fig. 4) bears three incised loops formed by from three to five continuous grooves, the loops gradually increasing in length. The free end of the shortest loop commences near one of the apices of the weapon, passes down the middle line for about one quarter its length, then turns to the left or concave side of the boomerang and is returned again to the apex, at this point rounding on to the convex side, which it follows to a point a trifle beyond the centre of the weapon. Here it turns to the left as far as the middle line, and is again returned in that plane until meeting with and joining the first bend, the loop curves on itself to the left and follows the concave margin throughout the remaining length of the weapon, i e., to the further apex, then returning on itself to the right, passes on to the convex margin, which it follows until coming in

---

* Proc. Linn. Soc. N.S Wales, 1894, ix. (2), p. 193.
† P. 237, t. 32, f. 1-3.
‡ Proc. Linn. Soc. N.S. Wales, l c., t. 15, f. 1.

contact with the first return of the second loop, again returns on itself to the middle line of the boomerang, pursues its course along that plane, and terminates as it commenced in a free end ; hence there are in this figure four turns to the left, and two to right. When there are more than three incised grooves, the additional ones are made by interpolation. Some of the inter-spaces of the loops are quite plain, one bears seven crosses in three and a half pairs, three others have continuous zig-zag incised lines, whilst outside the central loop on the convex side of the boomerang, the marginal space is occupied by a similar zig-zag, or almost festoon-like, figure of two incised grooves. One of the apices is similarly marked transversely, whilst the other is devoid of sculpture, but just within the return of the loop, and above the free end is a figure resembling an unsymmetrical letter W.

The length of this weapon across the curve is two feet four inches, the breadth two and a quarter inches ; and the weight ten and a half ounces It is from the collection of Mr. P. R. Pedley, and was obtained at St George on the Balonne River, a branch of the Maranoa River, in South-east Queensland.

The second boomerang (Fig. 3) differs from Fig. 4 only in detail. The loops are identical in number and execution, but at the returning points instead of four deflections to the left and two to the right, there are two and four respectively. The interspaces are also sculptured in the same manner, although not within corresponding loops. The apices on the contrary are differently marked, both bearing a diagonal of four incised lines, the spaces on either side carrying sharp v-shaped figures.

The length is two feet four inches; the breadth two and a half inches, and the weight eleven ounces. It is from the same locality and collection as the last.

The third weapon (Fig 2) resembles Fig. 3, except that only two loops have been incised, almost equally dividing the surface, with two deflections to the right and two to the left. Only one inter-space bears a single zig-zag line, the others are devoid of sculpture. At one end the loop is contiguous to the apex, at the other the free space beyond the return of the loops is occupied by sigmoidal figures of two incisions each, and a central gently lunate outline.

The length is two feet three and a quarter inches; the breadth two inches; and the weight nine ounces. It is from the same locality and collection as the two previous weapons.

The two succeeding boomerangs (Figs. 1 or 7) have disconnected loops, or rather half-loops placed back to back and touching in pairs. Cross bars are also present, but differ in the two weapons. In both the loops are formed of six undulating grooves, producing a figure along one margin of each weapon, then returning on itself, and proceeding along the other margin, leaving a wide space in the middle line. In Fig. 1 there are seven of these half-loops, and in Fig. 7 six. In Fig. 1 the apical half-loops are turned in opposite directions, and one is smaller than the other. That at one of the apices is cut off by a single incised transverse line, and is followed by two half loops abutting against one another, and again divided off near the middle of the weapons by another transverse incised line. Two further half-loops repeat the same order, separated by the third transverse incised line from the first large half-loop referred to as occupying one of the apical portions of the boomerang. The arrangement in Fig. 7 is practically the same, but in consequence of the penultimate apical half-loops being nearly of a size, the sculpture is almost bilaterally symmetrical. The cross-bars are only two, wide quadrangular spaces, vertically incised with close grooves. The interspaces between the two parts of each half-loop are occupied in the middle line of the weapon with a zig-zag figure of two incised lines, the angles of the zig-zag either continuous or broken. In Fig. 7 there is only one such figure, but in Fig. 1 three of the half-loops are infilled with an additional series of a single incision. Again in Fig. 7 an additional zig-zag line is represented immediately along the central convex edge of the weapon.

		Fig. 1.	Fig. 7.
Length...	...	2ft. 9in.	2ft. 6in.
Breadth	...	2¼in.	2½in.
Weight	...	12oz.	12oz.

Both boomerangs are from Angeldool, on the Narran River, near the Queensland border, and are from the collection of Dr. J. C. Cox.

The next weapon to be described (Fig. 6) is well ornamented with four parallel series of small conjoined ovals, extending nearly the entire length of the boomerang, the two nearest the convex margin being the shortest. This margin is also scalloped. The ovals are obliquely incised with single grooves not all in the same direction, but the scalloped edge is plain.

The length is two feet eight inches; the breadth two and a quarter inches; and the weight thirteen ounces It is from the same locality and collection as Figs. 1 and 7.

The original of Fig. 5 like that of Fig. 6 is a large boomerang, with the sculpture excellently done, consisting of a median line of six inequilateral rhombs, the intervening triangular spaces on each side being vertically incised with grooves. The surfaces of the rhombs are smooth, and devoid of sculpture, with the exception of the shaped nicks, in from one to four series in each rhomb, but too disjointed to assume a zig-zag pattern.

The length is two feet nine and a half inches; the breadth two and a half inches; and the weight thirteen and a half ounces. This example is also from Dr. Cox's Angeldool collection.

Fig. 8 represents a boomerang imperfect in itself, but exactly coinciding in its sculpture with one of those described by me from Norley, on the Bulloo River,* and therefore need not be described further. We have here either an example of wide distribution of a certain pattern of sculpture, or a case of a weapon passed on by barter. The specimen is again from Angeldool.

Deeply scalloped margins distinguish Fig. 12, the scalloping edged with a wide groove, and itself obliquely incised. The middle line or crown is quite smooth with the exception of a fluctuating or serpentine line of two grooves, fairly well coinciding in its fluctuations with the groove edging the scalloped figure on the

---

* Proc. Linn. Soc. N.S. Wales, 1894, ix. (2), t. 15, f. 2.

2

concave side of the weapon.   The immediate apex at one end is cross-incised, and bears a few irregular v-shaped nicks.

The length is two feet three and a half inches; breadth two and a quarter inches; and the weight twelve ounces.   It is from St. George, Balonne River (Mr. P. R. Pedley).

Fig 10 is again a bilaterally unsymmetrical boomerang as regards the incised sculpture.   There are three cross-bars formed of one obliquely cross-notched incised line.   One of these is near the centre, another half way between this and one of the apices, and the third at the apex referred to, thus dividing the surface into three unequal lengths.   The middle line bears acute small rhombs, extending throughout the two larger divisions.   On each side the line of rhombs are the usual rolling or fluctuating grooves four to five on either side; whilst the middle line of the division unornamented by rhombs, is occupied by similar grooves.   The apex at this end bears a transverse double zig-zag pattern, and a single similar series is intra-marginal on the convex side of the boomerang

Length two feet three and a half inches; breadth two and a quarter inches; and the weight eleven ounces.   This is a much shorter and more highly curved weapon.

St. George, Balonne River (Mr. P. R. Pedley).

The middle line of this boomerang (Fig. 11) instead of rhombs is ornamented by deeply incised rolling grooves.   Flanking these are two similar grooves, intra-marginal in position, and between the latter and both edges of the weapon are a series of crosses. There are four cross-bars, one towards each apex, and one on either side the centre of the weapon.   At the apices are broad semi-lunar transverse bands, both concave in the same direction, and vertically grooved.

The length is two feet four and a half inches; breadth two inches; and weight ten and a half ounces.

This boomerang is again from St. George, and in the collection of Mr. P. R. Pedley.   Like Fig. 10 it is a good deal curved.

Another very bilaterally unsymmetrical boomerang is represented by Fig. 9.   The principal sculpture consists of three ranges of

fluctuating grooves, four to six grooves in each range, one group
in the middle line, and one on either side, extending from apex to
apex, but twice interrupted by cross bars, that differ widely, however, from those figured on preceding weapons. That on one side of
the centre consists of two parallel grooves, united by transverse
incisions, the other near one of the apices of two such bands, somewhat separated from one another, the plain interspace carrying five
v-shaped figures placed transversely. On the concave side of the
boomerang, and along one part of the edge, is the ever-recurring
single zig-zag line, whilst between the fluctuating lines over the
general surface, either the same kind of incised sculpture or v-
shaped figures parallel to the longer axis of the weapon.

Length two feet five and a half inches; breadth two and a
quarter inches; and weight twelve and a half ounces. This
boomerang was received from Normanton, Gulf of Carpentaria,
by Mr. N. Hardy, to whom it belongs.

A very peculiarly ornamented boomerang is represented in Fig.
13 · Along the convex margin is a series of very deep scallops,
reaching transversely to near the middle line of the weapon, and
grooved parallel to its longer axis.   The middle or centre line is
occupied by a single zig-zag, and between this and the concave
edge are three deep and wide slightly fluctuating lines of two
grooves each. The whole produces a very marked pattern   The
apices in this weapon are very sharply pointed.

Length two feet six inches; width two inches; and weight ten
ounces.

From Angeldool, on the Narran River, in the collection of Dr.
J. C. Cox.

The last boomerang but two (Fig. 14) bears on each side of the
sculptured face long moderately deep festoons, five on either side,
and obliquely grooved, but not reaching to either apex.   The
middle line is occupied by five large ovals, so arranged that each
more or less falls into the space left between opposite re-entering
angles of the festoons.   These are also deeply and obliquely
grooved. Clear spaces are left at both apices, one containing two

and a half rhombs placed transversely, whilst at the other is an oblong enclosure, with two parallel zig-zags of a single line each.

Length two feet five inches; breadth two and a quarter inches; and weight eleven and a half ounces.

From Angeldool, on the Narran River, in the collection of Dr. J. C. Cox.

In the last specimen but one (Fig. 15) runs a sub-central longitudinal line of eleven large ovals, and along the concave and convex margins respectively rows of fifteen and eighteen narrower ovals. Intervening between the central row and that on the convex margin at one end of the weapon is an additional row of larger ovals, but this only extends for half the length of the weapon. At each end this larger row dies off into a single zig zag line, whilst between the sub-central line of ovals and that on the concave margin is another.   All the ovals are grooved obliquely.

Length two feet four inches; breadth two and a quarter inches; and weight eleven ounces.

Again from Angeldool, on the Narran River, and in the collection of Dr. J. C. Cox.

The last boomerang (Fig. 16) is figured with some hesitation, not as to the genuineness of the weapon itself, but of the carving; the former betrays nothing out of the common   The natural objects represented are a large fish in the centre, bounded by two incised lines, and filled in with single diagonal lines in two directions, producing a series of rhombs probably representing scales.   Following this, and in front of it, is a by no means bad representation of a bird; below this again four rhombs, one within the other, followed by a nondescript object, infilled with incised lines coincident with the outline; and finally at the apex a heart-shaped body.   It is the execution of the bird, with the appearance of the nondescript and heart-shaped bodies that might raise the suspicion that the carving on this weapon is not purely " blackfellow."   Birds, however, are by no means uncommon on the highly decorated dilly-baskets of North Australia, whilst the heart has on more than one occasion been observed amongst rock paintings, undoubtedly the work of the Aborigines.   The boomerang is the property of Mr. Norman Hardy, and is from Queensland.

Figs. 2 to 4 are obviously after the type of the ornamented boomerangs from the Bulloo River, figured by myself,* differing merely in minor details; the loop pattern is here paramount. I think it very possible also that the sculpture fore-shadowed on a boomerang from Queensland, figured by Smyth,† is only this pattern in an incomplete state. Knight figures‡ a boomerang exhibited at the Philadelphia International Exhibition, said in the same breath to be both from N.S. Wales and Victoria, and bearing those serpentine figures that are probably of the same nature.

Figs. 1 and 7.—The half-loops do not correspond to any previously published illustrations known to me. The weapon represented by Fig. 6 is to some extent allied in its pattern to another figured by Smyth,§ from Rockingham Bay, that from Coomooboolaroo given by Lumholtz,‖ and one of those from the Alligator River Tableland, figured by myself in the Macleay Memorial Volume,¶ except that Fig. 6 is wanting in the marginal festoon work and possesses an additional row of ovals. Fig. 15 also stands in much the same relation.

The pattern of the broken boomerang, Fig. 8, again corresponds to one from the Bulloo River.**

The remainder of the figures are not related to any published forms so far as I know. Broken zig-zag double lines, as in Figs. 1, 7, 8, 15, &c., are by no means uncommon on aboriginal weapons, whilst crosses are very uncommon (see Fig. 11). For instance a Bull-roarer, figured by Angas, from S. Australia, and called · *Wimmari*, is decorated in this manner.

---

* Proc. Linn. Soc. N.S. Wales, 1894, ix. (2), t. 15, f. 1.
† Aborigines of Victoria, 1878, i., p. 285, f. 37.
‡ Smithsonian Ann. Report for 1879 [1880], p. 227, f. 28, lower fig.
§ Smyth, loc. cit. p. 329, f. 112.
‖ Among Cannibals, 1890, p. 51, f. *b*.
¶ t 32, f. 3.
** Proc. Linn. Soc. N.S. Wales, 1894, ix (2), t. 10, f. 2.

On taking a general glance over the figures of these boomerangs one is struck with the limited number of designs that appear to have been used amongst the aboriginal artists, notwithstanding that so far as detail goes no two are precisely alike   The designs are confined to the loop, half-loop, rhomb, oval, cross, rectangular bars, and semilunate, festoon, and zig-zag patterns, with modifications of one or the other.   The chevron or herring pattern is also often met with.   Circles and spirals are conspicuous by their absence on boomerangs.   True it is the incised work of our Aborigines is devoid of that finish and delicacy of execution seen in the carvings of many other dark races—for instance, compare some of the beautiful designs employed by the Dyaks to ornament their wood work.   At the same time the incised patterns of our Aborigines have a character of their own not to be mistaken for those of any other race.

Whenever natural objects are represented they are always to a greater extent recognisable as such, and do not seem to be degenerate representations of a higher and more advanced art previously existing, the realism being maintained and not abandoned.   Writing on the "Decorative Art of Torres Straits," Professor A. W. Haddon says* :—" We see that the animals are always represented individually, and are not utilised for the purpose of making patterns, or of telling a story, or for conveying information "   At present there is no evidence to show that figures of the animate creation were otherwise used by our Aborigines on their boomerangs.

In the majority of instances the designs run parallel to the longer axis of the boomerangs, although not in all.   Etched designs appear remarkable by their absence on this class of weapon, so far as my observation goes, although they are met with on some Womerahs, neither does there seem to be that appreciation of the grotesque that forms so marked a feature in carvings from New Zealand and the Pacific Islands.   One very noticeable character exists throughout the whole series, without exception, the emargination and produced centre of all the apices.

---

* R. Irish Acad. Cunningham Mem. 1894. No. x. p 65.

ON A NEW GENUS AND SPECIES OF FISHES FROM MAROUBRA BAY

By J. Douglas Ogilby.

(Communicated by T. Whitelegge, F R.M.S.)

It is again my pleasing duty to record yet another new fish from Maroubra Bay, where it was obtained by Mr. Whitelegge early in February. The constant recurrence of new forms of animal life in this small bay, probably the only spot on the Australian coast which has been systematically and scientifically explored, is an additional proof, if one were needed, of how imperfect a knowledge of our littoral fauna we possess.

I am puzzled to know in what family this genus should be placed; a casual glance would indicate affinity to the *Apogonidæ*, but the absence of vomerine teeth and the number of the dorsal spines apparently deny it a resting-place among these little fishes, unless indeed it should be considered to be an aberrant Apogonid with sciænoid affinities.

Apogonops, gen.nov.

Body elongate-oblong and somewhat tapering posteriorly, compressed. Head large. Mouth rather large, with oblique cleft, the premaxillaries protractile and forming almost the entire anterior margin of the upper jaw; maxillary exposed, without supplemental bone; lower jaw the longer. Two nostrils on each side, the anterior rather the larger and situated much nearer to the eye than to the extremity of the snout Eye large. Preorbital entire; preopercle with a double ridge; the inner ridge entire, the outer with a few weak spines round the angle, opercle with two spines; the membranous portion produced and pointed, extending well beyond the lower spine, posttemporal spiniferous. Gill-membranes separate from the isthmus; gills four, a slit behind the fourth; seven branchiostegals; pseudobranchiæ present; gill-rakers moderate, rather slender. Narrow bands of villiform teeth in the jaws, vomer, palatines, and tongue edentulous. A single dorsal fin, deeply notched, with x 10 rays, the spinous portion longer than the soft, anal short, with iii 7 rays, the second spine strong and laterally grooved; ventrals inserted below the base. of the pectorals, close together, with a strong spine; pectorals pointed, with 14 rays, the second the longest and much stronger than the third; caudal emarginate, the peduncle long and strong. Scales moderate, cycloid, concentrically striated, very deciduous; head partially naked, soft dorsal and anal fins with a basal scaly sheath; no scaly process between the ventrals. Lateral line continuous, extending on the base of the caudal fin, the tube straight and simple, not quite reaching to the extremity of the scale.

Etymology :---Apogon; ὤψ, resemblance.

Distribution :--Coast of New South Wales.

APOGONOPS ANOMALUS, sp.nov.

D. x 10.   A. iii 7.   Sc. 55.

Length of head $2\frac{5}{6}$, depth of body $4\frac{1}{10}$ in the total length,[*] depth of head $1\frac{1}{2}$, width of head $2\frac{2}{3}$ in its length. Eye very large, situated on the upper half of the side of the head, its diameter one-third of the length of the head; snout obtuse, shorter than the eye; interorbital region flat, its width $3\frac{3}{6}$ in the length of the head   Maxillary not quite extending to the vertical from

---

[*] In this and all future papers the total length, as taken in connection with the comparative measurements, is the distance between the extremity of the jaws and the base of the caudal fin, unless special mention to the contrary is made; in giving the length of the fish this fin is of course included. In taking the measurement of the head the free opercular flap is not included, nor, unless definitely stated to the contrary, such portion of the lower jaw as may project beyond the upper.

the middle of the eye, its length half of that of the head; its distal extremity is expanded, two-fifths of the diameter of the eye in width, its posterior margin sinuous. The preorbital and the inner ridge of the preopercle are entirely unarmed, while the outer ridge has a few feeble spines at or near the rounded angle; lower opercular spine the longer; posttemporal with five spines. About 22 gill-rakers on the lower branch of the anterior arch. The dorsal fin originates above the base of the pectoral; the spines are rather weak; the first short, about one-third of the second and sub-equal to the eighth; the fourth spine is the longest, two-fifths of the length of the head and five-sixths of the anterior soft rays; the ninth spine is very short, and the tenth is intermediate in length between the sixth and seventh: the anal originates betneath the fourth soft ray of the dorsal; the first spine is very short and stout, the second much stronger, but not so long as the third, which is one-third of the length of the head, and not much shorter than the anterior rays: ventral not reaching to the vent, the outer ray the longest, four-sevenths of the length of the head: pectoral two-thirds of the head: caudal emarginate, the peduncle long and tapering, its depth immediately behind the dorsal fin 1¾, its least depth 2¾ in its length.

Brownish-green, the sides strongly tinged with yellow; thorax and abdomen silvery; upper surface of head bluish, the lips, interorbital region, and an angular band on the occiput darkest; opercle bluish. a series of five large olive brown spots along the side; lower side of tail with three groups of crowded brown specks; dorsal fin sparsely, caudal densely covered with similar specks, the latter with two large dark basal spots.

The single example collected measures 54 millimeters and is apparently full grown.

CATALOGUE OF THE DESCRIBED COLEOPTERA OF AUSTRALIA SUPPLEMENT, PART II.

By George Masters.

Issued separately as a Supplement to the Part.

ON THE OCCURRENCE OF CALLOSITIES IN *CYPRÆA* OTHER THAN *CY. BICALLOSA* AND *CY. RHINOCERUS;* AND ON THE OCCURRENCE OF A SULCUS IN TRIVIA.

By AGNES F. KENYON.

(Communicated by John Brazier, F.L.S.)

I have lately come across several specimens of different species of *Cypræa (helvola, tabescens, miliaris, erosa)*, which have the termino-dorsal arches adorned with callosities Though these do not occur in every specimen, still finding it in several specimens of the genus, it proves that it is not an abnormal incident; and therefore I think the circumstance deserving of being recorded

Cy helvola (callused variety) possessing a double or twin callosity at the posterior extremity; the callus is not so well defined anteriorly, though in some specimens well marked; extremities white.

Cy. tabescens (callused var.) : extremities with a callus more or less conspicuous, and in some instances furnished with two at the posterior extremity.

Cy. erosa (callused var.) : differing in no respect from the normal type except in having at both extremities more or less well defined callosities; some specimens bear double ones on the posterior terminal arch.

Cy miliaris (callused var.) : evidently a lighter variety, but bearing a well marked callus.

Cy. carneola (pustulated var.): I have several specimens of this species, in which the margins are pustulated; this I should say was rather a rare occurrence, though evidently not unique

Cy. lynx (pustulated var.): I have several specimens even more pustulated than those of *Cy. carneola.*

Cy. angustata (var.): I found at Flinders, Victoria, several specimens with the margins unspotted and dorsal surface uncoloured.

On the occurrence of a Sulcus in Trivia australis It is somewhat unusual to find any species of the genus *Trivia* with a dorsal impression or sulcus, as the authorities have agreed in defining them with none. I have, however, several specimens distinctly marked; also one in which the base is not white; and one which has only one spot at each end may be pronounced a Victorian variety of *T. napolini*, it having been found at Flinders, Victoria. I have also a pair of *T napolini* from West Australia with a distinct sulcus.

It will therefore be noted that some of the distinguishing marks of this genus are absent in these specimens

Mr. Hedley called attention to specimens of *Fiona marina*, Forskal, collected at Maroubra Bay, on February 9th, 1896, by Mr. T. Whitelegge, who first found the genus in Australia last year, the discovery being recorded in Proc. Malac. Soc. I. p. 333, footnote. The first examples found were swimming free, and were tinted that shade of dark blue common to *Ianthina*, *Glaucus*, *Porpita*, *Velella*, *Physalia* and other pelagic animals. In the present instance they were of a pearl-grey colour, and were sunk in deep grooves evidently gnawed by themselves in fragments of an indeterminate species of *Sepia* shell, upon which grew examples of *Lepas ansifera* about 10 mm. in length. With them were associated several masses of ova, resembling those figured by Bergh (Result. Camp. Scient. Prince Monaco, Fasc. iv. Pl. i. f. 16). In support of the suggestion that the coloration of these specimens was a protective adaptation to the colour of the Sepia, the molluscs, ova and cuttlebone were exhibited.

Mr. Hedley also reported that on March 8th last Mr. Whitelegge had further increased the list of Australian genera by the discovery of the specimens of *Firoloides desmaresti*, Lesueur, which were exhibited on behalf of the finder. Two males and three females were thrown by the waves on the sandy beach at Maroubra Bay, and were so little injured as to swim about actively for some hours in a vessel of sea-water. The species had been identified by the excellent figures in Pl. xvi of the "Voyage de la Bonite: Zoologie." The bibliography of this species brought down to a late date would be found in Challenger Reports, Vol. xxiii., Heteropoda, p 22. Like the preceding, this genus is not included in Prof. Tate's census (Trans. Roy. Soc. 1888, pp. 70-81), but an undetermined species of *Firoloides* had been recorded from Bass Straits by Dr. Macdonald (Trans. Roy. Soc. Edinburgh, Vol. xxiii., 1862).

Mr. Edgar R. Waite exhibited a large number of living young Green Tree Snakes *(Dendrophis punctulata)*, the property of Mr.

A. P. Kemp, of Kempsey. These snakes were hatched in captivity, the eggs having been obtained at Unkya, on the Macleay River. In a gully, at this place, individuals of the species were said to exist, not in scores, but in hundreds; and in view of the large number exhibited at the meeting the statement was by no means difficult of belief.

In illustration of Mr. Kenyon's paper, Mr. Brazier exhibited specimens of *Cyprœa helvola, C. tabescens, C. erosa, C. miliaris, C. lynx,* and *C. carneola,* all showing callosities; a colour variety of *C. angustata;* and examples of *Trivia australis* with a distinct dorsal sulcus, a character not in conformity with the generic definition.

Mr. Turner exhibited some well grown fruits of *Pyrus domestica,* L, the True Service Tree, from a garden at Camden, a species which, it is believed, has seldom been observed to fruit here.

WEDNESDAY, APRIL 29TH, 1896.

The following Meetings of the Society were held in the Linnean Hall, Ithaca Road, Elizabeth Bay, on Wednesday evening, April 29th, 1896.

ADJOURNED ANNUAL GENERAL MEETING.

The Hon. Treasurer read the report of the Auditors, who, after an examination of the books, vouchers, and securities, certified the accounts for 1895 to be correct.

On the motion of Mr. P. R. Pedley, the report was adopted.

SPECIAL GENERAL MEETING.

On the motion of the Hon. Dr. Norton, M.L.C , Hon. Treasurer, the following addition to Rule xxiii. was adopted :—

xxiii. *bis*—All moneys received on behalf of the Society shall be paid to an account in the name of the Society in the Commercial Bank of Sydney or such other Bank as shall be approved by the Council.

No moneys shall be drawn out of the said account except by cheque drawn by the Treasurer and countersigned by the Secretary and no claims on the Society shall be paid except by such cheques or out of petty cash from time to time authorized by the Council to be paid into the hands of the Secretary.

MONTHLY MEETING.

The President gave notice that upon requisition he convened a Special General Meeting to be held on May 27th, to take precedence of the Monthly Meeting. *Business :* Professor Haswell to introduce the subject of the establishment of a Biological · Station on the Society's grounds at Elizabeth Bay.

DONATIONS.

Zoologischer Anzeiger. Bd. xix. Nos. 496-498 (Feb.-March, 1896). *From the Editor.*

K. K. Zoologisch-botanische Gesellschaft in Wien—Verhandlungen. Jahrgang 1896. xlvi. Bd. 1 Heft. *From the Society.*

Société Hollandaise des Sciences à Harlem—Archives Néerlandaises. T. xxix. 4me et 5me Livs. *From the Society.*

Muséum d'Histoire Naturelle, Paris—Bulletin. Année 1895. Nos. 1 and 8. *From the Museum.*

Société Scientifique du Chili — Actes. T. iv. (1894) 5me Livraison. *From the Society.*

Field Columbian Museum—Historical Series. Vol. i. No 2 (May, 1895): Geological Series. Vol. i. No. 1 (Aug. 1895): Botanical Series. Vol. i. No. 1 (Aug. 1895). *From the Director.*

American Philosophical Society—Proceedings. Vol. xxxiv. No. 147 (Jan. 1895). *From the Society.*

Portland Society of Natural History—Proceedings. Vol. ii. 1895. Part 3 : The Portland Catalogue of Maine Plants. Second Edition. *From the Society.*

Academy of Science of St. Louis—Transactions. Vol. vi. No. 18: Vol. vii. Nos. 1-3 (Jan.-Feb., 1895). *From the Academy.*

U. S. Geological Survey—Bulletin. Nos. 118-122 (1894) : Monographs. Vols. xxiii. and xxiv. (1894): Fourteenth Annûal Report (1892-93). Parts i. and ii. *From the Director.*

Smithsonian Institution—Report of the U.S. National Museum, 1893. *From the Institution.*

.   Seven Pamphlets by Prof. J. F. James. (From the Journal of the Cincinnati Society of Natural History, July, 1884–July, 1894). *From the Author.*

American Museum of Natural History—Memoirs. Vol. i. Part ii. (Aug. 1895): Bulletin. Vol. viii. (1896), Sig. 1, pp. 1-16 (March). *From the Museum.*

Naturhistorisches Museum in Hamburg—Mitteilungen. xii. Jahrg. (1894). *From the Museum.*

Gesellschaft fûr Erdkunde zu Berlin—Verhandlungen. Bd. xxii. (1895), Nos. 4-6 : Zeitschrift. Bd. xxx. (1895), Nos. 2-3 *From the Society.*

Pamphlet entitled "Geogenetische Beiträge." By Dr. Otto Kuntze. *From the Author.*

K. K. Naturhistorisches Hof-Museum in Wien—Annalen. Bd. x. (1895), Nr. 1. *From the Museum.*

Verein fur vaterländische Naturkunde in Württemberg— Jahreshefte. li. Jahrg. (1895). *From the Society.*

Konigl. böhmische Gesellschaft der Wissenschaften in Prag— Jahresbericht fur das Jahr 1895: Mathematisch-Naturwissen- schaftliche Classe—Sitzungsberichte, 1894. *From the Society.*

American Geographical Society—Bulletin. Vol. xxvii. No. 4 (1895). *From the Society.*

Natural History Society of New Brunswick—Bulletin. No. xiii. (1895). *From the Society.*

Geological Survey of India—Records. Vol. xxix. (1896), Part 1 *From the Director.*

Société Impériale des Naturalistes de Moscou—Bulletin. Année 1895. No. 3. *From the Society.*

Perak Government Gazette. Vol. ix. Nos 4-6 (Feb.-Mar., 1896). *From the Government Secretary.*

Bureau of Agriculture, Perth, W.A.—Journal. Vol. iii. No. 6 (Mar. 1896). *From the Secretary.*

Pharmaceutical Journal of Australasia. Vol. ix. No. 3 (Mar. 1896). *From the Editor.*

Société d'Horticulture du Doubs, Besançon—Bulletin. Série Illustrée. No. 2. (Feb., 1896). *From the Society.*

Zoological Society of London—Abstracts, 18th Febry., March 3rd (and Rules for the Scientific Naming of Animals, &c.), and March 17th. *From the Society.*

Royal Society, London—Proceedings. Vol. lix. No. 354 (Feb., 1896). *From the Society.*

L'Académie Royale des Sciences et des Lettres de Danemark, Copenhague—Bulletin. Année 1895, Nos. 3-4. 1896, No. 1. *From the Academy.*

Marine Biological Association of the United Kingdom— Journal. N.S. Vol. iv. No. 2 (Feb., 1896). *From the Association.*

Royal Microscopical Society—Journal, 1896. Part 1 (Feb.). *From the Society.*

Societas Entomologica Rossica—Horæ. T. xxix. (1894-95). *From the Society.*

Seven Conchological Pamphlets. By Edgar A. Smith, F.Z.S., &c. *From the Author.*

"The Wealth and Progress of New South Wales, 1894." *From the Government Statistician.*

Department of Public Instruction, Sydney —Technical Education Series, No. 11—"Gems and Precious Stones." By H. G. Smith, F.C.S. *From the Curator, Technological Museum.*

Royal Society of Queensland— Proceedings. Vol. xi. Part 2 (1896). *From the Society.*

Naturwissenschaftlicher Verein in Hamburg —Abhandlungen. xiv. Band (1896) : Verhandlungen, 1895 (Dritte Folge, iii.). *From the Society.*

Société des Sciences de Finlande—Observations Météorologiques faites à Helsingfors en 1894. Vol. xiii. 1ʳᵉ Liv. *From the Society.*

Société Royale Linnéenne de Bruxelles—Bulletin. 21ᵐᵉ Année. Nos. 4-5 (Feb.-March, 1896). *From the Society.*

Museo di Zoologia ed Anatomia comparata della R Università di Torino—Bollettino. Vol. x. (1895), Nos. 210-220; Titlepage, &c.: Vol. xi (1896), Nos. 221-226 (Jan.-Feb.). *From the Museum.*

Royal Physical Society of Edinburgh—Proceedings, Session 1894-95. Vol. xiii. *From the Society.*

Australasian Journal of Pharmacy. Vol. xi. No 124 (April, 1896). *From the Editor.*

Johns Hopkins University—Circulars. Vol. xv Nos. 122-123 (Nov., 1895, Feb., 1896): Annual Reports. Tenth, and Twelfth-Twentieth (1885 and 1887-95). *From the University.*

American Naturalist. Vol. xxx. No. 351 (March, 1896). *From the Editors.*

Harvard College, Cambridge, Mass.—Bulletin of the Museum of Comparative Zoology. Vol. xxvii. No. 7 (Jan. 1896). *From the Director.*

Gordon Technical College, Geelong—The Wombat. Vol. i. No. 2 (April, 1896) : Annual Report, 1895. *From the College.*

Department of Agriculture, Sydney —Agricultural Gazette. Title Page and Index to Vol. vi. (Jan.-Dec., 1895): Vol. vii. Part 3 (Mar., 1896). *From the Hon. the Minister for Mines and Agriculture.*

Pamphlet entitled " Remarks on the Past, Present, and Future of the Australian Flora." By the Rev. W. Woolls, Ph.D., F.L.S. *From Mrs. Woolls.*

Archiv für Naturgeschichte lviii. Jahrgang (1892). ii. Bd. 3 Heft: lxi. Jahrg. (1895). i. Bd. 3 Heft. *From the Editor.*

Auckland Institute and Museum—Annual Report for 1895-96. *From the Institute.*

Woods and Forests Department, Adelaide, S.A.—Fourteen Annual Progress Reports (1881-95). *From the Conservator of Forests, Adelaide.*

Victorian Naturalist. Vol. xii. No. 12 (March, 1896). *From the Field Naturalists' Club of Victoria.*

THEORETICAL EXPLANATIONS OF THE DISTRIBUTION OF SOUTHERN FAUNAS.

By Captain F. W. Hutton, F.R.S., Hon. Memb. L.S.N.S.W.

On considering the present geographical distribution of land and purely fresh-water vertebrates the first and most obvious generalisation is that while the same or closely allied species are widely spread in the northern hemisphere—through Europe, Asia, and N. America—there is, in the southern hemisphere, a great difference between those inhabiting S. Africa, Australasia, and S. America. When we turn our attention to the marine vertebrates—including the migratory fishes which pass a part of the year in fresh water—we notice that the opposite is the case; for while closely related species are widely diffused in the southern hemisphere, the seals, whales, sea-birds and fishes of the N. Pacific differ considerably from those of the N. Atlantic. The reason for these peculiarities is, of course, the peculiar configuration of the land and sea, giving most of the land to the northern and most of the sea to the southern hemisphere; and a necessary conclusion is that the present configuration of the oceans and continents must have remained much as it is now for a very long time. Indeed oceans and continents could not have been widely different from what they now are ever since most of the present genera—and in some cases even families—of birds and mammals came into existence; for, if such had not been the case, we should not now find these genera and families isolated from each other by barriers of land in the northern, and of sea in the southern hemisphere. We may, therefore, safely infer that the physical geography of the earth has not altered greatly during the latter half of the Tertiary era.

But when we pass from the general aspect of the question to study the details, we find many exceptions (especially in the distribut on of t e land plants and land animals of the southern hemisphere) which do not bear out the conclusion forced upon us by the majority of the facts, and the question arises : Have these relationships been brought about by the former existence of more land in the southern hemisphere, or can they be explained without any such assumption ?

The first discussion of the question was by Sir Joseph Hooker, who, in 1853,* advocated a "larger and more continuous tract of land than now exists" in the Antarctic Ocean to explain the distribution of the flowering-plants of the Southern Islands. He assigned no date to this extension of land, but, no doubt, supposed it to be not very ancient.

In 1870, Professor Huxley, in his Anniversary Address to the Geological Society of London, said that the simplest and most rational mode of accounting for the differences between the mammalian faunas of Australia, S. America, and Arctogæa, as well as for the sudden appearance of Eutheria in the latter and in S. America, is the supposition that a Pacific continent existed in the Mesozoic era which gradually subsided, Australia being separated at the end of the Triassic period before the higher mammalia had come into existence. These Eutheria subsequently migrated into North and South America when the Pacific continent finally sank. He says :—"The Mesozoic continent must, I conceive, have lain to the east, about the shores of the N. Pacific and Indian Oceans, and I am inclined to believe that it continued along the eastern side of the Pacific area to what is now the province of Austro-Columbia, the characteristic fauna of which is probably a remnant of the population of the latter part of this period."†

In 1873 I proposed the following hypothesis to explain the complicated problem of the origin of the New Zealand fauna. An Antarctic Mesozoic continent which subsided in the upper Cretaceous period. During the Lower Eocene a second extension of land from New Zealand northwards so as to include New

---

* Flora Novæ Zealandiæ, Introduction, p. xxi.
† Quart. Journ. Geol. Soc. Vol. XXVI. p. lxiii.

Caledonia and part of Polynesia. Subsidence in the Oligocene and Miocene, followed by a third elevation in the Older Pliocene when New Zealand was connected with the Chatham Is , Auckland Is., and perhaps others to the south, but did not stretch north into Polynesia. This large island was broken up by subsidence during the Newer Pliocene.*

In 1874 Prof. A. Milne-Edwards presented to the Academy of Sciences, Paris, a report on the fossil birds of the Mascarene Islands showing that they were related to those of New Zealand. As an explanation, he supposed that land communication had formerly existed between these islands and New Zealand, which was also joined to some islands in Polynesia, while it remained separated from Australia. The connection with Polynesia was to explain the occurrence of *Rhinochetus* in New Caledonia and *Didunculus* in Samoa.

In 1876 Prof. H. N. Moseley supported Sir Jos. Hooker's theory of a former greater extension of land in the Antarctic Ocean†; and in the same year Mr. A. R Wallace published his "Geographical Distribution of Animals," which treats of the whole question.

In 1880 Mr. Wallace published " Island Life," in which he proposes the following hypothesis relating to Australia and New Zealand. During the Cretaceous period, and probably throughout a considerable portion of the Tertiary era, S. W Australia (including the southern part of S. Australia) was separated from Eastern Australia by a broad sea, which contained some islands in what is now Northern Australia. This western island had received its mammalia at an earlier epoch from Asia, and no mammals existed in Eastern Australia. New Zealand was connected with the northern part of Eastern Australia, the land forming a horse-shoe open towards the Tasman Sea. Probably the Bampton Shoal, west of New Caledonia, and Lord Howe's Island formed the western limits of this land; but it is possible, though hardly probable, that

---

* Trans. N.Z. Inst. Vol. v. p. 227, and A.M.N.H. Ser. 4, Vol. xv. p. 25.
  † Linn Soc. Journ. Botany, Vol. xv. p. 485.

it extended northward to the Kermadecs and even to Tonga and
Fiji. Whether it also extended to the Chatham Islands and
Macquarie Island we have, he says, no means of ascertaining, but
such is possible. Separation of New Zealand from Australia took
place at the close of the Cretaceous period, or in the early
Tertiary. At a somewhat later date a southern extension of
New Zealand towards the Antarctic continent seems probable
" as affording an easy passage for the numerous species of South
American and Antarctic plants, and also for the identical and
closely allied fresh-water fishes of these countries."*

In 1882 M. Emile Blanchard contributed a paper to the
Academy of Sciences, Paris, called "Proofs of the subsidence of
a Southern Continent during recent Geological Epochs."†

In 1884-5 I made a further contribution to the subject,‡ in
which I abandoned my former idea of a Mesozoic Antarctic
Continent, and substituted for it a Mesozoic Pacific Continent,
stretching, more or less completely, from Melanesia to Chili. I
still adhered to the other portions of my former paper, but laid
more stress than before on a greater extension of Antarctic
islands during the Older Pliocene.

In 1888 Dr. Theodore Gill published, in the Memoirs of the
National Academy of Sciences, Philadelphia, a paper called "A
comparison of Antipodal Faunas," in which he also advocated the
existence of "some terrestrial passage way" between Tasmania,
New Zealand, and South America, " at a time as late as the close
of the Mesozoic period. The evidence of such a connection
afforded by congeneric fishes is fortified by analogous representatives among insects, molluscs, and even amphibians. The

---

* Island Life, p. 455.

† See N. Z. Journal of Science, Vol. i., p. 251. In the same Journal
will be found a paper by Dr. H. Filhol on the Geological and Zoological
Relations of Campbell Island with the neighbouring Islands.

‡ Part I. in N. Z. Journ. Sci. Vol. ii. p. 1, and A. M. N. H. (5), xiii.,
425; Part II. in N. Z. Journ. Sci. Vol. ii. p. 249, and A. M. N. H. (5),
xv , 77.

separation of the several areas must, however, have occurred little later than the early Tertiary, inasmuch as the salt-water fishes of corresponding isotherms found along the coasts of the now widely separated lands, are to such a large extent specifically different."

In 1892 Dr. H. von Jhering published a paper in the Trans. N. Z. Inst. Vol. xxiv. "On the Ancient Relations between New Zealand and South America." He here supposes that during the whole of the Mesozoic era a continent—which he calls Archiplata—existed which included Chili and Patagonia and extended into the South Pacific. This gradually subsided, throwing off first the Polynesian Islands, then New Zealand, and finally New Guinea and Australia. All this took place before and during the Eocene period; after which Archiplata was joined to Archiguyana, which occupied the high lands of Brazil and Venezuela Dr. F. Ameghino has also, quite independently, advocated a Pacific Mesozoic continent to explain the relations of the Eocene marsupials of Patagonia to those of Australia, and Prof. Zittel has expressed a favourable opinion of this theory.[*]

In 1893 Dr. H. O. Forbes published a paper in the "Geographical Journal (Supplementary Papers ") called " The Chatham Islands : their relation to a former southern continent," in which he reproduced the old theory of an Antarctic continent, but made it last until late Pliocene times, when, he thinks, the Antarctic fauna and flora were driven north by the coming on of a glacial epoch. This continent is supposed to have been unconnected either with S. Africa or with W. Australia (which formed a large island); but sent out prolongations northward, (1) to Madagascar and the Mascarene Islands, (2) to Tasmania and E. Australia, thence through New Guinea and the Solomon Islands to Borneo and Sumatra, (3) to New Zealand, New Caledonia and Fiji; and (4) to S. America, reaching to beyond the Amazon.

In the same year Mr. C. Hedley published in the Proc. Linn. Soc. N.S.W. a short note advocating the existence during Mesozoic

---

[*] See Geol. Mag. New Series. Decade iii., Vol. 10, p. 512 (1893).

and early Tertiary times of a strip of land extending from S. America across the pole to Tasmania; New Zealand, in Tertiary times, reaching near this antarctic land without joining it. And in "Natural Science," he had a paper "On the Relations of the Fauna and Flora of Australia to those of New Zealand," in which he supports the idea of an ancient continent, or "Melanesian Plateau,"* which included the Solomon Islands, Fiji, New Hebrides, New Caledonia, Lord Howe Island and New Zealand, but was separated from Australia and New Guinea. No date is given to this island-continent, but it is supposed to be later than the "Australian Tertiary and Mesozoic beds"; later, therefore, than the Antarctic land.

In 1895, Mr. Hedley returned to the subject in a paper to the Royal Society of N.S.W. called "Considerations on the surviving Refugees in Austral Lands of ancient Antarctic Life." Here he advocates an Antarctic continent, which was a very unstable area, "at one time dissolving into an archipelago, at another resolving itself into a continent.", He thinks that snakes, frogs, monotremes and marsupials passed across this continent, from S. America to Tasmania, during a warm, Mid-tertiary period. He also now thinks that the southward extension of New Zealand, mentioned in his former paper, was synchronous with its northern extension to the Melanesian plateau, that is, it was late instead of early Tertiary date.

This short historical sketch will, I think, make it clear that a considerable amount of ingenuity has been expended in trying to solve the interesting problem of the distribution of southern faunas. The differences of opinion are due partly to some of the authors having taken only a small number of the known facts into consideration, and partly to constant additions to our knowledge, either by the discovery of new facts, or by the correction of old errors. No doubt our knowledge will still increase, but it seems hardly possible to make any more theories. The problem is a very intricate one, and we may be sure that the true solution is not simple.

---

* Called Antipodea by Dr. Forbes

It is evident that in any large district, like Australasia, there is no reason to suppose that the ancestors of the animals and plants now inhabiting it all came from the same direction or at the same time : consequently the first step to take is to try to separate the fauna and flora into groups which find their nearest relations in different directions. Thus in Australasia we have—

1. An Australasian fauna and flora which have no near relatives now living.

2. A northern fauna and flora related to the Oriental fauna and flora of the present day.

3. A south-tropical or sub-tropical fauna and flora whose nearest relations at present are either in S. Africa or in S. America north of 40° S. That the differences between these countries are far greater than their resemblances does not do away with the existence of these resemblances, but rather accentuates them. They are vestigial remains with all the importance that vestigial remains always possess.

4. A south-temperate or cold-temperate fauna and flora, with relations to plants and animals in Patagonia or Chili and the Antarctic Islands. This is usually called the Antarctic element.

Judging by the relative closeness of the relationship of these different faunistic elements to their foreign connections, we must conclude that the first and third are much older inhabitants of Australasia than the second and fourth. The second element, which is best developed in north-eastern Australia, presents no difficulty and everyone is agreed as to its origin. The fourth element, which is better developed in New Zealand than in any other part of Australasia, consists of marine animals with a few migratory fresh-water fishes and possibly some land mollusca and worms; and there is a general consensus of opinion that these spread by means of a greater development of land in the Antarctic region. This may have been as late as the Older Pliocene, but not later, as considerable changes have taken place in the animals since it occurred   Also, as pointed out in the first paragraphs of this paper, this land could not have been continuous between S.

America and Australasia, for in that case there would have been a far greater commingling of the land faunas and floras. It is the origin of the first and third elements which has given rise to such differences of opinion. These are developed far more strongly in Australia and Tasmania than in New Zealand; and the explanation of the third will probably explain the first also. I will, therefore, briefly review the three hypotheses (variously modified) which have been proposed.

1. The first explanation is that the different groups of animals and plants in question have migrated from the northern hemisphere into the southern by the present continents and have since then become extinct in the north. With regard to the South African connection, this explanation will be readily accepted. The fact that Proteaceous plants-·now almost confined to S Africa and Australia—were formerly abundant in Arctogæa is a proof, so far as they are concerned; and we may accept the same explanation for the occurrence of the Baobab-tree *(Adansonia)* in W. Australia and the Fern-bird *(Sphenœacus)* in New Zealand. This theory also explains the occurrence of the curious genus of wingless locusts —*Anostostoma*—in Madagascar and Australia and the connection of some birds of Madagascar and the Mascarene Islands with others of New Zealand and Polynesia. It will also explain the abundance of parrots in Australia and S. America, for these lived in Europe in the Miocene period, as well as the occurrence of tapirs and trogons in Central America and Malaya; for these, like the large carnivora, must have passed from one continent to the other by a northerly passage. Probably also it will explain the relation of the curassows of S. America to the megapodes of Australia and Polynesia, and the connection between the lower passerine birds of both continents, as these relationships are all very distant.

But, however this may be, there are certain facts of distribution which this theory cannot solve. A typical case is the distribution of the tree-frogs belonging to the genus *Hyla*. This contains 83 species in S. America, 28 in Australia, 17 in N. America, and one each in India, China, and Europe; while *Hylella* is found

only in Australia and tropical America. Again the fresh-water
tortoises belonging to the family *Chelydidæ* are restricted to
Australia and S America. The fresh water fish *Osteoglossum* is
represented by species in S. America, Queensland, and Borneo;
and the South American beetles are more closely related to those
of Australia and Africa than they are to those of N. America.
Indeed the connection between S America and Australia is so
marked in the *Buprestidæ* and *Longicornia* that Mr. Wallace,
who as a general rule strongly supports the northern route, says
that "there must probably once have been some means of communication between the two regions better adapted to these
insects than any they now possess." And as several of the
Eocene mammalia of Patagonia were closely allied to those now
living in Australia the evidence for a former land passage between
the two countries may be considered as conclusive. The northern
route therefore fails to give a full and satisfactory account of the
whole of the facts, and we must look to some other route to
supplement it   The portions of the faunas unaccounted for are
all old forms of life, and consequently we must conclude that the
means of communication used by them has been long ago destroyed;
for if not it would also have been used for modern groups.

2 Turning now to the proposed southern route by an Antarctic
continent, it has this in its favour that, as the greater extension
of Antarctic land in the late Tertiary era has been allowed, it is
not difficult to suppose that at a still earlier time that is in the
Mesozoic era, a large continent might have existed there. One
difficulty is in the climate. How could tropical, or sub-tropical,
snakes, insects, and fresh-water tortoises and fishes pass through
such high latitudes? The example of Greenland is pointed to,
but in Greenland the climate indicated is temperate only, not
sub-tropical or tropical. Again it is stated, in explanation, that
there is evidence of a much warmer climate having obtained in
the southern hemisphere in Miocene times than now. But this
appears to have been a period of depression throughout southern
Australasia, and it does not follow that the climate would be
equally mild when an Antarctic continent existed. I do not

think that the climatic objection is fatal, for we cannot tell what the climate may have been in the Jurassic and Cretaceous periods, but it is a difficulty, and I cannot go so far as Mr. Hedley, who supposes that venomous snakes, frogs, monotremes and marsupials passed round the head of a deep bight of the Pacific Ocean which "stretched within a few degrees of the pole."

A far greater difficulty remains for consideration, which is this: Aplacental Mammals — both Multituberculata and Polyprotodontia—existed in Europe and N. America in the Triassic and Jurassic periods, and these Polyprotodontia were, no doubt, the ancestors of the living Polyprotodontia of Australia. In the Eocene strata of Patagonia remains of a large number of Polyprotodontia have been found which are far more closely related to the Polyprotodontia of Australia than to the Mesosoic forms of Europe and N. America; consequently a direct land communication must have existed between these two southern countries. Now there is strong geological and palæontological evidence that no land ridge existed between N. and S. America during the Mesozoic and early Cainozoic eras; consequently we must assume that the southern forms migrated through the Malay Archipelago; and, if they went to Patagonia by means of an Antarctic continent, they must have passed through Australia. But mingled with the Eocene marsupials of Patagonia there are a number of Eutheria of typically South American character—*Edentata, Toxo-. dontia, Typotheria, Perissodactyla, Rodentia*, and even *Platyrrhine* monkeys—without any northern forms of *Artiodactyla, Carnivora*, or *Insectivora;* and it is hardly possible that these should have passed through Australia without leaving any record behind. This is, to me, a fatal objection to the theory of migration by means of an Antarctic continent.

3. The theory of the former existence of a South Pacific Mesozoic continent seems to be the only theory left, but it has been objected to both on account of the present depth of the ocean and because, it is said, no record has been left in the Polynesian Islands of the supposed passage of the plants and animals. Both these objections apply equally to the former

existence of an Antarctic continent. According to the latest maps the ocean south of Tasmania, and the Pacific below 45° S., are considerably deeper than the Pacific between 10° and 30° S , and the answer in both cases is that this continent existed a very long time ago. The answer to the second objection is that no record has been preserved of the fauna and flora on the Antarctic continent because of a change in climate, and in the Polynesian Islands because the continent disappeared entirely below the sea, the present volcanic and coral islands being merely outgrowths on its submerged back. But the statement that no record exists in the case of the Pacific continent is not quite correct, for the Iguanas of Fiji can hardly be explained in any other way.

The theory of a Mesozoic South Pacific continent not only explains the origin of the Australian and S. American marsupials, but also the almost simultaneous appearance of different Eutherian mammals in North and South America. We must suppose that this continent threw off first New Zealand, then Australia, then Chili, and finally disappeared under the waves. The reasons why we must suppose New Zealand to have been at one time attached to the continent are the existence in that country of *Sphenodon, Unio,* and *Astacidæ,* none of which are found in truly Oceanic islands*. At a later date, as I pointed out in my former papers, New Zealand must have formed part of a large island joined to New Caledonia, but not to Australia. This has lately been called Antipodea by Dr. Forbes, and the Melanesian Plateau by Mr. C. Hedley. Still later again, New Zealand must have stretched south and obtained its Antarctic fauna and flora from Patagonia through a number of islands.

From a biological point of view I see no reason to object to this theory. The objections are geological, and most geologists at the present day would, I think, say that the doctrine of the persistence

---

* It is also hardly possible to account for the distribution of frogs, slugs, wingless and feebly flying insects, earth-worms, myriapods, and fresh water animals generally, except by the supposition of land passage.

of continental and oceanic areas negatives it. This doctrine—which is not accepted by all geologists*—is founded on the undoubted fact that the principal mountain ranges in the northern hemisphere, and, perhaps, in Australia also, are formed of shallow water sediments representing all periods from the Silurian upwards; consequently land must have existed in their neighbourhood all that time; and from this it is inferred that the present oceanic areas have always been sea. The proof, however, is far from being complete, and no explanation has, as yet, been given either (1) of the remarkable submarine plateaux found in the basins of the S. Pacific and S. Atlantic Oceans; or (2) of the sudden irruption of mollusca, bony-fishes and dicotyledons into N. America during the close of the Cretaceous period, followed by a host of Eutherian mammalia in the Eocene; or (3) of the place of origin of the peculiar S. American mammalia. The former existence of a Mesozoic Pacific continent seems to me, as it did to Professor Huxley, the simplest explanation of all these difficulties; we can never expect to attain certainty in the matter, but I think that the weight of the evidence is in its favour.

---

* Gardner, Geol. Mag. 1882, p. 546 ; Hutton, N.Z. Journal of Science, Vol. I. p. 406 (1883) ; Blandford, Q.J.G.S. XLVI. Proceedings, p. 59 (1890); Oldham, Geol. of India, 2nd Ed. p. 211 (1893).

REPORT ON A BONE BRECCIA DEPOSIT NEAR THE WOMBEYAN CAVES, N.S.W.:

WITH DESCRIPTIONS OF SOME NEW SPECIES OF MARSUPIALS.

By R. Broom, M.D., B.Sc.

(Plates vi.-viii.)

About 18 months ago I discovered a small bone breccia deposit in the neighbourhood of the Wombeyan Caves. The deposit is situated in a small depression near the top of the hill above the present caves and no doubt is portion of the floor of an older cave whose walls and roof have long since been weathered away. The deposit consists of a rather hard light brown calcareous matrix containing imbedded in it innumerable small bones. In some parts the bones are almost all small and packed together so closely that there is very little matrix; in others the matrix is comparatively free from bones, only containing a few of the larger forms. As the deposit is unquestionably old and contains some forms new to science—two of which I have already described*— I have thought it well to give a detailed account of the forms found, as it will give a fair idea of the smaller animals living in later Tertiary times.

Macropus (Halmaturus) wombeyensis, n.sp

(Pl vi figs. 1-3).

Though the deposit is essentially one of small bones, there are a number of bones of a species of *Macropus*. Besides a number of vertebræ and long bones, I have succeeded in finding three imperfect fragments showing the upper molars, and four moderately well preserved lower jaws—two of which are presumably from the same individual. In size the form was apparently

---

* Proc. Linn. Soc. N.S.W. (2) Vol. x (Pt. iv. 1895).

about that of *Macropus ualabatus*, but the dental details are decidedly different. Of existing species the only one to which it comes at all close is *M. agilis;* but from this species it differs in the narrowness of the molars and in the jaw being considerably thinner. Among extinct forms the only ones approaching it in dental details and measurements are some fragmentary specimens from Queensland, referred to by De Vis.* Thinking my form might possibly belong to the same species as one or other of the fragmentary Queensland specimens, I submitted a specimen to Mr. De Vis, who kindly writes me as follows :—" I have compared the Halmaturus jaw with my types—it agrees with none of them. In size and general features it is like *H. agilis*, but appears to me to be quite distinct from that species." As my specimens thus appear to differ from all existing or previously observed extinct species, I have conferred on it the above distinctive name from the locality in which the form has first been observed.

In general form the lower jaw resembles that of the larger Wallabies; there is, however, a greater disparity between the anterior and posterior depth of the jaw than is usually the case in existing forms. The dental portion of the jaw is comparatively narrow—more so than in any of the existing larger Wallabies. The angle is more inflected than in the Wallabies, closely resembling the condition in the Rock-Wallabies. The premolar (p⁴) is well developed, rather narrow without internal cusp. It is slightly ridged, there being three very shallow vertical grooves. In the specimen figured (Pl. vi. fig. 3) there are also on the outer aspect two small horizontal furrows. The molars resemble very closely those of *M. ualabatus*—the crests being curved and the links well developed.

Though two specimens illustrate the palatal region, in neither are the teeth well preserved. The upper premolar (p⁴), however, appears to have had a moderate internal cusp. One point of great interest is the presence of large palatal vacuities. In this

---

* Proc. Linn. Soc. N S.W. (2) Vol. x. (Pt. i. 1895).

the form agrees with the smaller Wallabies and Rock-Wallabies and differs from the larger sorts.

Though the form thus equals in size the larger Wallabies, its affinities are probably more with the smaller sorts, and in some respects it seems to come very near to the Rock-Wallabies (*Petrogale*).

The following are some of the principal measurements :—

Depth of mandibles behind p^4 (4 sp.), 17, 18, 18·4, 18·4 mm.

    ,,   ,,   in front of m^4 (3 sp.), 15·4, 16·9, 16·9 mm.

Length of p^4 (2 sp.)... 6·8 mm. (worn), 7·4 mm. (unworn).

    ,,   m^1-m^2 (2 sp) .13·4, 13·5 mm.

            m^1-m^3 (2 sp)...21·8, 21·9 mm.

            m^2-m^4 (2 sp.)...25, 26 mm.

            m^1-m^4 (3 sp.)...29·2, 30·8, 31·4 mm.

    ,,   m^3-m^4 (3 sp.) ..17·8, 18·, 18·8 mm.

    ,,   p^4-m^4 (3 sp)...36·5, 37, 37·4 mm.

Width of m^3 (3 sp.). .5·7, 5·8, 5·8 mm.

Thickness of mandible below m^3, 9·3 mm.

POTOROUS TRIDACTYLUS, var. ANTIQUUS, n.var.

In the deposit are the remains of a small Potorous. Though not abundant a number of specimens have been obtained. As I have been unable to obtain a skull of the existing *Potorous tridactylus* I am in some doubt as to the exact position of the fossil form. *Potorous tridactylus*, as defined by Thomas, is apparently a very variable species, and it seems probable that the fossil form is but a variety. As regards the premolar of this species Thomas (Brit. Mus. Cat. Marsup.) says:—"P^4 very variable; in the large Tasmanian form ('*apicalis*') 7 or 8 millim. long, with four distinct grooves; in the smaller New South Wales examples, and in the still smaller Tasmanian form described as '*rufus*' 6 or 7 millim. long with only three grooves." In the fossil form the upper premolar measures 6·1 mm., but has four grooves. The three anterior grooves are well marked, but the fourth, though well marked at the edge, does not extend so far up the tooth as the others. In the deciduous p^3 there are but three grooves. In the lower p^4 there are four grooves, all well developed.

Dental Measurements.

Length of upper p^4 ...6·1 mm.

,,   ,, dp^4...3·4 mm.

,, m^1 ...4·8 mm.

,, m^2 ...4·9 mm.

lower p^4 ...5· mm.

BURRAMYS PARVUS, Broom.

(Pl. VII. figs. 1-2).

This most interesting little form which I recently described before this Society [*] occurs in the deposit pretty abundantly, but from its minute size and the obliquity of the large premolar it is difficult to extract perfect specimens. Since I described the form I have succeeded in discovering a few more points in its structure. In my paper on this species I expressed the opinion that it forms a connecting link between the Phalangers and the Kangaroos, finding in the large grooved premolars a relationship with the Rat-Kangaroos and in the entire masseteric fossa, and the small teeth between i^1 and p^4 an affinity with the Phalangers. No perfect specimen has yet been discovered of the upper jaw, but a few fragmentary specimens enable us to almost complete the dental formula. Within the upper large premolar and a little in front is a minute two-rooted premolar similar to p^3 in the lower jaw. In front of this is a very considerable diastema where the palate has a rounded edge somewhat like that in Macropus, and with apparently no anterior premolars. In front is a small but well formed canine implanted in the maxillary more after the manner of the small Macropods than of the Phalangers. The dental formula so far as known would thus appear to be, in the notation used by Thomas :—

$$\text{I } \frac{\cdots}{1\ 2\ 0} \quad \text{C } \frac{1}{0} \quad \text{P } \frac{0\ 0\ 3\ 4}{1\ 0\ 3\ 4} \quad \text{M } \frac{1\ 2\ 3\ 0}{1\ 2\ 3\ 4}$$

---

[*] "On a small fossil Marsupial with large grooved premolars." Proc. Linn. Soc. N.S.W. (2) Vol. x. (Pt. 4, 1895).

There appears to be no upper m⁴, while the rudimentary lower m⁴ is apparently variable. The dental formula shows much resemblance to that of *Hypsiprymnodon* as regards the upper teeth, but in the possession of the two small teeth between i¹ and p³ there is considerable difference in the lower jaw. As regards the number and arrangement of the teeth in the lower jaw the agreement with some of the smaller Phalangers is very marked; *Dromicia nana*, for example, having an entire dental formula almost exactly like that of *Burramys*. To *Dromicia nana* there is also a marked resemblance in the lower minute teeth and some resemblance in the molars.

A considerable fragment of the skull gives a fair idea of the outline, but adds little to the settlement of the affinities of the genus. The skull has been apparently sharp-snouted as in *Petaurus* or *Dromicia* The lacrymal foramen is placed distinctly. in front of and beyond the orbit. The infraorbital foramen is large, and placed in front of the large premolar—in this resembling the condition in the Phalangers and differing from the normal Macropod arrangement. The interorbital region of the skull is comparatively broad, but there is no distinct supraorbital ridge. The olfactory lobes of the brain have been well developed, and the whole brain appears to have been relatively large. The zygomatic arch passes out from the maxilla in the usual manner: it arises near the posterior part of the large premolar and is com paratively slender.

PETAURUS BREVICEPS, Waterh.

Some time ago I found an imperfect fragment of a lower jaw, with the roots of three teeth in position. Though the fragment was manifestly that of a Petaurus and in size agreed with *P. breviceps*, I hesitated to refer it definitely to that species on such imperfect evidence. Since then I have found a fragment of the cranium with the frontal bones almost perfect, and from the size and the formation of the supraorbital ridges, there is no doubt in referring the specimen to *P. breviceps*, and there is little doubt but that the lower jaw fragment also belongs to this species.

As these are the only remains found the species must have been very rare in the district at the time of the deposit.

At present the species is found in the district and may be regarded as not infrequent, though I am led to believe that 50 years ago it was very abundant, the present scarcity being due apparently to the havoc made amongst them by domestic cats.

PALÆOPETAURUS ELEGANS, Broom.

(Pl. VII. fig. 3).

This small Petaurus-like Marsupial I recently described[*] from some jaws and a well preserved specimen with the maxillary teeth.   Since then I have found besides numerous jaws a moderately good portion of the skull (Plate   , fig. 3) and a number of other fragments.   The frontal bones differ from those of *Petaurus*, and agree apparently with *Gymnobelideus* in being without supraorbital ridges; and the hinder part of the frontals is considerably broader and flatter proportionally than in *Petaurus*.   The snout though narrow appears somewhat broader than in *Gymnobelideus* judging by the figure.   In one of the type specimens the upper p^1 was found to be single-rooted, or rather its two roots were united together.   This, too, appears to be rather variable as in two other specimens one is found with the roots close together but distinct, while the other has the roots somewhat apart.   In all the observed specimens, however, p^3 is double rooted.

DROMICIA NANA, Desm.

One of the most interesting discoveries is that of *Dromicia nana*, of which I have found a large number of both lower and upper jaws.   There can thus be little doubt but that in later Tertiary times *Dromicia nana* was very common in New South

---

[*] "On a small fossil Petaurus-like Marsupial," Proc Linn. Soc. N.S. W. (2) Vol. x. (Pt. 4, 1895).

Wales. From the existing species being believed to be confined to New Guinea, Tasmania, and West Australia, Thomas regards it as practically certain that *Dromicia* existed in former times in Eastern Australia. The correctness of this conclusion is now established. The fossil form so far as known does not differ from the existing *D. nana*.

As regards the present distribution of this species Thomas considers it to be exclusively confined to Tasmania. In this, however, it is probable that he is in error. For though the species must be excessively rare in New South Wales it most probably still survives, as it is quite certain that it existed within very recent years. In the Grand Arch at the Wombeyan Caves there are near the entrance numerous ledges of rock frequented by Rock Wallabies, and on which the animals leave quantities of their excrement. Mingled with the dry and decomposing dung are to be found quantities of small bones—chiefly those of *Phascologale flavipes, Petaurus breviceps*, and of the Bush Rat (*Mus* sp.), but with also a few of *Pseudochirus peregrinus, Perameles obesula*, and of small birds and snakes While searching among these I discovered, to my surprise, two jaws of *Dromicia nana* in tolerably good preservation It is hard to say what may be the age of the bones, but as the ledge is quite exposed to atmospheric influences and as the bones show little sign of weathering, it cannot well be more than a very limited number of years since the animals died. Considering the wild region in which the caves are situated it is very probable that the species still survives in the district, though I have sought it in vain. On mentioning my discovery to Mr. J. J. Fletcher, he kindly called my attention to Krefft's discovery of *Dromicia unicolor* [= *D. nana*] at North Shore, Sydney, in 1863, and to the fact that Thomas regards the specimens as almost certainly Tasmanian specimens which had escaped from captivity. Such an explanation will not do for the recent bones found at the Wombeyan Caves, nor is such a theory now required for even Krefft's specimens, considering that in former times *Dromicia nana* was one of the commonest of New South Wales marsupials.

PSEUDOCHIRUS ANTIQUUS n.sp.

(Pl. VII. Figs. 4-6).

One of the commonest forms whose remains are found in the deposit is a species of *Pseudochirus*. In size and structure it much resembles the common ring-tailed Phalanger (*P. peregrinus*), but the careful study of a large series of specimens has satisfied me that the remains are those of a distinct and new species. In average size the teeth are appreciably larger than in *P. peregrinus*, yet on the whole the form comes nearer to that species than to either *P. cooki* or *P. orientalis.*

The following table illustrates the features so far as known and the points distinguishing the fossil form from *P. peregrinus*.

P. peregrinus.	*P. antiquus.*
Upper p^1 small, about 1 mm. in front of p^3	Upper p^1 moderate size, placed close to p^3
Length of m^1-m^3 — 11·2-12·6 mm.	m^1-m^3 in only three specimens, showing complete series— 12.7, 12·9, and 13· mm.
Cusps of upper and lower molars moderately developed	Cusps of upper and lower molars well developed
Post Ext. Cusp of upper m^1 (4 sp.) min. 1·7, max. 2·0, average 1·85	Post. Ext. Cusp of upper m^1 (5 sp) min. 2·1, max. 2·3, average 2·22
Ant. Int. Cusp of lower m^4 (3 sp.) min 1·6, max. 1·8, average 1·7	Ant. Int. Cusp of lower m^4 (3 sp.) min. 2·3, max. 2 5, average 2 4
Palate with a distinct lateral depression in region of p^3 and p^4	Palate moderately flat, no distinct lateral depression in region of p^3 and p^4
Angle of jaw produced well backwards.	Angle of jaw relatively small and passing backwards but a short distance

It is unfortunate that I have not succeeded in getting any specimens with the upper p¹ in position, and only one specimen (Pl. fig. 4) showing the socket. From this specimen the tooth appears to have been almost double-rooted and placed much closer to p³ than in *P. peregrinus*, and in this resembling more *P. cooki*.

PERAMELES WOMBEYENSIS, n.sp.

(Pl. VIII. figs. 1-8).

The above name I propose for a species of *Perameles* which must have been very common at the period when the bone deposit was formed. Though from the nature of the matrix I have been unable to develop a single perfect jaw, yet I have succeeded in finding sufficient fragmentary specimens to enable me to give almost all the important details of dentition. The species seems to have been a form a little larger than *P. obesula*, and to have resembled it in being short-nosed.

The upper incisor teeth are unknown, the premaxillary being absent from all the upper jaw specimens I have. The canine is moderately developed and rather larger and flatter than in *P. obesula*. P¹ is considerably larger than in *P. obesula*, and directed some. what forward. It is placed about 2 mm. behind the canine. P³ is about equal in size to p¹ and placed a little less than 1 mm. from it. It has a distinct anterior secondary cusp and a less marked posterior one. P⁴ is unknown. The upper molars resemble those of *P. obesula* in being soon worn down, and in old specimens leaving no trace of the cusps. In shape there is considerable agreement with those of *P. obesula*, the section of the worn tooth being quadrangular, with rounded angles. M⁴ is unknown.

The lower jaw bears more resemblance to *P. obesula* than any other existing form. The anterior edge of the coronoid process is straight and the process itself passes back obliquely. The angle seems less produced than in *P. obesula*, though it is possible a portion of the slender tip may have been broken off in the figured specimen (Pl. fig. 1). The lower incisors are absent, but in fig.

3 the anterior part of the jaw is seen. The canine appears to be small, though as the specimen figured (Pl. VIII. fig. 3) is from a young animal, the canine has probably not attained its full size. P¹ and p³ resemble the upper teeth in size, and are both furnished with small anterior and posterior secondary cusps. P⁴. is relatively large. Lower molars resemble those of *P. obesula*.

The following are some of the principal measurements :—

Height of canine	...	...	...	3·1 mm.
Length of p¹	...	...	...	2·8 mm.
„	unworn m¹ ...	...	...	4·0 mm.
„	worn m²	...	...	3·6 mm.
„	worn m³	...	...	3·4 mm.
Estimated length of unworn m¹-m³	...	11·3 mm.
Lower p³-m⁴, aged specimen ...	...	21·3 mm.
Estimated upper c-m⁴ ...	...	...	28-28·5 mm.

THYLACINUS CYNOCEPHALUS, Harris.

Of this species I have found two teeth—a perfect lower canine and a perfect lower premolar—but no bones.

PHASCOLOGALE FLAVIPES, Waterh.

This small pouched mouse is represented by a very large number of jaws and other remains. It appears to be the commonest species in the deposit with the exception of the Bush Rat. So far as I have been able to make out, the fossil animal in no way differs from the existing species. *Phascologale flavipes* is still found in the district, and though it is very rare if not extinct in the settled parts, in the wilder regions it is fairly common.

PHASCOLOGALE PENICILLATA, Shaw.

This species though met with is distinctly rare. I have only found one complete lower jaw, a fragment of a second, and two fragments of the upper jaw. The anterior premolars and canine are a trifle larger than in the recent skull in my possession (a female), but there is no doubt that the remains belong to the existing species. The form is still met with in the district, though by no means common even in the mountainous regions, while in most of the settled parts it appears to be extinct.

ECHIDNA sp.

(Pl. VIII. figs. 9-10). .

A number of bones of a large Echidna have been found, and which in all probability belong to the described form *Echidna oweni*, Krefft. The specimens are, however, too fragmentary to enable me to refer them definitely to this form. The remains comprise the greater portion of the left ilium, with a fragment of the sacrum attached, the lower portion of left femur, the articular head of the femur, two vertebral centra, and a number of fragments of long bones.

The femur differs in one or two respects from *E aculeata*. The constriction of the shaft immediately above the condyles is much less marked, and the shaft at this part is more flattened than in the common existing species, while the depression above the patellar surface is more marked and broader.

The ilium is very considerably stouter proportionately than in *E. aculeata*. From the union by complete anchylosis of two small fragments of the sacrum with the ilium it is evident that the extinct species agrees with the living in the complete anchylosis of the sacrum with the ilia.

Max. width across lower end of femur ...	32·5 mm.
„ „ „ in *E. aculeata* (adult male) ...	22·5 mm.
Oblique measurement from outer depression of shaft to inner condyle	26·4 mm.
Oblique measurement in *E aculeata*	17·8 mm.
Trans. measurement above patellar depression ...	24· mm.
„ „ „ in *E aculeata*	14·5 mm.

Besides the above forms there are a few remains too fragmentary for certain identification. Two fragmentary teeth probably represent *Thylacoleo*, while a detached molar belongs to a small species of *Macropus*. There are also innumerable remains of Bush Rats (*Mus* sp.) which I have not had an opportunity of identifying with certainty. Of birds there have been found the perfect cranium of one about the size of a Sparrow and some small bones, while of lizards there occur the remains of a moderate sized member of the *Scincidæ*.

CONCLUDING OBSERVATIONS.

Though a few of the forms found in the deposit are still surviving, the general character of the fauna is very different from that of recent times. With the exception of *Thylacinus*, the *Macropus* and the *Echidna*, the animals may almost all be classed as feeble and defenceless, and have apparently flourished owing to the absence or scarcity of natural enemies. *Dromicia*, *Palæopetaurus* and *Burramys* were probably all of very similar habits, the conditions suitable to the one being equally so to the others, while those inimical to any would probably tend to the destruction of all. The two species of *Phascologale*, though probably suffering from the same adverse condition which has destroyed the small Diprotodonts have been less affected and able to survive. The cause of the destruction of the smaller forms is probably to be found in the introduction into their midst of some common enemy. A glance at the recent fauna of the district suggests a not improbable explanation of the change. To-day the forms which may be said to be numerous are *Trichosurus vulpecula*, *Phascolarctus cinereus*, *Dasyurus viverrinus*, *D. maculatus*, and *Macropus ualabatus*. All these are absent from the deposit, and though their absence does not prove that they were not then in the district, it may safely be taken to indicate that they were at least rare. The absence of the common Phalanger for example could not have been due to unfavourable conditions, as the abundant remains of the species of Ring-tailed Phalanger show there must have been plenty of suitable trees. The conclusion thus seems probable that *Trichosurus* is a comparatively recent addition to the local fauna. If it could be proved that with it came the Dasyures we would have at once a satisfactory explanation of the disappearance of the small Diprotodonts. It is at present, however, impossible to say more than that at the time of the deposit Dasyures were absent or rare, that in more recent times they have become numerous in the district, and that their introduction or increase has been the probable cause of the destruction of the smaller forms. The fact of *Petaurus breviceps* having not only survived but increased, while the closely allied *Dromicia* has been all but exterminated, seems to suggest that the former with the parachute expansions was able to escape from some enemy to which *Dromicia* fell a prey. *Palæop-taurus*, if we may assume, as is quite probable, that it resembled *Gymnobelideus* in being without lateral expansions, would fall as easily a prey as *Dromicia*.

I must acknowledge my indebtedness to Mr. J. J. Fletcher, Mr. R. Etheridge, Junr., Mr. De Vis, and to my father for kind assistance they have rendered me.

EXPLANATION OF PLATES.

Plate VI.

Macropus wombeyensis.

Fig. 1.—Right jaw—nat. size.
Fig. 2.—Right lower teeth—nat size.
Fig. 3.—Lower p^4 ($\times$ 3).

Potorous tridactylus, var. *antiquus.*

Fig. 4.—Left upper molars ($\times$ 4·5).
Fig. 5. —P^3 (left upper ?) ($\times$ 4·5).
Fig. 6 —Left upper p (4·5).
Fig. 7.—Right lower p^4 ($\times$ 5).

Plate VII.

Burramys parvus.

Fig. 1.—Side view of skull of ($\times$ 3·4). The lower jaw is seen somewhat obliquely to represent its true side view when placed in the skull.
Fig. 2.—Upper aspect of fragment of skull ($\times$ 3·4).

Palæopetaurus elegans.

Fig. 3.—Upper aspect of fragment of skull (× 2).

Pseudochirus antiquus.

Fig. 4.—Upper premolars (× 3·6).
Fig. 5.—Lower m^3 (× 4).
Fig. 6.—Back part of lower jaw—nat. size.
Fig. 7—Exactly similar aspect of lower jaw of *Pseudochirus peregrinus.*

Plate VIII.

Perameles wombeyensis.

Fig. 1.—Back part of lower jaw with m^4 (× 2).
Fig. 2—Anterior part of upper jaw (× 2).
Fig. 3.—Inner view of anterior part of lower jaw of young—nat. size.
Fig. 4.—Inner view of adult lower jaw—nat. size.
Fig. 5.—Right upper m^1 unworn (× 4).
Fig. 6.—Left upper m^3 somewhat worn (× 4).
Fig. 7.—Inner view of lower m^4 (× 5·5).
Fig. 8.—Outer view of lower m^4 (× 5·5).

Echidna sp.

Fig. 9.—
Fig. 10.—

ON A *GALAXIAS* FROM MOUNT KOSCIUSKO

By J. Douglas Ogilby.

At the meeting of this Society in March, 1882 (Vol. vii. p. 107) the late Sir William Macleay read a paper descriptive of a species of *Galaxias* which had been forwarded to him by Baron von Mueller to whom examples had been sent by Mr. S. Findlay, who found them inhabiting the streams which drain the southern slopes of Mount Kosciusko and form a section of the watershed of the Snowy River; for this form he proposed, at the request of Baron von Mueller, the name of *Galaxias findlayi* in honour of its discoverer and collector.

With the exception of its inclusion in the "Supplement" to Macleay's "Descriptive Catalogue of Australian Fishes" there does not appear to be any further published information respecting the Kosciusko Galaxiid, nor do any specimens from that district seem to have been collected until the autumn of 1889, when a few examples were secured and brought to Sydney by Mr. Richard Helms on the occasion of his visit to that mountain, a short account of which is published in the Records of the Australian Museum, Vol. i. pp. 11-16. These specimens were also obtained from streams flowing into the Snowy River, and writing of their distribution Mr. Helms observes (p. 13):—"The absence of *Galaxias* at this elevation" (Wilkinson's Valley) "struck me as peculiar. It is, however, remarkable that on the Snowy River side these fishes are met with almost everywhere"

The paragraph from which this quotation is taken is not clear, but the most reasonable deduction from it is that, in Mr. Helms' opinion, Galaxiids were scarce or even absent on the Murrumbidgee slope.

Pressure of business prevented a full examination of these specimens being made at the time, and they were put on one side and neglected until the commencement of the current year, when Mr. Helms requested me to furnish him with a report on these fishes, and it was then discovered that owing to the changes which had taken place in the Museum and the consequent shifting of specimens from place to place the examples in question were not immediately forthcoming.

In default of these the next best thing to do was to endeavour to get other specimens from the same locality, and an opportunity for effecting this occurred through the visit in January last of the Rev. J. M. Curran and Mr. C. Hedley to Mount Kosciusko, and the writer thereupon called the attention of the latter gentleman to the subject in the hope of procuring a good working series for examination; however, the specimens thus obtained, two or three in number, were, on Mr. Hedley's return, handed to the authorities of the Australian Museum, and became, therefore, unavailable for the purpose required, which included such an exhaustive examination as the difficulty of determining the species of this intricate genus and the interest attaching to this particular form as an inhabitant of a greater altitude than is reached by any other Australian fish warranted.

In this unsatisfactory state our knowledge must again have been indefinitely left but that, the Rev. Mr. Curran having occasion to return almost immediately to Kosciusko, the writer took advantage of his going to request him to collect sufficient material to enable the complete examination which was deemed necessary to be made. So well was this request acceded to that on the return of that gentleman from his second trip I received a fine series numbering no less than sixteen individuals in perfect condition, and this collection was afterwards supplemented by a further contribution of eleven, and I take this opportunity of acknowledging my obligations and tendering my grateful thanks to that gentleman for the trouble which he took in procuring so fine a series of specimens.

A critical investigation of these examples reveals facts which greatly invalidate certain apparently well established characters which have hitherto been considered of sufficient importance to justify specific separation. As an instance, it will be remembered that the fishes of the genus *Galaxias* have naturally fallen into two groups, characterised—the one by a short, stout body, of which group *truttaceus* may be taken as typical, the other by a long, slender body, to which *attenuatus* and its allies are to be referred; yet in this one small species I am confronted with individuals varying from one-fifth to one-eighth in the proportionate measurement of depth to length, and with a corresponding difference in colour from a dull dark brown without or with but very slight indications of markings to bright golden beautifully blotched, spotted, or barred with black. These differences, however, great as they appear to a casual glance, are entirely attributable to the nature of the locality and the water which the individual fish inhabits, the stout, sombre-coloured form being found in the deep still pools and small subalpine tarns, the slender brilliant one in the more rapid gravelly or sandy shallows where it is exposed to the sunlight; but between these two limital forms every conceivable variation, both of contour and colour, may be found.

The distribution of *Galaxias*, comprising as it does the southern extremities of the three great continental areas which converge upon the Antarctic Circle, is unique among fishes, though the Marsipobranchians of the genera *Geotria* and *Caragola* and the recent members of the clupeoid genus *Diplomystus** somewhat

---

* The genus *Diplomystus* was instituted by Prof. Cope (Bull. U.S. Geol. Survey Terr. 1877, p. 808) for the accommodation of certain fossil forms of Tertiary Clupeids from the Green River portion of the Wasatch Beds, which are situated in the central region of the United States, where it is numerous both in species and individuals. Three recent species are known. two of which—*novæ hollandiæ* and *sprattellulæ*—belong to the fauna of south-eastern Australia, and the third (*Clupea notacanthus*, Gunth.) to that of Chile. Not being aware of its earlier severance by Cope, I proposed (Records Austr. Mus. ii p. 24, 1892) to separate, under the name *Hyperlophus*, all those Herrings in which a predorsal serrature was present, but, my attention being kindly drawn thereto by Dr. Smith Woodward, I used Cope's name for Valenciennes' *Meletta novæ-hollandiæ* in a subsequent work (Edible Fish. and Crust. N.S Wales, p. 184, 1893). At present, however, I am uncertain whether *Diplomystus* can properly be retained for these forms, as Dr. Eigenmann in 1891 diagnosed the family *Diplomystidæ*—of which presumably the central genus is *Diplomystus*—for certain South American Nematognaths, and I have not as yet been able to learn the date of this genus; if, however, it is prior to Cope's the clupeoid fishes must take the name *Hyperlophus*.

closely approach it, but in other biological Classes a much more intimate geographical relationship between these Regions may be discerned.*

Several theories have been enunciated to account for this singular distribution of a family of fresh-water fishes in such widely separated regions as western South America, south-eastern Australia, and South Africa. Apparently the most favoured of these theories, as it is also the most natural and the most strongly supported by recent facts, is that, at some remote period of the world's history, there existed a great austral continent, which has now largely disappeared beneath the surface of the ocean and which extended northwards on the one hand through Tierra del Fuego to the southern and south-western parts of South America, on the other through Tasmania to south-eastern Australia, and possibly also to New Zealand and South Africa

So far as Australia and America are concerned I see no reason to doubt that they were at one time connected at their southern extremities by a belt of land stretching across the south pole, and that the antarctic continent so formed enjoyed a mild and equable climate, and supported a large and varied flora and fauna, the remains of which are abundantly visible in both to the present day, but especially in Australia, where forms of animal life, elsewhere extinct or nearly so, still constitute characteristic features in the faunic aspect, among which may be mentioned the *Marsupialia* among Mammals, the Struthionids among Birds, certain Lizards such as *Chlamydosaurus*, and Fishes such as *Neoceratodus*.

With regard to the claims of New Zealand and South Africa to a post-mesozoic junction with Antarctica the testimony is by no means so convincing, in fact the weight of evidence clearly points to the conclusion that at no more recent time was there any intimate connection between them, while there are many indications that the distance separating them was not so wide as

---

* For references see Hedley, Proc. Roy. Soc. N.S. Wales, 1895, p. 3, note 1.

5

to preclude the possibility of many plants and animals finding their way across "either by flight or drift."*

In the case of *Galaxias* the ova might easily have been carried across on the feet or plumage of water-birds, or, as seems to me a more simple and natural solution, some individuals having been swept out to sea by floods in their native rivers, have survived the passage across the intervening belt of ocean and successfully colonised the shores to which they wandered †

GALAXIAS FINDLAYI.

Galaxias findlayi, Macleay, Proc. Linn. Soc. N. S. Wales, 1882, vii p. 107.

B. ix. D. 12-13. A. 11-12.‡ V. 9. P 16. C. 16. Vert. 37-38/23.

Body stout to slender, the head broad and depressed. Length of head $4\frac{3}{4}$ to $5\frac{1}{2}$, depth of body $5\frac{1}{3}$ to 8 in the total length; width of body equal to or a little less than its depth, $1\frac{1}{3}$ to $1\frac{3}{4}$, of interorbital region $2\frac{4}{5}$ to $3\frac{1}{5}$, diameter of eye 4 to $5\frac{2}{5}$ in the length of the head; snouth obtuse, from three-eights to three-fourths of a diameter longer than the eye, which is very small. Lips thick and fleshy; the maxillary reaches to the vertical from the middle of the eye or not quite so far; lower jaw included. Seven or eight gill-rakers on the lower branch of the anterior arch. Jaws with a single series of moderate hooked teeth of somewhat irregular size, palatines with a similar series along their inner border directed inwards and backwards, a series of five strong hooked teeth on each side of the tongue and a single median tooth in front; vomer toothless. Dorsal fin obtusely pointed or rounded,§

---

* Hedley, l.c. p. 6.

† For an analogous example of colonization see Ogilby, Proc. Roy. Dublin Soc. 1885, p 529, re *Coregonus pollan*.

‡ The small rod-like rays in front being variable in number are not included, the computation being made from the first normally articulated ray.

§ In the largest example all the fins are rounded except the caudal.

the space between its origin and the base of the caudal $2\frac{2}{3}$ to $2\frac{1}{4}$ in its distance from the extremity of the snout; the fourth and fifth rays are the longest, $1\frac{4}{5}$ to 2 in the length of the head; the base of the fin is $1\frac{1}{10}$ to $1\frac{1}{5}$ in its height and $1\frac{1}{2}$ to $1\frac{3}{3}$ in the space between its origin and that of the anal: the anal fin is similar in shape to and originates beneath the last fourth of the dorsal; the fifth and sixth rays are the longest, as long as or a little longer than the dorsal rays; its base is $1\frac{1}{10}$ to $1\frac{1}{5}$ in its height, and 1 to $1\frac{1}{4}$ in its distance from the caudal: ventral inserted nearer to the anal than to the base of the pectoral, not reaching to beneath the dorsal fin; the distance between its origin and the base of the caudal is $1\frac{1}{10}$ to $1\frac{1}{4}$ in its distance from the tip of the snout; the middle rays are the longest, $1\frac{1}{2}$ to $1\frac{3}{4}$ in the length of the head and 2 to $2\frac{1}{4}$ in the distance between its origin and the anal: pectoral cuneiform, $1\frac{1}{5}$ to $1\frac{1}{2}$ in the head and $2\frac{1}{8}$ to $2\frac{2}{3}$ in the space between its origin and the ventral: caudal slightly emarginate with the lobes rounded, $1\frac{1}{5}$ to $1\frac{1}{3}$ in the length of the head, the peduncle rather slender and compressed, its depth $2\frac{2}{5}$ to $3\frac{1}{5}$ in its length.

Colours variable: from dark greenish-brown above and yellowish-brown below, the sides with more or less distinct darker markings, which may take the form of irregular transverse bands, or of minute spots, which again may be concurrent so as to form blotches or may be distributed so as to almost obliterate the ground-colour, generally with a more or less well defined series of dark spots along the middle of the body, with the fins shading from yellowish-brown basally to orange distally; to golden with regular transverse bands or large blotches of a black or dark chestnut colour, with the fins yellow. Irides silvery.

In addition to the above, the Rev. Mr. Curran tells me that there is in the living fish "over the eye a crescent-shaped area coloured reddish like metallic copper"; that the opercles "are metallic gold and green," and that the sides are irradiated with "peacock hues." As to its habits he reports it as being "very sprightly and lively," and hiding cunningly under stones or in holes in the bank when pursued; also that it leaps to the fly, and can be easily caught in this way."   "I saw some stockmen amusing themselves in this manner, the whole outfit consisting of a piece of black thread, a bent pin, and a fly."

Distribution :—Streams and tarns on Mount Kosciusko and the neighbouring uplands, including the head waters of the Snowy River and its tributary, the Crackenback, where they were obtained by Messrs. Curran and Hedley.   Later on the former gentleman obtained specimens from the streams draining the northern and western slopes of Kosciusko and flowing into the Murrumbidgee.   Spawning in February

Eleven specimens measuring from 63 to 105 millimeters, were utilised in drawing up the above description.

Appended is a list of the species of *Galaxias* at present known, arranged in chronological order :—

1801 *alepidotus*, Forster, Bloch and Schneider, Syst. Ichth. p. 395, New Zealand.

1817. *truttaceus*, Cuvier, Règne Anim. ii. p. 283; Tasmania and Victoria.

1842. *fasciatus*, Gray, Zool Misc. p 73; New Zealand.

1842. *maculatus*, Jenyns, Zool. Beagle, Fish. p. 119, pl. XXII. f. 4; Patagonia, Tierra del Fuego.

1842 *alpinus*, Jenyns, l c. p. 121; Alpine Lakes of Hardy Peninsula, Tierra del Fuego.

1842 *attenuatus*, Jenyns, l c. pl. XXII. f. 5; New Zealand, Tasmania, Victoria, Falkland Islands, Western South America northwards to Peru.

1846 *scriba*, Cuvier and Valenciennes, Hist. Nat Poiss. xviii. p. 347; Port Jackson, New South Wales.

1864. *gracillimus*, Canestrini, Arch. Zool. Anat. e Fisiol. iii. p. 100, pl. IV. f. 2; Chili.

1866. *ocellatus*, McCoy, Intern. Exh. Ess. p. 14; River Yarra, Victoria

1866. *olidus*, Günther, Catal. Fish. vi. p 209; New Zealand.

1866. *kreffti*, Gunther, l.c. p. 211; New South Wales.

1866 *punctitus*, Günther, l.c. p. 213, New South Wales.

1866. *brevipinnis*, Günther, l.c.; New Zealand.

1867. *waterhousei*, Krefft, Proc. Zool. Soc. Lond p. 943; South Australia.

1869. *schomburgkii*, Peters, Monatsb. Ac. Wiss Berlin, 1868. p. 455; Adelaide, South Australia

1872. *rostratus*, Klunzinger, Arch. f. Nat. p. 41; Murray River.

1872. *versicolor*, Castelnau, Proc. Zool. Soc. Vic. i. p. 176, Marsh near St. Kilda, Victoria.

1872 *cylindricus*, Castelnau, l.c p. 177, Lower Yarra, Victoria.

1872. *delicatulus*, Castelnau, l.c. p. 178; River Yarra, Victoria

1872. *amœnus*, Castelnau, l.c.; River Yarra, Victoria.

1873. *ornatus*, Castelnau, l.c. p. 153; Cardinia Creek, Victoria.

1880. *campbelli*, Sauvage, Bull. Soc. Philom. (7) iv. p 229; Campbell Island.

1880. *coxi*, Macleay, Proc. Linn. Soc. N. S. Wales, 1880, v. p. 45; Mount Wilson, New South Wales.

1881. *coppingeri*, Gunther, Proc. Zool. Soc. Lond. p. 21, Alert Bay, Straits of Magelhaen.

1881. *planiceps*, Macleay, l.c. vi. p. 233; Rankin's Lagoon, Bathurst; New South Wales.

1881. *bongbong*, Macleay, l.c.; Mossvale and rivers at Bongbong; New South Wales.

1881. *nebulosa*, Macleay, l.c. p. 234; Long Bay, Sydney, New South Wales.

1882. *findlayi*, Macleay, l.c. vii. p. 107; Streams on Mount Kosciusko, New South Wales.

1882. *auratus*, Johnston, Proc. Roy. Soc. Tas. p. 131; Great Lakes, Tasmania.

1882. *weedoni*, Johnston, l.c.; Mersey River, Tasmania.

1882. *atkinsoni*, Johnston, l.c.; Pieman River, Tasmania.

1886. *kayi*, Ramsay and Ogilby, Proc. Linn. Soc. N. S. Wales (2) i. p 6; Fifth Creek, Adelaide, South Australia.

1888. *indicus*, Day, Fish. Ind. Suppl. p. 806, fig.; Littoral districts of Bengal and Madras.

1892. *nigothoruk*, Lucas, Proc. Roy. Soc. Vic. (2) iv. p. 28; Lake Nigothoruk, Gippsland, Victoria.

1894. *capensis*, Steindachner, Ichth. Beitr. (xvii.) p. 18; Lorenz River, South Africa.

There can be little doubt that many of the species, 32 in number, here catalogued are merely nominal, but, though detailed descriptions of each would doubtless prove of great assistance in indicating the various degrees of affinity which connect the local forms with their antarctic progenitors, it is plainly impossible to even approximately delimitate the species in a satisfactory manner, until a full series of each variety or subspecies shall have been brought together for examination. The local variations in the same form inhabiting the same little subalpine runlets is shown to be so great, as is manifest by the study of the present species, that the wonder is, not that so many nominal species have been created, but that there are not infinitely more.

This perplexing number of local varieties finds its analogue in the common Brook Trout of the North of Ireland,* where every angler is well aware that the Trout from each stream differ so greatly in outward appearance from those inhabiting the next,

---

* I only mention this locality because it was there that I observed the local differences in *Salmo fario*, but no doubt sportsmen from other districts can testify to the accuracy of the above remarks. Salmon also vary much in different rivers, and even when taken in nets set in the sea many miles from the mouths of the rivers in which they spawn, the individuals belonging to each water way can be unhesitatingly selected (*vide* Ogilby, Proc Roy. Dublin Soc. 1885, p. 526).

that, to any one who knows the waters, the fish from any given stream may be selected at a glance from those of a dozen other streams, but no one now-a-days would venture to assert that they were of different species, even were it not well known that on being transferred from one stream to another the colonists soon assume the characteristics of the local race.* These variations are attributable (in both genera, *Galaxias* and *Salmo*) to similar local causes, such as the depth, stillness or rapidity of the water, the quality and the supply of food, the character of the bottom, the composition of the water, &c.; indeed as to the latter trout taken from streams fed from limestone springs are as different from those residing in waters which have their origin in peat mosses as *Galaxias truttaceus* is from *G attenuatus.*

As to the affinities of the species, it is useless in the present state of our knowledge to attempt any generalisation, and it is only by obtaining a series of specimens from the localities whence they were originally described that such species as Castelnau's and (in a less degree) Macleay's can be with certainty identified; nevertheless the following corrections and suggestions may be of use :—

Galaxias olidus, Günth., doubtfully attributed by that author to Queensland, proves to be a New Zealand species, and must be erased from the number of Australian fishes.

Galaxias waterhousei, Krefft, is a variety of *G. attenuatus* according to Klunzinger, as is also *G. obtusus*, Klunz. (Sitzb. Ak. Wiss. Wien, 1879, lxxx. i. p. 412) I mention this latter fact because Lucas includes both *attenuatus* and *obtusus* in his "Census of Victorian Fishes, 1889";† although Klunzinger had himself pointed out his own error (*l.c.*), while he omits *truttaceus* which that author had received from "Port Phillip." G. *schomburghii*, Peters, and *G hayi*, R and O. are possibly varieties of *Waterhousei*.

---

* This does not apply with equal force to the anadromous Salmonids.

† Proc. Roy. Soc. Vic. 1889, pp. 15-47.

Galaxias nebulosa, Macleay, is probably the same as *G. scriba*, Val. The variation in the number of the dorsal and anal rays cannot be considered of any value in this genus if the small unarticulated anterior rays be included, the number of these being extremely inconstant; there is no other character of sufficient consequence to warrant their separation except the size of the eye, which is stated by Valenciennes to measure " two-fifths of the length of the head," a proportion which is quite unknown among the members of the genus, and is very unlikely to be correct. *G. rostratus*, Klunz., should also be compared with *scriba*.

Galaxias auratus, Johnston. Through the courtesy of Mr. Alexander Morton of the Tasmanian Museum, I have had an opportunity of examining two fine examples—225 and 185 millimeters in length—of the form inhabiting the Great Lakes, Tasmania, which lie at an altitude of 4000 feet above the sea level. These specimens I believe to be mere varieties of *G. truttaceus*, modified by their surroundings.

Galaxias indicus, Day. From the first I looked with distrust on the possibility of the occurrence of a species of this genus in Indian waters, and I am, therefore, pleased to find that Dr. Gill not only shares that distrust, but has had the courage to publish his disbelief (Nature, liii. p. 366). Should the species on further examination prove to be a true *Galaxias*, its presence in the Indian littoral would seriously interfere with the theory of an antarctic origin for these fishes.

It will be observed that no less than seven species (*truttaceus, attenuatus, ocellatus, versicolor, cylindricus, delicatulus,* and *amœnus*) are said to be resident in or in the immediate neighbourhood of the Yarra, and since the two first are well known to be of wide distribution and variable appearance I must be permitted to doubt the specific value of all or most of the residual forms, for none of which have their authors pointed out such distinctive characters as would enable one, from a study of the descriptions alone, to determine their specific value. Too much importance has evidently been placed by Australian authors (I might perhaps with equal truth say by all authors) on the shape and colour of these fishes, both of which characters I have shown above to be quite worthless in distinguishing the species.

Finally I am not satisfied, notwithstanding my scepticism with regard to the number of Australian species, to accept as proved the identity of the New Zealand and Tasmanian *attenuatus* with the Falkland Island and Peruvian form, referred to by Günther under the same name, nor am I prepared to go as far as Macleay in considering that "it is more than probable that they"—all the known forms of *Galaxias*—"are one and all only permanent local varieties of the same fish."

THE ENTOMOLOGY OF THE GRASS-TREES (*XANTHORRHŒA*).

By Walter W. Froggatt.

(Plate ix.)

Four species of *Xanthorrhœa* are recorded from the County of Cumberland, within the limits of which all my entomological specimens have been collected; as their general structure is similar, it is not surprising that the same species of insects are to be found frequenting all four alike.

At first sight a grass-tree might not appear to be a profitable field for investigation by the entomologist; yet whether alive or dead it is the home of a considerable number of interesting insects, some of which are born and die in it, while others are only passing visitors. A grass-tree presents three distinct parts, each with its special frequenters; first the stout cylindrical stem or trunk, generally two or three feet high, and consisting of a tubular sheath composed of the basal portion of the fallen leaves matted together into a solid ring, and thickly impregnated with the yellow resinous gum, and in which nothing lives; this encloses the caudex, composed of close fibrous matter, which in a living tree contains nothing, but after death it decays very rapidly, and soon becomes the abode of much insect life, for which the outer covering or sheath forms a protection. Secondly, there is the coarse grass-like foliage which is the resort of many small beetles, spiders, &c., which lurk about the bases of the stalks; it is also eaten by several beetles and is visited by others. Thirdly, the flower-stalk and scape which both alive and dead furnishes food or a home to certain beetles, bees, and ants.

As the grass-trees generally thrive best in poor sandy country covered with low scrub, great numbers are scorched up by the bush fires every season. It is in such burnt patches that most of the grass-trees examined by me occurred.

COLEOPTERA.

MICROPŒCILA BREWERI, Janson.

Larva about 1½ inches in length; white, rather elongate and cylindrical; head reddish-brown, rugose, rounded behind, slightly impressed in the centre with a wavy line running across on either side to the base of the antennæ; stout black jaws armed with three small blunt teeth; a broad elongate brown patch on either side of the first thoracic segment, above the first spiracle, legs long, covered with long ferruginous hairs; thoracic segments and first seven abdominal segments furrowed into three ridges covered with short dark spiny bristles, together with a transverse row of longer hairs across the tip, 8th segment smooth and shining, covered with scattered short spines, and tinged with blue from the internal food, the anal segment rounded at the tip.

Beetle 10½ lines in length, all the underside, legs, head, the centre of the thorax and elytra smooth, shining black, with a broad marginal band encircling the thorax and elytra deep orange yellow, sides of the wing-covers showing shallow punctured parallel striæ.

Near Hornsby I obtained a large number of larvæ early in July from a patch of dead grass-trees in which they were living in the rich black vegetable mould into which the inner portion of the caudex had been transformed by the action of the weather and their jaws. Towards the beginning of May they began to form earthy oval cocoons on the bottom of the tin, where they remained until the end of November, when the beetles began to come out.

The beetles are found with many others feeding upon the flowers of the dwarf Angophora.

CISSEIS 12-MACULATA, Fab.

I have never found the larva of this pretty little buprestid, and do not know anything about its life-history, but the beetle is common about Sydney in early summer, feeding upon the leaves, clasping the foliage with its legs, but dropping to the ground at the least alarm.

Beetle 5 lines in length, with the head bright metallic green, thorax and elytra of a much darker tint, the whole deeply and closely punctured; sides of the thorax ornamented with a pale buff patch on either side, with four transverse rows of the same coloured oval spots, the first and last containing two and the middle ones four each, undersurface of a bright metallic green, with a patch of buff below the hind legs, and at the margin of each abdominal segment.

TRIGONOTARSUS RUGOSUS, Boisd.

(Plate IX., figs. 1-3.)

Larva with smooth castaneous head; thoracic segments pale reddish-brown and not more than half as thick as the centre of the pale yellow abdominal ones, which are generally arched up behind the head; length in repose about an inch, but when moving about it extends its body half as far again; thoracic segments rather flattened upon the dorsal surface, with the abdominal ones of a uniform length and very much wrinkled, anal one terminating with two short truncate tubercles of a reddish-brown colour, with several smaller ones round them.

The larvæ bore holes into the fibrous caudex near the bottom of the trunk of the grass-tree, where they must feed for some time, as I have taken the larvæ, pupæ, and beetles in the same tree about the middle of the year.

Pupa 14 lines in length, white to dull yellow in colour; snout very thick, and curved straight down over the breast, both it and the rest of the head lightly clothed with stout bristles, which also extend over the sides of the thorax; wing-cases drawn round the shoulders, short, and rounded at the tips, and deeply and regularly striated; thoracic segments bearing a transverse ridge of coarse irregular spines across the centre of each segment except the anal one, which is ornamented with a crescent-shaped mark turning downwards, clothed with a few scattered hairs.

Beetle is 16 lines in length, stout and rather flattened on the back, of a uniform black colour, with the broad head and thorax finely rugose, the elytra being deeply ridged with regular punctured striæ. The curious form of the tips of the tibiæ which terminate in a long slender spine projecting beyond the tarsi enables it if touched to cling very tightly to anything when laid upon its back.

ACANTHOLOPHUS MARSHAMI, Kirby.

This is the common Amycterid about the neighbourhood of Sydney. Most of the members of this large genus live upon the grass, but this one climbs up the leaves of the grass-tree, and clinging round them gnaws pieces out

Beetle slightly under an inch in length, of a sooty-brown colour; the head stout, an angular spine on either side between the antennæ, a stout double pointed knob in front of each eye, and the antennæ and mouth parts hairy, thorax rather oval, flattened on the summit but very rugose, with three stout conical spines along the outer margins, and two irregular lines of shorter ones divided by the stout median suture, legs stout, with tibiæ and tarsi hairy; elytra broad, flattened on the summit, the sides transversely corrugated, the upper margins ornamented with an irregular line of large conical spines and numerous smaller ones covering the whole of the back; abdominal plates beneath covered with fine silvery scales or hairs.

TRANES sp.

Beetle 6 lines in length, all black; head small; snout long and stout, antennæ thick at the tip; thorax rounded in front, the sides flattened on the summit and thickly covered with fine circular punctures; legs short and strong; dark ferruginous, with the tarsi lighter coloured; elytra much broader than thorax, which is arched slightly in front, flattened on the back, and thickly ribbed with parallel deeply punctuate striæ.

This beetle is not very common; it occurs towards the base of the flower stalk and the young leaves. My specimens were obtained from trees at the Hawkesbury.

SYMPHYLETES SOLANDRI, Fabr.

The life-history of this fine longicorn is given by me in detail
in the Proceedings of this Society (Vol. ix. (2), p. 115, 1894).
Though not generally a very common beetle unless in an exceptional season, it is one that is very easily bred from infested
flowerstalks if kept in a box.

XANTHOLINUS ERYTHROPTERUS, Erichs.

(Plate IX., figs. 4-5.)

Larva slender, flattened, $7\frac{1}{2}$ lines in length, with the head,
prothorax, and legs ferruginous, the rest of the thoracic and all
the abdominal segments pale yellow, lightly fringed with hairs;
head longer than broad, rounded behind, and armed with long
slender black jaws; antennæ 4-jointed, 2nd and 3rd joints long,
slender, and swollen at the apex, 4th shorter and rounded at the
tip, prothorax rounded in front, truncate behind, both head and
thorax with a slight median suture; legs short and thick, with
slender tarsal claws; abdominal segments uniform with metathorax, the anal one tapering to the tip and armed with a slender
hairy appendage on either side.

Pupa is a tightly swathed ferruginous bundle, the thoracic
portion forming a roof-like covering over the turned down head,
the legs in front, the hind pair forming a rounded projection in
front of the upper abdominal segments, which are round and
cylindrical to the tip.

Beetle six lines in length, all smooth, shining, black, except
the wing covers, which are bright reddish-brown; head rounded,
much broader than the thorax, deeply impressed above the long
sickle-shaped jaws, and lightly fringed in front with reddish
hairs; antennæ with brownish pubescence, the terminal joint of
palpi ferruginous; thorax broadest in front, sloping on either side,
and rounded at apex, lightly fringed with blackish hairs; legs
short and spined, thickly covered with blackish hairs; elytra
finely punctuate, broadest at apex, truncate; abdomen rather

short, thickly fringed and lightly covered upon both sides with long blackish hairs; first four segments of uniform size, fifth nearly twice as wide and tapering to the small anal segment.

The larvæ are plentiful in spring between the sheath and the caudex, preying upon the many minute creatures attracted by the decaying matter. Like others of the *Staphylinidæ*, the beetles are very active, and are found in the same stumps with the larvæ; the pupa bred out in the Museum under glass in some damp earth.

HOLOLEPETA SI NENSIS, Marsham.

This is one of the commonest beetles found in the top of the decaying caudex, or between it and the outer sheath. Though I have examined great numbers of the stems at all seasons of the year, I have never come across the larval or pupal forms.

Beetle half an inch in length, smooth, shining black, broad and flat; the head armed in front with two curved stout pointed horns projecting in front of the eyes and touching at the tips, hollowed out in front at base of horns, with an excavation behind the eyes, and a small blunt spine on the side; thorax with a faint impressed line in the centre, and along the outer edges slightly pitted with small punctures; elytra without any punctures, but a slender purse-like cavity on either margin caused by the edge of the elytra turning upwards; chitinous plates covering the apex of the abdomen impressed with larger rounded punctures on their edges; underside except the central plate between the legs also finely punctured.

I have never collected this species any where else, though others in the north are often found crawling on tree trunks.

PLATYSOMA sp. ?

This beetle evidently passes through all its transformations in the decaying caudex, but after examining a great number of plants in all stages of decay, and at all seasons of the year, I have never been able to identify the larva, though once or twice I have found the pupa just ready to turn into the perfect insect,

from which it only differs in colour, being dull white. The beetles
are often very numerous, twenty or thirty being obtained from
one stump.

Beetle 1½ lines in length, broad and oval, black and shining;
head small, round in front, thorax smooth, truncate behind;
elytra smooth in the centre, with four very distinct striæ on each
side, and truncate at the apex; the tip of the abdomen sloping
downwards.

ALLECULA SUBSULCATA (?), Macl.

Larva is a typical heteromerous wire worm; slender, cylindrical,
smooth, and shining, about an inch in length, of a uniform ochreous
colour; head and tip of the abdomen ferruginous, and an apical
narrow band round the abdominal segments dark brown, head
small, rounded in front, with slender sickle-shaped jaws, short
antennæ, and long drooping palpi; legs are comparatively long,
with slender tarsal claws.

They are very active little creatures, living in the rich black
mould left by the decaying caudex; sometimes they are very
numerous, common in July and August.

Pupa pale yellow, short and angular, with the head drawn down
over the thorax, antennæ curling round under the fore legs, and
coming over the hind ones, labial palpi projecting over the fore
legs and showing the peculiar axe-shaped terminal joint; outer
edges of the abdominal segments flanged and finely serrate, the
anal one terminating in two fine spines, wing cases short and
wrinkled.

Beetle 7 lines in length, all black, except the last three joints
of antennæ and last two joints of the tarsi, which are pale
ferruginous; head and thorax closely and finely punctured;
antennæ 11-jointed, long, slender, and cylindrical, 2nd joint very
short, 3rd longest, apical joint of the labial palpi large and axe-
shaped; legs long, apex of tibiæ and the tarsi clothed with fine
reddish hairs; elytra rugose and deeply grooved with parallel striæ,
thickly and deeply punctured; all the ventral surface closely
punctured.

The beetles began to emerge from the earth, in which the larvæ had buried themselves, about the middle of November.

They are often found in the summer time hiding among the dead leaves among the bushes or clinging to the twigs.

HYMENOPTERA.

Lestis bombiliformis, Smith.

This beautiful carpenter bee forms its nest in the flower stalks of the grass-trees found about Sydney, after they have borne the flower and have become dry and hard. It begins by boring a circular hole, $3\frac{1}{2}$ lines in diameter, about three or four feet up the stalk, in towards the centre, when it turns downwards, excavating nearly all the pith out for a distance of about four inches down, then working upwards, so that the tunnel is about eight inches from end to end, with an average of half an inch in diameter. The cells are made about half an inch in length, with a ball of bee-bread and an egg deposited in the far end, each being partitioned off from the other by a stout pad or wad of triturated pith. I have never found the whole length of the chamber filled with bee larvæ, a space being usually left unoccupied in the centre.

Larva a dull white-coloured grub of cylindrical shape, attenuated towards both extremities, about half an inch in length when full grown. They can be found in all stages about November.

♂. Bee $7\frac{1}{2}$ lines in length, bright metallic green, with the face yellow, eyes brown; antennæ, ocelli, and mouth parts black, sides of the face, back of head, thorax and legs thickly covered with short golden yellow hairs, with three dark parallel bars of blackish hairs crossing the centre and on either side; above the wings clouded with brown, covered with fine brown spots over the marginal cells, and having fine metallic purple iridescence, upper surface of the abdominal segments finely rugose, without hairs; under surface covered with dark brown hairs, the tip with black.

♀. Bee 9 lines in length, of a brilliant metallic blue colour, with the abdominal segments showing coppery tints, face and

6

head behind the eyes covered with greyish white hairs, thorax, legs, and under surface of abdomen thickly clothed with black hairs except the sides of the anal segments, which are fringed with white hairs; wings darker than in the male

Mr. F. Smith gave a short account* of the habits of this bee, communicated to him by Mr. Ker, who stated that it inhabited the hollow stem of a Zamia or grass tree, the entrance to the tube being rounded like the mouth of a flute.

Dolichoderus doriæ, Emery.

These ants are very common about Hornsby, and are very fond of the sweet sugary lerp formed upon the leaves of the Eucalypts by the larvæ of several species of *Psylla*, so that where the lerp is plentiful the leaves are often covered with them, all intent upon the enjoyment of their sweet food.   They form their nest between the caudex and dry outer sheath of the dead and dry grass trees, often in such numbers that the cavity between the caudex and the outer mass is a living mass of ants.

Ant ♀, 4 lines in length, head and thorax black, very rugose, the latter armed with a pair of stout spines projecting in front of the prothorax, with a similar pair at the base of the metathorax, longer and pointing downwards; antennæ and legs ferruginous, the node short but stout, abdomen black, covered with a brownish pubescence, heart-shaped, hollowed out in front down the centre, with the outer margins rounded and forming regular rounded tips.

Iridomyrmex gracilis, Lowne

A small slender black ant that makes its nest in the dead flower stalks of the trees, hollowing out the interior in irregular parallel passages, a large nest of them often occupying the whole stalk.

♀. Ants are under 3 lines in length, pitchy brown, with very long slender legs covered with a very fine grey pubescence; head

---

* Notes on the Habits of Australian Hymenoptera, Trans. Ent. Soc. London, Vol. i. (2nd Ser.) p. 179, 1850

large, smooth, and shining, truncate at the base, and rounded towards the jaws; thorax narrow, smooth and shining; abdomen short, rounded and pointed towards the tip.

DIPTERA.

ORTHOPROSOPA NIGRA, Macq.

(Plate IX., figs. 6-8.)

Larva 8 lines in length, dirty white to brownish, rounded at the head, widest about the centre, tapering towards the tip of abdomen which is produced into a stout horny ochreous appendage truncate at the tip and armed at the base with a short fleshy spine on either side.

The maggots, frequently in great numbers, are found living in the slime and putrid water which accumulates between the outer shell and the caudex of the dead stem, about midwinter, numbers kept under observation remained about six weeks before changing into pupæ. The latter were simply the skin of the maggot hardened into a brown oval case covered with particles of earth attached to it, and the anal appendage shortened and retracted.

This handsome fly (one of the *Syrphidæ*) is 7 lines in length, shining black, with the antennæ and face bright yellow; thorax covered with a very short fine blackish down and ornamented with a pair of rounded naked black spots in the centre; wings slightly fuscous, legs black; abdomen stoutest at the base, rounded towards the tip.

ORTHOPROSOPA sp.

(Plate IX., figs. 9-11.)

Larva dirty white, 10 lines in length, but able to retract or extend its segments considerably; head rather truncate in front, with the sides round, narrow, with segments of uniform size, tapering towards the tip which is produced into a slender fleshy tail; two-thirds of the length of the whole of the body terminating in a slender horny tube or spine, truncate at the tip.

The larvæ live in the decaying wood and putrid water that has
accumulated between the caudex and the sheath, crawling about
mixed up with the maggots of the last described species, sometimes
in considerable numbers.   Specimens kept in a damp jar pupated
among the rotten wood at the bottom about three weeks after they
were taken.   Pupa case light brown, covered with bits of dirt;
the apex and sides rounded, oval, with the long slender anal
segment produced into a slender tube curving sharply round, and
retaining the anal tube at the tip.

Fly 5 lines in length, steely blue, thorax and abdomen smooth
and shining; face and antennæ covered with fine hairs, the latter
short with the last segment oval and flattened, ornamented with
a fine bristle; legs piceous, covered with fine hairs; wings hyaline,
very slightly clouded.

EPHIPPIUM ALBITARSIS (?), Bigot.

(Plate ix., figs. 12-13.)

Larva 8 lines in length, 2 in width, varying from greyish-
brown to black; head much narrower, slender, horny, broadest at
the base, sloping up to a truncate tip, with an eye-like spot on
either side, and several short bristles along the sides, the mouth
concave; thoracic and abdominal segments broad, convex on both
dorsal and ventral surfaces, the hind margin of the first five
sloping back, first arcuate behind the head, narrow, the following
ones gradually increasing in size to the fourth, and of a uniform
width to the ninth, tenth smaller, the last spatulate, with a round
impression on the dorsal surface; outer margins of each segment
fringed with two long bristles, a few scattered ones over the
dorsal surface.

The pupa undergoes its transformation in the larval skin, the
fly emerging from the base of the head.   They are plentiful in
decaying stems between the caudex and sheath, living among the
rotten matter, and are very sluggish in their habits.   Specimens
I collected remained among some rotten wood and mould for

about three months before the flies began to emerge about the end of September.

Fly varying from 4½ to 3 lines in length, all black except the white tarsi; head broad, rugose between the eyes; antennæ spindle-shaped, pointed towards the tips, standing straight out, without any terminal bristle; thorax rounded in front, broadest about the middle, finely granulated on the dorsal surface; scutellum almost square, the apical edge having a short spine on either side; legs stout; wings dusky, nervures black, the wings creased in the centre and folded down over the tip of the abdomen; the latter constricted at the base, large and round, finely granulated, with the apical segments turning downwards, and the extreme tip truncate.

This is a typical form of the family *Stratiomyiidæ*, and is, I believe, identical with Bigot's *C. albitarsis*, one of the few described Australian species

Another very pretty little fly also lives in the rotten caudex, the larvæ of which I have never observed, but have bred several from the pupæ, which are oval brown cases covered with particles of earth, the front broadest, with a cylindrical short truncate spine on either side, standing out like a little horn, the apical tip rather pointed.

The fly, which belongs to the family *Trypetinæ*, is often found upon the leaves, moving its wings up and down (as many members of this family do when resting), but is very hard to catch ; common in November.

Fly 3 lines in length; head black, narrow; last joint of the antennæ large and circular, terminated with a stout bristle; head and thorax hairy, the latter steely blue; scutellum large, yellow, with black markings on the apical edge which is truncate and fringed with hairs; legs long, pale yellow; wings hyaline, thickly mottled with irregular black blotches over the apical half, abdomen broad, heart-shaped, pale ochreous yellow, rounded on dorsal surface, with a curious imprinted brown mark in centre, thin and flat on the underside, tinged with black towards the tip, and tufted with silvery white hairs on the sides.

LEPIDOPTERA.

APHOMIA LATRO, Zeller.

Larva half an inch in length, dark brown to black upon the dorsal surface, with lighter parallel stripes down the centre of back, and along each side; head large, smooth, shining, and divided in the centre by a suture; prothorax rounded and large; other thoracic segments uniform with the abdominal ones; legs moderately stout, with small pointed tarsal claws; ventral surface pale yellow.

The larvæ live in small communities, feeding upon the scape of the flower stalk, gnawing up all the undeveloped buds, which become matted together with their loose web. They move about very rapidly, and pupate on the flower head, forming elongate white silken cocoons.

Pupa long and slender, reddish-brown, with the wing-cases curving round in front and covering the first five segments; a raised ridge running down the centre of back; anal segment armed with a number of short conical spines.

Moth $1\frac{1}{2}$ inches across the wings, which are long and slender, and rounded at the tips; creamy buff colour shot with fine black spots, and divided down the centre with a broad parallel stripe of white. Hind wings silvery grey, thickly fringed with long semi-opaque hairs along the tips and lower margin; body slender, apical segments darkest.

Mr. Ernest Anderson, who identified this species for me, says that it is common in Victoria, where it also feeds upon grass-trees and stems of rushes Bred in the Museum about the end of October, from infested flowers received from the Curator.

HOMOPTERA.

ASPIDIOTUS ROSSI, Mask.

The foliage is often quite discoloured with the number of black scales (adult females) infesting the leaves, often overlapping each other like a lot of oyster shells.

CHIONASPIS EUGENIÆ, Mask.

I found this scale very plentiful upon the leaves of a patch of grass-trees last March at Botany, but it is more generally found upon *Leptospermum*, *Melaleuca*, and *Eugenia*. The adult female coccids are pale yellow at the tip, with the long slender test pearly white, and are attached along the outer edge of the under-surface of the leaves.

EXPLANATION OF PLATES.

Trigonotarsus rugosus, Boisd.

Fig 1.—Larva (nat. size).
Fig 2.—Larva—front view of head (enlarged).
Fig. 3.—Pupa (nat. size).

Xantholinus erythropterus, Erichs.

Fig. 4.—Larva (enlarged).   The line beside shows the length.
Fig. 5.—Pupa (enlarged).   The line beside shows the length.

Orthoprosopa nigra, Macq

Fig 6.—Larva (enlarged).
Fig. 7.—Pupa (enlarged)
Fig. 8.—Fly (enlarged).

Orthoprosopa sp.

Fig 9.—Larva (enlarged).
Fig 10.—Pupa (enlarged)
Fig. 11.—Fly (enlarged).

Ephippium albitarsis (?), Bigot.

Fig. 12.—Larva (much enlarged).
Fig. 13.—Fly (enlarged).

NOTES AND EXHIBITS.

Mr. North exhibited the types of the new genus and species of birds obtained by the members of the "Horn Expedition" in Central Australia, and described by him in the July number of "The Ibis" for 1895, also more fully in the "Report of the Horn Scientific Expedition," Part II Zoology, just published. The genus *Spathopterus* formed for the reception of the Princess of Wales' Parrakeet is a most extraordinary one. The fully adult male, of which a beautiful specimen was exhibited, has the end of the third primary prolonged half an inch beyond the second and terminating in a spatulate tip. It is entirely different from the wing of any other bird found in Australia, but the peculiar terminations of the third primaries resemble somewhat the tail-like appendages to the lower wings of the Queensland butterfly *Papilio ulysses*. The new species comprised the following :—*Rhipidura albicauda, Xerophila nigricincta, Ptilotis keartlandi, Climacteris superciliosa, Turnix leucogaster,* and *Calamanthus isabellinus,* a sub-species of *C. campestris,* Gould.

Mr. Hedley exhibited on behalf of Mr. J. Jennings some living *Strombus luhuanus* from Vaucluse. As none had been observed alive for several years it had been feared that this interesting colony, the most southern recorded of this species, had become extinct, a fear happily now shown to be unfounded.

Mr. Rainbow showed a Sydney spider (*Celæria excavata,* Koch) which mimicks the excreta of a bird. Also examples of the egg-bags of the same species, which in appearance resemble the kernels of the Quandong *(Fusanus).*

Mr. Froggatt exhibited specimens of the insects frequenting the four species of *Xanthorrhœa* to be found in the County of Cumberland, together with drawings illustrative of the life-history of some of them. Also a living specimen of the "Thorny Lizard" *(Moloch horridus,* Gray), received by post from Kalgoorlie, W.A. Mr. Froggatt likewise communicated some observations on the habits of this specimen.

Mr. Pedley also exhibited a living specimen of *Moloch horridus* from West Australia.

Mr. Lucas showed a fossil fish in Wianamatta Shale from Marrickville.

WEDNESDAY, MAY 27TH, 1896.

The Ordinary Monthly Meeting of the Society was held in the Linnean Hall, Ithaca Road, Elizabeth Bay, on Wednesday evening, May 27th, 1896.

The President, Henry Deane, Esq, M.A., F.L.S., in the Chair.

Mrs. Agnes Kenyon, Richmond, Victoria, was elected an Associate Member of the Society.

The Special General Meeting, of which notice had been given, was postponed.

DONATIONS.

Pharmaceutical Journal of Australasia. Vol. ix. No. 4 (April, 1896). *From the Editor.*

Société d' Horticulture du Doubs, Besançon—Bulletin. Sér. Illustrée. No. 3 (March, 1896). *From the Society.*

U.S. Dept. of Agriculture — Division of Ornithology and Mammalogy—Bulletin. No. 8: Division of Entomology — New Series. Bulletin. No 3. Technical Series. No. 2. *From the Secretary of Agriculture.*

Asiatic Society of Bengal—Journal. N.S. Vol. lxiv. (1895). Part i. No. 3 ; Part ii. No. 3. *From the Society.*

Royal Society of Victoria—Proceedings (1895). Vol. viii. (New Series). *From the Society.*

Geelong Naturalist. Vol. v. No. 3 (April, 1896). *From the Geelong Field Naturalists' Club.*

K. K. Zoologisch-botanische Gesellschaft in Wien—Verhandlungen xlvi. Bd. Jahrgang 1896. 2 Heft. *From the Society.*

Bureau of Agriculture, Perth, W.A.—Journal. Vol. iii. Nos. 7, 8, 10 and 11 (March-May, 1896). *From the Secretary.*

Pamphlet entitled "Sur la Deuxième Campagne Scientifique de la Princesse Alice.' Par S. A. S. Albert 1ᵉʳ., Prince de Monaco *From the Author.*

Papuan Plants. No. ix.; Iconography of Candolleaceous Plants. First Decade (1892) By Baron Ferd. von Mueller, K.C.M.G, M. & Ph D., LL.D., F.R S. *From the Author.*

Museo de la Plata—Anales. i. (1890-91), Seccion de Arqueologia ii -iii. (1892); Seccion Geologica y Mineralogica. i (1892), Seccion de Historia General. i. (1892), Seccion Zoologica. i.-iii (1893-95); Paleontologia Argentina. ii.-iii. (1893-94): Revista T. i.-v. (1890-94). T. vi Primera Parte (1894). T. vii. Primera Parte (1895): Pamphlets entitled "The La Plata Museum" By R. Lydekker, B.A, F Z S.; and "Le Musée de La Plata" Par F. P. Moreno. *From the Director.*

Gordon Technical College, Geelong, Victoria—Annual Report for 1894. *From the Secretary.*

Journal of Conchology Vol. viii. No. 6 (April, 1896). *From the Conchological Society of Great Britain and Ireland.*

Entomological Society of London—Proceedings, 1896. Part i *From the Society.*

Muséum d' Histoire Naturelle, Paris - Bulletin. Année 1896. No. 1. *From the Museum.*

Zoologischer Anzeiger. xix. Band. Nos. 499-500 (March-April) 1896). *From the Editor.*

Societas pro Fauna et Flora Fennica—Acta. Vol. v. Pars iii , Vols. viii -x.; Vol xii. (1890-95): Meddelanden. 18-21 Haftet (1892-95): Herbarium Musei Fennici. Ed. 2. Pars ii. (1894). *From the Society.*

Naturhistorischer Verein der preussischen Rheinlande, &c. Bonn—Verhandlungen. Jahrgang li. Zweite Halfte (1894). *From the Society.*

Geological Survey of New South Wales—Records. Vol. iv. (1894-95), Title page, &c.; Vol v. Part i. (1896). *From the Hon. the Minister for Mines and Agriculture.*

Victorian Naturalist. Vol. xiii. No 1 (April, 1896). *From the Field Naturalists' Club of Victoria.*

Société Royale de Botanique de Belgique—Bulletin. Tome xxxiv. (1895). *From the Society.*

Perak Government Gazette. Vol. ix. Nos. 7-8 (March-April, 1896); Title page, &c., to Vol. viii. (1895). *From the Government Secretary.*

Geological Survey of India—Memoirs Vol. xxvii. Part i. (1895); Palæontologia Indica. Ser. xiii. Salt-Range Fossils Vol ii. Part 1; Ser. xv. Himalayan Fossils. Vol. ii. Trias, Part 2 (1895). *From the Director.*

Cincinnati Society of Natural History—Journal. Vol. xviii. Nos. 1 and 2 (April-July, 1895). *From the Society.*

Field Columbian Museum, Chicago—Zoological Series. Vol. i. Nos. 1-2 (Oct -Nov. 1895). *From the Director.*

Nova Scotian Institute of Science—Proceedings and Transactions. Session 1893-94. Vol. i. Second Series. Part 4 *From the Institute.*

Tufts College, Mass.—Studies. No. iv. (Sept. 1895) *From the College.*

New York Academy of Sciences—Transactions. Vol. xiv. (1894-95). *From the Academy.*

American Academy of Arts and Sciences—Proceedings. New Series. Vol. xxii. (1894-95). *From the Academy.*

Academy of Natural Sciences of Philadelphia—Proceedings, 1895. Part ii. (April-Sept.) *From the Academy.*

Boston Society of Natural History—Memoirs. Vol. v. Nos. 1-2 (July-Oct. 1895) : Proceedings. Vol. xxvi. Part 4 (1894-95). *From the Society.*

Rochester Academy of Science — Proceedings. Vol. ii. Brochures 3-4 (1894-95). *From the Academy.*

L' Académie Royale des Sciences, &c , de Belgique—Annuaire lx.-lxi. (1894-95) : Bulletins. 3me. Sér. Tomes xxvi.-xxix. (1893-95). *From the Academy.*

Gesellschaft fur Erdkunde zu Berlin—Verhandlungen. Bd. xxii. (1895). No. 7 : Zeitschrift. Bd. xxx. (1895). Nos. 4-5. *From the Society.*

Société Helvétique des Sciences Naturelles — 77me Session réunie à Schaffhausen (July-Aug. 1894): Actes et Compte Rendu : Mitteilungen der naturforschenden Gesellschaft in Bern, 1894. *From the Society.*

L' Académie Impériale des Sciences de St. Pétersbourg— Bulletin. T. xxxii. Nos. 1 and 4 (1887 and 1888); Nouvelle Série iii (xxxv.) Nos. 1-4 (1892-94) : Mémoires viie. Sér. T. xxxviii. Nos 9-14 (1892); T. xxxix. No. 1 (1891); T. xl. No. 1 (1892); T. xli. Nos. 1-7 (1892-93); T. xlii. Nos. 1, 3-9 and 10 (1894). *From the Academy.*

Four Excerpts from the " Report of the Horn Expedition to Central Australia   Part iii "—[Physical Geography, General Geology, Palæontology, Botany]. *From Prof. R Tate, F L.S.*

Australasian Journal of Pharmacy. Vol. xi. No. 125 (May, 1896) *From the Editor.*

L' Académie Royale des Sciences et des Lettres de Danemark, Copenhague—Bulletin, 1896. No. 2. *From the Academy.*

Société Royale Linnéenne de Bruxelles—Bulletin. 21me. Année. No. 6 (April, 1896). *From the Society.*

Department of Agriculture, Sydney—Agricultural Gazette. Vol. vii. Part 4 (April, 1896). *From the Hon. the Minister for Mines and Agriculture.*

Société Impériale Mineralogique, St. Pétersbourg—Verhandlungen. Zweite Serie. xxxiii. Band, i. Lief. (1895). *From the Society.*

American Naturalist. Vol. xxx. No. 352 (April, 1896) *From the Editors.*

Harvard College, Cambridge, Mass.—Bulletin of the Museum of Comparative Zoology. Vol. xxix. No. 1. *From the Curator*

Société Scientifique du Chili—Actes. T. v. (1895) 1re., 2me et 3me. Livs. *From the Society.*

Canadian Institute—Transactions. Vol. iv. Part 2 (Dec., 1895,: Archæological Report, 1894-95: Inaugural Address (Nov., 1894). By J. M. Clark, M.A., LL.B *From the Institute.*

American Museum of Natural History, New York.—Bulletin. Vol. vii. (1896). Sig. 3-4 (pp. 33-64). *From the Museum.*

Konink. Natuurk. Vereeniging in Nederl Indie—Tijdschrift. Deel lv. (1896): Supplement-Catalogus (1883-93) der Bibliotheek. *From the Society.*

OBSERVATIONS ON PERIPATUS.

BY THOS. STEEL, F.C.S.

The following remarks refer entirely to the ordinary New South Wales *Peripatus*, the form for which the name *P. Leuckarti*, var. *orientalis* has been proposed by Mr. Fletcher.*

For some years past I have taken a good deal of interest in this creature amongst other of the cryptozoic fauna of Australia; and having had numerous living specimens of all ages under constant observation in vivaria during a continuous period of over a year, I have thought that my observations would be of interest to naturalists.

In the course of a number of visits to the Moss Vale district during the summer of 1894-5, and again in 1895-6, I was successful in collecting a considerable number of specimens.

The most remarkable feature about my collection, apart from the unusually large number of individuals of both sexes secured, is the very interesting range of colour variation which it illustrates.

It is not my intention to enter into any details regarding classification or structure, but to give a statement of such facts in connection with the habits and life-history of the creature as I have observed ; together with a few details of the individual range of colour, and the relative proportions of the sexes in the specimens collected.

The summer of 1894-5 was remarkable, in the district above mentioned, for the abundance of various cryptozoic forms of life, particularly land Planarians, and the conditions seem to have been peculiarly favourable for *Peripatus*, judging by the number of individuals which I observed.

The total number of adults which I collected in the Moss Vale district during that summer was 579, of which 390 were

---

* P.L.S. N.S.W. (2 Ser.) Vol. x. 186.

females and 189 males; that is 67 per cent. of the former and 33 per cent. of the latter. Besides these a large number of young, ranging from newly born upwards, were noticed.

The summer of 1895-6 having been preceded by a prolonged spell of very dry weather, the organisms mentioned were found to be very scarce. Where in the previous summer I found hundreds of land Planarians, only scattered individuals of the more hardy and common species were to be met with, and it was only by diligent searching over a somewhat wide area that I was able to secure a very moderate number of Peripati. Particular spots which I specially remembered as being where I met with plenty of specimens in 1894-5, in 1895-6 I found to be quite deserted or only very sparingly populated by Peripatus, while the other usual forms of life—with the exception of ants and termites, which seem to flourish under any conditions—were equally scarce in proportion. This collection, though a good deal smaller, contained much the same relative proportions of males and females, and a similar range of colour variation, as that made in 1894-5.

When collecting in 1894-5, whenever I saw young Peripati under logs I made it a rule to replace them in the position in which I had found them; and as I noted numbers of these logs I was able to examine them again in 1895-6. In many cases where I had left large numbers of young of various ages I found on my second visit not a trace of any, and in others only a few.

My friend, Mr. C. Frost, F.L S., informs me that in Victoria, where the summer of 1895-6 was similar to that experienced in New South Wales, he found the land Planarians exceedingly scarce, and in some cases altogether absent, in districts such as Fern Tree Gully, which are known to be usually prolific in these forms of life.

Such dry conditions, and the attendant "bush fires," must cause an enormous mortality amongst these lowly creatures, and it is greatly to be desired that as much information about them as is possible should be gained, as many local forms are certain to be now rapidly approaching extermination.

In the favourable summer of 1894-5, the individual adult Peripati ranged very much larger in size than was the case in 1895-6. The dry conditions of the latter period appeared to have stunted the growth of the creature. In 1894-5 large numbers of females were 1½ inches in length when crawling, not counting the antennæ, and the males 1 inch; while in 1895-6 the longest female seldom exceeded 1 inch and males about ¾ inch. These are the dimensions when crawling naturally, and not when stretched to the fullest extent What became of the large sized individuals of 1894-5, I cannot say. They may have perished, or could they have shrunk in size as a result of the unfavourable conditions? Whatever may be the cause, their absence was very marked.

In his account of the Mammalia of the Horn Expedition,[*] Professor Spencer gives exceedingly interesting information on the effect of the prolonged spells of arid conditions on the bodily development of some of the mammals of that region; and of the remarkable manner in which, on the other hand, they respond to the more favourable state of matters when a wet period intervenes.

A somewhat analogous series of observations is quoted in *Nature* from *The Entomologist,*[†] in which Standfuss, of Zurich, has investigated the effect on the dimensions, and on the patterns and colours of the wings of certain butterflies; of the subjection of the eggs, larvæ and pupæ to various periods of exposure to different conditions of heat, cold, and moisture. Amongst other results arrived at was this, that the effect of abnormal heat on the larva was to hasten the development, but to cause a notable reduction in the size of the wings.

A very noticeable peculiarity was the intensely local nature of Peripatus. Considerable numbers would be met with in a very restricted area, and without any apparent cause none at all, or very few, would be found on precisely similar ground adjoining.

---

[*] Account of the Horn Expedition to Central Australia, Part 2, 1896.
[†] "Nature," Vol. liii., 540, April, 1896.

After a little experience I got to know the likely-looking parts, and even the most promising logs under which to search All the specimens were underneath logs, either on the ground or on the undersurface of the log, and in the cracks and crannies in the soil beneath the logs. Small easily rolled logs yield the best results for Peripatus as well as for land Planarians and the other creatures that live under them; large heavy ones lie too hard and close to the ground, and do not give the necessary room underneath.

The colours of the individuals were exceedingly variable. Adopting a similar method of comparison to that used by Mr. Fletcher* in his description of the collection made by Mr. Helms at Mt. Kosciusko, my specimens very naturally divide themselves into four groups:—a. Black or blue-black. b. Black, sparingly speckled with rufous brown. c. Rufous brown with black antennæ and with or without visible scattered black spots or specklings. d. Entirely rufous brown or red, including the antennæ, and without any visible black.

The relative numbers of individuals in each of these classes was:—

a. Black or blue-black	... ...	77½ per cent	
b. Black, speckled with brown	...	6½	,, ,,
c. Brown, black antennæ ...	...	10	,, ,,
d. Entirely brown ...	... ...	6	,, ,,

In the Mt. Kosciusko collection the proportion of entirely black individuals is very much smaller than the above, amounting to only about 9 per cent. of the whole, the greater number being dark, sparingly speckled with brown.

No specimens with antennæ and body both entirely brown are mentioned, and indeed, judging from the published descriptions and my own experience, this particular form appears to be much less common than the others. Such being the case, it may be well for me here to briefly describe those in my collection. To the naked eye or the microscope there is no trace of black visible. The lozenge-shaped pattern which has been so fully treated of by

---

* P.L.S. N.S.W. (2 Ser.) Vol. v. 471.

Fletcher and Dendy, while quite distinct, is not nearly so boldly outlined as is commonly the case in *P. oviparus*, Dendy; it is marked out by alternate light and dark areas of skin, the pattern being entirely due to differences in intensity of the brown pigment. This form of Peripatus is exceedingly beautiful; it is a very striking object, and from its bright colour, much more conspicious than its black brethren. When a number of specimens of the brown form are put in spirit together, I have noticed that the latter acquires a distinct brown tinge, which would show that the colour pigment, like that of land Planarians, is to some extent soluble in alcohol.

. Most if not all of the specimens which to the eye or the pocket lens appear quite black, under the microscope present numerous scattered skin papillæ and minute patches of the skin of a brown colour. The antennæ appear to be the last part to lose the black pigmentation or the first to gain it, whichever the case may be. It very commonly happens that the entire body may be brown and the antennæ alone black, and I have not observed a specimen having entirely brown antennæ which had black on any part of the body.

This recalls to my mind a matter in connection with dogs which I have noticed for many years, that they invariably have the tip of the tail white if there is white on any part of the body, and frequently the tail tip is the only white part.

It may also be noticed that in Peripatus the colour variations are pretty uniformly proportionately divided between the males and females.

The adult females are, in my experience, invariably larger than the males, the former being usually from ⅓ to ½ longer than the latter; and the females are also a good deal stouter in proportion to their size, the males being more slender.

The males are distinguishable under the microscope from the females by the white leg papillæ, when about 12 mm. in length, corresponding to about eight months old.

Judging from the rate of growth in captivity I think the females do not mature before they are over two years of age, and it would appear very probable to me that the young are not born until the mother is at least three years old.

In life both blades of the jaws lie with their convex edges outwards, the outer simple bladed jaw lying close up to the inner toothed one, with the points close together. When feeding the jaws are moved very rapidly, with a circular sweep.

I have counted the claw-bearing legs of several hundreds of specimens, and have found them invariably fifteen pairs, exclusive of the oral papillæ. In living individuals the narrow white line in the centre of the dorso-median furrow, described by Prof. Dandy in *P. oviparus*,* and by Mr. Fletcher in *P. Leuckarti*,† is very readily seen under the microscope in the dark coloured specimens, and can be distinctly observed in the light brown ones also, especially when it crosses patches of the darker brown. In young ones it is even more conspicuous than in adults. In adults a somewhat similar line lies at the bottom of the numerous horizontal skin furrows which cross the median line, and indeed wherever there is a furrow in the skin its course is more or less distinctly marked out by white.

These lines are well seen when the animal is extended in the act of crawling, but when it is at rest they are closed over by the skin folds.

The food of Peripatus consists of insects, wood lice, and suchlike. Termites are a favourite article of diet, and are eaten freely. All the soft parts are eaten, including the legs of small insects. The skin of the outer integument of such creatures as wood lice is scraped completely off. Its feeding, as one might expect from the nature of its jaws, is by no means confined to sucking the juices of its prey, but all parts, save the hard integument, are devoured. Of Termites only the hard part of the head is rejected, the remainder, including the antennæ, being entirely eaten.

---

* P.L.S.N.S.W. (2 Ser.) x. 196.

† Ibid. 183.

It is rather interesting to observe the behaviour of wood lice, the creatures with which I have most frequently fed my Peripati, when dropped into the vivarium. At first they scramble under the little pieces of rotten wood, under which the Peripati are lurking, but they very quickly appear to recognise the presence of an enemy and crawl out again, finally clustering together as far as they can get from their foes. Wood lice eat any sort of organic matter, vegetable or animal, and I have seen one biting and nibbling at a sickly Peripatus which was too weak to defend itself.

I have never observed Peripati eat one another; even when kept without food they do not attack each other or the young.

When feeding the movements of the animal are very graceful and deliberate. The antennæ are endowed with a high degree of sensitiveness, and are used by cautiously touching the insect, when so occupied being carried somewhat erect with the tips curved downwards From the manner of using them sometimes, by bending them round and over an object which is being examined, without touching it I think it is highly probable the antennæ are the medium of a sense analogous to that of smell.

In securing its prey Peripatus does not always use the slime secretion, but appears to resort thereto only when the insect which it is endeavouring to secure appears likely to escape, or when it struggles violently, or again when the animal is hungry and wants to make certain of the capture It then becomes animated, raises the front part of its body and ejects the viscid fluid from both papillæ simultaneously. The secretion is ejected with sufficient force to project it several inches. The slime appears to be of an albuminous nature. It is not at all acid or acrid, but is merely useful mechanically, through its tenacious stickiness. When freshly emitted it is rather liquid, but quickly toughens in the air. It is tasteless and has no effect when applied to a sensitive mucous surface of the human body. It mixes with water, but is at once coagulated and rendered insoluble by alcohol.

When the creature is alarmed by sudden exposure to light, the slime is often discharged, the object obviously being self-defence.

Peripatus is a very sociable creature. They do not molest one another, and love to crowd together in congenial lurking-places. I have often observed several of them around one insect feeding in perfect harmony.

Although they will readily feed on dead insects, I have not been able to induce them to eat raw or cooked meat. Occasionally one will after a long examination pull at the meat for a little while with its jaws, but very soon leaves it.

The skin is cast at apparently somewhat irregular intervals, but I have not observed how often   The earliest casting which I have noticed was in the case of young ones born in captivity, which shed the skin when between one and two weeks old. The skin splits along the median dorsal furrow, and is gradually worked off by expansive and contractile movements of the animal, the front end being first worked forward out of the skin and then the whole gradually crumpled off in a very perfect state, including that of the antennæ, feet, and appendages. The exuviæ are pure white, the colour pigment being situated entirely in the inner skin layer which remains

During the shedding of the skin, the operation is frequently assisted by the animal bending round and pulling at it with its jaws, and as soon as it is cast the skin is often eaten, being taken up by the mouth, worked about for a little while by the jaws, and then swallowed entire.

By watching the creatures I have been able to secure several specimens of the cast skins, and with a little careful floating on water have uncrumpled them and caused them to spread out to their full extent, when they form a very delicate and beautiful object Examples of these, both young and adult, are amongst the specimens exhibited. The young appear to be usually born fully extended, but at times doubled up in a thin membrane   I am not sure, however, that in the latter case the birth is not somewhat premature. However, the newly-born young soon crawl about, though they generally remain about the mother for several days. When born they are nearly white, but the colour pigment is plain on the antennæ and those parts of the skin which, in after life, are darkest.   I have frequently witnessed the natural birth of the young, and have succeeded in keeping them alive for over twelve months.   When newly born they are about 5 mm in length, without the antennæ, and from frequent measurements I have found the rate of growth during the 12 months which I had them under observation to be rather less than 1 mm. per month.

Pregnant females somewhat readily extrude the young when distressed by close confinement or uncomfortable conditions. Frequently soft adventitious eggs are laid   These bear no resemblance to those described by Dendy from *P. oviparus* * but are quite smooth and have a very flaccid thin envelope.   They very soon break up into a drop of turbid liquid.   My supposition is that they are merely ova which have escaped fertilization, and are thus making their natural exit from the body.

From my own observations I have seen the young born at all times, from the middle of November till the middle of March. Females which I had in captivity from January, 1895, began to give birth to young at the former date, and continued doing so for over a month, while specimens collected in December, January and February of different years, had young in the course of these and the following months.

So far as my observations go, the young follow the colours of the mother.   Mothers, in whom brown is the prevailing colour, have young of a similar character, and those that are black have dark progeny

I have never noticed the presence of external parasites of any kind on Peripatus.

During the colder months they become sluggish, and remain for considerable periods without eating, but in the warmer part of

---

* P L S N S W. (2 Ser) x. 195.

the year they move about very freely at night, crawling all over the accessible parts of the vivarium in which they are confined, and in the day time hiding away in crevices and beneath lumps of earth or pieces of wood.

The kind of vivaria in which I have been most successful in keeping my specimens alive, consist of ordinary glass jam jars having metal lids, which slip or screw on not quite air tight. These are filled with lumps of moist earth and odd pieces of rotten wood.   An arrangement such as this is very convenient for observation, and allows of taking out the contents when desired for examination, without injury to the specimens.

DESCRIPTIONS OF NEW AUSTRALIAN FUNGI.

By D. McAlpine, F.L.S.

No. I.

(Communicated by J. H. Maiden, F.L.S.)

Meliola funerea, n.sp.

(Plate x., figs. 1-6)

Amphigenous, but most developed on upper surface of leaf Spots velvety, funereal black, with hair-like pile, orbicular or irregular, usually confluent, $\frac{1}{4}$-$\frac{3}{16}$ inch or in a continuous mass $\frac{1}{2}$ inch or more, and very conspicuous

Mycelium of dark brown, thick-walled, septate, branched interwoven threads, about $8\frac{1}{2}$ μ dia., springing from deeper-seated, delicate, colourless hyphae, about 2 μ dia. Bristles on surface looking like masses of black hairs, rigid, sooty-brown, septate, curved, tapering to a point, generally about 11 μ broad.

Perithecia globose, apparently black but with a distinct purple tint, slightly warted, 310-350 μ diameter.

Asci generally 4-spored, ovate to fusoid, up to 90 × 45 μ. Sporidia brown or yellowish, sausage-shaped or elliptic, 3-septate, constricted, 54-62 × 18-20 μ.

On leaves of *Grevillea robusta*, Cunn., in March. Lismore, N.S.W. (Maiden).

The spots and patches are very conspicuous, often almost covering the pinnæ of the fern-like leaf, as well as the leaf-stalk. The sporidia are seen in the same perithecium at different stages of development, varying in colour from hyaline to grey, then yellowish, and finally brown.

Cyathus plumbagineus, n.sp.

(Plate x., figs. 7-12)

Peridium cartilaginous, campanulate, narrowing towards base, externally colour of substratum of dried cow-dung, rough, internally steel-gray, smooth, up to 9 mm. high, and 8 mm. across mouth, rigid when dry, flexible when moist, margin slightly revolute at maturity.

Peridiola or sporangia black-lead-like, discoid, irregularly oval in shape, surface slightly wrinkled, with distinct umbilicus, about 2 mm. dia., with white elastic cord stretching to 7 mm., and attaching it to inner wall of peridium Sometimes the sporangia are attached to outside wall of peridium.

Spores colourless, globose or sub-globose, 24 μ dia., or 24-27 × 21-24 μ, wall sometimes 3 μ broad.

Gregarious, in clusters on cow-dung in March. Near Mercey-road, Homebush, Sydney, N.S W. (Maiden).

The generic nature of this fungus is seen in the three-layered peridium shown in fig. 2, and in the sporangia being umbilicate in the centre of one side. The wall of the peridium is composed of three layers as seen in microscopic section, an outer dark brown layer about 56 μ thick, an inner paler brown layer about 34 μ thick, and a central layer comparatively transparent and loose in texture like a central medulla or pith about 112 μ thick The average thickness of the entire wall is about 200 μ.

Several species of this genus have been found on dung in Australia, but differ from this one in various respects.

C. baileyi, Mass , is externally tomentose and cinnamon colour, and the spores are only 18-20 × 15-16 μ.

C fimicola, Berk., is minutely velvety and umber-coloured, and sporangia are of the same colour, while C. fimetarius, DC., is tawny-rufous and externally velvety.

The specific name is given from the appearance of the sporangia

PHOMA STENOSPORA, n.sp.

(Plate XI., figs. 13-15.)

Spots small to largish, roughly oval, grey, with distinct reddish-brown margin.

Perithecia on upper surface, minute, black, punctiform, semi-immersed, globular to oval, opening by pore, 112-280 μ diameter.

Sporules hyaline, cylindrical, rounded at both ends, on short straight hyaline stalk, with 3 guttules, one at each end and another central or eccentric, $4 \times 1 \mu$.

On living leaves of *Notelœa longifolia*, Vent., in October. New South Wales (J. H. Maiden).

Before the sporules are expelled a yellow plug of matter is extruded, and then the sporules imbedded in a glairy substance.

EXPLANATION OF PLATES

Plate x.

Meliola funerea, n.sp.

Fig. 1.—Portion of upper and under surface of leaf, showing spots and blotches (nat. size).

Fig. 2.—*a*, bristle (× 115); *b*, portion of bristle showing septum (× 600).

Fig. 3.—Perithecium split and unsplit (352 μ and 310 μ in diameter), with stiff pointed bristles (× 65).

Fig. 4 —Asci with sporidia (× 600).   The sporidia were still pale in colour, and comparatively thin-walled

Fig. 5.—Asci with sporidia (× 265).   *a*, four sporidia dark brown in colour; *b*, pale yellow; *c*, greyish; *d*, *e*, hyaline.

Fig. 6.—Two groups of four fully developed sporidia (× 265).

Cyathus plumbagineus, n.sp.

Fig. 7.—Peridium (nat. size)

Fig. 8.—Section of wall of peridium (× 65)

Fig. 9 —Portion of middle layer of wall (× 600).

Fig. 10.—Sporangia (enlarged).

Fig. 11.—Section of sporangium (enlarged)

Fig. 12.—Spores (× 600).

Plate xi.

Phoma stenospora, n.sp.

Fig. 13.—Upper surface of leaf with perithecia (nat. size)

Fig. 14.—Perithecium with projecting yellow matter (× 115).

Fig. 15.—Sporules (× 1000)

DESCRIPTION OF A NEW SPECIES OF *ASTRALIUM* FROM NEW BRITAIN.

By Charles Hedley, F.L.S., and Arthur Willey, D.Sc.

(Plate XII.)

The following species was dredged up by one of us in Talili Bay, off the north-east coast of the Gazelle Peninsula, New Britain, in 30-40 fathoms on a shelly floor, in company with species of *Xenophorus*, *Ranella*, *Oniscia*, *Pleurotoma*, *Fusus*, *Nassa*, *Conus*, &c.

The entire material at our disposal consisted of some three dozen specimens, and was obtained in one haul of the trawl. The stages of growth exhibited ranged from young shells about 16 mm. in diameter, inclusive of spines, to adult shells of some 45 mm. in diameter.

This handsome shell is nearest allied to the well-known Japanese species, *A. triumphans*, from which it differs chiefly by a reduction of the peripheral spines in the adult and in the greater number of spines.

Adopting Pilsbry's classification as given in the Manual of Conchology, Vol. X., it should enter the sub-genus *Guildfordia* of Gray.

Description of Species.

ASTRALIUM MONILIFERUM, n.sp.

Shell.—Low, trochiform, imperforate.

Colour.—Light purplish beads on a ground of old gold, with a metallic lustre; paler below.

Whorls —Seven, inclusive of the embryonic portion of the shell; the upper whorls convex, the last whorl becoming distinctly concave towards the aperture.

Sculpture.—The first three whorls are comparatively smooth, with oblique wavy lines between shoulder and suture; they are angled at the shoulder by a ridge, which commences as a raised thread and at about the fourth whorl breaks up into beads. As growth proceeds, additional bead-lines are intercalated until they reach the number of 8 or 9 rows* on the last whorl, where the subsutural row is composed of large, somewhat oblique, transversely flattened, and closely appressed beads.

Below the subsutural row, the outer rows are placed closer together, the median ones further apart.

The impressed suture is sinuously wound, the spines of the preceding whorl being absorbed.

Periphery is set about in the adult with ten to twelve short forwardly directed, stout, compressed spines† of a maximum length corresponding to about one-third the width of the last whorl; but at the age of four whorls the periphery is armed with 11 closed tubular spines, as long as the whorl is wide.

Base is flattened, becoming convex towards the lower lip of the aperture, a double row of beads, about 50 in a row, forms the margin of the spiked periphery, within which occurs a wide shallow furrow, normally devoid of beads, but frequently containing one or even two intercalated rows, then three or, exceptionally, four rows of beads encircle a heavy boss of callus, excavated at the centre, proceeding from this boss a stout rib thickens the anterior margin of the lip.

Aperture.—Oblique, ovate, angled, and channelled at periphery, pearly within, and reinforced at the upper angle by a heavy

---

* Sometimes there is indication of a tenth row.

† Sometimes there are indications of as many as 14 spines. In the adult the peripheral spines may be locally quite suppressed.

buttress of callus.   A deep sinus is formed by the projection of a tongue of non-nacreous shell, as shewn in the figures accompanying this paper.

Operculum.—Slightly hollowed out on its external surface, very sharply angled on the distal margin, thick and regularly oval.

Dimensions of adult shell.—Height 26 mm., major diameter 45 mm. (maximum measurement), minor diameter about 39 mm.

———

EXPLANATION OF PLATE.

In both figures the buttress of callus is shown at the upper angle of the aperture.   In Fig. 2 only a portion of the bead-rows have been inserted; this specimen had four rows about the central callus, and a row of very small beads at the bottom of the submarginal furrow (indicated by the dark shading).   Finally, in Fig. 2, the non-nacreous tongue at the outer margin of the aperture, mentioned in the text, is indicated by the dotted line dividing it off from the nacreous portion of aperture.

ON A RARE VARIATION IN THE SHELL OF *PTEROCERA LAMBIS*, LINN.

By Arthur Willey, D Sc.

(Communicated by Jas. P. Hill, F.L.S)

(Plate XIII.)

With the view of ascertaining the nature of the variations which the shell of this common tropical species presented, I recently made a collection, amounting to 67 specimens, both from New Britain and from the Eastern Archipelago of New Guinea, the majority coming from the latter locality.

As might be expected from such a comparatively large series, variations of greater or less intensity were very numerous. I am indebted to Mr Charles Hedley for his kind assistance in arranging and classifying the collection.

As is known, Bateson (Materials for the Study of Variation. London, 1894) has divided variations into two main categories, namely, (1) Meristic variations, comprising numerical variations in members of a series, as the rings of an earthworm or, what concerns us at present, the digitations of *Pterocera*, and (2) Substantive variations, comprising variations in the form and bulk ("substance") of individual parts or regions.

My collection shows numerous substantive variations, the more striking of which relate to the curvature of the digitations, their lengths, the intervals between them, and to the extent to which the apical whorls of the shell are involved in, concealed by or fused with the posterior digitation   The last point is essentially co-terminous with the extent of the ascent of the last whorl upon the spire.

Excluding about 15 of the shells as being young, *i.e* , with unthickened outer lips, in the majority of the adult shells a greater or less number of the apical whorls are free. In two specimens only, that is to say in about 1 per cent. of the individuals, was the apex of the spire entirely fused with and, in one of them, deeply imbedded in the base of the posterior digitation. In the other shell the apex was not imbedded in the posterior digitation, but was applied very closely against it.

Pterocera also varies very much as to the stage of growth at which the deposition of callus on the outer lip of the shell takes place. As is known, this deposition of callus eventually leads to the complete closing up of the canals which, in the younger shells, passed from the mouth of the shell into the tubular digitations. This fact is analogous to what has been observed in some other of the lower animals, namely, that they can become sexually mature at very different sizes, and then cease to grow in linear dimensions

In the adult animal of *P. lambis*, therefore, the border of the mantle is not digitated.

We now pass on to the description of the rare variation referred to in the title of this paper.

Out of the whole collection only three specimens exhibited a variation in regard to the number of the labial digitations. In all cases the intercalated digitation occurred between the second and third normal digitations. Although small, its presence offered a striking contrast to the other shells. Of the three specimens exhibiting this variation, two (Figs. 1 & 2) came from New Britain. In both cases the rudimentary digitation was backed up by a definite ridge on the outer surface of the shell as is the case with normal digitations.

The third specimen, from New Guinea (Fig 3), presented a rather puzzling aspect. The intercalated digitation had a double character, and was not backed up by a prominent ridge on the outer surface. It appeared to have had a distinctly later origin than in the other two cases. Two furrows proceeded from it to the mouth of the shell, one being independent and the other produced by a bifurcation of the furrow belonging to the second normal digitation.

The constancy in the position of the above described rudimentary intercalated digitation in *P. lambis* should be emphasized.

It can be identified, I think, with absolute certainty, with one of the digitations of *P. millepeda*, Linn , namely, the fourth. I obtained four specimens of *P. millepeda*, which has nine labial digitations, from New Guinea. In two of these the fourth digitation was markedly smaller than any of the others, while agreeing in position with that above described in *P. lambis*. In fact, in *P. millepeda* the intercalated digitations are obviously the second and fourth, and probably the seventh.

It may also be remembered as indicating the significance of the appearance, by variation, of an extra digitation in *P. lambis*, that in *P. elongata*, Swainson, there are eight labial digitations, in *P. violacea*, Swainson, ten, and in *P. chiragra*, Linn., five.

— —

EXPLANATION OF THE FIGURES.

Fig. 1.—The canals leading into the tubular digitations are still open, the deposition of callus having only commenced.

Figs. 2 and 3. —The canals are closed up by callus, their previous existence being indicated by shallow furrows.

i.d., intercalated digitation.

The shell represented in Fig. 1 was the same in which the apex of the spire was imbedded in the posterior digitation as mentioned in the text.

Mr. Maiden exhibited specimens of the fungi described in Mr. McAlpine's paper.

Mr. Steel exhibited a fine series of beautifully preserved specimens of Peripatus from Australia, Tasmania, and New Zealand.

Mr. Froggatt exhibited living specimens ($\male$ and $\female$) of *Cœlostoma australe*, described in 1890 by Mr. Maskell in the Society's Proceedings (Second Series, v., 280). The male is a very beautiful and rare insect Six were taken, round the stump upon which the female was found, the first examples the exhibitor had ever seen.

Mr. Froggatt also exhibited a number of the larvæ of the Acacia Goat Moth [*Zeuzera (Eudoxyla) eucalypti*], victims of an attack of a fungoid growth allied to *Cordyceps*, and turned into "vegetable caterpillars," so called. Some of the specimens were cut out of the trunks of Acacias *(A. longifolia)* growing near Manly, in which they were found in the tunnels formed by the larvæ. Others were from larvæ taken alive and kept in breeding boxes; probably they had become infected previously, as after living for months they changed into similar hard masses. The late Mr. Olliff in one of his latest papers in the Agricultural Gazette upon Australian Entomophytes, in describing the hosts of *Cordyceps* says that it attacks only subterranean root-feeding larvæ, and never those of true wood borers, as so often stated by entomologists. The specimens exhibited bear out his statements, for the fungus concerned is a species without the projecting clubbed growth, which would be at a disadvantage in the confined tunnels of a wood-boring caterpillar. It may belong to the genus *Xylostroma*, which is often found in the centre of decaying trees.

The President exhibited a "Cotton-grass Snake" (*Typhlops* sp.) forwarded from Menindie, N.S.W., by Mr. A. G. Little.

8

WEDNESDAY, 24TH JUNE, 1896.

The Ordinary Monthly Meeting of the Society was held at the Linnean Hall, Ithaca Road, Elizabeth Bay, on Wednesday evening, June 24th, 1896

The President, Mr. Henry Deane, M.A., F.L.S., in the Chair.

The President announced that Professor Haswell would be glad to receive and forward contributions to the Huxley Memorial Fund.

The President also announced that Mr. Duncan Carson had presented to the Society his collection of British plants; but as the utilisation of such a collection was hardly within the scope of the Society's operations at present, the Council, with the donor's approval, was prepared to offer the same for distribution among Members desirous of supplementing their British collections.

DONATIONS.

Naturhistoriske Forening i Kjobenhavn—Videnskabelige Meddelelser for Aaret, 1895. *From the Society.*

Naturwissenschaftlicher Verein zu Bremen—Abhandlungen. xiii. Band, 3 Heft (1896); xiv. Band, 1 Heft (1895). *From the Society*

Bombay Natural History Society—Journal. Vol. x. No. 2 (March, 1896.) *From the Society.*

Perak Government Gazette. Vol. ix. Nos. 9-11 (April-May, 1896). *From the Government Secretary.*

Radcliffe Library, Oxford—Catalogue of Books added during the year 1895. *From the Radcliffe Trustees.*

Société Géologique de Belgique—Annales. T. xxiii. 1ʳᵉ Liv. (1895-96). *From the Society.*

K. K. Zoologisch-botanische Gesellschaft in Wien—Verhandlungen. xlvi. Band (1896), 3 Heft. *From the Society.*

Pharmaceutical Journal of Australasia Vol. ix. No. 5 (May, 1896). *From the Editor.*

Bureau of Agriculture, Perth, W.A.—Journal. Vol. iii. Nos. 9 and 12-14 (April-June, 1896). *From the Secretary.*

British Museum (Natural History)—Catalogue of Birds. Vols. xxv and xxvii. (1895-96) : Catalogue of Fossil Fishes. Part iii. (1895) : Catalogue of the Fossil Plants of the Wealden. Part ii. (1895) : An Introduction to the Study of Rocks (1896) : Guide to the British Mycetozoa (1895). *From the Trustees.*

Zoological Society of London—Abstract, April 21st, May 5th : Proceedings, 1895. Part iv : Transactions Vol. xiv. Part i. (April 1896). *From the Society.*

Royal Society, London—Proceedings. Vol lix Nos. 355-356 (March-April, 1896). *From the Society.*

Morphological Laboratory, Cambridge University — Studies. Vol. vi (1896). *From the Balfour Library.*

Zoologischer Anzeiger xix. Band. Nos. 501-502 (April-May, 1896). *From the Editor.*

Nederlandsche Entomologische Vereeniging—Tijdschrift. xxvi. Deel. Jaargang 1882-83. Afl. 1-2 : xxxvii. Deel Jaargang 1893-94 Afl. 1-4. *From the Society.*

Société Impériale des Naturalistes de Moscou—Bulletin. Année 1895, No. 4. *From the Society.*

Société des Naturalistes de Kieff—Mémoires. Tome xiii. Livs. 1-2 (1894) : Tome xiv. Liv. 1 (1895). *From the Society.*

Société d'Horticulture du Doubs, Besançon—Bulletin. Série Illustrée. No 4 (April, 1896). *From the Society.*

Zoologische Station zu Neapel—Mittheilungen. xii. Band. 2 Heft (1896). *From the Director*

Report on the Work of the Horn Scientific Expedition to Central Australia Part ii. Zoology : Part iii. Geology and Botany. *From W. A. Horn, Esq , per Professor Baldwin Spencer, M A.*

University of Sydney—Calendar, 1896. *From the Senate.*

L'Académie Royale des Sciences, Stockholm—Oefversigt. lii. Ärgangen (1895). *From the Academy.*

Victorian Naturalist Vol. xiii. No 2 (May, 1896). *From the Field Naturalists' Club of Victoria.*

Birmingham Natural History and Philosophical Society— Proceedings Vol. ix. Part ii (1895). *From the Society.*

Hooker's Icones Plantarum. Fourth Series. Vol v. Part iii. (May 1896). *From the Bentham Trustees.*

Société Royale de Géographie d'Anvers—Bulletin. Tome xx. 4me Fascicule (1896). *From the Society.*

Department of Agriculture, Sydney—Agricultural Gazette Vol. vii. Part 5 (May, 1896) *From the Hon the Minister for Mines and Agriculture.*

Museo di Zoologia ed Anatomia comparata della R. Università di Torino—Bollettino. Vol. xi. Nos. 227-242 (Feb.-May, 1896). *From the Museum.*

Royal Society of New South Wales—Journal and Proceedings. Vol. xxix. (1895). *From the Society.*

Australasian Journal of Pharmacy. Vol. xi. No. 126 (June, 1896). *From the Editor.*

Natural History Society of Montreal—Canadian Record of Science. Vol. vi. Nos. 3-7 (1894-95). *From the Society.*

Department of Agriculture, Brisbane—Botany Bulletin. No. xiii. (April, 1896). *From the Government Botanist.*

Museum of Comparative Zoology at Harvard College, Cambridge, Mass.—Bulletin. Vol. xxix. No. 2 (March, 1896). *From the Director.*

American Naturalist. Vol. xxx. No. 353 (May, 1896) *From the Editors.*

American Museum of Natural History, New York—Bulletin Vol. viii. Sig. 5 (pp 65-80—April, 1896). *From the Museum*

A NEW FAMILY OF AUSTRALIAN FISHES.

By J. Douglas Ogilby.

The family, of which the following diagnosis is given, is intended to accommodate those forms of percesocoid fishes in which, among other characters which separate them from the *Sphyrænidæ* and *Atherinidæ*, the first dorsal fin is composed of a single pungent and two or more flexible, unarticulated rays, and by the position of the anal fin, which is more elongated and advanced than in the typical Atherinids, and which on account of its anterior insertion pushes forward the position of the anal orifice and of the ventral fins so far that the latter become thoracic, and the family thus makes a distinct advance towards the more typical Acanthopterygians.

To Prof Kner and Dr. Steindachner, and subsequently to Count Castelnau, the claim of these little fishes to rank as a distinct family has commended itself. Prof. Kner, in 1865, alluded to the expediency of forming a family, *Pseudomugilidæ*, for the reception of certain small fishes, alleged to have been obtained by the collectors of the Novara Expedition at Sydney, and to which he gave the name of *Pseudomugil signifer*; he, however, gave no definition of the proposed family, though during the following year he, in conjunction with Dr. Steindachner, again makes incidental mention of the family while describing a closely allied genus, *Strabo*; these authors also neglect to formulate a diagnosis.

In 1873, Count Castelnau, after describing as new a genus which he named *Zantecla*, notices the differences in "its characters from all the families established till now," he being doubtless unaware of the previous discoveries of Drs. Kner and Steindachner; this author also places his genus "near the *Atherinidæ*," and considers that it "will be the type of a new family, which might be called *Zanteclidæ*." In the previous year the same author, after diagnosing a new genus as *Atherinosoma*, had suggested that it might prove necessary to form a new family for its reception, and again in 1875, having formulated yet another new genus under the name of *Neoatherina*, he returns to the subject and proposes "forming on it a family to be called *Neoatherinidæ*," which was also to contain the genus *Atherinosoma*.

We have, therefore, already three different families—*Pseudomugilidæ*, *Zanteclidæ*, and *Neoatherinidæ* – proposed for the reception of different genera of these fishes, for not one of which has any diagnosis been even attempted.

To prevent confusion with these older undefined names, it has appeared advisable to me to suggest a new name for the family, though for reasons which I give below I am constrained to make that genus typical, which from its slight specialization is the least suitable; nevertheless, since Dr. Gill has already formulated for certain of these fishes a subfamily of the *Atherinidæ* under the name *Melanotæniinæ*, I do not feel justified in proposing to change his name for the more suitable one of *Rhombatractidæ*.

There are several cogent reasons which point to this course as being the most fitting to pursue under the circumstances Taking Castelnau's proposed families first :—

The use of *Zanteclidæ* is precluded, its typical genus *Zantecla* being synonymous with and of later date than *Melanotænia*, and therefore inadmissible; while *Neoatherinidæ*, as well as being the last suggested name and belonging to a less distinctly specialized genus, is formed on a bastard title, the employment of which should be as much as possible deprecated, at any rate so far as the names of families are concerned; besides which it labours under the disability of having been associated by its author with a genus which undoubtedly belongs to the *Atherinidæ* proper.

My choice, therefore, is restricted to the use of *Pseudomugilidæ* –the only one of the three proposed names which in the author's opinion, is entitled to consideration—or to the substitution of *Melanotæniidæ*, and I believe that I am consulting the best interests of science by taking the latter course, for the following reasons :—

Pseudomugilidæ—also a bastard name, and therefore open to the same objection as *Neoatherinidæ*--is misleading, since the

genera which are here segregated have little in common with the
true Mugilids, but form conjointly a connecting link between the
percesocoid and acanthopterygian types, furthermore, *Pseudomugil*
is a small and obscure form, not ranking either in distribution or
importance with *Melanotænia* or *Rhombatractus*

I shall now proceed to give a diagnosis of the family, in which
I include five genera—*Neoatherina, Pseudomugil, Rhombatractus,
Aida,* and *Melanotænia*—which form a very natural group,
characterised by the structure of the first dorsal fin, the advanced
position of the ventrals, &c.

The metropolis of the family appears to be in north-eastern
Australia, where no less than four of the genera have their home;
thence it has spread northwards into the rivers of south-eastern
New Guinea, westwards to Port Darwin and the Victoria River,
south-westwards into the central districts of South Australia, and
on, in the aberrant *Neoatherina*, to Swan River, and finally southward to the Richmond and Clarence Rivers District of New
South Wales, and perhaps even as far as the Nepean watershed

MELANOTÆNIIDÆ.

Pseudomugilidæ, Kner, Voy. Novara, Fische, p. 275, 1865 (*no
definition*).

Pseudomugilidæ, Kner & Steindachner, Sitzb. Ak. Wiss. Wien,
liv. 1866, p. 372 (*no definition*).

Zanteclidæ, Castelnau, Proc Zool. & Acclimat. Soc. Vict ii.
1873, p. 88 (*no definition*).

Neoatherinidæ, Castelnau, Res. Fish. Austr. p. 32, 1875 (*no
definition*).

Melanotæniinæ, Gill, American Naturalist, 1894, p. 708.

Body rhombofusiform to elongate-oblong, more or less compressed. Mouth moderate, terminal, oblique. Two nostrils on each
side. Premaxillaries not protractile, forming the entire dentigerous
margin of the upper jaw; maxillaries narrow.   Gill-openings
wide; gill-membranes separate, free from the isthmus; five or six

(seven ?) branchiostegals; pseudobranchiæ present; gill-rakers short. Opercular bones entire; preopercle with a double ridge. Jaws and vomer toothed; palate with or without teeth; tongue smooth. Two separate dorsal fins; the first with a strong, acute spinous ray anteriorly, followed by two or more flexible, often elongate, unarticulated rays; the second with a similar strong spinous and several articulated and branched rays: anal similar to but more developed than the second dorsal: ventrals separate, thoracic, with one spinous and five soft rays: pectorals well developed, rounded: caudal emarginate, the peduncle stout. Body entirely scaly, the scales cycloid or ciliated, smooth; cheeks and opercles scaly; no scaly sheath to the vertical fins; no scaly process at the base of the ventrals; lateral line inconspicuous or absent Air-vessel present, simple Pyloric appendages wanting.

Small fishes from the fresh and brackish waters of tropical and subtropical Australia and southern New Guinea.

As indicated on a previous page I propose to associate in this group five genera, the diagnoses of which, so far as the scanty material available to me permits, will be found below, but unfortunately, from lack of specimens, I have not been in a position to personally examine any of these genera except *Rhombatractus*, of which a detailed description is given, the principal characters of the remaining genera being taken from the works of their respective authors.

NEOATHERINA.

Neoatherina, Castelnau, Res. Fish. Austr. p. 31, 1875.

Body subelongate, compressed, with the anterior portion of the back convex; snout pointed, rather projecting; mouth moderate and oblique, the upper jaw the longer. Teeth rather strong, in two series in the upper jaw, long and blunt anteriorly, triangular laterally; in the lower they are very numerous, in pavement form, with an external row of enlarged conical ones; anterior teeth in both jaws directed forwards; palate with several transverse series of strong teeth.* Two dorsal fins, well separated, the first formed of one rather long spine and of four much longer filamentary rays; the second dorsal long, composed of one spine and eleven rays : anal fin long, with one spine and seventeen strong, spine-like rays : ventrals inserted far behind the base of the pectorals, and very little in advance of the insertion of the first dorsal, with one spine and six† elongate rays : pectorals small, with twelve rays : caudal forked   Scales large, ciliated; cheeks and opercles scaly; lateral line indistinct.

E t y m o l o g y :—*véos*, new; *Atherina*.

T y p e :—*Neoatherina australis*, Castelnau, 1 c. p. 32.

D i s t r i b u t i o n :—Swan River, West Australia.

In the increased number of the ventral rays (if correct), the ciliation of the scales and the character of the dentition *Neoatherina* differs from all the other Melanotæniids, while it approaches *Pseudomugil* in the presence of a lateral line; its affinity, however, to the melanotænioid rather than to the atherinoid forms is shown in one character, incidentally alluded to by Castelnau in the following terms :— "The small specimen has a more elongate form; the upper profile being much less convex . . " This character was passed over as of little or no value by that author, probably because he was unaware of the sexual differences in form which are so strongly marked in his *Aristeus* (= *Rhombatractus*), but, in my opinion, it is significant of the systematic position of the genus, which, from the more backward insertion of the ventral fins, some authors might be inclined to retain among the true Atherinids.

---

* It is probable that, either through insufficient knowledge of the language or carelessness on the part of the author, there is some error in this sentence; either "vomer" should be substituted for "palate," or "longitudinal" for "transverse," probably the former.

† If this character be correct it is unique in the Percesocids.

PSEUDOMUGIL.

Pseudomugil, Kner, Voy. Novara, Fische, p. 275, 1865.

Body subelongate, compressed, with convex ventral profile; forehead broad and flat, snout short, with the mouth oblique; a band of acute teeth in both jaws; eyes large, preorbital smooth; two separate dorsal fins, the first with four or five flexible, unarticulated rays; scales large and cycloid, the lateral line little conspicuous. Air-vessel simple. Dorsal and ventral fins with elongate, filiform rays in the male. (*Kner*).

From the description of the only known species we also learn that the lower jaw projects slightly beyond the upper, the maxillary does not reach to the eye, and is almost entirely concealed beneath the preorbital; that the teeth in the jaws are small, acute, directed inwards, and arranged in a narrow band, the outer series being enlarged and almost caninoid, while there are no perceptible teeth on the palate.

The absence of palatine teeth, presence of an inconspicuous lateral line, and similarity in form of the sexes are the only important characters which are available for the separation of this from the succeeding genus, and it is quite possible that, when examples of the two can be compared, the line of demarcation will be found untenable, and *Rhombatractus* will have to merge in the older *Pseudomugil*.

Etymology :—ψεῦδος, false, *Mugil*.

Type.—*Pseudomugil signifer*, Kner.

Distribution :—York Peninsula. In the Voyage Novara it is alleged that the fishes from which Professor Kner's description was drawn up, were collected at Sydney, but this is manifestly erroneous, no member of the family being so far known with certainty to exist on the coastal watershed of our dividing range south of the Richmond and Clarence District, from whence the late Sir William Macleay described a species under the name of *Aristeus lineatus*. The locality here given is that from which Dr. Günther received his *Atherina signata*, which is said to be identical with Kner's fish.

RHOMBATRACTUS.

Aristeus (not Duvernoy) Castelnau, Proc. Linn Soc. N S. Wales, iii. 1878, p. 141.

Rhombatractus, Gill, American Naturalist, 1894, p. 709.

Body rhombofusiform or oblong, strongly compressed, with the dorso-rostral profile more or less emarginate, and the ventral profile convex; head small, the snout broad and depressed; mouth moderate, anterior, with oblique cleft, the lips thin; jaws equal or the lower a little the longer; premaxillaries not protractile, forming the entire dentigerous margin of the upper jaw, broad and projecting horizontally in front, narrow and oblique behind; maxillaries narrow, extending a little beyond the premaxillaries, entirely concealed beneath the preorbital except at the extreme tip. All the bones of the head entire, the preopercle with a double ridge Gill-membranes separate, entirely free from the isthmus; gill-openings wide; five branchiostegals; pseudobranchiæ present, gill-rakers widely separated, moderate, stiff, and serrulate. Jaws with a band of short, stout, conical teeth, which are more numerous in the lower, the outer series being much enlarged and recurved; vomer and palatine bones with narrow bands of small, stout, conical teeth, tongue toothless.* Two separate dorsal fins with v-vii, i 9-14 rays, the first not so long as the second and composed of one strong and a variable number of flexible, unarticulated, spinous rays, the second with a similar spine and several branched rays anal fin originating beneath the base of the first dorsal and more developed than the second, with i 17-21 rays : ventral fins close together, thoracic, inserted a short

---

* The teeth on the vomer and some or all of those behind the anterior series upon the horizontal portion of the premaxillaries are occasionally wanting in adult specimens, and are probably more or less deciduous with age.

distance behind the base of the pectorals, with a slender spinous and five soft rays : pectorals rather small, moderately pointed, with 13-15 rays, those in the upper half of the fin the longest, the upper ray simple and somewhat inspissate : caudal fin emarginate, with short deep peduncle. Scales large, cycloid, smooth, not deciduous, the posterior border being more or less truncated, especially on the tail; cheeks, opercles except the outer ridge of the preopercle, and occiput scaly, the rest of the head naked; dorsal and anal fins without a basal scaly sheath; no enlarged scales at the base of the first dorsal, pectoral, or ventral fins, and no scaly process between the latter; lateral line wanting, a series of large open pores from the maxillary symphysis along the lower border of the preorbital, passing upwards in front of and above the eye to the occiput, where it connects with a similar series extending from the mandibulary symphysis below the eye and round the naked outer preopercular surface. Vertebræ 33 to 37 (22 + 15 in *Rhombatractus fluviatilis*). Air-vessel large and simple. Abdominal cavity very large, extending backwards far beyond the vent, the intestines very long and convoluted.

Etymology :—ῤόμβος, rhomb; ἅτρακτος, a spindle; in allusion to its shape.

Type :—*Aristeus fitzroyensis*, Castelnau.

Distribution :—Fresh waters of Australia as far south as the 32nd parallel, and of southern New Guinea

The sexual differences are strongly marked in these fishes, both as regards the form of the body and the development of the fins.

In adult males the depth of the body is much greater than in females of the same age; for instance, in a series of specimens of *Rhombatractus fluviatilis*, collected from a single haul in Yulpa Creek, near Deniliquin, the depth of the males is from $2\frac{1}{2}$ to $2\frac{3}{4}$, of the females from $3\frac{1}{5}$ to $3\frac{1}{4}$ in the total length; this variation is entirely due to the slight development in the latter of the postoccipital convexity, which is so pronounced a character in the males, the rostro-dorsal contour in the females being gently and evenly arched from the extremity of the snout to the caudal peduncle.

The caudal peduncle in the male is a little deeper than long, in the female a little longer than deep.

The development of the dorsal, anal, and ventral fins shows similar sexual distinctions; thus, the flexible spines of the first dorsal, the posterior rays of the second dorsal and of the anal, and the outer rays of the ventral fins are prolonged into filaments in the males, while in females and immature males this character is inconspicuous or absent

Though not the oldest, this genus is by far the most important of the group, whether as regards its degree of specialization, area of distribution, or number of species

Up to the year 1878, when Castelnau first described this genus under the name *Aristeus*, all but one of the authors (Richardson, Gunther, Kner, and Steindachner), who had written on the fishes which are here collected together in one family, had recognised their affinity to the Atherinids, the exception being Dr. Peters, and though Castelnau himself, first in proposing to separate in a distinct family his closely allied genus *Zantecla* (= *Melanotænia*), which, as he says, "comes near the *Atherinidæ*," definitely gives in his adhesion to this view, and two years subsequently endorsed this recognition by proposing to separate from that family his two new genera, *Atherinosoma* and *Neoatherina*, which he coupled. notwithstanding their manifest differences, as *Neoatherinidæ*, he nevertheless, in spite of his acquaintance with two of the genera —*Melanotænia* and *Neoatherina*—and his acknowledgment of their connection with the true Atherinids, commits the extra, ordinary error of referring *Aristeus* to the *Gobiidæ*, a family with which it has not the slightest affinity, either in its external or its internal structure; this error is perpetuated by Macleay and others.

In 1886, in a paper on the fishes obtained by the collectors of the New South Wales Geographical Society's Expedition to New Guinea, I described two very distinct species from the Strickland River, substituting for *Aristeus* Peters' name *Nematocentris*, this being, so far as I knew at that time, the earliest attempt to remove Castelnau's genus to its true systematic position; however, as was kindly pointed out to me by Dr. Gill, Steindachner had previously recognised the close relationship of these two genera (Zoöl. Jahresb. 1879, p. 1061).

Mr. Zietz, the latest writer on the subject, who has followed Steindachner and me in making *Aristeus* synonymous with *Nematocentris*, refrains from enlightening us as to his views of the systematic affinities of this genus; two new species from Central Australia are described by this author, who places them (Horn Exped. Centr. Austr. pp. 178-9) between the Theraponids and the Eleotrine Gobiids, below which *Gobius* itself is ranked, thus securing so wide a margin for selection that we are left in doubt as to the family in which he is in favour of leaving it, though we would be justified in inferring that he considers Castelnau correct in allying *Aristeus*—and, therefore, by his own admission of the identity of the two genera *Nematocentris*—with *Eleotris*, since by no possibility could the percesocoid fishes be so placed.

Curiously enough Castelnau himself, in the same pamphlet in which the diagnosis of *Neoatherina* is published, described yet another new genus as *Aida*, of the close relationship of which to *Rhombatractus* I shall have something to say further on, and places it " with considerable doubt in the family of the *Percidæ*," that is to say, in that section of Günther's *Percidæ*, which we should now call *Apogonidæ* or *Chlodipteridæ*; there it is left without comment by Macleay.

Prior, however, to the publication of Castelnau's paper, Dr. Peters had already assigned to his genus *Nematocentris* a position near to the Apogons, although the species on which his diagnosis was formed had been described many years previously by Richardson as *Atherina nigrans*, and holds a place in Günther's Catalogue as *Atherinichthys nigrans*. Kner and Steindachner, however, in the same year point out the affinity existing between *Nematocentris* and the Atherinids, though none of these authors appear to have suspected the identity of their respective species with that of Richardson.

The above remarks will, however, suffice to show how diverse
the views of authors have been as to the position which these
fishes and their allies are entitled to hold in the ichthyological
system.

AIDA.

Aida, Castelnau, Res. Fish. Austr p 10, 1875.

Body very compressed; upper part of the head unequal; opening
of the mouth very oblique, almost perpendicular, opercle and
preopercle without teeth or spines, the first with a double edge.
Teeth fine, minute, disposed on one line, two very feeble canine
teeth in front of the upper jaw; a transverse line of teeth on the
palate. Two dorsal fins, the first composed of five spines, the
four last prolonged; the second with one spine and thirteen rays,
which increase in length backwards anal with two spines and
seventeen rays, formed like the second dorsal · ventrals inserted
behind the pectorals and united at their base, formed of one spine
and five rays. pectorals placed at about half the height of the
body, rather small. caudal bilobed. Scales rather large and entire
on their edges, the posterior part of the head and the opercle
covered with scales similar to those of the body; no lateral line.
*(Castelnau).**

E t y m o l o g y :—unknown.

T y p e :—*Aida inornata*, Castelnau.

D i s t r i b u t i o n .—Gulf of Carpentaria.

If an analysis be made of the differences between the
above description and that of *Rhombatractus*, it will be found
that they are but slight and such as, bearing in mind the care-

---

* With the exception of rearranging the sequence of the sentences and
of omitting some unnecessary words no change has been made in
Castelnau's own phraseology; and these transpositions have been undertaken merely to bring the above diagnosis into sequential accordance with
that of *Rhombatractus*, and so make the comparison of the two genera
easier for those who follow me in the study of these interesting forms.

lessness which characterises Castelnau's work, may be easily set aside or explained away; the main differences are as follows :—

(i.) *Gill-covers.*—Castelnau writes : "opercle and preopercle without teeth or spines, the first with a double edge." This is probably mere carelessness; by substituting "last" for "first" the description would be quite correct.

ii) *Dentition.*—By turning to the foot-note p. 124 my readers will find that I there suggest that certain of the teeth in *Rhombatractus* may be deciduous with age, and it is merely necessary to carry this deciduousness a little further to arrive at a dentition somewhat similar to that described by Castelnau.

(iii.) *Fin rays.*—"Anal with two spines." I do not think it necessary to attach much importance to this character, seeing that Castelnau was possessed of but one specimen from which to draw up his description. It may be taken for granted that in all these small fresh-water fishes the first soft ray is liable to take the form of an additional spine, and it would, of course, be but natural to describe this genus as having two anal spines if the diagnosis was taken from an example having this individual peculiarity.

As an instance of this tendency I may mention that when some years ago a species of *Ambassis* was present in great abundance in the Parramatta and George's Rivers, I noticed that in a number of specimens taken at random almost as many would be found having two rays in front of the second dorsal as those having one, and this increase was always coordinated with a corresponding decrease in the number of soft rays, thus plainly showing that this was not a structural character, but a simple, though common, variation caused by the calcification of the anterior soft ray.

That Castelnau on the one hand was either unaware of or paid no attention to this tendency to acanthination in fresh-water fishes, while on the other hand placing undue prominence on the presence of one or more additional spines, we know from his own writings and from his treatment of *Macquaria australasica*, of

9

which fish he makes, in a single paper (Proc. Zool. & Acclimat. Soc. Vict. i. 1872, pp. 57 & 61-64), no less than five new species, which he distributes in three different genera, two of which are described as new,* the principal reason given being the disagreement in the number of the dorsal spines; thus, referring to *Dules christyi*, he writes :--" It is so much like *Murrayia cyprinoides* in form that I should have thought it belonged to the same species had it not been for the difference in the number of the spines of the first dorsal." And in the diagnosis of *Riverina* the following passage occurs :—" This genus is very nearly allied by its form to *Murrayia*, but the dorsal has twelve spines." *Murrayia* has eleven spines and twelve rays, *Riverina* twelve spines and eleven rays.

(iv). *Lepidosis.*—Of the gill-covers only the opercle, according to Castelnau, is scaly; but even here by the simple substitution of "opercles" for "opercle" the diagnosis would be sufficiently close for that author.

I think, therefore, that it is quite possible that when Castelnau penned his description of *Aida* he had a specimen of *Rhombatractus* before him, and in any case, until I am satisfied that the differences relied on are constant and are supported by other structural characters, I am content to consider *Aida* a true Melanotæniid.

MELANOTÆNIA.

Melanotænia, Gill, Proc. Acad. Nat. Sc. Philad. 1862, p 280.

Nematocentris, Peters, Monatsb. Ak. Wiss. Berlin, 1866, p. 516.

Strabo, Kner & Steindachner, Sitzb. Ak. Wiss. Wien, liv. 1866, p. 372 (1867).

Zantecla, Castelnau, Proc. Zool. & Acclimat. Soc. Vict. ii. 1873, p. 88.

---

* These are *Dules christyi*, p 57 ; *Murrayia güntheri*, p. 61 ; *M. cyprinoides*, p. 62 ; *M bramoides*, p 63 ; and *Riverina fluviatilis*, p. 64.

Body fusiform, little compressed, with the dorso-rostral profile slightly curved; snout short, depressed, prominent; mouth small, with horizontal cleft. Opercle spineless; preopercle with a double ridge. Gills four; six branchiostegals; pseudobranchiæ present. Jaws, vomer, and palatines with a band of villiform teeth, the outer series in the former being enlarged, conical, and curved. Two separate dorsal fins, the first with one stout and four or five slender, flexible rays, the second longer, with one spine and nine to twelve articulated and branched rays: anal long, with a single stout spine: ventrals thoracic. Scales of moderate size, cycloid, with the margins feebly crenulated. No lateral line. Pyloric appendages in small number. Air-vessel simple.

Etymology :—μέλας, black; ταινία, a band.

Type :—*Melanotænia nigrans*, Gill, = *Atherina nigrans*, Richardson.

Distribution :—Fresh and brackish waters of northern and eastern Australia, extending southwards at least as far as the Richmond River District, and possibly further since, after describing *Aristeus fluviatilis*, Castelnau remarks :—" I have two specimens of this fish, one, two and a half inches long. It comes from the Murrumbidgee . . . . the other was found by Mr. Duboulay in Rope's Creek, and is three and a half inches long. It has a very feebly marked black longitudinal stripe on each side." This latter specimen is probably a *Melanotænia*, and the locality given would bring the range of that genus as far south as the metropolitan district.

It is much to be regretted that owing to the uncertainty which prevails as to the correct name of the genus which I have called *Rhombatractus* in this paper, I have been obliged to adopt as the sponsor of the family a genus which is distinctly less specialized and, in its little compressed, non-ventradiform body more closely approaches to exotic forms than the others. If I could have satisfied myself that future investigations would justify the separation of *Rhombatractus* from *Pseudomugil* and *Aida*, I should certainly have preferred to name the family *Rhombatractidœ*, that genus being the most highly specialized and most widely diffused of all the forms at present known.

In reference to the position which this family is entitled to hold in the system, I am unable to agree with those authors who would place it between the *Atherinidœ* and the *Mugilidœ*, much less with those who would associate it with the *Eleotrinœ* or the *Apogonidœ*, but though the position of these fishes near *Apogon* is untenable, it cannot be denied that there is considerable external resemblance between them and some Ambassids; in *Nannoperca*,[*] for instance, we find the same posterior insertion of the ventrals, reduced number of branchiostegal rays (six as in the Ambassids, not seven as in the Apogonids), absence or irregularity of the lateral line, and concavity of the dorso-rostral contour.

That, however, its affinities are distinctly percesocoid I believe that no one, who is acquainted with one or more of the various forms, and who has more than a superficial knowledge of fishes in general, will deny, and it is only, therefore, with regard to the degree of affinity which exists between it and the other Percesocids that I am at issue with those scientists who would make it a link between the Gray Mullets and the Atherines.

The forward position of the ventral fins, which is so characteristic of this family, marks a decided advance in the direction of the more typical Acanthopterygians, while the increased strength of the dentition clearly points to relationship with the *Sphyrænidœ*, in which family we find, in our *Dinolestes*, an example of the tendency towards an enlargement of the anal fin and consequent advancement of the position of the ventral fins.

It seems to me, therefore, that the most natural sequence in which to place the Percesocids with relation to other fishes would be as follows :—

---

[*] *Paradules*, Klunzinger (not Bleeker) and *Microperca*, Castelnau (not Putnam) are synonymous, and very closely allied to if not identical with *Nannoperca*; *Microperca yarræ = Paradules obscurus*

Suborder—S Y N E N T O G N A T H I.*

Suborder—P E R C E S O C E S.

Family—M u g i l i d æ.

      „   A t h e r i n i d æ.

      „   S p h y r æ n i d æ.

      „   M e l a n o t æ n i i d æ.

Suborder—A C A N T H O P T E R Y G I I.

Appended is a list of the Melanotæniids described up to the present time :—

1. *Neoatherina australis*, Castelnau, Res. Fish. Austr. p. 32, 1875. Swan River, West Australia.

2. *Pseudomugil signifer*, Kner, Voy. Novara, Fische, p. 275, 1865. Sydney, New South Wales.

3 *P. signata*, = *Atherina signata*, Gunther, Ann. & Mag. Nat. Hist. (3) xx. 1867, p. 64. Cape York, Queensland.

4. *Rhombatractus fitzroyensis*, = *Aristeus fitzroyensis*, Castelnau, Proc. Linn. Soc. N.S. Wales, iii. 1878, p. 141. Fitzroy River, Queensland.

5. *R. fluviatilis*; = *Aristeus fluviatilis*, Castelnau, l.c. Murrumbidgee River, New South Wales.

6. *R. rufescens*; = *Aristeus rufescens*, Macleay, Proc. Linn. Soc. N S. Wales, v. 1880, p. 625 [1881]. Rivers of Northern Queensland.

7. *R. lineatus*, = *Aristeus lineatus*, Macleay, l.c. p. 626. Richmond River, New South Wales.

8. *R. cavifrons*; = *Aristeus cavifrons*, Macleay, l c. vii. 1882, p. 70. Palmer River, Queensland.

---

* Possibly the Lophobranchiate fishes should intervene between the Hemirrhamphids and the Percesocids.

9. *R. goldiei*, = *Aristeus goldiei*, Macleay, l.c. viii. 1883, p. 269. Goldie River, New Guinea.

10. *R. perperosus;* = *Aristeus perperosus*, De Vis, Proc. Linn. Soc. N.S. Wales, ix. 1884, p. 694.

11. *R. nova-guineæ*, = *Nematocentris nova-guineæ*, Ramsay & Ogilby, Proc. Linn. Soc. N.S. Wales (2) i. 1886, p. 13. Strickland River, New Guinea.

12. *R. rubrostriatus;* = *Nematocentris rubrostriatus*, Ramsay & Ogilby, l.c. p. 14.   Strickland River, New Guinea.

13. *R. loriæ;* = *Aristeus loriæ*, Perugia, Ann. Mus. Genov. (2) xiv. 1894, p. 549.

14 *R. tatei;* = *Nematocentris tatei*, Zietz, Rep. Horn Exped. Centr. Austr. Zool p 178, f. 2, 1896.   Finke River, South Australia.

15. *R. winneckei;* = *Nematocentris winneckei*, Zietz, l c p. 179, f. 3.   Finke River, South Australia.

16. *Aida inornata*, Castelnau, Res. Fish. Austr. p. 10, 1875. Gulf of Carpentaria

17. *Melanotænia nigrans*, = *Atherina nigrans*, Richardson, Ann. & Mag. Nat. Hist. xi. 1843, p. 180.   Rivers of North Australia.   As before remarked (p. 131) the same species may range nearly as far southward as Sydney, but much confusion exists as to the members of this genus. Dr. Günther apparently is content to consider the four species identical, but I think that any such conclusion, based on the small material available to him, is hasty, and that judging by analogy with the allied genus *Rhombatractus*, the distribution of which is also wide but the species of which are known to be numerous, it is unwise to unite in one species all the black-banded forms from widely separated parts of the continent.

18. *M. splendida;* = *Nematocentris splendida*, Peters, Monatsb. Ak. Wiss. Berlin, 1866, p 516.   Fitzroy River, Queensland.

19. *M. nigrofasciata,* = *Strabo nigrofasciatus,* Kner & Steindachner, Sitzb. Ak Wiss. Wien, liv. 1866, pp. 373, 395,
pl. iii. f. 10, [1867], and lv. 1867, p 16. Brisbane and
Fitzroy Rivers, Queensland.

20 *M. pusilla,* = *Zantecla pusilla,* Castelnau, Proc Zool. &
Acclimat. Soc. Vict. 1873, ii. p 88. Port Darwin, North-
West Australia.

In the above list I have made no attempt to indicate the degree
of affinity between any of these species, but it is generally conceded that *Atherina signata,* Gunther, is identical with *Pseudomugil signifer,* and that *Nematocentris splendida,* Peters, and
Strabo nigrofasciatus, Kner & Steindachner, cannot be separated
specifically from *Melanotænia nigrans, Zantecla pusilla,* Castelnau,
is a good species in my opinion

It is, however, improbable that all the twelve described species
of *Rhombatractus* are tenable, but I trust soon to be in a position,
with the cooperation of other scientific societies and of individual
students, to publish in this Journal a monograph of the family
with original descriptions of all the species

DESCRIPTIONS OF TWO NEW GENERA AND SPECIES OF AUSTRALIAN FISHES.

By J. Douglas Ogilby.

Macrurrhynchus, gen.nov.

Body elongate, compressed, head moderate, the snout somewhat pointed, conical, deep, projecting, convex above; mouth small, prominent, subinferior, with transverse cleft; lips thin; dentigerous portion of the upper jaw slightly curved, of the lower semicircular; cleft of mouth extending to beneath the middle of the eye; nostrils superior, the anterior pair rather close together, about as far from the eye as from the tip of the snout; the posterior pair more widely separated, midway between the eye and the anterior nostril; no nasal nor orbital tentacles; eyes lateral; interorbital region moderate and flat   Gill-openings reduced to a small foramen in front of the upper angle of the base of the pectoral.   Teeth in a single series in both jaws, fixed, those of the upper well developed, laterally compressed, of rather unequal length; with the tips truncated and slightly bent backwards; of the lower smaller, more slender and crowded, and of equal length; upper jaw without, lower with an enormously developed tusk-like canine at the outer extremities of the series and fitting into a sheath in the upper jaw when the mouth is closed.   One dorsal fin, with the outer border entire, with xii 30 rays, the spines flexible, the spinous portion about half as long as the soft, all the rays of which are unbranched, the membrane of the last ray not extending to the caudal fin: anal fin originating beneath the commencement of the soft portion of the dorsal, with 30 soft rays, the tips of which are but slightly inspissate and free: ventrals in contact at their bases, inserted in advance of the base of the pectorals, with i 3 rays: pectorals small and rounded, with 12 equally developed simple rays: caudal emarginate, with the middle ray somewhat thickened.   No trace of a lateral line.

Etymology:—*Macrurus; ῥύγχος*, snout; in allusion to the form of the snout, which bears a marked resemblance to that of many of the *Macruridæ*, such for example as *Cœ'orhynchus australis*.

Distribution:—Western Pacific.

I would gladly have given to this genus the name *Aspidontus* of G Cuvier, but that I am unaware whether any diagnosis of that genus was ever published. Dr. Gunther apparently did not know of any such definition, and merely quotes Quoy & Gaimard for the name, making it synonymous with Ruppell's *Petroscirtes*.

MACRURRHYNCHUS MAROUBRÆ, sp.nov.

D. xii 30.   A. 30

Body of nearly equal depth throughout. Length of head $4\frac{2}{5}$, depth of body $6\frac{3}{4}$ in the total length; depth of head $1\frac{3}{4}$, width of head 2, of the flat interorbital region $3\frac{2}{3}$, diameter of the eye 4 in the length of the head; snout projecting, macruriform, with the profile convex, as long as the eye, the lower surface linear and oblique, as long as the upper. The posterior angle of the mouth extends to the vertical from the middle of the eye, the naked portion of the retangular cleft on each side as long as the entire dentigerous portion and $4\frac{1}{2}$ in the length of the head. Dorsal fin commencing immediately behind the posterior border of the preopercle, the distance between its origin and the extremity of the snout being five-sixths of the length of the head; the rays are of about the same length throughout, the middle ones being a little the longer, $2\frac{1}{4}$ in the length of the head: the anal originates a little behind the vertical from the last spinous ray of the dorsal and is considerably lower than that fin   the ventrals are composed of slender rays, three-sevenths of the length of the head· the pectoral fins are small, rounded, and symmetrical, their length five-eighths of that of the head: caudal fin small, slightly and evenly emarginate, $6\frac{1}{4}$ in the total length, its peduncle short and stout, with a depth of a half of that of the body.

Back olive green, lower half of the sides and the abdominal region silvery white washed with rose-colour; these tints are sharply defined, but from the lower border of the green numerous short vertical bars, as wide as the interspaces, extending downwards encroach on the sides, a narrow bright blue stripe extends backwards from the snout, above and in contact with the eye, along the side almost as far as the base of the caudal fin, about equally dividing the darker ground colour; they meet on the upper lip, where also they connect with a similar band which traverses the side of the snout, immediately below the rostral ridge, and is continued backwards below the eye to the opercles; a third stripe runs along the median line of the head to the dorsal where it is broadly forked, the branches being short; extremity of the snout orange on the lower surface, dorsal and anal fins silvery, with several broad dark vertical bands composed of numerous, closely set, blackish dots, and with a narrow marginal band of the same; ventral, pectoral, and caudal fins uniform grayish silvery, the latter with a dark band formed like those of the dorsal along the middle ray.

A single specimen was washed ashore during the month of May, on the beach at Maroubra, and was secured by Mr. White-legge, by whom it was presented to the Australian Museum, its length is 52 millimeters.

Petroscirtes tapeinosoma, Bleeker, and *P. rhinorhynchus*, Bleeker (Gunther, Fische d. Sudsee, p. 195, pl. cxv. D & E), would belong to this genus, as well as *Aspidontus tæniatus*, Quoy & Gaimard (Voy. Astrolabe, Poiss p 719, pl. xix f. 4).

DERMATOPSIS, gen.nov.

Body elongate and compressed, especially behind, head moderate, the snout short and blunt, mouth anterior and rather wide, with moderate cleft. Premaxillaries slightly protractile, forming the entire dentigerous portion of the upper jaw, maxillary narrow in front, greatly expanded behind, extending backwards well behind the eye, anterior border of the expanded portion bent downwards behind the premaxillary so as to form a strong, compressed, odontoid process. Nostrils lateral, widely separated, the anterior pair smaller than the posterior, surrounded by a skinny, vesicular lip. Eyes small and lateral, completely covered by similar skin. Opercles covered by a continuous skin; opercle with two strong spines, the upper of which pierces the skin. Gill-openings of moderate width, extending forwards to below the posterior border of the preopercle, isthmus wide, seven branchiostegals, no pseudobranchiæ. gill-rakers reduced to small, serrulate tubercles. Upper jaw with a band of villiform teeth and a single small, curved, canine-like tooth on each side of the symphysis; lower jaw with a narrow band of villiform teeth anteriorly, the inner series much enlarged and continued backwards along the sides in the form of a row of widely separated, curved, canine-like teeth; vomer with an angular series of small, acute, conical teeth, the posterior tooth on each side greatly enlarged, palatine teeth in a triangular patch anteriorly, small and conical, with a single central and three posterior basal enlarged ones, pterygoids and tongue smooth Anterior dorsal fin represented by a single spinous tubercle which does not pierce the skin; dorsal and anal fins low, separated from the caudal by a distinct interspace. ventral fins close together, inverted behind the isthmus, reduced to a slender filament, which is composed of two intimately connected rays: pectorals moderately developed, pointed, composed of twenty slender branched rays tail diphycercal, the caudal fin narrow and pointed. Scales small, deeply embedded, widely separated, head, except the snout, with scattered scales; vertical fins for the most part covered with skin, which is scaly like the body. A series of large pores along the outer border of the snout and preorbital, and a pair of similar pores at the angle of the preopercle, lateral line inconspicuous.

Etymology:—δέρμα, skin; ὄψις, eye.

Distribution:—Coast of New South Wales.

Apparently the dorsal tubercle represents the rudiments of a first dorsal fin, and its presence would, therefore, necessitate the removal of the genus from the *Brotulidæ* to the *Gadidæ*, a course

which I am very unwilling to take since in all other characters it
is a true Brotulid, in fact its affinity to *Dinematichthys* is so close
that its disassociation with that genus would be out of the
question, the dentition and the form of the maxillary being the
only prominent external differential characters.   I have not had
access to Dr Bleeker's paper diagnostic of *Dinematichthys*, and
am, therefore, unaware as to whether or not he notices any such
rudimentary first dorsal in that genus; certainly no other authors,
such as Drs Ayres, Gunther, Gill, and Jordan, who have made
personal examinations of the various species, have mentioned it.
It would be interesting if some scientist, possessed of a series of
that genus, were to investigate the matter with a view to detecting
the existence of the same structure in *Dinematichthys*, since,
should it be so discovered, the two genera would, I presume, have
to be removed from the *Brotulidæ*, or at least one of the structural
characters which separate that family from the *Gadidæ* would
have to be modified.   Perhaps Dr Jordan would examine one of
his examples of *Dinematichthys ventralis*, and let us know whether
any such rudiment is present

DERMATOPSIS MACRODON, sp nov.

D. 78.   A. 52.

Body elongate and compressed, the tail very strongly so, its
posterior portion tænuform.   Head moderate, with the cheeks
and opercles rather swollen, its length $4\frac{1}{2}$, the depth of the body
$6\frac{2}{3}$ in the total length; depth of the head $1\frac{4}{5}$, width of the head
$1\frac{6}{7}$, of the interorbital region $5\frac{3}{4}$, diameter of the eye 7 in the
length of the head; snout blunt, its profile linear and slightly
oblique, covered with a loose skin, three-fourths of a diameter
longer than the eye, interorbital region convex, the supraciliary
bones slightly prominent   Mouth rather large, its cleft extending
to the vertical from the middle of the eye; the premaxillaries are
very little protractile and form the entire dentigerous surface of
the upper jaw, they are moderately broad anteriorly, but are
slender and rod-like on the sides, maxillary narrow in front,
greatly expanded behind, its lower border curved downwards and

forwards so as to form a strong, compressed, tooth-like process, into the curved base of which the rounded distal extremity of the premaxillary fits, behind this process the maxillary bone forms a gentle and even arc, of equal width throughout, the extremity rounded and directed slightly upwards, the maxillary extends to about one diameter behind the eye, and its length from tip to tip is $1\frac{4}{5}$ in that of the head; the lower jaw is a little shorter than the upper, and is provided with an inferior low skinny flap, which extends entirely across its anterior border and is pectinated at the edge, the mandibular bone reaches as far back as the maxillary, along the inner surface of which it lies. The anterior nostrils are small and circular, and are situated rather close together on the edge of the maxillary and directly in front of the posterior pair, which is much larger and subtriangular, and opens immediately in advance of the eye, both are surrounded by a loose, skinny, vesicular lip, which entirely conceals the orifice Eye very small, entirely covered by loose skin. Opercle with a pair of stout, sharp spines; the upper one running in a horizontal direction below its upper border, the lower rising from the same base is directed downwards and a little backwards, both are entirely concealed beneath the loose skin, which is continuous across the gill-covers, with the exception of the extreme tip of the upper one which just pierces the skin. Twelve rudimentary, tubercular gill-rakers, each of them crowned with a few short acute serræ, on the lower branch of the anterior arch. The band of villiform teeth on the premaxillaries is broad in front, but rapidly decreases in width on the sides, about midway along which it ceases; on each side of the symphysis anteriorly is a small, acute, curved, canine-like tooth; the mandibulary band is much narrower than that of the premaxillaries, and does not extend so far laterally; there are no enlarged teeth anteriorly at the symphysis, but the inner series is considerably enlarged, conical, and acute, the lateral dentition consists of seven (or more) very strong, widely separated, caniniform teeth, which are curved backwards and inwards, the largest teeth being about the middle of the series; there is an angular ridge on the head of the vomer, which is armed with a single series of acute, conical, separated teeth, those at the apex and along the sides being of moderate size, while the posterior tooth on each limb is similar to the largest mandibulary teeth, and is directed backwards and slightly outwards; palatine teeth in an acutely triangular patch with the apex pointing forwards, and consisting of small, strong teeth, with a central and three basal enlarged and conical ones. The dorsal tubercle is situated immediately behind the base of the pectoral; it does not pierce the skin, but is distinctly perceptible to the finger-nail; the origin of the dorsal fin is above the middle of the pectoral, and rather more than a diameter of the eye behind the dorsal tubercle, its distance from the extremity of the snout is $3\frac{2}{3}$ in the total length, the rays are very slender and but little branched, of almost equal length throughout, those which are inserted somewhat behind the middle of the fin being a little the longest and about one-third of the length of the head: the anal originates beneath the commencement of the middle third of the dorsal, and is in all respects similar to that fin; the distance between its origin and the tip of the snout is as long as its distance from the base of the caudal fin: ventral inserted beneath the hinder margin of the preopercle, not quite so long, the pectoral half as long as the head caudal fin truncate at the base, not quite as long as the pectoral, with thirteen rays.

Reddish-brown, the upper surface of the head and the vertical fins rather darker; sides and lower surface of the head, the abdominal region, and the paired fins yellowish-brown.

The single example from which the diagnosis is taken was picked up dead, but in a perfectly fresh condition, on the beach at Maroubra by Mr. Whitelegge in May last, after a heavy gale, and measures 80 millimeters

From the small size of the eyes, and the fact of their being protected by a complete covering of skin, one is led to infer that in its natural state this fish is accustomed to burrow in the sand or mud for purposes of concealment, or perhaps as a means of seeking food, a similar protective eyelid is present in *Leme*. It is probably an inhabitant of the littoral zone or, at most, of shallow water in the neighbourhood of the shore.

ON THE AUSTRALIAN *CLIVINIDES* (FAM. *CARABIDÆ* .

(REVISION OF THE AUSTRALIAN SPECIES OF THE GENUS *CLIVINA* WITH THE DESCRIPTION OF A NEW GENUS, *CLIVINARCHUS*).

BY THOMAS G. SLOANE

The *Clivinides* form a division of the tribe *Scaritini* of world-wide distribution, but found most plentifully in the warmer portions of the globe; they are very plentiful in Australia

Following Dr. G. H. Horn's classification of the *Carabidæ*, their position will be as follows :—

Family CARABIDÆ.

Sub-Family CARABINÆ.

Tribe SCARITINI.

The *Scaritini* may be divided into two main divisions thus:—

Mentum broad and concealing at sides base of maxillæ. ..*Scaritides.*
Base of maxillæ not covered by mentum . . . *Clivinides.*

CLIVINIDES.

As represented in the Australian fauna, the *Clivinides* comprise the genera *Dyschirius, Clivina, Clivinarchus* and *Steganomma.* For the present I have to pass over *Steganomma* which is founded on a unique species, *S porcatum*, Macl , in the Macleay Museum, Sydney; it is very closely allied to *Clivina.*

For the purposes of the Australian fauna the genera *Dyschirius, Clivina* and *Clivinarchus* may be tabulated thus :—

Prothorax globose. .... *Dyschirius*
Prothorax not globose.
    Mesosternal episterna strongly impressed on each side of
      peduncle..... .... *Clivina.*
    Peduncle without lateral impressions   .... *Clivinarchus.*

Genus CLIVINA.*

Scolyptus, Putzeys (in part): *Ceratoglossa*, Macleay

The following features of universal application in the genus *Clivina* are extracted from Dr Horn's definition of the tribe *Scaritini*.†

Eyes not distant from mouth   Head with two supra-orbital setæ   Ligula small and prolonged, bisetose at tip, paraglossæ slender.   Palpi with penultimate joint bisetose in front ‡   Thorax with two lateral punctures   Body pedunculate, scutellum not visible between elytra   Sides of elytra narrowly inflexed, margin entire   Metasternal epimera distinct.   Posterior coxæ contiguous. Legs stout, the anterior femora especially stout.

To the universal characters given above I would add for the Australian species the following :--

Labrum usually truncate (sometimes the middle lightly advanced), gently declivous to anterior margin; five (rarely) or seven (normally) setigerous punctures above anterior declivity—the lateral puncture on each side larger than the others and the seta rising from it longer than the other setæ and erect (in species with only five setæ the one next to the lateral is wanting); anterior angles rounded, ciliate.   Mentum emarginate with a wide median tooth Clypeus with a seta on each side.   Vertex with a ridge on each side above supra-orbital punctures *(facial carina* —" carène

---

* Latreille, Consid. Gén. sur les Cr. et les Ins.

† Trans. Am. Ent. Soc. ix. 1881, pp. 119, 120.

‡ The following are Dr. Horn's words in reference to the palps of the *Scaritini* :—" Palpi moderate, terminal joint variable in form, shorter than penultimate *(Scarites)* equal or longer *(Clivinæ)*, the penultimate bisetose in front *(Clivinæ)* plurisetose *(Scarites)*." It is evident he only refers to the labial palps, but for all that the differences sought to be established cannot be maintained, for in his " group " *Clivinæ* some Australian species *(e.g., C. planiceps,* Putz.) have the penultimate joint of the labial palps evidently longer than the terminal, and in *Carenum* too the relative proportions of these joints varies.

oculaire " of Putzeys); a sulcus on inner side of each of the facial carinæ *(facial sulcus)*. Throat and temples normally rugulose; gular sutures wide apart; a short oblique ridge *(gular cicatrix)* extending inwards on each side of base of neck and dividing the gular and temporal regions. Prothorax and disc canaliculate, and normally with a transverse arcuate impression *(anterior line)* near anterior margin; a deep channel along each lateral margin, its course terminated before the posterior marginal puncture by a slight upward curve of the border at posterior angle. Body winged Peduncle with a concavity on each side (normally punctate) to receive intermediate femora. Elytra normally with seven punctate striæ and a lateral channel, third interstice with four foveiform punctures along course of third stria. Prosternum strongly bordered on anterior margin; the episterna normally overhanging on sides anteriorly—(the antennæ pass under the overhanging part of the sides when in repose). Metasternal episterna—with epimera—normally elongate and narrowed posteriorly, rarely short. Ventral segments transversely sulcate. Intermediate tibiæ with an acute spur on external side above apex, rarely at apex.

The features given above are normally present in Australian species of *Clivina*, therefore little, and often no use has been made of them in the descriptions which follow; but in all cases where any variation from the normal form has been observed it has been noted (except in the case of differences of the gular and temporal regions of the head, the gular sutures, the gular cicatrix and the anterior margin of the labrum), and where no allusion is made to any of the characters enumerated above in my descriptions of specimens before me, it is to be assumed that the form is normal.

The following characters seem to call for special notice, the more so because I have been compelled for the sake of descriptive exactness to adopt a new terminology for some features not hitherto used in diagnosing species of *Clivina*, and to vary some of the terms used by M Putzeys for certain features.

The head is longitudinally impressed on each side, the anterior part of each of these impressions usually forming a wide and

10

irregular depression of variable depth *(frontal impressions)*; the seta found on each side of the clypeus is situated in the frontal impression, often the puncture from which it rises is lost in the rugosity of the impression : from the frontal impressions the facial sulci extend backwards on each side of the face, and in some species (e.g., *C. obliquata*, Putz.) a short light internal impression extends from the anterior part of the facial sulcus obliquely inwards and backwards on each side of the face—the facial sulci may then be said to be *recurved* (this is a feature of evident classificatory importance)   The clypeus is large, usually not divided from the front between the frontal impressions, when it is so divided it is by a wide usually irregular impression. It is necessary for descriptive purposes to divide the clypeus into three areas, viz —(1) *The clypeal elevation* ("elévation antérieure" of Putzeys) being the raised part of the clypeus between the frontal impressions —(reference is usually made by me only to the shape of the anterior margin of the clypeal elevation), (2) *the median part* ("epistome" of Putzeys) being the central part of the clypeus in front of the clypeal elevation (usually I refer to the anterior margin only as the median part); (3) *the wings* ("petites ailes" of Putzeys) being the lateral parts of the clypeus (usually a finely marked suture is noticeable between the wings of the clypeus and the supra-antennal plates).   The form of the anterior margin of the clypeus varies greatly, these variations being important for grouping the species; among the Australian species there are three well marked forms of the anterior margin of the clypeus, of one or other of which all different forms may be considered as merely modifications; these are :—

(*a*) The median part projecting on each side beyond the wings, in which case it is *angular*, the lateral angles being more or less marked *(e g., C. angustula*, Putz).

(*b*) The median part in no way separated from the wings along the anterior margin *(e.g., C. australasiæ*, Bohem)

(*c*) The wings projecting strongly beyond the truncate median part *(e g., C. procera*, Putz).

The median part is often defined on each side from the wings by a ridge, more or less distinct (I have made but little use of this feature, though these ridges seem not without value for diagnostic purposes).

The supra-antennal plates ("grandes ailes" of Putzeys) are the "frontal plates" (Horn) of the head under which the antennæ are inserted.

The elytra have the striæ at the base either (a) *all free*, or (b) *the four inner free*, the fifth uniting with the sixth, or (c) *the three inner free*, the fourth uniting with the fifth at the base. These variations are of great classificatory importance and seem to offer the most reliable means of grouping the species into primary divisions. The first stria of the elytra rises in an ocellate puncture at the base, and in some species, especially the larger ones, the first and second striæ unite at the base; sometimes a short scutellar striole is very noticeable at the base of the first interstice (this is an important feature). The interstices vary, the eighth usually forming a narrow carina near the apex. A *submarginal humeral carina* is generally present at the humeral angles, when present it may vary in length and prominence and may be formed by the basal part of (a) the seventh interstice, (b) the eighth interstice, or (c) the seventh and eighth together. The position of the posterior puncture of the third interstice varies; but, though useful when comparing specimens, I have not used it in my descriptions.

The prosternum may be divided into the *pectoral part* and the *intercoxal part:* the point of union between these parts varying in width, five different degrees of width may be used, (a) *very wide* (C. procera, Putz, &c), (b) *wide (C. lepida,* Putz, &c.), (c) *narrow* (C. australasiæ, Bohem., &c), (d) *very narrow (C. obliquata,* Putz, &c), (e) *attenuate (C. melanopyga,* Putz., &c.). The difference in width of the intercoxal part anteriorly is of high classificatory importance and of the greatest assistance in arranging the Australian species. The pectoral part is sometimes margined on each side posteriorly by a prominent border, these may be termed *the pectoral ridges* (vide *C. lepida).* The

base of the intercoxal part may be either transversely sulcate
or not; this seems a useful feature for separating species.

The differences in the legs are of great classificatory importance,
but need no special note beyond attention being drawn to the
differences between the terms used by M. Putzeys in describing
the digitation of the anterior tibiæ and those adopted by me. M
Putzeys disregarded the external apical projection and only made
reference to the teeth on the outer side above the apex, while,
in conformity with the usage of writers on the *Carenides*, I
include the apical projection in counting the external teeth of
the tibia

I have made no use of the maxillæ; in all the species which I
have examined the inner lobe has been found to be hooked and
acute at the apex; this form I believe to be invariable among the
Australian species of *Clivina*, but Dr. Horn's drawings* of the
maxillæ of North American species show that sometimes the
inner lobe is obtuse at the apex

M. Putzeys reduced the genus *Ceratoglossa*, Macleay, to a
synonym of his genus *Scolyptus*, and, as far as the Australian
fauna is concerned, I would merge *Scolyptus* in *Clivina*.   There
is no doubt in my mind that the species placed by me in
the "*procera* group," several of which M. Putzeys put in
Scolyptus, are congeneric with *C. basalis*, Chaud., &c ; *C.
planiceps* (with allied species) might be thought to require a
different genus from *C. basalis*, but, if so, other species (e.g, *C
frenchi*, Sl) are equally deserving of separation from both *C
basalis* and *C planiceps*   On the whole I think the only course
is to place in the central genus *Clivina* all those Australian
species which have been put in *Scolyptus*, at least till someone is
prepared to give sound reasons for the generic separation of any
of them from the other species of *Clivina;* this I am not, at
present, prepared to do.

The first Australian *Clivina* to be described was *C. basalis* by
M. de Chaudoir in 1843, and this remained the only species

---

* Trans. Am. Ent. Soc. ix. 1881, pl. v.

known till 1858, when Bohemann described *C. australasiæ* from Sydney. In 1862 M. Putzeys published his "Postscriptum," in which he described four new Australian species. It may be noted that of these four species, all founded on unique specimens, three, viz., *C. elegans, C. attrata,* and *C. suturalis,* never seem to have turned up again; as will be seen from my notes on them, I suspect a possibility of the identity of two of them with subsequently described and known species. In 1863 Sir William Macleay described two *Scaritides* from N.S. Wales as *Ceratoglossa foveiceps* and *C rugiceps;* these are species of *Clivina,* but both have to be dropped out of the Australian list for reasons stated below. In 1866 Putzeys published a Revision of the Australian species of Clivina, including descriptions of thirteen new Australian species—these descriptions he afterwards embodied in the "Révision Générale." I do not think it will be easy, if indeed possible, ever to identify *C. juvenis, C. prominens,* and *C. verticalis.* In 1867 Putzeys published his "Révision Générale," describing four new Australian species; and also he received for description the whole of Count Castelnau's collection of *Clivinides,* among which he found fourteen species of *Clivina* from Australia to describe as new; of these I have been able to identify six. Between 1868 and 1873 Putzeys added three species to our list, all of which are known to me. After 1873 no more species of Australian *Clivina* were described till 1889, when the Rev. Thos. Blackburn described nine new species, and since that date he has described three additional species, bringing the number known from Australia up to fifty-two. I have now thirty-one to add, making a total of eighty-three species for Australia, a number which I expect to be largely augmented when the continent has been more carefully searched for these insects

A few words on size and colour in reference to distinguishing species of the genus *Clivina* from one another will not be out of place. M. Putzeys seems to have regarded slight differences in size as of more than legitimate value in determining closely allied species, *vide* his descriptions of *C. juvenis, C. lepida* and *C. rubripes,* which are not decidedly differentiated among themselves or from

C. australasiæ, by mere size, though it is made a point of the first importance in the original descriptions *

Occasional dwarfed specimens of probably most species of *Clivina* occur, which are so much smaller than the average size of their species that if only two specimens, one small and the other of normal size, were placed in anyone's hands for description they would more likely be regarded as different species than as representatives of the same species   It is only when we have before us a large series of specimens from one locality that we realise the amount of variation in size, and therefore in appearance, which may occur in a species of *Clivina*.   For instance, a specimen of *C. biplagiata* only 5 5mm in length is in my possession—7-7·5mm being the normal length of the species; and small specimens of some species, *e.g.*, *C. adelaidæ*, appear to the eye too narrow and light to be associated without hesitation with large specimens of the same species.

It appears to me that too much importance must not be attached to mere colour for distinguishing species; immature specimens are always more lightly coloured than those that are mature; and speaking as a practical collector I would call attention to the fact that several immature specimens will sometimes represent all those of a species taken at one time and place, in this way immature specimens may be considered as typical in colour of a species, and so confusion may arise   A good example of colour-differences in a single species is afforded by *C. sellata*, three specimens of which in my collection taken at the same time and place differ in colour as follows.   One, showing the mature colour of the species, has the head and prothorax black, the elytra reddish testaceous with a black dorsal spot, the second has the head and prothorax testaceous-red, the elytra testaceous with the place of the dorsal spot a little obscured; the third has the upper surface wholly testaceous, the elytra being paler than the head and prothorax.

---

* For a note by M. Piochard de la Brûlerie criticising M. Putzeys' work as an author of species, *vide* Ann. Soc. Ent. Fr., 1875, (3), v. p. 128

I have divided the Australian species of *Clivina* into thirteen groups; a synoptical view of these groups is given in the table below. The groups are formed in an arbitrary way, and no doubt their number might advantageously be reduced had I a surer knowledge of the affinities of the species.

Table grouping the Australian species of Clivina.

L **Elytra with striæ free at base.** (Submarginal humeral carina wanting).

    A. Facial sulci simple, clypeus emarginate; inter-coxal part of prosternum wide anteriorly. . *biplagiata* group.

    AA. Facial sulci recurved, clypeus with median part angular; inter coxal part of prosternum very narrow anteriorly ... . *cribrosa* group.

II. **Elytra with four inner striæ free at base, fifth joining sixth at base.** (Submarginal humeral carina normally present).

    B. Mandibles short.

        C. Clypeus with five triangular projections in front................... . ... *coronata* group.

        CC. Clypeus with median part more or less angular laterally.................................. *obliquata* group.

    BB. Mandibles long, decussating.

        D. Prothorax with border reaching base on each side. .... *planiceps* group.

        DD. Prothorax with border not reaching base *grandiceps* group.

III **Elytra with three inner striæ free at base, fourth joining fifth at base.** (Submarginal humeral carina usually well developed).

    E. Clypeus with median part more or less distinctly divided from wings along anterior margin (usually more prominent than wings).

        F. Anterior femora with posterior edge of lower side strongly dilatate in middle. ... . *punctaticeps* group.

        FF. Anterior femora not greatly dilatate on lower side.

            G. Head very wide across occiput, eyes not prominent.

H. Size small; prothorax longer than
broad, without anterior line.. . .. *blackburni* group.

HH. Size moderate; prothorax broader
than long, anterior line present. . *olliffi* group.

GG. Eyes prominent.

I. Prosternum with intercoxal part
attenuate.. .. . . *heterogena* group.

II. Prosternum with intercoxal part
narrow.. .. .. *bovillæ* group.

EE. Clypeus roundly emarginate, median part
not divided from wings. .... *australasiæ* group.

EEE. Clypeus deeply truncate-emarginate, wings
strongly advanced; (size usually large). . *procera* group.

Following M. Putzeys' example, I define each group as I come
to it.

I begin the descriptions of species by treating of two species,
viz , *C. attrata*, Putz., and *C obliterata*, Sl , which I have felt unable
to place in any of the thirteen groups into which I have arranged
the species of *Clivina* found in Australia   *C. attrata* may not be
an Australian species at all.   *C. obliterata* seems a species of
anomalous position, and, in view of its strong resemblance to *C.
australasiæ*, Bohem , even of doubtful validity.

C. ATTRATA, Putzeys.

Mém. Liége, 1863, xviii. p 54 , Stett Ent. Zeit 1866, xxvii.
p. 36; Ann. Soc. Ent. Belg. x. 1866, p. 179.

" Nigra, antennis brunneis, palpis pedibusque dilutioribus.
Mandibulæ latæ, breves.   Antennæ longæ, crassiusculæ.   Labium
[? labrum] bisinuatum.   Clypeus emarginatus, alis prominentibus.
Vertex 3-impressus denseque punctulatus Oculi prominentes postice
cupulati.   Pronotum subquadratum, antice subangustatum, basi
vix prolongata   Elytra elongata, basi truncata, punctato-striata,
striis apice evanescentibus, punctis maxime distinctis   Femora
antica subtus unidentata , tibiæ sulcatæ, extus unidigitatæ atque
unidentatæ; intermediæ calcaratæ.   Long. 11½, El. 6¼, lat. 3 mill."

The above is M Putzeys' original description, which he supplemented by a longer and more minute one in French, from which I take the salient features as follows :—*

The epistoma is widely emarginate, its angles are prominent and clearly separated from the wings which are rounded and a little more advanced The eyes are very prominent; posteriorly they are enclosed in the lateral margins of the head. The impression which separates the head from the neck is hardly distinct, especially in the middle. The striæ of the elytra are rather weak, but their puncturation is very distinct; they are less strongly impressed towards the external margin and hardly perceptible at the apex. The sixth interstice unites very indistinctly with the marginal border above the shoulder; not one of the striæ touches the base The anterior tibiæ have at the apex a rather short digitation and a large strongly marked tooth.

In his "Révision Générale" M. Putzeys forms a separate group (twenty-fifth) for *C. attrata,* and treats of it in the following terms : This species, unique up to the present, has so much resemblance to *C. australasiæ,* that at first sight it might be taken for a mere variety. The tooth of the mentum is longer, attaining the height of the lateral lobes. The mandibles are very short, broad, less arcuate, less acute, only carinate at the base. The prothorax is much more convex, hardly narrowed in front, almost square, with the sides rounded and the anterior angles very declivous. The elytra are truncate at the base, the shoulders marked, the striæ wider and more deeply punctate The fifth stria, and not the fourth touches the eighth interstice at the base The central carina of the prosternum is rather strongly narrowed between the coxæ, shortly and lightly canaliculate; the apex is oval, deeply foveolate on the base

* This revision being intended for the use of students in Australia, who often are unable to refer to the older (and scarce) literature of other countries, all M. Putzeys' species have been dealt with, and translations of his remarks (except Latin diagnoses) on all species that are unknown to the author have been given.

In regard to its habitat, the original description states that the author had seen only a single specimen which came from New Holland. The "Révision Générale" rather throws doubt upon this by saying that this insect, formerly received as coming from South America, appears rather to be Australian.

It may be noted that in his tabular view of the species of Clivina in his "Postscriptum," p 32, M. Putzeys gives as a distinguishing character of *C attrata*—eighth interstice not prolonged above the shoulder.

The species for which I propose the name of *C obliterata*, is an anomalous one among Australian species. It so closely resembles *C. australasiæ*, Bohem., as to seem merely a variety of that species; but as five specimens are before me, all agreeing in the basal characters of their elytra, I have felt compelled to regard it as distinct, and to place it with *C. attrata*, Putz. It requires more study, and should it prove to be a "sport" of *C. australasiæ*, of which there seems a possibility, it is a remarkable fact that the striæ free at the base should be accompanied by the total obliteration of the submarginal humeral carina

CLIVINA OBLITERATA, n.sp.

Facies as in *C. australasiæ*, only the elytra more truncate at base, with striæ free at base and submarginal humeral carina wanting; anterior tibiæ 3-dentate. Black, four posterior legs piceous. Only differing from *C. australasiæ* as follows ·—Head more evenly narrowed before eyes, (the sinuosity between the supra-antennal plates and wings of clypeus nearly obsolete), clypeus less deeply emarginate, the wings narrower; elytra with shoulders more marked (though rounded), more declivous, lateral border very fine, marginal channel very narrow behind and at shoulders, interstices flatter, eighth more finely carinate on apical curve, striæ lighter especially towards sides, fourth free, fifth hardly joining sixth at base, external teeth of tibiæ a little weaker. Length 9·5, breadth 2 6 mm.

Hab. : N.S. Wales—Carrathool, Mulwala (Sloane); Victoria (Kershaw).

The anterior margin of the clypeus is exactly as in *C. austral-asiæ*, emarginate with the wings not divided from the median part; the prosternum is exactly as in *C. australasiæ*. Apart from its smaller size, and the form of the clypeus and anterior tibiæ, this species seems to present a remarkable resemblance to *C. attrata*, Putz.

Biplagiata group.

Head wide, short, strongly and roundly angustate in front of eyes; clypeus deeply emarginate, median part not divided from wings. Elytra with striæ free at base; submarginal humeral carina wanting. Prosternum with intercoxal part wide anteriorly, sulcate on base. Anterior femora wide, lower side rounded, tibiæ 3-dentate.

Clivina biplagiata, Putzeys

Stett. Ent. Zeit. 1866, xxvii. p. 43, Ann. Soc. Ent. Belg 1866, x. p. 191.

Robust, convex. Black, with a reddish spot on each elytron just before apical declivity; anterior legs piceous, four posterior legs piceous red. Head wide, a shallow punctulate depression between clypeus and front; vertex smooth; clypeus deeply emarginate, wings small, not divided from median part; eyes prominent. Prothorax about as broad as long (1·8 × 1·75 mm.), widely convex, decidedly narrowed anteriorly; anterior angles very obtuse; basal curve short, rounded. Elytra convex, ovate, truncate at base, abruptly and deeply declivous to peduncle; striæ free at base, strongly punctate towards base, lighter and more finely punctate towards apex, seventh interrupted towards apex; interstices convex at base, depressed towards apex, eighth carinate on apical curve; submarginal humeral carina wanting. Prosternum with intercoxal part wide anteriorly, transversely sulcate on base; episterna finely transversely striolate. Anterior femora compressed, very wide, lower side rounded; anterior tibiæ 3-dentate. Length 7-7·8, breadth 2 mm. (One specimen in my collection only 5·5 mm. in length).

Hab · Queensland—Cape York (from Mr. French), Port Denison and Wide Bay (Masters); N.S. Wales—Sydney [common], Goulburn and Mulwala [rare] (Sloane); Victoria—Melbourne

An isolated and easily identified species. The red subapical maculæ of the elytra vary in size and brightness; in one specimen from Sydney in my possession they are wanting, the elytra being entirely black. I have not found any perceptible punctures on the prothorax as mentioned by Putzeys.

Cribrosa group.

Size moderate. Head short, wide and convex on occiput; clypeus with median part angular; facial sulci recurved; eyes depressed Prothorax short, parallel; anterior angles marked. Elytra with five inner striæ.free at base; submarginal humeral carina wanting. Prosternum with intercoxal part very narrow anteriorly, sulcate on base. Anterior tibiæ strongly 4-dentate.

The species known to me may be divided into sections thus :—

I. Clypeus with angles of median part obtuse .. .. $\begin{cases} C.\ cribrosa\ \text{Putz.} \\ C.\ boops,\ \text{Blkb.} \\ C.\ fortis,\ \text{Sl.} \end{cases}$

II. Clypeus with angles of median part prominent, dentiform. .... *C. frenchi*, Sl.

CLIVINA CRIBROSA, Putzeys.

Ann. Soc. Ent. Belg. 1868, xi p 20.

Robust, cylindrical, parallel. Head large, convex, coarsely punctate, eyes depressed; prothorax short, not narrowed anteriorly, anterior angles marked; elytra truncate on base, shoulders marked, striæ not deep, punctate, free at base; anterior tibiæ 4-dentate. Black (or piceous), legs reddish.

Head very convex, wide at base, sloping from vertex to anterior margin; vertex and occiput coarsely punctate, the punctures extending to middle of front· clypeus short; median part truncate (obsoletely emarginate between angles), angles prominent, short, triangular; wings wide, short, external angles wide, obtuse,

marked; a well marked sinuosity between wings and supra-
antennal plates, these wide, rounded externally; frontal impressions
wide, shallow, hardly marked; facial sulci hardly marked,
recurved part well marked; facial carinæ distant from eyes,
straight, carinate; eyes not enclosed behind. Prothorax broader
than long (1·3 × 1·4 mm.), very declivous to base; upper surface,
excepting basal declivity, densely and strongly rugulose-punctate;
sides parallel; anterior margin truncate, anterior angles marked,
very lightly advanced; posterior angles rounded; median and
anterior lines distinctly marked; lateral basal impressions obsolete.
Elytra a little wider than prothorax (3 2 × 1·5 mm); base truncate, deeply and abruptly declivous to peduncle; apex widely
rounded; striæ shallow, strongly punctate, entire, weaker near
apex, seventh weak, obsolete on apical curve; marginal channel
shallow in middle. Prosternum with intercoxal part very narrow
anteriorly, sulcate on base; episterna overhanging anteriorly, very
finely striolate near lateral margins. Anterior tibiæ wide, 4-
dentate; intermediate tibiæ with external spur distant from apex,
long, erect, acute.

Length 6-6·5, breadth 1·5 mm.

Hab. : West Australia—King George's Sound (Masters),
Beverley (Lea).

It greatly resembles *C. boops*, Blkb., some differences being its
smaller size, lighter form, the whole of the disc of the prothorax
strongly punctate, and the less strongly impressed elytral striæ.
The description given above is founded on specimens sent to me
by Mr. Masters; their colour is coal black; a specimen sent by
Mr. Lea is piceous; Putzeys gives the colour as piceous.

Note.—It is evident that Putzeys' measurements are incorrect;
the species is rather a stoutly built little one, and, even in the
most narrow species of *Clivina*, such a shape for the elytra as
"4½ × 1½ mm." would be unheard of.

CLIVINA BOOPS, Blackburn.

P.L.S.N.S.W. 1889 (2), iv. p. 719.

Very closely allied to *C. cribrosa*, Putz., which it exactly
resembles as to the head, shape of prothorax, elytra, legs, &c.; for

some apparent differences between them see description of *C. cribrosa* (*ante*, p. 157).

These species require careful study with large series of fresh specimens from different localities.

The dimensions of a specimen sent to me by Mr. Blackburn are: length 7; head 1·2 × 1·4; proth. 1·6 × 1·75; el. 4 × 1·9 mm.

Hab. : South Australia—Adelaide, Port Lincoln (Blackburn), Victoria—Melbourne (Kershaw).

CLIVINA FORTIS, n.sp.

Robust, cylindrical. Head punctate, large, wide and convex posteriorly, declivous in front, facial sulci recurved; prothorax broader than long, not narrowed anteriorly, striolate-punctate towards sides; elytra with striæ free at base, prosternum with intercoxal part very narrow anteriorly, sulcate on base, episterna hardly rugulose, very finely transversely striolate; anterior tibiæ 4 dentate. Black.

Head large, finely punctate on base of clypeus and middle of front; vertex and occiput very convex, not punctate; a wide shallow impression between clypeus and front: clypeus deeply declivous and rugose to median part, this narrow, strongly emarginate, its angles not marked, wings small, anterior margin sloping roundly and very lightly backwards from median part, supra-antennal plates rounded, bordered, divided from wings of clypeus by a light sinuosity, a submarginal ridge extending backwards from this sinuosity, facial sulci lightly impressed, recurved part elongate and very distinct ; facial carinæ short, strong ; eyes very depressed. Prothorax transverse (1 75 × 1·9 mm), widely convex, strongly declivous to base, smooth anteriorly, rugose-punctate towards sides of disc; sides parallel; anterior angles obtuse, but marked; posterior angles rounded, basal curve short, border narrow; median line strongly impressed; anterior line very lightly impressed Elytra wider than prothorax (4·2 × 2·2 mm), convex, parallel, truncate and abrupt at base, widely rounded at apex, striæ lightly impressed, entire, finely punctate; interstices lightly convex, eighth narrow (not carinate) on apical curve. Intermediate tibiæ wide, incrassate, about three small projections above external spur.

Length 7·8, breadth 2 2 mm.

Hab. : N.S. Wales (unique in Rev. T. Blackburn's Collection).

This species is closely allied to *C. boops*, Blkb., from which its most conspicuous differences are its larger size, more depressed eyes, and the obtuse anterior angles of the prothorax.

Note.—A specimen sent to me for examination by Mr. Masters, and ticketed Tasmania, only differs from the above in having the fine punctures of the head spread over all the posterior part; and the strong puncturation of the prothorax over nearly the whole of the disc, the angles of the median part of the clypeus a little marked, and the anterior angles of prothorax more prominent; I do not feel quite sure that it is conspecific with *C. fortis*, but am unable to regard it as distinct.

CLIVINA FRENCHI, n.sp.

Parallel, cylindrical. Head large, facial sulci recurved; prothorax broader than long, not narrowed in front; elytra with five inner striæ free at base, submarginal humeral carina obsolete; anterior tibiæ 4-dentate. Head, prothorax, and legs piceous (four posterior legs more lightly coloured than anterior); elytra brown.

Head large (1·7 × 1·8 mm.), wide behind eyes, convex, on upper surface a shallow puncturation, except on posterior part of vertex: clypeus not divided from front; median part truncate, its angles forming a strong triangular projection; wings about as prominent towards sides as the angles of median part, defined posteriorly by an oblique line, external angles rounded; lateral setigerous punctures large, placed behind angles of median part a little in front of the line defining the wings behind; supra-antennal plates large, projecting decidedly beyond wings of clypeus; facial sulci not clearly marked, turning inwards in front, an ill-defined short impression extending obliquely inwards and backwards from their anterior part on each side of vertex; facial carinæ short; eyes deeply embedded, hardly more prominent than supra-antennal plates; sides of head behind eyes finely and densely rugose-punctate; gulæ hardly rugulose. Mandibles short, flat. Mentum deeply and obliquely emarginate; lobes rounded at apex; median tooth broad, long, triangular. Prothorax a little broader than long (2·1 × 2·25 mm.), not narrowed anteriorly, convex, transversely striolate towards sides; anterior margin truncate; anterior angles lightly advanced, posterior angles rounded; basal curve short, border narrow; median line well marked, linear; anterior line variable (sometimes well marked, sometimes obsolete); lateral basal impressions usually well marked, elongate (reaching beyond middle of prothorax), rugulose   Elytra convex, a little wider than prothorax (5 × 2·5 mm.), parallel on sides, truncate at base, widely rounded at apex; striæ punctate for whole length, more lightly impressed towards apex; interstices lightly convex towards base, eighth not carinate at base, distinct and wide (not carinate) on apical curve.   Prosternum with intercoxal part attenuate anteriorly, transversely sulcate on base; episterna minutely shagreened, with fine wavy transverse striolæ.   Ventral segments smooth.   Anterior femora short, wide; anterior tibiæ 4-dentate, the upper tooth prominent, triangular; intermediate tibiæ with external spur long, acute.

Length 7 6-9, breadth 2-2 5 mm.

Hab. : North Queensland (from Mr. French); S. Australia— Lake Callabonna (Zietz).

The specimen of which the measurements are given in the description is 9 mm. in length.

Coronata group.

Size small.   Head depressed; eyes not prominent; clypeus with five triangular projections along anterior margin; supra-antennal plates also triangular in front.   Elytra with four inner striæ free, fifth joining sixth at base   Prosternum with intercoxal part attenuate anteriorly.   Anterior tibiæ 4-dentate

CLIVINA CORONATA, Putzeys.

Ann. Soc. Ent. Belg. xvi. 1873, p. 17.

Narrow, cylindrical. Clypeus with five prominent projections in front, prothorax parallel on sides; elytra parallel on sides, fifth stria joining sixth at base; prosternum with intercoxal part attenuate anteriorly; anterior tibiæ strongly 4-dentate. Testaceous, elytra more lightly coloured than head and prothorax.

Head depressed, lightly impressed, finely punctulate, frontal foveæ nearly obsolete, facial sulci obsolete, forming a wide shallow depression on each side of vertex; facial carinæ distant from eyes, feebly developed; supra antennal plates large, overshadowing the eyes at base, obtusely pointed in front; eyes not prominent. Prothorax rather longer than broad (1·25 × 1 2 mm), finely striolate near sides, lateral basal impressions elongate Elytra hardly wider than prothorax (2·7 mm. × 1·25 mm.), punctate-striate; striæ entire; interstices lightly convex, eighth marked on apical curve, submarginal humeral carina very fine and weakly developed Prosternum with episterna minutely shagreened, not transversely striolate Anterior femora wide, with lower edge rounded.

Length 5·2, breadth 1·25 mm.

Hab West Australia—King George's Sound (Masters)

This species is readily distinguished by the form of the anterior margin of the head with seven triangular projections. I have not found any perceptible punctures on the sides of the prothorax as mentioned by Putzeys. I have not been able to observe the base of the prosternum with accuracy in my specimen, so cannot say if it is transversely sulcate or not.

Obliquata group.

Size moderate or small. Front punctate, clypeus with angles of median part marked; facial sulci more or less recurved. Mandibles short. Elytra with four inner striæ free, fifth joining sixth at base; submarginal humeral carina present, not strongly developed. Prosternum with intercoxal part very narrow or

11

attenuate anteriorly, sulcate on base.   Anterior tibiæ 4-dentate
(the upper tooth sometimes feebly indicated or obsolete).

Table of Species.

I. Elytra punctate-striate.

  A. Unicolorous.

    B. Dorsal surface depressed.

      C. Prothorax as long as, or longer than broad.

        D. Size medium, fourth stria of elytra out-
turned at base . . . . *C. obliquata*, Putz.

        DD. Size small, fourth stria of elytra not out-
turned at base ...   .   *C. debilis*, Blkb.

      CC. Prothorax broader then long (none of the
elytral striæ outturned at base) . .   *C riverinæ*, Sl.

    BB. Form cylindrical.

      E. Anterior tibiæ 3 dentate, interstices of
elytra convex .. ..*C cylindriformis*, Sl.

      EE Anterior tibiæ 4-dentate, interstices of
elytra depressed . ..   . *C. obsoleta*, Sl.

  AA. Bicolorous.

      F. Elytra with basal part red, apical black   *C. melanopyga*, Putz.

      FF. Elytra reddish with a black sutural vitta *C. dorsalis*, Blkb.

      FFF. Elytra entirely ferruginous red .   *C. bicolor*, Sl.

II. Elytra with striæ simple . ... *C. denticollis*, Sl.

The members of this group which I do not know are *C. wildi*,
Blkb., evidently coming near *C. debilis; C. eremicola*, Blkb., allied
to *C. obliquata; and C. adelaidæ*, Blkb.

CLIVINA OBLIQUATA, Putzeys.

Ann. Soc. Ent. Belg. 1866, x. p. 188; and 1868, xi. p. 16.

Parallel, rather depressed.   Head widely convex, eyes not
prominent, front lightly punctate; facial sulci recurved; prothorax
depressed, parallel, not perceptibly narrowed anteriorly: elytra
parallel, punctate-striate; fourth stria outturned, but not joining

fifth at base; interstices lightly convex on basal part of disc, depressed posteriorly, eighth narrowly carinate at apex; submarginal humeral carina short, feebly carinate. Prosternum with intercoxal part small, very narrow anteriorly, sulcate on base; episterna minutely shagreened, the transverse striolæ hardly perceptible. Anterior femora wide, lower side rounded; tibiæ 4 dentate.

Head rather small; frontal impressions wide, well marked; clypeal elevation raised and prominent, clypeus divided from front by a shallow punctulate impression, depressed near anterior margin; median part emarginate truncate, its angles hardly advanced beyond wings, hardly marked; wings truncate, external angles marked, obtuse; supra-antennal plates large, projecting strongly and sharply beyond wings of clypeus, rounded and margined laterally; eyes lightly convex, not prominent, strongly enclosed behind. Prothorax rather longer than broad (1·75 × 1·7 mm.), sides widely and very feebly sinuate behind anterior angles; anterior margin truncate, anterior angles marked, obtuse Elytra elongate, very little wider than prothorax (3·8 × 1 75 mm); four inner striæ strongly impressed, fifth and sixth strongly impressed near base, becoming obsolete after anterior third, seventh entire, distinctly impressed; posterior puncture of third interstice near apex.

Length 7, breadth 1·75 mm.

Hab. . South Australia—Port Lincoln (Coll. Castelnau). (Two specimens were sent to me by Mr. Masters, ticketed South Australia.)

It appears probable that the identification of *C. obliquata* has been rendered difficult by a certain vagueness in Putzeys' description, *e g.*, when he says that *C. obliquata* may be distinguished at the first glance by its long, narrow and almost cylindrical elytra; this probably should be read as comparative to *C. melanopyga*, Putz.; the only other member of the group in which he placed *C. obliquata*, known to him, and of which he says the elytra are *elongate, almost cylindrical* (though, being a more than usually depressed species, I should not call them so); again,

though he places *C. obliquata* in a group characterised by the
fifth stria, not the fourth, reaching the eighth interstice, he says,
in the description, that the fourth *unites more or less distinctly
with the eighth at the base;* in *C obliquata* it turns out at the
base, but does not actually join the fifth.

CLIVINA DEBILIS, Blackburn.

P.L S N.S.W 1889 (2), iv. p 722.

Black, legs testaceous Narrow, elongate, subdepressed.
Clypeus with median part truncate, hardly distinct from wings,
its angles very weak, wings truncate, external angles squarely
obtuse, supra-antennal plates projecting strongly beyond wings of
clypeus. Prothorax quadrate (1 2 × 1 1 mm.). Elytra parallel
(2·8 × 1·3 mm); fifth stria joining sixth at base, seventh well
marked in all its course Prosternum with intercoxal part very
narrow anteriorly, transversely sulcate on base. Anterior tibiæ
narrow, 3-dentate (only an obsolete trace of an upper prominence).

Length 5, breadth 1 3 mm

Hab . South Australia—Adelaide, Port Lincoln (Blackburn).

Closely allied to *C. obliquata*, Putz , from which its small size
will at once distinguish it. The description above is founded
on a specimen for which I am indebted to Rev. T. Blackburn.

A specimen brought from Lake Callabonna (Central Australia)
by Mr A. Zietz, in 1893, differs slightly, being a little larger
(5·3 × 1 4 mm.), and having the prothorax with longer sides (basal
curve short), (1 4 × 1 2 mm.), the disc punctate near the sides; the
angles of the median part of the clypeus more prominent, the
" wings " more angulate, &c It may be a different, but closely
allied species ; to study it satisfactorily several specimens would
be necessary

CLIVINA RIVERINÆ, n.sp.

Wide, parallel, very depressed. Prothorax quadrate; elytra
punctate-striate, four inner striæ free at base; prosternum with
intercoxal part very narrow anteriorly, transversely sulcate on
base, anterior tibiæ 3-dentate. Black, shining, legs piceous

Head large (1·4 × 1·5 mm.), anterior part depressed; vertex wide, lightly convex, more or less punctate: clypeus declivous, divided from front by a wide—usually punctulate—depression; median part bordered, wide, lightly emarginate-truncate, its angles projecting obtusely beyond wings; these small, almost square, with external angle obtuse; supra-antennal plates large, bordered, projecting strongly and squarely beyond wings of clypeus, anterior angle obtuse, but marked; facial sulci deep, recurved part obsolete (sometimes feebly indicated); facial carinæ strong; eyes convex, rather prominent, lightly enclosed behind. Mentum wide, deeply and obliquely emarginate; lobes widely rounded at apex: median tooth triangular, acute. Prothorax depressed, quadrate (2 × 2·1 mm.), widest behind middle, very shortly declivous to base, a little narrowed anteriorly (ant. width 1 9 mm.); sides very lightly rounded; posterior angles rounded, not marked; basal curve short; anterior margin truncate; anterior angles wide, obtuse, a little prominent; border narrow; median and anterior lines strongly impressed; lateral basal impressions obsolete, or very lightly marked. Elytra depressed, hardly wider than prothorax (4 5 × 2·2 mm.), parallel, widely rounded at apex, truncate at base; striæ punctate, weaker towards apex, fifth and sixth obsolete except near base, seventh lightly marked, not punctate; eighth interstice narrow, subcarinate on apical curve; border narrow. Prosternum not protuberant; episterna finely shagreened, marked with wavy transverse lines. Anterior femora short, wide; anterior tibiæ strongly 3-dentate, a small triangular prominence above the upper tooth.

Length 7·2-8·6, breadth 2-2·7 mm.

Hab.: Victoria—Swan Hill (C. French); N.S. Wales—Urana District (Sloane—moderately plentiful on the edges of a large marsh 20 miles N.E. from Urana.)

Allied to *C. obliquata*, Putz., which it greatly resembles; it is a broader and more depressed species (being the most depressed Australian species), the prothorax is more transverse, being broader than long, and less parallel on the sides. The sub-marginal humeral carina of the elytra is very short and hardly carinate—it 'might be described as nearly obsolete. The specimen (δ) from which the measurements used in the description were taken was 8·4 mm. in length.

CLIVINA CYLINDRIFORMIS, n sp.

Narrow, cylindrical. Head with recurved facial sulci; prothorax as long as broad, longitudinally convex; elytra strongly punctate-striate, fourth stria free, lightly outturned at base, fifth joining sixth at base; prosternum with intercoxal part very narrow anteriorly; anterior tibiæ 3-dentate. Head, prothorax, and under surface of body piceous black; elytra piceous brown (piceous black near suture at beginning of apical declivity); under surface of prothorax piceous red, legs ferruginous.

Head convex (1 1 × 1·3 mm.); clypeus divided from front by a wide punctate impression, an elongate punctate depression in middle of front extending backwards from this impression; sides of head punctate behind eyes, the puncturation strong on each side above base of facial carinæ; median part of clypeus emarginate-truncate, bordered, its angles widely obtuse, hardly projecting beyond wings; these small, subrotundate in front with external margin widely rounded (their margin extends in a slightly uneven curve from median part to supra-antennal plates); supra-antennal plates large, explanate towards margin, projecting strongly and sharply beyond wings of clypeus, rounded on external margin: facial sulci strongly impressed, a short impression extending backwards from their anterior part on each side of vertex; facial carinæ strong, elongate, eyes convex, rather prominent, lightly enclosed behind; gulæ lightly striate on anterior part. Prothorax lævigate, convex, as long as broad (1·8 × 1·8 mm.), widest a little before the posterior angles, lightly narrowed anteriorly (ant. width 1·6 mm.); sides very lightly rounded; posterior angles not marked. basal curve rounded, anterior margin truncate; anterior angles subprominent, obtuse; border narrow on sides; median line linear, deep; anterior line obsolete; lateral basal impressions lightly marked. Elytra hardly wider than prothorax (4 × 1·9 mm), very convex, sides lightly rounded; base roundly truncate; seventh

stria not interrupted towards apex; interstices convex, eighth
narrow and distinct on apical curve; submarginal humeral
carina short and feebly developed; lateral border narrow.   Prosternum not protuberant, transversely sulcate on base; episterna
minutely shagreened, not transversely striolate.   Anterior femora
short, wide; lower side canaliculate, with posterior edge rounded.
   Length 7, breadth 1·9 mm.
   Hab. : Queensland—Gulf of Carpentaria (one specimen sent to
me by Mr. C. French).
   Differs from *C. obliquata*, Putz., in colour, facies, and the
3-dentate anterior tibiæ.

CLIVINA OBSOLETA, n.sp.

   Narrow, cylindrical.   Head wide; facial sulci obsolete; clypeus
with angles of median part projecting beyond the wings; eyes not
prominent; prothorax about as long as wide, very lightly
narrowed anteriorly; elytra parallel, fifth stria joining sixth at
base; prosternum with intercoxal part attenuate anteriorly;
anterior tibiæ strongly 4-dentate.   Ferruginous, elytra a little
more lightly coloured than head and prothorax.
   Head wide between eyes and across occiput; front finely, not
densely, punctate; vertex finely punctate on each side behind
facial carinæ; clypeal elevation truncate; median part of clypeus
depressed, defined on each side by a carinate ridge, truncate, its
angles projecting decidedly beyond wings in the form of obtuse
triangular teeth; wings small, concave, quadrate, external angle
marked; supra-antennal plates projecting beyond and divided
from clypeal wings by a sharp sinuosity; facial carinæ short,
weakly developed, eyes convex, not prominent, hardly at all
enclosed behind. Prothorax convex, smooth (except for a few transverse striolæ); anterior margin truncate; anterior angles obtuse,
feebly indicated; posterior angles widely rounded, basal curve
short, lateral basal impressions short, lightly impressed; median
line well marked; anterior line hardly marked.   Elytra long,
parallel (3·3 × 1·5 mm.), truncate and strongly declivous at base,
widely rounded at apex, very declivous to sides and apex; striæ

lightly impressed, entire, finely punctate; interstices not convex, eighth narrow near apex; submarginal humeral carina short, narrow, weak.   Prosternum with episterna minutely shagreened Anterior femora wide, lower side rounded; anterior tibiæ widely palmate, upper internal spine thick, curved, incrassate.

Length 6, breadth 1·5 mm.

Hab. · Queensland—Cape York (unique in the collection of the Rev. T. Blackburn).

This is an isolated species; in general appearance it is rather like *C. blackburni*, Sl , but its nearest ally known to me seems to be *C. frenchi*, Sl., which it resembles in its widely palmate tibiæ; in *C. frenchi* the upper internal spine of the anterior tibiæ is greatly developed, though not so thick as in *C. obsoleta*. I have placed it in the "*obliquata group*," because it has the elytra with the fifth stria joining the sixth at base, and has a submarginal carina at each shoulder.

CLIVINA MELANOPYGA, Putzeys.

Stett. Ent. Zeit 1866, xxvii. p. 41; Ann. Soc. Ent. Belg. x 1866, p. 187.

This species is at once distinguished from all other Australian species by its colour, its rather depressed form, and by having the four inner striæ of the elytra free at the base.   The following brief note will sufficiently characterise it.

Head, prothorax, undersurface and apical part of elytra black, elytra reddish on more than anterior half; legs piceous   Head, including clypeus, as in *C. obliquata*, Putz., prothorax quadrate (1·5 × 1·5 mm.): elytra depressed on disc (3 × 1 5 mm), punctate-striate, four inner striæ free, fifth joining sixth at base; submarginal humeral carina short, weakly developed; prosternum with intercoxal part attenuate anteriorly; anterior tibiæ 3-dentate, a fourth upper tooth feebly indicated.

Length 5·6-6·5, breadth 1·5-1 8 mm.

Hab. : N.S. Wales—Urana District (Sloane—one specimen), Victoria—Swan Hill (French), Melbourne (Kershaw); South Australia.

CLIVINA DORSALIS, Blackburn.

P L.S.N.S.W. 1889 (2), iv. p. 719.

Parallel, lightly convex.   Black; elytra red with a black sutural stripe (this stripe occupying only first interstice at base, widening posteriorly and extending over three inner interstices, not reaching apex); anterior legs ferruginous, four posterior testaceous   Front punctate, clypeus with median part lightly emarginate-truncate, its angles hardly marked, its wings small with anterior margin truncate, their exterior angles obtuse but marked, facial sulci recurved   Prothorax quadrate (1·2 × 1·2 mm), evenly and lightly convex, punctulate.   Elytra a little broader than prothorax (2 5 × 1·35 mm), widely rounded at apex, evenly and lightly convex; striæ strongly impressed, entire, punctate, fifth joining sixth at base.   Prosternum with intercoxal part attenuate anteriorly, transversely sulcate on base; episterna minutely shagreened, obsoletely transversely striolate.   Anterior tibiæ 4-dentate, the upper tooth very feeble.

Length 5, breadth 1·35 mm.

Hab. · Victoria (Kershaw); South Australia—Adelaide, Port Lincoln (Blackburn); West Australia—King George's Sound (Masters), Beverley (Lea).

This species agrees with M. Putzeys' original description of C. suturalis in every particular, except that from the group in which he placed C. suturalis it should have the fourth stria joining the fifth at the base, but he placed C. planiceps in the same group as also having the fourth stria joining the fifth at the base, which was incorrect, and it is impossible for me to avoid a suspicion that C. dorsalis, Blkb., = C suturalis, Putz. If so, Putzeys' description is erroneous, and nothing but an inspection of his type, or the discovery of a species coloured like C. dorsalis, and having the fourth and fifth striæ of the elytra confluent at the base, can now settle the point *

* See descriptions of C. suturalis and C. verticalis (post) for further remarks on this subject.

CLIVINA BICOLOR, n.sp.

Narrow, parallel, subdepressed. Head short, convex, facial sulci recurved, eyes not prominent: prothorax longer than broad, parallel on sides; upper surface densely and strongly punctate: elytra parallel, finely punctate-striate; four inner striæ free, fifth joining sixth at base; interstices depressed, eighth carinate at apex, and shoulders; anterior tibiæ 4-dentate. Elytra ferruginous-red; prothorax and head piceous, under surface piceous.

Head convex and smooth on vertex, a few fine punctures on anterior part of front: clypeus with median part truncate, its angles prominent, triangular; wings wide, subquadrate, hardly as advanced as angles of median part, external angles strongly marked, obtuse at summit, external margin straight; supra-antennal plates large, projecting sharply and strongly beyond wings of clypeus; facial carinæ hardly marked, eyes convex, not at all prominent, weakly enclosed behind. Prothorax longer than broad (1·2 × 1·1 mm.), lightly convex, lightly declivous to base; upper surface—excepting basal declivity and anterior collar—strongly punctate; sides parallel, a little narrowed at anterior angles; anterior margin truncate; anterior angles marked; lateral basal impressions lightly marked, elongate. Elytra very little wider than prothorax (2 5 × 1 25 mm.); sides subparallel (hardly rounded), a little narrowed to base; shoulders obtuse, but marked, base lightly emarginate behind peduncle, striæ entire, lightly impressed, finely punctate, seventh entire, interstices depressed; submarginal humeral carina long, narrow. Prosternum with intercoxal part cordate, narrow anteriorly, episterna sublævigate (very minutely shagreened).

Length 4·7, breadth 1·25 mm

Hab. : West Australia—King George's Sound (unique, sent by Mr Masters).

Allied to *C. dorsalis*, Blkb., from which its colour and the prothorax with the whole of the disc punctate at once distinguish it. the angles of both the median part and the wings of the clypeus are far more prominent than in *C. dorsalis*

CLIVINA DENTICOLLIS, n.sp.

Robust, lightly convex. Head depressed, transversely impressed posteriorly, eyes very large and convex; prothorax subquadrate, posterior angles marked, shortly dentate: elytra parallel, simply striate; four inner striæ free at base; a well marked striole at base of first interstice; submarginal humeral carina wanting: prosternum with intercoxal part canaliculate, wide anteriorly, transversely sulcate on base, episterna very finely transversely striolate, not overhanging in front; lateral cavities of peduncle punctulate: anterior tibiæ strongly 3-dentate; intermediate tibiæ not wide, external spur stout, acute, very near apex. Ferruginous, eyes black.

Head depressed, widely impressed across occiput; front depressed, rugulose; frontal impressions very shallow; facial sulci wide, shallow, nearly obsolete; vertex smooth, minutely punctulate, facial carinæ wide, short, lightly raised: clypeus with median part truncate, its angles small, obtuse, very lightly advanced; wings small, concave (less advanced than median part), external angles rounded; supra-antennal plates rather depressed, rounded externally, a strong sinuosity dividing them from clypeus wings; eyes very large, convex, prominent, projecting far beyond supra-antennal plates; gulæ smoother than usual, lightly punctate near eyes. Labial palpi stout, terminal joint stout, subfusiform (obtuse at apex). Prothorax broader than long ($1\cdot3 \times 1\cdot4$ mm.), lightly and evenly convex: disc covered with fine transverse striolæ; anterior margin truncate, vertical at sides of neck; anterior angles obtuse; sides evenly rounded; posterior angles marked by a short but decided dentiform projection; basal curve short; border narrow, lightly reflexed on sides, very fine (not reflexed) on sides of basal curve; median and anterior lines strongly impressed; lateral basal impressions wanting   Elytra much wider than prothorax ($3\cdot3 \times 1\cdot8$ mm.), lightly rounded on sides, widely rounded at apex; base truncate; striæ simple, entire, lightly impressed, fifth joining sixth at base, seventh entire; interstices depressed, eighth hardly carinate on apical curve. Anterior femora not channelled below, lower side not dilatate or rounded.

Length 6, breadth 1 8 mm.

Hab. : West Australia—N. W. Coast (?); (sent to me by Mr. C. French).

A remarkable and isolated species, not nearly allied to any other Australian species In facies it resembles *C. pectoralis*, Putz.; its head is much like that of *C. bovillæ*, Blkb., but the eyes are larger; the form of the clypeus is like that of the species of the "*obliquata group*", the intercoxal part of the prosternum is as wide as in typical members of the "*australasiæ group.*" Although I have placed it in the "*obliquata group*," it might well be regarded as the type of a new group, of which the characters would be those of the preliminary paragraph of the description above.

Planiceps group.

Size large. Mandibles long, decussating. Clypeus with median part truncate, wings wide, truncate, sharply advanced. Labrum truncate, 5-setose. Labial palpi with penultimate joint slender, longer than terminal. Elytra with four inner striæ free at base, fifth joining sixth; submarginal humeral carina present. Prosternum with intercoxal part very wide anteriorly, non-sulcate on base.

Table of species.

A. Anterior tibiæ 3-dentate   .   ...   *C. planiceps*, Putz.

AA. Anterior tibiæ 4-dentate   . ...

    B. Head rugulose ......···....................... . ... *C. quadratifrons*, Sl

    BB. Head smooth ..   .   . *C. carpentaria*, Sl.

C. crassicollis, Putz., allied to *C. planiceps*, is unknown to me

Clivina planiceps, Putzeys.

Mém. Liége, 1863, xviii. p. 42, *Ceratoglossa rugiceps*, Macl., Trans. Ent. Soc. N.S.W. 1863, i. p. 72; *Scolyptus planiceps*, Putz., Ann. Soc. Ent. Belg. 1866, x. p. 24.

A well-known species, which may be distinguished by the following note :—

Cylindrical. Black, under surface piceous, legs reddish or reddish piceous. Head large (2 3 × 2·3 mm.), depressed, rugulose; clypeus with wings strongly and obliquely advanced beyond the truncate median part. Prothorax longer than broad (3·5 × 3·3 mm), lightly narrowed anteriorly (ant. width 3 mm). Elytra parallel (7·6 × 3 5 mm.), crenulate-striate; four inner striæ free at base, fourth a little outturned at base, fifth joining sixth at base, eighth interstice distinct on apical curve; a submarginal carina at shoulders. Anterior tibiæ 3-dentate

Length 12 5-16·5, breadth 3-4 mm.

Hab N S. Wales—Murray and Murrumbidgee Rivers.

M. Putzeys in his "Postscriptum" places this species in a group characterised by having the fourth and fifth striæ confluent at base, he makes no reference to this feature in his description, nor does he remark on it in Stett. Ent. Zeit , nor in his " Révision Genérale," where he merely puts it in *Scolyptus*, and places *Ceratoglossus rugiceps*, Macl., as a synonym without comment. Rarely the fourth interstice does turn outwards at the base, and actually join the fifth; one such example is in my collection from Mulwala on the Murray, where this species is very common.

CLIVINA CRASSICOLLIS, Putzeys.

Scolyptus crassicollis, Putz., Ann. Soc Ent. Belg. 1866, x p. 25

The following is a translation of Putzeys' whole description (*sic*) of this species :—

Larger than *C. planiceps*, its elytra are proportionately more elongate, the prothorax is very noticeably more convex, more declivous particularly towards the anterior angles, the anterior margin is less emarginate.

Length 18, el. 9, breadth 4 mm.

New South Wales—two specimens.

The above is an example of the uselessness of some of M. Putzeys' descriptions; it might be founded on the large specimens

from the Gulf of Carpentaria mentioned below under *C. quadratifrons*, Sl.; but, if so, the description does not aid one in determining it, besides the inference is that the anterior tibiæ are
3-dentate as in *C. planiceps*.

CLIVINA QUADRATIFRONS, n.sp.

Robust, parallel, cylindrical. Head flat, rugulose; prothorax
about as long as broad; elytra with fifth stria joining sixth at
base, eighth interstice distinctly marked on apical curve, a well-
developed submarginal carina at shoulders; anterior tibiæ 4-
dentate. Black, under surface piceous, anterior legs reddish
piceous, four posterior legs and antennæ testaceous brown.

Head quadrate (2 × 2·1 mm.), flat, rugulose: clypeus not
divided from front; median part truncate; wings divided from
supra-antennal plates by a light linear impression, lightly and
obliquely advanced beyond median part, wide, truncate, external
angle marked, rounded; supra-antennal plates depressed, declivous
before eyes, divided from clypeal wing by a light sinuosity,
external margin sinuate; facial sulci lost in facial rugulosity,
facial carinæ distant from eyes, feebly developed; eyes convex,
prominent; orbits narrow, abruptly truncate behind eyes. Mandibles wide at base, decussating. Mentum concave; lobes rounded
at apex, lightly longitudinally striate; median tooth large, rounded
at apex. Prothorax of almost equal length and breadth
(3·6 × 3·5 mm), parallel on sides, very little narrowed to apex,
convex, roundly declivous to base; anterior margin truncate;
anterior angles obtuse; posterior angles not marked; border wide
and explanate near anterior angles, narrow backwards, not interrupted at posterior angles; median and anterior lines well marked,
lateral basal impressions short, shallow, subfoveiform. Elytra
parallel, cylindrical (8·5 × 4 mm.), truncate on base; striæ entire,
lightly crenulate, deeply impressed, becoming shallow towards
apex, first outturned to join second at base, fourth free at base,
interstices lightly convex, eighth forming a narrow carina on
apical curve; border narrow. Prosternum protuberant, intercoxal part very wide in front, widely and lightly channelled,

abrupt and non-sulcate on base; episterna covered with fine wavy
transverse striæ. Ventral segments smooth. Anterior femora
short, wide, compressed, lightly channelled below, posterior margin
of lower side wide in middle; tibiæ wide, palmate, external teeth
strong and close together, intermediate tibiæ wide, incrassate,
external edge arcuate above subapical spur, this strong, acute.

Length 13·5-16, breadth 3·3-4·2 mm.

Hab.. New South Wales—Urana District (Sloane); Victoria
—Mildura (French).

Note.—Two specimens have been sent to me by Mr. C. French
as coming from near Burketown on the Gulf of Carpentaria,
which, though appearing at first sight to be a different species
from *C. quadratifrons*, yet, on a close examination, reveal no
differences that I can see, except their larger size. I regard them
as merely the northern form of a widely distributed species
(dimensions, head 2·8 × 2·8 mm , prothorax 4·5 × 4·3 mm., elytra
10 × 4·6 mm.). It is possible this may be *C. crassicollis*, Putz.,
but it is not to my eye a more elongate and convex species than
C. planiceps; besides Putzeys' brief note (not a description) on
C crassicollis seems to infer only 3-dentate anterior tibiæ for
that species.

C. quadratifrons is closely allied to *C planiceps*, which it
resembles in size and appearance; but decided differences to which
attention may be directed are the shorter and more parallel
prothorax, the clypeus with the wings less advanced beyond the
median part, and the 4-dentate anterior tibiæ.

CLIVINA CARPENTARIA, n.sp.

Narrow, cylindrical. Head not rugulose; prothorax longer
than broad. elytra with striæ entire, fifth joining sixth at base;
interstices convex, eighth not visible on apical curve; ventral
segments rugulose laterally; anterior tibiæ 4-dentate   Black,
shining; legs piceous brown.

Head smooth, large, depressed (1·6 × 2 mm.); a shallow transverse line dividing clypeus from front, and a strong sulcus dividing
clypeal wings from supra-antennal plates ; clypeal elevation

well defined, almost semicircular: clypeus with median part
truncate, wings lightly and abruptly advanced beyond median
part, wide, flat, truncate, rounded at external angles and laterally,
supra-antennal plates depressed, declivous externally, lightly
rounded, narrowly margined, facial sulci short; supra-orbital
setæ placed near each eye in a short depression, upper edge of
this depression forming a thick round carina, lower edge forming
a narrow carina; eyes globose, very prominent, projecting strongly
from sides of head.   Mandibles large, wide at base, decussating
Mentum deeply and obliquely emarginate, median tooth wide,
short, lobes strongly striolate, rounded at apex.   Prothorax
lævigate, longer than broad (2·8 × 2 5 mm.), widest a little in
front of posterior angles, a little narrowed anteriorly (ant. width
2 25 mm), sides lightly and widely sinuate, posterior angles
rounded; anterior margin truncate, anterior angles obtuse, border
reflexed on sides, median and anterior lines strongly impressed,
lateral basal impressions wanting.   Elytra cylindrical, parallel,
hardly wider than prothorax (5·7 × 2 6 mm), base widely and
very lightly emarginate, shoulders obtuse, apex strongly declivous,
striæ strongly impressed, crenulate, interstices convex, seventh
and eighth uniting and forming a short carina at base; lateral
border narrowly reflexed.   Prosternum protuberant, intercoxal
part wide anteriorly, not transversely sulcate on base, episterna
finely rugulose and transversely striolate.   Ventral segments
smooth in middle, first and second strongly and closely longitudinally striolate, third striolate-punctate, fourth, fifth and sixth
rugulose-punctate at sides   Anterior femora short, wide, lightly
channelled below, posterior margin of lower side wide, anterior
tibiæ wide, palmate, three external teeth very strong and close
together.

Length 11, breadth 2·5 mm.

Hab. : Queensland—Gulf of Carpentaria (sent to me by Mr.
C French).

Grandiceps group.

Size large.   Head large, clypeus with median part wide,
rounded, a light wide sinuosity dividing it on each side from

wings; these very wide, rounded, hardly more advanced than centre of median part. Mandibles long, decussating, wide at base Labrum 5-setose. Palpi filiform; labial with penultimate joint slender, longer than terminal. Prothorax transverse, border not reaching base on sides of basal curve, anterior marginal puncture very near anterior angle. Elytra with four striæ free at base; submarginal humeral carina short, feebly developed. Prosternum with intercoxal part greatly narrowed (not attenuate) anteriorly. Anterior tibiæ 4-dentate.

Clivina grandiceps, n.sp.

Comparatively short. Head large, smooth, vertex convex; prothorax short, lateral border not attaining base; anterior tibiæ 4-dentate Black, shining; legs light piceous brown; palpi piceous.

Head large, transverse (2·4 × 2·9 mm.); vertex convex, lævigate: clypeus slightly rugulose, divided from front by a straight transverse impression (this impression hardly distinct in middle); anterior margin sinuate; median part lightly rounded in middle; wings large, wide, divided from median part by a light sinuosity, widely rounded in front and laterally, a little more prominent than median part, lateral setæ placed in a sharply defined foveiform puncture about middle of each wing; supra-antennal plates small, convex, divided from clypeal wings by a light sinuosity, roundly protuberant and margined laterally; facial sulci lightly impressed, two supra-orbital setæ on each side placed a considerable distance from eye in a deep groove, the lower as well as the upper edge of this groove carinate; eyes convex, projecting beyond supra-antennal plates; orbits enclosing eyes lightly behind, sloping obliquely to neck. Mandibles large, wide at base, decussating. Labrum large; anterior margin subrotundate (lightly truncate in middle), 5-setose. Mentum lightly and squarely emarginate, median tooth short, widely triangular; lobes rugulose, wide, obliquely truncate to apex on external side. Palpi filiform. Antennæ long, slender, not incrassate, first joint long (about as long as two succeeding ones). Prothorax short, transverse (2·2 × 2·9 mm.),

12

widest just behind anterior angles, convex, slightly depressed on each side of median line, abruptly declivous to base, sides parallel. anterior margin emarginate in middle; anterior angles obtuse, explanate: posterior angles wide, but marked, basal curve short, lateral border wide and reflexed on sides, interrupted and upturned at posterior angles just before posterior marginal puncture, thick and indistinct on anterior part of basal curve, obsolete on posterior part and not reaching base, border strongly reflexed and marginal channel wide on base: median and anterior lines strongly impressed, lateral marginal punctures large, anterior placed near anterior angle on the explanate border. Elytra convex, very little wider than prothorax (5 7 × 3·1 mm), hardly narrowed to base, wide at apex, sides lightly rounded, base truncate, shoulders rounded, striæ entire, crenulate, strongly impressed, weaker on apical declivity, fifth joining sixth at base, seventh obsolete on apical curve; interstices convex, eighth obsolete towards apex, submarginal humeral carina short, thick; lateral border wide, reflexed.   Prosternum with intercoxal part lightly concave, narrow (not attenuate) anteriorly, base abrupt, not transversely sulcate; episterna overhanging in front, transversely rugulose-striate.   Ventral segments smooth, excepting two basal ones lightly longitudinally striolate   Anterior femora light, lower side straight; anterior tibiæ 4-dentate, apex strongly outturned, external teeth wide apart, strong, triangular; external spur of intermediate tibiæ fine, acute

Length 10 5, breadth 3·1 mm

Hab   Queensland—Gulf of Carpentaria (one specimen, given to me by Mr. C French)

Punctaticeps group.

Size small. Facial sulci not recurved, clypeus with median part emarginate, its angles more or less marked   Elytra with fourth and fifth striæ confluent at base, seventh not interrupted at beginning of apical curve, submarginal humeral carina well marked, a distinct elongate striole at base of first interstice

Prosternum with intercoxal part attenuate anteriorly, sulcate on base. Anterior femora with posterior margin of lower side strongly dilatate in middle, tibiæ 4-dentate.

Table of species.

A. Form cylindrical, prothorax longer than broad { *C. punctaticeps*, Putz. { *C. tumidipes*, Sl.

AA. Form subdepressed, prothorax broader than long . .. . *C. lobipes*, Sl.

CLIVINA PUNCTATICEPS, Putzeys.

Ann. Soc Ent. Belg. 1868, xi. p. 18.

Closely allied to *C. tumidipes*, Sl., of which it seems the northern form, and from which it only appears to differ by its ferruginous colour; prothorax proportionately wider; elytra a little more deeply striate, the interstices more convex. The legs are similar in all respects.

I offer the following brief diagnosis founded on a specimen sent to me for examination by the Rev. Thos. Blackburn :—

Elongate, cylindrical. Head moderate; front punctulate; vertex coarsely punctulate in middle and posteriorly from side to side· clypeus with median part projecting strongly beyond wings, lightly emarginate, its angles prominent, triangular, wings small, rounded, strongly divided from median part and lightly from supra-antennal plates. Prothorax a little longer than broad (1·6 × 1·5 mm.), a little narrowed anteriorly (ant. width 1·3 mm.). Elytra oval (3·5 × 1·75 mm.), strongly punctate-striate; fourth stria joining fifth at base; a distinct striole at base of first interstice, the interstices convex, eighth well defined for whole length, carinate at base. Prosternum with intercoxal part attenuate anteriorly. Anterior femora thick, strongly and roundly dilatate on middle of lower side, anterior tibiæ 4-dentate.

Length 5 5-6, breadth 1·7-1·75 mm.

Hab. . Queensland—Cape York , Rockhampton (Coll. Blackburn, Macleay Museum).

CLIVINA TUMIDIPES, n.sp.

P.L.S.N.S.W. 1889 (2), iv. p. 720.

Elongate, parallel. Head punctulate anteriorly, eyes prominent, prothorax longer than broad, convex: elytra parallel, convex, punctate-striate; fourth and fifth striæ confluent at base; a short distinct submarginal carina at shoulder; an elongate fine striole at base of first interstice, anterior femora with posterior margin of lower side strongly and roundly dilatate, anterior tibiæ 4-dentate. Black, shining; under surface piceous; anterior legs piceous brown; four posterior legs, antennæ and palpi reddish testaceous.

Head moderate; front closely and finely punctate; vertex smooth (sometimes some fine punctures near posterior extremity of each facial carina): clypeus not divided from front; median part deeply and rather angularly emarginate, its angles obtuse, very lightly advanced beyond and hardly divided from wings; these small, hardly divided laterally from supra-antennal plates; lateral setæ of clypeus placed in a rugose depression at base of each wing, supra-antennal plates small, depressed; eyes globose, prominent, lightly enclosed behind; orbits abrupt behind. Prothorax smooth (sometimes a few transverse wrinkles on disc), longer than broad (1·7 × 1·5 mm.), widest near posterior angles, very little narrowed anteriorly (ant. width 1·4 mm.); sides lightly subsinuate behind anterior marginal punctures, decidedly narrowed from these to anterior angles; anterior margin truncate; anterior angles projecting very slightly: lateral basal impressions obsolete. Elytra narrow, parallel, hardly wider than prothorax (4 × 1·7 mm); base truncate; striæ entire, narrow, lighter towards apex, closely punctate, seventh strongly marked in all its course; interstices lightly convex, eighth well developed on apical curve. Prosternum with intercoxal part attenuate anteriorly, transversely sulcate on base; episterna overhanging greatly anteriorly, shagreened, transversely striolate. Ventral segments minutely shagreened under a strong lens. Anterior femora short, wide, compressed; anterior tibiæ with two strong external teeth and a short triangular prominence above apical projection; anterior trochanters projecting lightly and obtusely beyond base of femora.

Length 5-6·7, breadth 1·3-1·7 mm.

Hab.: N.S. Wales—Junee District, Urana District (Sloane); Victoria—Swan Hill (French); South Australia—Adelaide (Blackburn).

This species must be very closely allied to *C. emarginata*, Putz., but evidently differs in colour. I took it plentifully twenty miles north-east from the town of Urana on the margins of tanks dug to water sheep (the only permanent water), in the months of December and January, as many as 32 specimens were washed out of part of the muddy margin of one tank in less than half an hour.

CLIVINA EMARGINATA, Putzeys.

Ann. Soc. Ent. Belg. 1868, xi. p. 15.

"Nigra nitida, ore, antennis, pedibus, elytrorum basi apiceque externis testaceis. Clypeus emarginatus, alis subæqualis. Vertex antice profunde .et dense punctatus. Prothorax subquadratus, lævis. Elytra cylindrica, basi intus oblique truncata, humeris rotundatis. Femora antica extus in medio inferiore dilatata. Long 5¾, El 4, Lat. 1¼ mill."*

M. Putzeys supplemented this diagnosis by remarks which I translate as follows :—

This species forms a link between the twenty-seventh group in which the rounded wings of the epistoma extend considerably beyond the epistoma itself and the twenty-eighth,† in which the

---

* It is evident there is an error in these measurements; the length given for the elytra is certainly too great.

† By twenty-seventh and twenty-eighth groups M. Putzeys appears to have meant, on this occasion, the groups of which *C. nyctosyloides*, Putz., for which he formed a new twenty-seventh group in place of his old twenty-seventh, *C. procera* being transferred to *Scolyptus*, and *C. heterogena*, Putz., are respectively the types; but as on the following page he refers *C. heterogena* to a *thirtieth group* it is apparent that twenty-eighth is a mistake.

epistoma, more or less emarginate, has its angles prominent, extending beyond the wings, which are usually angular

In *C. emarginata* the epistoma is deeply emarginate; its angles are not more advanced than the wings, from which it appears to be separated by a depression which there is between them. The anterior elevation, broad, though but little raised, is strongly punctate the same as all the anterior part of the head; the puncturation almost disappears on the vertex, which is very convex and the fovea of which is shallow. The prothorax is almost square, just a little longer than broad; the sides are lightly narrowed at the anterior third, but then regain their width up to the anterior angles, which are obtuse and declivous. The surface is smooth, the median line is very deep from the base to the anterior line, one can hardly distinguish a feeble trace of the two lateral foveæ. The elytra are cylindrical, obliquely truncate, internally at the base, the shoulders are rounded; the striæ become hardly distinct towards the apex; they are strongly punctate The anterior femora are thick, their lower surface is dilatate externally so as to form a rounded prominence, but the trochanter projecting at the apex makes a prominent angle.

Australia   One specimen (Coll. Casteln.)

In facies *C. emarginata* must resemble *C tumidipes*, Sl., but it is differently coloured. The clypeus may resemble that of *C. lobipes*, Sl., but seems as if it should be not unlike *C. bovillæ*, Blkb. I should expect the tibiæ to be 4-dentate, and the prosternum with the intercoxal part narrow. Its colour should render its recognition easy. I have associated it with *C. adelaidæ* on account of the form of the anterior femora.

CLIVINA LOBIPES, n.sp.

Robust, parallel, subdepressed. Head short, wide, finely rugulose-punctate; prothorax subquadrate, punctate on disc; elytra punctate-striate, fourth stria joining fifth at base, prosternum with intercoxal part attenuate, transversely sulcate on base, episterna strongly rugose and transversely striolate; anterior femora lobate, tibiæ strongly 4-dentate. Reddish piceous; elytra lighter coloured than head and prothorax, with a dark piceous spot on posterior part of disc.

Head wide, depressed; front and clypeal elevation closely rugulose-punctate; a round fovea in middle behind punctate part, vertex wide, smooth; frontal impressions wide, shallow; facial sulci lightly impressed; clypeal elevation hardly raised: clypeus not divided from front; median part deeply emarginate, defined on each side by a slight ridge, not angulate laterally, wings small, not divided from median part, sloping roundly backwards to and divided from supra-antennal plates by a faint wide sinuosity; eyes prominent, hemispherical, lightly enclosed behind. Prothorax subquadrate ($1\cdot5 \times 1\cdot55$ mm.), lightly convex, coarsely punctate except on anterior part of disc and near sides; anterior margin truncate, angles obtuse, but marked; sides parallel, lightly and widely emarginate; posterior angles marked; basal curve sloping sharply to base on each side; median line deeply, anterior line lightly impressed. Elytra very little wider than prothorax ($3\cdot2 \times 1\cdot6$ mm.), convex—not cylindrical,—parallel on sides; base truncate; shoulders rounded, with border prominent; striæ entire, seventh not interrupted at beginning of apical curve; interstices lightly convex, eighth finally carinate at base, narrow and lightly carinate near apex. Anterior femora with lower side forming a wide round protuberance; external spur of intermediate tibiæ long, acute.

Length 6·3, breadth 1·6 mm.

Hab.: Queensland—King's Plains Station (28 miles S.W. from Cooktown; one specimen sent to me by Mr. N. H. Gibson).

It seems to be allied to *C. emarginata*, Putz., the clypeus and anterior femora are apparently similar, but *C. lobipes* is evidently a broader species, differing in having the prothorax not longer than broad, and roughly punctate on the disc. From *C. tumidipes*, Sl., and *C. punctaticeps*, Putz., species with lobate anterior femora, it is easily distinguished by its wider and less cylindrical form, shorter punctate prothorax, &c.

Blackburni group.

Size small, form cylindrical. Head large, convex; occiput short, wide; eyes not prominent; facial sulci recurved; clypeus with angles of median part very lightly advanced beyond wings, these with external angles rounded, but marked; supra-antennal plates projecting strongly beyond clypeus. Prothorax longer than broad, anterior line wanting Elytra with fourth and fifth striæ confluent at base Prosternum with intercoxal part attenuate anteriorly, sulcate on base Anterior tibiæ 4-dentate.

The facies of this species, the short wide head, the long narrow cylindrical prothorax and elytra, the non-prominent eyes, &c , have caused me to separate *C. blackburni* from *C. heterogena*, Putz., and form a distinct group for it

CLIVINA BLACKBURNI, n sp.

Narrow, parallel, cylindrical. Head large, facial sulci recurved, eyes very depressed; prothorax longer than broad, anterior line wanting. elytra lightly punctate-striate, fourth stria joining fifth at base, interstices depressed, eighth carinate at base, narrow and carinate on apical curve; anterior tibiæ 4-dentate. Piceous brown.

Head large, convex; vertex smooth; front finely punctate: clypeus not divided from front, declivous to median part; this depressed, truncate-emarginate, its angles projecting lightly and obtusely beyond wings, lateral ridges short, wide, distinct; wings subquadrate, with external angles rounded, supra-antennal plates long, lightly rounded externally, projecting sharply and decidedly beyond wings of clypeus, bordered; a longitudinal ridge extending backwards from base of clypeal wings; facial sulci lightly impressed, an elongate impression extending backwards from their anterior part, facial carinæ distant from eyes, short; eyes depressed, deeply set in head, hardly projecting; orbits forming a thick ridge above eyes, projecting sharply but lightly from head behind. Antennæ moniliform, incrassate; joints 5-11 very short, transverse, compressed Mentum deeply emarginate;

median tooth moderate, triangular, pointed. Mandibles short, thick. Prothorax smooth (a few light rugæ near sides), parallel, very little wider than head with eyes, longer than broad (1·4 × 1 mm.), roundly and strongly declivous to base; anterior margin truncate; base wide; basal curve short, rounded; posterior angles widely rounded; basal angles obtuse; median line well marked, linear. Elytra parallel, cylindrical (3 × 1·2 mm), truncate at base, widely rounded at apex; apical declivity roundly abrupt; striæ entire, lightly impressed, finely punctate; interstices not at all convex, posterior puncture of third much nearer apex than usual Prosternum with intercoxal part attenuate anteriorly, transversely sulcate on base; episterna obsoletely transversely striolate, overhanging anteriorly. Legs short; anterior femora short, thick, rounded on lower side; anterior tibiæ strongly 4-dentate; upper tooth short, triangular; posterior tibiæ short, incrassate.

Length 5 3, breadth 1·2 mm.

Hab : South Australia—Lake Callabonna.

A very distinct species; its narrow cylindrical shape, with the elytra shortly and widely terminated, give it a general resemblance to a member of the family *Bostrychidæ*.

CLIVINA OLLIFFI, n sp.

Robust, parallel. Head large; prothorax a little broader than long : elytra long, parallel; fourth stria joining fifth at base; submarginal humeral carina feebly developed; eighth interstice marked, but not carinate on apical curve; a well marked striole at base of first interstice Prosternum with intercoxal part attenuate anteriorly; transverse sulcus of base obsolete. Anterior tibiæ 4-dentate. Black; prothorax piceous black, anterior legs testaceous brown, four posterior legs testaceous.

Head large (1·3 × 1·5 mm), densely rugose-punctulate on gulæ and behind eyes; vertex convex, lævigate; front lightly impressed and punctulate in middle, lightly and widely impressed on each side (the impressions a little rugulose), clypeal elevation slightly raised, narrow, arcuate . clypeus wide, depressed; median part truncate, its angles small, triangular, projecting; wings strongly divided from median part, anterior margin sloping lightly forward to external angles, these prominent, obtuse at apex; supra-antennal plates depressed, very strongly divided from clypeal wings, prominent and rounded externally: eyes convex, not prominent, lightly enclosed behind; facial sulci obsolete; facial carinæ short, distant from eyes. Mandibles wide, short, lightly decussating. Labrum 5-setose. Mentum rugulose-striate Labial palpi slender, two apical joints of about equal length. Antennæ short, lightly incrassate. Prothorax a little broader than long (1·8 × 1·9 mm.), lightly convex, subdepressed along median line, lightly declivous to base, transversely striolate, lightly punctulate except near anterior margin on middle of disc and on basal declivity, sides parallel, not narrowed anteriorly, posterior angles rounded, not marked, anterior margin truncate on each side, emarginate in middle, anterior angles obtuse, border narrow, median line deeply impressed; anterior line well marked. lateral basal impressions hardly marked. Elytra hardly wider than prothorax (4 5 × 2 mm.), widest behind middle, subparallel on sides, very lightly rounded, a little narrowed to shoulders. disc subdepressed, sides and apex strongly and deeply declivous. base truncate, shoulders marked, striæ deep, except towards apex, strongly crenulate-punctate, seventh entire, interstices subdepressed, hardly convex, eighth convex, narrow (hardly carinate) on apical curve, greatly narrowed about basal fifth, shortly and feebly carinate at humeral angle, lateral channel shallow; posterior puncture of third interstice placed at extremity of third and fourth striæ. Prosternum protuberant; episterna shagreened, obsoletely transversely striolate, overhanging anteriorly. Anterior trochanters projecting strongly and obtusely at apex; femora wide, compressed, posterior edge of lower side rounded; tibiæ wide, palmate; external spur of intermediate tibiæ long, acute.

Length 8, breadth 2 mm.

Hab. : West Australia – Beverley (sent to me by Mr. A M. Lea).

A remarkable and isolated species, for which I have found it necessary to form a separate group. In general appearance, shape of head, prothorax, elytra, prosternum and legs it resembles the species of the *"cribrosa group"*; but the fourth stria is outturned to join the fifth at the base. The crenulations of the elytral striæ are deep and punctiform, and from them fine short transverse striæ are given off, causing the interstices to have an undulate appearance. The external angles of the clypeal wings are strongly marked and quite as advanced as (if not a little more so than) the angles of the median part; the anterior margin of the wings slopes inwards and thus causes the median part to project sharply forward on each side. The elytra are concave on the three inner interstices near the base, and have a distinct elongate scutellar striole.

I have named this species in memory of my friend Mr. A. S Olliff, late Government Entomologist for New South Wales.

Heterogena group.

Size small. Eyes prominent; clypeus with median part angular, the angles projecting beyond the wings, these angular laterally. Elytra with fourth and fifth striæ confluent at base, seventh not interrupted at beginning of apical curve; submarginal humeral carina present; no striole noticeable at base of first interstice. Prosternum with intercoxal part attenuate anteriorly, sulcate on base. Anterior tibiæ 4-dentate.

Nine species, viz., *C. angustula*, Putz., *C. australica*, Sl., *C. deplanata*, Putz., *C. difformis*, Putz., *C. flava*, Putz., *C. heterogena*, Putz., *C. odontomera*, Putz., *C. 'oodnadattæ*, Blkb, and *C. tuberculifrons*, Blkb., seem to belong to this group; of these I know only two, therefore do not attempt to tabulate them.

CLIVINA HETEROGENA, Putzeys.

Stett. Ent. Zeit. 1866, xxvii. p. 41; Ann. Soc. Ent. Belg. 1866, x. p. 189.

Although I have a suspicion that *C. heterogena* will ultimately prove to be identical with *C. angustula*, the evidence before me is insufficient to enable me to feel absolutely certain about this, I therefore append a translation of the description of *C. heterogena.*

The anterior elevation, well marked and rather short, is separated from the vertex by a punctate impression of but little depth, the summit of the head bears a wide longitudinal impression containing some large punctures; the punctures on each side near the eyes are of the same size.

The eyes, of which only half is distinct, are very prominent. The prothorax is square, a little sinuate on the sides, as broad in front as behind; all the surface, except the anterior part in the middle, is covered with very distinct punctures.

The elytra are very elongate [and] cylindrical; their rounded shoulders are reflexed; they are of a piceous brown, but their external border, the suture before and behind, and the shoulders are of a testaceous colour. The fourth stria turns out at the base and reaches the eighth interstice.

The under surface of the body is black; the legs, except the upper side of the femora, the palpi and the antennæ are testaceous. The anterior tibiæ have externally two very long teeth and a small not very distinct tooth.

Length 5½, El. 2¾, breadth 1½ mm.

Australia. One specimen belonging to M. de Chaudoir, who received it from M. Melly.

The specimen noted under form " e " of *C. angustula*, Putz, *(vide post)*, from Windsor, N S.W., agrees in all respects with the description of *C. heterogena*   If "e" be merely a form of *C angustula*, then that species must sink to a synonym of *C. heterogena.* but this is a point which, with the identity of *C. difformis*, Putz, and *C. odontomera*, Putz., cannot be determined till exhaustive series of specimens of *C. angustula* and allied forms, from various localities (including Rockhampton) on the east coast of Australia, have been examined

CLIVINA ANGUSTULA, Putzeys.

Stett. Ent. Zeit. 1866, xxvii. p. 42; Ann. Soc. Ent. Belg. 1866, x. p. 190.

Narrow, parallel, subcylindrical. Black, head and prothorax piceous black; elytra with suture and margins (excepting base) reddish, legs reddish, four posterior paler than anterior. Head wide, short before eyes, front and vertex punctate: clypeus divided from front by a wide shallow punctate depression; clypeal elevation prominent, widely rounded, a wide depressed space near anterior margin; median part emarginate-truncate, the angles lightly advanced beyond wings, obtuse, wings square, with external angles rounded, supra-antennal plates wide, rounded externally, projecting decidedly beyond clypeal wings; eyes prominent; facial sulci hardly impressed, facial carinæ narrow, well developed. Prothorax about as long as broad (1·3 × 1 2 mm) a little narrowed anteriorly (ant. width 1·1 mm.), convex, punctate; sides lightly and widely sinuate behind anterior marginal puncture. Elytra parallel (2·7 × 1·3 mm.), convex, punctate-striate; striæ entire; eighth interstice carinate at base and on apical curve. Prosternum with intercoxal part attenuate ·anteriorly; episterna rugulose and striolate. Anterior tibiæ 4-dentate.

Length 4·2-5·2, breadth 1-1·4 mm.

Hab : N.S. Wales—Clarence River, Windsor (Lea), Carrathool (Sloane); Victoria—Lillydale, Ferntree Gully (Sloane); South Australia (Blackburn).

The description given above is founded on specimens taken at Lillydale and Ferntree Gully, near Melbourne. Putzeys' description suggests the inference that the prothorax is not narrowed anteriorly, but in my specimens, which I have no doubt are *C. angustula*, Putz., the prothorax certainly is narrowed ; different specimens vary in degree in this respect, which I believe to be a sexual difference.

C. angustula seems to present considerable differences in colour and size;* its constant features are the puncturation of the head and prothorax, the form of the clypeus, the striation of the elytra, the anterior femora not dilatate on lower side, the trochanters prominent at base of femora, and the digitation of the anterior tibiæ.

I offer the following notes on some variations that have come under my notice :—

(1). A numerous series of specimens sent to me by Mr A. M. Lea, taken at Windsor, N S W., vary as follows —

Length 4·2-5·2, breadth 1-1·4 mm.   Colour (*a*) testaceous (immature); (*b*) ferruginous (slightly immature ?), (*c*) ferruginous with interstices 2-5 of elytra obscurely piceous on posterior part of disc; (*d*) ferruginous with interstices 2-5 wholly piceous except at apex; (*e*) head and prothorax piceous brown, elytra reddish with interstices 2-4 piceous black on posterior part of disc and apical declivity.

(2) Specimens from the Clarence River, also received from Mr Lea, are apparently narrower and more depressed, testaceous with posterior part, excepting apices of interstices 2-4, obscurely piceous.   This form seems a variety or closely allied species, but requires studying with more specimens than are available to me

(3). Specimens from Carrathool (Murrumbidgee River) have the elytra more depressed; one specimen (immature) is pale testaceous, the others are coloured as in the description above This form has also been sent to me by the Rev. T. Blackburn, from South Australia, it seems likely to be *C deplanata*, Putz

(4) A specimen has been sent to me by the Rev T. Blackburn, which cannot in any way be distinguished from " No. 3 " above, except by having the anterior femora with the lower edge forming a decidedly acute triangular projection about anterior third This might be *C. odontomera*, Putz , but I should be unwilling to

---

* *Vide* Ann Soc. Ent. Belg 1866, x. p 190, where seven varieties are noted by M. Putzeys.

separate it from " No. 3 " on a single specimen, and without a knowledge that the form of the lower side of the tibiæ was constant, especially seeing that gummed on the same card, and therefore presumably from the same locality, was a specimen exactly resembling it, but with femora as in *C. angustula*.

CLIVINA DEPLANATA, Putzeys

Ann Soc. Ent. Belg. 1866, x. p. 190

In his unsatisfactory note on this species all that M Putzeys has to say is that it is with hesitation he separates this species from C. *angustula*, which it resembles in every respect except that the prothorax is a little broader and especially decidedly flatter. The colour is as variable as in *C. angustula*. All the specimens seen came from Melbourne.

CLIVINA FLAVA, Putzeys

Ann. Soc. Ent. Belg. 1868, xi. p. 17.

"Testaceo-flava, capite prothoraceque obscurioribus. Caput in vertice late nec profunde foveolatum, parce punctulatum. Prothorax brevis subquadratus, angulis anticis deflexis, lateribus rectis, utrinque in medio praesertim punctatus. Elytra subcylindrica, basi truncata, humeris rotundatis, striis integris punctatis, interstitio 3º quadripunctato. Tibiæ antice latæ, apice longe digitatæ, extus bidigitatæ denticuloque superiore armatæ

"Long. 5½, El. 2¾, Lat. 1⅓ mill "

Putzeys' remarks on this species are very full. I select for translation those bearing on important features.

Of a testaceous red, with the head, prothorax, and apex of the mandibles of a clear brown. The epistoma is rather narrow, a little emarginate; its angles are prominent and project beyond the little wings, which are very definitely separated from them; the anterior elevation is hardly marked, glabrous, separated from the vertex by a deep irregular punctate impression

The vertex bears a longitudinal fovea, in the centre of which some large punctures are noticeable; the occiput and the sides of

the head alike bear some punctures   The eyes are very prominent and project decidedly beyond the large wings; the posterior
border extends over half their breadth.

The prothorax is almost square, a little broader than long, the
anterior margin is not emarginate; the sides are straight; the
anterior angles are obtuse, but depressed; the border widens a
little and forms a slight prominence at the posterior angles,
which are marked by a large puncture, the surface is very lightly
convex, the median line is wider and deeper anteriorly than
towards the base; each side of the prothorax is covered with
punctures, which are particularly distinct in the middle and do
not extend to the base; the two lateral impressions are oblong
and very lightly marked.

The elytra are a little wider than the prothorax, cylindrical,
truncate at the base; their shoulders are rounded; the striæ are
deep and very distinct for their whole length, punctate almost to
the apex; the interstices are lightly convex   The head is strongly
rugose beneath, the prothorax is much more finely rugose and
transversely striolate.   The abdomen is smooth.   The anterior
trochanters form a feeble prominence at the base of the femora,
the tibiæ are wide, strongly digitate externally, and sulcate on
upper surface, the intermediate tibiæ have three or four spiniform
bristles above the spur.

Hab.—Rockhampton (Coll. Castelnau; several specimens).

I have been unable to identify *C. flava* among the species I
have seen

Clivina difformis, Putzeys.

Ann. Soc. Ent Belg. 1868, xi. p. 19.

"Castanea, capite elytroque singulo in medio piceo, palpis,
antennis pedibusque brunneo-testaceis.   Prothorax elongatus,
antice angustatus obsolete punctulatus.   Elytra cylindrica, basi
truncata.   Tibiæ anticæ extus bidentatæ.

"Long. 5⅓, El 3½, Lat. 1¼ mill."

The following is a translation of Putzeys' remarks on this species, which is unknown to me :—

The vertex is punctate; it bears a lightly impressed oblong wide fovea, where the punctures are denser   The antennæ are thick, moniliform   The eyes are prominent, but greatly enclosed by the postocular tubercles.   The prothorax is longer than broad, narrowed in front, but particularly behind the anterior angles; these are lightly advanced: the posterior angles are distinct; the lightly convex surface bears some striolæ and some small scattered punctures.

The elytra are cylindrical, their base is truncate, but the shoulders are a little rounded, under a strong lens it is seen that the interstices are covered with small transverse undulations not close together.   The elytra are piceous, with all their margins (including the suture) of a rather clear brown.

The femora are narrow.   The anterior tibiæ, sulcate on upper side, have externally two very strong teeth.   The apical digitation is thicker, and one-half longer than the inner apical spine

Hab . Probably the north-west of Australia (Coll. Castelnau; a single specimen only).

CLIVINA AUSTRALICA, n.sp.

Narrow, parallel, subcylindrical.   Head short, convex; eyes large, convex, not prominent, facial sulci lightly recurved: prothorax parallel, longer than broad: elytra long, parallel; fourth stria joining fifth at -base; eighth interstice distinct on apical curve, submarginal humeral carina moderate, narrow; prosternum with intercoxal part attenuate anteriorly; episterna very finely striolate near lateral margins, overhanging anteriorly; anterior tibiæ 4-dentate.   Ferruginous.

Head sparsely covered with minute, nearly obsolete punctures: clypeus with median part wide, truncate (obsoletely emarginate between angles), angles obtuse, hardly prominent, wings small, hardly divided from but not so prominent as angles of median part, outer angles obtuse, external side straight, supra-antennal plates

13

projecting sharply beyond wings; recurved part of facial sulci well marked, lightly oblique; facial carinæ well developed, narrow; eyes very lightly enclosed behind Prothorax longer than broad (1·15 × 1 mm ', hardly narrowed anteriorly, declivous to base, transversely striolate near sides, anterior margin truncate, anterior angles marked, not prominent, posterior angles widely rounded; border narrow; median line strongly impressed; anterior line lightly marked; lateral basal impressions obsolete. Elytra hardly wider than prothorax (2·3 × 1·1 mm.), parallel, convex, widely rounded, and very declivous to apex; base lightly emarginate; shoulders rounded but marked; striæ lightly impressed, entire, finely punctate, seventh not interrupted near apical curve; interstices lightly convex on anterior part of disc. Anterior femora short, wide, intermediate tibiæ wide, external margin arcuate, external spur long, slender, acute

Length 4 3, breadth 1·1 mm

Hab. · N W. Australia (sent by Mr Masters.)

Allied to *C. angustula*, Putz., but distinguished by its more cylindrical form, impunctate prothorax, &c. The form of the clypeus is as in *C. dorsalis*, Blkb , but the outer angles of the wings are more rectangular It should resemble, judging from the description, *C. verticalis*, Putz., but is smaller, its prothorax is exceptionally long, and the outer angles of the wings of the clypeus should be more marked It is evidently distinct from *C. difformis*, Putz , attention may be directed to the following points of difference from Putzeys' description, the smaller size, different colour, eyes lightly enclosed in the weakly developed posterior part of orbits, anterior femora wide, tibiæ 4-dentate

Clivina odontomera, Putzeys.

Ann. Soc. Ent. Belg 1868, xi. p. 18

" Dilute brunnea. Caput undique grosse rugoso-punctatum Prothorax latitudine longior, convexus antice parum augustatus, parce punctulatus. Elytra subcylindrica, basi intus truncata, humeris rotundatis, striis integris punctatis, interstitio 3"

4-punctato. Femora antica subtus ante apicem dentata; tibiæ latæ, apice longe digitatæ, extus digitatæ [? bidigitatæ] denticuloque superiore armatæ.

"Long. 5, El. 3¼, Lat. 1¼ mill."

"Rockhampton (Coll. Castelnau)."

Appended is a translation of his further remarks on this species :—

It has the appearance of *C. punctaticeps*; however, the prothorax is more convex, narrower, particularly anteriorly; it is usually a little more distinctly punctate.

The epistoma is wider, more truncate, the head is covered with punctures [which are] much more numerous and almost rugulose. The anterior femora, less wide and less thick, have not beneath an inflation analogous to that of *C. lobata*, but they have, a little before the apex, a strong acute tooth, and the apex of the trochanters is equally raised in the form of a tooth.

It appears to me that *C. odontomera* must be allied rather to *C. angustula*, Putz., than to *C. adelaidæ*, Blkb.

Bovillæ group

Clypeus with median part and wings almost on same level, median part divided from wings on each side by a small triangular sinuosity. Elytra with fourth and fifth striæ confluent at base, submarginal humeral carina present. Prosternum with intercoxal part very narrow and canaliculate anteriorly, sulcate on base; pectoral ridges short, well developed Anterior tibiæ 4-dentate.

I do not feel sure that I am right in separating *C. bovillæ* from the "*heterogena group*"; this has been done on account of the different form of the intercoxal part of the prosternum. Probably the "*punctaticeps, blackburni, olliffi, heterogena*, and *bovillæ groups*" might with advantage be regarded as sections of one large group.

CLIVINA BOVILLÆ, Blackburn.

P L S N.S.W. 1889 (2), iv. p. 717.

Piceous brown. Robust, parallel. Head wide, depressed anteriorly; clypeal elevation prominent, convex, hardly arcuate.

clypeus widely depressed near anterior margin; median part wide, subtruncate (hardly emarginate), its angles obtuse, very lightly marked, hardly advanced beyond wings; these small, with external angles rounded; supra-antennal plates projecting sharply and decidedly beyond wings; facial carinæ wide, eyes prominent, enclosed behind. Prothorax convex, subquadrate (1·65 × 1·65 mm), lightly narrowed anteriorly (ant width 1·5 mm), sides hardly rounded (not sinuate), basal curve short. Elytra wider than prothorax (3·6 × 1·9 mm.), convex; sides parallel; lateral channel wide and strongly bordered at shoulders; striæ entire, deeply impressed, finely crenulate; interstices convex (depressed near apex), eighth narrowly carinate at base and apex. Prosternum with intercoxal part small, narrow and canaliculate anteriorly, base sulcate, pectoral ridges short, distinct, episterna coarsely rugulose Anterior femora wide, tibiæ strongly 4-dentate, the upper tooth small.

Length 6-6·8, breadth 1·7-1·9 mm.

Hab. · Northern Territory of S.A (Mrs. Bovill); West Australia; Queensland — Gulf of Carpentaria (received from Mr. French).

The position of *C bovillæ* is between *C. australasiæ*, Boh., and *C. heterogena*, Putz. The clypeus conforms nearly to that of *C. heterogena*, but the intercoxal part of the prosternum, though narrower than in *C. australasiæ*, is wider and does not form a narrow ridge, as it does in *C. heterogena.* It appears to be widely spread along the north coast of Australia, and judging from specimens in my possession varies considerably in facies, the form of the clypeus and the intercoxal part of the prosternum are its constant features. The description given above is founded on a type specimen kindly lent to me by the Rev T. Blackburn.

Clivina cava, Putzeys.

Stett. Ent. Zeit. 1866, xxvii. p. 38. Ann. Soc. Ent. Belg. 1866, x. p. 184; l.c. xi. p. 13.

Convex, parallel. Head wide, depressed, eyes prominent, prothorax subquadrate elytra with striæ entire, punctate, fourth joining fifth at base; submarginal humeral carina short, weak; interstices lightly convex, eighth carinate at apex: prosternum with intercoxal part angustate (narrow, but not attenuate) anteriorly, sulcate on base, episterna very finely transversely striolate, anterior tibiæ strongly 4-dentate. Ferruginous brown, legs testaceous.

Head with front and vertex depressed, finely but distinctly punctate; supra-antennal plates and wings of clypeus flat; clypeal elevation lightly raised, subtruncate (lightly rounded)· clypeus not divided from front, depressed near anterior margin, median part with margin lightly rounded; wings short, wide, strongly advanced beyond median part, external angles rounded but a little marked; supra-antennal plates projecting strongly and sharply beyond wings of clypeus; eyes prominent. convex, very lightly enclosed. Prothorax subquadrate (1·8 × 1 8 mm.), very little narrowed anteriorly (ant. width 1·65 mm.); disc smooth, basal declivity rugulose; sides subparallel, hardly rounded or sinuate; posterior angles rounded, but lightly marked; anterior margin truncate; anterior angles rounded, not marked; border reflexed, passing round anterior angles; median and anterior lines well marked; lateral basal impressions rather long, deep, narrow, punctulate. Elytra wider than prothorax (3·8 × 2 mm.), parallel on sides, widely rounded at apex; base truncate towards sides, emarginate in middle, shoulders rounded, seventh stria entire, not interrupted at beginning of apical curve. Anterior femora short, wide

Length 7, breadth 2 mm.

Hab. North-west Australia (two specimens sent by Mr. Masters), Queensland—Rockhampton (Putzeys, Coll. Castelnau).

The species on which the above description is founded agrees so well with Putzeys' description of *C. cara*, that I have little hesitation in regarding it as that species. The strongly 4-dentate anterior tibiæ associate it with *C. bovillæ*, Blkb., but the depressed head and the clypeus deeply truncate-emarginate, with wide wings isolate it from all other Australian species. I have not included it among the species of the "*australasiæ group*," but have felt unwilling to form a separate group for it, so have left it in an intermediate position between the "*borillæ*" and "*australasiæ groups.*"

Australasiæ group.

Mandibles short; eyes prominent, clypeus with anterior margin emarginate, wings widely rounded, not divided from median part. Elytra with fourth and fifth striæ confluent at base, submarginal humeral carina well developed; eight interstice carinate near apex.

The "*australasiæ group*" may be divided into four sections as shown in the following table :—

A Prosternal episterna more or less rugulose-striolate, not punctate.

   B. Prosternum with intercoxal part
        attenuate anteriorly, anterior
        tibiæ 4-dentate. ..   Section I. (Type *C. sellata*).

   BB. Prosternum with intercoxal part
        narrow anteriorly, anterior
        tibiæ with two strong external
        teeth and a slight prominence
        above apical projection ... Section II. (Type *C. australasiæ*).

   BBB. Prosternum with intercoxal
        part wide anteriorly, anterior
        tibiæ 3-dentate . . . . .. Section III. (Type *C. basalis*)

AA. Prosternal episterna punctate.. Section IV. (Type *C. pectoralis*).

Section I.

Table of Species known to me.

c. Bicolorous. . . .. *C. sellata*, Putz.

cc. Unicolorous.

   d. Anterior tibiæ 4-dentate . . . . *C ferruginea*, Putz.

   dd. Anterior tibiæ 3-dentate

      e. Black, convex, interstices of elytra convex   *C. occulta*, Sl.

      ee. Testaceous, depressed; interstices of elytra
         flat (size very small)   .. .. *C nana*, Sl.

It appears as though *C. suturalis*, Putz., *C. verticalis*, Putz, *C. dimidiata*, Putz., and *C. æqualis*, Blkb, should be placed in this section.

CLIVINA SELLATA, Putzeys

Stett. Ent. Zeit 1866, xxvii. p. 40; Ann. Soc. Ent. Belg 1866, x. p. 186.

Head and prothorax black; elytra testaceous, with a large black patch on posterior part of disc; four posterior legs testaceous, anterior legs ferruginous, under surface piceous. Narrow, cylindrical Front rugulose-punctate; vertex foveate in middle, clypeus with median part not divided from wings, lightly emarginate, clypeal elevation prominent, arcuate, a decided sinuosity between supra-antennal plates and wings of clypeus. Prothorax smooth (disc lightly transversely striolate and covered with scattered minute punctures), convex, rather longer than broad (1·35 × 1 25 mm), lightly narrowed anteriorly (ant width 1 mm.). Elytra convex, parallel (2 9 × 1·5 mm), strongly punctate-striate, striæ entire, fourth joining fifth at base, interstices convex, eighth distinct on apical curve, a submarginal carina at shoulder. Prosternum with intercoxal part attenuate anteriorly, transversely sulcate on base; episterna finely shagreened and transversely striolate. Anterior femora wide, compressed; tibiæ 4-dentate (upper tooth a small triangular prominence).

Length 4·3-5·5, breadth 1·25-1·5 mm.

Hab Queensland—Gayndah (Masters); N.S. Wales—Richmond River, Tamworth, Sydney (Lea), Narrandera, Carrathool, Mulwala, Junee (Sloane); Victoria—Melbourne (Kershaw); South Australia (Masters).

The characteristic features of this widely distributed species are the 4-dentate tibiæ, the attenuate intercoxal part of the prosternum, and the colour. Immature specimens are often taken of an entirely testaceous colour

CLIVINA FERRUGINEA, Putzeys.

Ann. Soc. Ent. Belg. 1868, xi. p. 14.

"Ferruginea. Caput in vertice foveolatum, parce punctulatum. Prothorax subquadratus, antice leviter angustatus, convexus, utrinque in medio et in foveis basalibus oblongis

punctulatus. Elytra subcylindrica, basi truncata, humeris sub-
rotundatis; striis integris punctatis, interstitio tertio quadripunctato. Tibiæ anticæ apice longe digitatæ, extus bidigitatæ
denticuloque superiore armatæ.

"Long. 6, El. 3, Lat. 1½ mill."

After the Latin diagnosis M. Putzeys has some remarks, of
which the following is a translation :—

The epistoma roundly emarginate and closely united to the
wings, which are rounded, classes the species very clearly among
those of the twenty-seventh [? twenty eighth] group.

It has a very great resemblance to *C. flava*, in which, however,
the epistoma is quite differently shaped; but the colour of the
elytra is the same as that of the head and prothorax; the
prothorax is less quadrate, more elongate, decidedly more
convex, the sides are less straight; the vertex is more convex,
less punctate, and the anterior elevation is less distinctly
separated by a transverse impression.

The episterna of the prothorax are hardly distinctly striolate
on their internal part.

Hab. : Rockhampton (Coll. Castelnau).

Specimens sent to me by the Rev. T. Blackburn as coming
from Cairns, North Queensland, agree with the description of
C. ferruginea, except in the following points :—size a little
smaller, prothorax smooth (a few very minute punctures are
discernible in and near the lateral basal impression with a very
powerful lens). The following brief diagnosis gives particulars of
some characters not mentioned by Putzeys.

Narrow, cylindrical. Head with a light lateral sinuosity
dividing the wings of the clypeus from the supra-antennal plates:
prothorax as long as broad (1·25 × 1·25 mm.), very lightly narrowed
anteriorly: elytra (3 × 1·35 mm.) with striæ entire, lightly punctate,
fourth joining fifth at base, eighth interstice carinate at
base and apex: prosternum with pectoral part protuberant:
intercoxal part small, attenuate anteriorly, sulcate on base:
episterna very finely rugulose and transversely striolate. Anterior

trochanters projecting beyond base of femora, these not dilatate on lower side; tibiæ 4-dentate.

Length 5·3, breadth 1·35 mm.

A specimen sent by Mr. Masters, as coming from N.W. Australia, cannot be separated from the specimens from Cairns.

CLIVINA OCCULTA, n.sp.

Narrow, convex. Head wide before eyes; prothorax narrow, convex; elytra strongly punctate-striate, fourth stria outturned and joining fifth at base; prosternum with intercoxal part small, attenuate anteriorly, sulcate on base; anterior tibiæ wide, strongly 3-dentate. Black, shining; antennæ ferruginous, legs reddish testaceous.

Head short, rather depressed, sparsely and coarsely punctate; vertex convex; frontal foveæ very wide: clypeus lightly declivous to anterior margin; median part truncate, not divided from wings; these oblique on inner side to median part, decidedly advanced beyond median part, widely and lightly rounded in front; supra-antennal plates wide, rounded externally, projecting lightly but decidedly beyond wings of clypeus; eyes prominent; orbits truncate behind. Prothorax small, narrow, hardly broader than long (1·4 × 1·42 mm.), a little narrowed to apex (ant. width 1·25 mm.), convex, strongly declivous to base; disc transversely striolate; sides widely and very lightly sinuate behind anterior marginal puncture; lateral basal impressions distinct, narrow, elongate-foveiform Elytra narrow (3·25 × 1·65 mm.), widest behind middle, same width as prothorax at base, truncate on base; striæ strongly impressed, entire, coarsely punctate (the punctures stronger than usual towards apex), seventh stria entire; interstices convex, depressed towards apex, eighth shortly carinate at base.

Length 6·2, breadth 1·65 mm.

Hab. : Queensland—Cape York (Coll. Blackburn; a single specimen).

This species must be associated with *C. sellata*, Putz., though the form of its clypeus is more that of the "*obliquata group*" than of *C. sellata*.   In general appearance it resembles *C. queens-landica*, Sl., and *C. dilutipes*, Putz   from *C. queenslandica* it may be distinguished by its more convex shape; clypeus with median part more truncate, the wings wider, concave, more decidedly advanced beyond median part and roundly subtruncate, elytra with striæ more coarsely punctate, prosternum with inter-coxal part attenuate: from *C. dilutipes* the wider and punctate anterior part of the head, the stronger external teeth of the anterior tibiæ, and the shape of the intercoxal part of the prosternum thoroughly differentiate it.

Clivina nana, n.sp.

Small, depressed, parallel.   Head wide, depressed; prothorax subquadrate; elytra lightly crenulate-striate, fourth stria joining fifth at base, interstices flat, eighth weakly carinate at base, finely and weakly carinate near apex; prosternum with intercoxal part narrow anteriorly; episterna minutely rugulose-striolate; anterior tibiæ wide, strongly 3-dentate   Testaceous, eyes black.

Head depressed, vertex roundly concave in middle, clypeal elevation well marked, lunulate   clypeus divided from front by a shallow depression, anterior margin subtruncate (hardly emargi-nate), wings small, not divided from median part, rounded laterally, divided from supra-antennal plates by a decided sinuosity; supra-antennal plates convex, prominent before eyes, extending obliquely backwards without interruption above eyes to form the wide facial carinæ; these reaching behind base of eyes; facial im-pressions wide, shallow, not sulciform, eyes depressed.   Prothorax depressed, about as long as broad (0 75 × 0·8 mm), very lightly narrowed anteriorly, disc obsoletely and minutely punctulate, sides roundly subparallel , basal curve short, lateral channel feebly marked, marginal punctures wide, shallow, the anterior distant from anterior angle, the posterior behind posterior

angle, not touching margin. Elytra very little wider than prothorax (2 × 0·9 mm.), depressed, sides parallel; base truncate.

Length 3 6, breadth 0·9 mm.

Hab : N.S. Wales—Tamworth (Lea).

An isolated species among those known to me, and the smallest
Australian Clivina yet described.

CLIVINA SUTURALIS, Putzeys.

Mém. Liége, 1863, xviii. p. 39; Stett. Ent. Zeit. 1866, xxvii.
p 40, Ann. Soc. Ent. Belg. 1866, x. p 186.

"Nigra, nitida, ore, antennis pedibus elytrisque testaceo-ferru-
gineis, hisce plaga suturali nigra ornatis. Clypeus truncatus
angulis elevatis prominulis. Vertex depressus, punctatus.
Pronotum subelongato-quadratum, punctatum, basi utrinque
longitudinaliter impressum. Elytra elongata subcylindrica,
profunde punctato-striata. Tibiæ anticæ extus obtuse bidentatæ."

"Long. 6, El. 3, Lat. 1⅓ mm."

M. Putzeys added to his Latin diagnosis a fuller description
in French; the following is a translation of the more salient
parts :—

The epistoma is almost truncate, bordered; its angles project
in the form of prominent teeth, the wings are hardly
distinct from the supra-antennal margins. The vertex is
flattened in the middle, irregularly foveolate and punctate; the
longitudinal carinæ of the sides of the head are very distinct and
straight, they do not become broader towards their source.

The prothorax is a little longer than broad, its sides are parallel;
the anterior angles are lightly rounded and very declivous; the posterior angles are only marked by the interruption of the marginal
border and by a piliferous puncture placed within it, all the
surface (except the margins) is covered with rather large punctures,
which are stronger and more numerous on the sides near the
basal foveæ; these are oblong, rather wide, but shallow.

The elytra are of the same width as the prothorax, elongate,
their sides are almost parallel; the base appears truncate and the

apex is rounded; they are strongly punctate-striate. The scutellar striole is oblique and short. The suture is occupied by a stripe of brown-black which, at the base, covers the first interstice, and becomes wider after the basal fourth without extending beyond the third interstice.

Hab. : Australia—Port Phillip; (one specimen).

In his "Révision Générale" the following is all that is said of this species :—

In a great many respects it comes very near *C. verticalis*; the prothorax has the same form, but it is less convex, longer and still more enlarged behind the anterior angles; it is covered with a very distinct puncturation. The epistoma has the external angle of its wings more marked, simply obtuse, and the wings are not separated from the posterior wings. The anterior elevation is less marked, the vertex has only some scattered punctures anteriorly. All the external teeth of the tibiæ are obliterated, which may well be only accidental

Length 5, El. 2½, breadth 1½ mm.

In spite of M. Putzeys' having placed *C. suturalis* in a section in which the fourth stria joined the fifth at the base,* I cannot help a suspicion that it did not do so, and that *C. suturalis* was founded on the same species that Mr. Blackburn has since named *C. dorsalis.*† The difference in the dimensions given in Putzeys' two descriptions, apparently founded on the same specimen, and the absence of any comment thereon are unsatisfactory.

Clivina verticalis, Putzeys

Stett. Ent. Zeit. 1866, xxvii. p. 40; Ann. Soc. Ent. Belg. 1866, x p. 186.

The following is a translation of M. Putzeys' whole description:—

It differs from the preceding [*C. sellata*] by its wholly testaceous colour, a little darker on the head and prothorax, the suture

---

* He placed *C. planiceps* in the same section, *vide ante*, p. 173.
† *Vide* description of *C. dorsalis, ante*, p. 169

is slightly brownish. The decided difference is found in the shape of the prothorax, which is almost square, as broad before as behind: the lateral margin is a little sinuate before the anterior angle. The elytra are more cylindrical, not at all narrowed behind. In all other respects it resembles *C. sellata.*

Length 5¼, El. 2¾, breadth 1½ mm.

Australia—(Coll. Chaudoir; two specimens).

I have an immature specimen of *C. dorsalis*, Blkb., from Victoria, which is wholly testaceous in colour, and I cannot help suspecting that *C. verticalis* has been founded on immature specimens and is in reality conspecific with *C. dorsalis.* In support of this suspicion it may be noted that the characters of the basal striæ of the elytra do not appear to have been taken into account by M. Putzeys at the time he described *C. verticalis;* under the circumstances there is nothing for it but to retain both names, but, if I am right in my suspicion as to their identity, a want of carefulness on the part of M. Putzeys has saddled the Australian list with at least one name for which no species is likely to be found in nature.

CLIVINA DIMIDIATA, Putzeys.

Stett. Ent. Zeit. 1866, xxvii. p. 39; Ann. Soc. Ent. Belg. 1866, x. p. 185.

The disposition of the colours is almost the same as in *C. basalis*, but the black part is not so large, very oblique from the lateral margin to the suture where it is prolonged beyond the middle of the elytra; the anterior colour instead of being a dull red is a light reddish testaceous; the legs and antennæ are also of a clearer tint. The eyes are less prominent; the prothorax is less narrowed in front, and less emarginate in the middle of the anterior margin; the elytra are shorter and narrower.

Length 7, El 3·5, breadth 1⅞ mm.

Australia —Melbourne (?) (Coll. Chaudoir; two specimens).

In addition we learn from the Révision Générale (p. 183) that the central carina of the prosternum is very narrow in *C. dimidiata.* It must greatly resemble *C. melanopyga*, Putz., and indeed on account of its having the intercoxal part of the prosternum very narrow, and from the fact that M. Putzeys in his memoir in the Entomologische Zeitung placed *C. melanopyga* in the same group as *C basalis*, taking no notice of the basal characters of the striæ of the elytra, I suspect that it is not unlikely to have been founded on specimens of *C melanopyga*, which, probably chiefly on account of their larger size, had been taken to belong to a distinct species.

SECTION II.

Table of Species.

f. Unicolorous.

  g. Size large. . . ..   *C australasiæ*, Bohem.

  gg. Size small   .   .   .   .   . *C queenslandica*, Sl.

ff. Bicolorous

  h. Black, with apex of elytra reddish   . *C leai*, Sl.

  hh. Elytra black, with a reddish vitta on each

    side .... *C. vittata*, Sl.

The species I do not know are *C. juvenis*, Putz., and *C. helmsi*, Blkb

CLIVINA AUSTRALASIÆ, Bohemann.

Res Eugen. Coleoptera, 1858, p. 8.

Robust. Head wide, punctulate on each side at posterior extremity of facial carinæ, prothorax not longer than wide, decidedly narrowed anteriorly (ant. width 2·15 mm); elytra strongly punctate striate, fourth stria outturned and joining fifth at base, interstices convex, eighth carinate at base and apex; anterior tibiæ strongly 3-dentate (hardly 4-dentate), inner apical spine (♂) not obtuse at apex. Black, antennæ, tibiæ and tarsi piceous.

Head large, wide before eyes, obliquely angustate, with a well marked sinuosity between supra-antennal plates and wings of clypeus; front and vertex rather depressed. clypeus obsoletely divided from front, anterior elevation arcuate; anterior margin wide, lightly and roundly emarginate, wings wide, concave, rounded, not divided from median part, supra-antennal plates broad, widely depressed near clypeus, facial sulci deep, parallel posteriorly. Prothorax lightly convex, of nearly equal length and breadth (2 55 × 2·6 mm); anterior angles rounded, bordered, lateral basal impressions obsolete, or very faint Elytra long, parallel (6 × 2·8 mm.), lightly convex, dorsal surface rather depressed; base truncate; marginal channel wide at humeral angle; striæ deep and strongly punctate on disc, becoming faint and finely punctate towards apex; interstices convex, except on apical declivity. Prosternum with intercoxal part narrow anteriorly, sulcate on base; episterna closely rugulose. Anterior tibiæ strongly 3-dentate, a sinuosity above upper large tooth causing a fourth tooth to be weakly developed. ♂ with anterior tibiæ hardly less strongly dentate than ♀; the inner apical spine longer and more curved, but not obtuse at apex.

Length 8-10·5, breadth 2·4-2·8 mm.

Hab. N.S. Wales, Victoria, and South Australia (widely distributed); Lord Howe Island (Macleay Museum); New Zealand (Broun).

The description given above is founded on specimens sent to me by Mr Lea, and taken by him at Windsor, near Sydney, the form found on the Murray and Murrumbidgee Rivers seems to vary a little from the typical form, being a lighter and more convex insect, but I cannot find any differences between them that are worth considering of even varietal value. The original description seems inexact in giving the shape of the prothorax as "*latitudine dimidio longior*," and the elytra, "*prothorace haud latiora.*" Sometimes the anterior part of the front is densely punctate, and often the punctures that are always present on the sides of the occiput, near the facial carinæ, extend across the occiput Specimens of *C australasiæ* from Lord Howe Island are in the Macleay Museum; they are probably identical with the species considered *C. vagans* by the late Mr A. S. Olliff (Mem. Aust. Mus. 1889). A specimen (♂) sent to me many years ago, from New Zealand by Capt. Thos Broun, under the name of

C. rugithorax, Putz., in no way differs from *C. australasiæ*, so it appears as if *C. rugithorax* should be regarded as a synonym of *C. australasiæ*.

Specimens only 8 mm in length are rarely found.

CLIVINA JUVENIS, Putzeys

Stett Ent Zeit 1866, xxvii. p. 37; Ann. Soc. Ent Belg. 1866, x p. 183

Subjoined is a translation of Putzeys' entire description. It seems quite useless as a means of identifying any species, and appears to be founded on an immature specimen. The question of whether, in spite of the differences given as distinguishing it from *C. australasiæ*, it may not be that species, I leave for him who can to decide.

Entirely of a slightly reddish testaceous colour Behind the anterior elevation of the front a wide deep impression is noticed. The impression of the vertex is short and less marked [than in *C. australasiæ*]. The prothorax is narrower, its anterior angles are less rounded; the elytra are a little shorter; the teeth of the tibiæ are finer.

Length 8, El. 4, breadth 2 mm.

Hab. : Melbourne (Coll. Chaudoir).

In addition to the particulars given above we learn from the Révision Générale that the base of the elytra is more distinctly truncate than in *C. australasiæ*.

CLIVINA QUEENSLANDICA, n sp.

Form light, rather depressed. Head wide, lightly punctate on vertex, prothorax depressed, elytra lightly striate, fourth stria joining fifth at base ; prosternum with intercoxal part rather wide in front; anterior tibiæ strongly 3-dentate. Black, shining (prothorax sometimes piceous black); legs piceous red, the four posterior lighter coloured than the anterior.

Head wide, subdepressed, front lightly punctate: clypeus not divided from front, lightly and widely emarginate, a wide depressed rugulose space along anterior margin; wings small, rounded, not divided from median part; clypeal elevation depressed, widely arcuate; a light sinuosity dividing wings from supra-antennal plates ; facial sulci lightly impressed, wide apart, parallel posteriorly; facial carinæ wide, depressed. Prothorax lævigate, subquadrate (1·7 × 1·7 mm.), narrowed anteriorly (ant. width 1·3 mm.); sides lightly rounded; lateral basal impressions distinct, short, narrow. Elytra a little depressed, very little wider than prothorax (3·5 × 1 8 mm.), very little narrowed to base, sides subparallel; shoulders rounded; striæ entire, lightly impressed, finely crenulate; interstices lightly convex on disc, eighth carinate at base and apex. Prosternum with base sulcate; episterna rugulose and transversely striolate Anterior tibiæ strongly 3-dentate, with a feeble projection above large teeth. ♂ with inner apical spine long, arcuate.

Length 6 2-7; breadth 1·65-1·9 mm.

Hab. : Queensland—Darling Downs District (Lau); South Australia—Lake Callabonna (Zietz).

This species is allied by the form of the anterior tibiæ in the ♂, and the shape of the head to *C. australasiæ,* Bohem , rather than to those species which resemble *C lepida,* Putz , in these respects, as *C. vagans,* Putz , and *C. dilutipes,* Putz . It is very like *C. dilutipes* in general appearance, but may be distinguished by having the head wider and punctate, eyes less prominent, prothorax more depressed, elytral striæ more finely punctate, prosternum sulcate on base, external teeth of anterior tibiæ stronger; it has even a closer resemblance to *C. occulta,* Sl., but differs in shape of clypeus, shape of prothorax, prosternum with the intercoxal part wider anteriorly, &c.

CLIVINA LEAI, n.sp.

Narrow, convex. Head depressed, wide before eyes; prothorax of equal length and breadth, decidedly narrowed anteriorly; elytra strongly punctate-striate, fourth stria outturned and joining fifth at base, a fine submarginal carina at shoulder; anterior tibiæ

14

strongly 3-dentate.   Black; elytra with apical third testaceous red, under surface piceous; anterior legs piceous brown, four posterior legs testaceous.

Head wide before eyes (1·2 mm. × 1·2 mm.), vertex with a few shallow rugæ, not punctate except finely on each side near extremity of facial carinæ: clypeus not divided from front, lightly and widely emarginate, anterior angles (wings) widely rounded; median part depressed, bordered, defined on each side by a short, narrow, longitudinal ridge; wings small, concave; clypeal elevation distinct, arcuate; supra-antennal plates rather depressed, large, wide, strongly rounded and bordered externally, projecting sharply and decidedly beyond wings of clypeus; facial sulci lightly impressed, facial carinæ short, wide; eyes convex, projecting slightly, deeply enclosed by supra-antennal plates in front, lightly enclosed behind; orbits abruptly constricted behind. Prothorax smooth (a few transverse striolæ on disc), as long as broad (1 8 mm. × 1·8 mm), widest a little before posterior angles, decidedly narrowed anteriorly (ant. width 1·5 mm.), basal curve short; border rather wide on anterior part of sides, median and anterior lines well marked; lateral basal impressions short, distinct.   Elytra convex, very declivous on sides, widest a little behind middle (4 × 2 1 mm.), a little narrowed to base; sides lightly rounded; base shortly truncate in middle, rounded on each side; humeral angles not marked; striæ deeply impressed on basal two-thirds, becoming faint towards apex, closely punctate; the punctures strong towards base, weaker towards apex.   Prosternum with intercoxal part narrow (not attenuate) anteriorly, transversely sulcate on base; episterna finely rugulose and transversely striolate.

Length 7-7·5, breadth 2·1 mm.

Hab. : Queensland—Pine Mountain (Masters); N.S. Wales—Clarence River (Lea); Central Australia (Horn Scientific Expedition).

The colour of the elytra, with the whole apical part testaceous red from just behind the third puncture of the third interstice, distinguishes this elegant species, which was first sent to me by Mr. A. M. Lea, after whom I have named it.

Var ? *C. apicalis.* A specimen sent to me by Mr. Masters, as coming from N W. Australia, differs from the type form of *C. leai* by being smaller; the head smooth; the prothorax a little shorter (1·5 × 1·6 mm.), more convex, more rounded on the sides, the lateral basal impressions obsolete, the striæ of the elytra deeper and more strongly punctate.

Length 6, breadth 1·7 mm.

It is probably a distinct species, but requires studying with a number of specimens before one; its general resemblance to *C. biplagiata*, Putz, is very noticeable.

CLIVINA VITTATA, n.sp.

Robust, convex. Front punctate-foveate, prothorax convex, broader than long (1·35 × 1·45 mm), lightly narrowed anteriorly (ant width 1·15 mm.) Elytra rounded on sides, widest behind middle, a little narrowed to base (3 mm. × 1 6 mm.), strongly punctate-striate; interstices convex, eighth narrowly carinate at base, and on apical curve. Prosternum with intercoxal part narrow (not attenuate) anteriorly, sulcate on base; episterna finely rugulose-striolate Anterior femora wide; tibiæ 4-dentate, the upper tooth very feeble. Piceous black; a reddish lateral vitta (interstices 5-7) on each elytron, not reaching apex; legs reddish piceous

Length 5 3, breadth 1·6 mm.

Hab · N. S Wales—Sydney (one specimen sent by Mr. Masters).

A second specimen, labelled Victoria, is in the collection of the Rev. Thos. Blackburn, who has kindly forwarded it to me for examination; it is smaller (4·3 × 1·2 mm.', and has the prothorax piceous red, but otherwise agrees with the type.

This species is allied to *C. sellata*, Putz, but, besides being differently coloured, it differs by its wider and more convex form; wider prothorax; elytra less parallel, more rounded on the sides, widest behind the middle and evidently narrowed to the shoulders, more widely rounded at apex; intercoxal part of prosternum wider anteriorly: the clypeus is very similar to that of *C. sellata*, but the wings are smaller and recede a little more at the sides, which causes the angles of the median part to be just the least indicated; the clypeal elevation is less prominent, and the head is less rugulose.

SECTION III.

Head with space between facial impressions smooth, usually convex; lateral sinuosity between supra-antennal plates and clypeus obsolete or hardly marked. Prosternum with intercoxal part wide anteriorly. Anterior tibiæ 3-dentate (in ♂ narrower, and with the teeth much less developed than in ♀); inner apical spine in ♂ longer than in ♀, curved and obtuse at apex, in ♀ pointed at apex.

Table of Species.

i. Bicolorous.

j. Elytra with basal part reddish, apical part black. . *C. basalis*, Ch.

jj. Elytra reddish, with a large discoidal plaga .... { *C. felix*, Sl.
                                                       C. eximia, Sl.

ii. Unicolorous.

k. Prosternum not transversely sulcate on base ...

l. Size medium, head narrow and obliquely angustate before eyes ... .

m. ♂ with external teeth of anterior tibiæ obtuse   *C. dilutipes*, Putz.

mm. ♂ with external teeth of anterior tibiæ slender and prominent   .... *C. angustipes*, Putz.

ll. Size large, head wide and roundly angustate before eyes ..   .. *C. simulans*, Sl.

kk. Prosternum transversely sulcate on base............

n. Elytra with sides very lightly or not percep-}   *C. vagans*, Putz.
tibly narrowed to base ....}   *C. lepida*, Putz.

nn. Elytra with sides strongly rounded, decidedly narrowed to base.... . ... *C. sydneyensis*, Sl.

Evidently *C. microdon*, Putz., *C. rubripes*, Putz., and *C. isogona*, Putz., come into this section.

CLIVINA BASALIS, Chaudoir.

Bull. Mosc 1843, iv. p. 733; Putzeys, Mém. Liège, 1863, xviii. p. 38.

Black, base of elytra red (the red part about one-third of elytra in middle of disc and sloping backwards to half the length on each side); legs reddish testaceous. Head smooth, convex, angustate with hardly a perceptible sinuosity on each side before eyes; clypeus not divided from front, anterior margin bordered, widely emarginate, anterior angles rounded. Prothorax convex, smooth, of almost equal length and breadth (1·8 × 1·7 mm.), narrowed anteriorly (ant. width 1 5 mm); sides lightly rounded; basal curve short; lateral basal impressions well marked. Elytra lightly convex, a little depresed on disc, lightly rounded on sides, not perceptibly narrowed to base (4 × 2 mm), strongly punctate-striate, the striæ entire, but weaker towards apex, fourth out-turned and joining fifth at base; five inner interstices convex towards base, becoming flat towards apex, eighth distinctly marked on apical curve; a submarginal carina at shoulder. Prosternum with intercoxal part wide anteriorly, transverse sulcus of base obsolete. Anterior tibiæ 3-dentate. in ♂ narrow, first external tooth strong, short, second shorter, projecting but little beyond margin of tibia, inner apical spine elongate, curved and obtuse at apex: in ♀ external teeth much stronger, inner apical spine slender and acute.

Length 5 75-7, breadth 1·6-2 mm.

Hab · N.S. Wales - Sydney, Tamworth (Lea), Junee, Narran-dera, Urana, and Mulwàla (Sloane); Victoria; South Australia.

A well known and easily identified species.

CLIVINA FELIX, n.sp.

Head and prothorax black; elytra reddish testaceous, with a large ovate black plaga on the posterior two-thirds of disc (not reaching margin), lateral margins and under surface piceous; legs, antennæ, and palpi testaceous.   Facies, head, prothorax, elytra, prosternum, and legs as in *C. basalix*, Chaud.

Length 6-7, breadth 1·5-1·9 mm.

Hab. : Queensland—Port Denison (Masters), N. S. Wales— Junee, Narrandera, Carrathool, Urana, and Mulwala (Sloane), Victoria; South Australia (Blackburn).

This species is rather common in Southern Riverina during the summer months.   It resembles *C. basalis* so closely that it may be taken for it at a casual glance, but the colour differentiates it, the black discoidal patch of the elytra in *C. felix* never reaches the margins (as it does in *C. basalis*), but is separated by the testaceous seventh and eighth interstices, on the average it is smaller than *C. basalis*; the only specimens more than 6 5 mm. in length that I have seen have been those from Port Denison. A specimen from Narrandera has the base of the elytra clouded with black.   From *C. sellata*, Putz., it differs by its larger size, less cylindrical shape, smooth head, intercoxal part of prosternum not attenuate anteriorly, anterior tibiæ 3-dentate, &c.

Clivina eximia, n. sp

Robust, broad, lightly convex.   Head as in *C. basalis*, Ch.: prothorax broader than long, basal curve short, lateral basal impressions strongly marked, elytra wide, parallel, truncate at base, punctate-striate, fourth stria outturned and joining fifth at base, interstices convex, eighth carinate at base and apex; anterior tibiæ 3-dentate, with a small protuberance above upper tooth. Head, prothorax, and a large dorsal plaga on elytra black; base (widely), sides, and apex of elytra reddish; under surface reddish or reddish piceous, antennæ, mouth parts, and four posterior legs testaceous, anterior legs reddish.

Head convex, smooth (vertex and front covered with minute punctures); lateral impressions light: clypeus not divided from front, wide anteriorly, and very lightly emarginate; wings small, rounded, not divided from median part; eyes convex, prominent, very lightly enclosed behind.   Prothorax transverse

(1·8 × 2 mm), lightly narrowed anteriorly (ant. width 1·7 mm.), convex, declivous to base, finely transversely striolate; sides hardly rounded (nearly straight); posterior angles rounded but marked; anterior margin lightly and widely emarginate; anterior angles obtuse, lightly marked; border narrow, not weaker on sides of basal curve; median and anterior lines strongly impressed; lateral basal impressions short, deep, narrow. Elytra wide (4·5 × 2 4 mm), lightly convex, subdepressed on disc, shortly declivous to peduncle, base truncate (a little roundly); shoulders rounded; striæ deep, strongly crenulate, becoming lighter towards apex, first stria curving in towards suture a little before base and turning out towards second at basal extremity; interstices convex, depressed posteriorly. Prosternum protuberant; intercoxal part wide anteriorly, sulcate on base; episterna strongly rugulose and transversely striolate.

Length 8, breadth 2·4 mm.

Hab.—North West Australia. (Two specimens sent by Mr. Masters.)

Closely allied to and resembling *C. felix*, Sl., in colour, but larger, wider, and more depressed. The discoidal black patch on the elytra is oval, and extends in its widest part over the four or five inner interstices.

CLIVINA MICRODON, Putzeys.

Ann. Soc. Ent. Belg. 1866, x. p. 183

Of a slightly duller testaceous colour than *C. juvenis*, the last half of the elytra even more obscure than the base. The antennæ are more slender. The anterior elevation of the head is not declivous and narrowed behind as in *C. juvenis*, where it has the shape of a horseshoe; the vertex has not a central fovea; the prothorax is a little flatter, wider, and the impressions of the base are more marked and rounded towards base. The anterior tibiæ have only two very short and triangular teeth above the apical digitation.

Length 7, El. 3¾, breadth 1¾ mm.

Hab.: Melbourne (Coll. Chaudoir ; two specimens.)

The above is a translation of the whole of Putzeys' description of *C. microdon*. I cannot help thinking that it looks not unlike a description founded on an immature specimen of *C. basalis*, Ch., (♂), discoloured with age.

CLIVINA DILUTIPES, Putzeys.

Ann. Soc. Ent. Belg. 1868, xi. p. 12.

It appears to me likely that M. Putzeys confused two species under this name, viz., the Victorian species which I consider *C. vagans*, Putz., and a species from the coastal districts between Sydney and Brisbane, to which I attribute the name *C. dilutipes*. It is to be regretted that M. Putzeys gave no indication of the differences which divided *C. dilutipes* from *C. vagans*, for it seems not unlikely that both may have been founded on the same species; however, as there appear to be two closely allied species, to either of which either name seems equally applicable, it is probably best to apply the older name, *C. vagans*, to the species which it strikes me as being most fitted to, and then to allot the later name to the remaining species. The resemblance between these two species is very great, the only points of difference apparent to me being that, in *C. dilutipes* the elytra are more deeply striate, with coarser punctures in the striæ, and the prosternum is not sulcate on the base. The following is a description of *C. dilutipes* :—

Narrow, cylindrical. Head small, smooth, lightly bi-impressed; prothorax convex, sides rounded; elytra narrow, strongly punctate-striate, fourth stria joining fifth at base; prosternum with inter-coxal part wide anteriorly, non-sulcate on base, anterior tibiæ lightly 3-dentate. Black (sixth and seventh interstices sometimes piceous red on anterior third), legs piceous (four posterior often testaceous).

Head small, narrowly angustate before eyes; front and vertex lightly convex between facial sulci; clypeus not divided from front, roundly emarginate; facial sulci lightly impressed, sub-parallel, hardly divergent posteriorly; eyes convex, enclosed behind. Prothorax as broad as long (1·75 × 1·75 mm.), convex, narrowed anteriorly (ant. width 1·35 mm.); lateral basal impressions short, linear, well marked. Elytra narrow (4 × 1·9 mm.), widest a little behind middle; sides subparallel, hardly narrowed to shoulders; base truncate; shoulders rounded, not marked; striæ strongly impressed, deeply punctate, lighter towards apex; interstices convex near base, depressed behind basal third, eighth finely carinate at base and near apex.

Length 6·5-7·5, breadth 1·8-2·2 mm.

Hab.: N. S. Wales—Windsor, Clarence River, and Tweed River (Lea); Queensland—Brisbane (Coates).

The specimens from the Tweed River and Brisbane are darker coloured and have a greater tendency to lose the piceous red patch on the anterior part of the sides than those from the Clarence River.

CLIVINA ANGUSTIPES, Putzeys.

Ann. Soc. Ent. Belg. 1868, xi. p. 12.

Narrow, elongate. Black; legs dark piceous; antennæ, palpi, and tarsi ferruginous. Head small, smooth, convex, narrow, angustate without any sinuosity before eyes; clypeus bordered, roundly emarginate; frontal impressions arcuate, deep; eyes convex, prominent. Prothorax longer than broad (1 75 × 1·7 mm), greatly narrowed anteriorly (ant. width 1·4 mm.), lightly rounded on sides, smooth, convex; anterior angles obtuse; median line lightly impressed; anterior line strongly impressed; lateral basal impressions short, linear, distinct. Elytra a little broader than prothorax (3·8 × 2 mm.), lightly convex, parallel on sides; base truncate; shoulders rounded; striæ moderate, becoming shallow towards apex, strongly punctate (the punctures very fine towards apex), first flexuous near base, fourth outturned and joining fifth at base; interstices lightly convex near base, depressed towards apex, eighth carinate near shoulders, narrowly carinate on apical curve. Prosternum without pectoral ridges; intercoxal part wide at base, angustate but remaining wide anteriorly, transverse sulcus of base lightly marked, sometimes obsolete; episterna rugulose and transversely striolate. Anterior tibiæ narrow,

3-dentate; apical digitation long, lightly arcuate, external teeth short, prominent; inner apical spine as long as apical digitation, truncate, not incrassate.

Length 6·5-7·5, breadth 1·9-2·2 mm.

Hab : West Australia—Swan River, Newcastle, and Donny-brook (Lea).

Very closely allied to *C. lepida,* Putz , with which it agrees in facies, the head is similar, the prothorax seems a little narrower and longer, the elytra present no differences   The reasons for regarding it as distinct from *C. lepida* are that the prosternum is without pectoral ridges, and not so decidedly (if at all) trans-versely sulcate on base, and, that the anterior tibiæ differ slightly, their external teeth being longer and more prominent, the apical digitation longer and less obtuse, and the inner apical spine not incrassate at apex.

CLIVINA SIMULANS, n sp.

Robust, elongate, parallel, subcylindrical   Head smooth; pro-thorax as long as broad, narrowed anteriorly, elytra with fourth stria outturned and joining fifth at base, eighth interstice shortly subcarinate at base, narrowly carinate near apex; anterior tibiæ 3-dentate, ♂ with external teeth much weaker than ♀, and with inner apical spine long, incrassate, obtuse Black, shining; anterior legs piceous brown; antennæ and four posterior legs ferruginous.

Head smooth, strongly roundly angustate before eyes: the lateral sinuosity between the wings of clypeus and supra-antennal plates hardly perceptible; front and vertex convex, lævigate : clypeus not divided from front, a wide depressed space near anterior margin; clypeal elevation raised, lunulate; anterior margin roundly emarginate, wings not divided from median part, small, external angles rounded.   Prothorax convex, almost equal in length and breadth (2 48 × 2·5 mm.), narrowed anteriorly (ant width 2mm.); lateral basal impressions elongate, decidedly impressed Elytra truncate oval (5·9 mm. × 2·8 mm.), convex; sides parallel; striæ strongly impressed, crenulate-punctate; interstices convex on disc, depressed towards apex, seventh and eighth uniting at base to form a short humeral carina; marginal channel narrowed at humeral angles. Prosternum with pectoral part protuberant; intercoxal part wide anteriorly, non-sulcate on base; episterna overhanging anteriorly, minutely rugulose and finely transversely striolate

Length 9·3-10·5, breadth 2·7-2·8 mm.

Hab N.S. Wales—Urana District (Sloane; common on the edge of the more permanent creeks and swamps).

This species resembles *C. australasiæ*, Bohem., so closely that it is impossible to distinguish them except by a close scrutiny. The head is smoother, it is not punctate as is always the case in more or less degree with *C. australasiæ;* the sinuosity between the supra-antennal plates and the wings of the clypeus is less marked, the antennæ are a little lighter and slightly less incrassate; the supra-antennal plates diverge from the head more gently before the eyes; the prothorax is more convex, more strongly narrowed in front, the lateral basal impressions more distinct, the elytra are more convex, the sides being more declivous from the fifth stria to the margin, the basal declivity is greater, the striæ a little more distinctly crenulate, the submarginal humeral carina shorter and less developed, the base of the prosternum is not sulcate, and the wavy rugulosity of the episterna is finer; the external teeth of the anterior tibiæ are weaker in both sexes (especially in ♂), the upper being smaller and less outturned, the upper internal spine is longer, straighter, more acute, the apical spine is lighter in both sexes, and in ♂ is obtuse at the apex (in *C. australasiæ*, though the inner apical spine is longer in ♂ than in ♀, it is bent and pointed at the apex).

CLIVINA VAGANS, Putzeys.

Stett. Ent. Zeit. 1866, xxvii. p. 38; Ann. Soc. Ent. Belg. 1866, x p. 185.

Narrow, convex. Head small, smooth; prothorax smooth, rather longer than broad; elytra narrow, prosternum with strong pectoral ridges, intercoxal part wide anteriorly, sulcate on base. Coal black, shining; legs black, four posterior tibiæ piceous.

♂. Head small, smooth; front and vertex lightly convex; clypeus not divided from front, lightly emarginate, wings not divided from median part; supra-antennal plates narrow, not divided from wings of clypeus by a lateral sinuosity; frontal foveæ small, shallow ; facial sulci lightly impressed, diverging lightly backwards; facial carinæ wide, depressed; eyes not prominent. Prothorax a little longer than broad (2 × 1·9 mm.), evenly convex, narrowed anteriorly (ant. width 1 6 mm); anterior angles lightly rounded, lateral basal impressions shallow, elongate, minutely punctate; median and anterior lines distinctly impressed. Elytra convex (4 × 2·2 mm.); sides lightly rounded, a little narrowed to base; shoulders rounded; base truncate, lateral channel narrow at humeral angles, striæ lightly impressed, finely punctate, first entire, others (excepting seventh) becoming obsolete on apical declivity; interstices lightly convex near base, flat on apical half, seventh carinate at base, eighth narrowly carinate near apex Prosternum with pectoral part flat, margined by strong carinæ, these oblique, but becoming parallel at anterior extremity, episterna finely rugulose and transversely striolate   Anterior tibiæ narrow; the apical projection short and but little outturned the external teeth feebly developed, the upper not projecting beyond edge of tibiæ; inner apical spine very long, curved, obtuse at apex.

♀. Anterior tibiæ wider, with strong external teeth, the upper lightly prominent, prosternum with pectoral ridges shorter and more feebly developed.

Length 6·5-7·75, breadth 1·8-2·2 mm.

Hab. : Victoria—Lillydale (Sloane).

It appears to me that this must be *C. vagans*, Putz.; it certainly should be the species he mentions as from Melbourne, at the end of his description, if so, the type specimen was a very small one, though one equally small has been sent to me by Mr. Blackburn It is very closely allied to *C. lepida*, Putz., of which it seems the Victorian representative; the more convex and less parallel elytra seem the most decided character distinguishing it from *C. lepida* The black legs seem characteristic of the typical form of *C. vagans*, but specimens sent me from Swan Hill by Mr. C. French have the four posterior legs testaceous. The black species allied to *C. lepida* require careful study with large series of freshly collected specimens from many different localities

CLIVINA LEPIDA, Putzeys.

Stett. Ent Zeit. 1866, xxvii. p. 38; Ann. Soc. Ent. Belg. 1866, x p. 184.

Narrow, parallel. Head small, smooth, prothorax convex, not broader than long, decidedly narrowed anteriorly (ant. width 1·7 mm.); elytra parallel on sides, punctate-striate, fourth stria out-turned and joining fifth at base. Prosternum with intercoxal part wide anteriorly, sulcate on base; anterior tibiæ 3-dentate; ♂ with teeth of the anterior tibiæ much weaker than in ♀, and with the inner apical spine stout, curved and obtuse at apex. Black, shining; four posterior legs testaceous red, anterior legs piceous

Head narrow, obliquely angustate, with hardly any trace of a lateral sinuosity on each side behind wings of clypeus, convex and smooth between facial impressions; clypeus not divided from front, anterior margin roundly emarginate, wings small, not divided from median part Prothorax rather longer than broad (2·2 × 2·15 mm.), sides lightly rounded, not sinuate behind anterior angles, anterior margin lightly emarginate behind neck; anterior angles obtusely rounded, median and anterior lines well marked; lateral basal impressions distinct, linear. Elytra very little wider than pro-thorax (4·5 mm. × 2·3 mm), lightly convex; sides parallel, not perceptibly narrowed to shoulders; base truncate; shoulders rounded; apical declivity lightly declivous; striæ more strongly marked and punctate on disc than towards apex; interstices convex towards base, depressed towards apex, seventh shortly carinate at base, eightly finely carinate near apex; lateral border narrow, hardly perceptibly wider posteriorly. Prosternum with pectoral ridges strongly developed; episterna finely rugulose and transversely striolate. Anterior femora dilatate, upper side very arcuate.

Length 7-8·5, breadth 2·1-2·3 mm.

Hab. : N S. Wales—Windsor (Lea); New Zealand (Broun).

This species is readily separated from *C. australasiæ*, Bohem., by its smooth head, narrower before eyes, by the weaker external teeth of the anterior tibiæ in both sexes (the fourth tooth is quite obsolete); and by the ♂ having the inner apical spine more curved and obtuse at apex. A specimen sent to me from New Zealand by Capt. T. Broun, under the name of *C. rugithorax*, Putz., is identical in every respect with the ♂ of *C. lepida*, it seems to have been confused with *C. australasiæ* by New Zealand coleopterists. I believe *C. lepida* is also found in Victoria and South Australia.

Var. ? *C. tasmaniensis*, Sl. Coal black, shining, legs black. Differing from *C. lepida* by its darker colour; more convex form; prothorax with lateral basal impressions feebly developed, shallow, short; elytra less parallel, more rounded on sides, striæ less strongly impressed.

Length 7 2-8, breadth 1 9-2·2 mm.

Hab. : Tasmania (sent to me by Mr. A. M. Lea, as from Tasmania).

It requires further study and comparison with *C vagans*, Putz.; it is doubtless the species that Mr. Bates considered *C. vagans* (Cist. Ent. ii. 1878).

CLIVINA SYDNEYENSIS, n.sp

Robust, convex. Head small; frontal sulci diverging backwards; prothorax of equal length and breadth; elytra oval, narrowed to base, fourth stria outturned and joining fifth at base; anterior tibiæ 3-dentate; the external teeth much weaker and the inner apical spine longer (obtuse) in ♂ than in ♀. Black; legs piceous red, anterior darker than four posterior.

Head small, smooth, narrow, convex; clypeus not divided from front, roundly emarginate; eyes not prominent. Prothorax convex, of equal length and breadth (1·9 × 1 9 mm.), decidedly narrowed anteriorly (ant. width 1·6 mm.); anterior angles lightly marked, obtuse; anterior margin lightly emarginate; lateral basal impressions shallow, linear (sometimes obsolete). Elytra oval (4 x 2·1 mm), convex, widest behind middle, sides rounded, decidedly narrowed to base; shoulders not marked, base rounded, striæ narrow, deep on disc, lighter towards apex; their puncturation fine, dense, interstices narrow, convex towards base, eighth finely carinate near apex, a short distinct submarginal carina at shoulder. Prosternum with intercoxal part wide anteriorly, sulcate on base; pectoral ridges well developed.

Length 6·5-8, breadth 1·8-2 2 mm

Hab. N.S. Wales—Sydney District (Sloane, Lea).

Very closely allied to *C. lepida*, Putz., but evidently a distinct species. The marked character distinguishing them is the shape of the elytra. In *C. sydneyensis* the elytra are more convex, more deeply and abruptly declivous on base, sides, and apex, the sides are greatly rounded and strongly narrowed to the base, the interstices are narrower and more convex, the fourth being much narrower at the base, the lateral border is wider on the sides, except near the shoulders. From *C. dilutipes*, Putz., which it resembles, it may be distinguished by the more rounded sides of the elytra, and by the presence of a sulcus on the base of the prosternum. From *C. vagans*, Putz., it is separated by the stronger striæ and more convex interstices of the elytra, &c. It appears to be one of the commonest species of Clivina in the neighbourhood of Sydney.

CLIVINA RUBRIPES, Putzeys.

Ann Soc Ent Belg 1868, xi. p. 13.

The following is a translation of Putzeys' entire note (it cannot be called a description) on this species :—

A little smaller than *C. lepida*. Very distinct by its legs entirely of a red testaceous colour; its prothorax wider, flatter, shorter, nearly quite square, scarcely a little narrowed to the anterior angles, which are a little more rounded; its elytra longer, and its shoulders more marked.

Length 8, El. 4¼, breadth 1¾ mm.

Hab. Rockhampton (Coll. Castelnau)

CLIVINA ISOGONA, Putzeys.

Ann. Soc. Ent. Belg. 1868, xi. p. 13.

"Fusca, elytris pedibusque 4 posticis fusco-testaceis. Clypeus vix emarginatus; vertex in medio oblonge profunde foveolatus et antice parum punctatus.   Prothorax quadratus parum convexus, sulco medio profundo, transversim undulatus neque punctatus. Elytra cylindrica, basi truncata, humeris rotundatis, profunde punctato-striata.   Tibiæ anticæ apice digitatæ, extus unidigitatæ, denticuloque superiore vix perspicuo armatæ.

"Long. 8, El. 3¼*, Lat. 1¾ mm."

I translate the remarks which follow, as under:—By its size and general appearance it comes near *C. rubripes*, but the elytra are a little longer and the shoulders less rounded; the prothorax is shorter, still less narrowed in front, a little less convex; the median line is more deeply impressed and the surface bears much more distinct undulate striæ; the two impressions of the base are less marked.

The vertex bears in the centre a deep oblong fovea which is preceded by some large scattered punctures.   The epistoma is much less emarginate and more strongly bordered in the middle; the antennæ are a little less thick.

The collection of M. de Castlenau contains a single specimen without exact locality; presumably from Melbourne.

SECTION IV.

Submarginal humeral carinæ of elytra nearly obsolete.   Prosternum with intercoxal part narrow anteriorly, sulcate on base; episterna punctate.   Ventral segments punctulate laterally.

---

* There is evidently a mistake in these figures; judging from the statement which follows that the elytra are longer than those of *C. rubripes* it is probable we should read 4½.

CLIVINA PECTORALIS, Putzeys.

Ann. Soc. Ent. Belg. 1868, xi. p 14.

Robust, convex; prothorax broader than long; elytra oval with base truncate, crenulate-punctate, fourth stria joining fifth at base, submarginal humeral carina hardly developed; prosternum with intercoxal part sharply narrowed, not attenuate anteriorly, sulcate on base, episterna finely punctate; anterior tibiæ strongly 3-dentate Head, prothorax, legs, suture and lateral margins of elytra reddish brown; elytra piceous brown.

Head not large, punctate between posterior extremities of supra-orbital carinæ; vertex and front convex· clypeus not divided from front, anterior margin widely emarginate, bordered; wings not divided from median part, widely rounded; supra-antennal plates convex, rounded externally, projecting . strongly and sharply beyond wings of clypeus; frontal foveæ large, wide; facial carinæ wide, merely a backward prolongation of the supra-antennal plates; facial sulci wide, divergent; eyes convex, not prominent, orbits prominent and convex behind Prothorax finely shagreened, convex, widest a little before posterior angles (1·3 × 1·35 mm.), narrowed anteriorly (1·1 mm.), sides short, evenly rounded, anterior margin emarginate; angles obtuse, posterior angles marked; median line strongly impressed· anterior line lighter. Elytra wider than prothorax (2 9 × 1 6 mm.), oval; shoulders rounded, not marked; striæ entire, deeply impressed, finely crenulate, seventh not interrupted at beginning of apical curve; a short distinct striole at base of first interstice, interstices convex, minutely shagreened, eighth broad, hardly carinate near apex. Intermediate tibiæ with external margin spinulose, the spine nearest the apex a little stronger than others.

Length 4·5-5·2, breadth 1·35-1·6 mm.

Hab.: Queensland — Rockhampton (Coll. Castelnau); N.S. Wales—Clarence River (Lea); West Australia (sent by Mr. French, probably from N.W. Coast).

15

A completely isolated species among the Australian members of the genus. The external spur of the intermediate tibiæ is very weak and situated not far from the apex.

The description given above is founded on specimens (♀?) from the Clarence River, sent to me by Mr. Lea, which, although appearing to differ slightly from M. Putzeys' description of *C. pectoralis* in having the puncturation of the head, prothorax, and prosternal episterna weaker, seems undoubtedly that species. One specimen (♂ probably), of which only the elytra now remain, is much smaller (4·5 mm.), differently coloured—the elytra being black, with the suture and lateral border reddish — the puncturation of the metasternum and ventral segments stronger, and the ventral segments foveate laterally. In the specimen described above, the puncturation of the prothorax is so obsolete as to require a powerful lens to distinguish it; the metasternum is finely punctate near the sides, also the episterna, and the ventral segments are without punctures or lateral foveæ. A specimen sent to me by Mr. French, as from West Australia, is of an entirely ferruginous colour.

Procera group.

Size large, or above the average. Clypeus truncate-emarginate (median part truncate, wings projecting strongly forward, and roundly obtuse at apex). Elytra with fourth and fifth striæ confluent at base, a submarginal carina at shoulder (sometimes feebly developed, *e.g.*, *C nyctosyloides*, Putz). Prosternum with intercoxal part very wide anteriorly, not sulcate on base. Anterior tibiæ 3-dentate, external teeth weaker in ♂ than in ♀; inner apical spine in ♂ long, curved, obtuse at apex.

Fifteen species are associated in this group; of these, twelve known to me, are tabulated below. The group could readily be broken up into seven sections represented by *C. procera, C. monilicornis, C. oblonga, C. regularis, C. nyctosyloides, C. mastersi*, and *C. marginata*. The species I do not know are *C. elegans*, Putz., *C. prominens*, Putz., and *C. obscuripes*, Blkb.

Table of Species known to me.

A. Lateral cavities of peduncle punctate or rugulose.

  B. Metasternal episterna elongate (metasternum between intermediate and posterior coxæ longer than posterior coxæ).

    C. Prothorax not longer than broad, mandibles short ... . *C. procera*, Putz.

    CC. Prothorax longer than broad, mandibles decussating (antennæ very short, moniliform) .... *C. monilicornis* Sl.

  BB. Metasternal episterna very short (metasternum between intermediate and posterior coxæ shorter than posterior coxæ).

    D. Head with a strong transverse occipital impression. .... *C. oblonga*, Putz.

    DD. Head without a transverse occipital impression (or at most only lightly indicated on sides).

      E. Head without a noticeable lateral sinuosity between supra-antennal plates and wings of clypeus. Prosternal episterna rugose on basal declivities...... . ...… *C. abbreviata*, Putz.

      EE. Head with a decided lateral sinuosity between supra-antennal plates and wings of clypeus. Prosternal episterna smooth on basal declivities *C. macleayi*, Sl.

AA. Lateral cavities of peduncle smooth.

  F. Prothorax not broader than long, normally narrowed anteriorly................................ .... *C. regularis*, Sl.

  FF Prothorax broader than long, greatly narrowed anteriorly.

    G. Mandibles short.

      H. Elytra with striæ deep, entire, strongly punctate, antennæ subfiliform, second joint decidedly longer than third...... } *C. nyctosyloides*, Putz. *C. interstitialis*, Sl.

      HH. Elytra smooth on sides and apex; antennæ filiform, third joint not shorter than second.

I. Striæ of elytra simple, interstices
    not convex ... . *C. mastersi*, Sl.

II. Striæ of elytra punctate, interstices
    convex on anterior part of disc.... *C. ovipennis*, Sl

GG. Mandibles long, decussating.

      K. Elytra with testaceous margin .. *C. marginata*, Putz.

      KK. Upper surface entirely black . *C. gracilipes*, Sl.

Clivina procera, Putzeys.

Stett. Ent. Zeit. 1866, xxvii. p. 34; Ann. Soc. Ent. Belg. 1866, x. p. 180; *Scolyptus procerus*, l c xi. p. 8.

A widespread and well known species; the following diagnosis will enable it to be identified :—

Elongate, parallel, subcylindrical. Black, shining; legs piceous. Head smooth, lateral margin sloping obliquely and evenly forward from a little before the eyes: clypeus not divided from front; median part truncate, wings strongly advanced, rounded at apex; facial sulci lightly impressed; eyes prominent, lightly enclosed behind. Mandibles short. Antennæ not short, submoniliform, lightly compressed Labrum 5-setose Prothorax subquadrate (4 × 4·1 mm.), lightly convex, narrowed anteriorly (ant. width 3·3 mm). declivous to base, anterior margin very lightly emarginate Elytra a little wider than prothorax (9·5 × 4 5 mm.), parallel, striæ crenulate, strongly impressed near base, becoming lighter towards apex and sides, fourth outturned and joining fifth at base, seventh interstice carinate at humeral angle, eighth very narrowly and lightly indicated (sometimes obsolete) near apex. Prosternum protuberant, intercoxal part very wide anteriorly, bordered on each side by a strong wide carina, vertical and non-sulcate on base; episterna covered with a faint wavy rugulosity. Lateral cavities of peduncle punctate. Metasternum longer between intermediate and posterior coxæ than length of posterior coxæ; episterna elongate. Anterior femora thick, not channelled below in ♂; tibiæ 3-dentate (much narrower and with external teeth ch weaker in ♂ than in ♀); inner apical

ut, incurved and truncate at apex.

ength 13·5-17, breadth 3·75-4·7 mm.

ab.: Queensland—Burketown District

(Coll. Castelnau); N.S. Wales—Murra

ers; Victoria; South Australia.

ote.—A specimen in the possession of M

win is of the following dimensions:

x 5·25 × 5·3, elytra 13·5 × 6, lengtl

est Clivina I have seen, but, beyond i

d, I cannot differentiate it from *C. proc*

Clivina prominens, Putz

Stett. Ent. Zeit. 1866, xxvii. p. 35; Ann.

p. 182; *Scolyptus prominens*, l.c. 1868, x

Putzeys' whole description is in three lin

'ery near *C. procera*, of which it is pe

is smaller; the prothorax is a little sl

Head short (1·6 × 1·8 mm.), wide before eyes; vertex and front smooth, wide, lightly convex; clypeal elevation prominent, rounded: clypeus divided from front by a strong transverse impression, depressed near anterior margin; median part truncate, bordered; wings strongly advanced, rounded externally, very obtuse at apex, oblique on inner side, supra-antennal plates wide, rounded externally, a light sinuosity dividing them from clypeal wings; eyes globose, prominent, projecting lightly beyond supra-antennal plates; orbits narrow and abruptly constricted behind; facial sulci diverging backwards from ends of clypeal suture, facial carinæ thick, prominent. Labrum 5-setose. Palpi stout; penultimate joint of labial about same length as terminal. Antennæ with second joint decidedly longer than third, joints 4-10 short, quadrate. Prothorax smooth, longer than broad (3 × 2·8 mm.), narrowed anteriorly (ant. width 2·3 mm.), very convex transversely, lightly convex longitudinally, very declivous to base; anterior margin subtruncate (lightly emarginate behind neck). anterior angles obtuse, hardly marked; posterior angles rounded: basal curve short; border narrow; median and anterior lines lightly impressed; lateral basal impressions distinct, round, foveiform. Elytra very convex, suboval (6 × 3 mm.), lightly rounded on sides, widely rounded at apex, very declivous to humeral angles, these rounded, striæ finely crenulate, strongly impressed on disc, weaker towards apex and sides, seventh hardly marked; interstices convex near base, becoming depressed towards apex, first of each elytron together forming a wide lightly raised sutural ridge, the four large punctures of third interstice stronger than usual. Prosternum protuberant, not canaliculate between coxæ or sulcate on base, episterna minutely shagreened and very finely transversely striolate. Anterior femora short, wide, compressed: anterior tibiæ wide, strongly 3-dentate; upper tooth prominent, triangular, inner apical spine long, curved, pointed. upper internal spine long, slender, acute; intermediate tibiæ wide, compressed, external spur strong, erect.

Length 9 5-11·5, breadth 2·6-3 mm.

Hab. : Queensland—Port Denison (Masters).

omalous species ; the arrangement of
he elytra and the form of the clypeus a
·a, Putz., and *C. abbreviata*, Putz.; prol
llied to *C. abbreviata*, Putz., than to a
o me, but the longer metasternal ep
its being put with that species.   The r
much shorter than in *C. procera*, being
'. *gracilipes*, Sl., *C. emarginata*, Putz., or
much narrower, especially in front, than

Clivina elegans, Putzeys.

Liège, 1863, xviii. p. 44; Stett. Ent.
n. Soc. Ent. Belg. 1866, x. p. 179.
, nitida, palpis tarsisque testaceis; labr
unneis.   Clypeus truncatus, alis angula
notum planiusculum, oblongo-subqua
n, a basi rotundatum, angulis posticis
ngato-oblonga, punctato-striata, inter
    Tibiæ anticæ sulcatæ extus for
calcaratæ.

little backwards.   The eyes are not very prominent, their posterior
third being embedded in the lateral margin of the head.   The
impression which divides the head from the neck is hardly marked
in the middle.

The prothorax is quadrate, a little longer than broad, narrowed
anteriorly, very rounded at the posterior angles, not much
prolonged posteriorly; the surface is lightly convex, the anterior
margin is widely emarginate; the angles are a little prominent;
the sides, cut obliquely for their first half, are regularly curved to
the base; the posterior angles form no prominence; only a large
internal puncture is seen above a tubercle, which does not project
beyond the marginal border.   The transverse anterior impression
is rather close to the margin; the longitudinal impression extends
a little past the first.   In the middle of each side of the prothorax, facing the posterior angles, a rather wide shallow fovea is
noticed, which extends forward in a straight impressed and more
marked line, reaching beyond the anterior third of the prothorax.

The elytra form a very elongate regular oval, their upper surface
is depressed longitudinally along the suture on the anterior third,
the striæ are punctate, but the interstices are not raised.   It is
a prolongation of the seventh interstice, which at the shoulder
unites with the marginal border; only the interstices 1-3 touch
the base.

The anterior tibiæ are wide, sulcate on upper side; externally
they have a rather long strong tooth, and above this a second short
and broad tooth.   The intermediate tibiæ are wide, spinose along
the posterior side, which is armed with a spur

Underneath all the body is covered with undulating transverse
striolæ, dotted with rather scattered punctures.

Hab. : Australia (one specimen).

In his " Révision Générale " M. Putzeys has formed a separate
group for *C. elegans*, of which he treats as follows, being a
translation of his remarks in the Entomolgische Zeitung :—

Twenty-sixth Group.

It has much resemblance to the twenty-seventh group
[*C. procera*].   It differs by its less shining colour, its darker

antennæ, its legs of a blackish-brown, its epistoma with less narrow wings, its thicker antennæ, its eyes enclosed on all sides, the anterior impression of the head a little deeper, its head more convex, very finely punctate, its prothorax more oval and more emarginate in front, its elytra more convex, of a very regular elongate-oval shape, its striæ deeper, the under surface of the prothorax finely striolate-punctate, and particularly by the metasternal episterna, which are short and square; the paronychium is a little longer.

The central carina of the prosternum is broad, canaliculate only between the coxæ.

M. Putzeys also says that he had possessed this insect a long time, and that it was given to him as coming from South America. As the greater part of its features show an affinity to the Australian species he adds that he suspects that this country may well be its true habitat.

The impression left upon my mind by a study of Putzeys' description, with specimens of *C. oblonga*, Putz., before me, is that it may well have been founded on a specimen (♀) of that species, and it is to be regretted that M. Putzeys when describing *C. oblonga* did not compare it with *C. elegans*. The only features that separate these species seem to be the punctate striæ and the interstices not raised, with the striolate-punctate under surface of *C. elegans*; however, a specimen of *C. oblonga*, referred to under that species as identical with *Ceratoglossa foveiceps*, Macl., (*vide* p. 235), presents elytral characters that might be described as are those of *C. elegans*. It is possible the fine punctures of the head and under surface may be a *post mortem* effect; still, as M. Putzeys regarded the species he named *C. oblonga* as undescribed, his opinion, must, I think, be upheld, though not without doubt on my part.

CLIVINA OBLONGA, Putzeys.

Scolyptus oblongus, Putz., Ann. Soc. Ent. Belg. 1873, xvi. p. 10; *Ceratoglossa foveiceps*, Macleay, Trans. Ent. Soc. N.S.W. 1863, .i. p. 73.

Robust, elongate-oval.   Head strongly transversely impressed
behind vertex; antennæ moniliform; mandibles short: elytra
oblong-oval; striæ deep, entire; lateral cavities of peduncle
punctate; metasternum and metasternal episterna short; anterior
tibiæ 3-dentate.   Black, shining; under surface minutely shagreened.

♀. Head smooth, narrowed to a neck behind eyes; lateral
margins sloping obliquely and evenly forward from a little
before eyes; a deep oblique impression dividing clypeus on
each side from supra-antennal plates—these impressions sometimes turning inwards and dividing the clypeus from the front at
each side: clypeus not divided from front in middle, convex,
declivous to anterior margin; this bordered, deeply truncate-
emarginate; wings concave, strongly advanced, widely rounded at
apex, sloping gently to median part on inner side; supra-antennal
plates large, convex, not divided from the wide convex facial
carinæ; facial sulci strongly impressed; eyes convex, deeply
enclosed in orbits; these large, strongly protuberant (about
two-thirds size of eyes) behind eyes; supra-orbital punctures
distant from eyes, temporal region strongly rugulose; gulæ
finely rugulose.   Antennæ stout, moniliform, incrassate: joints
5-10 short, strongly compressed.   Palpi with apical joint
thick, oval.   Prothorax smooth (faint transverse striolæ noticeable under a lens), a little longer than broad (3·7 × 3·5 mm.·,
narrowed anteriorly (ant. width 3 mm.), depressed, shortly
declivous to base, sides very lightly rounded; posterior angles not
marked; anterior margin emarginate, widely and obtusely truncate on each side of neck; border narrow, reflexed on sides;
lateral basal impressions weakly developed or obsolete; anterior
line deeply impressed.   Elytra a little wider than prothorax
(7·7 × 3·8 mm.), subdepressed; sides lightly rounded; base narrow
and submarginate between humeral angles; striæ deeply
impressed, entire (the inner ones often obsoletely crenulate),
fourth joining fifth at base, but not outturned; interstices
convex, eighth shortly carinate at base, narrowly carinate
on apical curve; border reflexed; lateral channel wide.   Pro-

sternum with intercoxal part channelled, wide anteriorly, almost vertical and non-sulcate on base; pectoral carinæ weakly developed, widely divergent anteriorly. Metasternum much shorter between intermediate and posterior coxæ than length of posterior coxæ. Legs in every way similar to those of *C. procera*.

Length 13·5-16, breadth 3·8-4·6 mm.

Hab.: N. S. Wales—Richmond River (Macleay), Narrara Creek (Sloane), Burrawang (Fletcher).

Allied to *C. abbreviata*, Putz., from which the strong transverse occipital impression, which is characteristic of *C. oblonga*, at once separates it.

The number of punctures on the third interstice of the elytra. varies from four to five; the posterior puncture in *C. oblonga* is deep and placed opposite the extremity of the fourth interstice, and is much nearer the apex than in any other of the large species of Clivina from Australia. The form of the apical extremities of the third and fifth interstices is worthy of note—these interstices are strongly raised and confluent at their apices, the apex of the fourth interstice terminating in a rather deep depression formed by this union of the third and fifth.

A specimen (♀) is in my collection which I have compared and found identical with the type of *Ceratoglossa foveiceps*, Macl. It is larger (16 × 4·6 mm.) and more convex than typical specimens of *C oblonga*, has the prothorax a little shorter (3·8 × 3 8 mm.), the striæ of the elytra distinctly crenulate, and the posterior large puncture of the third interstice a little further from the apex: but I cannot think it a different species. The name *foveiceps* was preoccupied in *Clivina* when Sir William Macleay bestowed it on his species; the later name *oblonga* therefore has to be adopted.

CLIVINA ABBREVIATA, Putzeys.

Scolyptus abbreviatus, Putz., Ann. Soc. Ent. Belg. 1873, xvi. p. 10.

This species agrees with *C. oblonga*, Putz., in most features; the head is similar, excepting that the transverse occipital impression is wanting; the metasternum and its episterna are similar; the legs are similar, but the external teeth of the anterior tibiæ are much stronger.   The following brief description will enable it to be recognised :—

Black, legs piceous, or reddish.   ♂. Prothorax as long as broad (3·1 × 3·1 mm), decidedly narrowed anteriorly (ant. width 2·4 mm.), lightly convex; sides lightly rounded; basal curve short; anterior margin emarginate, anterior angles lightly advanced, widely rounded. Elytra oval (6·5 × 3·4 mm,), striæ and interstices as in *C. oblonga*, eighth interstice feebly and shortly carinate near apex.   Prosternum as in *C. oblonga*, the pectoral carinæ more strongly developed.   Anterior tibiæ 3-dentate, the external teeth strong.   Under surface minutely shagreened.

Length 12·5-13·5, breadth 3·4-3·8 mm.

Hab : Queensland—Wide Bay District (Spencer, Masters)

Note —In the specimen before me, the third interstice has five punctures on each elytron, the three anterior ones not being placed quite similarly on each elytron. In *C. abbreviata* the posterior puncture is placed at the beginning of the apical declivity, not on the declivity at the junction of the third and fourth striæ, as in *C. oblonga*, Putz.

CLIVINA MACLEAYI, n.sp.

Short, robust, convex. Head convex, facial carinæ diverging strongly backwards, clypeus deeply truncate-emarginate; prothorax subquadrate, lightly narrowed anteriorly, elytra oval, strongly striate, fourth stria outturned and joining fifth at base, interstices equal, lightly convex, seventh forming a weak submarginal carina at shoulders, eighth obsolete on apical curve; lateral cavities of peduncle minutely shagreened, not punctate; metasternal episterna short; anterior tibiæ 3-dentate. Piceous brown, prothorax and upper part of head darker.

Head wide before eyes, abruptly constricted on sides behind eyes; front and vertex wide, convex; frontal impressions wide, shallow; clypeal elevation convex, declivous in front: clypeus divided from front by an irregular shallow impression, this impression obsolete in middle; median part not divided from wings, truncate; wings advanced, rounded at apex and externally, inner side gently oblique; supra-antennal plates short, wide, rounded externally, projecting strongly beyond clypeal wings; eyes deeply embedded in orbits behind, small, convex, hardly more prominent than supra-antennal plates; orbits projecting strongly from sides of head behind eyes; facial carinæ strongly developed, converging roundly in front and reaching clypeus. Mandibles short. Labrum 5-setose. Labial palpi stout; penultimate joint not longer than terminal; this thick, obtuse at apex. Antennæ short, moniliform; third joint shorter than second; joints 5-10 short, quadrate. Prothorax subquadrate (2·3* × .2·45 mm.), widest just before posterior angles, a little narrowed anteriorly (ant. width 2·15 mm.), convex, very declivous to base; sides lightly and widely sinuate, rounded to anterior angles, anterior margin widely and deeply emarginate; anterior angles distant from neck, obtuse but marked; posterior angles rounded, not marked, basal curve very short; lateral channel well developed; median line strongly impressed, reaching base; anterior line distinct, very near margin; border narrow, not upturned at posterior angles. Elytra oval (4·5 × 2·5 mm.), widest a little behind middle, sides strongly rounded; shoulders rounded; apex widely rounded; striæ deep, simple, seventh hardly less deeply impressed than others. Prosternum with intercoxal part wide anteriorly, non-sulcate on base; episterna very feebly transversely striolate, overhanging near anterior angles. Anterior femora short, wide, strongly arcuate above, rounded not channelled below; tibiæ rather wide, apex short, wide, curved, first external tooth wide, prominent, upper tooth wide, not prominent, inner apical spine thick, truncate, longer than apical digitation (as long as three basal joints of tarsus), upper internal spine finely

---

* This is the length in the middle; from anterior angle to base the length about equals the breadth.

acuminate; intermediate tibiæ with outer edge spinulose, the external spur prominent and placed considerably before the apex.

Length 9, breadth 2·5 mm.

Hab. : Queensland—Port Darwin, Roper River (sent by Mr. Masters).

A very distinct species, in general appearance much resembling a small species of *Promecoderus.* Its affinity is to *C. abbreviata,* Putz., but it differs greatly from that species by its smaller size, head much wider in front of eyes, more strongly rounded (a strong sinuosity behind wings of clypeus) to anterior angles, the facial carinæ long, incurved, forming a border to the inner side of the supra-antennal plates, eyes more deeply enclosed· in orbits, these more abruptly constricted behind; prothorax more quadrate, the sides sinuate, the basal curve still shorter; prosternum with intercoxal part not bisulcate, &c.

CLIVINA REGULARIS, n.sp.

Robust, parallel. Head as in *C. procera;* clypeus deeply emarginate-truncate; prothorax as long as broad, lightly narrowed anteriorly; elytra parallel, simply striate, striæ deep on disc, weak on sides, interstices convex on disc, eighth feebly indicated near apex, submarginal humeral carina short; prosternum with intercoxal part very wide anteriorly, episterna smooth; lateral cavities of peduncle deep, not punctate; metasternal episterna of medium length; metasternum between anterior and posterior coxæ not longer than posterior coxæ; anterior tibiæ 3-dentate Black.

Head smooth, large (2 x 2 2 mm.), convex, obliquely angustate before eyes, lateral impressions light; clypeal elevation convex: clypeus divided from front on sides, depressed along anterior margin; median part truncate; wings concave, strongly advanced, obtusely rounded at apex; eyes prominent, convex, enclosed by orbits. Prothorax as long as broad (2·9 mm. x 2·9 mm.), lightly narrowed anteriorly (ant. width 2·4 mm.), smooth, convex; sides nearly straight, obsoletely sinuate; posterior angles rounded, not marked, anterior margin widely and very lightly emarginate, anterior angles obtuse, but slightly prominent; median and anterior lines strongly impressed; lateral basal impressions elongate, very shallow.   Elytra truncate-oval (6·2 × 3 mm.), a little narrowed to base, very convex; sides rounded; apex widely rounded; base truncate; shoulders rounded, striæ obsoletely crenulate, four inner ones very strongly impressed, weaker towards apex, fifth, sixth and seventh successively weaker (seventh very faint); five inner interstices convex, seventh and eighth united at base and forming a short, rather broad and lightly raised carina at humeral angle.   Legs stout; anterior trochanters not projecting at base of femora; tibiæ with apical digitation short, thick, two external teeth short, thick, prominent, inner apical spine longer than apical digitation, obtuse at apex; external spur of intermediate tibiæ as in *C. australasiæ.*

Length 11·5, breadth 3 mm.

Hab. · New South Wales—New England.

Two specimens, both apparently ♂, were sent to me by Mr. Masters   A very distinct species—in general appearance it resembles *C. australasiæ,* Bohem., but the smooth prosternal episterna and peduncle, the emarginate-truncate clypeus, &c., show it to be allied to *C procera,* Putz, and *C. oblonga,* Putz.; probably its nearest ally is *C. monilicornis,* Sl, with which it is associated by the length of the metasternal episterna, but its antennæ, though moniliform, are longer; the head is larger, with wider supra-antennal plates; the prothorax is shorter, less strongly narrowed anteriorly, and without the rounded basal foveæ of *C. monilicornis;* the elytra are less convex   From *C. simulans,* Sl , it is readily distinguished by its thicker antennæ; the form of the clypeus. elytra more rounded on sides, the striæ not punctate; the prosternal episterna not rugulose on the basal declivities, &c.

CLIVINA NYCTOSYLOIDES, Putzeys.

Ann. Soc. Ent. Belg. 1868, xi. p. 10.

Oval, robust, convex.   Head large, eyes prominent, prothorax transverse, subtrapezoid, very convex, elytra oval, deeply punctate-striate, striæ entire, fourth joining fifth at base, interstices convex, eighth interrupted at beginning of apical curve, very narrowly carinate near apex, submarginal humeral carinæ obsolete; prosternum with intercoxal part very wide anteriorly, lateral cavities of peduncle smooth, wide, shallow; anterior tibiæ 3-dentate, external spur of intermediate tibiæ oblique and near apex. Black, legs piceous, antennæ and tarsi reddish. .

Head large (1·8 × 2·2 mm.), smooth between lateral impressions; a punctiform impression in middle between eyes; a strong lateral sinuosity between wings of clypeus and supra-antennal plates: clypeus not divided from front, depressed along anterior margin; median part truncate; wings concave, strongly advanced beyond median part, roundly obtuse, oblique on inner side; throat very convex, gulæ with a few faint wavy striolæ; eyes convex, prominent, enclosed on lower side posteriorly Labial palpi stout, penultimate joint about same length as terminal, this stout, fusiform, truncate.   Antennæ not long, lightly compressed, not incrassate; second joint decidedly longer than third.   Prothorax smooth, transverse (3·2 × 3 5 mm.), widest a little before posterior angles, greatly narrowed anteriorly (ant width 2 5 mm.), rounded on sides, evenly convex, gently and roundly, but deeply declivous to base; anterior angles obtuse ; posterior angles obtuse, but marked, border thick, widened at and passing round anterior angles; median line deeply impressed; anterior line distinct and near margin; lateral basal impressions wanting.   Elytra oval (7 5 × 4 mm.), convex, wide across base; shoulders rounded; apex widely rounded; striæ strongly impressed, entire, coarsely punctate, the puncturation strong on apical third, seventh hardly impressed, but distinctly indicated as a row of punctures; interstices convex for whole length, seventh wide and convex on apical curve, joining first at apex.   Prosternum with intercoxal part bisulcate, non-sulcate on base; episterna smooth (only some minute wavy transverse scratches), hardly overhanging anteriorly.   Metasternum a little longer between intermediate and posterior coxæ than length of posterior coxæ; episterna rather wide posteriorly Anterior femora compressed, tibiæ with apical digitation long, stout, strongly curved, first external tooth prominent, stout, obtuse, upper not prominent, upper internal spine slender, very acuminate: intermediate femora long; tibiæ with external spur a little above apex, pointing obliquely downwards.

Length 13, breadth 4 mm.

Hab.. Queensland—Rockhampton (Coll. Castelnau), Dawson River (Barnard).

M. Putzeys formed a separate group for the reception of this species, but I have placed it among the large assemblage of species which I term the "*procera group*," in which it is the representative of a distinct section. Putzeys describes the inner apical spine of the anterior tibiæ as equalling in length the apical digitation, not diminishing in width and truncate at apex in the ♂, and acuminate in the ♀; I only know the ♀, in which it does not actually equal the apical digitation in length.

The elytra (only) of a specimen are in my collection received from the late Mr. G Barnard from Coomooboolaroo, Dawson River, in which the fourth stria is free at the base.

CLIVINA INTERSTITIALIS, n.sp.

Oval, robust, convex. Head convex, eyes convex; prothorax transverse, subtrapezoid, longitudinally convex; elytra ovate, wide, deeply punctate-striate, fourth stria joining fifth at base, interstices very convex, eighth interrupted at beginning of apical curve, finely carinate near apex, submarginal carinæ of shoulders obsolete ; prosternum with intercoxal part bisulcate, wide anteriorly, non-sulcate on base, episterna smooth, not overhanging in front; lateral cavities of peduncle wide, very shallow, not punctate; metasternal episterna shorter than usual in genus; anterior tibiæ narrow, 3-dentate, apex long, wide, curved, external spur of intermediate tibiæ short, stout, nearer apex than usual. Black, antennæ and tarsi piceous.

Head not large (1·6 × 1·8 mm.), convex, smooth between facial impressions, obsoletely transversely impressed behind vertex; frontal impressions narrow, extending on to wings of clypeus; facial sulci linear, deep, divergent : clypeus with median part

16

truncate; wings concave, strongly advanced beyond median part, obtusely rounded anteriorly; gulæ convex, hardly at all rugulose. Labial palpi with penultimate joint stout, rather short, about same length as terminal; this wide and obtuse at apex. Antennæ with third joint shorter than second; joints 4-11 short, hardly compressed. Prothorax smooth, transverse (2 6 mm. × 2·9 mm.), widest a little before posterior angles, greatly narrowed anteriorly (ant. width 2 mm.), very convex, strongly and roundly declivous to base; sides rounded, anterior angles obtuse; posterior angles obtuse, but marked. basal curve short; border thick, wide and reaching neck at anterior angles. median line weak, anterior line strongly impressed, lateral basal impressions obsolete   Elytra ovate (5·5 × 3 5 mm), striæ deep, entire, very coarsely punctate on disc ; interstices subcarinate for whole length, narrow and more carinate on apical declivity.

Length 10, breadth 3·5 mm.

Hab. : Queensland—Cooktown (from Mr. French).

This species agrees in all points of structural detail with *C nyctosyloides*, Putz., of which it may possibly be a marked variety. though I regard it as a distinct species   The following differences from *C. nyctosyloides* may be noted; the smaller size; more convex form; more elongate head. prothorax more convex, narrower, more strongly narrowed anteriorly; elytra more convex, striæ deeper, interstices more convex, especially towards apex.

Clivina mastersi, n sp ,

Very large, robust, convex   Head as in *C. procera·* prothorax smooth, greatly narrowed anteriorly, convex, strongly declivous to base; basal curve short, rounded : elytra oval, smooth on sides and apex; five inner striæ impressed towards base, first only entire, fourth and fifth confluent at base; sixth interstice narrow, not carinate at humeral angle,[*] eighth not visible near apex:

---

[*] The weakly developed submarginal humeral carina is a continuation of the sixth interstice; it is very narrow and hardly raised.

prosternum with intercoxal part wide anteriorly, bisulcate between
coxæ, non-sulcate on base; episterna smooth, hardly overhanging
anteriorly; metasternal episterna short; lateral cavities of
peduncle feebly developed, impunctate : anterior tibiæ slender,
3-dentate; intermediate tibiæ narrow, external spur short, placed
at apex. Black, antennæ and tarsi piceous red.

♂. Head rather large (2·7 × 3 mm.), convex, smooth, obsoletely
and widely transversely impressed behind facial carinæ; sides
obliquely narrowed and widely sinuate before eyes. clypeus not
divided from front, declivous; median part wide, truncate; wings
narrow, impressed, strongly and obtusely advanced, facial impressions strongly impressed, sinuate; facial carinæ short, wide,
convex, not greatly raised; eyes prominent, strongly enclosed by
orbits on posterior part of lower side. Palpi filiform; labial with
penultimate joint not longer than terminal. Antennæ filiform,
third joint not shorter than second. Prothorax nearly as long
as broad (4·5 × 4·6 mm.), widest a little behind middle, greatly
narrowed anteriorly (ant width 3 5 mm), roundly and deeply
declivous to base; sides oblique, hardly rounded; anterior margin
lightly emarginate; anterior angles rounded; posterior angles
rounded, border thick, hardly reflexed on sides, weaker behind
posterior angles, extending round anterior angles to neck, median
line linear, distinct: anterior line lightly but decidedly impressed,
lateral basal impressions shallow, wide, distinct. Elytra oval
(10·5 × 5·5 mm.), convex; sides rounded; shoulders rounded, not
marked; striæ simple, four inner ones strongly impressed towards
base, first entire, joining second at base, others not reaching apex,
successively shorter, fourth not outturned at base, fifth inturned to
meet fourth at base, sixth and seventh obsolete; three inner
interstices lightly convex near base, sutural interstice of each
elytron separately convex on basal third, after that together forming a lightly raised sutural ridge; lateral border narrowly reflexed,
reaching nearly to peduncle at base. Anterior femora thick,
hardly compressed, lower side rounded; tibiæ slender, apical
digitation long, narrow, curved, obtusely pointed, first external
tooth prominent, triangular, second obtuse, feebly developed,

middle of lower side greatly raised and forming a prominent triangular tooth above upper internal spine, inner apical spine about as long as apical digitation, cylindrical, curved, obtuse, upper spine long, slender, very acuminate; four posterior legs light.

Length 19, breadth 5 5 mm.

Hab. : Queensland—Port Darwin.

A single specimen of this fine species was sent to me for description by Mr. G. Masters. Excepting a specimen sent to me by Mr. Masters as from Port Darwin, which I cannot separate from *C. procera*, Putz., this is the largest Clivina I have seen. It represents a distinct section, its nearest ally being *C. ovipennis*, Sl., which agrees with it in facies, and in form of metasternal episterna and legs.

CLIVINA OVIPENNIS, n.sp.

Elongate-oval, robust, convex. Head obsoletely impressed on each side behind vertex; prothorax greatly narrowed anteriorly · elytra oval, smooth on sides and apex; four inner striæ deeply impressed and coarsely punctate on basal half; eighth interstice obsolete on apical curve; a very feebly developed submarginal carina at shoulder · prosternum with intercoxal part bisulcate, very wide anteriorly, non-sulcate on base; episterna smooth, not overhanging anteriorly (the inflexed margins of the pronotum projecting a little at the anterior angles): lateral cavities of peduncle smooth ; metasternum short : anterior tibiæ obtusely 3-dentate; external spur of intermediate tibiæ narrow, short, placed at apex. Black, shining; under surface and femora dark piceous , four posterior tibiæ and tarsi clear brown; antennæ ferruginous.

Head not large (2 × 1 9 mm.), smooth, convex, lateral margin sloping obliquely forward from a little before eyes : clypeus not divided from front, not bordered on anterior margin; median part wide, truncate ; wings not divided from the supra-antennal plates, concave, narrow, strongly advanced, rounded at apex; supra-antennal plates narrow, convex; facial sulci strongly impressed, facial carinæ raised; eyes globose, prominent; orbits feebly developed behind eyes. Mandibles short. Antennæ stout, long, subfiliform; third joint not shorter than second; joints 5-10 oblong, hardly compressed. Prothorax smooth, of equal length and breadth (3·5 mm. × 3·5 mm.), widest a little before posterior angles, greatly narrowed anteriorly (ant width 2·6 mm.), convex, roundly and deeply declivous to base; sides rounded; posterior angles rounded; anterior margin lightly emarginate, angles rounded; basal curve short; border narrow, reflexed on sides, extending round anterior angles to neck; median line lightly impressed; anterior line strongly impressed; lateral basal impressions lightly marked, elongate Elytra oval (8 × 4·1 mm.), strongly and evenly convex; a wide smooth space on sides and apex, base truncate between shoulders; humeral angles rounded off, not the least marked; striæ deeply impressed and strongly punctate on basal half of disc, first entire, joining second at base, none of the others attaining apex, successively shorter towards sides, fourth joining fifth but not outturned at base, first interstice of each elytron together forming a convex ridge for whole length of suture, interstices 2-4 convex towards base, flat on apical half, 6-8 not divided from one another, sixth finely carinate at base, border reflexed, reaching very nearly to peduncle. Metasternum and its episterna short (distance between intermediate and posterior coxæ a little shorter than length of posterior coxæ). Ventral segments smooth. Anterior femora stout, not channelled below; tibiæ narrow, first external tooth short, wide, projecting, second a mere obtuse prominence, inner apical spine very long, narrow, truncate.

Length 14, breadth 4·1 mm.

Hab.: North Queensland. (A single specimen given to me by Mr C. French).

The type specimen is evidently the ♂. *C. ovipennis* is allied to *C. mastersi*, Sl., which it resembles in general appearance; the chief differences being its smaller size; prothorax slightly shorter and more narrowed in front; elytra with deeper and strongly

punctate striæ on the basal part of disc, the interstices much
more convex, the suture not impressed near the base, &c.

Clivina marginata, Putzeys.

Scolyptus marginatus, Putz., Ann. Soc. Ent. Belg. 1868, xi. p. 8.

♂. Black; sides of elytra for posterior two-thirds, (excepting
border) apex and legs testaceous red; antennæ and palpi testaceous.
Robust, convex. Head smooth, convex, not transversely impressed
behind vertex; front depressed : clypeus not divided from front,
median part wide, truncate; wings shortly but decidedly advanced,
widely rounded at apex; frontal impressions lightly impressed.
facial carinæ feebly developed.   Mandibles long, decussating.
Palpi long, filiform; penultimate joint of labial rather longer than
terminal, of maxillary as long as terminal.   Antennæ filiform,
third joint not shorter than second.   Prosternum a little broader
than long (3·8 × 4 mm), greatly narrowed anteriorly (ant. width
3·1 mm), smooth, convex, roundly and deeply declivous to base.
basal curve short; sides hardly rounded; anterior margin lightly
emarginate, anterior angles obtuse; posterior angles rounded, but
marked; border extending round anterior angles; median line
lightly impressed; anterior line strongly impressed; lateral basal
impressions distinct, wide, shallow. Elytra wide, oval (8·8 × 5 mm),
five inner striæ strongly impressed, lightly crenulate, first entire,
others obsolete near apex, fourth a little outturned and joining
fifth at base, sixth lightly impressed except near base, seventh
only indicated by a row of fine punctures, five inner interstices
very convex at base, becoming more and more depressed towards
apex, two inner ones together forming a sutural ridge, three
lateral ones confluent except at base, seventh narrow, subcarinate
at shoulders, eighth feebly indicated near apex by a very narrow
carina.   Prosternum with pectoral ridges well developed, intercoxal part very wide, not narrowed anteriorly, non-sulcate on base,
episterna not overhanging anteriorly, covered with wavy tran-
verse striolæ.   Lateral cavities of peduncle well developed, smooth
Metasternal episterna not long, wide posteriorly.   Legs light
anterior femora long, thick, not compressed, rounded on lower

side; tibiæ 3-dentate, narrow, apex short, lightly curved, first external tooth short, triangular, prominent, upper feebly developed, middle of lower side of tibia forming a ridge and ending in a strong triangular tooth near upper internal spine; inner apical spine about twice as long as apical digitation, thick and very obtuse at apex, upper spine slender, finely acuminate; four posterior legs long, light; intermediate tibiæ narrow, external spur very near apex, short, oblique.

Length 15·5, breadth 5 mm.

Hab. : Queensland—Port Denison (Masters).

The description given above is founded on a specimen kindly lent to me by Mr. Masters. This species may be considered the type of a separate section consisting of *C. marginata* and *C. gracilipes*, Sl. The following will be the characteristic features of this section ·—Mandibles decussating; clypeus with median part truncate, the wings shortly but decidedly advanced; antennæ filiform, third joint as long as second : palpi long, filiform, the labial with the penultimate joint longer than the terminal; maxillary with penultimate joint about as long as terminal, prothorax widest near posterior angles and greatly narrowed anteriorly, posterior angles marked; prosternum wide between the coxæ, the sides not overhanging in front ; metasternal episterna shorter and much wider than in *C. australasiæ*, Bohem., but longer than in *C. oblonga*, Putz.; legs light, external spur of intermediate tibiæ small and placed almost at apex, the tarsi long, slender.

CLIVINA GRACILIPES, n.sp.

Elliptic-oval. Head small; mandibles decussating, labial palpi with penultimate joint long, slender : prothorax subtrapezoid; elytra widely ovate, crenulate-striate; fourth stria joining fifth at base, seventh obsolete; eighth interstice shortly carinate at base, not indicated on apical curve. prosternum with intercoxal part bisulcate, very wide anteriorly; lateral cavities of peduncle smooth, shallow: legs light; anterior tibiæ narrow, 3-dentate, intermediate tibiæ narrow, external spur short, oblique, very near apex.

Black, under surface piceous black; legs, antennæ and palpi testaceous.

Head small (1·5 × 1·5 mm.), convex, smooth; a shallow almost obsolete fovea in middle of vertex; lateral margins sloping obliquely and roundly forward from a little before eyes: clypeus not divided from front, lightly emarginate-truncate; median part wide; wings small, not divided from supra-antennal plates, lightly advanced, rounded at apex, sloping very gently on inner side to median part; supra-antennal plates small, rather depressed; facial sulci lightly impressed, parallel; facial carinæ wide, not greatly raised; eyes large, convex, prominent, lightly enclosed behind. Mandibles rather long, decussating, wide at base, narrow and acute at apex. Mentum deeply emarginate; median tooth very wide, short, obtuse. Palpi slender; penultimate joint of maxillary nearly as long as terminal, of labial longer, terminal joint fusiform. Antennæ filiform, very lightly incrassate; second and third joints of about equal length. Prothorax smooth, broader than long (2·8 × 2·9 mm.), widest considerably before posterior angles, greatly narrowed anteriorly (ant. width 2·2 mm.), convex, strongly declivous to base; sides rounded; posterior angles lightly marked; base of disc curving gently between posterior angles; anterior margin truncate; anterior angles widely obtuse, finely bordered; border narrow, fine on basal curve; median and anterior lines well marked; lateral basal impressions lightly marked, rather long. Elytra ovate, much wider than prothorax (6·5 × 4 mm.), lightly and evenly convex, rounded on sides, narrowed to apex; humeral angles not marked; base very lightly emarginate in middle; striæ crenulate, 1-5 deeply impressed on basal half, becoming faint towards apex: interstices convex on disc, minutely shagreened under a strong lens; border reflexed, reaching base of fourth interstice; marginal channel wide. Prosternum not protuberant, abrupt and non-sulcate on base; pectoral ridges short, hardly carinate; episterna minutely rugulose. Metasternum shorter than usual, distance between intermediate and posterior coxæ equal to length of posterior coxæ; episterna considerably longer than broad. Legs light: anterior femora compressed, not stout, not channelled below; anterior tibiæ narrow, apex long, outturned, external teeth small, prominent; posterior tibiæ light, a little incrassate, not arcuate.

Length 11, breadth 4 mm.

Hab.: Queensland—Gulf of Carpentaria (a single specimen given to me by Mr. C. French, as from the Burketown District).

CLIVINARCHUS, n.gen.

Head with frontal region a little raised above occipital region, clypeus with median part angulate.

Mandibles short; upper surface depressed; outer margin obtusely angled near basal third.

Mentum deeply emarginate; lobes widely rounded at apex; median tooth long, obtusely pointed, keeled, projecting forward as far as lobes. Submentum large, projecting strongly and vertically from throat; a ridge vertically raised from throat, extending between submentum and base of orbits and defining suborbital channel behind.

Palpi: Labial with penultimate joint short, stout (about as long as terminal), bisetose, terminal joint stout (stouter than penultimate), truncate (hardly narrowed) at apex; maxillary stout, penultimate joint short, conical, terminal joint compressed, oval, obtuse at apex.

Antennæ short, stout; four basal joints cylindrical, first stout not elongate, second not long (but longer than third) joints 5-11 short, compressed, decidedly separated from one another, apical joint obtuse.

Prothorax longer than wide, convex, not declivous to base; a raised declivous "collar" (or wide border) along anterior margin.

Elytra very long, cylindrical, punctate-striate; fourth stria sharply outturned and joining fifth at base; no submarginal carina at shoulder; third interstice 4-punctate.

Prosternum with pectoral part not protuberant, intercoxal part wide anteriorly, non-sulcate on base; episterna over-hanging along anterior half, smooth—a few faint transverse striolæ perceptible with a lens.

Mesosternum smooth, without a lateral impression on each side of peduncle to receive intermediate tibiæ.

Metasternum large, long, transversely striolate on each side; episterna very long and narrow.

Legs : Anterior tibiæ wide, 3-dentate, apical projection short, strong, external teeth short, wide at base, the edge of the tibia triangularly excised above upper tooth so as to form a fourth small non-projecting tooth, inner spines long; intermediate tibiæ with two short prominent triangular external teeth, the anterior at the apex, the upper a little distance above the apex.

Peduncle wide.

Body winged.

This genus is thoroughly distinct from *Clivina*. Evident differences that may be noted are . its very elongate form, wide peduncle without lateral cavities, the raised and declivous collar along anterior margin of prothorax and the bidentate intermediate tibiæ. The formation of both the upper and lower surfaces of the head is also very different. There are two supra-orbital punctures, and two prothoracic marginal punctures as in *Clivina*.

CLIVINARCHUS PERLONGUS, n.sp.

Very elongate, narrow, cylindrical. Head, prothorax and under surface piceous black; elytra reddish brown; anterior legs and antennæ reddish piceous; palpi and four posterior legs piceous red.

Head (with eyes) broader than long (2·3 × 2·6 mm.); clypeal suture, facial sulci and facial carinæ lost in rugulosity of anterior part of head; this rugose part raised and sharply defined posteriorly between base of eyes; frontal impressions wide, shallow, irregular, rugose: clypeus with median part divided from wings by a carinate ridge, widely and squarely emarginate, its angles porrect, projecting strongly forward in a triangular prominence; wings small, angular, anterior margin truncate and about on a level with margin of median part; supra-antennal plates short, wide, projecting sharply and widely beyond wings of clypeus, external angles widely rounded; eyes large, globose, prominent, lightly enclosed.   Prothorax cylindrical, parallel, very widely and lightly sinuate on each side, longer than broad (4 × 3 mm.), lightly convex longitudinally, lightly transversely striolate (the striolæ wavy and more strongly impressed near sides); anterior angles very obtuse, rounded from anterior marginal puncture to neck; posterior angles rounded, not marked; basal curve short; base wide; border narrow and reflexed on sides, a little upturned at posterior angles, wide on base, very wide and declivous along anterior margin; marginal channel obsolete on sides.   Elytra narrow, cylindrical (10·5 × 3·5 mm.), shortly, not vertically, declivous to base; shoulders rounded, not marked; striæ entire, closely and strongly punctate, the punctures becoming finer from base to apex; interstices hardly convex; three posterior punctures of third interstice on apical half; marginal channel narrow, not deep, lightly punctate.   Anterior legs stout; femora thick, compressed, posterior edge of lower side roundly and widely dilatate; intermediate tibiæ incrassate, external edge arcuate, spinose, bidentate.

Length 18, breadth 3·5 mm.

Hab. : Queensland (sent to me by Mr. C. French as coming from the Gulf of Carpentaria, opposite Wellesley Islands).

Distribution of the Australian Clivinides.

I have thought that a few notes on the geographical distribution of the Clivinides in Australia may be not without interest, though the observations I can offer on the subject must be very defective owing to the scantiness of my knowledge of the range of the various species.   The only parts of the continent that have been tolerably well searched for these insects seem to be the

Sydney coastal district; the Melbourne district; the southern parts
of South Australia, where the Rev. T. Blackburn has collected;
and a part of inland New South Wales lying between Narrandera,
on the Murrumbidgee River, and Mulwala on the Murray, over
which I have collected, though not with sufficient care.   Good
collections have also been made by Mr. Masters at Port Denison
and Gayndah in Queensland, and at King George's Sound; by
Mr. Froggatt at King's Sound; and by Mr. Lea at Tamworth in
New South Wales   No use can be made by me, from want of
accurate knowledge, of the collections from Melbourne, South
Australia, Gayndah and King's Sound.

The Clivinides are a well defined division of the subfamily
Scaritini.   They reach their greatest development in the warm
parts of the earth, and it is, as might have been expected, in
tropical Australia that they are most numerous and show the
greatest diversity of form.   All the Australian genera, viz,
Dyschirius, Clivina, Steganomma, and *Clivinarchus* have representatives in tropical Queensland, the two last being peculiar to that
region.

Dyschirius (5 species) seems spread over the continent.

Clivina (83 species) has representatives wherever there is
water of any permanence all over Australia.   The following are
a few remarks on the dispersion of the thirteen groups into which
I have divided the Australian species :—

(1) *C. biplagiata* extends over eastern Australia from the Gulf
of Carpentaria to Melbourne.

(2) The "*cribrosa group*" (4 species) is typically a western
and southern one.   *C. frenchi* from Central Australia and
Queensland is not closely allied to the other three species.

(3) The "*obliquata group*" (11 species) has its headquarters
in the southern and western parts of the continent.   The two
species, *C. cylindriformis* and *C. obsoleta*, from tropical Queensland, are both isolated species, not closely allied to one another or
to any of the other members of the group.

(4) *C. coronata* is from south-western Australia.

(5) The "*planiceps group*," though spread from the Gulf of Carpentaria to Bass Strait, is probably of tropical origin; it has not yet been reported from the western half of the continent.

(6) *C. grandiceps* is from the neighbourhood of Burketown on the Gulf of Carpentaria.

(7) The "*punctaticeps group*" (4 species) is evidently a tropical group with one species, *C. adelaidæ*, in the Murray River watershed.

(8) *C. blackburni* is from Lake Callabonna in Central Australia.

(9) *C. olliffi* is from West Australia.

(10) The "*heterogena group*" (9 species) has representatives already reported from most parts of Australia, though none is yet known from West Australia, south of the tropics.

(11) *C. bovillæ* seems to have a wide distribution along the northern coastal region.

(12) The "*australasiæ group*" (27 species) is spread over the whole continent. I have further divided it into four *sections*, of these—*Section I.* (type, *C. sellata*, Putz.—8 species) apparently belongs to eastern Australia, and seems to be of tropical origin. *Section II.* (type *C. australasiæ*, Bohem.—6 species) is of eastern origin, though now found over the greater part of the continent; it also has a representative in New Zealand and Lord Howe Island. *Section III.* (type *C. basalis*, Ch.—12 species) is spread over all Australia and Tasmania, and has a species in New Zealand. *Section IV.*, founded for *C. pectoralis.* is undoubtedly a tropical type.

(13) The "*procera group*" (15 species) has its headquarters on Eastern Australia. It may be divided into seven sections, of which six have representatives in tropical Australia.

The members of the genus *Clivina* are strong fliers; often in summer evenings they may be noticed flying to the lamps in lighted rooms. All the species are found in damp ground near the margins of rivers, marshes, ponds, or, indeed, any tolerably permanent water; their habits are fossorial. Some species may be found all the year round, though more rarely in the winter

when they hibernate, hidden in the earth, often away from the
immediate proximity of water   During floods they may be taken
plentifully in the débris drifted along by the swollen streams
Owing to their habits it is evident that their dispersion may be
aided by streams, and there seem no reasons, except those of
climate and food-supply, why a species having once gained a
footing on any watershed should not spread along all the streams
of such watershed.

    With the insufficient data at my command no conclusions or
inferences of any practical worth in regard to the distribution of
the Australian species of *Clivina* can be attempted; but the
following suggestions may be offered —(1) The sameness in
climate will have permitted a wide range for species from east to
west   (2) The number of different species may be expected to be
greater on the coastal side of the mountain ranges owing to the
greater number of separate river systems   (3) The large area
included in the watershed of each of the two great river systems
which collect the waters flowing from the inland slopes of the
dividing ranges of Eastern Australia, from the boundary between
The Northern Territory of South Australia and Queensland to
Western Victoria, viz., the Barcoo watershed and the Murray
watershed, will have been conducive to a wide range for the
species found in the areas of these river systems.   There certainly
seems to have been a migration from tropical Queensland towards
South Australia by way of the Barcoo watershed, and thence
into Victoria and New South Wales by way of the Murray and
its tributaries, this is evidenced by the range of *C. procera, C
quadratifrons*, and *C. felix,* while *C. australasiæ, C. basalis, C.
sellata, C. angustula* and *C adelindæ* are species that evidently
have had their distribution helped by the Murray river-system.

    In conclusion, attention may be drawn to the great scarcity of
the *Clivinides* in New Zealand (only two species) in comparison
with their great development in tropical Queensland as offering
some evidence against an actual land connection in former
geological times between New Zealand and North Eastern
Australia.

The following lists of species give those known to me as coming from (1) Tropical Queensland, (2) the Sydney district, (3) the part of New South Wales between the Murray and Murrumbidgee Rivers along the 146th parallel of longitude (Riverina), (4) South West Australia.

Tropical Queensland.	Sydney.	Riverina.	South-west Australia.
C. biplagiata ...	C. biplagiata	C. obliterata . .	C. cribrosa... .. ..
C. frenchi	C. angustula	C. biplagiata... ...	C. coronata ....
C. cylindriformis	C. sellata	C. melanopyga .	C. dorsalis .... .
C. obsoleta	C. australasiæ ...	C. riverinæ.. .	C. bicolor.. . .
C. quadratifrons.	C. vittata	C. planiceps. .. .	C. olliffi ....
C. carpentaria ...	C. lepida	C. quadratifrons	C. angustipes .
C. grandiceps.. ..	C. dilutipes.	C. tumidipes .	....
C. puncticeps	C. sydneyensis ..	(C. angustula*).	....
C. lobipes	C. basalis.	C. sellata	....
C. fava	C. oblonga.........	C. australasiæ .	....
C. odontomera ...	....	C. vagans..	....
C. borillæ...	....	C. simulans..	....
C. cava.	....	C. basalis.........	.. ..
C. occulta...	....	C. felix	....
C. ferruginea.....	:.....................	C. procera.... .	....
C. felix	....	....	....
C. rubripes........	....	....	....
C. procera	....	....	....
C. monilicornis ..	....	....	....
C. nyctosyloides..	....	....	....
C. interstitialis ..	....	....	....
C. ovipennis.......	....	....	....
C. marginata..	....	....	....
C. tenuipes	....	....	....

The following is a list of the authors who have dealt with the nomenclature of the Australian Clivinides, with references to their papers :—

CHAUDOIR. Carabiques Nouveaux. Bull. Mosc. 1843, xvi. p. 733.

BOHEMANN. Eugenies Resa, Coleoptera, 1858.

---

* I have not found *C. angustula* further east than Carrathool, on the Marrumbidgee River, 32 miles east from Hay.

PUTZEYS, JULES.   Postscriptum ad Clivinidarum Monographium
atque de quibusdam aliis. (Mense Novembris 1861.)* Mém. Soc. Roy. Sc. Liège, 1863,
xviii. pp. 1-78.

——————— Révision des Clivinides de l'Australie. Stett.
Ent. Zeit. 1866, xxvii. pp. 33-43.

——————— · Révision Générale des Clivinides. Ann. Soc.
Ent. Belg. 1867, x. pp. 1-242.

——————— Supplément à la Révision Générale des Clivinides. l.c. 1868, xi. pp. 5-22.

——————— Deuxième Supplément à la Révision Générale
des Clivinides. l.c. 1873, xvi. pp. 1-9.

MACLEAY, WILLIAM.   On the *Scaritidæ* of New Holland. Trans.
Ent. Soc. N.S.W. 1863, i. Part 1, pp. 71-74.

BLACKBURN, THOS.   Notes on Australian Coleoptera, with Descriptions of New Species, Part iv. Proc.
Linn. Soc. N.S.W. (2). iv. 1889, pp.
717-722.

——————— Coleoptera (of Elder Exploring Expedition).
Trans. Roy. Soc. S.A. (1892), xvi. p. 22.

——————— Notes on Australian Coleoptera, with Descriptions of New Species, Part xv. Proc.
Linn. Soc. N.S.W. 1894 (2) ix. pp. 86-88.

My thanks are due to friends who have helped me by the
gift and loan of specimens, viz., to Mr. C. French, Government
Entomologist of Victoria, for his generosity in giving me specimens
of a great many new and rare species; to Mr. G. Masters, Curator
of the Macleay Museum, Sydney, for sending me for examination

---

* I believe this memoir appears in Mém. Liège, Vol. xviii., but my
separate copy bears the following date, " Leodii, 1862," so that it was
evidently published in 1862.

a splendid collection of 120 specimens, representing 40 different species, of which 7 were new, and for the gift of many rare specimens; to the Rev. T. Blackburn, of Adelaide, for loan of specimens of new and rare species, and for the gift of specimens of various species, to Mr. A. M. Lea, of the Bureau of Agriculture, West Australia, for generously placing his whole collection of species taken by him in New South Wales at my disposal, and for specimens from West Australia; and to Mr W. Kershaw, of Melbourne, for some Victorian specimens.

ON THE BAG-SHELTERS OF LEPIDOPTEROUS LARVÆ OF THE GENUS *TEARA*.

By Walter W. Froggatt.

(Plate xiv.)

In many parts of the Australian bush one frequently comes across brown liver-coloured silken bags of an irregular funnel-shape, spun round a stout twig enclosing several others, and frequently a few leaves, all matted together and rough on the inner surface, but smooth and regular on the outside. They vary in size from 3-8 inches in diameter at the broad end, which may be quite open or loosely covered with a few silken strands; upon examination, if freshly constructed, they will be found full of very hairy caterpillars mixed up with their castings and moulted skins.

When they have served their purpose, and are abandoned by the full grown caterpillars, they will remain for a considerable time, a solid mass of skins and castings, compact and firm, protected by the strong silken coverings. These curious structures are woven round the twigs by the gregarious larvæ of several different species of moths belonging to the genus *Teara* (Family *Liparidæ*). They are constructed for shelter during the day, and are not used for pupating purposes. Hiding therein during the day, the caterpillars issue forth at dusk, feeding all night over the tree and returning to cover at daybreak. When moving about they travel in procession. The first large nest I came across I carried home, and was very much surprised next morning to see a string of large hairy caterpillars stretching right across the roof of the tent; they had emerged from the nest in the night, but were unable to find their way back.

Some twenty species of the genus, which is peculiar to Australia, have been described; most of them are short thickset moths with feathery antennæ, and the tip of the abdomen bearing a tuft of fine hairs. Our commonest species, *Teara tristis*, is generally very slow and sluggish in its habits, and is usually found clinging to low bushes.

I have, during the last season, been fortunate in breeding out one of our largest species, which spins a somewhat different form of shelter, which is described below with the life-history of the species.

TEARA CONTRARIA, Walker.

The larva, when full grown, is two inches in length, of a uniform thickness, with the head ferruginous, rounded on summit and sides, a pale median suture running into the triangular clypeus; labium and jaws small; all the head thickly covered with long reddish-brown hairs standing out in front. Thoracic and abdominal segments black across the centre, which is raised into a row of large tubercles, out of which spring a number of long fine white and reddish-brown hairs; between the segments thickly covered with small white spots, from each of which springs a short black hair. Under side pale ochreous yellow, with a double row of dark ferruginous tubercles tufted with reddish-brown hairs; legs ferruginous, black at the tips, covered with short reddish hairs; tubercles on the 1st and 2nd abdominal segments, and claspers upon the following segments covered with stout reddish-brown hairs.

The larvæ live in communities of a hundred or more, forming a felted silken bag or net of a dark reddish-brown colour on the sheltered side of the tree trunk, close to the ground, under which they hide during the day, half buried in the cast skins and excreta which accumulate beneath. They crawl up the tree at dusk, feeding upon the foliage, and returning to their retreat at daylight. In April last a clump of very fine wattles (*Acacia pro-minens*) were completely defoliated by them near the Penshurst railway station. Every other tree had a large bag at the foot of its trunk, while branches and trunk were festooned with strands of dirty yellow silk down to the top of the bag.

About fifty specimens of nearly mature larvæ were collected and placed in a large glass jar in the Museum, where they remained huddled together in a hairy mass, unless disturbed, when they would all set off in a procession round the walls of their prison, one behind the other, often keeping it up for hours together.   In about a fortnight they began to burrow into the loose sand at the bottom of the jar, constructing soft felted cocoons out of the hairs upon their bodies   The pupæ were stout and short, smooth, shining, of a reddish-brown colour, with the anterior portion small and the tip of the abdomen curved upwards.   The first moths emerged about the end of September, and the last two months later; but from the fifty specimens not more than eight moths were obtained

The moths vary considerably in size, the male about 2 inches across the wings, and the female often over $2\frac{1}{2}$ inches; they are of a general dark brown colour, with a small oval white spot in the centre of the forewings; and a very small and indistinct one in the hind ones   The head and thorax are thickly clothed with long brown hairs, bright yellow and lance-shaped at the tips; the upper surface of the abdomen is covered with bright reddish-orange barred with black at the apex of each segment, and tipped with hairs of the same colour.   The moths are very difficult to breed, those mentioned being the first I have obtained in four seasons   Mr. E Anderson, of Melbourne, to whom I am indebted for the identification of the moth, tells me that he knows no other instance of success in breeding them, though the larvæ are common in Victoria and New South Wales.

EXPLANATION OF PLATE XIV
Teara contraria, Walk.

Fig 1 —Larva.
Fig 2 —Pupa in cocoon.
Fig. 3.—Moth.
Fig. 4. – Rough sketch showing bag shelter formed at the base of a tree stem.
Figs. 5-6 —Forms of bag shelters made by larvæ of *Teara* spp

NOTE ON THE OCCURRENCE OF DIATOMACEOUS EARTH AT THE WARRUMBUNGLE MOUNTAINS, NEW SOUTH WALES.

By T. W. Edgeworth David.

(Plates xv.-xvii)

I.—*Introduction*

Deposits of diatomaceous earth have been recorded as occurring in New South Wales at the following localities:—Barraba (between Tamworth and Bingara), the Lismore District; the Richmond River, the Tweed River, Cooma; Newbridge, and the Warrumbungle Mountains. The deposit near Barraba has been described by Mr E F. Pittman, the Government Geologist, in general terms.*

Mr Pittman states that the diatomaceous earth is capped by basalt, and attains a thickness of about 8 feet, having a layer of coarse sand (2 inches thick) about 3 feet from the top. The infusorial earth rests on a bed of sandy mudstone, about 1 foot in thickness, under which is an impure infusorial deposit containing rolled pebbles and fragments of imbedded lava, pointing to the fact that volcanic eruptions were common at the time of its deposition. Finally, an overwhelming flow of lava filled up what was, doubtless, during the Miocene epoch, a lake, and it now forms an elevated tableland. As far as I am aware, this is the only reference to the mode of occurrence of diatomaceous earth in New South Wales. Descriptions have been given by other observers of hand specimens of the diatomaceous earth.

Ann. Rept. Dep. Mines, 1881, pp. 142-143. By Authority Sydney, 1882.

In 1888 Professor Liversidge published an account of *Tripoli or Infusorial Earth*,* from Barraba.

He states that the "tripoli" at Barraba is made up almost entirely of the remains of Diatoms resembling *Melosira*. The same author refers to a deposit (*op. cit.* p. 194) of "cimolite" from the Richmond River. There can now be little doubt that this material, described as "a very white and porous hydrous silicate of alumina,† often sent down to Sydney as meerschaum," must graduate into a clayey diatomaceous earth, as Diatoms in some numbers have been observed by me in a similar rock from the same locality. Professor Liversidge gives analyses of the rocks from both the above localities.

Mr. R. Etheridge, Junr., has published a short description of some hand specimens of the diatomaceous earth from the Warrumbungle Mountains, and also of similar specimens respectively from the Lismore District, Tweed River, and Richmond River Districts.‡

He refers the barrel-shaped Diatoms, so conspicuous in these deposits, to *Melosira*, and notes the association with them of spicules of freshwater sponges.

Last September Judge Docker and the author were afforded an opportunity, through the kindness of Mr. W. L. R. Gipps, of Bearbong Station, of examining the deposit of diatomaceous earth in the Warrumbungle Mountains.

II.—*General Geological Features of the District.*

In the neighbourhood of the diatomaceous earth deposit there are two formations represented:—(1) The Permo-Carboniferous Coal-measures, and (2) Trachyte lavas, dykes, and tuffs,

---

* The Minerals of New South Wales, &c.   By A. Liversidge, M.A., F.R.S. p. 177.   Trubner & Co. London, 1888.

† Ann. Rept. Dep. Mines, for the year 1887, pp. 165-166.   By Authority Sydney, 1888.

‡ Ann. Rept Dep. Mines, for the year 1888, p. 190.   By Authority. Sydney, 1889.

with which last are associated the deposits of diatomaceous earth, and a seam of lignite. It is not my intention here to attempt to give a detailed description of that grand chain of trachytic volcanoes, of which the Warrumbungle Mountains form a not insignificant portion. Suffice it to say that they are the wrecks of large volcanoes; and their cores of coarsely crystalline trachyte, which have cooled deep down in the volcanic chimneys, now rear themselves skywards as gigantic monoliths, between 3,000 and 4,000 feet above the sea, and over 2,000 feet above the surrounding plain, ringed round with alternating beds of coarse trachyte tuff and lava.

The chain extended probably from at least as far south as the Canobolas, near Orange, northwards, perhaps, with intervals, to the Glass-House Mountains on the coast north of Brisbane, a distance of nearly 400 miles. As the diatomaceous earth deposits are interstratified with the trachytes it is obvious that any evidence which throws light upon the age of the trachytes has an equally important bearing upon the question as to the age of the diatomaceous earths.

As shown on Plate xv., accompanying this paper, there is clear evidence to show that the trachytes have intruded the Permo-Carboniferous Coal-measures in this neighbourhood The latter consist of sandstones, quartzites, cherts containing well preserved specimens of *Glossopteris*, finely laminated black shales, and at least one seam of coal, over 6 feet in thickness. The coal has been calcined by the trachyte dykes, and at the extreme right of the section, beds of trachyte tuff are seen resting, with strong uncomformity, on the Permo-Carboniferous strata. Obviously then the eruption of the trachytes was later than Permo-Carboniferous time.

At several localities in the Warrumbungle Mountains the trachyte series is seen to overlie sandstones, which are almost certainly of Triassic age, and in this case the trachytes would be proved to be Triassic or Post-Triassic.

If now the chain of trachytic volcanoes be followed up into Queensland, and traced north of the Glass-House Mountains, it may be noted that near Port Mackay trachyte lavas and tuffs are abundantly interstratified with rocks of the Desert Sandstone Series, the age of which is Upper Cretaceous *

It is unlikely that these extensive eruptions took place in Lower Cretaceous time, as that was a period of prolonged subsidence, and Mr. R. L. Jack has commented on the fact that in Queensland, at any rate, no lavas nor tuffs have as yet been noted in the Rolling Downs Series (Lower Cretaceous). As regards the downward limit in time of these eruptions, it is improbable, therefore, that it was earlier than Upper Cretaceous

As regards the upward limit, the following considerations suggest themselves :—It is improbable that the Warrumbungle trachyte volcanoes, at the time they were active, were far distant from the sea. They are now over 300 miles inland from the Pacific, but during the Lower Cretaceous epoch the waters of the inland sea, which, at that time, must have extended from the Gulf of Carpentaria to the Australian Bight, must very nearly have washed the bases of the Warrumbungles. In Upper Cretaceous time elevation took place, and marine conditions were largely replaced in Central Australia by shallow lacustrine conditions. There is no evidence to show that marine conditions obtained within a hundred miles of the Warrumbungles in Tertiary time   On physical evidence therefore it might be inferred that the age of the trachyte series might be placed at the close of the Cretaceous, or at the commencement of the Eocene periods. There is also some palæontological evidence in support of this supposition, as will be stated in the next division of this paper.

III.—Details of the Diatomaceous Earth Deposit.

The deposit makes two distinct outcrops at the bottom of the shallow valley or gully through which flows Wantialable Creek.

---

* "Geological Features and Mineral Resources of the Mackay District." By A. G. Maitland. By Authority. Brisbane, 1889. Also see Geology and Palæontology of Queensland and New Guinea. Jack & Etheridge, Junr. Text. pp. 546-547. 1892.

As shown in the upper section on Plate xvi. a sheet of trachyte at least 20 feet thick caps the ridge overlooking Wantialable Creek. Below this is a thickness of about 30 feet of trachyte tuff varying in texture from fine to coarse. A remarkable rock succeeds which I have termed a silicified trachyte tuff, 1½ ft. to 2 ft. thick. This rock has already been ably described by Mr. G W. Card,* the Mineralogist to the Geological Survey of the Department of Mines.

Underlying this is another also very remarkable bed of trachyte tuff, almost exclusively composed of translucent crystals of sanidine, from a fraction of an inch up to ½ an inch in diameter. The crystals exhibit their usual tabular habit, the clinopinacoid faces being extensively developed. The bed being only loosely coherent, the rain washes quantities of the larger sanidines out of it, and forms with them miniature snow-white talus slopes.

Next follows the bed of diatomaceous earth, 3 feet 9 inches thick, then come 19 feet 3 inches of strata, chiefly trachyte tuffs, resting on the surface of a sheet of vesicular trachyte. Half-a-mile higher up the creek, the lower section shown on Plate xvi. may be studied. It resembles the section above quoted, but in addition fossil leaves occur on a horizon immediately above and intimately associated with the diatomaceous earth, as was shown me by Mr. W. L. R. Gipps. We had here the good fortune to discover a fossil leaf fairly well preserved in the fine tuff, which Mr. R. Etheridge, jun, and Mr. W. S. Dun, Assistant Palæontologist to the Geological Survey, identify as *Cinnamomum Leichhardtii*, Ettingshausen. (See Plate accompanying this paper). This leaf is elsewhere in Australia associated with Eocene deposits.

The age therefore of the Diatoms and of the freshwater sponge spicules associated with them at this spot may, I think, be provisionally set down as early Eocene or late Cretaceous.

I have purposely abstained from attempting a detailed description of the different species of Diatoms and sponges represented

---

* Records Geol. Surv. N.S. Wales. Vol. iv. Pt. iii. pp. 115-117. Plate x. By authority Sydney. 1895.

in this deposit, as I understand that this is a work which has already been commenced by Mr. W. S. Dun and Mr. G. W. Card, and an interesting paper from them on this subject may shortly be expected. I would merely add that *Melosira* appears to greatly predominate among the Diatoms, but not to the entire exclusion of other forms. The sponge spicules are acerate or fusiform, slightly arcuate, and some are thorny, but the majority smooth

I should like to emphasise the fact that hitherto all our diatomaceous earths in New South Wales have been found in association with volcanic rocks, and I would venture to suggest that this association is probably far from accidental. The superheated water flowing from hot springs and from the lavas themselves during the trachytic eruptions would be certain to carry more or less silica in solution, and its high temperature, combined with its dissolved silica, would probably render it a very favourable medium for the development of Diatoms to the exclusion of most other kinds of plant. While some species of Diatoms flourish luxuriantly in the cold waters of the Antarctic Ocean, others may be found equally flourishing in the hot and highly mineralised waters of geysers. For example, Mr. H. N. Moseley* has described the occurrence of Diatoms near the Boiling Springs at Furnas, St. Michael's, Azores, and their neighbourhood.

Mr. Moseley states (*op. cit.* p. 322) "The *Chroococcus* [*Botryococcus Braunii*, Ktz , as would appear from the footnote. T. W. E. D.] was not so abundant in the samples of incrusting matter in this hot spring as in those from the spring at Furnas. Amongst the green matter are a few skeletons of *Diatomaceæ* (a *Navicula*), but these are very probably derived from a cool spring, situate just above the sulphur spring, the water of which mingles with that of the sulphur spring, and indeed appears to supply a large share of the water of most of the hot springs, the water being merely heated and impregnated with various

---

* Journ. Linn. Soc. Bot. Vol. xiv. p. 322.

minerals by the discharge of steam and various gases from apertures in the several basins into which it finds its way. . . . The small cool spring above referred to contains abundance of *Naviculæ* and other Diatoms, such as those met with amongst the green matter growing in very hot water." He also observes (*op. cit.* p. 323), "In this water, which was too hot to bear the finger, the same *Chroococcus* as observed at the springs near the lake was abundant," etc. . . . "A little lower down in a small pool of hot mud and water, so hot that the finger could only be borne in it for a short time, grows a sedge . . . and an abundant growth of algæ, *Chroococcus, Oscillatoriæ* [*Tolyphothrix* f. Archer. T.W.E.D] and some Diatoms with endochrome complete."

The temperature of the springs in the lake of Furnas is quoted (*op. cit.* p. 324), f. Hartung* as from 78° to 190° Fahr. The water in which the *Chroococcus* grew is estimated to have had a temperature of 149° to 158° Fahr., and that in which the sedges grew of 113° to 122° Fahr. Mr. W. T. Thiselton Dyer, in notes on Mr. Moseley's collections (*op. cit.* p. 326), states that in the collection submitted to him "from among the sedges at Furnas in very hot water" he identified a number of Diatoms, which he specifically names. He adds that they were not numerously represented, however, and says (p. 327), "These are all forms of common occurrence, and seemed in no way affected by the high temperature of the water." A useful bibliography of references to the vegetation of hot waters is contained in Ninth Report, Geol. Sur. U.S.A. 1887-88, pp. 620-628. It is noted (*op. cit.* p. 625, quoted from Manual of Geology, by James D. Dana, 3rd ed., 1880, p. 611) that "Mr. James Blake found diatoms in water having a temperature of 163° F. at Pueblo Hot Springs, Nevada." It is also stated (*ibidem*), "At the Mammoth Hot Springs, Dr. F. V. Hayden observed the occurrence of pale yellow filaments about the springs and the green confervoid vegetation of the waters, as well as the presence of diatoms in the basins of the main springs, two species of the latter, *Palmella* and *Oscillaria*,

---

* "Die Azoren," Leipzig, W. Englemann, 1860, p. 173.

being recognized by D Billings." . . . (*Op. cit.* p. 627) "The extreme temperature at which vegetation has been observed is 200° F, recorded by Prof. W. H. Brewer at the California Geysers "

It is clear therefore that Diatoms are capable of flourishing in the waters of hot springs, the water of which must necessarily be more or less highly mineralised, though apparently they do not flourish in water at so high a temperature as that in which some algæ, such as the *Oscillatoriæ*, can flourish. The fact must not be forgotten that spicules of *Spongilla* are at the Warrumbungle Mountains associated with the Diatoms, and obviously if the Diatoms flourished in hot water the Sponges must have existed under similar conditions.

Animal life was well represented in the neighbourhood of Furnas by Rhizopods, but no mention is made of freshwater sponges

It is at all events certain that at the Warrumbungle Mountains the Diatom *Melosira* and a variety of *Spongilla* occur in association with trachytic lavas and tuffs of early Tertiary, possibly of late Cretaceous Age.

EXPLANATION OF PLATES.

Plate xv.

Section showing junction between the Trachyte Volcanic Group of the Warrumbungle Mountains, and the Permo-Carboniferous Coal Measures in a tributary of Uargon Creek, Wollongulgong, near Tooraweena, N S.W.

Plate xvi.

Upper Figure.

Section in Wantialable Creek, near Tooraweena, Warrumbungle Mountains, showing intercalation of Diatomaceous Earth in the Trachyte Series

Lower Figure.

Section in Wantialable Creek, near Tooraweena, Warrumbungle Mountains, showing Diatomaceous Earth in association with *Cinnamomum Leichhardtii*

Plate xvii.

Cinnamomum Leichhardtii, Ettings.

On behalf of Mr. F. M. Bailey, Government Botanist of Queensland, the Secretary exhibited an interesting collection of botanical specimens specially brought together to illustrate the plants of Queensland which are known to possess active or medicinal properties. As such it might be considered to illustrate a later edition of the knowledge summarised in a paper by the exhibitor "On the Medicinal Plants of Queensland" in the Society's Proceedings for 1880 Vol. v. First Series, p. 4).

On behalf of Dr. Broom, the Secretary exhibited specimens illustrative of the fossil Marsupials from a bone-breccia deposit near the Wombeyan Caves, described at the Meeting of April 29th, 1896.

Mrs. Kenyon sent for exhibition, and contributed a note upon, specimens of varietal forms of *Cypræa*

Mr Darley exhibited a specimen of rock from Newcastle bored by specimens of *Pholas*, with examples of the molluscs *in situ*. Also from the roof of a building in Sydney a piece of sheet-lead which had been perforated by Termites

Mr. Steel showed an elegant fungus, probably *Polyporus portentosus*, Berk, from Bundanoon

Mr. Froggatt exhibited drawings and specimens of the larva, pupa, moth, and bag-shelters of *Teara contraria* from Penshurst, near Sydney; in this locality during April many trees of *Acacia prominens* were completely defoliated by the caterpillars, the shelters being placed at the foot of the trees. Also the more substantial silken shelter of a species from Kalgoorlie, W.A.; and a series of specimens of the commoner species of the genus occurring in New South Wales.

The President exhibited a rare and remarkable spider, *Actinopus* sp., forwarded by Mr. A. G. Little, Railway Surveyor, Menindie, N.S.W. This is apparently the first recorded occurrence of the genus in Australia. In respect of the length of the palpi and the shortness of the abdomen it appears to come nearest to *A. longipalpus* from Brazil.

WEDNESDAY, JULY 29th, 1896.

The Ordinary Monthly Meeting of the Society was held at the Linnean Hall, Ithaca Road, Elizabeth Bay, on Wednesday evening, July 29th, 1896.

The President, Mr. Henry Deane, M A., F.L.S., in the Chair.

Mr. J. Douglas Ogilby, Livingstone Road, Petersham, was elected a Member of the Society.

DONATIONS.

Natural History Society of Montreal—Canadian Record of Science. Vol. vi. No. 8 (1896). *From the Society.*

American Geographical Society—Bulletin. Vol. **xxviii.**, No 1 (1896). *From the Society.*

Zoological Society of Philadelphia—Twenty-fourth Annual Report of the Board of Directors (1895-96). *From the Society.*

Société d'Horticulture du Doubs, Besançon — Bulletin. Nouvelle Série. Nos. 43, 55, and 58 (1894-95): Série Illustrée. No 5 (May, 1896). *From the Society.*

Muséum d'Histoire Naturelle, Paris—Bulletin. Année 1895, Nos. 4-5, and 7. *From the Society.*

La Faculté des Sciences de Marseille—Annales. Tome v. Fasc. 4; Tome vi. Fasc. 1-3. *From the Faculty.*

Cambridge Philosophical Society—Proceedings. Vol. ix. Part ii. (1896). *From the Society.*

Royal Microscopical Society—Journal, 1896. Part 2 (April). *From the Society.*

Société Belge de Microscopie—Bulletin. Tome xxii. Nos. 5-7 (1895-96). *From the Society.*

Geological Society, London — Quarterly Journal Vol lii. Part 2 [No. 206] (May, 1896). *From the Society.*

Zoologischer Anzeiger. xix. Bd. Nos. 503-505 (May-June, 1896). *From the Editor.*

K. K. Zoologisch-botanische Gesellschaft in Wien—Verhandlungen. xlvi. Bd. (1896), 4 u. 5 Hefte. *From the Society.*

Verein für naturwissenschaftliche Unterhaltung zu Hamburg—Verhandlungen, 1894-95. ix. Band. *From the Society.*

Department of Mines and Agriculture, Sydney — Annual Report for the year 1895: Agricultural Gazette. Vol. vii. Part 6 (June, 1896). *From the Hon. the Minister for Mines and Agriculture.*

Australasian Association for the Advancement of Science — Report of the Sixth Meeting held at Brisbane, January, 1895. *From the Association.*

Pharmaceutical Journal of Australasia Vol. ix. Nos. 6-7 (June-July, 1896). *From the Editor.*

Department of Agriculture, Brisbane—Bulletin. No. 8, Second Series (1896). *From the Secretary for Agriculture.*

Bureau of Agriculture, Perth, W.A.—Journal. Vol. iii. Nos. 15-17 (June, 1896). *From the Secretary.*

University of Melbourne—Examination Papers—Matriculation. May, 1896. *From the University.*

Asiatic Society of Bengal—Journal. Vol. lxiv. (1895), Part i. No. 4; Part ii, Title page and Index: Vol lxv (1896), Part ii. No. 1: Proceedings 1895, Nos. ix.-x. (Nov.-Dec), 1896, No i (Jan.)· Annual Address. By A Pedler, F R.S etc (Feb., 1896). *From the Society*

Zoological Society of London—Abstract, May 19th, June 2nd and 16th, 1896. *From the Society.*

Madras Government Museum—Bulletin. No 4 (1896). *From the Superintendent.*

Museo Nacional de Montevideo—Anales iv (1896). *From the Museum*

Perak Government Gazette. Vol ix. Nos. 13-14 (June, 1896). *From the Government Secretary.*

Société Royale Linnéenne de Bruxelles—Bulletin. 21me Année, No 7 (May, 1896) *From the Society.*

Sociéte Hollandaise des Sciences à Harlem—Archives Néerlandaises. Tome xxx. 1re Livraison (1896). *From the Society.*

Victorian Naturalist. Vol. xiii. N [illegible] June, 1896). [illegible] *the Field Naturalists' Club of Victo*[illegible]

Entomolo[illegible] [illegible]iety of Lon[illegible] [illegible]nsactions, 1896. Part ii (June). *F*[illegible] *Society.*

Nederlandsche Entomoï ïsche Vereeniging—Tijdschrift voor Entomologie. xxxviii. Deel. Jahrgang 1894-95. Afl. 2, 3, and 4 *From the Society.*

Société des Sciences de Finlande—Oefversigt. xxxvii. (1894-95). *From the Society.*

Royal Society of South Australia – Transactions. Vol xvi. Part. iii. (June, 1896). *From the Society.*

Zoological and Acclimatisation Society of Victoria—Thirty-second Annual Report (Feb., 1896). *From the Society.*

Three Pamphlets entitled "Report of the Research Committee appointed to collect Evidence as to Glacial Action in Australasia" By Professor R. Tate, Mr. W. Howchin, and Professor T. W E. David (1895) ; "Address by the President : Section of Geology and Mineralogy"—Aust. Assoc. for Adv. Sci. By Professor David (Brisbane, 1896); "Antarctic Rocks," &c. (1895) *From Professor T. W. E. David B.A., F.G.S.*

Australasian Journal of Pharmacy. Vol. xi No 127 (July, 1896) *From the Editor.*

... aturalist. Vol. xxx. No. 354 (June, 1896) *From ...*

... of Compa... Zoology at Harvard College, Cambridge ...—Bulletin ... xxix. No. 3 (April, 1896). *From ... rator.*

...S. Dept. of ...—Division of Entomology—Bulletin ...anical Series ...). *From the ... ary of Agriculture.*

Comité ...St. Pétersbourg ... T. xiv. Nos. 6-9; T. xv ... Mémoires. ... 2 (1894). *From the Society.*

... Australia ... gy. By Rev. Thomas

Two Pamphlets entitled " Further Coccid Notes, etc.;" and " Contributions towards a Monograph of the *Aleurodidæ*, a Family of Hemiptera-Homoptera." By W. M Maskell. (From Trans. N.Z. Inst. Vol. xxviii. [1895]). *From the Author.*

Indian Museum, Calcutta—Natural History Notes. Series ii.· No. 18 [? 19] (1895). *From the Museum.*

Geelong Naturalist. Vol. v. No. 4 (July, 1896). *From the Geelong Field Naturalists' Club.*

L'Académie Royale des Sciences, etc. de Danemark, Copenhague. —Bulletin, 1896. No. 3. *From the Academy.*

APPENDIX TO THE AUSTRALIAN CLIVINIDES
(FAM. *CARABIDÆ*).

By Thomas G. Sloane.

The Clivinides of King's Sound and its Vicinity.

When the late Sir William Macleay described the *Carabidæ* collected by Mr. W. W. Froggatt in the vicinity of King's Sound in 1887,* he passed over the *Clivinides*, merely remarking that the collection contained seventeen species.† During a visit to Sydney, after completing the "Revision of the Australian *Clivinides*," I was able, through the courtesy of Mr. Masters, Curator of the Macleay Museum, to examine the *Clivinides* from King's Sound, and as the collection seems a representative one the following report on it will not be without interest.

The following is a list of the species :—

Clivina riverinæ, Sl. ? (var. ?)
C. denticollis, Sl.
C. quadratifrons, Sl.
C. punctaticeps, Putz.
    var sulcicollis, Sl.
C. australica, Sl.
C. bovillæ, Blkb.
C. cava, Putz.

Clivina sellata, Putz.
    var. inconspicua, Sl.
C. ferruginea, Putz
C. australasiæ, Bohem. ? (var. ?)
C. eximia, Sl.
C. leai, Sl.
    var. apicalis, Sl.
C. procera, Putz. (var.)‡

C. froggatti, n.sp.
Dyschirius macleayi, n.sp.

---

* P.L.S.N.S.W. 1888, iii. (2) pp. 446-458.

† l.c. p. 462.

‡ It is the large species mentioned under *C. procera* (vide supra, p. 229) as being from Port Darwin; and though probably distinct from *C. procera*, Putz., seems to offer no characters to distinguish it from that species except its large size.

My examination of this collection leaves the impression on my mind that all the specimens are not actually from King's Sound, but that some, as *C. procera* and *C. quadratifrons*, may be from Port Darwin or some other more easterly port of call, at which Mr. Froggatt may have touched.

CLIVINA RIVERINÆ, Sloane.*

The single representative of this species seems to agree with typical specimens in everything excepting colour.   It is brown with the elytra ferruginous.

CLIVINA PUNCTATICEPS, Putzeys (var. SULCICOLLIS).

A species which is plentifully represented in the collection agrees with *C. punctaticeps*, Putz., in respect of the head, elytra. prosternum, and eyes, but differs by having the prothorax shorter and rather more convex, the median line more deeply impressed, the basal curve shorter, the base more deeply and abruptly declivous, the marginal channel across the base much wider and deeper.   It may be a distinct species, though it seems probable that *C. punctaticeps* will be found to be a widely spread species varying sufficiently to take in this form as a variety.   The following is a brief description :—

Narrow, parallel, convex.   Piceous red, elytra with first stria of each elytron usually dark piceous, this sutural infuscation often spreading over the first three interstices above the apical declivity. a very distinct crenulate striole at base of first interstice; anterior femora lobate on lower side, anterior tibiæ 4-dentate.

Length 5·5-7, breadth 1·45-1·75 mm.

The characteristic feature of this variety is the wide deep channel of the base of the prothorax which interrupts the marginal border at each side, and prevents it from actually joining the basal border, as is usual in *Clivina*.

---

* *Vide supra* p. 164.

CLIVINA SELLATA, Putzeys (var. ?·INCONSPICUA).

A small *Clivina* represented by seven specimens (two immature) is among those from King's Sound. It agrees so closely with *C. sellata*, Putz, that I have placed it under that species as a variety; the only differences I can find are that it seems a smaller insect, and apparently the black dorsal spot on the elytra is quite wanting; however, I cannot separate immature specimens from immature specimens of *C. sellata*. It is quite likely that when this form is better known it will come to be regarded as a species distinct from *C. sellata*, and it is with this impression in my mind that I give it a varietal name, for I feel that it would be misleading to extend the range of *C. sellata* to King's Sound on the specimens before me.

The following description will suffice for its recognition :—

Ferruginous. Parallel, convex. Head short, vertex with a rounded punctate impression: clypeus emarginate, median part not divided from wings, these small, rounded, a strong sinuosity dividing them from supra-antennal plates. Prothorax about as long as broad (1·1 × 1 1 mm.), decidedly narrowed anteriorly. Elytra punctate striate, fourth stria joining fifth at base, seventh entire. Prosternum with intercoxal part attenuate anteriorly. Anterior tibiæ 4-dentate.

Length 3·7-4·2, breadth 1-1·15 mm.

CLIVINA AUSTRALASIÆ, Bohemann ? (var. ?).

A large black species is plentifully represented in the King's Sound collection. In general appearance it exactly resembles *C. australasiæ*, Bohem., the only noticeable differences that I can see being, the head less punctate and more roundly angustate before the eyes, the legs lighter coloured, the inner apical spine of the anterior tibiæ longer and more obtuse at the apex in the ♂. Some specimens have the clypeus more deeply emarginate than others.

Length 8-9·5, breadth 2·4-2·7 mm.

CLIVINA FROGGATTI, n.sp.

Robust, convex. Head short, wide, clypeus truncate-emarginate, prothorax subquadrate, with all its angles rounded; elytra oval, seventh and eighth interstices uniting at base to form a short, not strong, marginal carina, eighth interstice indicated by a fine carina near apex; prosternum with intercoxal part wide anteriorly, non-sulcate on base; episterna very finely shagreened, finely transversely striolate; metasternum, between intermediate and posterior coxæ, about as long as posterior coxæ, episterna sub-elongate; anterior tibiæ 3-dentate. Black, shining, legs and antennæ reddish piceous.

Head transverse, convex, anterior part rugulose, vertex wide, clypeal elevation arcuate, clypeus irregularly divided from front, deeply and widely truncate-emarginate, wings advanced, small, obtusely rounded, concave, gently oblique on inner side, supra-antennal plates convex, rounded externally, bordered, divided from wings of clypeus by a light sinuosity; facial sulci deep and divergent posteriorly; frontal impressions strongly marked, irregular; facial carinæ short, wide, prominent, supra-orbital punctures distant from eyes, set in a longitudinal groove, lower edge of this groove carinate, eyes globose, prominent, lightly enclosed behind; orbits abruptly constricted behind eyes. Antennæ moniliform, short, incrassate. Prothorax rather broader than long (2·2 × 2·25 mm.), widely convex; sides parallel, strongly and roundly narrowed in front of anterior marginal puncture, anterior margin lightly emarginate in middle, anterior angles obtusely rounded; posterior angles rounded, basal curve short, border wide, reflexed, median line well marked; anterior line strongly impressed; lateral basal impressions obsolete. Elytra oval (4·3 × 2·35 mm.), convex; sides strongly rounded, shoulders rounded, striæ deeply impressed, strongly crenulate except towards apex; interstices convex near base, depressed on apical declivity, lateral border strongly reflexed near shoulders; lateral channel wide

Length 7·2-8·5, breadth 2-2·3·5 mm.

Four specimens, the one measuring 7·2 mm. in length is, judging from the other three, an unusually small specimen.

Closely allied to *C. macleayi*, Sl., but differing in having the eyes more prominent and spherical, the facial sulci shorter, less arcuate and less convergent in front, the frontal foveæ deeper, the prothorax more convex, the sides not sinuate and much more strongly rounded to anterior angles, the anterior margin less emarginate, the anterior angles obtusely rounded and less marked; the elytra with distinctly crenulate striæ, the eighth interstice indicated near apex; the metasternum longer and with a deeply impressed channel near external margin, the metasternal episterna a little longer and with a strongly marked channel near inner margin, the colour deep black.

DYSCHIRIUS MACLEAYI, n.sp.

Robust, convex. Head strongly depressed between eyes, front carinate in middle, clypeus deeply and roundly emarginate with prominent lateral angles; elytra convex, basal part—in front of testaceous fascia—strongly punctate-striate (eight rows of punctures); anterior tibiæ 3-dentate. Head piceous black; prothorax shining bronzy-black; elytra ferruginous with a bronzy tinge, a wide testaceous fascia across apical third, legs, antennæ and under surface of prothorax reddish, body reddish piceous.

Clypeus declivous, anterior margin roundly emarginate, lateral angles advanced, obtuse at apex; supra-antennal plates large, quadrate, bordered, projecting widely and sharply beyond clypeus, declivous on inner side, anterior angles obtuse, anterior margins oblique; front depressed, a longitudinal carina in centre, two transverse impressions on each side between central carina and supra-antennal plates; vertex convex, smooth; supra-orbital carinæ well developed, thick; eyes globose, prominent   Prothorax globose, lævigate, a light transverse impression near anterior margin, median line wanting: marginal channel of base punctate. Elytra rounded on sides; shoulders rounded; striæ consisting of rows of deep coarse punctures, first stria only reaching apex, a short deep stria near margin on each side of apex; interstices convex on basal part of disc, third, fifth and seventh bearing some setigerous punctures; apical part of elytra smooth excepting for these punctures; marginal channel narrow on sides, stronger and more deeply impressed behind shoulders.   Anterior tibiæ with apical digitation long, arcuate; two upper teeth successively shorter, well developed, prominent, acute.

Length 4, breadth 1·15 mm.

Evidently allied to *D. torrensis*, Blkb., but differing in colour, and apparently in the sculpture of the head.

Note —It seems worthy of notice that there are eight striæ on each elytron of this species, the eighth stria consists of three or four punctures, and rises where the marginal channel narrows behind the shoulders.   *D. zonatus*, Putz., a specimen of which I have seen in the Macleay Museum, has only seven striæ on each elytron (the normal number among the *Clivinides*), and has the marginal channel wider and more punctate.

DESCRIPTION OF A NEW SPECIES OF *ABLEPHARUS* FROM VICTORIA, WITH CRITICAL NOTES ON TWO OTHER AUSTRALIAN LIZARDS.

By A. H. S. Lucas, M.A., B.Sc., and C. Frost, F.L.S.

ABLEPHARUS RHODONOIDES, sp.nov.

Snout broad, obtuse; rostral projecting. Eye incompletely surrounded with granules. Nasals large, forming a short suture behind the rostral; frontonasal much broader than long, forming a broad straight suture with the frontal; prefrontals widely separated, as long as the fronto-prefrontal suture; frontal large, longer than the frontoparietals and interparietal together, nearly as long as its distance from the nuchals, in contact with the the anterior supraoculars, three supraoculars, second largest; five supraciliaries; frontoparietals united; interparietal distinct; parietals about twice as broad as long, forming a suture behind the interparietal; three or four pairs of nuchals; five upper labials, fourth below the eye; five lower labials. Ear-opening minute, distinct. Body much elongate, scales in over sixty transverse series between axilla and groin, arranged in twenty longitudinal series; dorsals largest, laterals smallest. Two enlarged præanals. Limbs short, tridactyle, widely separated when adpressed; the fore limb shorter than the distance from the end of the snout to the ear-opening, hind limb a little shorter than the distance from the end of the snout to the shoulders; length of outer toe twice the length of the middle, four times that of the inner toe Tail almost as long as head and body.

Colour —Greyish above, each of the dorsal scales with a black central streak, forming four longitudinal series; a black lateral band from the nostril through the eye. Tail brownish. Under-surfaces yellowish.

19

```
         ...            „
Fore limb     ...    4·5 „
Hind limb     ...    9·5 „
Tail (reproduced)    35  „
```

Locality.—Mildura, Victoria. Two specimens obtained by favour of Rev. Walter Fielder.

Remarks.—This species is allied to *A. greyi*, Gray, by the head-scaling, but in habit resembles *A. lineatus*, Bell, and *A. muelleri*, Fischer. It differs from *A. lineatus* in head-scaling, in number of digits, and in the number of longitudinal series of body scales; and from *A. muelleri* in the head-scaling. The genus *Ablepharus* is characterised by its snake-like absence of movable eyelids, and the three species, *A. muelleri*, *A. lineatus*, and *A. rhodonoides*, show a further approach to the snake type in the reduction in size of the limbs and in the number of the digits.

It is convenient here to add remarks on two other lizards.

(1) *Ablepharus greyi*, Gray.

Within the year, Mr. H J. McCooey obtained specimens of an *Ablepharus* in the Boggabri District, which he subsequently described in a country paper. He has been good enough to forward examples to us. They do not differ in any particular from *A greyi*, Gray, which was first described from W. Australia, and was obtained by the Horn Expedition from the Centre. The species is thus one of those which is characteristically found in the interior regions of scanty rainfall.

(2) *Hemisphœriodon tasmanicum*, L & F.

After carefully examining a larger series of *Homolepida casuarinœ*, D & B., from New South Wales, and a series of examples kindly forwarded to us by Mr. A. Morton of the Hobart

Museum, we have come to the conclusion that our specimens described from the St. Clair Lake, Tasmania, in the P.L.S.N.S.W. 1893, p. 227, as *Hemisphœriodon tasmanicum*, are only among the numerous varieties of *Homolepida casuarinœ*, D. & B. Our chief reason for including the apparently new species under the genus *Hemisphœriodon* was the relatively large size of one of the teeth in each side of each jaw.

The genus *Hemisphœriodon* was separated off from *Hinulia* in 1867 by Peters. It is still considered, and we think rightly, as distinct from *Lygosoma*, in which *Hinulia* and *Homolepida*, with others, are included by Boulenger (B.M.C.)

The synonymy of *Homolepida casuarinœ*, D. & B., then consists of *Omolepidota casuarinœ*, Gray, *Cyclodus casuarinœ*, Dum. et Bibr., *Homolepida nigricans*, Peters, 1874, *Lygosoma muelleri*, Peters, 1878, and *Hemisphœriodon tasmanicum*, L. & F., 1891

Hemisphœriodon is separated from *Homolepida* thus :

In *Hemisphœriodon* (1) the pterygoid bones are separated on the median line of the palate, the palatal notch extending anteriorly to an imaginary line connecting the centre of the eyes; (2) lateral teeth with rounded crowns, one on each side of each jaw enormous, the others small.

In *Homolepida (Omolepidota)* (1) the pterygoids are usually in contact anteriorly, the palatal notch not extending forwards to beyond the centre of the eyes, (2) the maxillary teeth conical or obtuse, subequal.

In *H. tasmanicum (casuarinœ)* (1) the palatal notch extends forward to the hind border of the eye; (2) lateral teeth with rounded crowns, one on each side of each jaw much larger than the others, relatively as much larger as in young *H. gerrardii*. Thus this species may be claimed on the first ground by *Homolepida (Lygosoma)*, and on the second ground by *Hemisphœriodon*. Large individuals approach *H. gerrardii* to some extent also in habit. On the whole, pending a more satisfactory classification of the subgenera of *Lygosoma*, it is probably best to leave this variable form under the designation *Lygosoma (Homolepida) casuarinœ*.

DESCRIPTIONS OF NEW SPECIES OF AUSTRALIAN COLEOPTERA.

By Arthur M. Lea.

Part III.

TENEBRIONIDÆ.

Pterohelæus Darwini, n.sp.

Elliptic, convex, subnitid Piceous; under surface piceous-brown. Head minutely punctate; prothorax and elytra with very minute punctures, the latter with very feeble traces of striæ towards the base; under surface and legs with very minute punctures, those on the legs more distinct; abdomen feebly longitudinally strigose. Apex of tibiæ and tarsi with dense, reddish-brown, short setæ.

Head large, clypeus broad, very feebly emarginate, sides oblique, not at all reflexed, its suture with epicranium indistinct except at sides; feeble trace of a groove between eyes. Prothorax transverse, *at base wider than elytra;* margins flat, moderately wide, widest at base; angles acute, posterior slightly projecting on to prothorax, anterior passing eyes; disc from almost every direction without trace of median line. Scutellum widely transverse, feebly raised. Elytra *soldered together,* narrowing from base to apex, margins narrow, flat and feebly raised about the middle. *Wings rudimentary.* Legs moderate, three basal joints of anterior tarsi dilated (especially in ♂), 4th joint very small, the two apical slightly longer than the three basal, intermediate longer, two apical shorter than three basal; basal joint of posterior very long Length 16, width (at base of prothorax) 9½ mm.

Hab —Dongarra, West Australia (two specimens received from Mr. G. W. Ward).

A peculiar looking species, which I look upon as the most interesting in the whole subfamily. It evidently belongs to

Pterohelæus, but has rudimentary wings and elytra soldered together. The wings are gauzy, the veins connecting them with the metanotum are strong but short and abruptly terminated, the wings elsewhere without venation; near the termination of the veins they suddenly contract in width, thence parallel almost to apex, which is truncate. Length 6, width near base 2, width in middle ¾, longest vein 1½ mm.

I have examined *Helæus echinatus*, *Saragus rudis* and *Sympetes undulatus*, and find that in all three the metanotum is degraded, soldered to the elytra; and there are but the veriest rudiments of wings. Compared with the metanotum of *P. bullatus* or of *P. convexiusculus*, that of the present species differs in being much more transverse; the apex of a groove in a line with the scutellum marking the apex of a triangular extension, whilst in the two species named the metanotum is parallel; at the base in *Darwini* the angles of the scutellar groove are strongly rounded off, and —with another elevation—enclose a transverse pointed areolet; in *bullatus* and *convexiusculus* the angles are right angles and enclose a feeble slightly convex depression, the outer edge of which is not ridged; the groove in *Darwini* has a strong flattened ridge extending its whole length, in *bullatus* there is a faint trace of ridging, and none in *convexiusculus*.

PTEROHELÆUS BROADHURSTI, n.sp.

Convex, shining, glabrous. Reddish-brown, margins paler; under surface of head and mandibles piceous. Head densely and rather minutely punctate; prothorax with very minute punctures; each elytron with about seventeen rows of small punctures, and a short sutural row, sterna minutely punctate, abdomen very minutely punctate, and feebly longitudinally strigose.

Clypeus convex, its suture with epicranium distinct, both with reflexed sides; a shallow and moderately distinct impression between eyes; antennæ reaching intermediate coxæ, 3rd joint scarcely as long as 4th-5th combined. Prothorax widely transverse, with very feeble trace of median line, base sinuate, margins wide, very feebly raised at borders, anterior angles rounded, posterior acute, slightly recurved. Scutellum transverse, semi-circular, in some lights appearing feebly strigose. Elytra twice as long as head and prothorax combined, margins wide on basal half, narrowing thence to apex. Legs moderate, 1st joint of anterior tarsi scarcely as long as the rest, of intermediate distinctly shorter, of posterior as long as basal joint   Length 16, width 10 mm

Hab —Pelsart Island (Houtman's Abrolhos), W.A

In size and shape much the same as *confusus*, Macl. I have named this species after Mr. F C. Broadhurst, through whose kindness I was enabled to visit this interesting group of islands.

PTEROHELÆUS ABDOMINALIS, n sp

Oblong-elliptic, slightly convex, feebly shining, glabrous. Piceous-black, under surface and legs paler, margins, tibiæ and palpi piceous-red.   Head and prothorax densely minutely and obsoletely punctate, the former densely and minutely granulate at base, scutellum impunctate, each elytron with about eighteen rows of small punctures, becoming obsolete towards apex, under surface irregularly and feebly punctate, metasternum obliquely, the abdominal segments longitudinally strigose; legs minutely punctate.

Head wider across clypeus than the length to base of eyes, clypeus feebly convex in the middle, apex feebly emarginate, sides slightly raised, its suture almost obliterated.   Prothorax convex, with a feeble trace of a median line, deeply and semi-circularly emarginate in front, sinuate at base, anterior angles somewhat rounded, posterior acute, anteriorly feebly margined, lateral margins broad, slightly reflexed.   Elytra convex, parallel-sided to one-third from the apex, a little wider than prothorax at base, about twice as long as head and prothorax combined, not once and one-half as long as wide, margins broad, feebly reflexed, much narrowed from apical third to apex, a very feeble costa traceable from base to a little beyond the middle. Length 20, width 12 mm.

Hab.—Northam, W.A. (Master Percy Snelling).

From the description of *P. dispar*, the above species differs in being larger, its head decidedly broad in front, and the elytral suture slightly raised; my specimen is minus antennæ and tarsi.

PTEROHELÆUS TRISTIS, n.sp.

Oblong-elliptic, slightly convex, feebly shining. Piceous-black; prothoracic margins, tarsi, antennæ and palpi obscure reddish-piceous. Elytra with a few scattered short brownish hairs, scarcely visible to the naked eye; under surface with extremely minute and sparse pubescence. Head densely, minutely and irregularly punctate, and densely and minutely granulate at base; prothorax minutely and not so densely punctate as head, but in addition with extremely dense and almost microscopic punctures; scutellum extremely minutely punctate; elytra striate-punctate (in about eighteen rows), the striæ irregular at both base and apex, the punctures obsolete towards apex; under surface of head feebly granulate; prosternum sparsely and obsoletely, metasternum and abdominal segments distinctly punctate, the three basal segments of the latter feebly longitudinally strigose.

Head subquadrate; clypeus truncate, almost flat, its suture only visible at sides; antennæ flattened and widening to apex, reaching intermediate coxæ. Prothorax slightly convex, broadly transverse, median line unmarked, deeply emarginate in front, margins moderately broad, base feebly bisinuate, posterior angles acute. Scutellum transversely triangular. Elytra convex, parallel-sided to one-third from apex, as wide as prothorax at base, scarcely twice as long as wide, about once and one-half as long as head and prothorax combined, margins very narrow, feebly reflexed near base. Length 20, width 9 mm.

Hab.—Mt. Barker, W.A. (obtained under bark of a dead tree).

This species belongs to the 3rd subsection of Sir Wm. Macleay's second section of the genus; from either *P. parallelus* or *P. cereus* (the only two species belonging to the subsection from W.A.), its size will at once distinguish it. I do not know any species which it closely resembles.

P. PARALLELUS, Brême; Mast. Cat. Sp. No. 3756.

Hab.—Bunbury, W.A.

P. BULLATUS, Pasc.; Mast. Cat. Sp. No. 3742

Hab.—N.S.W., W.A.

P. CEREUS, Macl.; P.L.S.N.S.W. 1887, p. 545.

Hab.—Beverley, W.A.

P. CONVEXIUSCULUS, Macl.; l.c. p. 549.

Hab.—Cootamundra, N.S.W.

P. GLABER, Macl.; l.c. 547.

Hab.—Inverell, N.S.W.

P. HIRTUS, Macl.; l.c. p. 532.

Hab.—Forest Reefs, Sydney, N S.W.

P. ASELLUS, Pasc.; Mast. Cat. Sp. No. 3740.

Hab.—Tweed and Richmond Rivers, N.S.W.

P. LATICOLLIS, Pasc.; l.c. No. 3750.

Hab.—Forest Reefs, N.S.W.

P. CONFUSUS, Macl.; l.c. No. 3743.

Hab.—Armidale, N.S.W.

HELÆUS FULVOHIRTUS, n.sp.

Oval, shining. Piceous-brown, margins brownish-red, under surface brown. Elytra with four rows of long recurved brownish-red hair placed in small tufts. Prothorax and under surface with minute punctures, a minute hair arising from each. Margins very minutely granulate.

Antennæ reaching intermediate coxæ, 3rd joint longer than 4th-5th combined Prothorax with wide margins raised at an angle of about 45°, feebly curved at outer edge, the right side crossing the left in front of head, its point obtuse, posterior angles very slightly projecting on to prothorax; disc with a short narrow carina, nowhere angular or pointed. Scutellum transversely cordate, with a semicircular row of shallow irregular foveæ. Elytra widest behind the middle, margins at base raised at about 45°, becoming less towards apex, their outer edge more noticeably curved than in prothorax. Four basal segments of abdomen irregularly impressed at sides. Legs long, claw joint of anterior tarsi almost as long as the rest combined, of intermediate as long as basal joint, of anterior not as long as basal joint. Length 20, width 14mm.

Hab.—Dongarra, W.A. (Mr. G. W. Ward).

The small size of this species will serve to distinguish it from those of its congeners possessing hairy elytra; from the description it appears to be closest to *H. Kirbyi*.

HELÆUS GRANULATUS, n.sp.

Piceous-brown ; antennæ piceous-red. Head with shallow, moderately dense punctures ; prothorax covered with small, regular, feebly shining granules, margins feebly punctate and very feebly granulate. Elytra feebly striate-punctate, punctures almost obsolete, each bearing a minute erect bristle, seen from above the bristles appear to be all of the same height, but when viewed from behind there are seen to be five rows, between each of which are two rows of almost microscopic setæ; epipleuræ rather strongly and irregularly punctate; under surface with minute punctures and pubescence.

Head feebly grooved between eyes; antennæ reaching intermediate coxæ, 3rd joint longer than 4th-5th combined. Prothorax—including margins—subtriangular, not once and a quarter as wide as long, margins feebly curved, moderately wide, at base depressed, the posterior angles slightly projecting on to elytra, anterior angles subtruncate, right crossing left; disc with a raised shining carina continuous from head almost to base, near base descending at an angle of about 80°. Scutellum feebly raised, widely transverse. Elytra with suture carinate, each with a shining costa on 4th interstice terminated at posterior declivity; margins moderately wide at base, suddenly narrowed and then feeble to apex. Legs moderate, claw joint of anterior tarsi thick, longer than the rest combined, of intermediate as long, and of posterior not quite as long. Length 10, width 6½ mm

Hab.—Mullewa, W.A.

Described from a specimen taken alive; in two found dead (one of which measures 14 × 8 mm.) the elytral punctures are noticeable to the naked eye, and the setæ are sparse and minus the five more elongate rows. The species appears to be closest to *falcatus* from South Australia, from the description of which it differs in not having the anterior angles of prothorax acutely pointed, the elytra dull, and narrow margins without granules.

HELÆUS ECHIDNA, White ; Mast. Cat. Sp. No. 3771.

Sir William Macleay's description of this species is somewhat misleading, as he fails to mention the two tubercular spines on the prothorax, and that the sutural rows of spines terminate before the apex of the elytra. The species is readily identifiable by the figure accompanying the original description.

SYMPETES ACUTIFRONS, n.sp.

Broadly ovate, feebly shining. Piceous-brown, margins testaceous, their edges brown, apices of abdominal segments tinged with testaceous Elytra with very minute, pale, depressed setæ, under surface with moderately dense and very short pubescence. Head densely and irregularly punctate; prothorax minutely, its margins more noticeably punctate; elytra with dense and rather minute punctures, their epipleuræ very distinctly punctate; under surface minutely punctate.

Clypeus convex, its disc within a circular depression; a distinct shallow impression between eyes. Prothorax widely transverse, base trisinuate, irregularly transversely impressed in middle and more feebly towards sides, a feeble median carina becoming feebly

pointed at base; margins wide, edges recurved; anterior angles
acute, produced almost to apex of head, posterior sharp and
strongly curved. Scutellum widely transverse. Disc of elytra
very much wider than that of prothorax, bulged before middle,
constricted near apex, suture strongly raised, interstices irregular,
feebly raised; margins wide, their edges recurved. Legs moderately
short, claws long. Length ♂ 16, ♀ 17, width ♂ 12, ♀ 13½ mm.

In the male the margins are proportionately broader than in
the female, and they are also reflexed.

Hab.—Geraldton, W.A.

SYMPETES UNDULATUS, n.sp.

♂. Shining, subparallel. Reddish-brown, margins paler;
antennæ ferruginous. Upper surface with very minute setæ,
more noticeable on head and margins than elsewhere. Elytra
densely and irregularly punctate, abdomen densely and minutely,
the margins and sterna more coarsely punctate.

Head not projecting beyond prothorax; clypeus wide, perfectly
straight in front, very feebly convex, notched at the sides; eyes
scarcely visible; antennæ thin, joints 1st-7th cylindrical, 8th
pear-shaped, 9th-11th circular. Prothorax almost thrice as wide
as long, disc depressed on each side of middle, at sides and base;
margins each wider than disc, each forming the fourth segment
of a circle, anterior angles almost right angles, not at all produced,
posterior feebly curved and scarcely acute. Scutellum widely
transverse. Disc of elytra as long as prothorax is wide, ovate-
elliptic, suture strongly raised, each with six or seven feeble
irregular costæ, the alternate ones stronger; margins waved, in
middle almost as wide as each elytron, distinctly wider elsewhere,
edges scarcely recurved and very little darker. Legs long and
thin. Length 18, width 13 mm.

♀. Differs in being broader and more rounded; a more distinct
transverse impression at base of prothorax, the anterior angles
feebly produced; disc of elytra broadly ovate, and, except at base,
much wider than margins, outer edges of margins below level of

suture (in ♂ they are higher than the sutural crest), widest about middle (in ♂ the elytra are widest near base, the margins at the middle being slightly inwardly compressed); punctures of epipleuræ coarser   Length 17½, width 14 mm.

Hab.—Geraldton and Walkaway, W.A.

A rather fragile-looking species, having somewhat the appearance of an *Encara*; the clypeus is straighter than in any species of the subfamily with which I am acquainted.   When viewed against a light the margins appear to be thickly impressed with somewhat angular punctures   I have seven specimens under examination, two of which (sexes) measure but 16 mm.

SYMPETES DUBOULAYI, Pasc.; Mast. Cat. Sp. No. 3798.

This species was evidently unknown to Sir Wm. Macleay, as he simply quoted Pascoe's description, and allowed it to remain in *Saragus*.   Mr. Champion has since (Trans. Ent. Soc. 1894, p. 384) referred it to its correct genus.   The species is moderately common along the coastal regions from Swan River to Geraldton.   The posterior angles of the prothoracic, and the anterior of the elytral margins are turned down, a most unusual character in the family.

S. MACLEAYI, Pasc.; Mast. Cat. Sp. No. 3789.

Hab.—Northam, W.A.

S. TRICOSTELLUS, White; Mast. Cat Sp No. 3825

Hab.—Swan River, W.A.

SARAGUS STRIATIPENNIS, Macl., P.L.S.N S.W 1887, p. 668.

Hab. —New South Wales.   Widely distributed

S. RUDIS, Macl , l.c. p. 659

Hab.—New South Wales.   Widely distributed

S. LÆVICOLLIS, Oliv., Mast. Cat. Sp. No. 3807

Hab.—New South Wales.

specimen trom M. Kosciusk- .... ag
ascoe's description, and which is certai
ecimen is a male and has faint traces o
veral males of *sulcicollis* in my possess'
e head broader, the upper part of the e
nd more coarsely granulate; the protho
nvex, with the margins deflexed, a much
each side at base; elytral epipleuræ
here they are smaller; prosternal keel br
nd parallel; intercoxal process depress
ɔdominal segment smaller, with the 5th
.her but less noticeable differences.
ɪbescence is natural, and not due to ab
ɾidently very rare, and my specimen is th

APASIS PUNCTICEPS, n

♂. Elongate, slightly convex, shining.
ppery reflection, tarsi and palpi piceous.
attered reddish hairs (not always prese
biæ and the tarsi with dense short browni
ebly pubescent. Head distinctly a

suboval, striate, the 4th and 6th interstices slightly the widest, the sutural marked by irregular punctures   Under surface more shining than upper.   Femora stout; two small spurs at apex of tibiæ; anterior tarsi dilated.   Length 22, width 7 (vix) mm.

♀. Differs in being a little larger and duller, antennæ shorter and thicker, femora thinner, and the anterior tarsi no wider than the others.

Hab.—Mt. Kosciusko (Mr. W. E. Raymond)

Through the kindness of Mr. G. Masters I am enabled to compare the above with *A. Howitti*, from which it differs in being larger, the head distinctly punctate and less shiny, antennæ shorter and thicker (in both sexes), palpi much darker in colour; the prothorax is decidedly transverse (in *A. Howitti* it is—if anything—a little longer than wide); the scutellum is a little broader, the scutellar stria more distinct, and the other striæ are somewhat different at the apex.

MELANDRYIDÆ, ANTHICIDÆ, MORDELLIDÆ.

A paper by Mr. Champion (Trans. Ent. Soc. Lond. 1895), and two by myself (P.L.S.N S.W , 1894, and 1895) have clashed; and unfortunately several of the names proposed for species in the above families will have to rank as synonyms   The synonymy will be treated of by Mr Champion, but I would here like to offer a few brief remarks on three of the species described by me.

DIRCÆA LIGNIVORA, P L S N.S.W (2) x. 1895, p 266.

This species is very close to *venusta*, Champ., nevertheless I am satisfied that it should be considered as distinct   From *venusta* it differs in being narrower (♂♀); the thorax is much darker; apical macula on each elytron sublunulate (in that species it is dumb-bell shaped), basal macula much smaller and more rounded, not continued to lateral margins, and without a small spot or paler marking behind it, there are also several other but less noticeable differences

ANTHICUS EXIGUUS, P.L.S.N S.W. (2). ix. 1894, p. 616.

This name having been used by Mr. Champion for an American species, I propose to alter the name of the Australian species to *rubriceps*.

MORDELLA WATERHOUSEI, P L.S.N.S.W. (2), x. 1895, p. 300.

As Mr. Champion (Trans. Ent. Soc. 1895, p. 267) has substituted the name of *Waterhousei* for *obliqua*, Waterh., my name must fall; I therefore propose to alter the name of the Australian species to *Caroli*.

CURCULIONIDÆ.

AMYCTERIDES.

DIALEPTOPUS ECHINATUS, n.sp.

Narrow, deep, elongate-elliptic, subopaque. Piceous; prothoracic crests, elytral tubers and legs dull red; antennæ reddish-piceous. Rostrum and space about elytral suture with long blackish setæ; apex of prothorax with short setæ; head with very short depressed pubescence above and below eyes, a patch of whitish scales between eyes; prothorax with sparse elongate and rather small scales at sides; ocular lobes fringed with silvery white setæ, disc of elytra and tubercles with whitish scales variegated with pale brown along suture; lateral punctures filled with whitish-yellow scales; apical segment of abdomen with elongate setæ, and a spot of whitish scales.

Rostrum irregularly punctate, grooved in the middle, the ridges, together with those formed by scrobes, forming the letter M Prothorax with an elevated transversely granulate ridge on each side of middle, the ridges not conjoined at apex but separately overhanging head, the depression between the ridges deepest near apex, becoming shallower and with scattered granules near base; an oblique ridge formed by two irregular rows of granules from base to middle of ocular lobes, a few scattered granules below; there is also a very short intermediate basal ridge of obsolete granules.

Elytra narrow, with two distinct rows of sharp conical tubercles united at base and projecting on to prothorax; the outer row contains six to ten and the inner slightly more tubercles; there is also a short sutural row of from three to five smaller tubercles, commencing at about the middle and terminating at summit of posterior declivity; space between tubercles irregularly punctate; sides with four rows of large punctures, two of which are marginal; posterior declivity with small granules and punctures; apices rounded, very feebly emarginate.   Sterna sparsely punctate and with irregular depressions.   Two basal segments of abdomen with irregular depressions and ridges, all irregularly and (especially the apical) coarsely punctate at sides, a few feeble punctures across the middle; apical segment with a distinct circular squamose fovea in its middle.   Legs long, setose; femora moderately stout; anterior tarsi with an elongate pad on each side, the rest not padded.   Length 17, rostrum 2½; width 6 mm.

Hab.—Geraldton and Mullewa, W.A.

I have two specimens, one of which is almost scaleless and has the elytral extension larger, more obtuse and more obtusely granulate than in the other.   The species, on account of the number of rows and sharpness of its elytral tubercles, should be very distinct from any previously described.   The number of the tubercles in each row is never to be depended upon, as in most of the species I have examined they vary in number even on the same specimen.

DIALEPTOPUS LONGIPES, n.sp.

Narrow, deep, elongate-elliptic, subopaque.   Black; elytral tubercles dull red, legs piceous, antennæ black.   Rostrum and apex of prothorax with short blackish setæ; muddy-brown scales on head between eyes, and very small and indistinct muddy scales on prothorax and elytra.   ·

Rostrum almost impunctate, otherwise as in the preceding. Prothorax as in the preceding except that the crests become united at extreme apex and overhang the head as one, the lateral oblique ridge being more pronounced and less granulate.   Elytra narrow, with two rows of elongate triangular tubercles conjoined at base and projecting on to prothorax, the outer row containing four distinct tubercles and the inner three to five, becoming carinate towards base; punctures forming two sutural rows, two rows between tubercles and five larger and lateral rows, two of which are marginal, and one irregular touching outer row of tubercles; posterior declivity punctate and not granulate, apex narrowly and deeply emarginate and separately sharply mucronate. Sterna sparsely punctate. Abdomen irregularly and somewhat obsoletely punctate at sides, suture between 1st and 2nd segment deep and very distinct at sides, 2nd obliquely scratched, apical ridged in the middle and depressed on each side. Legs long, thin and setose, tarsi not padded. Length 12½, rostrum 1¾; width 4½ mm.

Hab.—Bridgetown, W.A.

D. sepidioides, Pasc., is a species larger than, but intermediate in shape between this and the following species, from either of which it may be distinguished by its much larger sutural punctures. The abdomen also is different from that of either of them.

DIALEPTOPUS SORDIDUS, n.sp.

Deep, opaque, moderately broad. Black, apical tubercles on elytra almost black, the rest entirely so. Rostrum and apex of prothorax with short blackish setæ. Muddy scales on head between eyes, at base of prothorax, and rather densely covering elytra; under surface (except apex of abdomen) glabrous.

Rostrum sparsely punctate, a shallow parallel-sided groove extending its entire length. Prothoracic crests as in the preceding, except that at apex they are more visibly united, oblique ridge feebly granulate, intermediate ridge more distinct than in either of the preceding and more obsoletely granulate. Elytra ovate, two rows of elongate triangular tubercles conjoined at base and projecting as a granulate extension on to prothorax, outer row composed of four distinct tubercles, the inner of three or four distinct only towards apex, space about suture and between rows

20

of tubercles irregularly punctate and obsoletely granulate, six lateral rows of punctures of which only one is distinctly marginal, the upper row irregular and touching tubercles, posterior declivity irregularly punctate and obsoletely granulate; apex semicircularly emarginate and each obtusely mucronate. Sterna sparsely punctate. Two basal segments of abdomen with shallow irregular impressions, except at sides of suture where they are distinct, 2nd segment irregularly feebly obliquely ridged at apex, apical segment with an outer row of coarse punctures, middle with a foveate elevation. Legs moderately long, thin, tarsi not padded. Length 13, rostrum 1¾; width 5½ mm.

Hab.—Swan River, W.A.

LÆMOSACCIDES.

LÆMOSACCUS ARGENTEUS, n.sp.

Entirely black. A median stripe on prothorax, a short oblique spot on each elytron conjoined at base (lying on the 1st and 2nd interstices, the two conjointly subobcordate), a small spot on each side of apical abdominal segment, clothed with silvery-white scales; a few whitish scales at apex of elytra, on sterna, and between eyes.

Eyes large, almost touching; rostrum long, shining, cylindrical, feebly curved, punctate* at base and apex, almost impunctate in middle; 1st joint of funicle nearly twice the length of 2nd, club almost as long as funicle. Prothorax subquadrate, slightly narrowed in front, and with a feeble median impression, base with an impression on each side. Scutellum transversely triangular. Elytra with angles slightly rounded at base and apex, each feebly convex in consequence of a sutural depression, interstices flat, feebly granulate. Abdomen, with meso- and metasternum,

---

* In the species here described I have not considered it necessary to give the puncturation of any parts but the rostrum, as it is much the same in all and therefore of little use for identification.

strongly convex. Femora edentate, 3rd tarsal joint small.
Length 2½, rostrum ⅘; width 1⅓ mm.

Hab.—Gosford, N.S.W.

The silvery scales on prothorax and about the scutellum (itself
nude), and the entirely black colour of this rather pretty little
species are its chief distinguishing features.

LÆMOSACCUS PASCOEI, n.sp.

Entirely black. A patch of yellowish pubescence about the
scutellum, extending on to the 1st and 2nd interstices to about two-
fifths from apex, and a much shorter distance on 3rd, the whole
forming an obtuse V; base of pygidium with silvery pubescence,
its apex nude ; sides of prothorax, sides of sterna and abdominal
segments with pale yellow and moderately dense pubescence, rest
of under surface with sparser and lighter coloured pubescence;
legs (except tarsi) glabrous.

Eyes very large, almost touching; rostrum short, thick, compressed, opaque, grooved, feebly bent and coarsely punctate; 1st
joint of funicle thicker and but slightly longer than 2nd. Prothorax with a short feeble irregular carina; on each side of middle
a large circular shallow impression feebly open towards apex.
Scutellum small, triangular, nude. Elytra about once and one
quarter as long as wide, interstices irregular. Anterior legs
moderately long; femora very minutely dentate ; 3rd tarsal joint
moderately bilobed, claw joint rather small. Length 2½, rostrum
½, width 1 mm.

Hab.—Clifton, N.S.W.

LÆMOSACCUS CARINICOLLIS, n.sp.

Black; legs (femora occasionally piceous) and antennæ dull red,
club darker. Above with dull orange-coloured and rather long
pubescence as follows—on the head between eyes, on prothorax
at sides and angles (becoming elongate spatulate scales lower
down) and a stripe continued from head, at middle of base a
patch parallel at commencement but becoming bilobed at the
middle (scarcely cordate in shape), on elytra irregularly X-shaped

and sparse at sides and apex. Pygidium with sparse greyish scales. Beneath with yellowish moderately elongate scales, sparsest down the middle. Legs somewhat densely pubescent.

Eyes very large, depressed, rostrum long, shining, distinctly curved, widening to apex, in ♂ densely punctate at base and apex, sparsely punctate in the middle and with oblong punctures at sides, in ♀ more regularly and sparsely punctate, 1st joint of funicle once and one-half as long as 2nd. Prothorax with a shallow longitudinal impression at apex, and a circular one on each side of middle; carina raised, shining, distinct, continuous from before the middle almost to base. Scutellum triangular, subcordate. Elytra moderately long ($3\frac{1}{3}$ × $2\frac{1}{2}$ mm.), parallel-sided, interstices flat, granulate Pygidium obsoletely carinate. Anterior femora with a small tooth moderately distinct in ♂, smaller in ♀; 3rd tarsal joint large, padded beneath with silvery hair, punctate above. Length 6, rostrum $1\frac{2}{3}$; width $2\frac{1}{2}$, range of variation $4-6\frac{1}{4}$ mm.

Hab.—Mt. Kosciusko (Raymond); Queanbeyan, Tamworth, Forest Reefs, Cootamundra, N.S.W. (Lea): Benalla, Vic. (Helms). Common on freshly felled Eucalypts.

The shining prothoracic carina and long curved rostrum should render this species easy of identification. The pubescence on the upper surface varies from a pale to a dark orange colour, the scutellum is always bare, the pattern on the prothorax, though always constant, varies in dimensions, on the elytra the pubescence occasionally almost covers the entire surface, a small transverse space close to the apex and several very small spots being left.

var. OCCIDENTALIS.

Differs only from the above by its much smaller size; by the pubescence of prothorax continuous across apex, that on the elytra forming a transverse H, and continuous across apex, and with the interstices somewhat smooth Length 3, rostrum $\frac{2}{3}$; width $1\frac{1}{2}$ mm.

Hab --Champion Bay, W.A.

LÆMOSACCUS CRUCICOLLIS, n sp.

Black, funicle piceous-black. Pale yellow or whitish pubescence at angles of prothorax, four elongate spots at base, apex and across middle, which if united, would form an inverted cross; elytra with a patch about scutellum, from the shoulders oblique to about the middle, then feebly widening for a short distance and terminated about the apical 4th, apex slightly pubescent, 6th-8th interstices slightly pubescent at apical third, and 8th-9th behind shoulders; under surface with moderately dense pubescence at sides, sparser and greyer in the middle.

Eyes large, almost touching. Rostrum moderately long, curved, shining, cylindrical, rather finely punctate. First joint of funicle thick, transverse, distinctly longer than 2nd; club as long as funicle. Prothorax bulged out in the middle, a longitudinal impression at base and apex, and a transverse one on each side of middle. Scutellum rounded, shining. Elytra moderately long, interstices transversely granulate. Pygidium carinate. Under-surface strongly convex; intermediate segments of abdomen with very distinct sutures. Femoral tooth very small, claw joint of tarsi moderately prominent   Length $3\frac{1}{2}$, rostrum $\frac{2}{3}$; width $1\frac{1}{3}$ mm. Range of variation very slight.

Hab—Clifton, Galston, Forest Reefs, N.S.W.

In build resembling *carinicollis*, but somewhat narrower, and without the shining prothoracic carina so distinct in that species. In one specimen I possess the patch of elytral pubescence is much smaller, it only extends to about the basal third, with a few spots about the apical third near the suture, and two very small spots on the 8th interstice.

LÆMOSACCUS FUNEREUS, Pasc.; Mast. Cat. Sp. No. 5325.

I have a male insect from Armidale which agrees very well with Mr. Pascoe's description of this species, except that the rostrum and legs (tarsi excepted) are black; but as both these are liable to sexual variation of colour, and Pascoe's specimen may have been a female, I have considered it inadvisable to describe it as new. Length $3\frac{1}{2}$, rostrum $\frac{1}{2}$; width $1\frac{3}{4}$ mm.

LÆMOSACCUS DUBIUS, n.sp.

♀? Black, antennæ red, club and tarsi reddish-piceous. Under surface and legs microscopically pubescent.

Eyes large, distinctly but not widely separated. Rostrum straight, moderately elongate, shining, cylindrical, sparsely punctate.   Antennæ long, scape almost straight, thin but thickened at apex; 1st joint of funicle large, twice as long as 2nd; club large, almost as long as funicle   Prothorax with a longitudinal impression feeble in the middle, much stronger towards apex, causing the surface near it to appear raised, each side of base with an oblique elliptic and distinct impression   Scutellum transverse.   Elytra wide, rather coarsely granulate, separately convex, 4th interstice widest.   Pygidium large, without trace of carina.   Legs moderately long, anterior femora with a very small basal tooth, the intermediate with a larger, sharper and more median tooth, claw joint distinct.   Length 5⅓, rostrum 1; width 2⅓ mm.

Hab.—Braidwood, N.S W.

This species also almost fits Mr. Pascoe's description of *funereus*, but as it was obtained in a mountainous district much farther south, and both species cannot be *funereus* (which evidently belongs to the group about *subsignatus, carinicollis, narinus, &c*) I have given it a name   From the specimen mentioned above as possibly *funereus* it differs in being considerably larger, without trace of pubescence on the upper surface, longer and straighter scape, darker tarsi, longer claw joint, and in several other details which may possibly be sexual.

LÆMOSACCUS COSSONOIDES, n.sp.

♂? Dull red; club and under surface piceous. Sparse and somewhat elongate yellowish pubescence on head, prothorax (a small spot on each side at base and apex nude) and elytra; the latter with a bare transverse space about the middle (continued towards base at sides) and a spot occupying the 5th-6th interstices near apex, pygidium densely covered with whitish scales, under surface rather densely (sparser in middle) covered with elongate scales or short greyish-yellow pubescence, legs with short pubescence

Eyes large, somewhat flat, widely separated.   Rostrum rather short and flat, widening to apex, densely and rather finely punctate, feebly curved.   First joint of funicle large, twice as long as

2nd, club not as long as joints 2nd-7th. Prothorax rounded, a feeble longitudinal impression down middle, and a feeble transverse one near apex. Scutellum small, round, not in a depression Elytra nearly once and one-half as long as wide, convex, interstices flat, very minutely granulate, those near the suture wider than towards the side. Pygidium with traces of a longitudinal carina Legs short, anterior femora with a moderately large basal tooth, 3rd tarsal joint deeply bilobed, but not much wider than 2nd, claw joint long, very distinct. Length $4\frac{3}{4}$, rostrum 1; width $1\frac{3}{4}$ mm.

Hab.—Sydney, N.S.W.

The elytra are more convex, with the interstices more feebly granulate than is usual in the genus. A slight resemblance to some of the broader species of *Cossonus* has suggested the specific name.

LÆMOSACCUS COMPACTUS, n.sp.

♂. Black; antennæ (club piceous) and tarsi dull red. Above and below with very sparse greyish pubescence.

Eyes widely separated. Rostrum short, thick, straight, opaque, coarsely punctate and grooved for its entire length Antennæ short; scape not twice the length of 1st joint of funicle, club large, compact. Prothorax rounded, a feeble carina at base, on each side of which is an almost circular and very distinct impression. Scutellum transverse, placed in a sutural depression. Elytra slightly longer than wide; interstices broad, coarsely granulate. Pygidium feebly carinate. Legs short; anterior femora with a minute tooth; claw joint very distinct. Length $1\frac{2}{3}$, rostrum $\frac{1}{3}$ (vix); width $\frac{3}{4}$ mm.

Hab.—Sydney, N.S W.

A small, dumpy, and rather strongly marked species, the size of which should alone be sufficient to render its identification easy.

LÆMOSACCUS FESTIVUS, n.sp.

Black; antennæ, tarsi and apex of tibiæ dull red. Golden yellow pubescence on prothorax at sides and apex, and encroaching

on the base, leaving a large discal patch nude; elytra with a transverse patch at base narrowing and then slightly widening to the
middle, behind it at a third from apex a small patch, and between
these on 5th-7th interstices another small patch, the whole
enclosing (to the naked eye) an elliptic bare space; pygidium and
apical segment of abdomen with sparse greyish scales; under
surface bare.

Eyes moderate, approximate. Rostrum short, straight,
cylindrical, shining, almost impunctate, scape short, curved; 1st
joint of funicle enlarged, not once and one-half the length of
2nd; club large. Prothorax subquadrate, a distinct impression
on each side at base, a median line invisible from most directions.
Scutellum small, subtriangular, not in a depression. Elytra somewhat convex, about once and one-third as long as wide, interstices narrow, transversely granulate. Pygidium with a short
moderately distinct carina. Femora edentate, claw joint small,
partially concealed. Length $2\frac{1}{3}$, rostrum $\frac{1}{2}$; width $\frac{4}{5}$ mm.

Hab --Tamworth, N.S.W.

A prettily marked little species but with no distinct structural
features.

LÆMOSACCUS OBSCURUS, n. sp.

♂. Black. Golden yellow pubescence forming a small spot at
base and apex of prothorax, a moderately long scutellar patch,
oblique from shoulders to about basal third, thence parallel to and
very slightly widening at the middle; pygidium with greyish
pubescence; under surface and sides of prothorax with greyish-
yellow pubescence.

Eyes large, almost touching. Rostrum short, straight, opaque,
slightly widening to apex, rather flat and densely punctate.
Antennæ inserted at basal two-fifths; 1st joint of funicle slightly
thickened, not much longer than 2nd; club large. Prothorax
rounded, a depression at base, apex and on each side of middle.
Scutellum small, elongate, shining, not depressed. Elytra not
once and one-quarter as long as wide, shoulders oblique, inter-

stices flat, moderately wide, transversely granulate. Propygidium large, pygidium small, feebly carinate. Anterior legs moderately long, femora edentate; tarsi narrow, 3rd joint deeply but not very widely bilobed, padded with silvery hair beneath, claw joint small, moderately distinct. Length $2\frac{1}{4}$, rostrum $\frac{1}{2}$ (vix), width $\frac{5}{6}$ mm

♀. Differs in having the rostrum shining, much less densely punctate and subcylindrical; club smaller; pubescence paler and sparser.

Hab.—Tamworth and Armidale, N.S.W.

One of the few species in which the scutellum is not situated at the base of a sutural depression; it is rather obscure and may cause some trouble to identify, though evidently distinct from any other known to me. From the preceding it differs in colour of tarsi and antennæ, markings on prothorax and elytra, slightly longer claw joint, and has a more angular outline.

LÆMOSACCUS ATER, n.sp.

♀. ? Black; antennæ (club piceous) and tarsi red. A few short yellowish hairs about base and across apical third of elytra; pygidium and sterna with sparse and very minute scales.

Eyes large, not widely separated. Rostrum short, straight, shining, cylindrical, finely punctate. Scape short, feebly curved; 1st joint of funicle large, the rest indistinctly jointed, club scarcely as long as funicle. Prothorax convex, a short distinct impression at apex, a feeble impression on each side at base, and a feeble impression almost at sides in middle. Scutellum small, within a feeble depression Elytra moderately long, interstices rather narrow, convex, transversely granulate. Pygidium very feebly carinate. Anterior femora edentate, claw joint small, moderately distinct. Length $2\frac{1}{5}$, rostrum $\frac{1}{2}$; width $\frac{4}{5}$ mm.

Hab.—Tamworth, N.S.W.

I have two specimens, both apparently females. The claw joint, though small, is not so minute as in *cryptonyx* and a number of other species.

LÆMOSACCUS VARIABILIS, n.sp.

♂. Head, base of rostrum, prothorax (apex tinged with red), scutellum, pygidium, under surface and base of femora piceous-brown or black; rest dull red, sides and base of elytra sometimes tinged with piceous.   Under surface and sides of prothorax microscopically and very sparsely pubescent.

Eyes moderately large, prominent, subapproximate.   Rostrum short, thick, curved, coarsely punctate, the two colours separated by a raised and triangular emargination, base feebly grooved Antennæ short, 1st joint of funicle thick, club almost as long as funicle.   Prothorax with an almost obsolete median and punctate carina, each side of base with a distinct transverse impression, and an almost invisible depression on each side of middle Scutellum small, elongate, depressed.   Elytra noticeably wider than prothorax, shoulders produced, oblique, apex feebly rounded, suture depressed, more distinctly towards scutellum, interstices narrow, strongly (for the genus) convex.   Pygidium feebly punctate.   Basal segment of abdomen with a shallow but distinct impression in its middle at suture with 2nd   Anterior legs moderately long, femora edentate, claw joint very small, scarcely extending beyond lobes of 3rd.   Length 2, rostrum $\frac{1}{2}$ (vix), width $\frac{3}{4}$; range of variation $1\frac{4}{5}$-$2\frac{1}{2}$ mm.

♀. Differs in having the prothorax (except for a piceous tinge about the basal impressions) red, without carina, and with a distinct median line; rostrum longer, thinner, smooth, almost entirely red, and much less densely punctate, abdomen narrower and more convex, and anterior femora shorter.

Hab.—Forest Reefs and Queanbeyan, N.S.W.

The entire absence of pubescence on the upper surface, with the colour of this species, and the peculiar rostrum of the male (appearing fractured in the middle) should render this species easy of identification, though the following one strongly resembles it.

LÆMOSACCUS VENTRALIS, n.sp.

♀. ? Dull red, head, scutellum, extreme base of pygidium, meso- and metasternum (except their sides) piceous-black.

Differs from the ♀ of preceding (which it strongly resembles) in having the rostrum a little broader and shorter, the prothorax with a feeble longitudinal impression with a feeble transverse impression crossing its middle, shorter legs, and femora with a very small tooth. Length 2, rostrum ½; width 1 mm.

Hab.—Swan River, W.A.

In all the numerous specimens of *variabilis* I have examined the abdomen and pygidium are entirely black, and neither of the sexes possesses a femoral tooth; in my specimen of the above the tooth, though small, is distinct and would seem to imply specific value.

LÆMOSACCUS RUFIPENNIS, n sp.

Black; elytra (except sides and apex), antennæ (club tinged or not with piceous), and tarsi dull red; apex of prothorax and knees occasionally tinged with red. Pygidium with silvery scales; punctures of under surface each with a small whitish scale.

Eyes large, approximate. Rostrum short, straight, shining, cylindrical, very finely and sparsely punctate. Scape short, distinctly curved; 1st joint of funicle large, twice the length of 2nd; club not as long as funicle. Prothorax rounded, a longitudinal impression very distinct at apex, feebly or not at all continued to base, base with an almost obsolete or moderately distinct impression at base, traces of a transverse impression on each side of disc. Scutellum small, round, situate in a depression. Elytra about once and one-third as long as wide, conjointly feebly convex towards apex, separately towards base, interstices narrow, strongly convex, very minutely granulate, the fifth with several (usually three) transverse and distinct granulations towards its apex. Pygidium densely punctate and with a shining impunctate longitudinal carina. Femora with a small tooth, 3rd tarsal joint moderately bilobed, claw joint small but distinct. Length 2¾, rostrum ⅗ (vix); width 1; range of variation 2½-3⅓mm.

Hab.—Tamworth, Forest Reefs, N S.W

At first sight resembling *variabilis*, but at once separated from that species by its perfectly straight rostrum. If, in the five

specimens of this species I have under examination, both sexes
are present, the difference is but slight: those I take to be males
have a slightly larger club and broader elytra, the prothorax
always entirely black, and the tarsi feebly tinged with piceous

LÆMOSACCUS INSTABILIS, n.sp.

♂. Black; antennæ and tarsi pale red, rostrum piceous, its apex
sometimes dull red, tip of femora and tibiæ and extreme apex of
elytra tinged with red   Pygidium and under surface almost nude.

Eyes large, prominent, almost touching.   Rostrum straight,
short, shining, perfectly cylindrical, with feeble elongate punctures.
Antennæ short, scape very short, inserted at eyes, almost geni-
culate, 1st joint of funicle large, transverse, distinctly wider than
scape, rest of the joints short, thick, their combined length not
equalling club.   Prothorax with bulged sides, much more strongly
punctate than usual in the genus, with a distinct longitudinal
furrow extending its entire length, a small and distinct impression on each side of middle.   Scutellum small, circular, within a
depression.   Elytra about once and one-third as long as wide,
feebly curved inwardly behind the shoulders, interstices narrow,
convex, transversely granulate.   Pygidium feebly carinate, seen
from the head appearing minutely mucronate.   Anterior femora
long, strongly toothed, tibiæ short, 3rd tarsal joint wide, claw
joint small but moderately distinct.   Length $2\frac{1}{8}$, rostrum $\frac{2}{7}$ (vix),
width $\frac{4}{7}$; range of variation 2-$2\frac{1}{4}$ mm.

♀. Differs in being slightly larger on an average; rostrum dull
red, tinged with piceous across its middle or apex; thorax tipped
at apex with red, elytra either entirely red or red with the sides
and apex black, sometimes with a transverse band at apical third
and piceous along suture, sometimes with four red spots (two near
apex and two near base), and occasionally with only two dull red
spots near the base; tibiæ and apical third of femora red; the
rostrum is slightly longer and narrower.

Hab.—Tamworth, Sydney, N S.W.

The short antennæ inserted so close to the eyes as to leave no visible space between them, the strongly bent scape, the unusually large 1st joint of funicle, and the distinct median groove on the prothorax render this species—despite the variable colour of the females—perhaps the most distinct of any in the genus. Resembling *variabilis* at first sight, the straight rostrum alone would distinguish it; the preceding species (which it resembles in miniature) has the antennæ inserted about the basal third.

LÆMOSACCUS RUFIPES, n.sp.

♀. Black; rostrum, antennæ and legs red. Pygidium feebly squamose at base.

Eyes moderately separated. Rostrum short, straight, shining, cylindrical, finely punctate. Antennæ inserted moderately close to eyes; scape short, curved, not twice the length of 1st joint of funicle, club very small. Prothorax rounded, a feeble impression at apex, continued but very feebly to near base, base with a sub-elliptic impression on each side. Scutellum small, triangular, scarcely in a depression. Elytra parallel-sided, about once and one-half as long as wide, interstices narrow, convex, scarcely granulate. Pygidium not carinate. Anterior femora with a small but rather distinct tooth, claw joint very small. Length 2⅛, rostrum ½ (vix); width ⅘ mm.

Hab.—Sydney, Galston, N.S.W.

An elongate parallel-sided species, somewhat resembling *instabilis*, but without a distinct median prothoracic line, and the antennæ not inserted at extreme base of rostrum though closer to it than usual. I have two specimens, both females.

LÆMOSACCUS GIBBOSUS, Pasc.; Mast. Cat. Sp. No. 5326.

This species was described from a male specimen; the female was described as *L. magdaloides* by the same author. I think it probable that the sexes of other species have received separate names. Of the above I have a pair taken *in cop.* The rostrum and colours of the legs are often subject to sexual variation; in some species the eyes are much closer to each other in the male than in the female, and the length of the anterior femora occasionally varies.

LÆMOSACCUS QUERULUS, Pasc., Mast. Cat. Sp No. 5334.

Mr. Pascoe has described only the female of this species; the male differs in having the rostrum thick, compressed, opaque, narrowing to apex, coarsely punctate and grooved for its entire length, or sometimes even carinate. I have numerous specimens from various parts of New South Wales and Swan River, the size ranges from 3 to 6 mm.; the elytral fasciæ are variable both in size and completeness; *L. narinus*, Pasc., is possibly a black variety.

LÆMOSACCUS AUSTRALIS, Boisd.; Mast. Cat. Sp. No. 5318.

I do not know how this species crept into the Catalogue, as Boisduval described it from New Guinea; and neither Pascoe nor Bohemann (the only two who have described Australian *Læmosacci*) mentions it as coming from Australia, though Pascoe compares several species with it.

LÆMOSACCUS CRYPTONYX, Pasc.; Mast. Cat. Sp. No. 5321.

In this species the clothing varies from pale yellow to dark orange; the size also is slightly variable. I have specimens from Bridgetown to Swan River.

LÆMOSACCUS DAPSILIS, Pasc.; Mast. Cat. Sp. No. 5322.

Mr. Pascoe doubtfully records this species from South Australia. I have specimens from Queanbeyan and Forest Reefs, N S W. The ♂ differs from the ♀ in being smaller, with a shorter and thicker rostrum, and the antennæ inserted much nearer the base than in the ♀.

L. ELECTILIS, Pasc.; Mast. Cat. Sp. No. 5323.

Hab —Whitton, N.S.W.

ʟ NARINUS, Pasc.; l.c. No. 5330.
Forest Reefs, Queanbeyan, N.S.W.

. NOTATUS, Pasc.; l.c. No. 5331.
Tweed River, N.S.W.

ʟ OCULARIS, Pasc.; l.c. No. 5332.
Forest Reefs, N.S.W.; Darling Rang

L. SUBSIGNATUS, Bohem.; l.c. No. 5336.
Tasmania (Simson's No. 2566).

ʟ SYNOPTICUS, Pasc.; l.c. No. 5337.
Forest Reefs, N.S.W.

e following tabulation of species kno
d as far as possible all characters .
n, where I do not know both sexes.

more or less noticeably curved.
rax with a distinct circular or elliptic impres
sion on each side at base.
almost touching.
othed above... ....

Prothorax with shining carina. . . . . *carinicollis*, n.sp.
Prothorax without shining carina.

Anterior femora edentate. .. . *synopticus*, Pasc

Anterior femora with small tooth.

Prothoracic impressions pubescent. *crucicollis*, n.sp.

Prothoracic impressions impubescent . *Pascoei*, n.sp.

Rostrum straight.

Scape inserted at extreme base of rostrum .. *instabilis*, n.sp.

Scape not inserted at extreme base of rostrum.

Form short and thick

Size very small . . . . *compactus*, n.sp.

Size larger.

Prothorax without basal impressions. . *dapsilis*, Pasc.

Prothorax with basal impressions.

Elytra more or less red. . . .. *querulus*, Pasc.

Anterior legs moderately long.

Anterior femora reaching apex of rostrum *longimanus*, Pasc.

Anterior femora not reaching apex of
rostrum. .... *subsignatus*, Boh.

Anterior legs short.

Feebly pubescent above . ... *funereus*, Pasc.?

Glabrous above.............................. *dubius*, n.sp.

Form rather elongate and subcylindrical.

Elytra and prothorax with distinct pubescence
forming patterns.

Claw joint moderately distinct . *obscurus*, n.sp.

Claw joint almost concealed.

Anterior tibiæ red . . .. *cryptonyx*, Pasc.

Anterior tibiæ piceous-black. . . . . *festivus*, n.sp.

Upper surface glabrous or feebly pubescent.

Elytra red... *rufipennis*, n.sp.

Elytra black.

Rostrum and femora red . ... *rufipes*, n.sp.

Rostrum and femora black.

Prothorax with a circular impression on
each side of disc . . . . .. *ater*, n.sp.

Prothorax without circular impression. . .. *gibbosus*, Pasc.

COSSONIDES

MASTERSINELLA, n.g.

Head small   *Eyes* small, prominent, coarsely granulate *Rostrum* cylindrical, parallel, elongate   *Antennæ* thick, funicle 8-jointed, club 3-jointed   *Prothorax* distinctly widest behind, distinctly longer than wide.   *Scutellum* small, distinct   *Elytra* slightly wider than prothorax, subcylindrical, apex acuminate. *Anterior coxæ* subapproximate, tibial hook sharp, very distinct, tarsi pseudo tetramerous.   *Body* fusiform, strongly sculptured, glabrous.

The eight-jointed funicle renders this genus at once distinct from any recorded by Mr. Wollaston, though, had specimens been before him, he might have considered it necessary to form a special group (as in *Notiomimetules*) to receive it. So far as I am capable of judging, its nearest Australian ally (although possessing a five-jointed funicle) appears to be *Microcossonus* (of which a species is herein recorded from New South Wales). Consequently I propose to treat it as an aberrant form belonging to the *Pentarthrides*.

MASTERSINELLA 8-ARTICULATA, n.sp

Dull red; rostrum and base of prothorax feebly tinged with piceous. Legs with feeble greyish pubescence. Head impunctate, a few coarse punctures between eyes, rostrum with coarse scattered punctures densest towards apex; prothorax with regular shallow punctures. elytra striate-punctate, the punctures large, shallow, approximate, tinged with piceous, interstices smooth   Under side of head feebly transversely strigose, sterna and alternate portions of abdomen with large shallow punctures

Rostrum once and one-half as long as head, feebly equally dilated towards apex, 1st joint of funicle wider than long, narrow at base, apex truncate, rounded outwardly, inwardly excavated. Prothorax subconvex, not once and one-half as long as wide, sides rounded, apex narrowed and feebly constricted, base feebly bisinuate. Elytra slightly wider than prothorax, parallel-sided to apical third.   Meta- twice as long as mesosternum, the two combined as long as abdomen.   Third tarsal joint strongly bilobed, entirely concealing true 4th joint except from below.   Length to eyes $1\frac{3}{4}$, rostrum $\frac{1}{3}$, width $\frac{1}{2}$ (six) mm.

Hab — N. Queensland (Mr. G. Masters), Barron Falls (Mr A. Koebele)   " In decaying timber."

HEXARTHROIDES, n g.

Head rather small.   *Eyes* small, prominent, coarsely granulate. *Rostrum* subcylindrical, parallel.   *Antennæ* moderately slender, funicle 6-jointed; club 3-jointed.   *Prothorax* widest across middle, longer than wide   *scutellum* almost invisible   *Elytra* subcylindrical, parallel, apex acuminate.   *Anterior coxæ* subapproximate, tibial hook distinct; tarsi pseudo-tetramerous, 3rd joint moderately bilobed.   *Body* elongate, narrow, strongly sculptured, feebly pubescent.

Although possessing a six-jointed funicle, I think this genus should go in with the *Cossonides* as limited by Mr. Wollaston. he himself places *Hexarthrum* (also with a six-jointed funicle) with them, and the present genus certainly cannot be placed with the *Onycholipides*.   I possess no Australian genus with which it can be satisfactorily compared, and from *Hexarthrum* it appears to differ widely.

HEXARTHROIDES PUNCTULATUM, n sp.

Narrow, subconvex.   Piceous-black; eyes brown, antennæ dull red, base of femora, apex of tibiæ and the tarsi tinged with red. Punctures with microscopic sparse pubescence, longest beneath Head feebly transversely strigose at base, it, the rostrum and prothorax with coarse dense punctures, elytra striate-punctate, the punctures coarse, approximate , under surface with strong regular punctures, head almost impunctate, and microscopically granulate, intermediate abdominal segments feebly and sparsely, apical more densely and strongly punctate, femora shallowly punctate and strigose.

Rostrum parallel-sided, except for a feeble dilatation to receive the antennæ.   Prothorax very feebly constricted near apex, and

with the head and rostrum elongate pear-shaped. Elytra much
wider than prothorax at base, but not much wider than across its
middle, parallel-sided to near apex, interstices very narrow.
Abdomen a little longer than meso- and metasternum combined.
Length to eyes $2\frac{1}{5}$, rostrum $\frac{1}{3}$; width $\frac{1}{2}$ mm

Hab—Galston, N.S.W.

MICROCOSSONUS PANDANI, n sp.

Subconvex. Dull red, antennæ and under side of head paler
Legs with feeble scattered pubescence Head both above and
below feebly transversely strigose, rostrum with shallow
punctures, prothorax with shallow, almost regular punctures
Elytra striate-punctate, the punctures large, shallow, approximate,
under surface with scattered large shallow punctures, and minutely
irregularly transversely or obliquely strigose, femora feebly
strigose

Scape feebly curved, slightly longer than the rest of antennæ,
1st joint of funicle longer than 2nd-3rd combined Prothorax
feebly constricted near apex, which is decidedly narrower than
base, base very feebly trisinuate Elytra feebly and equally
diminishing to apical third. Length to eyes $1\frac{1}{2}$, rostrum $\frac{1}{3}$ (vix),
width $\frac{3}{4}$ mm.

Hab.—Tweed and Richmond Rivers, N S.W.

Between decaying portions of the trunks and in old nuts of
Pandanus sp The species is moderately common and I have
taken both larvæ and pupæ, specimens of which are now in the
collection of the Department of Agriculture of New South Wales

STEREOBORUS LAPORTEÆ, n.sp

Cylindrical, shining, glabrous. Black or piceous-black, or piceous-
brown. Head, rostrum and prothorax densely punctate, elytra
punctate striate, the punctures large, subquadrate, interstices
convex, very sparsely punctate, under surface sparsely, sides of
sterna more densely punctate

Head broad; eyes indistinct, rostrum very broad, not much
longer than wide, feebly decreasing to apex, a feeble impression

between antennæ; antennæ short, scape curved, as long as funicle.
Prothorax slightly narrowed in front, as long as head and rostrum
combined, without trace of median line   Scutellum small, transverse.   Elytra parallel to near apex, suture slightly convex.
Sutures of intermediate abdominal segments very deep   Legs
short, anterior tibiæ fossorial.   Length 5, rostrum ½, width 1¼ mm

Hab —Clarence River, N.S.W

Numerous specimens taken from partly decayed trunks of the
large stinging tree *(Laportea gigas)*   The great number of closely
allied genera described by Mr. Wollaston renders satisfactory
determination of any but those with strongly marked features
somewhat difficult, and as this and the following species are at
least very close to *Stereoborus* (a species of which has already been
recorded from Australia) I have considered it advisable to place
them in that genus

STEREOBORUS INTERSTITIALIS. n sp

Elongate-elliptic, subconvex, shining, glabrous   Black, antennæ
and tarsi piceous.   Head (except base) and rostrum densely
punctate, prothorax less densely; elytra striate-punctate,
punctures moderately large, approximate, interstices flat, feebly
but distinctly punctate. sterna with moderately large regular
punctures, smaller on abdomen

Head wide, eyes moderately distinct, a small fovea between
them, rostrum short, broad, feebly dilating to apex, slightly
curved; antennæ inserted nearer base than apex of rostrum; scape
curved, as long as funicle and club combined: club short, obovate.
Prothorax constricted near apex, widest behind middle, with
feeble trace of median line   Scutellum small, transverse. Elytra
decreasing almost from base to apex, striæ deep at base, much
shallower towards apex. suture flat   Intermediate segments of
abdomen small, suture deep, apical segment feebly depressed in the
middle   Legs long. anterior tibiæ subfossorial.   Length 4¾,
rostrum ¾: width 1½ mm

Hab —Tweed River, N S W   Obtained under rotten bark

STEREODERUS MACLEAYI, n.sp.

Cylindrical, highly polished, glabrous. Black, antennæ piceous-red. Head and rostrum almost impunctate, mouth parts with long reddish hair, prothorax with sparse distinct punctures, sparsest towards base; elytra with regular rows of small distinct punctures, interstices flat, not punctate.

Head large, thick; eyes lateral, indistinct, a very feeble impression between them; rostrum very short, wider than long, antennæ inserted about middle of rostrum, scape very short, widening to apex, feebly curved. Prothorax about once and one-third as long as wide, feebly constricted near apex, which is slightly emarginate at its middle, and almost as wide as base. Scutellum distinct, subquadrate, within a depression. Elytra parallel to near apex, with an indistinct sutural stria. Intermediate segments of abdomen short, their sutures deep and wide. Legs very short, tibiæ strongly fossorial. Length $4\frac{1}{5}$, rostrum $\frac{1}{2}$; width $1\frac{1}{3}$, rostrum $\frac{3}{4}$ mm.

Hab.—Cairns, N.Q. (Macleay Museum).

Except for the shape of the prothorax this species agrees with Mr Wollaston's diagnosis of the genus *Stereoderus*, the base of the rostrum has three small tubercles immediately behind the long reddish hair with which the mouth is fringed.

COSSONUS INTEGRICOLLIS, n.sp

Broad, depressed, feebly shining. Head and prothorax black, elytra and scutellum dull brownish-red, the former tinged with piceous towards apex; under surface, legs and antennæ piceous-brown. Rostrum with dense small punctures, prothorax with large regular punctures except at apex where they are smaller, each elytron with about twelve rows of large, subquadrate punctures; interstices scarcely visibly punctate, about as wide as punctures; under surface densely punctate, punctures of sterna (especially of pro- and mesosternum) stronger.

Eyes lateral, distinct; rostrum narrow at base, suddenly widening to insertion of antennæ, parallel thence to apex; antennæ inserted about middle of rostrum, scape straight, as long as funicle, club short, obovate. Prothorax subconical, median line invisible on apical half, carinate towards base, base bisinuate. Scutellum small, distinct, circular, within a depression. Elytra wider than prothorax, parallel to apical third, interstices flat, scarcely raised (except posteriorly) Abdomen with a feeble depression at middle of 1st and 2nd segments, apical as long as two intermediate combined Legs long, femora (especially anterior) thickened   Length 4½, rostrum 1 (vix), width 1¾ mm.

Hab —Forest Reefs, N.S W.

Crawling over fences and logs at night time.

Cossonus impressifrons, n sp

Elongate, depressed, feebly shining, glabrous Piceous-black, under surface (except prosternum), legs and antennæ reddish-brown Head and rostrum densely punctate, the prothorax less densely but more strongly; elytra striate-punctate, punctures large, subquadrate, interstices scarcely visibly punctate, pro- and mesosternum with dense coarse punctures, on the mesosternum and two basal segments of abdomen they are smaller and somewhat irregular, intermediate segments sparsely punctate, apical densely and strongly

Head with a moderately large distinct fovea between eyes, rostrum moderately narrow at base, widening to insertion of antennæ, parallel thence to apex, flat, a groove continuous from ocular fovea almost to middle, where it distinctly terminates, from thence at the sides a feeble impression, scape straight, thickening to apex, as long as funicle, club obovate, as long as four preceding joints of funicle Prothorax with feebly bulged sides, an impunctate elevation extending almost from apex to base, with a depression on each side of it. Scutellum distinct, obtriangular, a feeble impression in its middle. Elytra wider than prothorax, feebly decreasing to near apex, alternate interstices feebly raised, all flat and rather narrow A depression extending from base of 1st to apex of 2nd abdominal segment. Legs moderately long, femora (especially anterior) thickened. Length 6¼, rostrum ⅔; width 1⅔ mm.

Hab.—New South Wales (probably from Sydney).

COSSONUS PRÆUSTUS, Redt., Mast. Cat. Sp. No. 5620.

Hab —N.S.W.; widely distributed.

PENTAMIMUS RHYNCHOLIFORMIS, Woll., l.c. No. 5615.

Hab.—Donnybrook, W A. In flowering stems of *Xanthorrhœa.*

P. CANALICULATUS, Woll ; l.c. No. 5614 .

Hab.—Tasmania (Macleay Museum).

ISOTROGUS BILINEATUS, Pasc ; l.c No 5621

Hab.—Cairns, N.Q. (Macleay Museum)

DESCRIPTIONS OF SOME NEW ARANEIDÆ OF NEW SOUTH WALES. No. 6. ·

By W. J. Rainbow.

(Plates XVIII -XX)

Family EPEIRIDÆ.

Genus N e p h i l a, Leach

Nephila ornata, sp.nov.

(Plate XVIII. figs. 1, 1a, 1b.)

♀. Cephalothorax 5 mm long, 4 mm. broad, abdomen 7 mm long, 4 mm broad

Cephalothorax dark mahogany brown, thickly clothed with silvery white hair, *caput* elevated, rounded on sides and upper part, deeply compressed at junction of cephalic and thoracic segments, two coniform tubercles at posterior extremity of cephalic segment *Clypeus* broad, moderately convex; a deep transverse groove at centre, indented laterally, indentations bare, transverse groove sparingly clothed with hoary pubescence.

Eyes glossy black, the four central eyes are seated on a moderately convex eminence and form an almost quadrangular figure, of these the two comprising the front row are somewhat closer together than the hinder pair; the lateral eyes are much the smallest of the group, and are placed obliquely on small tubercles, but are not contiguous.

Legs long, slender, yellow-brown, a few fine yellow hairs; *tarsi* dark brown. Relative lengths 1, 2, 4, 3, of these the second and fourth pairs are almost equal, and the third much the shortest.

Palpi rather short, somewhat darker than the legs, rather thickly clothed with short dark hairs.

Falces dark brown, conical, smooth, inner margin fringed with dark hairs, fangs much darker; the margins of the furrow of each falx armed with a row of three strong teeth.

Maxillæ dark at base; apex shiny, pale yellowish.

Labium longer than the base is broad, base and apex similar in colour to maxillæ

Sternum shield-shaped, straw colour, with small dark patches laterally.

Abdomen oblong, sinuous in outline, moderately convex, projecting over base of cephalothorax; superior surface dull yellowish, dark at anterior and posterior extremities, clothed sparingly with short silvery hairs, ornamented with a few dark spots, and from near the centre to anterior extremity with a network pattern of dark lines; sides and inferior surface dark brown, ornamented with a network of pale yellowish and uneven lines.

Epigyne a transverse oval, dark brown eminence, posterior lip more strongly elevated and convex than the anterior

Hab — Sydney.

(*Contribution from the Australian Museum.*)

NEPHILA PICTA, sp.nov.

(Plate XIX. fig. 1.)

♀. Cephalothorax 6 mm. long, 5 mm. broad; abdomen 11 mm. long, 7 mm. broad.

Cephalothorax shiny black, thickly clothed with silvery hairs; *caput* arched, clothed with silvery hairs, a few black shiny patches devoid of hairs; junction of cephalic and thoracic segments clearly defined; two shiny black coniform tubercles at base of cephalic eminence. *Clypeus* broad, slightly arched, clothed with silvery hairs, normal grooves distinct, black, shiny, and devoid of hairs; deeply indented at centre. *Marginal band* narrow, fringed with hoary hairs.

Eyes black; the four central eyes are seated on a moderately convex eminence, and form an almost quadrangular figure, the lateral pair are much the smallest, and are placed obliquely on small tubercles, but are not contiguous.

Legs long, slender, black, with broad yellow annulations, *trochanters* and *femurs* of first 2 pairs and femurs only of third and fourth pairs furnished at lower extremities with long black hairy plumes. *tibial joints, metatarsi* and *tarsi* black.

Palpi long, black, clothed with long black hairs or bristles

Falces black, arched in front, slightly divergent, a few short black hairs on inner margins, a row of three teeth on each margin of the furrow of each falx wherein the fang lies when at rest, fangs black

Maxillæ club-shaped, arched, outer margins black, inner margins shiny, yellowish.

Labium conical, rather longer than broad, black at base, shiny and yellowish at apex

Sternum cordate, longer than broad, surface uneven, black, with four small yellow lateral patches, a broad transverse curved yellow band at anterior part, and a small yellow patch at posterior extremity

Abdomen ovate, projecting over base of cephalothorax, superior surface sparingly pubescent, olive-green, spotted with yellow and ornamented with a network pattern of tracery, and two rather large yellow spots at centre, sides similar in colour to superior surface, inferior surface dark, ornamented with a broad wavy transverse yellow band situated just below epigyne, besides this there are three other transverse yellow lines seated lower down, the first of which is curved in a posterior direction, and the two others forward.

Epigyne dark, strongly arched, concave within.

Hab —Condobolin, N.S W.

Type specimen in the collection of the Australian Museum, to the Trustees of which Institution I am indebted for the privilege of describing it

Genus E p e i r a, Walck.

Epeira picta, sp nov.

(Plate xviii. figs. 2, 2a)

♀. Cephalothorax 3 mm. long, 2 mm. broad; abdomen 5 mm. long, 5 mm. broad.

Cephalothorax pale yellow. *Caput* elevated, rounded on sides and upper part, a few short fine pale yellow hairs in front and at sides. *Clypeus* broad, strongly convex; normal grooves indistinct. *Marginal band* narrow.

Eyes black; the four intermediate ones seated on a somewhat quadrangular protuberance, forming a square or nearly so, of these the pair comprising the first row are separated from each other by a distance equal to their individual diameter, those of the second by about one-half, and each row is separated from the other by about the diameter of one eye, lateral pairs much the smallest of the group, placed obliquely on small protuberances, and almost contiguous.

Legs moderately long and strong, pale yellow, armed with strong black spines, and sparingly clothed with short fine yellow hairs, relative lengths 1, 2, 4, 3.

Palpi short, pale yellow, clothed with fine yellow hairs, considerably longer than those of the legs.

Falces pale yellow, strong; the margins of each falx armed with a row of three teeth; fangs yellowish-brown.

Maxillæ pale yellow, arched, inner margins thickly fringed with yellow hairs.

Labium concolorous, broad at base, strongly arched, one-half the length of maxillæ.

Sternum cordate, yellowish-green, truncate in front, bare and uneven

Abdomen broad, ovate, overhanging base of cephalothorax strongly convex, green colour, with two large yellow spots, edged with dark brown towards anterior extremity; contiguous to each of these there is a much smaller yellow spot edged with dark brown; towards posterior extremity there is a network of fine dark and uneven lines, sides of a somewhat darker green than superior surface, underside olive green.

Epigyne an elevated eminence, the two openings, though sensibly separated, are connected at anterior part with a pale yellowish curved bar, immediately above the curved bar mentioned there is another bar larger, stronger, and much more arched than the first mentioned

Hab —New England District.

EPEIRA SIMILARIS, sp.nov.

(Plate XVIII. fig 3.)

♀ Cephalothorax 3 mm long, 2 mm. broad, abdomen 5 mm. long, 5 mm. broad

Cephalothorax pale yellow. *Caput* elevated, rounded on sides and upper part, a few short fine pale yellow hairs in front and at sides   *Clypeus* broad, strongly convex, normal grooves indistinct. *Marginal band* narrow

Eyes, legs palpi, falces, maxillæ, labium and *sternum* similar to *E ficta*

Abdomen broad, ovate, overhanging base of cephalothorax, moderately convex, green, with a broad transverse irregular patch of dull white towards anterior extremity, and which is broadest laterally, there are two large dark brown unevenly formed lateral patches so situated as to be surrounded by portions of the white patch referred to, in addition to these there are two small median depressions or dents, the depths of which are of a dark brown colour, from about the centre to the posterior extremity there is a network of fine uneven lines, sides green; under side dull green.

Epigyne an elevated eminence, the two openings more widely separated than in *E. ficta*, and not connected at anterior part, as in that species, with a curved bar; above the openings, and slightly overhanging them, there is a large strong arched bar as in the former species.

Hab.—New England District.

EPEIRA WAGNERI, sp nov

(Plate XIX. figs. 2, 2a, 2b, 2c, 2d)

♀. Cephalothorax 5 mm. long, 4 mm. wide; abdomen 6 mm. long, 5 mm. wide.

Cephalothorax yellow-brown. *Caput* elevated, rounded on sides and upper part. *Clypeus* broad, convex, normal grooves indistinct; a deep transverse cleft at centre. *Marginal band* narrow, black.

Eyes black; the four central eyes forming a square or nearly so, front pair separated from each other by about one eye's diameter, second pair by a distance equal to about three-fourths of their individual diameter; lateral pairs seated obliquely on tubercles, much the smallest of the group.

Legs long, strong, clothed with short black hairs and spines; coxæ pale straw colour, *trochanters* with lower half pale straw colour, the remainder reddish-brown, *femurs*, *tibiæ* and *tarsi* reddish-brown Relative lengths 1, 2, 4, 3.

Palpi long, similar in colour and armature to legs.

Falces reddish-brown, shiny, inner margin fringed with short hairs, the outer margin of the furrow of each falx armed with three teeth, and the inner two, fangs strong, dark brown.

Maxillæ yellow-brown, convex exteriorly, a thick fringe of short black hairs on inner margins, a few long black ones on the outer margins

Labium broad, half the height of maxillæ, rounded off at apex

Sternum shield-shaped, dark brown, lighter at the middle; surface uneven.

Abdomen oblong, convex, slightly projecting over base of cephalothorax; upper surface mottled yellow and brown; at anterior extremity two large dark and brown patches laterally; four rather deep indentations at the centre; a large leaf-like design, darkest at its outer edges, runs the entire length of the upper surface; sides mottled dark brown and yellow, with green markings, inferior surface yellowish, with dark brown patches.

The males of this species are pigmies in comparison to the females, but are exactly like them in colour and formation. The sexes pair during January and February, and live together in the same nest during that period. A more detailed account of their nidification, &c., will be found in another part of this paper. I have much pleasure in dedicating this species to my esteemed contemporary and correspondent, Professor Waldemar Wagner, of Moscow, who has published an admirable work, " L'Industrie des Araneina," in the " Mémoires de L'Académie Impériale des Sciences de St. Pétersbourg vii⁰ Série Tome xlvii No. 11."

Hab —Sydney.

Family LYCOSIDÆ

Genus D O L O M E D E S, Latr.

DOLOMEDES NEPTUNUS, sp.nov

(Plate XVIII., figs. 4, 4*a*.)

♀ Cephalothorax 4 mm long, 3 mm. broad, abdomen 3 mm. long, 5 mm. broad.

Cephalothorax pale yellowish, strongly convex, clothed with pale yellowish pubescence, normal grooves and indentations indistinct. *Marginal band* broad.

Eyes black; front row smallest of the group, and slightly procurved, middle eyes somewhat larger than their lateral neighbours, all equidistant, eyes of second row large, separated by a space equal to once their individual diameter; third row large, separated from each other by four diameters.

Legs strong, moderately long, pale yellowish; clothed with yellowish pubescence, and short, strong black spines. Relative lengths 4, 1, 2, 3.

Palpi moderately long, similar in colour and armature to legs.

Falces slightly divergent, strong, pale yellowish, clothed with pale yellowish pubescence, arched in front; a row of three black teeth along the margins of the furrow of each falx, those on the underside seated much nearer to the apex than those of the upper margin, fangs long, dark brown

Maxillæ long, arched in front, inclining inwards, thickly clothed with pale yellowish pubescence.

Labium half as long as maxillæ, coniform, arched in front, pale yellowish, thickly clothed with yellowish pubescence.

Sternum elliptical in outline, dark brown, shiny, clothed with yellowish pubescence.

Abdomen oblong, pale yellow, slightly projecting over base of cephalothorax, clothed with yellowish pubescence, and ornamented with dark brown spots, flecks, and at posterior extremity a rectangular figure, sides and inferior surface pale yellowish with yellow pubescence.

Epigyne a curved transverse slit.

Hab —The shores of Port Jackson.

DOLOMEDES SPINIPES, sp.nov.

(Plate XVIII., fig. 5)

♀. Cephalothorax 3 mm. long, 2 mm. broad, abdomen 4 mm. long, 2 mm. broad

Cephalothorax pale yellowish, convex, clothed with coarse yellowish hairs, normal grooves and indentations indistinct *Caput* elevated, rounded on sides and upper part, shiny, a few long coarse hairs at sides and in front. *Marginal band* broad.

Eyes black, front row smallest of the group, slightly procurved, middle eyes somewhat larger than their lateral neighbours, all equidistant, eyes of second row large, separated by a space equal to once their individual diameter, third row same size as those of the second, but separated from each other by four diameters.

Legs moderately long, strong, yellowish, thickly clothed with coarse yellowish hairs, and on upper sides of trochanters and femurs short, strong black spines, on the under sides of these joints long, strong black spines, *tibial and tarsal joints* furnished above and below with long, strong black spines   Relative lengths 1, 4, 2, 3.

Palpi moderately long, similar in colour to legs, clothed with long, coarse yellowish hairs

Falces slightly divergent, strong, pale yellowish, clothed with pale yellowish hairs, longest on the inner margins, arched in front, a row of three black teeth on each margin of each falx, fangs long, strong, dark brown

Maxillæ pale yellowish, long, arched in front, clothed with long, coarse, pale yellowish hairs

Labium pale yellowish, shiny, half as long as maxillæ, broad, rounded off at apex, a few long yellowish hairs, a thick fringe of long hairs at under side of apex

Sternum shield-shaped, pale yellowish, thickly clothed with long yellow hairs

Abdomen oblong, ovate, moderately convex, slightly projected over base of cephalothorax. superior surface, sides and inferior surface pale yellowish, thickly clothed with long, coarse, yellow hairs.

Epigyne a curved transverse slit, the curvature directed forwards

Hab —The shores of Port Jackson

Family MYGALIDÆ

Genus ACTINOPUS, Klug.

ACTINOPUS FORMOSUS, sp.nov.

(Plate XX)

♂. Cephalothorax 4 mm. long, 5 mm broad, abdomen 4 mm. long, 2 mm broad at base, 4 mm at posterior extremity

Cephalothorax broad.   *Caput* broad, high, strongly arched, truncate in front, bright red, junction of cephalic and thoracic segments sharply defined.   *Clypeus* broad, blue-black, moderately convex, normal grooves and indentations fairly distinct   *Marginal band* broad.

Eyes arranged in three groups; central pair dark, shiny, seated on a slightly raised dark brown eminence, and separated from each other by a space equal to once their individual diameter; lateral eyes in groups of three, each group forming a triangular figure; the front lateral eyes are sensibly the largest of the eight; the inner eyes of the triangular figures are the smallest of the group, and are of an opaline tint with black rings.

Legs long, strong, shiny, dark brown, almost black, furnished with rather long, fine black hairs, and few short stout spines. Relative lengths 1, 2, 4, 3.

Palpi long, strong, similar in colour to legs, and furnished with long black hairs; fifth joint much the strongest; copulatory organs tinged with red, directed backwards, spiral at base, tapering, and terminating with a long strong spine, the spine directed outwards in a horizontal position.

Falces long, strong, bright red, strongly arched, divergent at apex, where they are furnished with long coarse black hairs; fangs long, shiny, reddish-brown.

Maxillæ red, long, broad at base, tapering outwards to a point, arched in front, inner margins clothed with long coarse black and white hairs or bristles

Labium red, strongly arched, longer than broad, conical, fringed with black hairs at apex.

Sternum somewhat elliptical, red in front, darker laterally; dark brown, with reddish-brown lateral indentations towards junction with abdomen; a deep indentation in front under labium.

Abdomen triangular, slightly projecting over base of cephalothorax, broadest at posterior extremity; dark brown, nearly black, thickly clothed with long coarse hairs; a long, rather deep indentation runs down the abdomen from near its anterior to the posterior extremity, where it is slightly indented; sides and inferior surface similar to superior.

Hab—Menindie, N.S.W.

This species is the first of its genus recorded from Australia, and is consequently of more than ordinary interest. The spider was captured by Mr. A. G. Little, Railway Surveyor, Menindie. I

22

am indebted to Mr. Henry Deane, M.A., for the privilege of describing this species.

Of the eight species described in the present paper, five of them (*Epeira ficta*, *E. similaris*, *E. wagneri*, *Dolomedes neptunus.* and *D. spinipes*) are especially interesting from the fact that they, in common with hosts of other animals, are protected from the raids of predatory foes either by colouration or mimicry   Rambling along our sea beaches certain small spiders are occasionally found lurking amidst the masses of small and broken shells denoting high water mark, and corresponding so accurately in colour to the sea-wrack referred to, that it is utterly impossible to detect them unless they are in motion. and not only is this so, but their habit of feigning death, upon the approach of what they suspect to be danger, adds greatly to the deception. Of these, *Dolomedes neptunus* and *D. spinipes* are instances in point.

One day last summer, while helping my boys to gather some shells at Taylor Bay, Port Jackson, I discovered one of the spiders referred to (*D. neptunus*).   In endeavouring to catch it, it eluded me in the manner described, and so success-fully that it was only by probing the shells and pebbles until my forceps touched "something soft" that I succeeded in making my capture   Throughout the entire range of natural history there is no chapter more replete with interest than the marvellous provision of Nature in clothing her subjects, not otherwise protected, with colours identical with their. surroundings, thus enabling them not only to baffle the vigilance of their foes, but also by natural disguises to aid them in successfully stalking their prey. Numerous and extraordinary are the disguises assumed, and although many have been recorded and described, much work yet remains to be done.   This will require the exercise of much patient observation and labour, and will be of immense value to science.

Many spiders that are exceedingly conspicuous while resting in their webs are practically hidden from view when sheltering among leaves and twigs, the hues of which harmonise exactly with their own. All shades of green, brown, and grey are found among arboreal individuals. Mr. Arthur Lea gave me a number of spiders collected by him both in the New England district and at Queanbeyan, among which there are examples, not only coloured like withered leaves, but some are green and marked with mock-holes (as in *Epeira ficta*), and others with discoloured patches on their surface, having the appearance of leaves attacked by some insect (as in *E. similaris**) Quite a host of examples, both of spiders and beetles, whose colouration is protective, may be obtained by shaking a branch of any shrub over an inverted, open umbrella. Among the species whose haunts are confined to the ground, and those that ramble among rocks, the same rule obtains, the former harmonising with the colour of the soil, while the latter reflect not only the various tints of the rocks, but frequently mimick the lichens growing upon them.

Mr. C. M. Weed says that the Ash-Grey Harvest Spider, *Phlangium cinereum*, Weed, "is pre-eminently what may be called an indoor species. It abounds especially in sheds, out-houses, and neglected board piles, being rarely found . . . in the open field. Its colour especially fits it for crawling over weather-beaten boards, making it inconspicuous against such a background. During the day it is usually quiet, but at dusk and on cloudy days it moves about quite rapidly."†

Governed by the law of natural selection, the tints of animals frequently undergo certain modifications in order to suit them to altered conditions of surroundings‡. In tracts of bush that have been visited by fire, we find specimens so closely resembling the

---

* Writing upon the subject of his observations at Pera, Mr. H. W. Bates observes :—"The number of spiders ornamented with showy colours was somewhat remarkable. Some double themselves up at the base of leaf-stalks, so as to resemble flower-buds, and thus deceive the insects on which they prey." "The Naturalist on the River Amazon," p. 64.

† "American Naturalist," xxvi. p. 33.

‡ See Wallace's "Tropical Nature," pp. 167-172, for some interesting facts under this head ; also paper by Mr. R. Meldola, on "Variable Protective Colouring in Insects." Proc. Zool. Soc. Lond. 1873, p. 153.

charred branches or bark that when motionless it is utterly impossible to perceive them.* In some species the modification is very gradual, while in others the change is more rapid. An American author, Mr. J. Angas† states that when he placed a white variety of what he terms the "little flower spider" on a sun-flower it became quite yellow in from two to three days.

The habit of lying motionless when alarmed is common among sedentary spiders, such as the *Epeiridæ* and *Theridiidæ;* but it is badly developed in some and entirely absent in others of the jumping and swift-running species. Among the orb weavers the *Gasteracanthidæ* are singularly and effectively protected against the raids of insectivorous birds. Resting in the centre of their orbitular snares, fully exposed, the need of a protective armature is obvious, and this is afforded by their hard, horny and spiny abdomens. Likewise, the spines of *Acrosoma*, rendering the spiders similar in appearance to thorny leaves, knots of shrubs, acacias, &c , are also protective, and make these animals decidedly objectionable to insectivorous birds and reptiles. As in the case of the *Gasteracanthidæ*, the spiders of the genus *Acrosoma* also construct their webs in exposed situations, and sit fearlessly in the centre of the snares as though conscious of their security from attack.

In many instances specimens, when viewed in the cabinet, would not be likely to suggest the idea that their form and colouration are protective, yet when observed in the midst of their natural surroundings the fact that such is the case is forced upon the observer Again, some specimens lose their natural colours when placed in spirit. This is the case with *Epeira*

---

* Mr A T Urquhart in an interesting paper observes that "The generality of spiders found amongst burnt manuka, before it has become bleached, have the brownish-black colour of their environment, which causes them to be almost imperceptible at a very short distance."—"On the Protective Resemblances of the *Araneidea* of New Zealand," Trans. N.Z. Inst. Vol. xv. 1882, p. 175.

† "American Naturalist," xiv. p. 1010.

wagneri. Attus volans, Camb., on the other hand, redisplays all its brilliancy when taken out of the tube and the spirit has evaporated from its body.

The long attenuated bodies of the *Tetragnatha*, of which *T. cylindrica*, Koch, and *T. lupata*, Koch, each found in the vicinity of Sydney, are admirably adapted for concealment. These spiders when alarmed seek refuge upon the stems or branchlets of shrubs, and so closely do their tints agree with their surroundings that detection is exceedingly difficult *Epeira higginsii*, described and figured by Koch, and recorded by that eminent author from Darling Downs, but whose range extends far south of Sydney, is a singularly interesting example as far as its form is concerned; but in addition to that, its colouration and powers of mimicry are admirably adapted as a shield and protection. When disturbed it runs out of its snare to one of the supporting lines or guys, and there remains suspended, with its legs doubled up, the exact imitation, both in form and colour, of an autumn leaf. Writing to me upon the subject of protective colouration in spiders, my esteemed correspondent and contemporary, H. R Hogg, Esq., M.A., of Cheniston, Upper Macedon, Victoria, says :—" With regard to the protective colouring of spiders, I have frequently been asked if they have not sometimes the power of changing colour like chameleons in accordance with their surroundings. I must confess that all I have seen tends to show exactly the opposite, and that while many, if not most, are paler in their earlier stages, they get darker as they grow older. This is especially noticeable in laterigrades. The colouring matter of spiders, both in skins and hairs, is of a particularly lasting character, and even in spirits takes a long time to fade,* so that it would probably take a good many generations to alter the generally characteristic colouring of different species so as to conform to particular soils or vegetation. At the same time I

---

* I have spirited numerous specimens of *E. wagneri*, and not one retained its bright green and yellow colours two or three hours after emersion.—W.J.R.

have found a delicately-tinted green *Epeïra* on the similarly coloured green leaf of a lily, and a friend recently told me he had found a very brightly coloured yellow spider (which he did not bring me) on a yellow Cosmos flower."

Not only do spiders, in addition to colouration, possess the faculty of mimicry as a protection against birds, reptiles, &c., but their cocoons in some instances are also protected. The cocoon of *Epeira herione*, Koch, is made of withered leaves closely bound together, and suspended to one of the supporting lines or guys above the orbitular portions of the mesh, and looks more like a discoloured mass of rubbish rather than a nest containing eggs Writing "On the History and Habits of the *Epeira Aurelia* Spider,"[*] Mr Frederick Pollock remarks :—" The favourite haunt of *E aurelia* is the prickly pear—a plant from which the cocoon can scarcely be distinguished in colour, and so close is the resemblance that the first time I saw one of these cocoons, I could hardly believe that it was not a withered piece of the cactus " Anton Stecker also records a case of protective resemblance in the nest of an *Epeira* at Sokna (Tripoli),[†] covered with *débris* and the elytra of beetles, &c, and Odewahn [‡] obtained at Gawler (South Australia) some globular spiders' cocoons, found on branches of trees, and resembling the fruit of *Leptospermum*, the spiders of which were hanging near them, and resembled the excrement of some bird in appearance, a wonderful form of mimicry to which I shall presently have occasion to refer.

In *Cyrtarachne caliginosa*, recently described and figured by me,[§] we have, indeed, an extraordinary form It is well known that hairy caterpillars are exceedingly distasteful to birds: consequently it is only reasonable to assume that the long hairs upon

---

[*] Annals and Magazine of Nat. Hist. 3rd series, Vol. xv., p 459; June 1, 1865.

[†] Mittheilungun der africanischen Gesellschaft in Deutschland, ii pp. 78-80

[‡] Proc. Ent. Soc. 1864, p. 57.

[§] P L S N S.W. Vol. ix. (2nd series) pp 154-157; pl x. figs. 2, 2*a*, 2*b*

...t move w .. e jarred .. ｰ ..., ,

f movement when transferred to the cy

of *C. multilineata* were also described

ect galls. *Epeïra wagneri* is a comm

nd Sydney. It is brightly coloured wi

rs admirably adapted for concealment ᴠ

ʳeb, and seeks shelter among the coa

do when alarmed. It is chiefly interesti

f its web and leaf nest. The web iˌ

shape does not form a complete orb.

es from which the mesh depends, a

nd obliquely, and from the centre of th

e directed. The irregular lines at the

somewhat resemble the architecture o

The leaf-nest is placed at the base fro

d in this, during the period of mating

other periods the female is the only t

ᵐonly used is that of a Eucalypt, whi

red shape according to the leaf use

arrow leaf is rolled spirally, and a br

the edges being tightly bound down w

localities where Eucalypts are not abundant, other leaves are used, and those of *Lantana camara* are not uncommon.

The interior of these nests is beautifully lined with silk.   The cocoon is attached and suspended among the supporting lines on one side of the web; it consists of a Eucalypt leaf doubled over so that the tip and base nearly meet   The eggs are deposited inside the folded leaf, and then it is sealed up firmly and tightly, the female mounting guard during the period of incubation.   At Waterfall and Fairfield, I have met with another species of *Epeira* (at present undetermined) that constructs a mesh and makes a leaf-nest like the one just described.

Among the *Thomisidæ* there are some interesting examples of protective colouration and mimicry.   Two spiders found within the vicinity of Sydney, but whose range extends both to the northern and southern colonies, namely *Celænia excavata*, Koch, and *Thlaosoma dubium*, Cambr., mimick the excreta of birds.

When awaiting their prey these spiders lie on their backs, and in this position their appearance suggests that of a bird's dropping, the denser part of the body on the underside being of a chalky colour, spotted and streaked with dark markings; then, too, the legs, owing to their colour and being closely pressed up to the body, add greatly to the deception.   In addition to all this a little loose silk is spun over a portion of the surface of a leaf, in the centre of which the spider lies; this completes the deception as it resembles the more liquid portions of the fæces running off the leaf, and thickening at the edge as it trickles over.   The deception is just as complete as could well be imagined   No one looking at either one or the other of these spiders in the situation described would ever imagine, unless previously aware of the fact, that an animal lay before them patiently awaiting the descent of some unwary insect in quest of food, yet such is the case.   These spiders hold themselves in position by inserting the strong spines with which their legs are armed, under the loose silk referred to.   *C. excavata* makes a nest of dead, brown leaves; the cocoons or egg-bags vary in number.   Mr. F. A. A. Skuse recently showed me a living

specimen that had been forwarded to the Australian Museum
from Cavendish, in the Western District of Victoria; it was a
female, and was mounting guard over exactly one dozen egg-bags.
The cocoons are spherical, uniform in size, somewhat brittle, and
in appearance resemble the kernels of the Quandong (*Fusanus
acumina'nu*). Mr. H. O. Forbes, F.R.G S.,* discovered a like
case of mimicry in Java, but his book is so well-known that it
would be superfluous here to recapitulate the facts as communicated
by him It need only be noted, therefore, that the species discovered by him formed the type of a new genus, *Ornithoscatoides*,
Camb. Mr. G. F. Atkinson also notes a case of mimicry† by a
small spider of this family—*Thomisus aleatorius*, Hentz. This
species is very common on grass, to the summit of the culms
of which it climbs, where, clinging with its posterior legs
to the stem and its anterior legs on each side approximated
and extended outwards, it thus forms an angle with the
stem, strikingly similar to that formed by the spikelets.
The genus *Stephanopis*, Cambridge, is another group of remarkable spiders. By the form and arrangements of their legs, which
are laterigrade, they can move forwards, backwards, or in a lateral
direction with facility. They are generally found lurking under
loose bark, or among the rugulosities of trees. Their colour and
rugged appearance—closely resembling bark—not only shield
them from the raids of enemies, but aid them in the capture of
prey, which they take either by stealth or pursuit. The colouration and ornamentation of the genus *Cymbacha* are also protective.
These spiders also have laterigrade ambulatory limbs. They are
found in similar localities to the *Stephanopis*. *C. festiva* and *C.
saucia* are found both in Queensland and New South Wales, and
each has been found in the vicinity of Sydney. While upon the
Laterigradæ, I must not omit to mention those of the genus
Voconia, Thor. These huge uncanny spiders are common enough

---

* A Naturalist's Wanderings in the Eastern Archipelago, pp. 63-65, and
a figure.
† American Naturalist, xxii. pp. 545, 546.

in the bush around Sydney, as well as in the interior. If a piece
of loose bark be stripped off the trunk of a tree, or from a decaying log, several of them may be seen scampering off with great
rapidity. Representatives of this and allied genera are also to
be found lurking under stones These spiders have large, flat,
hairy bodies, and remarkably long legs, and so are well adapted
to the situations in which they are found, while their general
dull colour harmonises to a nicety with their surroundings.
Although the superior surface of the abdomen of some of these
spiders is ornamented to a certain degree, their appearance nevertheless is hardly such as could be expected to inspire confidence
Bushmen have a deep-seated horror of them, and state that the
results of their bite is not only painful, but exceedingly dangerous.
V. immanis, *V. dolosa*, and *V. insignis*, each of which is described
and figured by Koch in his admirable work, " Die Arachniden
des Australiens," are to be found in the bush, not only in the
vicinity of Sydney, but also at Brisbane and Rockhampton. In
a small collection forwarded to me by Dr. Roth, from Winton,
Central Queensland, there were specimens of *V immanis* and *V.
dolosa*, which, he informs me, he captured in his house.

The obnoxious odours and flavours of some insects, as in those
butterflies of the *Heliconii* and *Danavlæ*, render them safe from
the raids of natural enemies. Thus Mr. Belt, in his delightful
work,* states that when he tried to feed his pet monkey with
some of the former, though he (the monkey) would take them
when offered, and sometimes smell them, he would invariably roll
them up in his hand, and drop them quietly again in a few
minutes; also, whenever he placed any of the *Heliconii* in the
web of a species of *Nephila*, the spider would drop them out,
although another species of *Araneidæ* seemed fond of them.

It has been observed by naturalists working in different parts
of the world that some species of *Attidæ* are remarkable for their

---

* " The Naturalist in Nicaragua," pp. 316, 317.

mimicry of ants. Bertkau* has recorded the fact from Prussian-Rhineland and Westphalia; Walsh,† from Bengal; Bates,‡ and Peckham,§ from the United States; Belt,‖ from Nicaragua; Mansel Weale,¶ from Africa; Rothney,** from Barrackpur; besides other authors. The ants that are chiefly mimicked by spiders are those that live on trees or shrubs. Owing to their powers of biting, their acrid secretions which they can eject to a considerable distance at an approaching enemy, the obnoxious odours emitted, their dwelling in communities, and fighting battles in a united body for the common good, they are admirably protected from birds and small animals that prey upon insects. This being so, those spiders that mimick them and wander about their haunts must enjoy an almost absolute immunity from dangers that beset solitary wanderers. The *Attidæ* do not spin webs for the capture of prey, but take their victims by stealth, stalking them, and springing upon them from behind. So great is the resemblance of these *Attidæ* to the ants that experienced collectors viewing them when alive are frequently deceived.†† Not only does the colour harmonise with that of the insect mimicked, but the

---

* "Ameisenahnlichkeit unter Spinnen," &c, Verhand. des naturhist. Vereines der Preussischen Rheinlande und Westfalens (Bonn), xliii (1886), pp 66-69. Bertkau also notes in the same paper that certain *Drassidæ* mimick ants, more particularly the genera *Phrurolithus* and *Micaria*. Among the *Thomisidæ* and *Epeiridæ*, he observes, this kind of mimicry is unknown; but the *Theridiidæ* furnish a beautiful example in *Formicina mutinensis*. On elms infested by *Lasius* and *Formica* a species of *Lasæola* occurs, the male of which alone resembles ants.

† Journal of the Asiatic Society of Bengal, 1891, No. 1, pp. 1 4

‡ Trans. Linn, Soc. Vol. XXIII.

§ Papers of the Nat Hist. Soc. Wisconsin, 1892, pp. 1-83.

‖ "Naturalist in Nicaragua," p 314.

¶ "Nature," Vol. III p 508.

** Journal of the Bombay Nat. Hist. Soc Vol. v. p 44.

†† Mr. W. W. Froggatt informs me that a small black *Chalcid* on the tree trunks at Mosman's Bay mimicks a small jumping spider, and was taken by him as a spider.

contour of the body and the manner of carrying the first pair of legs, so as to appear like antennæ, and which, ant-like, they keep in motion when running about, make the deception complete All observers, whose works I have consulted, with the exception of Dr. E. G Peckham, are unanimous in their testimony as to the manner in which these ant-mimicking *Attidæ* carry the first pair of legs Of those species I have observed mimicking ants each carried the first pair of legs in imitation of antennæ. But Dr. Peckham says that an American species *(Synageles picata)* "holds up its second pair of legs to represent antennæ " Tull Walsh considers that this peculiarity of habit may be accounted for by a difference in the relative lengths of the legs, although another American species *(Synemosyna formica)* observed by Peckham[*] to use its second pair of legs in imitation of antennæ has the usual formula of legs—4, 1, 3, 2.

Tull Walsh in an interesting paper[†] says :—" I have noticed that the spiders are probably protected from birds and other enemies by their resemblance to ants, but there can be no doubt that frequently they also thereby gain another very considerable advantage The ants with which these spiders most do congregate are fairly omnivorous feeders, but show a decided preference for sweet juices often to be found exuding from trees, fruit, or flowers To these juices come also flies, small beetles and other insects which form the natural prey of the spiders, and which do not, under the circumstances, particularly fear the ants Thus while the flies are sucking up sweetness in company with the ants, the spider is no doubt able under its disguise to approach near enough to make a spring upon the unsuspecting victim, and to fix his sharp falces into its body. As regards the ants themselves, they do not seem to take any notice of the spiders, and do not apparently attack them." It would be absurd to suppose that spiders delude the ants by their disguise, on the contrary, it is

---

[*] " Protective Resemblance in Spiders." Papers of the Nat. Hist. Soc. Wisconsin, 1892, pp. 174-76.

[†] Journal of the Asiatic Soc. of Bengal, 1891, No. 1, p. 4.

more reasonable to assume that the disguise is solely for the purpose of shielding them from the attacks of insectivorous foes and
enabling them to stalk their prey. So far as these spiders are
concerned (the ant-like *Attidæ*), the ants have little to fear from
them; and, although I have watched closely on numerous
occasions, I never yet saw an ant attacked by a spider. Indeed,
their natural ferocity, hardness of body, and faculty of combining
to withstand assault, would tend to show that spiders were more
likely to be attacked by ants than that the ants would be
attacked by spiders. This view was held by Mr. Belt, who
observed :—" The use that the deceptive resemblance is to them
has been explained to be the facility it affords them for approaching ants on which they prey. I am convinced that this explanation
is incorrect so far as the Central American species are concerned.
Ants, and especially the stinging species, are, so far as my
experience goes, not preyed upon by any other insects. No
disguise need be adopted to approach them, as they are so bold
that they are more likely to attack a spider than a spider them.
. . . Their real use is, I doubt not, the protection the disguise
affords against insectivorous birds. I have found the crops of
some humming birds full of small soft-bodied spiders, and many
other birds feed on them. Stinging ants, like bees and wasps,
are closely resembled by a host of other insects; indeed, whenever
I found any insect provided with any special means of defence, I
looked for imitative forms, and was never disappointed in finding
them."[*] Among the Australian *Attidæ* that mimick ants are
Synemosyna lupata, Koch, recorded from Port Mackay, *Leptor-
chestes striatipes*, Koch, and *L. cognatus*, Koch. These two latter
species occur in the vicinity of Sydney. I have in my possession,
from various parts of New South Wales, several undetermined
species of *Attidæ* that mimick ants, and which will hereafter
provide material for description.

The late Mr. F. A. A. Skuse informed me of a remarkable
example of the mimicry of a dipterous insect by a spider
(undetermined, but probably an *Attid*) that came under his notice

---

[*] " Naturalist in Nicaragua," pp. 314, 315.

at Thornleigh.   Both spider and fly were equal in size, small, and
brightly coloured, the thorax bright red, and the abdomen bright
green; the tips of the tarsi of the spider were white like the tips
of the wings of the fly, and each were found on the bracken
(*Pteris aquilina*, var. *esculenta*).   When in want of a meal the
spider throws up two legs on each side of its body, loops them
together by hooking the tarsi, and beats the air vigorously, the
result being that the light striking through the loops gives the
appearance of a pair of bright transparent wings in rapid motion,
and the fly, evidently convinced that it is one of its friends,
alights, only to fall a victim to a remorseless enemy.   Mr. Skuse
also informed me that the spider in question is capable of jumping
a considerable distance—not less than six inches, and that when
in the air it has the appearance as if flying.*

Summary —Now it has been abundantly proved by Poulton,
Beddard, Wallace, Darwin, and others, that colouration and
mimicry in animals play an important and essential part either
for *protection* against natural enemies, as a *warning* to others, or
attraction for prey, and the more they are studied, and their life
histories investigated, the more clearly do we understand why the
tints of some animals are so bright and glaring, and others so
dull and sombre   After much patient work and investigation,
and the collection of a vast array of facts such as I have
enumerated, but which included observations from a far wider
field in animated Nature, Wallace divided living organisms into
five groups in his classification of " Organic Colours,"† namely:—

Animals.
1.—Protective colours
2.—Warning colours
   (*a*) of creatures specially protected.
   (*b*) of defenceless creatures mimicking *a*.
3.—Sexual colours.
4.—Typical colours.

---

* *Attus volans*, Camb , the " Flying Spider," which so far has only been
found at Sydney, is small and exceedingly bright.

† "Tropical Nature," p. 172.

Plants.—5.—Attractive colours.

For the purposes of this paper it will suffice to divide the *Araneidæ* into two groups, namely:—

1.—*(a)* Protective colouration, and *(b)* formation. .

2.—Spiders that mimick: *(a)* animate and *(b)* inanimate objects, and *(c)* whose colours are attractive.

Protective Colouration and Formation.—In the course of my remarks, I have drawn attention to the fact that certain spiders are protected by the uniformity of their colouration to surrounding objects. Thus we have seen that while the colour of one spider harmonises with that of the small and broken shells on our sea-beaches, another group *(Stephanopis)* finds shelter by its close resemblance to the bark of trees; then again, there are others whose physical formation is protective, and of such are the genera included in the subfamily of *Gasteracanthidæ*, whose hard, horny, and generally spiny epidermis make them anything but tempting morsels for insectivorous birds.

Spiders that mimick animate and inanimate objects, and whose colours are attractive.—This group contains those spiders whose protection is secured, or who capture their prey by the mimicry of animate and inanimate objects, and in this class we have the extraordinary case of mimicry reported by Mr. Skuse, in which, by the elevation of one pair of legs on each side of its body, looping them together by the tarsi, and beating them rapidly up and down, a certain species of spider, in addition to its colouration, adds that of the mimicry of a pair of wings, and thus secures as prey a certain dipterous insect. Again, there is the no less wonderful mimicry by certain spiders, even to the most minute detail, of birds' droppings—a form of mimicry that not only secures them from the raids of their common enemies, but also attracts those insects upon which they prey.

Conclusion.—Taken collectively, these facts add an important link to the great chain of evidence upon which the law of natural selection is based and built. Much more might be added, but sufficient has been given to illustrate the great truths comprised in that law. I am indebted to my colleague, Mr Edgar R Waite, for the admirable coloured drawing of *Actinopus formosus*, which has been reproduced in Plate xx.

EXPLANATION OF PLATES

PLATE XVIII.

Fig. 1. —*Nephila ornata* ♀.
Fig 1a.—   ,,   ,,   abdomen in prɔfile.
Fig. 1b —   ,,   ,,   Epigyne.
Fig. 2. —*Epeira ficta* ♀.
Fig. 2a.—   ,,   ,,   Epigyne.
Fig. 3. —   ,,   *similaris* ♀.
Fig. 4. —*Dolomedes neptunus* ♀.
Fig. 4a.—   ,,   ,,   eyes.
Fig 5. —   ,,   *spinipes* ♀.

PLATE XIX.

Fig. 1. —*Nephila picta* ♀.
Fig. 2. —*Epeira wagneri* ♀.
Fig. 2a.—   ,,   ,,   Folded eucalypt leaf nest
Fig. 2b. — ,,   ,,   Rolled eucalypt leaf nest
Fig 2c. —   ,.   ,,   Folded leaf (*Lantana camara*) nest
Fig. 2d.—   ,,   ,,   Leaf of a eucalypt folded over to form cocoon

} Entrance to each nest shown at A.

PLATE XX.

Fig. *Actinopus formosus* ♂ (× 3)

A NEW GENUS AND THREE NEW SPECIES OF MOLLUSCA FROM NEW SOUTH WALES, NEW HEBRIDES, AND WESTERN AUSTRALIA.

By John Brazier, F.L.S., C.M.Z.S., etc.

*Clathurella (?) Waterhouseæ, n.sp.

Shell fusiformly turreted, moderately solid, yellowish white, with a zone of double blackish brown nodes or spots on the last whorl, similar blackish markings being occasionally apparent here and there on the base and upper portion of the whorls; whorls 9, the three apical quite smooth, the others slightly convex, longitudinally ribbed and crossed with transverse spiral striæ, becoming sharply and prominently nodulous upon the ribs; spire sharp, apex light brown; aperture ovate, columella somewhat straight, white, canal short, outer lip more or less broken, barely showing any posterior sinus.

Long. 13, diam. 4½; length of aperture 5 mm.

Hab.—North Head of Botany Bay, New South Wales (*Mrs. G J Waterhouse*).

I place this pretty little species provisionally in *Clathurella* as the outer lip is broken, showing a very small sinus, the centre of the last whorl with two rows of black nodes on the ribs terminating on the second whorl above the suture; three similar rows on the base but not so clear and distinct, large blackish brown spots below the suture; the remaining whorls with a single row of blackish brown nodes above the suture with the spots here and there below. This interesting species was found by Mrs G J. Waterhouse and her sons on June 11, 1896, under a large stone at Botany North Head; the specimen was in the possession of a

---

* This species must now be referred to *Cantharus* A perfect adult specimen from Port Jackson, west side of Vaucluse, recently found by my son and myself, has the outer lip crenulated, thickened externally and denticulated within. Long. 15 ; diam. 5½ ; length of aperture 6 mm.—25 xi. 96.

23

hermit crab; the suture of the third whorl has been perforated by a *Nassa* or *Natica*.

Type in the Waterhouse Collection.

CONUS KENYONÆ, n.sp.

Shell solid, oblong, coronated ; spire very little raised, apex obtuse, whorls 6, with white nodes, the interspaces with yellowish brown spots, spirally sulcated at the lower part with 7 rather narrow grooves, the upper being the finest; colour cream yellow with snow white flexuous streaks and blotches in the centre; columellar base dark brown, ornamented with snow-flake spots; lip straight, somewhat thickened, interior of the aperture white.

Long. 43; diam. maj. 24, aperture 39 mm.

Hab.—Shark's Bay, W.A (*Mr. Podesta*).

The unique specimen of this new cone is slightly sea-worn but quite distinct from any of the species known to me. The upper half of the shell is quite smooth, the lower part having 6 or 7 rather narrow spiral grooves, and the centre ornamented with snow white flexuous streaks and blotches.

I have seen a second specimen formerly for many years in the collection of the late Mrs Brazier, which differs very much, both in colour and markings. I define it under a new varietal name.

CONUS KENYONÆ var. ARROWSMITHENSIS, var.nov.

Spire more raised, apex pinkish, less obtuse; colour flesh tinge, ornamented in the centre with somewhat broad white arrow-shaped markings, with the points to the right, spirally sulcated with 4 rather narrow but deep grooves rather wide apart, with two others below close together, columella tinged with violet; base tipped with brown intermingled with snow white spots, interior of aperture very light violet; lip thickish, straight.

Long. 36; diam. maj. 21; aperture, 28 mm.

Hab.—Arrowsmith Isl., Marshall Islands (*J. B.*, 22, ix.1872).

Types in the Kenyon Collection.

KENYONIA, g.n.

Shell subcylindrical, smooth ; spire much elevated ; whorls tabled at the suture, each whorl being connected with small curved shelly plates numbering about forty-four, giving the edge of the shoulder the appearance of being coronated with triangular pointed nodes; outer lip sinuous, forming an oblique posterior deep narrow sinus.

This is connected with *Conus* and *Pleurotoma* and may be placed under the former genus for the present until the animal is known.

KENYONIA PULCHERRIMA, n.sp.

Shell subcylindrical, rather thin, smooth, sometimes marked with faint slightly curved longitudinal lines of growth; whorls 8, tabled at the suture, each one being connected with small curious shelly plates that look like small deep pits when the shell is looked at end-on from the apex, giving the edge of the shoulder a coronated appearance, with triangular pointed nodes; last whorl more than half the length of the whole shell, ornamented with longitudinal reddish brown streaks and blotches, some of a zig-zag pattern, the three upper or apical flesh colour, smooth, outer lip sinuous, having an oblique posterior deep narrow sinus; columella straight, interior of aperture white.

Long 28, last whorl 17, the others 12, diam. maj 10 mm

Hab —New Hebrides (*A. F Kenyon*)

This very pretty shell Mrs Kenyon showed me some three years ago when in Sydney; she now writes (19 5/96):—"The curious shell I now send I used to think was a Cone. I do not think any more have been or are likely to be found I got it from a man who with his family had been over ten years resident in Fiji and the New Hebrides. The natives used to collect and bring him shells. There were some hurricanes during their residence, after which they used to pick up shells. I have had it in my possession about three years."

The shell being thin, I should take it to be a deep water species. The very curious little curved shelly plates at the suture make it coronated with small triangular shaped nodes; in places the suture is canaliculated and small rough shelly plates stand up somewhat like a minute roadway.

Mr. Baker contributed the following Note on a new variety of *Acacia decurrens*, Willd., a flowering specimen of which was exhibited :—*A. decurrens*, var *Deanei*, a shrub, from 3 to 5 ft, hoary, pubescent, the extremities of the branches silvery white; branches and branchlets terete, occasionally slightly ribbed by faint decurrent lines from the base of the branchlets. Pinnæ 6 to 12 pairs, leaflets 15 to 25 pairs, oblong, obtuse, 1 to 2 lines long, 1-nerved, minutely pubescent Glands regularly occurring along the rachis, one under each pair of pinnæ. Flower-heads small, few, in axillary racemes or forming a loose terminal panicle. Flowers not numerous, about 20 in a head, small, 5-merous. Calyx turbinate, broadly lobed. Petals minutely pubescent. Pod about 4 inches long and 3 lines broad, much contracted between the seeds Seeds oblong, arillus club-shaped, gradually tapering off into a short, straight funicle

Hab —Gilgandra, N S W. (Mr. Henry Deane)

This variety differs from the *A. decurrens* var. *normalis*, of Bentham, (1) in *not* having the strongly decurrent lines of that variety, in fact, the branches and branchlets are all but terete, and in that respect resemble *A. decurrens* var *mollis*. (2) in having shorter and broader leaflets ; and (3) in the narrower pod It resembles this variety in having only one gland between the individual pairs of pinnæ. Its greatest affinity is with *A. decurrens* var. *mollis*, of Bentham, resembling that variety in the terete branches, shape of pinnules and leaflets, but instead of the young shoots being golden yellow in colour they are silvery white, as in *A. dealbata* (from which species it differs in the size and shape of pod as well as the seed). The glands are also fewer in number than in *A. decurrens* var. *mollis*, there being only 1 to each pair of leaflets, and also the pods are longer and broader and more varicose than in that variety. *A. decurrens* var *mollis* flowers in December and this variety flowers in June In regard to *A. decurrens* var. *pauciglandulosa*, var. *Leichhardtii*, and vars *a* and *β* of Maiden (Ag. Gaz. N S.W. v , 607), its varietal differences are too well marked to need enumeration.

Mr. Edgar R. Waite exhibited a female Pouched Mouse and her eight young ones, *Phascologale flavipes*, Waterhouse ; and contributed the following note on the nidification of this species. So little has been recorded of the breeding habits of the pouched mice that the following extract from my note book dated November 23rd, 1893, and referring to the examples now exhibited, may be of interest. The mice were obtained at Berowera Creek, an arm of the River Hawkesbury. Clambering up a rocky slope, I noticed that one of the weathered holes, so common in the sandstone boulders of the district, was crowded with dry leaves. The hole was in a vertical face of the boulder about four feet from the ground, and as the leaves, all of Eucalypts, were regularly placed in a compact mass, I began to poke them out When a hat-full had been removed a rustling was heard within, and further leaves were cautiously withdrawn. A little snout and a pair of sparkling eyes appeared for a moment, and while removing more leaves, of which there seemed to be no end, the owner rushed out and was climbing up the perpendicular face of the rock when secured It was a half grown *Phascologale flavipes*, and as the hole was evidently not merely a retreat but probably contained an actual nest, I continued to remove the leaves. Scutterings within indicated that the occupants were in some number The nest was finally reached and contained two young ones the size of the one first caught. It was composed entirely of Eucalypt leaves and was completely domed over, but fell to pieces when handled, as the leaves were not secured together in any way. A larger, and evidently the mother mouse, came to the opening for an instant unaccompanied : almost immediately she reappeared and left the hole, this time with some young ones clinging to her back. Although thus heavily weighted she nearly escaped me. She ran under a horizontal slab of rock and clung like a fly, back downwards. When secured it was found that she had four young ones clinging to her, which together must have equalled more than her own weight. On removing the mouselings it was seen that each had a tuft of fur in its mouth, showing how they had retained their hold. I now had the mother and seven young ones and on feeling in the hole, which received my arm nearly to the elbow, I secured an eighth. The everted pouch exposed eight teats, so that the mother had her complement of young.

Although constantly stated that no true pouch exists in members of the *Phascologale*, this is scarcely correct. When very young the offspring are completely hidden by the outer wall of the pouch closing over them. As they increase in size the mouth dilates and no longer conceals the young. Mr. Oldfield Thomas* does not admit Krefft's statement that this species is provided with 10 teats † Although 8 is the usual number, I have examined several females with 10 teats, and there is one preserved in the Australian Museum with not only 12 teats, but also a young one on each teat. As far as can be judged without spoiling the exhibit, this animal does not otherwise differ from typical examples. It would therefore appear that in the *Dasyuridæ*, or at least in *Phascologale*, the number of mammæ is not such a constant character as has been insisted upon, or three otherwise similar species would have to be admitted ; characterised by the possession of 8, 10, and 12 mammæ respectively.‡

Mr. Rainbow showed the spiders described in his paper, with drawings of the same.

Mr. Lucas exhibited a specimen of the lizard described in the paper by Mr. Frost and himself

---

* B.M. Catalogue, Marsupialia, 1889, p. 289.

† Trans. Phil. Soc N.S Wales, 1862, p. 10.

‡ When writing the foregoing, I overlooked the fact that Prof. Spencer had already drawn attention to the variability in the number of teats in members of the smaller *Dasyuridæ ;* (Report of the Horn Expedition, Zoology, p. 42), and that Mr. J. J. Fletcher had previously exhibited a specimen of *Phascologale flavipes* with nine mammary fœtuses *in situ.* (Proc. Linn. Soc. N. S. Wales (2), I. p. 164.)

Mr. Brazier sent for exhibition specimens of the shells described in his paper, namely, a new species of *Clathurella* (?) [*Canthurus*—see p. 345] from the North Head of Botany Bay, a new Cone from West Australia, and a remarkable Shell from the New Hebrides for which a new genus is proposed.

The President exhibited three albums of mounted specimens of Western Australian wild flowers.

WEDNESDAY, AUGUST 26th, 1896.

The Ordinary Monthly Meeting of the Society was held at the Linnean Hall, Ithaca Road, Elizabeth Bay, on Wednesday evening, August 26th, 1896

P. N. Trebeck, Esq., J R, in the Chair.

Mr George William Card, A R.S M , A R.C S , Curator and Mineralogist, Geological Survey of New South Wales, and Professor Richard Threlfall, M.A., Sydney University, were elected Members of the Society

DONATIONS.

University of Melbourne—Examination Papers—Final Honour, Degrees, etc. (Feb., 1896). *From the University.*

Perak Government Gazette. Vol. ix. Nos. 15-17 (July, 1896) *From the Government Secretary.*

Johns Hopkins University Circulars. Vol. xv. Nos. 125-126 (May-June, 1896). *From the University.*

Bureau of Agriculture, Perth, W.A.—Journal Vol. iii. Nos. 18-20 (July-August, 1896) *From the Secretary.*

Zoologischer Anzeiger. xix. Bd. Nos. 506-508 (June-July, 1896). *From the Editor.*

Zoological Society of London—Proceedings, 1896. Part i. (June). *From the Society.*

Société d'Horticulture du Doubs, Besançon — Bulletin. Série Illustree. No 6 (June, 1896). *From the Society.*

Department of Agriculture, Brisbane — Bulletin. Nos. 9-10. Second Series (Feb.-May, 1896). *From the Secretary for Agriculture.*

Victorian Naturalist. Vol. xiii. No. 4 (July, 1896). *From the Field Naturalists' Club of Victoria.*

Pamphlet entitled "The Geological Structure of Extra-Australian Artesian Basins." By A. G Maitland, C.E., F.G S, Brisbane, 1896. *From the Geological Survey of Queensland.*

Journal of Conchology. Vol. iii. No. 7 (July, 1896) *From the Conchological Society of Great Britain and Ireland.*

Royal Society of Tasmania—Papers and Proceedings for 1894-1895 : Pamphlet entitled "The Health of Hobart." By R M. Johnston, F.L.S. (1896). *From the Society*

Société Royale Linnéenne de Bruxelles — Bulletin. 21me Année, No. 8 (July, 1896). *From the Society.*

Royal Microscopical Society—Journal, 1896. Part iii. (June). *From the Society*

New Zealand Institute—Transactions and Proceedings. Vol. xxviii. (1895). *From the Institute.*

Australian Museum, Sydney — Report of the Trustees for the year 1895. *From the Trustees.*

Public Library, Museums, and National Gallery of Victoria — Report of the Trustees for 1895. *From the Trustees.*

Agricultural Gazette of New South Wales. Vol. vii. Part 7 (July, 1896). *From the Hon. the Minister for Mines and Agriculture.*

Société Scientifique du Chili—Actes. Tome v. (1895). 4me Livraison. *From the Society.*

American Museum of Natural History, New York—Bulletin. Vol. viii. (1896), Sigs. 7-9 (pp. 97-144) [June]. *From the Museum.*

Museum of Comparative Zoology at Harvard College, Cambridge, Mass.—Bulletin.   Vol. xxix. No. 4 (June, 1896).   *From the Curator.*

U.S. Department of Agriculture —Division of Ornithology and Mammalogy—North American Fauna, No. 11 (June, 1896). *From the Secretary of Agriculture.*

American Geographical Society—Bulletin.   Vol. xxviii., No. 2 (1896).   *From the Society.*

American Naturalist.   Vol. xxx. No. 355 (July, 1896).   *From the Editors.*

Naturwissenschaftlicher Verein zu Elberfeld—Jahres-Berichte viii. Heft (1896).   *From the Society.*

Societas Entomologica Rossica — Horæ.   T. xxx. Nos. 1-2 (1895-96).   *From the Society.*

K. K. Zoologisch-botanische Gesellschaft in Wien—Verhandlungen.   Jahrgang 1896.   xlvi. Bd.   6 Heft.   *From the Society.*

Royal Dublin Society - Transactions (Series ii.)   Vol. v. Parts 5-12 (Aug., 1894-Jan., 1896): Vol. vi. Part 1 (Feb., 1896): Proceedings.   (N.S.) Vol. viii. Parts 3-4 (Aug., 1894-Sept., 1895). *From the Society.*

L'Académie Royale des Sciences, Stockholm — Handlingar. xxvii. Bd. (1895-96).   *From the Academy.*

Australasian Journal of Pharmacy.   Vol. xi. No. 128 (Aug. 1896).   *From the Editor.*

Pamphlet entitled "Synoptical List of Coccidæ," (1894).   By W. M. Maskell.   *From the Author.*

ON THE AUSTRALIAN *BEMBIDIIDES* REFERABLE TO THE GENUS *TACHYS*, WITH THE DESCRIPTION OF A NEW ALLIED GENUS *PYRROTACHYS*.

By Thomas G. Sloane.

In the present paper I have placed in the genus *Tachys* all the Australian Bembidiides which have the anterior tibiæ decidedly oblique above the apex on the external side; normally also a striole is present on the apical declivity of each elytron, but this character is not invariable.

The most important contribution to the knowledge of the Bembidiides of Australia is Sir William Macleay's notice and descriptions of seventeen species from Gayndah, all of which he referred to the genus *Bembidium*.* I have seen the types of Macleay's species in the Australian Museum, Sydney. Three of them, viz , *B. amplipenne, B. bipartitum* and *B. sexstriatum*, I am unable to deal with, as I do not possess specimens ; and, not residing in Sydney, I cannot see the types at present. Specimens of the eleven species to which the remaining fourteen must be reduced are in my possession. Nine are dealt with in the present paper; the tenth is *Bembidium jacksoniense*, Guér., = *B. subviride*, Macl., the eleventh, *Bembidium gagatinum*, Macl., is not a Bembidiid at all, but a Harpalid which may be referred, at least tentatively, to the genus *Thenarotes*.†

---

* *Vide* Trans. Ent. Soc. N.S W. 1873, ii. pp. 115-120.

† *Bembidium flavipes*, Macl., is a synonym of *B. gagatinum*, Macl., being founded on an immature specimen; the species, which extends as far south as the Murray River, may be known in future as *Thenarotes gagatinus*, Macl.

The principal features used in the synoptic table of species which follows seem to divide the species here placed in *Tachys* into distinctive groups that are readily separated from one another; indeed the most important of these groups are apparently so distinct that they might be removed from *Tachys* altogether and formed into separate genera; but to do this would require a fuller knowledge than I possess of the genera now regarded as capable of maintenance among the Subulipalpi, and of the system adopted in classifying them. The minor features used in the table for separating closely allied species from one another are not perhaps always the best that could have been chosen, though they have seemed to me to be so.

The following species of *Tachys*, described by the Rev. Thos. Blackburn, are unknown to me in nature, and, for that reason, have not been included in the table, viz., *T. baldiensis*, *T. infuscatus*, and *T. adelaidæ*.

Genus TACHYS.

Owing to the variable number of striæ on the elytra among the species of the genus *Tachys* (the full number is eight striæ and a marginal channel, but this only occurs in *T. yarrensis*, Blkb., among the species known to me) the ordinal number to indicate the stria next the marginal channel would vary, and as this stria seems a feature of great classificatory importance it becomes needful to use an unvarying term for it. I therefore call it the *submarginal stria*. The interstice between the submarginal stria and the marginal channel I call the *lateral interstice*.

Table of Species known to me.

I. Elytra with submarginal stria well marked.
   A. Prothorax with a submarginal lateral carina
      near base.
      b. Upper surface shagreened and finely
         punctulate (unicolorous). . ... *T. brunnipennis*, Macl.
      bb. Upper surface shagreened, impunctate
         (bicolorous).. . . .. *T. ectromioides*, Sl.
   AA. Prothorax without a submarginal lateral
      carina near base.

C Marginal channel of elytra simple,
ateral interstice convex.

Prothorax without a dentiform pro-
ection on sides before base.

. Elytra sexstriate on each side of
suture.

f. Elytra quadrimaculate, fifth stria
reaching border of base .... *T. buprestioides*, Sl.

ff. Elytra bimaculate, fifth stria not
reaching base..................... ...... *T. froggatti*, Sl.

ee. Elytra quinquestriate on each side
of suture

g. Elytra quadrimaculate. .. ... *T. striolatus*, Macl.

gg. Elytra bimaculate .... *T. bipustulatus*, Macl.

* *eee.* Elytra bistriate on each side of
suture.. .... *T. curticollis*, Sl.

eeee. Elytra unistriate on each side of
suture . ... *T. iaspideus*, Sl.

dd. Protborax with a dentiform projection
on sides a little before base.

h. Elytra bistriate on each side of
suture .... *T. spenceri*, Sl.

hh. Elytra unistriate on each side
of suture. .. .. *T. bis'riatus*, Macl.

CC. Marginal channel of elytra punctate,
lateral interstice depressed.

i. Elytra with eight punctate striæ on
each (seventh as well marked as
others' . .. . *T. yarrensis*, Blkb.

ii. Elytra with seventh stria obsolete.

j. Lateral basal foveæ of prothorax
concave, bordered by the widely
upturned lateral border.

k. Elytra sexstriate on each side of
suture, lateral margin of prothorax with one setigerous
puncture anteriorly .... *T. monochrous*, Schaum.

---

* *T. orensensis*, Blkb., (a specimen of which I received from Mr. Blackburn while this
paper was in the press) belongs to section "*eee*" For some differences between it and *T
curticollis*, see description of the latter (*post*, p. 364)

kk. Elytra quinquestriate on each side of suture, margin of prothorax plurisetose near anterior angles .... *T. seticollis*, Sl.

jj. Lateral basal foveæ of prothorax concave, divided from lateral border by a raised space .... *T. flindersi*, Blkb.

jjj. Posterior angles of prothorax forming the apex of a triangular marginal process.

    l. Colour piceous red, elytra with testaceous ante-apical maculæ *T. semistriatus*, Blkb.

    ll. Colour black .. .. *T. habitans*, Sl.

II. Elytra with submarginal stria obsolete on sides.

M. Form short, very convex; prothorax not perceptibly narrowed to base; elytra lævigate, unistriate on each side of suture... .. *T. ovatus*, Macl.

MM. Form varying, prothorax evidently narrowed to base.

N. Head impunctate, frontal impressions deep, oblique (converging anteriorly); tnird interstice of elytra bipunctate on disc.

    o. Elytra with six rows of strong punctures on each side of suture. .. . *T. mitchelli*, Sl.

    oo. Elytra with three or four punctulate striæ on each side of suture......... *T. australicus*, Sl.

NN. Head punctate, frontal impressions long, deep, narrow, parallel.

    p Surface of prothorax impunctate *T. leai*, Sl.

    pp. Surface of prothorax minutely and rather densely punctulate *T. murrumbidgensis*, Sl.

NNN. Head impunctate, frontal impressions wide, shallow.

    q. Each elytron bipunctate on disc, recurved striole of apex obsolete *T. captus*, Blkb.

    qq. Each elytron unipunctate on disc

        r. Elytra more or less distinctly striate on disc, recurved striole of apex well marked.

* s. Discoidal puncture of elytra
placed a little before middle
nearer suture than margin.
 t. Elytra depressed, sides
  parallel; prothorax piceous
  black .... *T. uniformis*, Blkb.
 tt Elytra lightly convex, sides
  rounded ; prothorax testaceous.
  u. Elytra with strongly impressed punctulate striæ
   on disc, base testaceous
   (a wide black fascia across
   middle of elytra)............. *T. atriceps*, Macl.
  uu. Elytra with faintly im-
   pressed striæ on disc,
   middle of base piceous *T. lindi*, Blkb.
 ss. Discoidal puncture of elytra
  placed about anterior third,
  nearer margin than suture *T. transversicollis*, Macl.
rr. Elytra lævigate, nonstriate,
 recurved striole of apex
 obsolete .. . . *T. macleayi*, Sl.

· · · TACHYS BRUNNIPENNIS, Macleay.

T. (Bembidium) brunnipennis, Macl., presents the characteristic
features of *Tachys*, viz., the anterior tibiæ oblique above apex on
external side, and the elytra with the sutural stria recurved at
apex, the recurved apical striole is very near the margin, and is
divided from the submarginal stria by a narrow subcarinate
interstice.

Hab.: Queensland—Cairns (Froggatt), Port Denison and
Gayndah (Masters).

TACHYS ECTROMIOIDES, n.sp.

Oval, subdepressed. Prothorax transverse, much wider at base
than apex, posterior angles rectangular : elytra oval, lightly

---

* *T. similis*, Blkb , (a specimen of which was received from Mr. Blackburn too late to be
worked into the table) comes into section "*s.*" It resembles *T. uniformis*, Blkb., in facies,
but differs in colour.

convex, finely striate; third stria more strongly impressed on
apical declivity and joining sutural stria at apex; submarginal
stria faintly impressed, very near margin.   Head dark piceous,
labrum testaceous; prothorax piceous brown, lateral margin and
middle of base testaceous; elytra testaceous, a very wide dark
piceous fascia across disc considerably behind base, apex widely
piceous; legs testaceous, antennæ infuscate, basal joints testaceous.

Head depressed, hardly impressed laterally; a feeble oblique
ridge on each side near eyes; clypeal suture finely impressed;
clypeus bifoveolate; eyes large, convex.   Antennæ filiform, not
long.   Prothorax transverse (0·65 × 0·85 mm.), widest about
anterior third, roundly declivous to lateral margin anteriorly;
sides strongly rounded to apex, straight posteriorly and hardly
narrowed to base; anterior margin truncate; anterior angles not
marked; basal angles rectangular, acute; base lightly and roundly
produced backwards in middle; lateral border reflexed, reaching
to sides of head at apex ; lateral channel wide, narrowed to
anterior angles; median line deep, a strongly marked arcuate
transverse line defining basal part of prothorax; a lightly carinate
longitudinal submarginal ridge near each basal angle   Elytra
oval, convex, much wider than prothorax (2 × 1 3 mm.); sides
rounded; shoulders rounded; five inner striæ lightly impressed,
finely crenulate, sixth and seventh obsolete; interstices depressed,
first narrow on apical declivity, second and third ampliate on
apical declivity, third with two small setigerous punctures—the
anterior just before, the posterior just behind discoidal piceous
fascia; lateral interstice very narrow, not convex, having four
setigerous punctures behind shoulders and about same number
towards apex: base not bordered; lateral border narrow, reflexed.
forming a very slight prominence at humeral angles   Anterior
tibiæ incrassate, oblique above apex on external side; a short
acute spur above obliquity.

Length 3, breadth 1·3 mm.

Hab.. West Australia—Donnybrook (Lea; Coll. Lea, unique).

I am not sure that I am right in putting this species in the
genus *Tachys;* no allied species is known to me; though I have

placed it with *T. brunnipennis*, Macl., in the table of species at p. 356; this has only been done on account of the submarginal carina near basal angle of prothorax, and not because I have thought there is any close affinity between these species. In general appearance it has a resemblance to a Lebiid of the genus *Surothrocrepis* or *Ectroma*. If the ground colour of the elytra be considered piceous, then the base (widely), the margin and a narrow fascia just above the apical declivity would be described as testaceous; the dark-coloured parts of the elytra do not anywhere reach nearer the sides than the submarginal stria.

TACHYS BUPRESTIOIDES, n.sp.

Robust, oval, convex. Head wide; prothorax transverse, wider across base than apex. elytra ovate, six inner striæ strongly impressed on each elytron; lateral stria and marginal channel strongly impressed, interstice between them convex. Bronzed black, each elytron with an elongate macula behind shoulder and a reniform macula on apical third testaceous, legs (excepting coxæ) testaceous, antennæ infuscate, under surface piceous, apical segments of abdomen reddish.

Head convex, finely shagreened, lightly bi-impressed, clypeal suture finely and distinctly marked; eyes large, convex, not globose. Maxillary palpi with penultimate joint elongate, thick, incrassate, setose; apical joint very small. Prothorax transverse, widest at anterior marginal puncture, sides strongly rounded on anterior two-thirds, lightly narrowed posteriorly, straight before base; anterior margin emarginate; anterior angles obtuse but marked, basal angles rectangular; base truncate on each side, roundly produced backwards in middle, border narrow, reflexed; median line very lightly impressed; a straight transverse line near base, this line strongly impressed in middle; lateral basal impressions short, placed at each side of rounded middle part of base Elytra wider than prothorax, convex; sides rounded; shoulders rounded; striæ simple, only first reaching apex, first, second and fifth reaching base, second, third and fourth extending past posterior margin (between macula and suture) of ante-apical

24

macula, fifth and sixth not extending past anterior margin of
ante-apical macula, fifth reaching basal border, sixth not reaching
base, seventh obsolete (only noticeable under a lens on black part
of space between sixth and eighth); lateral stria deeply impressed,
curving towards margin posteriorly; inner interstices convex;
submarginal interstice very convex, bipunctate near base and at
beginning of apical curve; lateral border extending on to base as
far as fifth stria. Anterior tibiæ shortly oblique above apex on
external side; a short acute spur above obliquity.

Length 3·1, breadth 1·3 mm.

Hab. : King's Sound (Froggatt; Macleay Museum).

Allied to *T. striolatus*, Macl, but larger and broader; the prothorax is more transverse and wider across the base, less rounded
on the sides, the anterior angles more strongly marked; the post-
humeral macula of the elytra is elongate; there are six (not five)
striæ on each elytron, the first, second and fifth striæ reaching the
base. The whole of the dark part of the elytra, excepting the
sides, is strongly striate, the third and fourth striæ do not reach
quite to the base, but there is not the wide lævigate basal space
that is so noticeable in *T. striolatus*.

TACHYS FROGGATTI, n.sp.

Robust, oval, convex. Head wide, lightly bi-impressed; prothorax transverse, wider across base than apex; elytra ovate, six
inner striæ strongly impressed, lateral stria and marginal
channel strongly impressed, interstice between them convex.
Black, head and prothorax with a greenish tinge; each elytron
with a testaceous macula about posterior third; under surface
piceous, legs (excepting coxæ) testaceous, antennæ infuscate
towards apex.

Head, prothorax and legs in every way resembling those of *T.
buprestioides*, Sl. Elytra similar to those of *T. buprestioides* in
shape, sides and apex; striæ hardly so deep, first entire, second
and third reaching almost to base, 4-6 rising at a considerable
distance from base on one level.

Length 2·2-2·6, breadth 1-1·15 mm.

Hab. : King's Sound (Froggatt; Macleay Museum).

Differs from *T. buprestioides* by its smaller size, by the absence of the post-humeral maculæ of the elytra, and by the fifth stria not reaching the base. It is closely allied to *T. bipustulatus*, Macl., from which it differs by having six (not five) striæ on each elytron and the striæ reaching nearer the base—especially the three inner ones.

TACHYS STRIOLATUS, Macleay.

T. (Bembidium) striolatus, Macl., has been redescribed and placed in *Tachys* by the Rev. Thos. Blackburn.*

Habits :—Riparian, running beside the margins of streams, or on sandy margins of pools, during summer months.

Hab. : Queensland—Gayndah (Masters); N.S. Wales—Narrandera and Mulwala (Sloane); Victoria—near Bright (Blackburn).

TACHYS BIPUSTULATUS, Macleay.

T. (Bembidium) bipustulatus, Macl., agrees in all points of structural detail and in striation of elytra with *T. striolatus*, Macl.

Habits :—Riparian ; two specimens occurred to me on the muddy edge of pools in Houlaghan's Creek near Junee.

Hab. : Queensland—Gayndah (Masters), N S Wales—Forest Reefs (Lea), Junee District (Sloane).

TACHYS CURTICOLLIS, n.sp.

Oval, convex. Prothorax transverse, evidently a little wider across base than apex, posterior angles rectangular and acute; elytra lævigate on disc, bistriate on each side of suture, lightly bipunctate near second stria. Black, or piceous black ; each elytron with a dull reddish spot near shoulder and another at beginning of apical declivity; legs pale testaceous.

---

* *Vide* P.L.S.N.S.W. 1891 v. (2), p. 785, and Trans. Roy Soc. S. Aust. 1894, xviii. p. 139.

Head smooth, frontal impressions long, straight, diverging backwards, extending forward to labrum; eyes prominent, hemispherical. Prothorax lævigate, convex, short, transverse, widest just behind anterior marginal puncture; basal part defined by a transverse impression, sides lightly rounded anteriorly, gently narrowed to base, meeting base at right angles; base sloping lightly forward on each side to posterior angles; lateral border reflexed, becoming wider towards base; median line obsolete, a flattened depressed space near each basal angle, a light transverse linear impression (hardly punctulate) connecting the lateral basal depressions. Elytra much wider than prothorax, oval, truncate at base (shoulders rounded), convex, declivous to base; striæ simple, first entire, second as strongly impressed as first, not reaching base or apical declivity, a deep lateral stria besides marginal channel on each elytron. Anterior tibiæ oblique above apex on external side, a spiniform spur above obliquity.

Length 2, breadth 1 mm.

Hab.: N.S. Wales—Tweed River (Lea; March, 1892), Cootamundra District (Sloane).

At a casual glance this species might be taken for a small form of *T. bistriatus*, Macl, but it differs decidedly from that species by having a second stria outside the sutural one extending from the anterior discoidal puncture to the apical declivity and by the shape of the prothorax, which is much wider at the base and has the basal angles rectangular, the sides not having a prominent angular projection above the base as in *T. bistriatus*. It is somewhat like *T. ovensensis,** Blkb., from which it differs by having a post-humeral reddish spot on each elytron; by the form of the frontal impressions which are further from the eyes narrow, and extend obliquely forward till they reach the anterior margin of the clypeus; by the sides of the prothorax being less rounded on the sides and wider at the base.

---

* In *T. ovensensis*, Blkb., the head and prothorax are similar in shape, &c., to those of *T. striolatus*, Macl.

TACHYS IASPIDEUS, n.sp.

Elongate-oval; prothorax transverse (not short); elytra lævigate, each elytron unistriate near suture and with recurved stria of apex distinct. Shining, polished, reddish or reddish brown; elytra lighter coloured than prothorax near base, almost black across middle and near apex, a large yellowish-red spot behind posterior discoidal puncture on each elytron

Head smooth, convex, lightly bi-impressed between eyes; the impressions short, not extending to clypeus; eyes large, convex. Prothorax small, transverse, a little wider than head, widest a little before middle, lightly narrowed to base, convex, lævigate, not declivous to middle of base, not transversely impressed across base, sides lightly rounded, gently narrowed (not sinuate) to posterior angles; apex and base truncate; posterior angles obtuse, not prominent; border narrowly reflexed; median line wanting; a lightly marked wide oblique impression at each basal angle. Elytra much wider than prothorax, suboval, convex, a little depressed on disc; base subtruncate; humeral angles rounded, apex narrowly rounded; one simple stria on each side of suture, one deep lateral stria besides the marginal channel on each elytron; lateral interstice convex and depressed posteriorly; lateral margin interrupted just behind shoulders causing the margin of the humeral angles to project slightly; two punctures placed longitudinally on disc of each elytron.

Length 2·8, breadth 1·3 mm.

Hab. : N.S.W.—Inverell, Tamworth (Lea)

This species exactly resembles *T. spenceri*, Sl., in shape and appearance; the marked features distinguishing it from that species are (a) the absence of any projection at the basal angles of the prothorax, and (b) the elytra having only one stria on each side of the suture, not two as in *T. spenceri*. The penultimate joint of the maxillary palpi is large and pyriform, the apical joint a mere short spike. The general colour is like that of polished yellowish-brown jasper.

TACHYS SPENCERI, Sloane.

Habits:—Found under stones besides edge of water (Spencer).
Hab.: Central Australia—Larapintine Region (Spencer); West Australia—King's Sound (Froggatt).

TACHYS BISTRIATUS, Macleay.

T. (Bembidium) bistriatus, Macl. (= *Bembidium convexum,* Macl.), has a short recurved striole on the middle of the apex of each elytron; the posterior angles of the prothorax form a small triangular prominence on the sides a little before the base itself. I have carefully compared the types of *Bembidium bistriatum,* Macl., and *B. convexum,* Macl., with one another and find them one species.

Hab.: Queensland—Gayndah (Masters); N.S. Wales—Tweed and Clarence Rivers (Lea).

TACHYS YARRENSIS, Blackburn.

Habits:—Found under logs and débris in very damp situations.
Hab.: Victoria—Upper Yarra (French); N. S Wales—Mulwala and Urana (Sloane), Tamworth (Lea).

TACHYS MONOCHROUS, Schaum.

No doubt remains in my mind, after comparing specimens of *Bembidium punctipenne,* Macl., with the description of *Tachys monochrous,* Schaum, but that the species are synonymous.

Habits:—Found under logs in very damp situations.
Hab.: Victoria—Lilydale (Sloane); N. S. Wales—Windsor and Tamworth (Lea), Ourimbah (Fletcher) ; Queensland—Gayndah (Masters).

TACHYS SETICOLLIS, n.sp.

Oval, robust. Prothorax lightly transverse, strongly narrowed to base, basal angles rectangular, margin plurisetose near anterior angles: elytra widely ovate; five rows of punctures and a submarginal stria on each elytron; recurved striole of apex distinct, rather short; two fine setigerous discoidal punctures on third interstice. Clear ferruginous red, subtestaceous above apical declivity of elytra; legs testaceous; antennæ ferruginous, basal joint testaceous.

Head convex; front widely but rather deeply bi-impressed; eyes large, convex. Antennæ short, stout, filiform (reaching back a little behind base of prothorax). Prothorax broader than long, widest a little before anterior third, evidently narrower across base than apex; disc lightly convex, rather depressed in middle, lightly declivous to basal area; sides strongly rounded anteriorly, shortly, strongly and roundly narrowed to anterior angles, strongly sinuate posteriorly, meeting base at right angles; anterior margin truncate; anterior angles not marked; base widely truncate in middle, oblique on each side; basal angles prominent, acute, basal area depressed, well marked, extending to lateral border at each side, defined anteriorly by a strongly marked transverse punctate impression; lateral border very narrow on rounded part of sides, thick and strongly reflexed near basal angles; median line very lightly impressed on disc; four or five setigerous marginal punctures between anterior third and anterior angles. Elytra widely ovate; base roundly truncate; humeral angles not marked; sides rounded; first stria entire, punctate for more than half its length, simple posteriorly; striæ 2-5 consisting of rows of closely set strong punctures extending from base to lighter-coloured lævigate apical part of elytra; submarginal stria punctate; lateral interstice not convex; marginal channel closely punctate; the punctures from shoulder to apical curve each bearing a long seta.

Length 2·25, breadth 1 mm.

Hab.: North West Australia—King's Sound (Froggatt; Macleay Museum).

Allied to *T. monochrous,* Schaum, but differing by its shorter, wider, and rather less convex form; the prothorax wider, more strongly narrowed to base, disc flatter and less strongly declivous to base, margin plurisetose behind anterior angles; elytra shorter, wider, less convex, five- (not six-) striate.

TACHYS FLINDERSI, Blackburn.

T. flindersi, Blkb. = *Tachys* (*Bembidium*) *rubicundus*, Macl.,
I have no doubt about the correctness of this synonymy,
Macleay's name was used in the genus *Tachys* as long ago as 1850,
therefore the later name must be adopted.*

Habits :—Found under logs and stones in very damp situations.

Hab. : Queensland—Gayndah (Masters); N.S. Wales—Tam-
worth (Lea), Sydney and Wagga Wagga (Sloane), Victoria—
Upper Ovens River (Blackburn), Lilydale (Sloane); Central
Australia (Spencer); West Australia—Darling Ranges (Lea).

TACHYS HABITANS, n.sp.

Oval, convex.   Prothorax convex, subcordate : elytra oval,
convex, six rows of punctures on basal part; apex lævigate; submarginal stria indicated, punctate; lateral interstice very narrow,
not convex; recurved striole of apex well marked.   Black, shining,
legs piceous, mandibles piceous brown.

Head convex, smooth; front widely bi-impressed anteriorly.
Prothorax small, lævigate, widest rather before middle, not
narrower across posterior angles than across apex; sides strongly
rounded on anterior two-thirds, shortly sinuate before posterior
angles; anterior margin truncate; anterior angles not marked;
posterior angles prominent, acute, base lightly oblique on each
side behind posterior angles; lateral border narrow, reaching sides
of head, median line obsolete; a lightly marked impunctate impression across base near margin; lateral basal foveæ round, deep,
placed near margin at basal angles.   Elytra much wider than
prothorax; six rows of punctures and a submarginal stria on each
elytron; first stria entire, finely and closely punctate on disc,
simple posteriorly, others (consisting of rows of punctures) not
reaching base, fifth and sixth short (sixth sometimes consisting
of only two punctures); third interstice with two fine setigerous

---

* *Vide* P.L.S.N.S.W. 1894, ix. (2) p. 90, for a note by the Rev. Thos.
Blackburn on this subject.

punctures, the anterior hardly noticeable among basal puncturation, the posterior on lævigate portion of elytra a little before apical declivity ; external margin of apical striole carinate ; marginal channel finely punctate; border passing round humeral angle on to base as far as fourth stria.

Length 2, breadth 0·8 mm.

Hab.: West Australia—Darling Ranges, Bridgetown, Pinjarrah (Lea).

Allied to *T. semistriatus,* Blkb , but differing in colour; its more elongate shape; the prothorax with posterior angles more prominent and explanate; the elytra proportionately narrower, less strongly punctate, with fewer punctures in the rows, especially the fifth and sixth.

TACHYS OVATUS, Macl.

T. (Bembidium) ovatus, Macl., = *Bembidium bifoveatum,* Macl.; I have seen the types and find these two species synonymous. It has a distinct recurved striole at apex of each elytron. Though usually of a pale testaceous colour, a specimen that is subpiceous has been sent to me by Mr. A. M. Lea, as coming from the Tweed River.

Habits : —Under stones in very damp situations.

Hab. : Queensland—Gayndah (Masters); N.S. Wales—Tweed River, Clarence River, Inverell, Tamworth and Sydney (Lea).

TACHYS AUSTRALICUS, n.sp.

Robust, very convex. Prothorax convex, transverse, rounded on sides, a little wider across base than apex; elytra very convex, lightly striate near suture, sides smooth. Head and prothorax red or testaceous red, eyes black, elytra piceous or piceous black.

Head smooth, convex; front with two rather wide nearly parallel impressions ; space between these impressions convex. Prothorax smooth, transverse, convex ; sides strongly rounded without any sinuosity before posterior angles, oblique to base on each side behind posterior angles ; basal area short, convex, defined by a strong transverse impression; posterior angles not prominent, their summit acute; lateral basal foveæ obsolete. Elytra wider than prothorax, oval, very convex, declivous to peduncle, truncate on base; shoulders rounded, not marked; two, or at most three, lightly impressed striæ near the suture, first entire, lightly punctulate on disc, others only marked on disc (not reaching base), lightly punctulate; space between striæ and margin smooth and without discoidal punctures; recurved striæ of apex obsolete; marginal channel not deep along sides; three strong punctures near margin behind each shoulder, and two strong submarginal foveiform impressions on apical third.

Length 1·7, breadth 0·75 mm.

Hab.: N.S. Wales—Tweed River, Windsor (Lea).

The affinity of this little species is to *T. mitchelli,* Sl., from which it differs by its smaller size; dark coloured elytra; shorter and less oblique frontal impressions; elytra with only two or three striæ next the suture marked, the remaining part smooth (the striæ are linear and hardly punctulate, not rows of punctures as in *T. mitchelli*), &c.

TACHYS LEAI, n.sp.

Elongate-oval; prothorax convex, transverse, subcordate, narrower between posterior angles than at apex; elytra depressed, truncate at base, finely punctate-striate. Black, shining; legs and under surface piceous brown; antennæ piceous brown at base, infuscate towards apex.

Head convex, smooth; front and vertex minutely punctulate; front bi-impressed, the impressions long, straight, deep, hardly diverging backward, extending forward to base of labrum; space between frontal impressions convex; clypeal suture obsolete; clypeus declivous to labrum; eyes convex, not very prominent. Prothorax a little wider than head, transverse, widest a little before the middle, lightly narrowed to base, convex, lævigate; anterior margin truncate; base truncate across peduncle, a little oblique on each side behind posterior angles; sides unequally rounded on anterior three-fourths, shortly sinuate before posterior angles; these acute, prominent, placed a little before the actual base; a well marked transverse impression extending across base just behind posterior angles and defining the basal part; median line very lightly impressed. Elytra wider than prothorax (1·5 × 1 mm.), depressed on disc; sides lightly rounded; base truncate, hardly emarginate, shoulders rather prominent, rounded; five finely punctulate lightly impressed striæ on each elytron (exclusive of marginal channel), first entire, flexuous (approaching suture) near base, second almost equally impressed as first on disc, obsolete towards base and apex, third and fourth much more lightly·impressed, not extending towards base beyond anterior discoidal puncture, fifth strongly impressed on anterior fourth near each shoulder, obsolete for remainder of its course; scutellar striole wanting; interstices flat, fourth with two discoidal punctures, the anterior at about one-fourth the length of elytra from base, the other a little behind middle on course of third stria; third interstice very finely punctulate on apical declivity, marginal channel deeply impressed along sides, three or four rather strong punctures behind the shoulders; apical declivity with two oblique impressions on each side, the external strongly impressed near the margin (extending round the apex to join the sutural stria), the inner short, placed closed to the external one.

Length 2·4, breadth 1 mm.

Hab.: N.S. Wales—Tamworth (Lea).

Sent to me by Mr. A. M. Lea, to whose generosity I am indebted for a specimen, and to whom I dedicate it

In all details of structure this species resembles *T. murrumbidgensis*, Sl., from which it differs by its larger size, wider and more convex shape, impunctate prothorax, black colour, &c. These two species form a well marked group among the Australian Bembidiides, and it is evident they can only provisionally be considered congeneric with such species as *Tachys monochrous*, Schaum, *T. flindersi*, Blkb., &c.

TACHYS MURRUMBIDGENSIS, Sloane.

Hab.: N.S. Wales—Narrandera (Sloane), Tamworth (Lea).

TACHYS CAPTUS, Blackburn.

Habits :—Found under sticks and stones in damp situations.

Hab.: South Australia—Port Lincoln, Adelaide (Blackburn);
N.S. Wales—Mulwala, Urana, Narrandera and Junee (Sloane),
Windsor and Tamworth (Lea).

TACHYS UNIFORMIS, Blkb.

Hab. : South Australia—Adelaide and Port Lincoln (Blackburn); West Australia—Beverley (Lea).

TACHYS ATRICEPS, Macleay.

Habits :—Found under logs in damp places near water.

Hab. : Queensland—Gayndah (Masters); N.S. Wales—Carrathool, Narrandera and Mulwala (Sloane); King's Sound (Froggatt).

TACHYS LINDI, Blackburn.

Among the Bembidiides from King's Sound, in the Macleay
Museum, the commonest species is one that I take to be *T. lindi*,
Blkb. (var.) It differs from a type specimen of *T. lindi* received
from Mr. Blackburn by being smaller (length 2·5 mm.) and of a
lighter build. *T. lindi* seems to be a variable species in size and
colour marks; its constant characters appear to be *(a)* a more or
less testaceous macula behind the shoulder and another towards
the apex of each elytron, *(b)* the anterior discoidal puncture of
the elytra placed about the middle of their length on the course of
the third stria. Many of the specimens from King's Sound
(evidently immature) have the elytra almost wholly testaceous
with variously pláced cloudy dark marks.

Hab. : South Australia—Port Lincoln District (Blackburn);
N.S. Wales—Windsor (Lea); West Australia—Swan River and
Beverley (Lea); Variety? King's Sound (Froggatt).

TACHYS TRANSVERSICOLLIS, Macleay.

The colour varies from pale testaceous (immature specimens) to
testaceous with the disc of the elytra infuscate, or even the whole

of the elytra infuscate; the head is blackish in mature specimens; the elytra are usually iridescent; the discoidal puncture on each elytron is situated along the fifth stria, considerably before the middle,—this is a constant character and valuable as an aid in the recognition of this species; the striæ of the elytra are faint and become obsolescent after the third.

Habits :—Found under sticks or stones near water in very damp situations.

Hab. : Queensland—Gayndah (Masters), Brisbane (Coates); N. S. Wales—Clarence River (Lea), Junee, Carrathool, Urana, and Mulwala (Sloane).

TACHYS MACLEAYI, n.sp.

Oval, subdepressed, lævigate. Head large, wide between eyes, prothorax subcordate ; posterior angles strongly marked, acute ; base (behind posterior angles) narrower than apex: elytra smooth, widely and lightly convex; two discoidal punctures on each elytron. Head piceous, prothorax obscure testaceous; elytra black with a large quadrate spot at shoulder, and a smaller round spot above apical declivity on each elytron pale testaceous; legs pale testaceous; antennæ pale testaceous with joints 3-6 infuscate.

Head lightly and widely bi-impressed between eyes. Antennæ filiform, long, slender. Prothorax lightly transverse, widest at anterior marginal puncture, angustate posteriorly; sides strongly rounded anteriorly, decidedly sinuate before posterior angles , anterior angles rounded; posterior angles triangular, prominent, acute, basal angles rounded; lateral border narrowly reflexed, reaching to sides of head; median line distinct; a well marked impunctate transverse line defining basal part of prothorax and reaching sides behind posterior angles. Elytra much broader than prothorax, wide between shoulders; base lightly rounded and margined on each side of peduncle; humeral angles obtuse; sides rounded, narrowed rather obliquely to apex; each elytron obtusely rounded at apex; three faint substriate impressions at apex of each elytron ; anterior discoidal puncture just behind humeral macula, posterior puncture in middle of subapical macula; border finely reflexed, extending from peduncle to apex; three or four setigerous punctures near margin behind shoulders, three foveiform submarginal impressions towards apex of each elytron.

Length 3, breadth 1·25 mm.

Hab. : King's Sound (Froggatt; Macleay Museum).

I know no Bembidiid closely allied to *T. macleayi;* its affinity is probably with *Bembidium bipartitum*, Macl., a species I have never critically examined. The legs and antennæ are long, the antennæ reaching back as far as the posterior maculæ of the elytra; the elytra are smooth without a submarginal stria on sides, and the marginal channel is not impressed.

PYRROTACHYS, n.gen.

Form parallel, depressed.

Head setigero-punctate, strongly constricted behind eyes; frontal impressions arcuate, extending backwards behind eyes.

Mandibles long, prominent, decussating.

Palpi with penultimate joint lævigate, swollen; terminal joint elongate, cylindrical.

Antennæ long, light, compressed, not narrowed to apex; terminal joint long, oval.

Prothorax setigero-punctate.

Elytra setigero-punctate, substriate, without striole on apex or submarginal stria on sides; margin not interrupted posteriorly by an internal plica.

Anterior tibiæ elongate; external side hardly oblique above apex; a short acute spur a little above apex externally.

Apparently this genus represents a distinct group among the Subulipalpi The absence of the slightest interruption of the margin of the elytra towards the apex or of any sign of a plica on the inner side of the elytra near the margin seems an important character. Dr. G. H. Horn, in his definition of the Bembidiini, makes "the margin interrupted posteriorly and with a distinct internal plica "* an important feature of the tribe.

PYRROTACHYS CONSTRICTIPES, n.sp.

Elongate, parallel, depressed Mandibles long, decussating; labrum deeply emarginate; antennæ with all the joints pubescent; prothorax transverse, narrowed to base; elytra pubescent, finely striate. Ferruginous; head reddish, eyes and adjacent parts black; elytra more obscurely coloured than prothorax, fuscous along suture and towards apex; legs testaceous; under parts of head and prothorax reddish, of body fuscous; antennæ testaceous at base, infuscate towards apex.

Head strongly bi-impressed; vertex convex, finely punctulate; impressions curved, diverging anteriorly and posteriorly, extending to sides of head behind eyes; front depressed between impressions; spaces between impressions and eyes convex, projecting sharply at base beyond sides of head; eyes prominent. Labrum large, deeply emarginate, a transverse linear impression a little before base. Prothorax depressed, transverse, widest at anterior marginal puncture, lightly narrowed to base, evidently wider at apex than base; sides very lightly rounded, shortly sinuate before posterior angles; anterior margin truncate; anterior angles obtuse; base truncate, a little oblique on each side behind posterior angles; these prominent, obtusely dentiform; a short depressed space along base, the impression defining this space curved forward in middle; median line well marked, not reaching either margin. Elytra narrow, a little wider than prothorax (1·7 × 1 mm.), depressed, parallel on sides, truncate at base, widely and evenly rounded (without sinuosities) at apex; whole upper surface covered with a short pubescence; striæ very fine, not perceptible after the fourth; marginal channel hardly impressed along sides, marked and punctate near shoulders; interstices flat;

---

* *Vide* Trans. Am. Ent. Soc. 1881, ix. p. 133.

three discoidal punctures on each elytron placed as in *Tachys leai*, S1.

Length 2·25-3, breadth 0·75-1 mm.

Hab. · N S. Wales—Tamworth (Lea).

The description is founded on a specimen of the largest size Mr. Lea regards the smaller specimens as representing a different species from the larger ones, but I have been unable to follow him in this; though, as the collector of a large number of specimens and a careful observer, his opinion in this matter should outweigh mine.

Appendix.

Specimens of a new species of *Tachys* were received from Mr. A. M. Lea after the completion of my notes on the genus, and too late to enable it to be put into its proper place according to the table of species given on p. 356; however its affinities will be found indicated in the note following the description below.

TACHYS OLLIFFI, n sp.

Robust, oval, convex; prothorax rather short, subcordate; five discoidal punctate striæ, a finely punctate submarginal stria and a well marked apical striole on each elytron.   Black; legs, upper side of mandibles and basal joint of antennæ testaceous; antennæ and palpi fuscous.

Head lævigate, convex; front lightly bi-impressed.   Antennæ short, stout.   Prothorax lævigate, transverse, subcordate; sides rounded, narrowed to base, shortly subsinuate before basal angles, anterior angles rounded, not marked; posterior angles subrectangular, slightly obtuse at summit, lateral border reflexed; lateral basal impressions wide, deep, short, extending to margin at basal angles; a light impression across base between lateral foveæ; median line obsolete.   Elytra a little wider than prothorax, convex, declivous on base; sides lightly rounded, first stria entire, punctate anteriorly; striæ 2-5 consisting of rows of punctures on disc becoming successively shorter (the punctures fewer, larger and more distant from one another in fourth and fifth rows);

submarginal stria not impressed on sides, but consisting of a row of fine punctures near margin; marginal channel hardly impressed, finely punctate; submarginal interstice depressed on sides; third interstice with two fine setigerous punctures.

Length 2·2, breadth 0·85 mm.

Hab. : N. S. Wales—Forest Reefs.

Allied to *T. flindersi*, Blkb., from which it differs by its more convex shape, its colour; its prothorax with the sides less strongly sinuate posteriorly, the base narrower, the basal foveæ deeper, the basal angles less prominently acute, &c. The prothorax appears to the eye of about equal width at base and apex.

Named in memory of Mr. A. S. Olliff, late Government Entomologist for New South Wales.

Note—With reference to my paper " On the Australian *Clivinides* (Fam. *Carabidæ*) " in the preceding Part of the Proceedings, attention is called to the following :—

CORRIGENDA.

Page 150, line 14—for *C. adelaidæ* read *C. tumidipes.*

Page 171, line 20—*for* clypeus *read* clypeal.

Page 180—omit line 2.

Page 181, line 5—*omit* South Australia *et seq.*

Page 182, line 27—for *C. adelaidæ* read *C. tumidipes.*

Page 195, line 18—for *C. adelaidæ*, Blkb., read *C. tumidipes*, Sl.

Page 253, line 7—for *C. adelaidæ* read *C. tumidipes.*

Page 253, line 27—*for* on *read* in.

Page 254, line 29—for *C. adelaidæ* read *C. tumidipes.*

Page 255, line 31—for *C. tenuipes* read *C. gracilipes.*

25 .

TWO NEW SPECIES OF *PROSTANTHERA* FROM NEW SOUTH WALES.

By R. T. Baker, F.L.S., Assistant Curator Technological Museum, Sydney.

(Plates xxii.-xxiii.)

Prostanthera discolor, sp.nov.

(Plate xxii.)

A tall slender shrub, 6 to 9 feet high, branches terete, branchlets only slightly angular; branches, branchlets, and calyx very hoary; branchlets slender and often nodding.

Leaves quite glabrous, lanceolate or oblong-lanceolate, obtuse, narrowing into a petiole 2 to 3 lines long, $\frac{3}{4}$ to over an inch long and $1\frac{1}{2}$ to rarely 3 or 4 lines broad, flat, entire, light underneath, dark coloured above, the midrib very prominent on the under side, particularly towards the petiole, but impressed above.

Flowers small in terminal compact heads or racemes, floral leaves reduced in size and very deciduous. Pedicels short, about half the length of the calyx. Calyx striate, very hoary pubescent particularly towards the base, the lips "blue," $1\frac{1}{2}$ lines long, both lips entire, the upper one slightly longer than the lower, and not so broad. Corolla about twice as long as the calyx, minutely pubescent, the lower lip exceeding the others. Stamens not exserted. Anthers mostly without any appendage to the connective, in fact, only rarely present, quite glabrous.

Hab.— At the foot of Cox's Gap Road, Murrumbo, Goulburn River, N.S.W. (R.T.B.)

As the anther appendages are only rarely met with in this species it might perhaps be placed in the Section Klanderia of Bentham's Table of the species of this genus, but the corolla is not similar in shape to that described under this group, and as this latter feature is such a well marked characteristic I have preferred to refer the species under consideration to the Section Euprostanthera. I am influenced in such a decision by its mode of inflorescence as well as by the fact that one or two species with only rudimentary appendages are already included in this Section.

In the species of Section Klanderia of *Prostanthera* the corolla tube is so very distinctive, being "narrow at the base, usually incurved and dilated upwards, the upper lip erect, concave or arched, the lower lip shorter or at any rate not longer and spreading," whilst in this species the corolla tube has the lower lip longer than the lobes, is not incurved or narrow at the base,—points that would not justify its being classified with this Section.

Neither can it be included under any of the species enumerated under Bentham's Series Convexæ and Subconcavæ, as all those have axillary flowers and anthers with one appendage about twice as long as the cell.

Of the species described under Euprostanthera it most resembles *P. rotundifolia* and *P. violacea* in its close terminal racemes, but differs from them in the form and size of its leaves, shape of corolla, and, of course, virtual want of anther appendages.

It also differs from *P. incana*, *P. hirtula*, and *P. denticulata* in its leaves being perfectly flat, also in inflorescence, indumentum, and absence of anther appendages, and for the same reason it is excluded from *P. rugosa*, *P. marifolia*, *P. rhombea*, *P. spinosa*, *P. cuneata*, *P. linearis*, *P. phylicifolia*, *P. decussata*, and *P. empetrifolia*.

Its greatest affinity is perhaps with *P. incisa* and *P. Sieberi*, but its leaves are so distinctly or uniformly entire that I prefer to regard it as a connecting link between those two species and *P. rotundifolia*. From the description of *P. incisa* one might be led to think it was that species, but when specimens of each are placed side by side the differences are very marked.

From the above considerations I conclude that in botanical sequence it should come after either *P. incisa* or *P. Sieberi*, and be followed by *P. rotundifolia*.

PROSTANTHERA STRICTA, sp.nov.

(Plate XXIII.)

A densely bushy shrub, drying black, with hirsute, terete branches and branchlets.

Leaves petiolate, lanceolate, sometimes broadly so, decussate, obtuse, entire, the margins recurved, scabrous-hispid above, bullate-rugose, dark coloured on the upper surface, whitish underneath, 4-9 lines long, 2-3 or even 4 lines broad, the midrib and lateral veins prominent underneath and impressed above, giving the surface a bullate appearance

Flowers opposite, in pairs in terminal compact cylindrical spikes or racemes, occasionally leafy at the base. Pedicels slender, above 1 line long. Bracts linear-subulate, almost as long as the calyx. Calyx $1\frac{1}{2}$ to 2 lines long, strongly ribbed towards the base, hirsute, glabrous inside except towards the mouth, where it is hoary pubescent, lips of about equal length and orbicular when surrounding the fruit. Corolla not twice the length of the calyx, glabrous, the lower lip longer than the other lobes. Anthers with one appendage exceeding the cell, the other adnate and shorter.

Hab.—Mt. Vincent, near Ilford, Mudgee Road, N.S W.

The compact terminal spikes or racemes give the plant a distinctive appearance, and by this mode of inflorescence it naturally falls into Bentham's Series R a c e m o s æ.

Its nearest ally in that Series is perhaps *P. denticulata*, that species resembling it somewhat in its leaves but not in inflorescence, indumentum, disposition of leaves, or anther appendages.

Its leaves bear a strong likeness also to those of *P. rugosa* and *P. marifolia*, but the attachment is quite different, and it differs also from these two species in its terminal inflorescence. It is also a much more rigid shrub than *P. marifolia*. The arrangement of its leaves would incline one from a casual examination to designate it *P. decussata*,—a Victorian species found on the rocky summits of the McAlister Range and Mt Mueller, with leaves narrower and smaller and not rugose, and with an inflorescence, which is axillary, and a transverse downy curved line inside at the base of the upper lip of the calyx,—characters absent in my species.

Following the classification of Bentham, I have placed this species in the Series R a c e m o s æ from its terminal spikes; and in botanical sequence after *P. denticulata*, having greatest affinity with that species, whilst resembling and possessing also some of the characters of *P. rugosa* and *P. marifolia*.

EXPLANATION OF PLATES.

Plate xxii.

Prostanthera discolor.

Fig. 1.—Twig showing inflorescence.

Fig. 2.—⎫
Fig. 3.—⎭ Individual flowers (enlarged).

Fig. 4—⎫
Fig. 5—⎬ Stamens, back and front views (enlarged).
Fig. 6—⎭

Fig. 7—Pistil and ovary.

Plate xxiii.

Prostanthera stricta.

Fig. 1—Twig with inflorescence.

Fig. 2—Individual flower (enlarged).

Fig. 3—Calyx showing bracts (enlarged).

Fig. 4—⎫
Fig. 5—⎬ Stamens with appendages (enlarged).

Fig. 6—Calyx with seeds (enlarged).

Fig. 7—Seed (enlarged).

EUCALYPTS AND LORANTHS IN THE RELATIONS OF HOST AND PARASITE: AND AS FOOD PLANTS.

By J. J. Fletcher.

The object of this paper was to introduce a discussion of the question whether, as has been stated, certain Loranths may be said to mimick Eucalypts.

NOTES AND EXHIBITS.

Mr. Rainbow exhibited a spray of Silver wattle (*Acacia dealbata*) with hymenopterous galls simulating the appearance of Lepidopterous larvæ. The specimen was procured by Mr. Affleck, M.L.A., at Bundarra, N.S.W.

Mr. Baker exhibited specimens of the plants referred to in his paper.

Mr. Froggatt exhibited a collection of Australian Coccids comprising representatives of thirty genera and ninety species, and including a number of rare species described by Mr. Maskell in some of his recent papers on this family. Among the rarer species of note were *Ceronema banksiæ* found upon *Banksia serrata*, *Aspidiotus pallens* on *Macrozamia*, *Mytilaspis spinifera* upon *Acacia pendula*, *Eriococcus spiniger* and *Ctenochiton eucalypti* upon *Eucalyptus*; also the well known St. José scale (*Aspidiotus perniciosus*) upon an apple bought in a Sydney fruit shop.

WEDNESDAY, SEPTEMBER 30TH, 1896.

The Ordinary Monthly Meeting of the Society was held at the Linnean Hall, Ithaca Road, Elizabeth Bay, on Wednesday evening, September 30th, 1896.

The President, Mr. Henry Deane, M.A., F.L.S., in the Chair.

Mr. Gilbert Turner, The Ridges, Mackay, Q , was elected a Member of the Society.

DONATIONS.

Melbourne University—Calendar, 1897. *From the University.*

Pamphlet entitled "Interzoœcial Communications in Flustridæ." By A. W. Waters, F.R.M.S., F.L.S. *From the Author.*

Pharmaceutical Journal of Australasia. Vol. ix. No. 8 (Aug., 1896). *From the Editor.*

Australasian Journal of Pharmacy. Vol. xi. No. 129 (Sept., 1896). *From the Editor.*

Zoological and Acclimatisation Society of Victoria—Twenty-fifth and Thirty-first Annual Reports (1888 and 1894). *From the Society.*

Museum of Comparative Zoology at Harvard College, Cambridge, Mass.—Bulletin. Vol. xxix. Nos. 5-6 (July, 1896). *From the Director.*

Linnean Society, London—Journal—Botany—Vol. xxx. No. 211 (Sept., 1895); Vol. xxxi. Nos. 212-217 (Nov., 1895—July, 1896) : Journal—Zoology—Vol. xxv. Nos. 161-162 (July, 1895 —Feb., 1896); General Index to Volumes i.-xx. (1838-90): Proceedings, Session 1894-95 : List, 1895-96 *From the Society.*

Linnean Society of London—Transactions. Second Series—Zoology. Vol. vi. Parts 4-5 (Feb -June, 1896) : Second Series—Botany. Vol. iv. Parts 3-4 (Dec., 1895—March, 1896): Vol. v. Parts 2-3 (Oct., 1895—May, 1896). *From C. Hedley, Esq., F.L.S.*

Royal Society, London—Proceedings. Vol. lix. No. 357 (June, 1896). *From the Society.*

Royal Irish Academy—Proceedings. Third Series. Vol. iii. No. 5 (May, 1896). Transactions. Vol. xxx. Parts xviii.-xx. (March-April, 1896): List of Members, 1896. *From the Academy.*

Nederlandsche Entomologische Vereeniging—Tijdschrift voor Entomologie. xxxix. Deel. Afl. 1-2 (June, 1896). *From the Society.*

Société d'Horticulture du Doubs, Besançon—Bulletin. Série Illustrée. No. 7 (July, 1896). *From the Society.*

Perak Government Gazette—Vol. ix. Nos. 18-19 (July-Aug., 1896). *From the Government Secretary.*

Department of Mines and Water Supply, Victoria—Annual Report, 1895. *From the Secretary.*

Field Columbian Museum, Chicago—Anthropological Series. Vol. i. No. 1 (Dec., 1895). *From the Director.*

Chicago Academy of Sciences—Bulletin. Vol. ii. No. ii. (1895): Thirty-eighth Annual Report (1895). *From the Academy.*

American Philosophical Society—Proceedings. Vol. xxxiv. Nos. 148-149 (July-Dec., 1895). *From the Society.*

American Museum of Natural History, N. York—Bulletin. Vol. vii. (1895): Vol. viii. (1896). Sigs. 10-12 (pp. 145-192— Aug.). *From the Museum.*

New York Academy of Sciences—Annals. Vol. viii. Nos. 6-12 (Nov., 1895): Memoir i. Part 1 (1895). *From the Academy.*

Academy of Natural Sciences of Philadelphia--Proceedings, 1895. Part iii. (Oct.-Dec.) *From the Academy.*

California Academy of Sciences—Proceedings. Second Series. Vol. v. Part i. (1895). *From the Academy.*

U S. National Museum—Annual Report, 1893 : Proceedings. Vol. xvii (1894): Bulletin. No. 48 (1895). *From the Director.*

Naturforschende Gesellschaft zu Freiburg, i. B.—Berichte. ix. Band. 1-3 Hefte (June, 1894—Nov., 1895). *From the Society.*

Gesellschaft fur Erdkunde zu Berlin—Verhandlungen. Bd. xxii. (1895), Nos. 8-10. *From the Society.*

Colonial Museum and Geological Survey of New Zealand— Sixteenth, Seventeenth, Nineteenth, and Twenty-fifth Annual Reports on the Colonial Museum and Laboratory (1882-91): Reports of Geological Explorations during 1881, 1882, and 1883-84: Broun's Manual of the New Zealand Coleoptera. Parts iii. and iv. (1886). *From Professor T. J. Parker, D.Sc , F.R.S.*

Royal Geographical Society of Australasia, Queensland Branch —Proceedings and Transactions. 11th Session (1895-96). Vol. xi. *From the Society.*

Geological Society of London—Quarterly Journal. Vol lii. Part 3 (No. 207; Aug, 1896). *From the Society.*

Société Néerlandaise de Zoologie—Compte-Rendu des Séances du Troisiéme Congrès International (Sept., 1895). *From the Society.*

Bureau of Agriculture, Perth, W.A.—Journal. Vol. iii. Nos. 21-22 (Aug.-Sept., 1896). *From the Bureau*

Hooker's Icones Plantarum. Fourth Series. Vol. v. Part iv. (July, 1896). *From the Bentham Trustees.*

Verein für Erdkunde zu Leipzig—Mitteilungen, 1895 : Wissen-schaftliche Veroffentlichungen. iii. Bd. 1 Heft (1896). *From the Society.*

Naturwissenschaftlicher Verein des Reg.-Bez., Frankfurt a. O. —Helios. xiii. Jahrg. Nos. 7-12 (Oct., 1895—April, 1896): Societatum Litteræ. 1895. Jahrg. ix. Nos. 10-12 (Oct.-Dec.): 1896, Jahrg. x. Nos. 1-6 (Jan.-June). *From the Society.*

Department of Agriculture, Sydney—Agricultural Gazette. Vol. vii. Part 8 (August, 1896). *From the Hon. the Minister for Mines and Agriculture.*

Pamphlet entitled "The Submarine Leakage of Artesian Waters." By R. L. Jack, F.G.S., F.R.G.S. (July, 1896). *From the Author.*

American Naturalist. Vol. xxx. No. 356 (August, 1896). *From the Editors.*

U S. Department of Agriculture—Division of Entomology— Bulletin. No. 31 (1893): Division of Ornithology and Mammalogy —Bulletin. Nos. 5 and 7 (1895). *From the Secretary of Agriculture.*

Zoological Society of London—Proceedings, 1896. Part ii. (Aug.) *From the Society.*

Geological Survey of India — Records. Vol. xxix. Part 3 (1896). *From the Director.*

Indian Museum, Calcutta—Natural History Notes. Series ii. No. 10 (Sept., 1894). *From the Museum.*

Zoologischer Anzeiger. xix. Band. Nos. 509-510 (Aug., 1896). *From the Editor.*

Archiv für Naturgeschichte. lxii Jahrg. (1896). i. Band. 1 Heft. *From the Editor.*

L'Académie Impériale des Sciences de St. Pétersbourg—Annuaire du Musée Zoologique. 1896. Nos. 1-2. *From the Academy.*

Royal Microscopical Society—Journal, 1896. Part 4 (August). *From the Society.*

Sociéte Hollandaise des Sciences à Harlem—Archives Néerlandaises. T. xxx. 2ᵐᵉ Liv. (1896). *From the Society.*

Société Royale Linnéenne de Bruxelles—Bulletin. 21ᵐᵉ Année. No. 9 (Aug., 1896). *From the Society.*

Royal Society of South Australia—Transactions. Vol. xx. Part i. (Sept., 1896). *From the Society.*

BRANCHII.

By J. Douglas Ogilby.

In the present paper I have endeavoured to reduce to some appearance of order the history of the Australasian Lampreys and such meagre and for the most part inaccurate literature as appertains thereto. It is undeniable that some such work had become necessary owing to the diversity of the views held by the various writers who have approached the subject, and which culminated in the recognition by Sir William Macleay of four genera and six species, two of the former and an equal number of the latter having been founded on ammocœtal or immature individuals; this list I have found it necessary to reduce to three genera, each of which is represented by a single species.

The first author to whom the honour of recording the existence of a hyperoartian Marsipobranchiate in the southern hemisphere is due is Sir John Richardson, who, under the name of *Petromyzon mordax*, described and figured a species in the Ichthyology of the Erebus and Terror; six years later Dr. John Edward Gray published a "Synopsis of the *Petromyzonidæ*" in the Proceedings of the Zoological Society of London, in which Richardson's species is made the type of a new genus *Mordacia*, while for a closely allied form from the rivers of Chile a second genus, *Carayola*, is proposed. Besides these the same paper contains descriptions and figures of two other austrogæan genera, namely, *Geotria*, founded on a specimen picked up on the beach in Hobson's Bay (*see p.* 425) by Mr. R. A. Pain, and by him forwarded to the British Museum; and *Velasia*, the type of which was a Chilian specimen in the collection of the same institution.

In a series of three papers (1857-1863) Philippi gave some particulars as to the Chilian Lampreys, and described two new species as *Petromyzon anwandteri* and *acutidens*; these papers appeared in Wiegmann's Archiv.

In the Annals of the Museum of Buenos Aires for 1868 Burmeister described a very curious form under the name of *Petromyzon macrostomus*, and as this Lamprey has no place in the Australasian fauna it may be dismissed here with the remark that it forms the type of a genus *Exomegas*, Gill (*see p.* 425), and is very rare, only two examples being known to science, the first having been picked up in the streets of Buenos Aires, and the second collected in the Bay of Monte Video.

Two years subsequently to the publication of Burmeister's paper the eighth volume of Dr. Gunther's Catalogue of Fishes appeared, and his treatment of the conclusions of previous authors is, to say the least of it, revolutionary; as a commencement *Mordacia mordax*, Gray, from Tasmania, *Caragola lapicida*, Gray, *Petromyzon anwandteri*, Philippi, and *P. acutidens*, Philippi, all three from Chile, are associated under the common name *Mordacia mordax*, though the author had at his disposal only Dr. Gray's two original specimens, one of which was in a notoriously bad condition; even the selection of the generic name was unfortunate, *Caragola* having a slight precedence over *Mordacia*, and though, for reasons hereafter stated, I have adopted the name *Mordacia*, it is not to be expected that all other authors will be equally complaisant,[*] and we shall, therefore, be cumbering our pages with a dual synonymy, one school of writers adhering to *Mordacia* while the other as strenuously upholds the claims of *Caragola*; all which confusion would have been avoided by the initial attention to the strict rules of nomenclature. Continuing, Dr. Gunther united Gray's *Geotria* and *Velasia*, a conclusion which is not borne out by a more careful examination of the two forms, and announced the occurrence of the latter in New

---

[*] Eigenmann & Eigenmann in "A Catalogue of the Fresh-water Fishes of South America" (Proc. U.S. Nat. Mus. xiv. 1891, p. 24) call the Chilian species *Caragola mordax*, thus possibly further confusing the synonymy as it is very unlikely that the Australian and Chilian forms are identical, and in view of my own discoveries in regard to the marked differences between *Velasia* and *Geotria* it is at least possible that both *Caragola* and *Mordacia* may be valid.

Zealand waters, determining the species found there as Gray's
chilensis, in which identification also I am not prepared to follow
him; he also records under the same name a Lamprey from
"Swan River," but whether this is the well known river of West
Australia or some other does not appear (*see p.* 419). In the
following year the same author described a new species of *Geotria*
from Tasmania as *G. allportii*, a proceeding which appears
unnecessary.

With this description the history of the Australasian Lampreys
as species, so far as exotic writers are concerned, ceases, with the
exception of two notices by Dr. Klunzinger of the occurrence of
Mordacia mordax in the estuary of the Murray in 1873, and of
Geotria australis as far west as King George's Sound in 1880.

With the cessation of outside interest in our Lampreys and
the conclusion of the British Museum Catalogue, an unwonted
and most gratifying activity on the subject of our fishes began to
be manifested by Australian writers, and among the rest the
Lampreys came in for their full share of attention.

The year 1872 is memorable for the production of two important essays, one of these being "The Fishes of New Zealand" by
Capt. Hutton, to which was appended a short account of the
edible species from the pen of Dr. Hector; the other, and in
many respects the more important of the two, was contributed by
Count Castelnau to the Proceedings of the Zoological and
Acclimatisation Society of Victoria under the title of "A Contribution to the Ichthyology of Australia." Both these authors,
and indeed all subsequent Australasian writers, accept Dr.
Gunther's synonymy without comment or protest.

In the first essay alluded to only Gunther's *Geotria chilensis*[*]
is mentioned, his description being copied direct from that author's
work, with the addition of certain rivers specially referred to as
being frequented by that species. And, as it must be done sooner
or later, I may as well take this opportunity of entering a strong

---

[*] *G. australis* was added to the New Zealand fauna in the following year
by Capt. Hutton.

protest against the practice which is so prevalent among writers on our fishes of copying the descriptions and remarks from the British Museum Catalogue without any attempt being made to test their accuracy, and by so doing perpetuating error, creating confusion, and indefinitely postponing the dawn of that accurate knowledge of our native fauna which every admirer of the marvellous products of our country must ardently desire.

Very different, however, is Count Castelnau's contribution; in it we find by far the best account of two of our species as yet published, and though in the case of one of them the author had evidently determined the species wrongly, this does not detract from the value of his remarks, while the very accuracy of his description has enabled me to correct his error without difficulty, a thing which would have been impossible had he also been content to be a mere copyist. Following his usual practice he has, however, given generic and specific names to two individuals, one of which was an ammocœte while the other had only just passed through its metamorphosis and assumed the habits and in part the dentition of the adult. Count Castelnau's long experience should have taught him to avoid this pitfall. His paper, therefore, increased the number of Australasian species to six, distributed among four genera, and at this they have been left up to the present time by all writers, even Sir William Macleay republishing without comment the descriptions of these nominal species in his Catalogue of Australian Fishes, where, at least, we might have expected that some effort would have been made to correct the errors of his predecessors.

I append here in parallel columns the names of the species as given by Macleay and those which I recognise as valid in the following paper :—

Mordacia mordax	*Mordacia mordax.*
Neomordacia howittii	
Geotria chilensis	*Velasia stenostomus.*
Yarra singularis	
Geotria australis	
Geotria allporti	*Geotria australis.*

In connection with the reinstatement of Gray's *Velasia* I wish to call the attention of those who may have the opportunity of examining this genus and *Geotria* during the ammocœtal stage and immediately after the metamorphosis has taken place, to the significance of the dental furrows in the latter genus; from the examination of the adult it appears to me that the evolution of the laminæ in *Geotria* will prove to be materially different from that which holds good for *Velasia*.

Finally, it is hoped that the present paper will not only throw some light on the affinities of these various forms, but also induce some of our southern naturalists to spare time for the study of these interesting animals, of whose life history much still remains to be learnt.

Class MARSIPOBRANCHII.

The Myzons.

Skeleton membrano-cartilaginous; skull imperfectly developed, not separate from the vertebral column, which consists of a stout notochord enveloped in a fibrous sheath; neural cartilages present, small; hæmal sheath present in the caudal region only.   Lower jaw, ribs, limbs, shoulder-girdle, and pelvic elements wanting Gills six or more on each side, represented by fixed sacs and destitute of branchial arches.   Mouth suctorial and subinferior, more or less circular, with or without lips.   Nasal aperture single. Eyes present or absent.   Vertical fins present, usually continuous around the tail, supported by feeble rays, which are rarely articulated or branched; no paired fins.   Skin naked.   Heart without arterial bulb.   Air-vessel absent.   Alimentary canal straight, little dilated, without pyloric appendages, pancreas, or spleen.   Generative openings peritoneal.

Etymology :—μαρσίπιον, a pouch; βράγχια, gills, in allusion to the sac-like formation of these organs.

Distribution :—Seas and rivers of the temperate regions of both hemispheres, no species having as yet been discovered either in high polar latitudes or within the tropics.

Geologically the Cyclostomes date back to the Lower Devonian.

Up to the present time but little has been definitely proven with regard to the degree of relationship which exists between the Marsipobranchiates on the one hand and the more recently and highly developed Teleostomes on the other, but the preponderance of evidence tends to show that the former are the survivors of a very primitive type of the Chordates, the oldest living representatives of which are to be found among the *Heptatrematidæ*.

The *Marsipobranchii* are divisible into two Orders, which may be briefly characterised as follows :—

Nasal duct tube-like, penetrating the palate; mouth without lips; eyes wanting; snout with barbels ... ...
$$\text{HYPEROTRETI}^{*}$$
Nasal duct a blind sac, not penetrating the palate; lips and eyes present; no barbels ... ...
$$\text{HYPEROARTII}[\dagger]$$

The first of these Orders contains two Families, the *Heptatrematidæ* and the *Myxinidæ*, the members of which are variously known as Hag-Fishes or Borers; they are small, colourless, more or less parasitic, marine animals, living at a moderate depth, and wholly carnivorous. In places where they are common they do no inconsiderable damage to the fishermen by destroying the hooked fishes, into whose body they burrow and upon whose tissues they feed internally. They inhabit nearly all the seas of temperate regions, and three genera, *Polistotrema*,[‡] *Heptatrema*,[||] and *Myxine*[§] have been differentiated.

---

[*] ὑπερώα, palate; τρητός, perforated.

[†] ὑπερώα, palate; ἄρτιος, entire.

[‡] *Polistotrema*, Gill, Proc. U.S. Nat. Mus. 1881, p. 30. Type, *Gastrobranchus dombey*, Lacépède. πολύς, many; ἱστός, vertical; τρῆμα, a perforation; in allusion to the increased number of external gill-openings.

[||] *Heptatrema*, Dumeril, ? Diss. Poiss. Cyclost. Type, *Petromyzon cirrhatus* (Forster), Bloch & Schneider. ἑπτά, seven; τρῆμα, a perforation; = *Bdellostoma*, Müller, Abh. Ak. Wien, 1834, p. 79 (1836).

[§] *Myxine*, Linnæus, Syst. Nat. i. 1758. Type, *Myxine glutinosa*, Linnæus. μυξῖνος, a slimy fish, from μύξα, slime; so named on account of the excessive amount of slime secreted by the mucous sacs of these animals, which is so great that the exudation from a single living example is sufficient to gelatinise a pailful of water.

26

So far, however, no Hyperotrete can be satisfactorily recorded as having occurred within our limits, but *Heptatrema cirrata*, being an inhabitant of the New Zealand seas, may occur or be represented by an allied form on our coast.*

The following synopsis will serve to show the most obvious characteristic of the three genera.

Eleven or more branchial apertures on each side; the base of the tongue situated between the seventh and eighth pair of branchiæ   ...   .
POLISTOTREMA.

Six or seven branchial apertures on each side; the base of the tongue situated between the anterior pair of branchiæ   ...
HEPTATREMA.

A single branchial aperture on each side   ...   .
MYXINE.

In all probability each genus is represented by a single valid species only; sexually they are hermaphrodite, but the ova and sperm attain maturity in each individual at a different period, the ripening of the latter taking place earlier in life than that of the former.

Order HYPEROARTII.

THE LAMPREYS.

Body anguilliform, naked, compressed or subcylindrical in front, compressed behind; mouth subcircular or oval, suctorial; lips present, usually fringed, but without barbels; nostril at the upper surface of the head, the nasal duct a blind sac, not penetrating the palate. Eyes present, small. Branchial apertures seven on each side, situated behind the head, the inner branchial ducts terminating in a common tube. Teeth cuticular, horny, simple or multicuspid, resting on soft papillæ, those immediately above and immediately below the opening of the œsophagus more or

---

* Krefft indeed (Australian Vertebrata, p. 779) gives, under the heading of *Bdellostoma cirrhatum*, the locality " New Zealand and Australian Rivers "; but this is obviously a mistake and refers to one of the Lampreys.

less specialised. Dorsal fin more or less deeply divided by a notch, the posterior portion usually continuous with the caudal. Intestine with a rudimentary spiral valve. Eggs small, fertilised after extrusion. Sexes separate.

Etymology:—ὑπερώα, palate; ἄρτιος, entire : in reference to the non-perforation of the palate by the nasal duct.

Distribution :—Seas and rivers of the temperate zones of both hemispheres.

All the Lampreys are subject to a metamorphosis; during the earlier stage of their existence, when they are known as *ammocœtes*, the eyes are in a rudimentary condition and they are entirely without teeth, their food consisting solely of vegetable substances gathered from the mud in which they live.

These ammocœtes are not unfrequently found of an equal or even larger size than individuals of the same species in which the eyes and teeth have already undergone development, this being due to arrested growth of these organs on the part of the individual

Several distinct genera, such as *Ammocœtes*, *Scolecosoma*, &c., have been constituted for the inclusion of these immature forms.

The suctorial disk which is so characteristic of the Lampreys is useful to them in various ways; it serves as an instrument by means of which they are able to adhere to rocks, piles, sunken logs and the like, and so resist the force of the current and escape the necessity for such continuous and violent muscular exertion as would be imperative in an animal possessed of such feeble swimming powers; by it they are able during the spawning season to remove stones and similar obstructions from that portion of the river bed which has been selected as suitable to the formation of the nesting-place or "redd," and, after the task of depositing the ova has been completed, to replace the stones, and so minimise the danger to which the eggs would be exposed in the event of the occurrence of heavy floods during the period of incubation; and finally, by it they are enabled to attach themselves to the substances which form their food.

Up to the year 1894 ichthyologists were content to segregate the various species of Lampreys in a single family, to which the name *Petromyzontidæ* had been given by Risso as early as 1826 (*Eur. Mérid. iii. p. 99*), the title being altered six years later by Bonaparte (*Saggio, &c. p. 41*) to the more correct orthographic reading *Petromyzonidæ*. So long ago, however, as 1882 Dr. Gill (*Proc. U.S. Nat. Mus v. p 524*) proposed to separate the genus *Mordacia* (= *Caragola*) from the remaining *Hyperoartii* in a sub-family *Caragolinæ*. In the volume of the same periodical for 1894 (*p. 109*) the same author went a step further and raised his *Caragolinæ* to family rank under the name *Mordaciidæ*, he having in the meanwhile become reconciled to the use of *Mordacia.*

In this later paper the author, in support of the proposed family, pertinently remarks :—"It behooves those who may object to these families to consider why the character used to distinguish them should not be of equal value with the union or separation of the lower pharyngeal bones and like modifications generally used."

As Dr. Gill's contention appears to me to be perfectly sound, I have accepted the families as here defined by him.

Analysis of the Families of the Hyperoartii.

Two distant lateral tuberculigerous laminæ developed from the upper arch of the annular cartilage ..   ..

MORDACIIDÆ.

A single median tuberculigerous suproral lamina developed from the upper arch of the annular cartilage   ...   .

PETROMYZONIDÆ.

There is one other character separating these two families, namely. the labial fringes, which taken in conjunction with the more perfect dentition of the former, appears to me worthy of special notice, all the *Petromyzonidæ* are provided with a more or less con-spicuous fringe of papillæ around the outer rim of the suctorial disk, which fringe is rudimentary in *Mordacia*. If we look upon these papillæ as having developed from the oral barbels of the more ancient *Hyperotreti*—and in so doing I scarcely think that we are assuming too much—it follows that both in this character as well as in the dentition the Mordaciids have attained to a higher degree of development than the Petromyzonids.

MORDACIIDÆ.

Caragolinæ, Gill, Proc. U.S. Nat. Mus. v. 1882, p 524.

Mordaciidæ, Gill, Mem. Nat. Acad. Sc. vi. p. 129, 1893 (*no definition*) *and* Proc. U.S. Nat. Mus. xvii. 1894, p. 109.

Two distant lateral tuberculigerous laminæ developed from the upper arch of the annular cartilage. Labial fringe rudimentary. Other characters similar to those of the Order.

One genus only.

D i s t r i b u t i o n:—Seas of South-eastern Australia, Tasmania, and Chile; entering fresh waters for the purpose of breeding.

MORDACIA.

Caragola, Gray, Proc. Zool. Soc. London, 1851, p. 239.

Mordacia, Gray, *l.c.*

Body elongate and slender, subcylindrical in front, the tail and a part of the body compressed; head small, oblong, attenuated, and somewhat depressed, with slightly pointed snout, suctorial disk moderate, oval, subinferior, extending backwards to the orbital region, with a well developed simple external lip, between which and the rim of the disk is inserted a regular series of short papillæ; rim of disk thin, forming a free, simple, cutaneous flap behind; surface of disk feebly plicated on its outer, smooth on its inner moiety. No gular pouch.* Branchial orifices small and subcircular, with a low raised rim and a well developed valve inserted anteriorly. Maxillary dentition consisting of two subtriangular plates, each of which is provided with three strong, sharp, hooked cusps, arranged in the form of a triangle; mandibular plate low and crescentic,

---

* The Chilian *Mordacia* is said by Philippi to be occasionally provided with a gular sac; this has never been observed in the Australian species, and is most unlikely.

cuspidate; disk with three strong unicuspid teeth anteriorly, the basal pair followed by two or three similar teeth, the sides and hinder portion with a series of broad tri- or bicuspid lamellæ; a row of small teeth inside the rim of the disk; tongue with two pairs of narrow multicuspid plates inserted on its dorsal surface and a finely cuspidate transverse plate below. Dorsal fin originating a short distance behind the middle of the body, divided into two portions (in the adult) by a short interspace, the anterior small, the posterior much larger and more or less continuous with the caudal, which is free or nearly so. Tail moderate, the vent situated well behind the middle of the second dorsal fin. No conspicuous series of pores on the head or body.

Etymology:—*Mordax,* voracious.

Type:— *Mordacia mordax,* Gray = *Petromyzon mordax,* Richardson.

Distribution:—South-eastern Australia, Tasmania, and Chile.

The absence of this genus from the New Zealand fauna when contrasted with its South American range is somewhat remarkable.

With regard to the propriety of retaining the generic name *Mordacia* for these Lampreys in place of *Caragola,* which both by a slight priority and by a more accurate diagnosis is fairly entitled to selection, I cannot do better than to quote the opinions of Drs. Gill and Boulenger as follows :—

The former remarks (*Proc. U.S. Nat. Mus. 1894, p. 109*):— " In 1882 I used in preference the first name (*Caragola*) based on a perfect individual. I have since been led to believe that the precedence of one name by such a little margin as *Caragola* has over *Mordacia* has no value, and that aptness of diagnosis, however desirable, is not necessary to procure priority, and I have, therefore, followed Dr. Günther in accepting the name *Mordacia* instead of *Caragola.*"

Dr. Boulenger writes (*in lit.*): " I cannot agree with you that the fact of one name appearing before another in the same book constitutes priority, and it would be a pity to alter the well known name *Mordacia* to *Caragola.*"

It is only in deference to the opinions as expressed above, of two so eminent scientists, that I have decided to adhere to the more generally accepted name *Mordacia;* nevertheless it is due to myself to say that the substitution of that name for *Caragola* is distinctly repugnant to me; so long as the rule remains in force, which provides that the earliest name, all other requirements having been complied with, shall take precedence, I cannot coincide with the contention that the accident of two names being published in the same volume, or even, as in this case, on the same page of the same volume, can under any circumstances justify our rejection of the earlier in favour of the later name; by so doing we are assisting to open a rift which may in course of time imperil the stability of the entire fabric; while the plea that a name should be retained because it is better known is sentimental and unsound, and therefore unworthy of consideration.

As is the case with all the Lampreys the dentary plates are provided with a horny covering, which may easily be removed in layers, but except for the necessary decrease in size both of plate and cusps consequent on the removal of each separate layer, no alteration in their appearance is noticeable, unless the entire corneous lamina be lost, and the underlying papillary prominence be thus exposed to view.

Gray's description of *Mordacia* was based on a specimen from Tasmania, the dentition of which was imperfect through the loss of the corneous lamellæ of many of the plates, while his type of *Caragola* was a Chilian example in which the lamellæ were intact; the diagnosis of *Caragola* is therefore more correct; surely an additional argument for the retention of that name.

Some interesting remarks on the pineal eye in this Lamprey, from the pen of Prof. Baldwin Spencer, will be found in the Proceedings of the Royal Society of Victoria, Vol. ii. 2nd Series, p. 102, 1890.

MORDACIA MORDAX.

Petromyzon mordax, Richardson, Voy. Erebus & Terror, Ichth. p. 62, pl. xxxviii. ff. 3-6*, 1845.

Mordacia mordax, Gray, Proc. Zool. Soc London, 1851, p. 239, pl. iv. f. 6†, *and* Catal. Chondropt. p. 144, pl. i. f. 6, 1851; Günther, Catal. Fish. p. 507, 1870; Klunzinger, Arch. f. Natur. xxxviii 1872, p. 45, *and* Sitzb. Ak. Wien, lxxx. i. 1879, p. 429 (1880); Castelnau, Proc. Zool. & Acclim. Soc. Vict. i. 1872, p. 229, *and* Edib. Fish. Vict. p. 17, 1873; Macleay, Proc. Linn. Soc. N. S. Wales, vi. 1881, p. 382; Johnston, Proc. Roy. Soc. Tas. 1882, p. 141 (1883), *and* 1890, p. 39 (1891); Stephens, Proc. Linn. Soc. N.S. Wales (2) i. 1886, p. 506; Lucas, Proc. Roy. Soc. Vict. (2) ii. 1890, p. 46.

Short-headed Lamprey.

Disk oval, its width when fully expanded somewhat less than its length, its posterior margin reaching to or nearly to the level of the eyes. Eyes conspicuous, the nasal tube opening a little in advance of their anterior margins. The distance between the extremity of the snout and the nasal opening is 21 to 26½ in the total length and 1¾ to 2 in that preceding the first branchial orifice, which is situated a little nearer to the last orifice than to the tip of the snout; the space between the last‡ orifice and the extremity of the snout is 6⅓ to 6⅘ in the total length. Maxillary plates widely separated; each plate is armed with three strong, acute cusps, the tips of which are directed slightly backwards; they are arranged in the form of a triangle, having the apex in

---

* Richardson's figure is unreliable, being taken from a specimen in which the lateral corneous lamellæ had been lost, a single papillary prominence alone being left to represent each plate.

† Copied from Richardson.

‡ In Richardson's description this measurement is erroneously given as the space in front of the *first* gill-opening.

front, the anterior cusp being rather stronger than the basal pair; mandibular plate with nine cusps, the last but one (rarely the last two) on each side much enlarged, the median one generally so; the discal dentition consists of three strong teeth anteriorly, the basal pair being on a line with the inner borders of the maxillary plates; they are similar in shape and arrangement to each triad of maxillary cusps, but differ in being entirely disconnected, though contiguous, at their bases; behind these a series of broad sharply-ridged lamellæ extends backwards along the sides of the disk close to the gular cavity and is continued behind the mandibular plate; each lamella is furnished with a strong cusp near its inner extremity and a smaller one at its outer, the lateral ones having a supplementary cusp outside and partially behind the inner cusp; between the discal lamellæ and the rim of the disk there is a row of small, sharp, hooked teeth, tongue with two narrow elongate plates arranged along each side of its dorsal aspect; the anterior pair are almost parallel, the distal extremity, however, being curved outwards and backwards, and armed with seven or eight fine subequal cusps and an enlarged terminal cusp, while on the linear portion seven cusps are present, the middle ones being the longest and the terminal one small; the posterior and outer pair of plates are inserted obliquely, with the convergent ends in front and in contact with the middle of the base of the inner plates; each is furnished with from twelve to fourteen fine cusps, which gradually decrease in size from the front; the ventral surface is armed at the base with a deep, transverse, V-shaped plate, the apex of which is radical: the outer border of each limb forms a deep concavity, which terminates in a stout, hooked cusp, outside the base of which the plate is curved inwards and backwards, both the recurved portion and the limb itself being armed with comb-like cusps, two or three of which on either side of the apex, are somewhat enlarged. The vent is situated beneath or a little in advance of the commencement of the last third of the second dorsal fin; the length of the tail is $6\frac{2}{3}$ to $7\frac{1}{4}$ in the total length. The distance between the origin of the dorsal fin and the tip of the tail is $1\frac{1}{4}$ to $1\frac{2}{3}$ in its distance from the extremity of the snout ; the anterior portion of the fin is small and evenly convex, and passes imperceptibly into the dorsal integument at both ends, the length of its base is from 1 to $1\frac{1}{3}$ in the interspace between the two divisions of the fin and $2\frac{1}{4}$ to $2\frac{1}{2}$ in the base of the second portion, which is connected with the caudal fin by a more or less conspicuous rayless membrane ; * the lower lobe of the caudal is more developed than the upper, to which it is joined round the extremity of the tail by a membrane similar to that which connects it with the dorsal.   Head and body without conspicuous pores.

In the ammocœte both the dorso-caudal and the intercaudal membranes are well developed and the dorsal is continuous, but in large examples the intervening membranes have entirely disappeared.

In the Nepean specimen (125 millimeters) the dorsals are con-nected by a low cutaneous fold, as also are the second dorsal and caudal, the fold in this case being almost as high as the latter fin but rayless; the lower lobe of the caudal extends forwards to the vent, and there is also a distinct fold for a considerable distance in front of the vent , the maxillary teeth are as large as in the adults.

Upper surfaces rich olive brown, the sides golden brown, lighter below ; lower surface of the head and the throat silvery , fins greenish yellow

Castelnau's description of the colours, taken from a recent specimen, is as follows :—

" Bluish gray, darkest on the back, head yellowish; eye silvery, first dorsal gray ; second bordered with pink, its posterior part black ; caudal black, with a pink margin."

The earliest intimation of the occurrence of a Lamprey in the Australasian Colonies is to be found in the Ichthyology of the Erebus and Terror, where Sir John Richardson describes this species from a Tasmanian example, without, however, separating it from the arctogæan genus *Petromyzon,* six years later, however,

---

* In large examples even this disappears.

Dr. Gray, when engaged on his Catalogue of Chondropterygians rightly removed the Tasmanian species from that genus under the name *Mordacia*, and further proposed for a very similar Chilian Lamprey the name *Caragola lapicida*, the generic differences relied on being due to the defective dentition of the former.

In 1863 Philippi (*Wiegm. Arch. p. 207, pl. x. f. b.*) described and figured a Chilian species under the name of *Petromyzon anwanlteri*, and in the following year (*l.c. p. 107, and Ann. & Mag. Nat. Hist. 3rd. ser. xvi. 1865, p. 221*) described yet another species from the same territory as *P. acutidens*.

All these various forms, Tasmanian and Chilian, were united together by Dr. Gunther in 1870 under the common name *Mordacia mordax*, a conclusion which—seeing that he had but a single example from each so widely separated locality, and that one of these (the Tasmanian) was admittedly in bad condition— is so manifestly inconsiderate that I prefer to regard the Chilian form distinct from that described by Richardson until conclusive evidence to the contrary shall have been brought forward.*

B r e e d i n g :—The habits of the Short-headed Lamprey during the breeding season are quite unknown, but it is not probable that they differ in any marked degree from those of the more carefully studied arctogæan species.

In the typical genus *Petromyzon* the eggs are minute, of spherical form, and number many thousands; the ova and sperm fall first into the body cavity and are emitted from thence through the abdominal pores; each ovum is enclosed in a delicate gelatinous membrane ; fertilization takes place in the water after extrusion; and the eggs arrive at maturity simultaneously after the lapse of about a fortnight.

An interesting account of the spawning habits of a species of *Petromyzon* is given by Prof. McClure and Dr. Strong, from

---

* "Less confusion arises from calling them "—*i.e.*, species from remote districts—" different until shown to be the same, than from calling them alike until shown to be different " (*David S. Jordan, in lit.*).

observations made by them in the neighbourhood of Princeton, New Jersey.

According to these authorities the eggs are deposited in shallow and clear water, so that the movements of the animals may readily be followed; the breeding season is in spring and the Lampreys remain upon the spawning grounds for two or three weeks; the nests are scattered thickly about the gravelly shoals, often only a few feet apart.   Each nest is occupied by several males and but a single female, which is conspicuous on account of its greater size.* When engaged in the act of spawning the Lampreys press together and cause a flurry in the water at the moment when the eggs and milt are in process of emission. Three or more layers of eggs are thus deposited, each layer being covered by a thin sheet of sand or gravel, the parents always returning to the same nest.   When all the ova have been deposited, the nest is strengthened by a dome-like mass of pebbles and stones which the Lampreys carefully drag to the spot; the nest is thus marked out as well as protected, and is said to be made use of during the ensuing season.†

The suctorial disk is used to keep the parents in position during the period of the emission of the spawn.

U s e s :—All the Lampreys are esteemed as food, and there is no reason to believe that the present species differ in this respect from the others; in fact, Castelnau distinctly states "that they are good food."

D i s t r i b u t i o n :—South-eastern Australia and Tasmania.

Athough long known from the neighbouring colonies of Tasmania and Victoria no record of the occurrence of a Lamprey in New South Wales waters had been furnished up to 1886, when the late Prof. Stephens exhibited a young example of this species

---

* Other observers insist that only one pair frequent each nest.

† See Bashford Dean, Fishes Living and Fossil, p. 182: consult also Gage, Lake and Brook Lampreys of New York, in Wilder, Quarter-Century Book, pp. 421-493, 1893.

at the May meeting of the local Linnean Society; this specimen, which is in the Macleay collection at the Sydney University, was obtained from the Nepean River, near Camden, but though efforts have since been made to obtain other examples in the same district they have hitherto resulted in failure.

Additional and reliable evidence of its presence in the Hawkesbury watershed has, however, been afforded by Mr. J. P. Hill, of the University, who informs me that a friend of his is acquainted with this Lamprey and has caught it in the Wollondilly by the following ingenious method :—a pickle bottle is baited with a piece of raw meat and, a string having been tied round its neck, is sunk in a likely spot; the animals enter the bottle to feed, and on perceiving the motion consequent on its periodical withdrawal, attach themselves thereto by means of the suctorial disk, and are found enclosed when the bottle is drawn out upon the bank.

There can be little doubt that its presence has been overlooked in the southern rivers of New South Wales, such as the Towamba, Bega, Clyde, Shoalhaven, and others, and that when opportunity has been afforded for a thorough investigation of the fresh-water fauna of the colony, this and many other species which are now considered rare will be found to be comparatively plentiful.

The earliest published record of the occurrence of this Lamprey on the mainland is that of Dr. Klunzinger in 1872 (*Arch. f. Natur. p. 45*), and consists of the curt notice " *Mordacia mordax*, Rich. Murray River. 12 Cm." We learn by a note (*l.c. p. 17*) that all the species sent to Klunzinger from the Murray River were taken near its mouth, and this therefore is the most westerly point from which I have been able to ascertain its presence.

During the same year in which Klunzinger's paper appeared Count Castelnau contributed to the Proceedings of the Zoological and Acclimatisation Society of Victoria a more full and interesting account of this Lamprey than any of his predecessors; his examples were collected in the lower portion of the Yarra, where he considered them to be common. He remarks that " their motions are very rapid; they are very voracious and pursue any object in the water, and they adhere to it with an extraordinary and ferocious tenacity."

From the above quotation one gathers that prior to 1872 these Lampreys were not only common in the Yarra, but that it was an easy matter to study their habits there; how different it is at the present day may be judged from the following :—" *Mordacia* seems sporadic and very rare generally; we got a few floating dead during the summer before last in the tidal Yarra " (*T. S. Hall, in lit. July, 1896*).

In his Catalogue of Tasmanian Fishes (*Proc. Roy. Soc. Tas. 1882, p. 141*) Mr. R. M. Johnston records this Lamprey as "abundant at certain seasons, clinging to the sides of perpendicular rocks under mill-shoots, Cataract Gorge, North Esk, Launceston;" and again (*p. 62*) speaking generally of the Tasmanian species, " the Lamprey, though abundant in some rivers, seems not to be in favour in the market, as they are rarely seen there." Notwithstanding this alleged abundance I have found it impossible to obtain a single specimen from the island.

Total length to 450 millimeters.

Type in the British Museum.

In the preparation of this article I have been able to examine seven specimens having a length of from 125 to 418 millimeters; four of these were collected in the lower Yarra, and were kindly forwarded to me by Sir Frederick McCoy (1) and Mr. T. S. Hall (3) of the Melbourne University; two are in the Macleay Museum, from the Nepean River and Tasmania respectively, and the seventh, also from the Yarra, belongs to Mr. J. P. Hill of the Sydney University.

For the opportunity of seeing two ammocœtes I am also indebted to the kindness of the latter gentleman, to whom they were given by Prof Baldwin Spencer.

PETROMYZONIDÆ.

Petromyzontidæ, Risso, Eur. Mérid. iii. p. 99, 1826.
Petromyzonidæ, Bonaparte, Saggio, &c. p. 41, 1832.

A single median tuberculigerous suproral lamina developed from the upper arch of the annular cartilage. Labial fringe more or less conspicuous. Other characters similar to those of the Order.

Seven recent genera are recognised as valid.

Etymology:—πέτρος, a stone; μυζάω, to suck; in allusion to the habit of clinging to stones and other substances by means of the oral disk.

Distribution:—Seas and fresh waters of the temperate and subtropical regions of both hemispheres, four genera belonging to the arctogæan and three to the austrogæan fauna, two of which latter inhabit Australian waters.

Analysis of the Australasian Genera.

Body elongate and slender; head small; suctorial disk very small, longer than broad, extending backwards midway to the eye; outer lip present, continuous behind; surface of disk plicated; no gular pouch; dental plates smooth; discal teeth approximate; ventribasal plate of tongue usually tricuspid; origin of first dorsal on the middle third of the body; head and trunk with conspicuous series of open pores, forming on the latter a well-marked lateral line ... ...
VELASIA, p. 407.

Body rather short and stout; head large; suctorial disk very large, broader than long, extending backwards more than midway to the eye; outer lip rudimentary; surface of disk smooth; gular pouch present; dental plates grooved; discal teeth widely separated; ventribasal plate of tongue bicuspid; origin of the first dorsal on the last third of the body; no series of pores on the head or trunk ... ...
GEOTRIA, p. 420.

VELASIA.

Velasia, Gray, Proc. Zool. Soc. London, 1851, p. 142.

Geotria, part. Gunther, Catal. Fish. viii. p. 508, 1870.

? *Neomordacia*, Castelnau, Proc. Zool. & Acclim. Soc. Vict. i. 1872, p. 232.

Body elongate and slender, strongly compressed; head moderate, oblong, attenuated and depressed, with narrow rounded snout; suctorial disk small, oval, subinferior, extending backwards about midway to the orbital region, with a smooth free outer lip, along the inner border of which a regular series of short, simple, distant papillæ is inserted anteriorly and laterally; on the rim of the disk is a second series of broad, profusely fringed, foliaceous papillæ, which is continued entirely round the hinder margin of the disk where it is widely separated from the external lip, surface of disk traversed by numerous series of closely set cutaneous ridges arranged more or less obliquely. No gular pouch. Branchial orifices moderate and slit-like, with distinct functional valves in front and behind, the latter fringed. Maxillary dentition consisting of a single transverse, crescentic, quadricuspid plate, the outer cusps being smooth and much larger than the inner pair, their extremities entire; mandibular plate low and crescentic, strongly cuspidate; disk with an inner series of moderate, diversely shaped teeth, from the bases of which radiate series of small, contiguous, graduated teeth, which are embedded in the hinder margin of the discal ridges; no subsidiary teeth behind the mandibular lamina; tongue with a single large plate, smooth on its outer, tricuspid on its inner margin, along either side of its dorsal surface; below with a strong, transverse, basal plate, provided with three (sometimes two*), slender acute cusps directed forwards. Two well developed dorsal fins, the anterior inserted far behind the middle of the body, the posterior much the larger and separated from the caudal by a moderate interspace; caudal fin well developed, continued around the extremity of the tail by a low, rayed membrane. Tail long, the vent situated below the origin of the second dorsal fin. Head with series of small, open pores; a series of widely separated pores along the middle of the trunk and along the bases of the fins.

---

*When the median cusp is absent the remaining two are widely separated at the base, not contiguous as in *Geotria*.

Etymology :—Unknown.

Type :— *Velasia chilensis*, Gray.

Distribution:—Coasts and rivers of south-eastern and southern Australia; ?South-western Australia; Tasmania; New Zealand; Chile.

VELASIA STENOSTOMUS.

Geotria chilensis, part., Günther, Catal Fish. viii. p. 509, 1870.

Geotria australis, (not Gray) Castelnau, Proc. Zool. & Acclim. Soc. Vict. i. 1872, p. 227 (1873) *and* Edib. Fish. Vict. p. 17, 1873; Lucas, Proc. Roy. Soc. Vict. (2) ii. 1890, p. 47.

Geotria chilensis, Hutton, Fish. N. Zeal. p. 87 *and* (Hector) p. 132, 1872 *and* Trans. N.Z. Inst. v. 1872, p. 271, pl. xii. f. 139 (1873) *and* viii 1875, p. 216 (1876) *and* xxii. 1889, p. 285 (1890); Macleay, Proc. Linn. Soc. N.S. Wales, vi. 1881, p. 384; Sherrin, Handb. N.Z. Fish. p. 36, 1886; Gill, Mem. Nat. Ac. Sc. Washingt. vi. p. 110, 1893 (not *Velasia chilensis*, Gray).

? *Petromyzon* sp., Kner, Voy. Novara, Fisch. p. 421, 1865.

? *Yarra singularis*, Castelnau, Proc. Zool. & Acclim. Soc. Vict. i. 1872, p. 231 (1873); Macleay, l.c. p. 385; Lucas, l.c.

? *Neomordacia howittii*, Castelnau, l.c p. 232; Macleay, l.c. p. 383; Lucas, l.c. p. 46.

Narrow-mouthed Lamprey.

Disk oval, its width when fully expanded less than its length, its posterior margin reaching backwards midway to the vertical from the middle of the eye. Eyes rather inconspicuous, the nasal tube opening between their anterior margins. The distance between the extremity of the snout and the nasal opening is $16\frac{2}{3}$ to $17\frac{1}{4}$ in the total length and $1\frac{3}{5}$ to $1\frac{3}{4}$ in that preceding the first branchial orifice, which is situated a little nearer to the last branchial orifice than to the tip of the snout; the space between the last orifice and the extremity of the snout is $5\frac{2}{5}$ to $5\frac{1}{2}$ in the total length. Maxillary plate smooth; the inner cusps triangular

27

and acute, the notch between them deeper than those which
separate them from the lateral cusps, which are much longer and
broader, with the inner border acute and convex, the tip pointed,
and the outer border obtusely rounded and almost linear, not
separated by a groove from the basal portion of the plate,
mandibular plate with eleven short, blunt cusps, the outer one at
each side and the median one inappreciably larger; inner series of
discal teeth large, triangular and acute in front, broad and
chiselled on the sides and behind ; the middle teeth behind the
maxillary plate are as large as the lateral ones; these teeth are
twenty-six in number, and the anterior pair correspond to the
inner maxillary cusps ; in front of the interspace between the
anterior pair a series of five teeth, which gradually decrease in
size from within, extend in a straight line to the outer rim of the
disk, from each of these a curved series of similarly developed
teeth radiates outwards and backwards on either side, the disk is
armed laterally with similar series of graduated teeth, each row
corresponding to one of the enlarged inner teeth and being so
strongly bent backwards towards the outer margin as to assume
a subconcentric appearance, the surface of the disk is divided into
series of low dermal ridges, on the inner posterior border of which
the teeth are embedded, these ridges are set so close together
that the teeth of one ridge overlap the succeeding ridge ; behind
the mandibular plate there are no teeth outside of the circumgular series, the tongue is armed with a single pair of dorso-
lateral plates, each of which is swollen and entire on its outer
border and bears on its inner three strong acute cusps, the
anterior of which is the smaller, the others being subequal.
the transverse ventribasal plate is strongly tricarinate on
its inner surface, each of the carinæ being produced into a
long, slender cusp, the tips of which are acute and slightly
curved upwards, the median cusp is as long as the outer pair
The vent is situated a little behind the origin of the second
dorsal, the length of the tail is 4 to $4\frac{3}{8}$ in the total length   The
distance between the origin of the first dorsal fin and the tip of
the tail is $1\frac{1}{2}$ to $1\frac{2}{3}$ in its distance from the extremity of the snout,

both dorsal fins rise gradually from the dorsal integument in front but terminate in a distinct though short posterior border; the outer border of the first dorsal fin is convex, its apical portion being situated somewhat in advance of the middle of the fin, and the length of its base is a little more than the interdorsal space and $1\frac{1}{2}$ to $1\frac{3}{5}$ in the base of the second, the outer border of which rises somewhat abruptly to above the origin of the median basal third and slopes gradually downwards from thence to its junction with the short posterior border, the anterior border being linear or somewhat convex; its height at the apex is one-third to one-half more than that of the first dorsal; the length of the tail behind the second dorsal is 1 to $1\frac{1}{2}$ in the base of that fin, which is entirely separated from the caudal by an interspace equal to about half the length of the latter fin; the caudal lobes are equally developed and are connected round the extremity of the tail by a low rayed membrane. A series of open pores extends from the throat along the rostral canthus to the antero-superior angle of the eye, where it curves downwards, and ultimately encircles three-fourths of the orbital ring, from the postero-superior angle of which it slopes backwards and downwards in the direction of the first branchial orifice; there is a short series of similar pores above and behind the posterior angle of the closed disk, and a few others along the lower surface of the head, the lateral line is indicated by a series of pores which extend along the middle of the sides of the trunk, and there are similar series along each side of the bases of the fins.

Back dark slate-colour, belly and the greater portion of the sides bronze, the line of demarcation well defined especially on the tail; head dark gray above, silver gray on the sides and below, the latter colour extending backwards along the branchial region; fins yellowish, broadly margined with slate-colour

The following is Castelnau's description of the colours in the fresh example :—:

" Dark blue on the back, silvery on the sides and belly, on the middle of the back, a little before the insertion of the first dorsal, begins a space of brilliant green, which extends to the tail; fins red, bordered with black."

Capt. Hutton describes the species as having " a broad band of green down each side of the back, the median line and the whole of the lower surface being pale brownish-white."

The brilliant green stripe on each side of the back appears, therefore, to be very distinctive of this Lamprey when alive or recently killed as compared with the uniform black or dark brown of the upper surface of *Geotria australis*.

It will be seen from the synonymy that I have included both of Castelnau's new species as synonyms of *Velasia stenostomus*, though from the size of the specimens, the insufficiency of the descriptions and the destruction or loss of the type,* it will always be impossible to say whether I am justified in my conclusions or, indeed, to what species his immature and ammocœtal forms should be united.   If, however, the types are extant and on examination show that my identification is correct in one or other instance, Castelnau's name must necessarily have priority over mine.

Yarra singularis.

The following are the points in Castelnau's description which induce me to believe that his *Yarra singularis* is founded on an ammocœte of the Narrow-mouthed Lamprey.   No generic diagnosis of *Yarra* was attempted by its author.

(1). "The body is elongate, being twenty-three times as long as high."

This character might apply with almost equal force either to this species or to *Mordacia mordax*; but when these two Lampreys (in the adult state) are laid side by side it will be seen that *Velasia* is noticeably the more slender of the two.   This character could not possibly apply to *Geotria*.

---

* These types may possibly be in the Paris Museum, where a part at least of Castelnau's collection is said to have gone.

(2). "The upper lip is flat and considerably prolongated over the buccal aperture."

This inferior position of the disk is also true of *Mordacia* and *Velasia*, but not of *Geotria*.

(3). "The lateral line is well marked in all the length of the body."

In my two adult examples of the Narrow-mouthed Lamprey there is a conspicuous series of open pores down the middle of each side of the body, homologous to the lateral line in the true fishes, in neither of the other genera is there any trace of such line.

(4). "There is only one dorsal, which begins at about two-thirds of the length of the body and is joined with the caudal and anal."

The posterior position of the origin of the dorsal fin is a distinct character of the Australian Petromyzonids, and entirely precludes the possibility of this example being a larval *Mordacia*, in which genus the fin commences in the adult at no great distance—one-fourth to two-fifths—behind the middle of the body, and it is not conceivable that the permanent anterior portion of the fin should develop after the metamorphosis has taken place, rather than that it should be isolated by the absorption of the intervening membrane. The want of accuracy in the expression "about two-thirds" makes it impossible to judge absolutely between the claims of *Velasia* and *Geotria*, but the balance is somewhat in favour of the latter, in which the insertion of the dorsal fins in the adult is distinctly more posterior than in the former.

The continuity of the two dorsal fins and of the second dorsal with the caudal is merely indicative of the ammocœtal character of the individual, as also is the absence of eyes and teeth.

Two other characters in Castelnau's description apparently favour the claims of *Geotria*; namely, that the body "is entirely divided in annular rings" and that "the skin of the throat is rather extensible."

Taking into consideration the small size and imperfect development of the specimen, I do not consider that these characters can be held to equal in importance the tenuity of the body and the presence of the lateral line.

Castelnau's reason for rejecting this ammocœte as the larval form of a *Geotria* seems to be mainly based on the fact that he had previously received " a very young individual, only three inches long, having exactly the same form, the same dimensions, and the same dentition " as the specimen of *Geotria australis* from which his description and measurements of the adult were drawn up, and which I shall show further on to have been in truth a *Velasia stenostomus*. His words are :—" I should have thought this might be the first state of *Geotria*,* but we have just seen that I had a still smaller specimen of this which has entirely the form of the adult "

That the length of the unique example of *Yarra singularis* was " four and three-eighth inches," or one and a-half times the length of the perfectly formed individual mentioned above, is not sufficient reason for denying its identity with the ammocœte of *Velasia*; the difference in size is capable of explanation in at least two ways, thus :—On the one hand the smaller specimen which, having developed teeth, must have passed the ammocœtal stage, may possibly have been the young of the true *Geotria australis*, which, as we shall subsequently show, occurs also on the Victorian coast, while on the other hand the metamorphosis may in this individual case have been retarded from some cause, or at all events incomplete.

Neomordacia howittii.

In his diagnosis of *Neomordacia* Castelnau relies for the validity of his genus on the following unstable character :—

It " has no first dorsal, or rather has only one dorsal, separate and rather distant from the caudal."

---

* Lege, *Velasia*   Castelnau does not appear to have ever seen a true *Geotria australis.*

The uninterrupted connection of the dorsal fin is of course only significant as showing the immaturity of the individual, and is, therefore, of no value as a generic character; this last sentence, however, is sufficient to separate the species from *Mordacia*, in which at all ages the dorsal and caudal fins are more or less distinctly united, and in examples up to 125 millimeters are conspicuously so.

The presence of "fringes round the mouth" is also peculiar to *Velasia* and *Geotria*, the external lip and discal rim of *Mordacia* being almost smooth.

The tenuity of the body and the absence of dilatation in the head are, however, characters which belong to *Velasia* as opposed to *Geotria*, and I have, therefore, decided to associate Castelnau's *Neomordacia howittii* with *Velasia stenostomus*.

Returning to the adult Lamprey, my reasons for considering that Castelnau's specimen was *Velasia stenostomus* and not *Geotria australis* as determined by him, will be found below, the more important points of that author's description being taken seriatim.

(1). "The maxillary lamina is formed of four teeth, the exterior of which are flat lobes, and the two interior ones long, conical, pointed teeth."

This gives a fair description of the maxillary cusps of *Velasia* in which the inner cusps are as described and the outer are simple and smooth, while in *Geotria* the inner cusps are lanceolate and the outer notched and grooved.

(2). "Suctorial teeth in numerous transverse series, those situated backwards larger than the others."

The number of the series of discal teeth in *Velasia* and *Geotria* is about the same, but from the great expansion of the disk in the latter they appear to be much less numerous than in the former, to which, therefore, the wording of Castelnau's paragraph would more naturally point; in *Velasia* too the posterior discal teeth are as large as the inner lateral ones, while in *Geotria* they are minute.

(3). " Lingual teeth two in number, straight, strong, and conical."

Without a re-examination of the specimen it is impossible to say whether there were in fact only two ventribasal cusps, as the third one might have been overlooked, either through careless or defective examination as is the case with the specimen most kindly forwarded to me from the British Museum in which the median cusp is as fully developed as either of the lateral ones, sometimes, however, it is absent as in Mr. Hill's specimen, but in that case the bases of the lateral cusps are widely separated.

(4). "The distance between the two dorsals and the base of the caudal is a little more than the diameter of the mouth."

It appears to me that this character in itself indubitably proves the identity of Castelnau's Lamprey with *Velasia* as will be seen by the following measurements taken from my own specimens :—In my Tasmanian type of *Velasia stenostomus* the longitudinal (longer) diameter of the closed suctorial disk is 16 millimeters and the dorso-caudal interspace – which is, I presume, what Castelnau intends—is 15; in *Geotria australis* on the contrary the longitudinal (shorter) diameter of the expanded—and, therefore, further shortened—disk is 27 millimeters and the dorso-caudal interspace only 12, or less than a half.

(5). " The diameter of the mouth is equal to half the distance from the end of the snout to the anterior edge of the eye."

This applies much more closely to the small-mouthed *Velasia* than to the large-mouthed *Geotria*, in which the disk extends full two-thirds of the preorbital portion of the head.

(6). The colours are those of *Velasia*.

In the table of measurements given by Castelnau we also find corroborative evidence of the correctness of my views, while on the other hand certain of the dimensions given are curiously subversive of those views but more in the direction of *Mordacia* than of *Geotria*. The following table has been drawn up for comparison, the measurements in columns 1, 3, and 4 being taken from specimens in my collection, while those in column 2 are from

Castelnau's figures, the circumference of the body being omitted as unnecessary.

	1	2	3	4
	V. stenostomus.	*G. australis, Cast. = V. stenostomus.*	*G. australis.*	*M. mordax.*
i. Total length (in millimeters)	468	513	375	413
ii. Muzzle to centre of eye to total length .	$14\frac{2}{3}$	$18\frac{1}{4}$	$8\frac{1}{5}$	$19\frac{3}{5}$
iii. Muzzle to first branchial orifice to total length	10	$13\frac{3}{5}$	$6\frac{1}{4}$	$12\frac{1}{2}$
iv. Muzzle to last branchial orifice to total length	$5\frac{1}{2}$	$6\frac{1}{2}$	$3\frac{1}{4}$	$6\frac{7}{10}$
v. Origin of first dorsal to tip of tail to its distance from tip of snout	$1\frac{1}{2}$	$1\frac{3}{5}$	$2\frac{1}{8}$	$1\frac{3}{5}$
vi. Interdorsal space to first dorsal .	$1\frac{1}{8}$	$1\frac{1}{15}$	$1\frac{7}{8}$	1
vii. Base of first dorsal to that of second	$1\frac{1}{2}$	$1\frac{1}{10}$	$1\frac{3}{10}$	$2\frac{1}{8}$
viii. Dorso-caudal interval to caudal	$2\frac{1}{8}$	$1\frac{3}{8}$	$2\frac{1}{8}$	$2\frac{5}{8}$
ix. Tail to total length	$4\frac{1}{2}$	$3\frac{9}{10}$	$5\frac{9}{10}$	$7\frac{1}{4}$

Of all these measurements only one (vii.) of Castelnau's shows a nearer approach to my *Geotria australis* than to *Velasia stenostomus*, while the two most important (v. and ix.) distinctly support the latter.

The three measurements connected with the head (ii. to iv.) are so curiously similar to those of my *Mordacia* that I cannot refrain from conjecturing that Castelnau had an example of each species (*Velasia* and *Mordacia*) before him, and somehow got the dimensions mixed; and if further evidence is necessary as to the probable truth of this conjecture, I may mention that in the table of measurements of *M mordax* given by Castelnau (*l c. p. 230*) the distance between the extremity of the snout and the centre of the eye is contained $14\frac{1}{5}$ times in the total length, or nearly the same as that in my *V. stenostomus*. In the same table the length of the first dorsal is erroneously given as 6½ inches; this is an evident *lapsus calami* for 1½ inches.

Taking all the characters which I have referred to above, for or against, together I consider that I am quite justified in my association of Castelnau's species with *Velasia stenostomus*.

Petromyzon sp.

Kner's description of the ammocœte from the Waikato River, New Zealand (*Voy. Novara, Fisch. p. 421*) gives no characters on which any accurate judgment as to its relationship can be based; the remark, however, that "the cavity of the suctorial disk is closely beset with papillæ" is clearly more indicative of affinity to *Velasia* than to *Geotria*. Günther is, therefore, probably right in conjecturing that "it is perhaps the young state of" *Geotria chilensis* (= *Velasia stenostomus*).

There is, however, one other character given by Kner which puzzles me; he says:—"The large triangular nostril lies nearly above the margin of the sucking disk in the middle of the forehead." Now in none of the species is the nostril situated "in the middle of the forehead," though it is of course placed on the middle longitudinal line of the head between or nearly between the anterior borders of the eyes; again the posterior margin of the suctorial disk does end beneath the middle of the forehead, *i.e*, of the preorbital space, in *Velasia*, but not beneath the nostril; it ends beneath the nostril only in *Mordacia*, which genus is not found in New Zealand; if it were I should unhesitatingly consider this little animal to be the larval form of the latter genus.

B r e e d i n g :—As with *Mordacia mordax* nothing definite is known of the propagation of this species, but it is worthy of note that such ammocœtes as have hitherto been recorded were all obtained in tidal waters, and as before their metamorphosis these animals remain buried in the mud, it would appear that the adults do not necessarily seek fresh water before depositing their spawn, nor is the purity of the element requisite to the development of the ovum.

Uses:- That these Lampreys were a common and favourite article of food among the Maoris we gather from the New Zealand writers; Sherrin tells us that " they are greatly esteemed by the Natives, who call them *Piharau* and used to pot them in large quantities. Maori chiefs, as well as Henry I., have died from a surfeit of Lampreys, the chiefs having the pick of large catches of all kinds of fish set apart for them."

Further on he writes :—" It is necessary to bear the construction of the mouth of the Lamprey in mind to understand what the Natives mean when they say they see them 'sucking their way up a waterfall in streams in hundreds at a time.' When thus found a net is placed at the foot of the fall, and the fish being detached fall into the net and are thus captured. They are also often found in their eel-weirs. They ascend the Waikato (and probably other rivers) when the whitebait is also ascending. If cooked in a hangi they have to be eaten with care, and a certain fluid they contain, the Natives say, must be expressed, or its effect will be similar to that induced by the eating of a certain kind of shark —the loss of the gourmand's skin. Cooked as Europeans would cook them, this apprehension would not be entertained."

Dr. Hector also writes :—" Most of the New Zealand rivers are visited early in summer by shoals of Lampreys, which are stated to be excessively delicate and well flavoured."

At the time this was written the occurrence of *Geotria australis* in New Zealand was unknown, nevertheless as the statement was based on hearsay evidence it must be taken as referring to both species.

Distribution:—Coasts and rivers of Victoria, South Australia, Tasmania, and New Zealand; ? West Australia.

In New Zealand the Whanganui, Waikato, and Raiwaka Rivers are specially referred to; the species inhabits, therefore, both the North and the South Islands.

I have placed a note of interrogation against the West Australian distribution usually accorded to this species on the strength of the British Museum Catalogue, in which it is recorded from "Swan River;" though without doubt the West Australian river is the most widely known, the name itself is so little distinctive that I am inclined to believe that some stream, possibly in Tasmania, where it has now been proved beyond question to occur, is intended.

Type in my possession.

Total length to 550 millimeters.

Three specimens have been available to me in the preparation of this description; for the first I am indebted to the authorities of the British Museum, who, on learning that I was working at the Australian Lampreys, with great kindness sent me one of the New Zealand examples recorded in Dr. Gunther's Catalogue as *Geotria chilensis,* while a second example from the same Colony was lent to me by Mr. J. P. Hill, only the anterior half of this individual having been preserved; the third was forwarded to me from Tasmania by Mr. Morton and measures 468 millimeters.

GEOTRIA.

· *Geotria,* Gray, Proc. Zool. Soc. London, 1851, p. 238

Body rather short and stout, strongly compressed; head large, oblong, with broad, rounded snout, suctorial disk very large, elliptical, subterminal, extending backwards more than half way to the orbital region, without free external lip, its rim thick and fleshy, and bearing on its inner margin two series of fringed, foliaceous papillæ, the hinder margin of the disk is low and bears a single series of similar but much enlarged papillæ, surface of disk smooth. Gular pouch present. Branchial orifices large and slit-like, with a rudimentary valve in front and behind. Maxillary dentition consisting of a single transverse, crescentic, quadricuspid plate, the basal portion divided from the cusps by a deep groove, outer cusp notched at the extremity; mandibular plate low and crescentic, smooth or feebly cuspidate; disk with an inner series of rather large, diversely shaped teeth, from each of which radiates a series of small, distant teeth; the series are curved obliquely backwards and widely separated; a single transverse series behind the mandibular plate, the median teeth the smallest; tongue with a single large plate, smooth on its outer, quadricuspid on its inner margin, along either side of its dorsal surface ; below with a strong, transverse, basal plate, provided with two stout, sharp cusps directed forwards. Two well developed dorsal fins separated by a moderate interspace, inserted on the posterior third of the body, the second entirely disconnected with the caudal and not much larger than the first; caudal fin moderate, continued around the extremity of the tail by a low rayed membrane. Tail short; the vent situated below or nearly below the origin of the second dorsal fin. Head and body without conspicuous series of pores.

Etymology :—Unknown.

Type:—*Geotria australis*, Gray.

Distribution :—Coasts and rivers of Southern Australia, Tasmania, and New Zealand; Chile and the Argentine Republic.

Dr. Günther, in the course of some remarks on *Geotria australis*, writes thus :—

"Philippi (*Wiegm. Arch. 1857, p. 266*)* has described a Lamprey from Chile under the name *Velasia chilensis*; the example was provided with the sac at the throat and the description agrees with *Geotria australis*; so that we must assume either that this latter species occurs not only in Australia but also in Chile, or that *Velasia chilensis* at a certain stage of development is provided with a gular sac. If the latter be the case the specific distinction of the two species would be questionable " (*Catal. Fish. viii. p. 509*).

From the above quotation it is evident that some species of Lamprey provided with a gular sac inhabits the rivers of Chile, and if I am correct in attributing that character to *Geotria* alone, it follows that the genus is represented there; but I cannot agree with Dr. Günther that the species is necessarily identical with *G. australis*, and much less that the latter species is indistinguishable from *Velasia chilensis*.

---

* I am unable to refer to a copy of this publication.

The function of the extraordinary pouch with which the members of this genus are furnished is quite unknown, nor have any observations as yet been made showing whether its presence is in any way connected with age, sex, or season.

GEOTRIA AUSTRALIS.

Geotria australis, Gray, Proc. Zool. Soc. London, 1851, p. 238, pls. iv. f. 3 & v., *and* Catal. Chondropt. p 142, pls. i. f. 3 & ii. 1851; Gunther, Catal. Fish. viii. p. 508, 1870; Hutton, Trans. N.Z. Inst. v. 1872, p. 272, pl. xii. f. 139a (1873) *and* xxii. 1889, p 285 (1890); Klunzinger, Sitzb. Ak. Wien, lxxx. i. 1879, p. 429 (1880); Macleay, Proc. Linn. Soc. N.S. Wales, vi. 1881, p. 384; Sherrin, Handb. N.Z. Fish. p. 56, 1886, Gill, Mem. Nat. Ac. Sc. Washingt vi. p. 110, 1893.

Geotria allporti, Günther, Proc. Zool. Soc. London, 1871, p 675, pl. lxx, Macleay, l.c. p. 385; Johnston, Proc. Roy. Soc. Tas. 1882, p 141, *and* 1890, p. 39.

Wide-mouthed Lamprey.

Disk elliptical, its length when fully expanded I$\frac{1}{4}$ in its breadth and 1$\frac{2}{3}$ to 1$\frac{3}{5}$ in the space between its anterior margin and the eye. Eyes conspicuous, the nasal tube opening between their anterior borders The distance between the extremity of the snout and the nasal opening is 7$\frac{1}{2}$ to 8$\frac{3}{4}$ in the total length, and 1$\frac{2}{3}$ to 1$\frac{1}{2}$ in that preceding the first branchial orifice, which is situated much nearer to the last than to the tip of the snout. The space between the last branchial orifice and the extremity of the snout is 3$\frac{1}{4}$ to 3$\frac{4}{5}$ in the total length. Maxillary plate grooved; the inner cusps are lanceolate and strongly keeled; they are entirely distinct from one another, the notch between them being as deep as those which separate them from the lateral cusps, which are much longer and broader, and are divided into two subequal portions by the prolongation of the basal groove. the free edge of the inner portion is strongly compressed, sharp.

and entire,* the remainder of the cusp being swollen and the tip
obtusely pointed; the external portion is broader than the inner
and is more or less truncated; it is as long as or shorter than the
inner portion; mandibular plate with ten cusps, the outer one on
each side acute and directed inwards and backwards, the other
short and blunt, sometimes rudimentary ; the inner series of
discal teeth are enlarged, triangular and acute in front, broad
and chiselled on the sides, those behind the mandibular plate
growing gradually smaller towards the middle; these teeth are
twenty-eight in number and the anterior pair correspond to the
inner maxillary cusps; in front of the interspace between the
anterior pair is a series of six teeth, which gradually decrease in
size from within and extend in a straight line to the rim of the
disk; from these and from the enlarged circumgular teeth extend
curved series of graduated teeth; these series are widely separated
from one another and the teeth themselves are not in contact
basally; there are no small teeth behind the postmandibular
series; the tongue is armed with a single pair of dorso-lateral
plates, each of which is deeply grooved near its outer border,
which is strongly convex, blunt, and entire, while the inner
border is quadricuspid, the anterior cusp being only about half
the length of the other three, which are subequal in size; the
transverse, ventribasal plate is also grooved round the base of the
cusps, but is otherwise smooth ; the cusps are two in number,
long, acute, and directed outwards and slightly upwards; there is
a minute median basal cusp behind the plane of the functional
pair. The vent is situated beneath the origin of the second
dorsal; the length of the tail is $5\frac{1}{3}$ to $6\frac{3}{8}$ in the total length. The
distance between the origin of the first dorsal fin and the tip of
the tail is 2 to $2\frac{1}{4}$ in its distance from the extremity of the snout;
both dorsal fins rise gradually from the dorsal integument in

---

* Gunther describes the outer cusps in *G. allporti* as being "finely
serrated on the inner margin," but there is no trace of any such serrature
in either of my specimens, though they agree perfectly in the transversely
plicated body.

front, but terminate in a distinct though short posterior border; the outer border of the anterior fin is evenly convex, its apical portion being situated above the middle of the base of the fin, and the length of its base is from one-fourth to three-fifths in the interdorsal space and $1\frac{3}{10}$ to $1\frac{2}{5}$ in that of the second dorsal, the outer border of which is also convex throughout, its apex being a little behind the commencement of the median third; its height at the apex is one-fifth more than that of the first dorsal; the length of the tail behind the second dorsal is a little more, equal to, or a little less than the base of that fin, which is entirely separated from the caudal by an interspace, which is equal to about two-fifths of the length of the latter fin, the caudal lobes are subequal in height, but the lower extends forwards much further than the upper; they are connected together around the extremity of the tail by a low rayed membrane. Head and body without series of conspicuous pores. Skin transversely plicated.

Black or dark brown above, lighter below; upper surface of head with a bluish, sides of head with a bronze tinge; lower surface of head, throat, and pouch grayish-white.

B r e e d i n g :—Unknown.

U s e s ·—Similar to the other species.

D i s t r i b u t i o n :—Having already shown that Castelnau's *Geotria australis* belonged in truth to the preceding species we are now reduced to a bare statement of the habitat of this Lamprey in so far as it can be separated with certainty from that of *Velasia*.

Gray's type specimen is said by Dr. Günther to have come from the "Inkarpinki River, South Australia"; but I have not succeeded in finding the locality of any river with such a name, and it must be remembered that throughout the British Museum Catalogue "South Australia" is used to denote our entire southern sea-board, and not restricted in the territorial sense which is customary here; this, however, is in this case of little consequence, as Count Castelnau informs us that Mr. Pain, by whom the specimen was forwarded to the British Museum, had personally assured him that he "picked it up on Brighton Beach, Hobson's Bay" As, however, Klunzinger records this Lamprey from King George's Sound it may be presumed that the species visits the rivers of our southern coast in greater or less numbers during the spawning season.

Under the name *allporti*, Johnston describes the Pouched Lamprey as being "not uncommon in fresh water, Derwent, North Esk, St. Leonards."

From New Zealand I can find no record except that of Capt. Hutton, who claims to have received it from Stewart Island.

Total length to ·500 millimeters.

Type in the British Museum, as also is that of *allporti*.

Only two specimens were available to me for examination, for both of which I have to thank Mr. Alexander Morton, to whose generous assistance I am greatly indebted for this opportunity of establishing the position of our Australian Hyperoartians on a more stable basis than they have hitherto enjoyed. Both my examples were collected in Tasmania and measure respectively 325 and 375 millimeters.

In order to render this paper as perfect as the means at my disposal permit I append the following brief diagnosis of the third austrogæan genus as given by its author.

EXOMEGAS.

Exomegas, Gill, Proc. U.S. Nat. Mus. v. 1882, p. 524.

"Discal teeth in concentric series, the outer containing the largest teeth (about 24 on each side); lingual teeth three, large, pointed, and curved, the median smallest, all standing on the same base."

Etymology: ἔξω, without; μέγας, large, in allusion to the enlarged size of the outer discal teeth.

28

Type:—*Exomegas macrostomus*, Gill = *Petromyzon macrostomus*, Burmeister.

Distribution:—Atlantic coast of South America (Argentine Republic); very rare.

For further information concerning this form consult Burmeister, Anal. Mus. Buenos Aires, pt. 5, 1868, Act. Soc. Palæont. p xxxvi., and Berg, Anal. Mus. La Plata, 1893.

ON THE BOTANY OF RYLSTONE AND THE GOULBURN RIVER DISTRICTS. PART I

By R. T. Baker, F.L.S., Assistant Curator, Technological Museum, Sydney.

The area of the colony treated of in this paper comprises the eastern divisions of the Counties of Phillip and Roxburgh,—a portion of New South Wales, which I believe has not previously been explored botanically.

The northern boundary of this area is the Goulburn River, which rises on the eastern slope of the Dividing Range, a few miles south-west of the town of Ulan, and flows easterly in a tortuous course, eventually joining the Hunter River a little south of Denman. It runs mostly through precipitous and mountainous sandstone ridges, and consequently is subject to inundations, — flood-marks being found at a considerable height above its ordinary level. The country between the river and the Dividing Range consists principally of mountain ranges, with occasional patches of good soil, derived from the disintegration of basalt from the volcanic outcrops, approximating in area about 1000 sq. miles. It is sparsely populated, there being only about half a dozen small Public Schools scattered throughout this large area. Settlements are therefore few and far between, and consequently much of the indigenous flora remains, so that it is a splendid country for botanising.

The eastern boundary is formed by Widdin Brook, a stream which rises in Corricuddy Mountain and flows north into the Goulburn River, and the main Dividing Range as far south as Capertee.

The western boundary is the Barrigan Ranges and a line drawn from these south through the town of Cudgegong to Ilford.

The main Dividing Range divides the district into the eastern and western watersheds.

The western slopes of the Range are much more fertile, and as settlements are more frequently met with, it will be easily understood that the indigenous vegetation has been considerably cleared.

The geological formation of the Main Range is the Hawkesbury sandstone (Triassic), which extends in outcrops down to and over the Goulburn River. Interspersed with the sandstone are basalt and the Tomago Series, which extend inland to beyond Dubbo, a fact that may account for the occurrence on the eastern watershed of several western species.

The sandstone of the Range is succeeded on the western slopes towards the Cudgegong River by the Newcastle Series, and next we have the Upper Marine Series, followed by Silurian, with outcrops of granite, quartz porphyries, felsites and limestones.

I have not been able to obtain any authentic records of any trips made by previous botanical collectors, but judging by the references to localities in the " Flora Australiensis," I am under the impression that until visited by me this country was botanically a *terra incognita*. A. Cunningham must have been on the outskirts, for in the " Flora Australiensis " (Vol. i. p. 443) under *Cryptandra buxifolia*, Fenzl, the locality is given as " Rocky Hills on the meridian of Bathurst, on the parallel of 30° 50'; Mount Yongo, on the route to Hunter's River and Goulburn River, *A. Cunningham*." This naturalist collected on the north-west branches of the Hunter River, so that the Census now offered fills the gap between his collectings and those of others on the Blue Mountains.

Mr. A. G. Hamilton's Mudgee Census includes the flora of the country to the east of the southern half of this district

My collections were made during the months of September, October, November and December, in the years 1895, 1893, 1892 and 1890 respectively, the actual collecting days being about 50.

Summarising the results of my expeditions, I find over 350 Species were collected representing 56 Natural Orders. Of these 10 were new species, and 7 have already been described, viz.—

Acacia Muelleriana, J.H M. et R.T.B., *Helichrysum tesselatum*, J.H M. et R T.B.; *H. brevidscurrens*, J.H.M. et R.T B.; *Daviesia recurvata*, J.H.M. et R.T.B.; *Isopogon Dawsoni*, R.T.B., *Prostanthera discolor*, R.T.B.; *P. stricta*, R.T.B.

Three species new to the Colony were also found, viz. :— *Eucalyptus trachyphloia*, F.v.M., *Grevillea longistyla*, Hook., *Loranthus Bidwillii*, Benth. The range of other forms hitherto regarded as inland species, has been extended to the eastern watershed.

The following is a list of the Natural Orders, with the number of species collected : —

RANUNCULACEÆ ..	...	1	COMPOSITÆ	... ...	32
DILLENIACEÆ ...	...	4	STYLIDEÆ	... ...	1
VIOLARIEÆ ...	...	1	GOODENIACEÆ ...	...	9
PITTOSPOREÆ ...	...	5	CAMPANULACEÆ...	...	3
CARYOPHYLLEÆ ...	...	1	EPACRIDEÆ	... ...	14
MALVACEÆ ...	...	4	JASMINEÆ	... ...	1
STERCULIACEÆ ...	...	3	APOCYNEÆ	... ...	1
LINEÆ	...	1	ASCLEPIADEÆ ...	...	1
GERANIACEÆ ...	...	2	LOGANEÆ	... ...	1
RUTACEÆ ...	...	9	GENTIANEÆ	... ...	2
OLACINEÆ ...	...	1	BORAGINEÆ	... ...	2
STACKHOUSIEÆ ...	...	1	SOLANEÆ	... ...	6
RHAMNEÆ ...	...	3	SCROPHULARINEÆ	...	2
SAPINDACEÆ ..	...	4	MYOPORINEÆ ...	...	5
LEGUMINOSÆ ...	...	58	LABIATÆ ...	...	8
ROSACEÆ ...	...	3	MONIMIACEÆ ...	...	1
SAXIFRAGEÆ ...	...	2	LAURINEÆ ...	...	2
DROSERACEÆ ...	...	1	PROTEACEÆ ...	...	27
MYRTACEÆ ...	...	38	THYMELEÆ ...	...	4
UMBELLIFERÆ ...	...	1	EUPHORBIACEÆ ...	...	5
ARALIACEÆ ...	...	1	URTICACEÆ ...	...	1
LORANTHACEÆ ...	...	4	CASUARINEÆ ...	...	3
RUBIACEÆ ...	...	5	SANTALACEÆ ...	...	7

CONIFERÆ	... ...	2	JUNCCAEÆ	... ...	4
CYCADEÆ	... ...	1	CYPERACEÆ	... ...	5
ORCHIDEÆ	... ...	5	GRAMINEÆ	... ...	6
IRIDEÆ ...	... ...	1	LYCOPODIACEÆ ...	...	1
LILIACEÆ	... ...	6	FILICES	... ...	6

Excluding new species, the next most interesting finds were:—

(a) *Pomaderris philicifolia*, Lodd., a species only recorded from this Continent from the "banks of subalpine streams under the Australian Alps, descending into the plains of Gippsland on the Hume and Murray Rivers, *F.v.Mueller*." It also occurs in Tasmania, and abundantly so in the northern island of New Zealand.

(b). *Eucalyptus trachyphloia*, F.v.M.

(c). *Loranthus Bidwillii*, Benth.

(d). *Grevillea longistyla*, Hook.

All these three species occur in Northern Queensland, and one would hardly have expected to have found them at Murrumbo, as they have never been collected in this Colony before.

I paid particular attention to the Acacias and have endeavoured to elucidate some of the difficulties surrounding the classification of the numerous species of this genus. Some points, I regret, still remain unsettled from want of perfect material; for instance, the occurrence in this Colony of *A. ixophylla* is still, I think, an open question; and the fruits obtained were not sufficiently mature for me to speak with any certainty, for as far as I was able to judge they differed entirely from those described by Bentham.

A. crassiuscula, Wendl., and *A. lunata*, Sieb., are also species I hope to deal with in a future paper, as the specimens collected were not altogether satisfactory.

To the Eucalypts I gave perhaps more attention than even the Acacias, as the late Dr. Woolls and Mr. A. G. Hamilton have already described the Mudgee representatives of this genus, and I look on my notes as the connecting link between the Mudgee and Coast Floras. I was surprised to find *E obliqua* so far north, as it has previously only been recorded from southern New South Wales, although it was recently found at the National Park by Mr. F. Williams. The shape of the fruit in the northern specimens differs entirely from the southern form, as will be observed in the notes under this species.

Other Stringybarks dispersed throughout the district are *E. eugenioides, E. macrorrhyncha,* and *E. capitellata.* Three species of Ironbarks were met with, but they were not plentiful. The most valued timber is perhaps "Slaty Gum," *E. polyanthema,* var ; *glauca,* var.nov.; and I consider it a distinct gain to the botany of the Colony to have the correct botanical sequence of this valuable tree made clear. *E. albens,* Miq , is a tree also valued for its durable timber. *E. globulus* occurring at Nulla Mountain is also worthy of note.

My list of grasses is poor, as most of my specimens were lost in transit.

I have followed Bentham and Hooker's classification.

I desire to tender my sincere thanks to Mr J. Dawson, of Henbury, Rylstone, Surveyor for the District, for his invitations to, and hospitality in, his several camps, from which I was enabled to reach without any expense what would otherwise have been inaccessible country, and I must also mention his kindness in placing at my disposal men, horses, and buggies in order to make my collections complete. He himself is no mean collector, for I am indebted to him for some valuable botanical material and specimens.

I must also acknowledge my indebtedness to Mr G. Harris, of Mount Vincent, near Ilford, for his kindness while staying at his homestead during my visit to the district in 1893, for it was from there I made my collections of the flora on the watershed of the Turon and Capertee Rivers.

Class I. **DICOTYLEDONS.**

Sub-class I.   POLYPETALEÆ.

Series 1. **Thalamifloræ.**

RANUNCULACEÆ.

CLEMATIS ARISTATA, R. Br.   Barrigan Ranges; only a few plants
seen, not in flower.

C. GLYCINOIDES, DC.   The most common Clematis in the
district; September and October.*

C. MICROPHYLLA, DC.   Murrumbo and Talooby; September.
I have placed my specimens provisionally under this
species as they very closely resemble it in form of leaf,
but differ in having anther appendages.

RANUNCULUS LAPPACEUS, Sm.   Murrumbo; September (flower and
fruit).

DILLENIACEÆ.

HIBBERTIA BILLARDIERI, F.v.M., var OBOVATA, Benth.   Murrumbo;
October.

H. ACICULARIS, F.v.M.   Only found on the barren sandy
soil at the top of the Gulf Road; leaves very rigid and
pungent-pointed, quite a distinct variety from the coast
form; November.

H. DIFFUSA, R Br., var. DILATATA, Benth.   An exceedingly
narrow leaved form of this variety is found over nearly
the whole district; November.

---

*Throughout this paper, unless otherwise stated, references of this kind
are used to denote the months in which the species were found to be
flowering.

H. LINEARIS, R. Br., var. ? OBTUSIFOLIA, Benth.   Murrumbo
Gate, growing amongst the Ironbarks, *E. sideroxylon.* It
seems to agree better with this doubtful variety of
Bentham than any other described *Hibbertia.*   I cannot
bring myself to regard it as a variety of *H. linearis,* as
an examination of the anthers shows it to have no
affinity with the type of *H. linearis,* which has 15-20
stamens, while the Murrumbo specimens have from 60
to 70.   If my specimens are this variety, then I think
the specific name of *H. obtusifolia,* DC. (Syst. Veg. i.
429), should stand.

VIOLARIEÆ.

HYMENANTHERA DENTATA, R. Br.   On the western watercourses
of the main Dividing Range at Carwell, near Rylstone,
and on the eastern watershed on the banks of the
Goulburn River, near Murrumbo; September.

PITTOSPOREÆ.

PITTOSPORUM PHILLYRÆOIDES, DC.   Near the summit of Range
on the right bank of Bylong Creek, near Bylong;
September.   Never before recorded so far east, being
strictly a dry country plant.

BURSARIA SPINOSA, Cav.   All over the district and in some
localities a perfect pest.   At the foot of the Barrigan
Ranges is a variety with very long leaves (2"), and
almost spineless.

MARIANTHUS PROCUMBENS, Benth.   Rare; October.

CITRIOBATUS MULTIFLORUS, A. Cunn.   Barrigan Ranges.

CHEIRANTHUS LINEARIS, A. Cunn.   Near Rylstone; rare; December.

CARYOPHYLLEÆ.

STELLARIA PUNGENS, Brongn.   Exceedingly common on sandstone
ridges.   Mt. Vincent; November.

MALVACEÆ.

SIDA CORRUGATA, Lindl., var. ORBICULARIS, Benth.   Not common; only one plant seen, and that at Murrumbo.   This is considered an inland species, with the exception of a specimen obtained at Broadland, on the Hawkesbury River, by Robert Brown; and its presence now on the Goulburn supplies the connecting link with the dry country varieties, October (flower and fruit).

ABUTILON TUBULOSUM, Hook.   Bylong; the most southern locality recorded; September.

HIBISCUS STURTII, Hook.   Rare, October (flower and fruit).

STERCULIACEÆ.

STERCULIA DIVERSIFOLIA, G. Don.   "Kurrajong."   On most of the ridges in the district; November and December.   These trees are never cut down, as the foliage is eaten by stock during times of drought.   A peculiar fact in connection with this species was related to me by Mr J. Dawson, surveyor for the district.   He states that when a living tree of any other species is blazed and marked the sapwood and bark eventually grow over the marks, and after a few years no trace of the cicatrices can be seen on the tree, but if the bark and alburnum be removed then the whole lettering, &c., is almost as clear as on the day it was originally cut.   I have seen marks after 6 inches of the outer growth had been removed as plain and distinct as when chiselled 36 years previously. With the "Kurrajong" (*S. diversifolia*) the survey mark is never covered by the alburnum or bark, and always remains on the surface to whatever size the tree may increase in girth

RULINGIA RUGOSA, Steetz.   Murrumbo; the most easterly habitat recorded; October.

R. PANNOSA, R Br.   Goulburn River.

Series II. Disciflorae.

LINEÆ.

LINUM MARGINALE, A. Cunn. Goulburn River; September.

GERANIACEÆ.

GERANIUM DISSECTUM, Linn. Talooby and Murrumbo; October.

ERODIUM CYGNORUM, Nees. Murrumbo; in fruit in October.

RUTACEÆ.

ZIERIA ASPALATHOIDES, A. Cunn. Murrumbo; October. The two previous recorded localities for this Colony are Wellington and Hunter River.

Z. CYTISOIDES, Sm Mt. Vincent and Rylstone; October and November.

BORONIA ? MOLLIS, A. Cunn. Bylong; the most northerly record if these specimens are those of *B. mollis*; November

B. ANEMONIFOLIA, A. Cunn., var. ANETHIFOLIA, Benth. Murrumbo, not common; September.

PHEBALIUM DIOSMEUM, A. Juss. Goulburn River; October and November.

P. GLANDULOSUM, Hook. Only found at one spot, at the foot of Cox's Gap (Murrumbo side). I have my doubts about placing the specimens under this species, but do so as they come nearer it than any other N.S. Wales species. It resembles the Western Australian *P. tuberculosum* in the leaves being channelled above and the margins scarcely, or not at all, recurved, and the flowers are in sessile umbels exceeding the last leaves; a showy shrub; height about 10 feet; September and October. Since writing the above, Mr. Dawson has found it at Kerrabie. Flowers on filiform pedicels.

P. SQUAMULOSUM, Benth.   Common on all the sandstone
ranges from Rylstone to Goulburn River; and it is
perhaps the most conspicuous shrub in the month of
September, when it is in full flower.   Height from 12-
20 feet, the coast representative rarely exceeding a
dozen feet.

PHILOTHECA AUSTRALIS, Rudge.   On sandstone ridges.   Most of
the specimens incline to Sieber's *P. Reichenbachiana*,
but as the leaf varies in nearly every plant, I have
placed them all under Rudge's species as suggested by
Baron von Mueller, September to November.   Petals
white or pink, as distinct from the mauve colour of the
coast plants.

OLACINEÆ.

OLAX STRICTA, R. Br.   Murrumbo, October (flower and fruit).

STACKHOUSIEÆ.

STACKHOUSIA MONOGYNA, Labill.   On moist damp flats; September
and October.

RHAMNEÆ.

ALPHITONIA EXCELSA, Reissek.   "Red Ash;" under the crown
of the high rocks on the banks of the Goulburn River
(Macdonald's Flat), Murrumbo; only small trees seen,
in early fruit; bark white and smooth.

POMADERRIS LANIGERA, Sims.   Kelgoola and Barrigan Ranges,
Bylong; September.

P. PHYLICIFOLIA, Lodd.   Only found at one spot, viz., about
two miles from the foot of Cox's Gap, Murrumbo side,
on the banks of a creek, and only one shrub seen, and
that about 5 feet high.   I first collected it in bud in
October, 1893, and again visited the same tree in
September, 1895, but was unfortunate in again finding
it only in bud,—owing no doubt to the severe drought
from which the whole Colony was suffering at that time

I look on this specimen as a particularly interesting find from the fact that it has only previously been recorded from this Continent from the "banks of subalpine streams under the Australian Alps," so that now this new locality brings its range very much farther north. It occurs abundantly in the northern island of New Zealand, and also in Tasmania I have compared this northern form with New Zealand and subalpine specimens, and it differs little from them. It has fewer leaf scars on the stems, and less numerous leaves; its height is also a little greater.

P. BETULINA, A. Cunn. Mount Vincent, near Ilford; November

SAPINDACEÆ.

DODONÆA TRIQUETRA, Wendl. Bylong, Murrumbo; in fruit in September.

D. ATTENUATA, A. Cunn. Mount Vincent, near Ilford; November.

D. CUNEATA, Rudge. Murrumbo; in fruit in October. Rylstone; in fruit in December.

D PINNATA, Sm. Barrigan Ranges; September

Series III Calyciflorae.

LEGUMINOSÆ.

OXYLOBIUM TRILOBATUM, Benth. Murrumbo; on sandflats near Goulburn River, and Kelgoola.

MIRBELIA GRANDIFLORA, Ait Kelgoola, September.

GOMPHOLOBIUM UNCINATUM, A. Cunn. Bylong Ranges; November. The pedicels are longer and the flowers larger than those described by Bentham (Fl. Aust. ii. 46), but I do not think it can be referred to any other species.

G. HUEGELII, Benth. A few miles west of Rylstone; October.

DAVIESIA CORYMBOSA, Sm., var. LINEARIS, Lodd.   A very narrow
   leaved form found at Talooby, October.

   D. LATIFOLIA, R. Br.   Mount Vincent, near Ilford and
      Talooby.   It is called "Native Hops" on account of
      the bitter principle contained in its leaves.   In full
      flower in October, and in fruit in November and Decem-
      ber.

   D. GENISTIFOLIA, A. Cunn.   Only seen in one locality, viz.,
      Murrumbo; September and October; mostly on grassy
      levels.

         var. COLLETIOIDES, Benth.   Kelgoola; source of the
            Cudgegong River.

   D. RECURVATA, J.H.M. et R.T B.   Bylong; November.

PULTENÆA SCABRA, R. Br., var. MONTANA, Benth.   Camboon and
   Talooby; October.

   P. SCABRA, R. Br., var. MICROPHYLLA, var.nov.   Bylong,
      November.   As my specimens possess smaller leaves than
      any described specimens, and are much shorter (about 2″)
      than the type, I propose to designate it a new variety.

   P. MICROPHYLLA, Sieb.   Portland and Camboon; October.

   P. TERNATA. F.v.M.   Only found in one spot, on the Range
      west of Murrumbo Plains; September to December.

DILLWYNIA FLORIBUNDA, Sm., var. SERICEA, Benth.   Murrumbo
   and Talooby; October.

   D. ERICIFOLIA, Sm., var. PHYLICOIDES, Benth.   Common on
      sandstone ridges.

BOSSIÆA MICROPHYLLA, Sm.   Rylstone and Camboon; October.

   B. BUXIFOLIA, A. Cunn.   Camboon; October.

TEMPLETONIA MUELLERI, Benth.   Murrumbo; September.

HOVEA LINEARIS, R. Br.   Murrumbo; September (flower and
   fruit).

H. HETEROPHYLLA, A. Cunn. Kelgoola; in flower in September: at Talooby in fruit in October.

H. LONGIFOLIA, R. Br., var. LANCEOLATA, Benth. Found throughout the whole district under shelving rocks; flowers blue, not showy; in flower in September, and in fruit in December.

H. LONGIFOLIA, R. Br., var. PANNOSA, Benth. Murrumbo and Mount Vincent; September. This is a very marked variety compared with the previous one, the leaves being smaller and the petioles shorter; tomentum on the underside of the leaves, branches and petiole, dense, woolly, and rusty-coloured.

LOTUS AUSTRALIS, Andr. Camboon, Bylong, and Murrumbo; November.

SWAINSONIA MICROPHYLLA, A. Gray. Bylong; September.

S. GALEGIFOLIA, R. Br. Throughout the district, in flower and fruit in November; eaten by cattle.

GLYCINE CLANDESTINA, Wendl. Talooby; October.

DESMODIUM VARIANS, Endl. Bylong; October to November.

GLYCINE TABACINA, Benth. Murrumbo; in fruit in October.

KENNEDYA MONOPHYLLA, Benth. Murrumbo; October (flower and fruit); fairly common; Cox's Gap, with leaves large and stipules persistent.

*MEDICAGO DENTICULATA, Willd. Murrumbo; in fruit in October.

CASSIA EREMOPHILA, A. Cunn. In flower at Bylong in September; Murrumbo; in fruit in October.

C AUSTRALIS, Sims. Not common; Bylong and Murrumbo; October to December.

---

* Introduced.

ACACIA LANIGERA, A. Cunn.   Henbury and Rylstone , in flower
in September and in fruit in December.   The first
authentic pods of this species were obtained from this
locality (P.L.S.N S.W. 2nd Ser. Vol. x.)

A. JUNIPERINA, Willd.   Murrumbo, Road to Goulburn River,
September.

var. BROWNII, Benth.   Barrigan Ranges.

A. ARMATA, R. Br.   Cox's Gap, Murrumbo; September.

A. VOMERIFORMIS, A. Cunn.   Rare; Kelgoola; September.
In the specimens collected there is a peculiar recurved
point or hook instead of the gland usually found on the
phyllodia of this species.

A. UNDULIFOLIA, A. Cunn.; var. SERTIFORMIS, Benth.; and var.
. DYSOPHYLLA. Benth.   Both forms are met with over the
whole district on sandstone ridges ; var. *sertiformis* is
most abundant in the Capertee Valley, but is found
interspersed with var. *dysophylla* at Camboon, Bylong,
and Murrumbo.

A. VERNICIFLUA, A. Cunn.   Between Rylstone and Mount
Vincent, September.

A. PENNINERVIS, Sieb.   This giant *Acacia* is found throughout the whole district, on both sides of the Dividing
Range.   It is known as "Blackwattle," and its
bark is valued for its tannin properties.   At the head
of the Capertee Valley some trees attain a height of
from 50 to 70 feet or more.   It is not recorded for
Mudgee in Mr. A. G. Hamilton's Census, although it
occurs plentifully not very far east of that town.

Of all the wattles known to me I think I can justly
assign the first place to this one for possessing the most
numerous varietal forms.

Bentham only gives one variety, viz., *falciformis,* under
which he includes *A. falciformis,* DC., and *A. astringens,*
A. Cunn.

At least three distinct forms are to be found in this district, viz. :—

1. Var. *normalis.*—Phyllodia lanceolate-falcate, obtuse or acuminate, thinly coriaceous, 3 to 5 inches long and 1 inch broad, 1-nerved and prominently penninerved, the margins nerve-like, and almost always with a short secondary nerve terminating in a gland a short distance from the base. Pod several inches long and 1in. broad, firm, margins parallel, often glaucous. A tree, up to 60 or 70 feet high. It is the bark of this tree that is highly prized for tanning.

2. Var. *lanceolata.*—A tall shrub : branchlets thin, angular, phyllodia uniformly lanceolate, narrowed at both ends, secondary nerve very indistinct ; always narrower than in var. 1. Pod much lighter in colour than any of the other forms, about ½ in. broad and 6 to 9 in. long.

3. Var. *glauca.*—A shrub of a few feet in height, branchlets red, terete, much stouter than in other varieties. Phyllodes broadly obtuse, glaucous, coriaceous, central nerve and margins very prominent, the gland rarely present, 3 to 5 inches long, 1 to 2 inches broad. Pod thickly coriaceous, 2 to 4 inches long, under one inch broad. Seed mostly orbicular.

(Mr. E. Dawson collected the whole series of pods and flowers upon which these remarks are based.)

A. NERIIFOLIA, A. Cunn. Talooby and Murrumbo, on sandstone ridges. Appears to have no local name At Murrumbo Gate there are a few fair-sized trees, measuring 18 inches in diameter and 20-30 feet in height; September.

A. GLADIIFORMIS, A. Cunn. Rylstone; September.

A. HAKEOIDES, A. Cunn. Talooby ; the nearest locality to the coast yet recorded for this dry country species; September.

A. SUBULATA, Bonpl.   Quite local; only found at Murrumbo
Gate, growing amongst Ironbarks, *E. sideroxylon*. A
tall, graceful shrub, with long pendulent green branches.
September and October.   The first recorded pods of this
species were obtained from this locality (P.L.S N S W.
2nd Ser. Vol viii)

A. ?CRASSIUSCULA, Wendl.   A common wattle throughout
the district; flowers in October and September and fruits
in December.   I have preceded the name with a query
as I have never seen an authenticated *A. crassiuscula,*
but as I am acquainted with almost every other species of
Acacia found in New South Wales I cannot place my
specimens under any other than this one   The fruit
does not agree with Bentham's description, but perhaps
his were wrongly matched.   It attains almost the size
of a young tree.

A. NEGLECTA, J.H.M. et R.T B.   Perhaps the most common
of all the *Acacias* found on the sandstone ridges and
ranges.   This is considered by some as *A. lunata*, but
the pods are entirely different from those described by
Bentham (B. Fl. Vol. ii. p. 373).

A. HOMALOPHYLLA, A. Cunn.   "Yarran."   Talooby ; never
recorded so far east before.

A. IXIOPHYLLA, Benth.   I have obtained only young pods of
this plant, so cannot speak with certainty as to its
identity; and yet if it is not this species I do not know
what other it can be, as its phyllodes are the most viscid
of all the *Acacias* known to me   It is by far the most
common wattle throughout the district of Bylong,
growing under the hills in dense, almost impenetrable
masses, and is in fact quite a pest.   It attains a height
of about 15 feet; September and October.

A. ? sp nov.   Rylstone, September   This I regard as a new
species, but have not been able to obtain the pods.   It

is a low shrub of a few feet, with long linear plurinerved phyllodes and short axillary racemes, with very few flowers in the head.

A. MELANOXYLON, R. Br. Only small trees seen; foot of Barrigan Ranges, Mt. Vincent and Kelgoola. The timber is not valued; August.

A. IMPLEXA, Benth. Barrigan Ranges; in early fruit.

A. LONGIFOLIA, Willd. (a). Var. *Bylongensis*, var nov. This is quite a distinct variety from any described by Bentham (B. Fl ii. 398). The length of the phyllode has already been recorded (P.L.S.N.S W. 2nd Ser. Vol. viii. p. 311). The racemes are shorter and more compact than the type and other known varieties, resembling in some respects those of *A. doratoxylon*; in fact it might be looked upon as an intermediate form between these two species. Gulf Road and Camboon.

(b). Var. TYPICA, Benth. This variety is found on the Barrigan Ranges.

A. DORATOXYLON, A. Cunn. "Hickory." At Murrumbo, on the ranges on the right bank of the Goulburn River. It also probably extends to the Hunter River, as a specimen of "Hickory" timber from that locality, which I have compared with the Murrumbo "Hickory," is exactly identical. I consider the finding of this species here of some importance, as it has only previously been recorded in this Colony from the interior, as the "Spearwood of certain tribes." Height generally from 15-30 feet; diameter up to 1 foot; in flower in September and in fruit in November and December.

A. CUNNINGHAMII, Hook., and also var. LONGISPICATA, Benth. Cox's Gap; September. I am indebted to Mr. J. Dawson for the pods of this *Acacia*. They hardly agree with any previous descriptions. Bentham had only unripe pods as he mentions (B. Fl. ii. p. 407), and

from the pods now in my possession I am inclined to
think his were not properly matched.   Mr. Dawson's
specimens of fruit are attached to twigs, with the
phyllodes, and stout, strongly 3-angled stems and
early flowers, so that there can be no doubt about
their identity.   They are not "long . . . . very
flexuose or twisted," but are straight or slightly curved,
2-3 inches long, under 2″ broad, valves *thin*, convex over
the seed   Seeds small, oblong, longitudinal, funicle at
first straight and filiform, and gradually thickening into
3 or 4 folds under the seed.

A. DISCOLOR, Willd.   Kelgoola, September, rare

A. DEALBATA, Link.   Occurs throughout the district from
Rylstone to the head of the Cudgegong ; September
Its bark is never used as a tan, the inhabitants having
found out the superior tanning properties of the black
wattle*(Acacia penninervis,* Sieb.).   An interesting feature
of this *Acacia* here, is that the plants on the ridges
have short leaflets, 2-3‴ long, and the whole tree is
glaucous, whilst the plants growing on the plains and in
gullies have linear leaflets, 4 to 6 lines long, and
glabrous; and the tree could very easily be mistaken for
A. *decurrens,* var. *normalis,* but for the pods.

A. MUELLERIANA, J H M. et R.T.B   Foot of Murrumbo
Ranges and Road to Cox's Gap, Murrumbo, in flower in
August, in fruit in December.

ROSACEÆ.

RUBUS PARVIFOLIUS, Linn.   Murrumbo, and on the banks of a
creek at Kelgoola.

ACÆNA OVINA, A. Cunn.   Kelgoola.   Only a few specimens seen.

SAXIFRAGEÆ.

CERATOPETALUM APETALUM, D. Don.   In the gullies at the head
of the Cudgegong River.   Vernacular name " White-
wood ;" timber used for lining boards of houses.

SCHIZOMERIA OVATA, D. Don. Gullies at the source of the Cud gegong River.

DROSERACEÆ.

DROSERA PELTATA, Sm. Camboon.

MYRTACEÆ.

CALYTHRIX TETRAGONA, Labill. Camboon and Murrumbo; in flower and fruit from September to December.

BÆCKEA CUNNINGHAMII, Benth. Found on the eastern and western slopes of the Dividing Range at Murrumbo toward the Goulburn River and Camboon, respectively. This is the first time it has been recorded on the eastern watershed, October.

LEPTOSPERMUM FLAVESCENS, Sm., var. GRANDIFLORUM, Benth. Bylong; November.

L. SCOPARIUM, R. & G. Forst. Sandy flats towards the Goulburn River; in fruit in September.

L. ARACHNOIDEUM, Sm. Camboon; in fruit in October.

L. LANIGERUM, Sm. Camboon; in fruit in October. I am not altogether certain about my determination in this case, as I failed to gather the flowers. The leaves are almost pungent-pointed and the fruits large. It is probably Bentham's variety (d) of this species.

L. PARVIFOLIUM, Sm. Camboon and Murrumbo; September and October. The Murrumbo specimens are characterised by an almost glabrous calyx, with triangular persistent lobes.

CALLISTEMON SALIGNUS, DC., var. ANGUSTIFOLIA, Benth. Murrumbo; October. I also collected a large-leaved variety at the same place.

ANGOPHORA INTERMEDIA, DC. Found mostly on the alluvial flats; very abundant at Bylong; February. This is a good fodder tree in time of drought. It is also an excellent shade tree for cattle. The timber is of very little value, but works up well in small cabinet work.

EUCALYPTUS STELLULATA, Sieb.   On the hills overlooking the
Capertee Valley.   Small trees with lead-coloured bark.

E. AMYGDALINA, Labill.   Known locally as " Peppermint ; "
rare; only one tree seen, at Kelgoola on the Cudgegong
River, in bud and mature fruit in September.

E. OBLIQUA, L'Her.   " Stringybark."   Gulf Road.   This
species has never been found so far north before.   The
fruits differ from those figured as *E. obliqua* in Hooker's
·" Flora of Tasmania " (i. 136, t. 28), and also from the
delineation in Baron von Mueller's ' Eucalyptograhia " In
both instances the fruits are shown with a contracted,
countersunk rim, but in my specimens *the fruits are
hemispherical, with a flat, broad truncate rim.*   The
shape of the leaves corresponds in every particular with
all the descriptions and figures published of *E. obliqua.*

A microscopial examination of the anthers showed
them also to agree with Bentham's description (B. Fl.
iii. p. 204).

This form of *E obliqua* is evidently peculiar to New
South Wales, as it has also been found near the National
Park (F. Williams).

This species probably occurs also at Mudgee, although
not collected by Hamilton (P.L.S.N.S W. 2nd Ser.
Vol. ii. p. 279)

E. CAPITELLATA, Sm.   Found throughout the whole district
in both basaltic and sandstone country.   From Rylstone
to the Goulburn River it goes by the name of " White
Stringybark," the same as *E. eugenioides* ; in fact, the
settlers look upon them as one and the same species, but
on the watershed between the Capertee and Turon Rivers
it is called "Silvertop" and "Messmate,"—rather unfortunate terms and not mentioned here to be perpetuated,
but only as a warning, as it is now generally decided by
botanists to reserve those terms for *E. amygdalina.*

The large-fruited form, the same as that found on "North Shore, *Woolls*" (B. Fl. iii. 206) predominates. The smaller-fruited forms are occasionally met with, and as *E. eugenioides* is also to be recorded from here, I should like to venture the opinion that this latter species should be merged into *E. capitellata* or *vice versa*, and the two regarded as extreme forms of the same species. Bentham places *E. eugenioides*, Sieb , as a variety of *E. piperita*, but there appears to me very little connection except in the matter of bark.

The type fruits of this species resemble the fruits of *E. eugenioides* in every particular except size, and the smaller varieties cannot be distinguished from those of *E eugenioides* ; in fact, they are the *E. eugenioides* of some authors.

E. MACRORRHYNCHA, F v.M.   "Red Stringybark."   This is considered the best stringybark in regard to durability of timber, and is highly prized   It occurs only on the western slopes of the ranges; November and December.

E. PIPERITA, Sm. "Blackbutt." Mount Vincent, near Ilford.

E. SIDEROXYLON, A. Cunn , var. PALLENS, Benth.   "Ironbark."   This variety previously had been recorded only from one locality, New England (C. Stuart).   Its southern extension must now be brought to the Murrumbo Plains, where it is the only Ironbark.   The buds are smaller than the typical Liverpool and Parramatta specimens of *E. sideroxylon,* and very much resemble those of *E. paniculata*.   The blue glaucous leaves contrasting with the black bark give certain patches of bush a very pretty appearance.   The timber is not considered of any value.   Flowers profusely from September to December.

E. MELLIODORA, A. Cunn.   "Yellowbox."   Throughout the district, mostly on flats.   Timber very durable, but difficult to obtain in any size, as most of the trees have a tendency to barrel in the trunks.

As A. Cunningham, C Moore, and F. v. Mueller each record a different bark (B. Fl. iii. 210), I may mention here that in all instances I found the bark "furrowed and presistent," and its inner surface, when freshly cut from the tree, has a very yellowish appearance as well as the exposed sapwood, hence its local name.

E. HÆMASTOMA, Sm , var. MICRANTHA. "Brittle Gum." Camboon, on the western slope of the Range, and Mount Vincent, near Ilford.

E. POLYANTHEMA, Schau. "Red Box," "Slaty Gum" There are three distinct varieties of this species to be found in the district.

(*a*). In the neighbourhood of Rylstone it goes by the name of "Red Box," and the timber is considered of no value whatever   The trees are of no great height, have a dirty scaly bark at the butt but smooth otherwise, and are found on poor sandstone country. The leaves are uniformly oval, on fairly long petioles, veins oblique, marginal one removed from the edge, under three inches long, and glaucous on both sides , flowers small , in flower in December ; fruit turbinate, under two lines long in diameter.

(*b*). At Camboon, 7 miles north of Rylstone, there is a variety with smooth bark, long lanceolate leaves green on both sides, the veins oblique, the marginal one *close* to the edge, the petiole long, sometimes twisted; flowers larger than in previous variety, outer stamens sterile , fruits turbinate, 3 lines in diameter, rim thin and notched and similar to the coast *E. polyanthema* ; flowers in October; timber good.

(*c*). On the eastern slope of the Dividing Range and extending to the Goulburn River there is to me the most important variety known as "Slaty Gum." The trees are large, with very straight barrels, and the timber is highly valued and considered equal to if not superior to Ironbark. The bark is smooth, with a silvery sheen. The leaves differ from those of the two other varieties in being much narrower and glaucous, the venation being the same as in the Camboon variety. The flowers are the smallest of the three varieties, the stamens are all fertile as in the first variety, the fruits glaucous, 1 line in diameter.

I was at first inclined to consider these as three disdinct species (being so looked upon by the residents), but a microscopial examination of the anthers proved them identical. The anthers are cylindrical, "truncated, opening by terminal pores" in each variety, and as faithfully figured by Baron von Mueller in his "Eucalyptographia." There is evidently an error in Bentham's description of the anthers (B. Fl. iii. 214).

In closing these remarks I would like to point out that the New South Wales *E. polyanthema* differs considerably in the character of its bark from the Victorian form, which has "an ashy-grey, persistent, rough and furrowed bark" (F.v.M., B. Fl. iii. 213), while all the trees seen by me, and I have collected from the coast to the western slope of the Dividing Range, are smooth-barked. The leaves of the Sydney *E. polyanthema* are much larger and more ovate than any of the three varieties above enumerated.

E. HEMIPHLOIA, F.v.M. "Box." Throughout the district on the flats. It is not by any means the fine upstanding tree growing on the coast near Parramatta.

It was found in flower at Bylong and Murrumbo in October. Mr. A. G. Hamilton gives the flowering time at Mudgee, 40 miles east, as April and May,—an evidence of the uncertain times of flowering of Eucalypts.

I have kept this species apart from the following, as I consider them quite distinct when the following differences are taken into account, namely :—size, shape, and venation of leaves; size of flowers and fruits ; and shape of anthers, which in this case resemble those of " Slaty Gum."

E. ALBENS, Miq. (E. HEMIPHLOIA, var. ALBENS, F.v M.) " Box; " " White Box." Bentham considered this " a very distinct species " (B. Fl. iii. p 219), but Baron von Mueller has placed it as a variety of *E. hemiphloia* When seen growing in juxtaposition with *E. hemiphloia* its characteristic differences are very marked. According to Baron von Mueller it has a dull green, persistent bark, but I have always found it with a whitish, persistent chequered bark, somewhat approaching *E. hemiphloia*, from which it also differs in the larger, angular, sessile calyx (nearly 9''' long), larger fruits, and in " the foliage being usually glaucous or almost nearly white." Anthers globular, opening at the side by almost circular pores, connective much developed.

The timbers of the two species are of equal merit. It is always found growing under the Ranges on both banks of Bylong Creek, and gradually ascending them till meeting the " Slaty Gum," *E. polyanthema;* September and October.

E. SIDEROPHLOIA, Benth. " Ironbark." On the sandstone ranges at Murrumbo, and only represented by poor specimens of this grand forest monarch of the coast, in fruit and bud in September.

E. CREBRA, F.v.M. Found throughout the district as the most common of all the " Ironbarks," and the only one valued for its timber, the others never appearing to grow to any size. Shingles that had lain on the ground exposed to the weather for over five years, were as good as when first split. None of the Ironbarks are considered equal in durability to " Slaty Gum " timber.

Splendid forests of this grand timber are being ring-barked by the selectors. The flowers are very much sought after by bees, and are their standby during times of drought when other flowers are scarce, September.

E. GLOBULUS, Labill. A small-fruited variety occurs at Nulla Mountain, 24 miles east of Rylstone.

E. DEALBATA, A. Cunn. " Sallow." I am not at all certain that my diagnosis in this instance is correct, but I place the specimen collected at Ganguddy Creek, 18 miles east of Rylstone, provisionally under this species.

E. VIMINALIS, Labill. Found throughout the district on low levels; known under several vernacular names such as "White Gum," " Swamp Gum," " River Gum," " Brittle Gum;" timber not used.

E. TERETICORNIS, Sm. " Red Swamp Gum ;" " Red Gum." Throughout the district on flats. A profuse flowerer during October, November and December. It is the common form with a long operculum. I am inclined to place this and the preceding species under one name.

E. STUARTIANA, F.v.M. " Woolly Butt." At Mount Vincent, near Ilford, and Ganguddy Creek, timber worthless.

E. PUNCTATA, DC. Kelgoola, at the source of the Currajong River. The dark copper-coloured foliage of this tree makes it very conspicuous amongst other Eucalypts of the bush in this locality, where it goes by the local name of " Ironwood." At Mount Vincent, near Ilford, it is known as " Red Gum "

E. GUNNII, Hook f. Occurs on both sides of the Dividing Range. Known as " Mountain Gum " at Kelgoola, but has no vernacular name at Murrumbo.

E. TRACHYPHLOIA, F.v.M. Only found at two places, Cox's Gap and Murrumbo Gate. It has not been recorded from any other locality in this Colony, and is known only

from the Burnett River, Queensland ("Bloodwood")
Kino exudes very freely. Timber hard, colour of Spotted
Gum; not used.   In fruit in September and October.

E. EUGENIOIDES, Sieb. "White Stringybark." Found on
the watershed between Capertee and Turon Rivers, and
also on the Barrigan Ranges, probably throughout the
whole district.   (See remarks under *E. capitellata*)

EUGENIA SMITHII, Poir.   Occurs plentifully in the gullies at the
extreme head of the River Cudgegong, and known as
" Lilly Pilly."

UMBELLIFERÆ.

ERYNGIUM ROSTRATUM, Cav.   Rylstone; in fruit in December

ARALIACEÆ.

ASTROTRICHA LEDIFOLIA, DC.   The narrow-leaved variety was
found at Camboon, in flower in October; and the broader
leaved form with narrower panicles at Bylong.

Sub-class II. MONOPETALÆ.

LORANTHACEÆ.

LORANTHUS BIDWILLII, Benth.   Only at one locality, Cox's Gap,
on *Callitris* sp.   Previously recorded only from Wide
Bay, Queensland.

L. CELASTROIDES, Sieb.   Rylstone; in fruit in December

L. PENDULUS, Sieb.   A long-leaved variety, the leaves
measuring sometimes over a foot.   Mount Vincent and
Camboon; in flower in November and in fruit in October

NOTOTHIXOS CORNIFOLIUS, Oliv.   Bylong.   On *Sterculia diversifolia*, G. Don; September and October.

RUBIACEÆ

CANTHIUM OLEIFOLIUM, Hook.   Collected when in flower at
Karrabie, by Mr. J. Dawson, L.S., and by me near the
Goulburn River, Murrumbo, but not in flower or fruit
Previously recorded only from the interior of the Colony

**COPROSMA HIRTELLA, Labill. Mount Vincent, near Ilford; November.

**OMAX UMBELLATA, Soland. Camboon; October.

**ASPERULA CONFERTA, Hook. Camboon; October.

**GALIUM GAUDICHAUDI, G. Don. Camboon. October.

COMPOSITÆ.

**OLEARIA RAMULOSA, Benth., var COMMUNIS, Benth. The common New England form, "with glabrous glandular achenes."

**VITTADINIA AUSTRALIS, A. Rich. Camboon; October.

V. AUSTRALIS, A. Rich., var. DISSECTA, Benth. Murrumbo; October.

**BRACHYCOME STURTII, Benth. Camboon; October.

B. GRAMINEA, F.v.M. Talooby, Murrumbo; October.

B. LINEARIFOLIA, DC. Camboon; October.

B. MULTIFIDA, DC. Murrumbo; October.

**SIEGESBECKIA ORIENTALIS, Linn Murrumbo; October.

**ECLIPTA PLATYGLOSSA, F.v.M Bylong; November.

**CRASPEDIA RICHEA, Cass. Murrumbo; October.

**CASSINIA ? LEPTOCEPHALA, F.v.M. In bud in November.

**IXIOLÆNA LEPTOLEPIS, Benth. Murrumbo; the most easterly locality recorded; generally regarded as an interior species; September.

**ODOLEPIS ACUMINATA, R. Br. Camboon; bracts very acuminate in my specimens; October.

**EPTORRHYNCHOS SQUAMATUS, Less. Talooby; October.

**ELICHRYSUM SCORPIOIDES, Labill. Common; some specimens measure 2 feet in height; October.

H. BRACTEATUM, Willd. A tall perennial of 2 feet, with long linear leaves; Murrumbo; October.

H. APICULATUM, DC. Throughout the district; September to December.

H. SEMIPAPPOSUM, DC., and var. BREVIFOLIUM, Sond. De Candolle considered this variety as a distinct species (*H. microlepis*, Prod. vi. 195). I was at first inclined to agree with his view, but I have since found it growing from the root or base of the stem of the typical form, thus proving what Bentham suspected (B Fl iii. 625)), that there is only one species. The two forms on the same stem make a unique herbarium specimen.

H. DIOSMIFOLIUM, Less. Throughout the district; October to December. Quite like the Sydney form.

H. BREVIDECURRENS, J.H.M. et R.T.B. Murrumbo; October.

H. TESSELATUM, J.H.M. et R.T.B. Murrumbo, and the hill overlooking Bylong on the east of Torrie Lodge.

H. CUNNINGHAMII, Benth. Barrigan Ranges, Bylong; September. I have placed my specimens under this species, although they differ from Bentham's description in having leaves over 1 inch long ($\frac{1}{2}$" Benth.) and 3 to 6 florets (3 Benth.)

HELIPTERUM ANTHEMOIDES, DC. Murrumbo; November.

H. INCANUM, DC. Common everywhere; October to December.

H. DIMORPHOLEPIS, Benth. Fairly common in places at Murrumbo and Camboon; September.

GNAPHALIUM LUTEO-ALBUM, Linn. Murrumbo, October.

ERECHTITES ARGUTA, DC, var. DISSECTA, Benth. Camboon, October.

Also a variety which is not "scabrous, with crisped hairs," and is without toothed auricles; flower heads not dense.

E. QUADRIDENTATA, DC. Camboon; October.

SENECIO LAUTUS, Sol. Murrumbo, Talooby and Mt. Vincent; October.

S VELLEIOIDES, A. Cunn. Talooby, Bylong Creek; October.

CYMBONOTUS LAWSONIANUS, Gaud. Camboon, October.

MICROSERIS FORSTERI, Hook. Not very common, only found at Murrumbo, September.

STYLIDEÆ.

STYLIDIUM LARICIFOLIUM, Rich. Camboon.

GOODENIACEÆ.

GOODENIA BARBATA, R. Br. An undershrub; on the eastern and western slopes of the Dividing Range at Camboon and Bylong respectively. This is its most northern locality; October and November.

G. DECURRENS, R. Br. Bylong Ranges; November.

G. OVATA, Sm. Bylong, under the shelter of rocks, mostly in moist situations; November. These specimens are *G. acuminata*, R. Br., placed under the above species by Bentham. The leaves are uniformly broadly lanceolate, denticulate, 1-1½ inches long, non-viscid and hoary on both sides.

G HETEROPHYLLA, Sm. Camboon; October.

G. PINNATIFIDA, Schlecht. Murrumbo; October and September.

G PANICULATA, Sm. Murrumbo; October.

SCÆVOLA MICROCARPA, Cav. Bylong Ranges; November.

DAMPIERA BROWNII, F.v M. Cox's Gap; September and November.

D ADPRESSA, A. Cunn. Murrumbo ; the most easterly recorded locality; October.

CAMPANULACEÆ.

ISOTOMA AXILLARIS, Lindl.   Bylong Ranges; November.

I. FLUVIATILIS, F.v.M.   Bylong; November.

WAHLENBERGIA GRACILIS, A.DC.   Everywhere; November.

EPACRIDEÆ.

STYPHELIA LÆTA, R.Br., var. ANGUSTIFOLIA, Benth.   At Bylong
and Murrumbo on the sandy flats and sandstone ridges.
Bentham (B. Fl. iv. p. 147) queries the colour of the
flowers, but in every instance I found them red. I have
never found this variety near Sydney.

S. LÆTA, R. Br., var. GLABRA, var.nov   I am in doubt
about the specimens placed here under a new variety,
but I prefer this to proposing a new species. The
flowers are red, the sepals acute, and the leaves narrow-
lanceolate,— characters not included under Bentham's
description of the species; Camboon; October.

ASTROLOMA HUMIFUSUM, Pers.   "Groundberry." Everywhere,
apparently in flower and fruit all the year round   At
Murrumbo it is quite an erect shrub; from 1-2 feet high.

MELICHRUS URCEOLATUS, R. Br.   The specimens found on the
western slope at Rylstone and Mt Vincent correspond
to A. Cunningham's M. medius; while those on the
eastern slope at Murrumbo to his M. erubescens. I
think they are good species, but as Bentham has placed
them under M. urceolatus, R.Br., I have followed
his classification.   September; in fruit in November and
December.

BRACHYLOMA DAPHNOIDES, Benth.   Only seen at Murrumbo, but
no doubt common.

LISSANTHE STRIGOSA, R. Br.   Murrumbo; September and October.

LEUCOPOGON LANCEOLATUS, R. Br.   Kelgoola; September.

L. MICROPHYLLUS, R. Br. Kelgoola; September.

L. VIRGATUS, R. Br. Camboon; October.

L. MUTICUS, R. Br. Camboon, Bylong Ranges; flowers and fruit in November.

L. ESQUAMATUS, R. Br. Very common on sandstone ridges; September and October.

EPACRIS RECLINATA, A. Cunn. Kelgoola, Camboon and Talooby; October. This is its most northern limit.

E. PULCHELLA, Cav. Only found on one patch of sandstone at Kelgoola.

DRACOPHYLLUM SECUNDUM, R. Br. Bentham notes under this species (B. Fl. iv. 263) "the filaments are represented in the Bot. Mag. [t. 3264] as free: I have always found them adnate to the corolla-tube." In the specimens collected at Kelgoola the anthers were free.

JASMINEÆ.

NOTELÆA MICROCARPA, R. Br. On the summit of the Dividing Range at Mt. Vincent, near Ilford; November. This is the most southerly locality for it yet recorded.

APOCYNEÆ.

LYONSIA EUCALYPTIFOLIA, F.v.M. Bylong; the most easterly locality in this colony yet recorded for it.

ASCLEPIADEÆ.

MARSDENIA SUAVEOLENS, R. Br. Murrumbo.

LOGANIACEÆ.

LOGANIA FLORIBUNDA, R. Br. Common throughout the district. It is of a lighter green than the coast variety, and also does not dry so black; September.

GENTIANEÆ.

SEBÆA OVATA, R. Br. Camboon; October.

ERYTHRÆA AUSTRALIS, R. Br. Camboon; October.
30

BORAGINEÆ.

MYOSOTIS AUSTRALIS, R. Br.   Only on the western slope of Dividing Range at Rylstone; December.

CYNOGLOSSUM AUSTRALE, R. Br.   " A tall, erect coarsely-hirsute plant."   Murrumbo; October.

SOLANEÆ.

SOLANUM STELLIGERUM, Sm.

S. VIOLACEUM, R. Br.   On the eastern slope of the Dividing Range from top of the Gulf to Murrumbo; in flower and fruit in October and November.   It differs from the ordinary *S. violaceum* in having broader calyx-lobes.

S. VIOLACEUM, R. Br., var. VARIEGATA, var nov.   I found this specimen growing between the bark and sapwood of *Angophora intermedia* on the Gulf Road.   The white markings gave it a very attractive appearance, and when first approaching it I thought I had got something new.   I propose to call it a variegated form of *S. violaceum.*

S. AMBLYMERUM, Dun.   Talooby; October.   Bentham suggests that this may prove to be a variety of *S. violaceum*, but after comparing specimens of both I think they are distinct species.

S. CAMPANULATUM, R. Br.   Murrumbo Ranges; October (flowers and fruits).

S. CINEREUM, R. Br.   Murrumbo; October; rare

SCROPHULARINEÆ.

GRATIOLA PERUVIANA, Linn.   " Brooklime."   In creeks at Mount Vincent, and Kelgoola.

EUPHRASIA BROWNII, F. v. M.   Throughout the district on damp ground; September.

MYOPORINEÆ.

MYOPORUM ACUMINATUM, R. Br., var. ANGUSTIFOLIUM, Benth.
Rylstone and at the foot of the Bylong Ranges. In
flower in September, and in fruit in November.

M. DESERTI, A. Cunn. Rylstone and Murrumbo. I do not
think it has been recorded further east than these two
localities. Bentham (B. Fl. v. p 5) in his description
of this species gives the number of stamens as five,
whilst I found only four in my specimens ; September
and October (flowers and fruits)

M. PLATYCARPUM, R. Br. Murrumbo; October. This species
has previously been recorded only from the dry interior,
i.e., Murray and Darling Rivers.

EREMOPHILA LONGIFOLIA, F.v.M. On the western slopes of the
Ranges to the east of Bylong Creek. This is the most
easterly locality yet recorded; September.

LABIATÆ.

SCUTELLARIA MOLLIS, R. Br. Camboon. This is its most northern
locality recorded; October.

PROSTANTHERA PRUNELLOIDES, R. Br. Murrumbo Ranges ;
October. A beautiful shrub, the profusion of large
white flowers making it most attractive.

P. DEALBATA, R.T.B. At the foot of Cox's Gap, Murrumbo
side; September.

P. STRICTA, R.T.B. Mount Vincent, near Ilford; November.

P. EMPETRIFOLIA, Sieb. Murrumbo; October.

WESTRINGIA LONGIFOLIA, R. Br. Murrumbo; October ·and
November.

TEUCRIUM CORYMBOSUM, R. Br., var. MICROPHYLLUM, var.nov.
Murrumbo; October.

AJUGA AUSTRALIS, R. Br. This species grows very luxuriantly at Bylong, reaching sometimes 3 feet in height. A pink form was found at Murrumbo.

Sub-class III. MONOCHLAMYDEÆ.

MONIMIACEÆ.

DORYPHORA SASSAFRAS, Endl. In the sassafras gullies at the source of the Cudgegong River.

LAURINEÆ.

CASSYTHA PUBESCENS, R. Br. Camboon; October (flowers and fruits).

    C. MELANTHA, R. Br. Murrumbo; October (flowers and fruits).

PROTEACEÆ.

PETROPHILA PULCHELLA, R. Br. On sandstone country; September (fruits).

ISOPOGON PETIOLARIS, A. Cunn. Bylong Ranges; October.

    I. DAWSONI, R.T.B. Murrumbo, on the summit of the Ranges on the north of the Murrumbo Plains. The original specimen upon which this species was founded was not a true representative. A second visit to this locality revealed a much larger shrub than formerly described. It is at least 20 feet high, in fact the tallest of Isopogons in Eastern Australia. The flowers are also very showy and attractive, and as it flowers freely it presents quite a picture, and is well worthy of cultivation; September.

CONOSPERMUM TAXIFOLIUM, Sm., var. LANCEOLATUM, R. Br. Bylong and Murrumbo; October and November. This variety was collected on the Hunter River by Robert Brown.

PERSOONIA CHAMÆPITYS, A. Cunn. At the top of the Gulf Road, on the loose sandy flat; October.

P. LINEARIS, Andr. The most common of all Persoonias, on worthless sandy ground and rocks; September (fruits).

P. RIGIDA, R. Br. Near the Goulburn River, Murrumbo.

P. CURVIFOLIA, R. Br. Only found on the western watershed, *i.e.*, at Camboon. This is therefore its most easterly habitat yet recorded; October (fruits).

P. ? OBLONGATA, A. Cunn. Not common ; on sandstone country at Kelgoola.

P. ? CUNNINGHAMII, R. Br. I have placed my specimens provisionally under this species as I was only able to obtain them in fruit. It differs from Bentham's description of *P. Cunninghamii* in having reflexed hairs on the branches, pedicels *not* glabrous nor slender, and a pubescent ovary, veins of leaf fairly prominent; Bylong Ranges.

GREVILLEA MUCRONULATA, R. Br. A small shrub occurring only at Murrumbo, and having "leaves rounded at the ends and shortly mucronate." This was the form found by A. Cunningham on the Hunter River (B. Fl. v. p. 443), and is made the type of the species by Bentham (*loc. cit.*); September and October.

G. LONGISTYLA, Hook. On the Ranges on the north side of Murrumbo Plains. The specimens obtained are referred to this species on the authority of Baron F. v. Mueller, who, in giving his reasons, says that Bentham's description of this species is incorrect as regards the length of pedicel, style, &c. My specimens differ from those described by Bentham in the length of the pedicels, which are under 6 lines, whereas Bentham gives 2-4 inches; the leaves are all under 1 line in width, whereas Bentham gives 2 lines ; they are linear, pinnatifid or

divided into long linear segments. It is a very showy
shrub and worthy of cultivation, its large beautiful
crimson flowers and long linear leaves having a very
pleasing effect. It is considered the prettiest shrub in
the bush at Murrumbo, where it was first found in this
Colony by Mr. J. Dawson, of Rylstone.

G. PUNICEA, R. Br.   Kelgoola.

G. SERICEA, R. Br.   Murrumbo; September and October.

G. TRITERNATA, R. Br.   On the road to Macdonald's Flat,
Murrumbo; September and October (flowers and fruits).

G. RAMOSISSIMA, Meissn.   Camboon and Rylstone; October.

HAKEA MICROCARPA, R. Br.   Throughout the district both in
grassland and sandy flats; October to December (flowers
and fruits).

H. DACTYLOIDES, Cav.   On the eastern slope of the Dividing
Range, at the top of the Gulf, Cox's Gap and Murrumbo;
October.

LOMATIA ILICIFOLIA, R. Br.   Fairly common on sandstone ridges
at Kelgoola.

L. LONGIFOLIA, R. Br.   Kelgoola; September (fruits).

BANKSIA MARGINATA, Cav.   A fair-sized tree at Mount Vincent,
near Ilford; also occurs at Kelgoola.

B. SERRATA, Linn. f. var. HIRSUTA, var. nov.   Only one clump
of this species seen, and that on one of the ranges at
Kelgoola.   The leaves are larger than those of Sydney
specimens, and covered on both sides with long white
hairs, which are also found on the branches.   This
species has never been recorded so far west before.

THYMELEÆ.

PIMELEA GLAUCA, R. Br.   Talooby.   Specimens differ from the
type in having the persistent portion of the perianth
glabrous; October.

P. COLLINA, R. Br. Camboon. The specimens are evidently the *P. Cunninghamii* of Meissn., which Bentham doubtfully places as a variety of *P. collina* (B. Fl. vi. 17); October.

P. LINIFOLIA, Sm. Everywhere; October to December.

P. CURVIFLORA, R. Br. A small delicate plant a few inches high. In flower at Murrumbo in October.

P. HIRSUTA, Meissn. A variety of this species with crowded, oval-shaped leaves was found at Murrumbo ; October. This is the most northern locality recorded for it.

EUPHORBIACEÆ.

PORANTHERA CORYMBOSA, Brongn. Top of Gulf Road and Murrumbo; September to November.

P. MICROPHYLLA, Brongn. Camboon; October.

BEYERIA VISCOSA, Miq. Murrumbo, on the banks of the Goulburn River; October (fruits).

BERTYA GUMMIFERA, Planch. Banks of Goulburn River, Murrumbo; September.

AMPEREA SPARTIOIDES, Brongn. Mount Vincent, near Ilford. Male plants.

URTICACEÆ.

FICUS SCABRA, Forst. Murrumbo; rare.

F. (STIPULATA) PUMILA, L. On the left hand side of the Gulf Road.

CASUARINEÆ.

CASUARINA STRICTA, Ait. This species occurs at Murrumbo, on the north-western slope of one of the ranges bounding the southern side of the Murrumbo Plains, and also on the side and summit of Bald Hill, Camboon These are the most northern localities recorded for this species, Mt. Dromedary in the south being the previous northern limit. It is mostly a swamp species; height 30 to 40 feet; in fruit in November and December.

C. SUBEROSA, Ott. et Dietr.   The only species of sheoak seen at Kelgoola, not very common.

C. DISTYLA, Vent.   On the hills on the left bank of Bylong Creek at Talooby, and Murrumbo.   A shrub of about 10 feet high.   In flower and fruit in October and ·November.   This is the most northern locality recorded for this species.   It differs from the coast form in having slender branches and much more elongated fruits.

SANTALACEÆ.

CHORETRUM SPICATUM, F.v.M.   Camboon (western watershed) ; October.   If this is a correct diagnosis this brings the range of the species very much further east than previously recorded.

C. LATERIFLORUM, R. Br.   Kelgoola; September.

C. CANDOLLEI, F.v.M.   Murrumbo ; September (flowers), October (fruits).

OMPHACOMERIA ACERBA, A.DC.   Mount Vincent, near Ilford.

EXOCARPUS CUPRESSIFORMIS, Labill.   "Native Cherry."   Barrigan Ranges.

E. STRICTA, R. Br.   Goulburn River; September.

Sub-class IV.   GYMNOSPERMÆ.

CONIFERÆ.

CALLITRIS CALCARATA, R. Br.   "Black Pine."   Talooby and Bylong.

C. COLUMELLARIS, F.v.M.   "White Pine "   Bylong.

CYCADEÆ.

MACROZAMIA SPIRALIS, Lehm.   On the foot and brow of most of the hills at Bylong; in fruit in September.

Class II. MONOCOTYLEDONS.

ORCHIDEÆ.

DENDROBIUM TERETIFOLIUM, R. Br. Kelgoola.

CYMBIDIUM SUAVE, R. Br. Mostly in the forks of dead standing timber ("Box," "White Box," and "Apple Tree") at Bylong and Talooby.

DIURIS AUREA, Sm. Murrumbo; September.

    D. SULPHUREA, R. Br. Talooby; September.

CALADENIA CARNEA, R. Br. Barrigan Ranges; September.

IRIDEÆ.

PATERSONIA SERICEA, R. Br. Murrumbo; October and September.

LILIACEÆ

GEITONOPLESIUM CYMOSUM, A. Cunn Rylstone; September (fruits).

BULBINE BULBOSA, Haw. Common throughout the district ; September to November.

ANGUILLARIA DIOICA, R. Br. Common; October.

JUNCACEÆ.

XEROTES LONGIFOLIA, R Br. Barrigan Ranges and Kelgoola.

    X. MULTIFLORA, R. Br. Camboon.

    X. FILIFORMIS, R Br. Bylong and Camboon.

XANTHORRHŒA HASTILIS, R. Br. Rare; found only on the sandy flats towards Goulburn River, Murrumbo; September.

CYPERACEÆ

SCHŒNUS ERICETORUM, R. Br. Murrumbo; September.

GAHNIA ASPERA, Spreng. Murrumbo; September.

    G. PSITTACORUM, Labill., var. (?) OXYLEPIS, Benth Kelgoola
31

CAUSTIS FLEXUOSA, R. Br.   Kelgoola.

CAREX PANICULATA, Linn   Talooby; October.

GRAMINEÆ.

ANTHISTIRIA CILIATA, Linn. fil.   Murrumbo; not common.

DANTHONIA SEMIANNULARIS, R. Br.   Throughout the district.

STIPA SETACEA, R. Br.   Rylstone.

* KOELERIA PHLEOIDES, Pers.   Murrumbo.

* FESTUCA RIGIDA, Mert. and Koch.   Murrumbo.

*CERATOCHLOA UNIOLOIDES, DC.} This American grass was found
    at Murrumbo

Class III. ACOTYLEDONS.

LYCOPODIACEÆ.

AZOLLA RUBRA, R. Br.   Very plentiful on Budden Creek.
    During the drought of 1895 it was the only green feed
    available for cattle, which seem to eat it with great
    relish.

FILICES.

TODEA BARBARA, T. Moore.   Rare, only found at Camboon, which
    locality would probably be its western limit ; in fructi-
    fication in October.

ADIANTUM AETHIOPICUM, Linn.   Barrigan Ranges

    A. FORMOSUM, R. Br.   Barrigan Ranges.

PTERIS AQUILINA, L.   Mount Vincent, near Ilford.

POLYPODIUM SERPENS, Forst.   Found in dense masses on the
    surfaces of rocks in the gullies.

NOTHOLÆNA DISTANS, R. Br.   Found on the eastern and western
    watersheds at Camboon and Murrumbo respectively.

---

* Introduced.

NOTE ON CYPRÆA ANGUSTATA, GRAY,

Var. *subcarnea*, Ancey.

By C. E. Beddome.

This variety of this species measures, from the syphonal end to the posterior apertural notch, 24 mm.; it is 16 mm. wide and 12 high, *i e.*, from the base to the most prominent part of the dorsum. It is therefore in all specimens I have seen a shorter, broader, and a more depressed shell than the type. Of a uniform pale flesh colour on the dorsal surface, without any indications of darker coloured bands or zones so frequently found in specimens of this species; base almost white from end to end, along the aperture, but approaching the thickened porcellanous sides of the base it shades off to a duller flesh colour than on the dorsum. This lateral intensified coloration continues forwards and backwards to the ends round which it is uninterruptedly continued with a dense porcellanous deposit, which characteristically separates the ventral from the dorsal aspects; this lateral thickening is sub-angulated, projecting beyond the surface with a slight upper recurved margin causing it to be shallowly channelled, most marked on the peristome, which is also less uneven than in most samples of *C. angustata;* in many forms of the latter the elevated surface points correspond with elevated ridges, which can be seen and felt distinctly running across the dorsum of the body whorl. I notice this character most marked in the zoned varieties of the species; they are less marked in this variety. Showing through the thickened porcellanous margin 8 to 10 small dark chocolate coloured round spots exist on each side, but are only hazily defined.

The aperture is proportionally wider than in the type form and rather more bent towards the left posterior end. The peristome

margin of the aperture is wider and more bent towards the left
than in typical forms such as I have, by me dredged alive in
Hobart Harbour on Coral; it has from 20 to 22 teeth, quite white,
inclined forwards, blunter, and spread outwardly more over the
base than in the typical specimens; in the latter forms the teeth
are sharp pointed, projecting into the aperture, and have a rusty
tinge.

On the left columellar margin there are 20 small white teeth
pointed directly across the aperture scarcely extended over the
base surface, but are seen extended down into the curved edge of
the columellar margin as it enters the cavity of the shell.   The
base, unlike the typical *angustata*, is densely porcellanous and
white; as a rule in the type it has a bluish tinge, whiter towards
the channelled ends of the aperture.

There is an absence of the dark colorations on either side of
the dorsal aspect of the anterior channel edges so characteristic
of the type forms, and this syphonal channel is not so produced
or notched, being obliterated by the more callous margin of this
form being continued directly round the ends   The dark zoned
specimens from the Derwent waters have many marginal spots,
at least 30, and although the angulated margins which separate
the base from the dorsal surface are decidedly thickened, they do
not round off the chanelled ends of the aperture as in this variety.

Hab.—Blackman's Bay, Derwent River, and Brown's River
beaches, Hobart Harbour, Tasmania (dredged).

The type specimens are in my private collection.   I have preferred to consider it only a varietal form in deference to my
esteemed friend Mr. Ancey, who named it from specimens I sent
him many months ago

[SEPTEMBER 30TH, 1896, *contd.*]

THE SOOTY MOULD OF CITRUS TREES: A STUDY IN POLYMORPHISM.

(*Capnodium citricolum*, n.sp.)

By D. McAlpine.

(*Communicated by J. II. Maiden, F.L.S.*)

(Plates XXIII.-XXXIV.)

CONTENTS.

This disease has been known for a long time, chiefly in Southern Europe, and now wherever Citrus trees are grown. It has had various common names in different countries, such as "Morfea," "Fumago," "Nero" in Italy; "Russthau or Sootdew" in Germany; "Sooty Mould" in Florida; and "Fumagine," "Black Mildew,"

32

"Black Blight" among ourselves. It is also often called "Smut" from its appearance, but does not belong to that division of Fungi which includes the true Smuts or *Ustilagineæ*. And the scientific names applied to it have been equally varied, for it assumes a variety of different forms to which different names have been given. In fact this "Sooty Mould" affords a very good illustration of what has been called Polymorphism—the same fungus appearing under different guises at different stages of its development, and it is this feature which will receive special attention here

In order to prove the fact of polymorphism it would be necessary to sow pure cultures and watch the development of the different forms under strictly test conditions, for otherwise the forms found together might be really different, and constitute merely a case of association. It is quite conceivable that the exposed surface of an Orange or Lemon leaf might be invaded by a fungus forming a dense felt by the intertwining of its filaments, and this would entangle, like a spider's web, any other spores wafted thither, so that a small community of organisms might be established, not necessarily genetically connected.

Instead of making artificial cultures, however, I have simply examined a number of specimens under natural conditions from different parts of this colony, as well as New South Wales and South Australia, carefully noting the forms found in association; and when I find a series of forms regularly occurring and constituting this "Sooty Mould," no matter what colony the specimens come from, I am led to the conclusion that they form links in a chain of successive or contemporaneous forms of the same fungus. And I am strengthened in this belief by experiments made by Zopf[*] and others on closely allied species. "Zopf studied his plants chiefly in pure cultures on microscopic slides in nutrient saccharine solutions of various degrees of concentration, and ascertained the agreement of the cultivated forms with those which occur in nature."

---

[*] N. Act. Leop. xl. 1878.

As already stated I have examined specimens from the three colonies of Victoria, New South Wales, and South Australia during the months of July and August. In Victoria I selected specimens from an orange tree in my own garden at Armadale; from another garden at Kew, a suburb of Melbourne; from the Royal Horticultural Gardens, Burnley; from a few other gardens, and from lemon trees grown on a large scale at Doncaster. The results obtained have been compared with those of South Australian and New South Wales specimens, and there is no doubt but the same fungus is common to all. The chief results will now be given from each district separately, to see how far similar forms are associated together in widely separated districts.

There is not only variety in the number of forms met with, starting with the gonidial and ending with the perithecial stage, but also in the different organs, and I have endeavoured to give some idea of this by representing variations in the characters of the self-same organs.

VICTORIAN SPECIMENS.

Doncaster specimens.—Doncaster is situated about 10 miles from Melbourne, where there is a well-known orchard with 23 acres mostly under lemon-trees, and in some situations and on certain trees there was abundance of the "Sooty Mould" The variegated lemon supplied the material, and as there was a greater variety of reproductive bodies met with than in any of the other specimens, it will be convenient to begin with it and give a general description of the fungus. It occurs on the living leaves particularly on the upper surface, but it may also appear more or less on the under surface. It is also on the branches as well as on the fruit, usually the upper or stem end as the fruits hang down. It forms black soot-like incrustations, often covering the entire upper surface of the leaf and peeling off in flakes. It is entirely superficial, not penetrating the tissues in any way, and therefore does not act as a parasite. There are all sorts of gradations in the nature and extent of the fungus. It may appear at first just like a sprinkling of dust on the leaf (in fact growers do confound it with dust), then of a dark muddy grey, peeling off as a thin papery layer, and finally as a sooty crust, soiling the fingers when rubbed. At times there is a considerable admixture of dust with the filaments, and then it is usually checked in its development. The depth of the colour is evidently largely influenced by the amount of more or less colourless and coloured hyphæ respectively, both of which are usually always present.

Fungus described.—When examined under the microscope it is seen to consist of a network of filaments and the reproductive bodies which they bear. These filaments are colourless or pale green, and darkly coloured, but there is a gradual transition from the one to the other. The thin-walled colourless filaments generally form a network in contact with the leaf, but they intermix with the thick-walled coloured filaments, and the more or less colourless may gradually become coloured, while the coloured may produce a colourless portion. When further developed, however, the colourless and the coloured hyphæ are distinctly seen.

Mycelium.—At an early stage the surface of the leaf shows numerous more or less colourless hyphæ creeping over it, and there are two kinds which may be distinguished—(*a*) closely septate, copiously branched hyphæ, in contact with each other and intermixing, so that a close-set pavement of cells is formed resembling a parenchymatous layer. The walls of these cells may become gelatinous, and thus not only stick together, but attach themselves more firmly to the epidermis of the leaf; and (*b*) at other times only creeping, colourless or pale green hyphæ are seen, very distantly septate and with their walls very uneven, as if thereby better able to adhere to the leaf. Even at this early stage there are abundance of colourless or pale green gonidia scattered about, which will be referred to subsequently.

When further developed the dark coloured hyphæ arise, and now there are the two kinds plainly discernible. The more or less colourless hyphæ are branched, septate, forming moniliform or elongated joints with mottled and usually vacuolated contents. The moniliform hyphæ averaged $3\frac{3}{4}$ μ in breadth, and the other, which were often of considerable length, $5\frac{1}{2}$ μ. Elongated and moniliform joints might occur in the same filament, but there were distinct, delicate, moniliform hyphæ and stouter hyphæ with elongated joints.

The dark coloured hyphæ are generally greenish-brown to dark brown, closely septate, either sparingly or copiously branched, thick-walled, bulging joints, often with oblique or longitudinal septa, $9\frac{1}{2}$-13 μ broad. The filaments often consist of several celled joints, and deeply constricted, so that their connection with each other is slight. The branches are very rigid, as may be seen when they are rolling about in a current, and the filaments anastomose as well as branch.

Reproductive bodies.—There is great variety in the mode of reproduction, and as this forms the distinguishing feature of the fungus it will be necessary to describe the different kinds with some fulness. The different forms are so unlike each other that the earlier mycologists assigned them to different form-genera, but they are now known to be stages in the life-cycle of the same fungus. The highest form or *Perithecium* will be described last, and this will enable us to fix the scientific position of the fungus.

(1) *Gonidia.*—These are produced in great abundance both by the colourless and coloured hyphæ, and no doubt contribute materially by their germination to weaving a web of hyphæ of firm texture. It will be convenient to consider them as produced by the colourless and coloured hyphæ.

(a) The gonidia produced by the colourless hyphæ at their tips are either colourless or pale green, and very varied. Some are in moniliform chains like a *Torula*, others spherical or oval and pale greenish, $7\frac{1}{2}$-13 × $3\frac{3}{4}$-$7\frac{1}{2}$ μ. Some are uniseptate and constricted at septa, 11-19 × $5\frac{1}{2}$-11 μ, others biseptate, about 24 × 8 μ.

A quadrate 4-celled body is very common, producing three radiating filaments, and bearing gonidia.

(*b*) The dark coloured hyphæ bear gonidia similarly coloured or a little paler, and are usually elliptical and uniseptate. They are very variable in size, $7\frac{1}{2}$-$16 \times 5\frac{1}{2}$-$8\frac{1}{2}\mu$. They are also in moniliform chains like a *Torula*, so that this form arises both from the transformation of the colourless and coloured filaments.

It has been shown by Zopf[*] that the ordinary joints of the dark coloured hyphæ are capable of germinating when detached.

(2) *Gemmæ.*—This is a convenient name for clusters of cells which detach themselves and reproduce the fungus. Detached portions of the coloured filaments, consisting of several joints and rounded at the ends, are very common. Also irregular groups of brown cells, which germinate and grow. Just as the genus-name of *Torula*, Pers., was applied to the moniliform chains of reproductive bodies, so the genus-name of *Coniothecium*, Corda, was given to the irregular groups of cells capable of germination. This form-genus would be represented both by the colourless quadrate bodies already referred to and the brown irregular clusters

There are also green mulberry-like clusters of cells which are capable of germination and are really gemmæ, but they naturally belong to the next form.

It will readily be seen that between the *Torula* and *Coniothecium* forms there is no sharp line of demarcation. In the *Torula* chain a cell may divide in the different directions of space, and thus pass into the other form.

The multiplication of the fungus is so far amply provided for by means of gonidia, gemmæ and detached joints or mycelia, and even these may pass, according to Zopf, into resting states, if the supply of food slowly diminishes. But while the fungus might multiply abundantly by means of the above-mentioned forms alone, there are various other reproductive bodies to be noticed, so that its rapid spread and extensive diffusion need not excite surprise.

---

[*] *L.c.* p. 13.

(3) *Glomeruli.*—I apply this term to pale or dirty green, or even brownish capsules, generally more or less spherical or hemi-spherical, and imbedded in and surrounded by the hyphæ. They are very common, and vary considerably in size from 75 to 470 'μ in diameter. The surface is raised into minute rounded elevations, a structure easily accounted for on crushing and examination. They are often arranged in groups or in chains, and then they become somewhat polygonal from pressing against each other.

These capsules burst readily when ripe, and are found to consist of an outer green layer and inner colourless contents. The outer layer is composed of numerous clusters of green cells, each like a miniature mulberry, and measuring about 22 μ in diameter, hence the mammillated appearance of the surface. These clusters act like gemmæ and reproduce the disease on another Citrus-leaf, according to Penzig * Inside this green shell are innumerable spherical, hyaline cells, large and small, imbedded in a gelatinous mass. They are either solitary or attached to each other by slender necks. The contents are turbid, with a relatively large vacuole, and while the larger are from 12-13 μ in diameter, the smaller are from 5-8 μ in diameter.

This has been assigned to the form-genus *Heterobotrys*, Sacc., and it is also found in connection with the "Sooty Mould" in Italy.

Penzig† describes and figures it as a stage in *Meliola penzigi*, Sacc , as a third conidial form, hitherto known as *H. paradoxa*, Sacc. It is interesting to observe that it is a different form of it we have in Australia, as the following account of the Italian form by Penzig will show (for the translation of which I am indebted to Dr. Gagliardi). He says :—" *H. paradoxa*, Sacc., appears to the naked eye as a small black globe, one-third of a millimetre in diameter, closely imitating the form of a perithecium. In fact, when we examine this small globe under the microscope, we can

---

* Annali di Agricoltura, p. 322, 1887.

† *L.c.* p. 321, and Atlas Pl. xxiv. fig. 4.

distinguish a parietal and a central part; but the parietal is not of solid structure, parenchymatous, as it consists of a number of dark coloured glomerules, just like those described as belonging to the second conidial form.   In the centre of this pseudo-perithecium we find innumerable spherical cellules, large, discoloured, with delicate walls, and one or two small guttules in the interior, isolated or united by a very narrow ligature.   The peripheric glomerules, as well as the central cellules, may reproduce, on germination, the 'morfea' on another leaf of a Citrus-plant." This is rather an economical form of reproductive-body, since the capsule itself, as well as its contents, is utilised in this way.

The Heterobotrys-stage is found both in Italy and Australia, with differences in detail, and it is conclusively proved, chiefly from the New South Wales specimens, that it is derived from the colourless or pale green filaments of the fungus.   The coloured hyphæ give rise to several other reproductive bodies, which are generally recognised as of three kinds—Spermogonia, Pycnidia and Perithecia—but when a number of specimens are examined it is not always easy to assign the forms met with to these three categories.   In the present instance, if we compare the forms with those of allied and known species such as *Capnodium salicinum*, Mort., there is no difficulty with the perithecia from their containing Asci, nor with the regular pycnidia and their septate stylospores or pycnospores, but there is a residue of forms which cannot, with any show of consistency, be all considered as spermogonia.   And the settlement of the question is not rendered easier by the fact that one branch of the pycnidium in *C. salicinum* may produce spermatia and another branch pycnospores.[*] There are at least three sufficiently distinct kinds with unicellular spores, and although we have not applied the test which De Bary lays down, that spermatia differ from spores in being incapable of germination, still the one which approaches nearest to the general type of a spermatia-bearing organ will be reckoned as such.

---

* Sorauer's Pflanzenkrankheiten, p. 336, 1886.

One of these three will be regarded as a spermogonium and the other two as gonidial receptacles or pycnidia, so that there will be three forms of pycnidia distinguished — (1) what may be called the *Antennaria-form*, with colourless, oval, unicellular spores; (2) the *Cerato-pycnidial-form*, with colourless, rod-like, unicellular spores ; and (3) the *Pycnidial-form* proper, with coloured, pluricellular pycnospores.

(4) *Spermogonia.*—The so-called spermogonia with spermatia occur in great abundance along with the other forms. They were so named by Tulasne, but as no male sexual function has been demonstrated here, the name is a misnomer, but it may be retained for distinction' sake. De Bary, however, considers spermatia to be non-germinating gonidia, and that might serve to distinguish them.

The spermogonia are dark coloured bodies, usually green by transmitted light, oblong, ovate or oval in shape, rounded and smooth at the free end, with irregularly netted surface. They vary in size from 62-190 by 37-77μ.

The spermatia are hyaline, rod-like, minute, $4\text{-}5\frac{1}{2} \times 1\text{-}1\frac{1}{2}$ μ.

(5) *Antennaria.*—These are dark green or brownish bodies, variable in shape and size, which may be swollen and flask shaped, with a short neck, or elongated oval or hemispherical, and opening irregularly at the apex. The contained spores are quite distinct from those of any of the other reproductive bodies, and I have utilised the genus-name of Antennaria, which is now generally regarded as a stage in the development of *Capnodium*. They are generally in clusters, dark green in colour, with decidedly marked walls, from 75-122 by 70-112 μ. Sometimes they are about as broad as long.

The spores are hyaline, oval to ovate, with granular contents and 2-5-guttulate, imbedded in mucilage, $5\frac{1}{2}\text{-}6\frac{1}{2} \times 2\frac{1}{2}\text{-}5$ μ, average $5\frac{1}{2} \times 4$ μ. Their size, shape and nature of contents distinguish them from the spermatia.

(6) *Cerato-pycnidia.*—I use this name for pale green, greenish-brown to dark brown, often swollen and curved, irregularly

shaped and sometimes branching pycnidia. They are distinct in
appearance and contents from the two preceding forms, and may
be very common.

They are so varied in character that it is difficult to describe
them generally, but a special form may be selected, as in
fig. 6a. It is an elongated, irregularly shaped body, the lower
three-fourths of a pale green colour with a tinge of yellow,
and the upper fourth of a decidedly darker tint. The upper
fourth is slightly swollen and tapering towards the free end, with
a round opening at the very apex, and contains the spores.

The lower portion tapers towards the base and bulges on one
side towards the centre, after which it narrows into the upper
portion. It is enveloped by and has hyphæ growing out from it,
while the upper fourth is bare. The wall is faintly marked out
into small irregular areas. The size is $240 \times 75\ \mu$, and the
terminal smooth portion is $66 \times 56\ \mu$. There is no decided line
of distinction between the upper and the lower portion, only the
darker colour is confined to the upper portion.

Other specimens are common enough, which are just straight
or curved cylindrical bodies, branched or unbranched, sometimes
swollen at the base, and generally becoming paler in colour
towards the tip. They may reach a length of $530\ \mu$, and narrow
down to a breadth between $20\text{-}30\ \mu$. The wall is evidently composed of elongated, jointed filaments, arranged end to end. The
spores escape by the opening at the apex, and are hyaline, rod-
like, rounded at the ends, minute, imbedded in a gelatinous matrix,
$4\text{-}6\frac{1}{2} \times 1\text{-}2\ \mu$, average about $5\frac{1}{2}\text{-}6 \times 1\frac{1}{2}\text{-}2\ \mu$. It will be observed
that the spores resemble spermatia closely, but the capsule is
different.

(7) *Pycnidia.*—These are not quite so common as the preceding
in the specimens examined by me, but they are plentiful enough.
They are generally somewhat flask-shaped or bottle-shaped bodies,
branched or unbranched, dark coloured but often pale green
towards the top, with walls resembling those of the preceding,
and mouth usually fringed with hairs. There is considerable
variety in the shape. It may be elongated and cylindrical, or

gradually tapering towards mouth, or swollen just below the opening. It may also be of a bright leek-green or greenish-brown or dark brown. The hairs fringing the mouth are simply tapering continuations of the cells of the walls, which are hyaline instead of being coloured. The pycnidia are sometimes very long, attaining a length of 670 μ.

The pycnospores are olive-green, pale yellowish-brown or yellowish. They are also colourless, but probably they pass from colourless to green, then to brown on maturity, like the sporidia They are ovate to oval, or even cylindrical, generally 3-(sometimes 2- or 4-) septate, slightly constricted at the septa, and sometimes longitudinally divided, $15\text{-}22\frac{1}{2} \times 5\frac{1}{2}\text{-}9\frac{1}{2}$ μ, average about $19\text{-}20 \times 7\frac{1}{2}$ μ. As already noticed, one branch may produce spermatia and the other pycnospores. I have observed no connection between spermogonia and pycnidia in their contents, but between the spermatia and the spores of cerato-pycnidia there is a close agreement.

(8) *Perithecia.*—They occur in large numbers at various stages of development, but none were found naturally opened. They are upright and deeply imbedded in the coloured hyphæ, so that their black-looking, rounded, upper portion is only distinctly seen. When crushed, the thick tough wall, as seen by transmitted light, is regularly of a characteristic sea-green or sage-green colour, and with a decided net-like surface.

They are oblong to oval or variously shaped, smooth in the upper portion, but often with adhering hyphæ in the imbedded portion, and varying in size from $112\text{-}250 \times 52\text{-}112$ μ.

The asci are hyaline, cylindrical-clavate in shape, sub-sessile, with rounded apex, 8- 6- 4-spored, and ranging from $49\text{-}81 \times 15\text{-}20\mu$. The fully mature asci average $70\text{-}80 \times 19\text{-}20$ μ.

The sporidia when mature are brown, oblong, sometimes a little fusoid, generally obtuse at both ends, constricted about the middle, 5-6-septate, often with longitudinal or oblique septa, arranged mostly in two ranks, but occasionally in three, and averaging $21\text{-}24 \times 8\frac{1}{2}\text{-}9\frac{1}{2}$ μ.

broade_ par .

The asci and paraphyses arise a
short chains of colourless cells.

Asci were met with in various st
sporidia pass through different co
contents of the ascus are finely
filling the interior and having a sn
centre.   Then the differentiation o
colourless sporidia takes place.   A
very pale green tint, and finally l
longer fill the ascus, as the space be
and the outer wall of the ascus may

It is worthy of note that these cl
to *green* and from green to brown
of the sporidia may turn out to be
genus *Capnodium*.   At any rate
Meliola I found the sporidia to pas
from yellow to· brown;* and in *Pl*
are first hyaline, then yellowish, an

Only a few mature sporidia wer
perithecia met with had opened the

which are often branched, and usually opening at the apex with a large fringed orifice. These are seated upon and amongst a dense subiculum of closely jointed or moniliform black hyphæ, so as to form large velvety patches, and are possibly, in some instances, the more complete developments of mould belonging to the genus *Fumago*." The accompanying figure of *Capnodium elongatum*, B. & D., with the spores leaves no doubt as to the pycnidium being meant. The pycnospores have a certain resemblance to the sporidia, but the latter have more septa, and of course are contained in asci (figs. 1-12).

Armadale Specimens.—Abundant examples were met with in my own garden, but only immature forms of perithecia were found. One side of the solitary orange-tree was decidedly less attacked than the other, and it was the most exposed and that which received most of the sun, the sheltered side receiving less of the sun being by far the worst.

Colourless and coloured hyphæ similar to the preceding were met with, and gonidia, gemmæ, glomeruli and antennaria forms.

Mycelium and Gonidia.—On the surface of a leaf only slightly attacked, numerous colourless to pale green creeping hyphæ were found, very irregular in outline, with very few septa and averaging $3\frac{1}{2}$ μ in diameter. Also numerous similarly coloured, oval to elliptic, continuous or uniseptate, and slightly constricted gonidia. The colourless hyphæ were generally branched, septate, thin-walled, and either with elongated or moniliform joints, and the gonidia were continuous, uni- or bi-septate. The dark coloured hyphæ were generally closely septate and constricted at septa, branched, thick-walled, and stouter than the colourless. The gonidia were usually uniseptate or in moniliform chains.

Gemmæ —The colourless and dark brown clusters of cells were met with germinating, also the mulberry-like clusters of green cells.

Glomeruli —These were in great abundance, and showed the green clusters of cells composing the wall, and the large and small colourless cells inside imbedded in mucilage, and often connected by an isthmus.

Antennaria-forms.—These were associated with the glomeruli, and seemed to be the most plentiful of all. They were imbedded in clusters among the hyphæ and emitted the colourless spores in great abundance, which remained in masses around the irregularly opening mouth.

No pycnidia were met with, although carefully looked for on a large number of leaves.

Perithecia.—Only immature forms were found of various sizes and at different stages of development. The only one figured (fig. 21) was of fair size ($150 \times 112 \mu$) dark coloured and oval in shape. On pressure the net-like areas of the wall were very distinct, and by transmitted light were either sea-green to sage-green or brownish. It contained numerous oil-globules and a few asci with paraphyses. The immature asci were shorter and narrower than the average ($39 \times 9\frac{1}{2} \mu$) and showed finely granular colourless contents within an inner envelope, and there was a small oval spot towards the centre. In some cases division of the contents had begun, and probably there were some mature forms of perithecia, but I did not happen to come across them (figs. 13-21).

Kew Specimens.—The specimens from Kew did not show very advanced stages. There were colourless to pale green hyphæ, bearing their unicellular or bicellular or simple gonidia, together with Torula-like chains and the quadrate gemmæ. The origin of these latter bodies was very clearly seen. A single cell might germinate and produce hyphæ in one or more directions, or it might divide into two and ultimately into four, each cell giving rise to a filament, but usually one stopped short, so that there were three radiating filaments.

There were also greenish-brown to brown hyphæ with their gonidia and gemmæ and detached joints. Sometimes the coloured hyphæ passed into colourless portions. The glomeruli and spores were also met with, and these, together with the quadrate gemmæ were very characteristic (figs. 22-25).

Burnley Specimens.—The specimens from the Royal Horticultural Gardens, about three miles from Melbourne, showed the ordinary colourless and coloured hyphæ, together with glomeruli, and pycnidia (principally pycnidia), were in great abundance, and seemed to be the prevailing form. There were also immature forms of perithecia, but not as yet in great quantity. The pycnidia varied in colour from leek-green when unopened to yellowish-brown when opened, and the specimen figured (fig. 28) was 526×122 μ. The pycnospores were generally pale green in colour, but sometimes brownish, and the average size was $19 \times 8\mu$. (figs. 26-30).

Other Victorian Specimens.—A few other specimens were obtained from Brighton and Elsternwick, suburbs of Melbourne.

The Brighton specimens were particularly rich in cerato-pycnidia and the antennaria (figs. 31-35), while the Elsternwick specimens showed abundance of pycnidia (figs. 36-37).

SOUTH AUSTRALIAN SPECIMEN.

An orange-leaf was forwarded by Mr. Quinn, Inspector under the Vine and Fruit Diseases Act, with the "Sooty Mould" upon it, but not very largely developed.

There were the colourless and coloured hyphæ, gonidia and gemmæ and abundance of glomeruli. The colourless hyphæ were septate, branched, with moniliform or elongated joints, and averaging $3\frac{1}{2}$-$4\frac{1}{2}$ μ broad.

The brown hyphæ were septate, sparingly branched, and varied in breadth from $4\frac{1}{2}$-$7\frac{1}{2}$ μ.

The gonidia were similarly coloured and usually simple.

The gemmæ were either clusters of dark brown cells or the green mulberry masses derived from the glomerules. None of the colourless quadrate bodies were met with.

The glomeruli were usually of a yellowish-green to pale green colour, and either isolated or in group.

The presence of brown gemmæ and glomeruli was the predominating feature (figs 38-39).

NEW SOUTH WALES SPECIMENS.

The specimens sent through the courtesy of Mr. Maiden, Govt. Botanist, from trees in the Botanic Gardens, Sydney, were badly infested with scale, but very little of the "sooty mould." There was also upon the scale a considerable quantity of a parasitic fungus known as *Microcera coccophila*, Desm.

In some cases on the upper surface of the leaf there was a very thin stratum of a mud colour, of just sufficient consistency to hold together when peeled off, but no more. It was evidently largely composed of fine dust, and scattered over it were little dark punctiform bodies, very variable in size when looked at with a magnifying glass.

Under the microscope it was seen to consist of a network of colourless hyphæ, and numbers of the spherical or irregularly shaped bodies we have already called glomeruli.

There were very few traces of the greenish-brown hyphæ developed, as the dust had evidently kept the fungus in check.

The colourless or very pale green hyphæ were closely septate, copiously branched and densely crowded so as to form a pavement of cells. The hyphæ were either moniliform or with longer or shorter joints, and bore various gonidia. The diameter of the hyphæ varied considerably, but the broadest was from 6-7$\frac{1}{2}$ μ, and narrowest about 4 μ.

The glomeruli were exceedingly numerous, scattered or in clumps, and were yellowish-green to pallid or even brownish. They varied considerably in shape from spherical to hemispherical or oval, and in size some measuring 250 μ or $\frac{1}{4}$ mm. in diameter. The mulberry-like green clusters and the contents were similar to those already described.

No other reproductive bodies were found.

Even in cases where to the naked eye there is nothing but a patch of dust on the leaf, there are the colourless hyphæ forming a close network of cells, and their gelatinous coating causes the dust to adhere.

As the result of the examination of a large number of specimens I find that the colourless filaments are the earliest formed, branching and intertwining so as to form a close network, adherent to the surface of the leaf.

And of the special reproductive bodies, the glomeruli originate from the colourless hyphæ, appearing in abundance when no other is present Even when the brown filaments are formed, the glomeruli are seen to be surrounded and not produced by them, as they leave a perfect cavity among the filaments, with the clear colourless layer at its base.

The remaining reproductive bodies are formed from the coloured hyphæ, and apparently appear in the following order when not developed simultaneously :—spermogonia, antennaria, cerato-pycnidia, pycnidia and perithecia.

This specimen served a very useful purpose in determining the origin of the coloured from the colourless hyphæ. At first nothing was observed but colourless hyphæ and numerous glomeruli, and from the constancy of this appearance I was inclined to the opinion that the colourless hyphæ with their reproductive bodies formed an independent fungus, afterwards overlaid by another fungus. But on further search, I found coloured hyphæ arising from the continuation of the colourless hyphæ, and thus the connection was established (figs. 40-44).

General development of sporidia —Taking an ascus in the young condition and when only about half the size of the adult form, it is found to be filled with finely granular protoplasm, only the short stalk being without it, and there is a minute, slightly oval primary nucleus in the centre (fig. 21).

When further grown the protoplasm recedes from the top, enveloped in its own membrane, and gradually gets further and further away, until in the mature form it may be 9 μ from the top of the ascus. It divides meanwhile into the sporidia, which soon acquire a distinct outline and a few septa. There is usually a slightly knobbed pedicel projecting from the top of the topmost sporidium when immature, apparently indicating a contracted portion of the protoplasmic membrane (fig. 12).

The contents of the at first colourless sporidia soon change into a pale green, increase in size and develop more septa (fig. 10).

This colour next changes to greenish-brown and finally a decided dark-brown like the mycelium, which is the mature form (fig. 12).

33

Alongside of each other in the same perithecium the three different coloured stages may be seen, but the sporidia in any individual ascus are all of the same colour.

When treated with potassium-iodide-iodine, the contents of the colourless sporidia immediately assumed a beautiful bright canary-yellow tint, but the rest of the ascus remained perfectly hyaline, showing that the epiplasm or glycogen-mass is not present as in Discomycetes, which gives a reddish- or violet-brown reaction. The green and the brown coloured sporidia were unaffected by this reagent. The contents of the paraphyses were also coloured bright canary-yellow, suggestive of their being simply sterile asci. The number of sporidia in each ascus is typically 8, but 4, 5 and 6 were also met with.

Characteristic Distinctions of the Special Reproductive Bodies.

1. *Glomeruli.*—They are generally of a dirty green colour, but may be pallid or greyish, or even brownish, apparently by coatings of dust, &c., and are more or less spherical or hemispherical in shape. They always originate from the colourless or pale green hyphæ, and are the first-formed of the special reproductive bodies. The covering is composed of clusters of mulberry-like green cells, and some of the hyaline cells in the interior are connected with each other by narrow joints. They vary considerably in size, reaching nearly $\frac{1}{4}$ mm. in diameter, and their shape, colour, wall and contents readily distinguish them from others.

2. *Spermogonia.*—The spermogonia resemble somewhat the antennaria in appearance, but differ in contents, while they resemble the cerato-pycnidia in contents, but differ in appearance. They vary considerably in shape and size, and it is difficult to distinguish them from the smaller forms of cerato-pycnidia, but the latter are usually elongated and slender, and have elongated regular cells composing wall, while the former have a net-like surface.

The spermatia so closely resemble cerato-pycnospores that they cannot be distinguished from each other.

3 *Antennaria.*—The spores here are the characteristic feature. They are simple, oval to ovate, with granular contents, and usually 2-guttulate, so that they are distinct from any of the others. The capsules are too variable in shape and size to be relied on for distinction, and they have a net-like surface like the preceding form, but they are often borne laterally on a filament.

4. *Cerato-pycnidia.*—When fully developed they are distinguished from the preceding forms by being very much elongated and often branched, and the regular pattern of their walls, and from the pycnidia proper by the naked, round or oval mouth-opening, but mainly by their contents. The simple, hyaline, rod-like minute spores distinguish the two forms at once.

5. *Pycnidia.*—The pycnidia proper, as already indicated, are distinguished by their usually fringed mouth opening and the coloured tri-septate pycnospores.

6. *Perithecia.*—The perithecia are distinguished from all the others by containing asci accompanied by paraphyses. They sometimes closely resemble spermogonia, although I was generally able to distinguish them by their sea-green or sage-green colour. However, with the exception of the glomeruli, the various reproductive bodies are so variable in size, shape and colour, that the nature of the contents must always be relied upon for final determination.

Connection with scale or other insects.—It is generally believed that this fungus is a saprophyte, since it does not penetrate the leaf in any way, and consequently does not extract nourishment from it. It must live at the expense of something else, and this is supposed to be the honey-dew secreted by certain insects, and associated with which it is invariably found. As a matter of fact I have never found "Sooty Mould" without the accompaniment of scale insects, and they secrete a sweet fluid known as honey-dew. Maskell, in his work on New Zealand Scale Insects, writes :—
"In many cases they exude, in the form of minute globules, a whitish, thick, gummy secretion, answering probably to the 'honey-dew' of the Aphididæ. This secretion drops from them on to the plant, and from it grows a black fungus, which soon

gives an unsightly appearance to the plant. This fungus or 'smut' is an almost invariable indication that a plant is attacked by insects, and may, indeed, give a useful warning to tree-growers." The occurrence of the fungus on the upper surface of the leaf may be variously accounted for. The upper surface is most readily moistened; the rain and dew are longer retained in the channel over the midrib at the tip. But the main reason evidently is that the honey-dew is dropped there by the coccids generally found on the under surface of the leaves. In the absence of honey-dew the fungus might grow on the accumulations of the excreta of insects, &c., but the general rule is that the fungus follows in the wake of insects, and to get rid of the one you must also get rid of the other

Since writing the above I have received a note from J. G O. Tepper, F.L.S., Adelaide, in which he shows how the destruction of honey-eating birds may affect the prevalence of this disease. He says:—"Regarding the 'Sooty Mould' and its prevalence *now* in many localities, it may be mentioned that it appears to have been practically absent, when nature was less disorganised by man, and for a very simple reason. It being due to the sugary exudations of scale insects, &c., coating the trees, its abundance depends upon that of its producers, and this upon the reduction of the sugar-loving, brush-tongued parakeets and other birds which formerly abounded so greatly. These I have often observed myself busy in the *early morning* among the foliage of gums, &c, upon which the honey-dew appeared. Later in the day the ants occupied these in overwhelming numbers, and drove the birds away, protecting the insects and cleaning the foliage.

"Now many plants have developed special *organs* to attract the ants as protectors against birds and animals which feed upon foliage, flowers or unripe fruit, and though rendering service to the plants by reducing superfluous quantities of either, and securing thus the greatest perfection of that remaining (also controlling other insect life), the birds constantly tend to overdo the work at certain critical periods. As our Eucalypts, &c, and many intro-duced plants have no such organs, they make use of the scales, aphides, &c., to secure indirectly the protective services of the ants, wherever there were birds, &c., available to keep the former under control within safe limits. Therefore the reduction of the birds, &c., by man, stimulated the limitless increase of the scales, aphides, psyllids, aleurodids, &c., and at the same time also the numbers of the ants, which helped to clean away the exudations of those of their pets left by the birds, &c., were greatly diminished. Hence excess of honey-dew insects and of their produce, which is naturally availed of by the low fungoid germ which, under normal conditions, had to be satisfied with the 'crumbs' left by the higher agents "

There is here a somewhat complex relation between the different forms of life used by the plant for protective purposes, and if one of the checks is withdrawn or diminished, the balance is disturbed and disorder ensues.

1. The *Scale* or other insects are used indirectly to attract the ants by their sweet secretions.

2 The *Ants* like a standing army protect the foliage against the attacks of leaf-eating animals

3 The abundance of honey-eating *Birds* is necessary to keep the scale or other insects within reasonable bounds.

4 The reduction of these birds by man tends to favour the increase of the scale insects and their produce

5. The scale and other insects now get the upper hand, and the ants protecting the insects also favour their increase.

6. The consequence is superabundance of honey-dew, and this is taken advantage of by the germs of the fungus to spread and multiply

Thus the destruction of the honey-eating birds has brought about an increase of the honey-dew and of the "Sooty Mould" which lives upon it, so that it is not only insectivorous birds which ought to be protected for the benefit of the grower.

It is interesting to observe the appearance of other checks to the spread of the scale or other insects. Here there are two parasitic fungi found respectively on the red and the white orange scale, *Microcera coccophila*, Desm., and *M. rectispora*, Cooke. In

Florida *Aschersonia tahitensis*, Mont., has been found attacking and destroying the larvæ and pupæ of the "Mealy Wing" (*Aleyrodes citri*, R. and H.), and bids fair to be of great use in combating the pest. This latter fungus has also been met with in Queensland on the foliage of a large climber, but no mention is made of its connection with scale or other insects.

Effect on trees —This fungus does not produce any marked injury to the tree at first, as when the "sooty mould" is removed from a leaf the surface beneath is often as green and glossy as a healthy one. The injury is rather of a mechanical nature, and, combined with the scale insects sucking the juices of the plant, there is often considerable damage done. The fungus will interfere with the process of assimilation, by preventing the access of light and the escape of watery vapour and other gases Indirectly this will hinder the growth of the tree and affect the production of bloom and of fruit. The leaves are less able to stand the effects of drought or other unfavourable conditions, and if the young fruit is attacked by it its development is hindered and it generally remains insipid.

Treatment.—It will be evident from the preceding remarks that the only sensible treatment will be to get rid of the lion's provider; and whatever insect provides the pabulum for the fungus to flourish on, should be dealt with. Mr. French, the Government Entomologist of Victoria, informs me that the principal scale insects attacking the Citrus leaves infested by "sooty mould" are the red scale of the orange (*Aspidiotus coccineus*, Gennad.) and the black scale (*Lecanium oleæ*, Bernard), and for these the treatment he recommends is the kerosene emulsion or resin wash. In a pamphlet issued this year by the U S. Department of Agriculture on "The principal diseases of Citrous fruits in Florida," by W. T. Swingle and H. J. Webber, spraying with resin wash or fumigation with hydrocyanic acid is said to be very effective.

In the course of this investigation I found a fungus-parasite on the scale insects on leaves with "sooty mould" from N.S.W This fungus, already known in Europe and hitherto only met with in Queensland, might become a useful ally in the treatment of scale insects, and so I have written a short paper upon this particular form. (*Vide* Appendix, p. 498.)

The fungus itself might be directly treated, but the only sure way is to get rid of the cause of the trouble, viz., the insects

The following is the formula recommended for the resin wash:—

Resin	... 20	lbs.
Caustic soda (98%)	... $4\frac{1}{4}$	,,
Fish oil (crude) ...	... 3	pints
Water to make ...	... 15	gallons.

This is a stock preparation, and when required for use one part thoroughly stirred is added to nine parts of water

Scientific Description.

CAPNODIUM CITRICOLUM, n.sp.—**Citrus Capnodium.**

Forming black soot-like incrustations, peeling off as a thin membrane, often covering entire surface of leaf. Colourless or pale green hyphæ creeping, copiously branched, septate, up to $6\text{-}8\frac{1}{4}$ μ. broad, intertwining and forming a pavement of cells, giving rise to ascending, short, simple, septate branches, bearing colourless or pale green gonidia, continuous, uni- or bi- septate, spherical, oval or elliptical, slightly constricted, smaller $7\frac{1}{2}\text{-}9\frac{1}{2} \times 4\text{-}5\frac{1}{4}$ μ, larger $11\text{-}24 \times 5\frac{1}{2}\text{-}11$ μ; or in moniliform chains.

Coloured hyphæ greenish-brown to dark brown, closely septate, deeply or slightly constricted, sparingly or copiously branched, rigid, $9\frac{1}{2}\text{-}11$ μ broad, bearing similarly coloured gonidia, usually elliptical, uniseptate, $7\frac{1}{2}\text{-}16 \times 5\frac{1}{2}\text{-}8\frac{1}{2}$ μ.

Perithecia intermixed with spermogonia, antennaria, cerato-pycnidia and pycnidia, sea-green to sage-green appearing black, oblong to oval or variously shaped, rounded and smooth at free end, with net-like surface, $112\text{-}250 \times 52\text{-}112\mu$.

Asci cylindrical-clavate; sub-sessile, apex rounded, 8- 6- or 4-spored, $70\text{-}80 \times 19\text{-}20$ μ.

Sporidia brown, oblong, sometimes a little fusoid, generally obtuse at both ends, constricted about the middle, 5-6- septate, often with longitudinal or oblique septa, arranged mostly in two ranks but occasionally in three, averaging $21\text{-}24 \times 8\frac{1}{2}\text{-}9\frac{1}{2}$ μ.

Paraphyses hyaline or finely granular, elongated-clavate, as long as asci and $9\frac{1}{2}$ μ broad towards apex.

Torula-, Coniothecium-, and Heterobotrys-stages occur.

On living leaves of orange and lemon, particularly on upper surface, also on branches and fruit; all the year round. Victoria, New South Wales, South Australia, Queensland.

There has been a considerable difference, and I might even say change of opinion, as to the true nature and scientific position of the fungus causing the "sooty mould" on Citrus trees. Probably it is due to different fungi in different countries; but as far as I have examined specimens in Australia, they all seem to be referable to the same fungus. Now what is this fungus? Having obtained the various stages of it and abundance of the highest or perithecial stage, there is plenty of material for coming to a definite conclusion.

Meliola penzigi, Sacc., is now recognised as the common "sooty mould" in Europe and America, but the globular perithecia, and the hyaline to brown sporidia 11-12 × 4-5 μ, distinguish it.

Meliola citri, Sacc., causes the disease known in Italy as "mal di cenere," on account of the ashy-grey crust formed by it; but apart from that, the bay-brown perithecia and hyaline sporidia do not agree with this one.

Meliola camelliæ, Sacc., has also been found on the leaves and branches of Citrus trees, but the absence of paraphyses distinguish it at once.

Capnodium citri, Berk., and Desm., has been determined by Dr. Cooke as being found on Citrus leaves in Victoria, but he had no asci and no ascospores to guide him in his determination. The published descriptions are so meagre, in the absence of the most important reproductive organs, that it is rather difficult to get distinctive characters for this species. The original description by Berkeley and Desmazières* mentions the peridia as being elongated, mostly acuminate, conical or lageniform, and the

---

* Journ. Hort. Soc. Vol. iv. p. 252 (1849).

sporidia as minute, oblong. Then Thuemen* speaks of the perithecia with net-like surface and oblong, very small, bright brown, 2-3-septate spores escaping by a pretty large opening at the apex. Next, Saccardo† describes the perithecia as elongated, often fusoid, ⅓ mm. high, and spermatia as 7 μ long. As no asci were found, it is doubtful if the bodies referred to were really perithecia, but the 2-3-septate sporidia of Thuemen are very different from the 5-6-septate sporidia of the present form.

Capnodium salicinum, Mont., has been determined by Farlow on orange leaves in America, and there is considerable resemblance in many points, but the asci and sporidia show marked distinctions. The asci measure 40-45 × 24 μ, while here they are on an average 70-80 × 19-20 μ, or nearly double the length. Then the sporidia correspond well in size in both cases, but instead of being tri-septate here, they are 5-6-septate.

Evidently, although the "sooty mould" is so common in Australia wherever Citrus fruits are cultivated, it has not yet been scientifically determined, and I propose naming it *Capnodium citricolum*.

Polymorphism.—Polymorphism literally means many forms, and has reference to the various forms assumed by fungi, especially in their reproductive bodies, in the course of their development. But the change of form may be accompanied by a change of host, and this is distinguished as heterœcism, or there may even be a desertion of the host, and then it is termed lipoxeny. The change of form referred to here occurs consecutively or simultaneously on the same individual, and all the changes were found even on a small portion of the same leaf.

In the present instance there are two different kinds of hyphæ associated—the thin-walled, colourless or slightly coloured hyphæ; and the thick-walled, distinctly coloured hyphæ—and each has its own reproductive bodies.

---

* Die Pilze—Fungi pomicoli, p. 53 (1885).
† Syll. Fung. I. p. 78 (1882).

The colourless hyphæ produce gonidia, gemmæ and glomerules; and the coloured hyphæ produce gonidia, gemmæ and the special reproductive bodies known as spermogonia, pycnidia and perithecia.

Detached portions of the hyphæ in both are able to reproduce the fungus, but that need not be specially considered here.

The starting point is with the colourless hyphæ producing gonidia, gemmæ and glomerules; and the final stage is with the coloured hyphæ producing perithecia.   The various reproductive bodies of both the colourless and the coloured hyphæ were found respectively in close contiguity, leaving no doubt as to their genetic connection, and the real point at issue is, do the coloured hyphæ grow out of the colourless, or is it simply a case of association ?   Fortunately, in the specimens from New South Wales, the hyphæ were nearly all colourless or pale green, and it was only very occasionally that a brownish filament was seen.   However, in some instances, the pale green or colourless fundamental hyphæ with projecting colourless filaments was observed to gradually pass into a pale brown shade, and from these cells the brownish and comparatively thick-walled hyphæ arose.   So that the colourless hyphæ may pass into the coloured, and since the various reproductive bodies may arise from the same or adjoining hyphæ there is genetic connection and not merely association throughout the different stages of this fungus.   The forms assumed by the different reproductive bodies are very varied and almost defy general description, so that I have drawn a number of the different shapes in order to give some idea of the wonderful wealth of variety occurring among them.   Besides I have only specially examined this fungus during the winter months, and it remains to be seen what are the prevailing forms at other seasons of the year.   I hope to examine it monthly, as it occurs with us all the year round, but at present at least seven stages or reproductive phases in the development-cycle of this fungus are known—(1) Gonidial and gemmal stage ; (2) Glomeruli stage (*Heterobotrys*); (3) Spermogonial stage; (4) Antennularia stage;

(5) Cerato-pycnidial stage; (6) Pycnidial stage; and (7) Perithecial stage.

My best thanks are due to all those who kindly supplied me with specimens for this investigation, viz.:—Messrs. Carson, Kew; Hunt, Elsternwick; Maiden, Sydney; Neilson, Burnley; Quinn, Adelaide; Turner, Brighton; and Williams, Doncaster.

EXPLANATION OF FIGURES.

(All the figures are magnified 1000 diameters unless otherwise indicated.)

PLATE XXIII., FIGS. 1 a-b ; FIG. 2 ; FIGS 3 a-g ; FIGS. 4 a-d.

Doncaster specimens—

Fig. 1.—Colourless hyphæ and gonidia.

Fig. 2.—Colourless quadrate gemma with three radiating hyphæ and bearing gonidia.

Fig. 3.—Coloured hyphæ, moniliform and otherwise, bearing gonidia (fig. c × 540).

PLATE XXIV., FIGS. 4 e-g ; FIGS. 5 a-c ; FIGS. 6 a-o.

Fig. 4.—Spermogonia with spermatia and pattern of wall (fig. a × 540 ; figs. b and e × 145 ; fig. f × 540).

Fig. 5.—Antennaria-form with spores and pattern of wall (fig. a × 270).

PLATE XXV., FIGS. 6 p-r ; FIGS 7 a-h.

Fig. 6.—Various forms of cerato-pycnidia with spores; the origin is shown in two instances from basal cells (fig. a × 270 ; fig. c × 540 ; fig. e × 540 ; figs. g-h × 270 ; figs. i-m × 145 ; fig. n × 270 ; fig. o × 145 ; fig. p × 145 ; fig. q × 270).

Fig. 7.—Various forms of pycnidia, showing in some cases fringed opening (figs. a-d and f-h × 145 ; fig. e × 270).

PLATE XXVI , FIG. 8 ; FIGS. 9 a-g.

Fig. 8.—Various forms of pycnospores—mature and immature ; two colourless forms at upper right-hand with finely granular contents.

Fig. 9.—Various forms of perithecia, some of them just peeping out from mass of hyphæ ; and pattern of wall (figs. a, c, f, and g × 540 ; fig. b × 270 ; figs. d and e × 145).

PLATE XXVII., FIGS. 10 *a-d*; FIGS. 11 *a-b*; FIGS. 12 *a-f*.

Fig. 10.—Asci with paraphyses, one with basal cell to left (figs. *a-d* × 540).

Fig. 11.—Two sporidia detached.

Fig. 12.—Asci containing 4-8 sporidia; the first contained colourless
sporidia, the next two pale green sporidia, and the remainder
were brown and mature, only the last one of the group being
colourless : paraphysis (fig. *f*) also shown.

PLATE XXVIII , FIGS. 13 *a-p*.

Armadale specimens—

Fig. 13.—Colourless hyphæ showing their varied forms, together with
gonidia, continuous or 1- to 2-septate (figs. *d* and *n* × 540).

PLATE XXIX., FIGS. 14 *a-b*; FIGS. 15 *a-m*; FIG. 16; FIGS. 17 *a-b*; FIGS. 18 *a-c*.

Fig. 14.—Quadrate colourless gemmæ (fig. *b* × 540).

Fig. 15.—Various forms of coloured hyphæ and gonidia (fig. *a* × 540).

Fig. 16.—Greenish-brown cluster of cells germinating.

Fig. 17.—Mulberry-like gemmæ.

Fig. 18.—Spores isolated and connected, large and small.

PLATE XXX., FIGS. 19 *a-l*; FIG. 20; FIGS. 21 *a-c*; FIGS. 22 *a-i*.

Fig. 19.—Antennaria-forms with spores and portion of netted wall (figs.
a-d × 540 ; figs. *e-i* and *k* × 270).

Fig. 20.—Immature form of antennaria (× 540).

Fig. 21. –Immature perithecium (fig. *a* × 145) and asci, showing origin of
latter from chain of colourless cells.

Kew specimens –

Fig. 22. –Colourless hyphæ and gonidia.

PLATE XXXI., FIG 23 (ten figures); FIG. 24 (six figures) ; FIGS. 25 *a-b* ; FIG.
26 ; FIGS 27 *a-c* ; FIGS. 28 *a b*; FIGS. 29 *a-b*; FIG. 30 ; FIGS. 31 *a-b*.

Fig. 23.—Quadrate gemmæ with triradiate hyphæ shown to originate from
a single cell.

Fig. 24.—Brown hyphæ and gonidia.

Fig. 25. –Glomerulus (fig. *a* × 270) and spores.

Burnley specimens—

Fig. 26.—Quadrate colourless gemmæ (× 270).

Fig. 27.—Pycnidia and pycnospores (fig. *a* × 52; fig. *b* × 97).

Fig. 28.—Pycnidium (× 145) and pycnospores more enlarged.

Fig. 29.—Wall of pycnidium formed of elongated, filamentous cells (fig. *a* near the top; fig. *b* lower down).

Fig. 30.—Green filaments of walls passing into colourless fringe at mouth.

Brighton specimens—

Fig. 31.—Quadrate gemmæ (× 540).

PLATE XXXII., FIGS. 32 *a-b;* FIGS. 33 *a-g;* FIG. 34; FIGS. 35 *a-b.*

Fig. 32.—Antennaria (× 145) and spores.

Fig. 33.—Cerato-pycnidia and spores (figs. *a, b, d,* and *e* × 145; figs. *c, f,* and *g* × 270).

Fig. 34.—Cerato-pycnidium conical and bullet-shaped (× 540).

Fig. 35.—Elongated jointed filaments composing wall of cerato-pycnidium, sometimes long and slender, sometimes short and stout.

Elsternwick specimens—

Fig. 36.—Quadrate gemma (× 540).

Fig. 37.—Upper portion of pycnidium and pycnospores (× 540).

South Australian specimens —

Fig. 38.—Dark brown gemmæ (figs. *b* and *c* × 540).

Fig. 39.—Glomeruli (× 145).

PLATE XXXIII., FIGS. 40 *a-d;* FIG. 41; FIGS. 42 *a-b;* FIGS. 43 *a-b*

New South Wales specimens—

Fig. 40.—Branching and gonidia-bearing colourless hyphæ.

Fig. 41.—Colourless and coloured cells and hyphæ. The colourless gradually pass into the pale brown towards the right, and produce thick-walled hyphæ, shown darker in colour.

Fig. 42.—Quadrate gemmæ (× 540).

Fig. 43.—Glomeruli, in chains and in groups (fig. *a* × 145; fig. *b* × 52).

PLATE XXXIV. (upper division of Plate), FIGS. 44 *a-h.*

Fig. 44.—Outlines of various isolated glomeruli (fig. *g* × 145).

Note.—The following are the magnifications assigned to Zeiss's Oculars and Objectives :—

$$Oc. \ 2. \ Obj. \ A = 52.$$
$$,, \ 4. \ ,, \ A = 97.$$
$$,, \ 2. \ ,, \ C = 145.$$
$$,, \ 4. \ ,, \ C = 270.$$
$$,, \ 2. \ ,, \ F = 540.$$
$$,, \ 4. \ ,, \ F = 1000.$$

APPENDIX.

MICROCERA COCCOPHILA, Desm.—**Coccus-loving Microcera.**

(Plate xxxiv., lower division of Plate.)

Minute, deep brick-red tubercles, rounded or flattened and disc-like on surface, usually in small groups, visible to the naked eye, hard and horny when dry, with short stem-like base.

Hyphæ at base of gonidiophores hyaline, septate, closely compacted, 3-4 μ broad.

Gonidiophores tufted, filiform, elongated (at least 280μ), septate, sometimes slightly constricted at septa, rose-pink in mass, with finely granular, and often vacuolated contents, 4-4$\frac{1}{2}$ μ broad.

Gonidia same colour as gonidiophores to hyaline, curved, elongated, usually blunter at free end than attached end, with finely granular, nucleated contents, variously septate, continuous up to 8-septate, average 5-6, size from tip to tip of curve and not actual length 75-103 × 5$\frac{1}{2}$-8$\frac{1}{2}$ μ.

Parasitic on Red Scale of Orange and Shaddock (*Aspidiotus coccineus*, Gennad.). July, August, &c Botanic Gardens, Sydney, New South Wales (Maiden).

In the original description the gonidiophores are given as 2$\frac{1}{2}$ μ thick and the gonidia as hyaline, acute at each end, 3-5-septate and 4-5 μ broad.   This European species has only hitherto been found in Queensland, where F. M. Bailey, the Colonial Botanist, observed it on a Coccus infesting the Lemon.   Mr. Tryon also refers to it in his "Report on Insect and Fungus Pests" as one of the natural enemies of the Red Orange Scale; and Mr. French, the Government Entomologist here, in his "Handbook of Destructive Insects," calls special attention to it as a possible auxiliary in keeping down the Red Scale, and possibly other scale insects.

So far it has not been met with in Victoria, but I hope to test its efficacy on the Orange Scale shortly,

It is closely allied to *Fusarium*, but the small tubercles differ and it is believed to be a conidial condition of *Sphaerostilbe*.

EXPLANATION OF FIGURES.

Microcera coccophila, Desm.

Fig. 1.—Gonidiophores and gonidia (× 527).

Fig. 2.—Gonidia with from 3-8 septa (× 1000).

Mr. Henn exhibited a collection of 43 species of Mollusca of the Family *Rissoridæ*, collected by himself in Port Jackson. The following, which are found also in Tasmania, are now for the first time recorded from Port Jackson :—*Rissoina elongata*, Petterd; *R. Budia*, Petterd, *R. spirata*, Sowerby; *R elegantula*, Angas; *Rissoia cyclostoma*, Ten-Woods; *R. Maccoyi*, Ten.-Woods, *R. Petterdi*, Brazier (=*pulchella*, Petterd). No less than sixteen species are apparently new; and Mr. Henn promised a paper dealing with them at a future date, after he had compared them with the Rissoiidæ of the neighbouring colonies. He also exhibited specimens of *Stylifer Lodderæ*, Petterd, and *Haminea cymbalum*, Q. and G, found by Mrs. Henn at Long Bay in October, 1893; *Turbonilla erubescens*, Tate; *Crosseia labiata*, Ten.-Woods; and *Zevlora Tasmanica*, Ten.-Woods, found by himself in shell sand at Middle Harbour, all previously unrecorded from New South Wales.

Mr. Edgar R. Waite contributed the following note on

The Range of the Platypus.

Mr. Oldfield Thomas (Brit Mus. Cat. of Marsupialia, p. 390) gives the northern range of the Platypus (*Ornithorhynchus anatinus*) as "southwards of 18° S. lat.," and quite recently Prof. W. Baldwin Spencer (Horn Expedition Report. Summary, p. 179) writes of the "absence of Platypus in the north-east," and evidences this as assisting the conclusion that the primitive Monotreme fauna entered Australia from the south.

While agreeing with Prof Spencer's inferences, it will be useful to point out that the northern range of the Platypus is more extensive than has hitherto been believed.

Some little time ago, on this question being raised, a letter was addressed to one of the Australian weekly newspapers ("The Bulletin "), and several replies were received. While some of the correspondents detail habitats further north than has been previously recorded, others give occurrences within the latitude above quoted, but at the same time supply localities whence the Platypus was not previously known. Such letters, together with information privately received, are therefore also reproduced, and I have inserted, within brackets, the latitude of the localities recorded.

The latitude of Trinity Bay (16° 45' S.) is the most northern limit of which I have record, and is supplied by two independent correspondents as follows :—

(1) "There are plenty of Platypi along from Mareeba to Kuranda in the Barron River, which runs into Trinity Bay north of the 27th [misprint for 17th] parallel. There's even a creek here named Platypus Creek.—*R. W. H., Cairns*"

(2) "The Platypus certainly lives a long way north of the Tropic of Capricorn. Years ago they were plentiful in the Barron (16° 45' S.) just above the falls, and I believe they can be found right along the North Queensland coast I have seen them both in the Herbert (18° 33' S) and Burdekin (19° 45' S.) and their tributaries, but mostly above the range On one occasion I saw one killed in Gowrie Creek, Lower Herbert District, where alligators [*Crocodilus porosus*] are quite plentiful. —*O K, Ravenswood.*"

Three other habitats are given below, which although further south than the Barron River, are yet a long way north of the 18th parallel. One of these observations (No 3) is peculiarly interesting, as it extends the range into the Gulf of Carpentaria, at a point very much further west (140° 56' E.) than any previous record from Northern Australia, and is thus the most northwesterly habitat at present known.

(3) "I have myself shot Platypi at Herberton (17° 25' S), and have met a Mr. Walcott, of Tenterfield, who has two Platypi shot or trapped in the Norman River, Normanton (17° 28' S., 140° 56' E.). While Normanton is no further north than Herberton, the above goes to show that the Platypus is to be found over a larger area than hitherto believed.—*Medicus, Drake, N.S.W.*"

34

(4) Mr. W. W. Froggatt informs me that he has obtained the Platypus on the Wild River (17° 45' S.).

I am indebted to Mr. Ernest Favenc for the following note :—

(5) " The highest point north, in Queensland, that I have seen the Platypus is on the head of the Broken River, a tributary, or rather a main tributary, of the Bowen River. The head of the Broken River is amongst the high ranges at the back of Port Mackay, and up there the river is permanently running and descends through a succession of gorges to the lower part, which is sandy. The country is peculiar in every way, and more resembles Southern Queensland than it does the general run of the country about there   The latitude is about 21° S   There are no crocodiles up there, but plenty in the Bowen River "

The following letter supplies localities which although well within the known area of distribution, are definite, and therefore worthy of record :—

(6) " Quite recently a son of Mr. John McPherson, of Rook-wood, killed a Platypus in Melaleuca Creek, where they are said to exist in numbers   Melaleuca Creek (23° 34' S.) runs into the Fitzroy about 20 miles from where the Platypus was killed. There are no alligators, so far as I am aware, in the creek, though they are fairly plentiful in the Fitzroy. The locality I refer to is due west of Rockhampton.—*J.T.S.B.*, *Rockhampton.*"

The known range of the Platypus, in time, has recently been extended by Mr. W. S. Dun, as detailed in an article in the Records of the Geological Survey of N.S. Wales (1895. iv. p. 123).

After the note was read, Mr. J. J. Fletcher drew my attention to the fact that the Platypus had been previously recorded from the Normanton District by Capt. W. E. Armit (Jour. Linn. Soc. Zoology. xiv. p. 413).

Mr. Froggatt exhibited an Arachnid from the New Hebrides, belonging to the genus *Thelyphonus* (Fam. *Phrynidæ*); and a very fine specimen of the Bag-shelter of a moth (genus *Teara*) from Quirindi, N.S.W.   Also, on behalf of Mr. Lyell, of Gisborne, Victoria, who was present, specimens of the rare butterfly

Ialmenus myrsilus, Doubl., bred by Mr. Lyell. Also, for Mr. Maiden, a bunch of curious horn-like galls (Fam. *Cynipidæ*) upon the twig of a Eucalypt.

Mr. R. T. Baker exhibited specimens of a Morell, *Morchella conica*, Pers., from Moonbi Plains, Tamworth, N.S.W., found by Mr. D. A. Porter: also a fossil leaf and some fossil wood from Wyrallah, Richmond River, the venation of the leaf is beautifully preserved, its characters being highly suggestive of *Eucalyptus*.

Mr T. Whitelegge exhibited a rare and curious Isopod, *Amphoroidea australiensis*, originally described from N.S. Wales by Dana in 1852, since when it appears to have escaped notice. The specimen exhibited was obtained on seaweed at Maroubra Bay last June, when alive it was bright olive-green, and of a similar tint to the seaweed to which it was adhering.

Baron von Mueller contributed the following

Notes on Boronia floribunda, *Sieber.*

In the earlier part of this century (during 1823) the Bohemian botanist, Franz Wilhelm Sieber, formed extensive collections of herbarium plants in the vicinity of Port Jackson and on the Blue Mountains; and although his stay in Australia lasted only seven months, and was limited to N.S. Wales, he extended largely our knowledge of the indigenous flora there, more particularly through the distribution of typic specimens, quoted in De Candolle's Prodromus and in other descriptive works. These records have had significance up to the present day, as will be instanced by one of Sieber's Boronias, namely, *B. floribunda*, which Professor Ignatius Urban, of Berlin, some few years ago, on a re-examination of this plant in Sieber's published set, restored to an independent specific position, Bentham in the Flora Australiensis having regarded it as having arisen from dimorphism. Authentic specimens from Sieber were not available in Melbourne when the first volume of the Flora became elaborated, and thus *B. floribunda* remained to be considered a mere state of *B. pinnata*, until the distinguished Berlin phytographer opened up this question anew, but I placed after his observation *B. floribunda* already into full specific rank in the Second Census of Australian Plants (p 18). Sprengel's diagnosis of this plant published in 1827 is very brief and applied as well to some forms of *B. pinnata* as to *B. floribunda*, the main distinctions not being given, namely, the much reduced size of four of the stamens and the short style with much dilated stigma. It was only recently that my attention from Prof. Urban's indications was directed to this subject, when Miss Georgina King, the zealous amateur lady naturalist of your colony, forwarded splendid specimens of *B. floribunda* to me from the Hawkesbury River, her plant proving to be the genuine one of Sieber. Unlike *B. pinnata*, which abounds in many places of four of the Australian colonies, the *B. floribunda* seems restricted to N S Wales, and I have it even from your territory only from Mrs. Capt. Rowan, the celebrated flower paintress, who sent it mixed with *B. pinnata* from the vicinity of Botany Bay, irrespective of the sendings of Miss King, and I have Sieberian specimens in the collections of Drs. Steetz and Sonder Thus it remains to be ascertained what are the geographic areas of *B floribunda*, and this might largely be settled at once by a re-examination of Sydney herbaria The specific validity of *B floribunda* will likely be affirmed still further by a search for the ripe fruit, which as yet is to me entirely unknown, good characteristics being derived from pericarp and seeds of many Boronias.

Mr. Ogilby contributed a note pointing out that there are *two genera* of recent rough-backed Herrings in our waters, both of them generically distinct from *Diplomystus*, which may be briefly characterised as follows :—

a. Maxillaries narrow, $3\frac{1}{2}$ to 4 in the diameter of the eye Jaws, palatines, and tongue toothed. Eight branchiostegals. Dorsal inserted well in front of the middle of the body, anal moderate, its base as long as its distance from the caudal, ventrals inserted beneath the anterior third of the dorsal. Scales with smooth posterior border ...   .

Potamalosa.

Fresh-water Herrings, represented by a single species, the "Australian Shad," *Potamalosa novæ-hollandiæ* (Cuvier and Valenciennes), Ogilby.

a'. Maxillaries broad, $2\frac{1}{3}$ to $2\frac{1}{2}$ in the diameter of the eye. Teeth entirely absent. Four branchiostegals. Dorsal inserted behind the middle of the body; anal rather long, its base much more than its distance from the caudal ; ventrals inserted in advance of the dorsal. Scales pectinated ...
Hyperlophus.

Marine Herrings, represented by a single species, the "Rough-backed Sprat," *Hyperlophus sprattellides,* Ogilby.

Dr. Cox exhibited some fine living specimens of *Terebratulina cancellata,* Koch, attached to a stone, which he had recently dredged off Forster, Cape Hawke, a new habitat which he thought well worthy of record. Besides the Brachiopods, Dr. Cox stated that he had also dredged the rare *Trigonia Strangei,* and he thought that the locality mentioned was the most northern at which this rare shell had been taken. Dr. Cox also exhibited a fine specimen of *Myochama Woodsi,* Petterd, from the Derwent River, Tasmania.

Professor David contributed the following note "On a remarkable Radiolarian Rock" from Tamworth, N S.W. :—"On September the 10th, in company with Mr. D. A. Porter, I observed the occurrence of a remarkable radiolarian rock on the Tamworth Temporary Common. Of this rock a hand specimen and section prepared for the microscope are now exhibited. The section is an opaque one prepared by cementing a slice of the rock about one-tenth of an inch thick on to an ordinary glass slip with Canada balsam and then etching its upper surface with dilute Hydrochloric Acid. The rock being partially calcareous, probably an old radiolarian ooze, the lime filling in the delicately latticed shells and interstices between the spines of the radiolaria is dissolved out, and the siliceous shells of the radiolaria become exposed to view. Some of them are exquisitely preserved for

Palæozoic radiolaria. The rock of which they constitute by far the larger proportion weathers into a brown pulverulent friable material like bath brick. The unweathered portions are dark bluish-grey and compact. The radiolaria appear to be chiefly referable to the porulose division of the Legion *Spumellaria*. This discovery confirms the previous determinations by me of radiolarian casts in the rocks of the New England district, and of the Jenolan Caves, N.S. Wales. The geological age of the formation in which this rock occurs is probably either Devonian or Lower Carboniferous, as *Lepidodendron australe* appears to occur on a horizon not far removed from that of this radiolarian rock. The Moor Creek limestone, near Tamworth, I find also contains numerous radiolaria. I propose to offer a paper on this subject at the next meeting of the Society."

WEDNESDAY, OCTOBER 28TH, 1896.

The Ordinary Monthly Meeting of the Society was held at the Linnean Hall, Ithaca Road, Elizabeth Bay, on Wednesday evening, October 28th, 1896.

The President, Mr. Henry Deane, M.A., F.L.S., in the Chair.

The President formally announced the death, on the 10th inst, of Baron von Mueller, who was one of the first two Honorary Members of the Society to be elected (Jan. 22nd, 1876).

On the motion of Mr. J. H. Maiden, F.L.S., it was resolved that :—

(1) The Members of this Society desire to express the profound regret with which the tidings of the decease of Baron von Mueller have been received; and at the same time to place on record their high appreciation of the Baron's life-work, which has in so eminent a degree contributed to the advanced state of our knowledge of the Flora of Australia.

(2) A copy of this resolution be forwarded to the surviving sister of the late Baron with an expression of the Society's sympathy in her bereavement.

The President read a letter from the Royal Society of Tasmania offering to co-operate in any movement to raise some appropriate Memorial of the late Baron von Mueller.

Pharmaceutical Journal of Australasia Vol. ix. No. 9 (Sept., 1896). *From the Editor.*

Indian Museum, Calcutta—Natural History Notes. Series ii. No. 23 (1896): Materials for a Carcinological Fauna of India. No. 2 (1896). *From the Museum.*

Perak Government Gazette—Vol. ix. Nos. 20-21 (Aug.-Sept, 1896). *From the Government Secretary.*

Société d'Horticulture du Doubs, Besançon—Bulletin. Série Illustrée. No. 8 (August, 1896). *From the Society.*

K. K. Zoologisch-botanische Gesellschaft in Wien—Verhandlungen. xlvi. Band (1896), 7 Heft. *From the Society.*

Société des Sciences de Finlande—Observations Météorologiques faites à Helsingfors en 1895 : Observations Météorologiques, 1881-90. Tome Supplémentaire : Pamphlet entitled "Météorologie et Magnétisme Terrestre." *From the Society.*

Marine Biological Association of the United Kingdom—Journal. New Series. Vol. iv. No. 3 (August, 1896). *From the Association.*

Bureau of Agriculture, Perth, W.A.—Journal. Vol. iii. No. 23 (Sept., 1896). *From the Secretary.*

Zoologischer Anzeiger. xix. Band. Nos. 511-512 (Aug.-Sept., 1896). *From the Editor.*

Department of Agriculture, Victoria—Guides to Growers. Nos. 27-28 (Aug.-Sept., 1896). *From the Government Entomologist.*

American Naturalist. Vol. xxx. No. 357 (Sept., 1896). *From the Editors.*

Johns Hopkins University Circulars. Vol. xv. No. 124 (March, 1896). *From the University.*

La Faculté des Sciences de Marseille—Annales. Tome vii. (1896). *From the Faculty of Science.*

Manchester Museum, Owens College—Report for the year 1895-96. *From the Museum.*

Archiv für Naturgeschichte. lvii Jahrg. (1891). ii. Band. 1 Heft: lx. Jahrg. (1894). i. Band. 2 Heft: lxi. Jahrg. (1895). ι. Band. 1-2 Hefte : Register, 26-60 Jahrg. (1895). *From the Editor.*

Senckenbergische Naturforschende Gesellschaft, Frankfurt a. M. —Abhandlungen. xix Band. 1-4 Hefte (1895-96) : xxii. Band. and Supplement (1896). *From the Society.*

"A Statistical Account of the Seven Colonies of Australasia." Sixth Issue (1895-96). By T. A. Coghlan. *From the Author.*

Geological Survey of India—Palæontologia Indica. Ser xvi. Vol ι. Part i (1895). *From the Director.*

Department of Agriculture, Sydney—Agricultural Gazette. Vol vii. Part 9 (Sept., 1896). *From the Hon the Minister for Mines and Agriculture.*

Melbourne Exhibition—Handbook to the Aquarium, Museum, etc 2nd edition (1896) *From the Exhibition Trustees.*

Pamphlet entitled "Description of a Collection of Tasmanian Silurian Fossils, &c." By R. Etheridge, Junr. (1896). *From the Royal Society of Tasmania.*

Royal Society of Edinburgh—Proceedings. Vol. xx (1893-95): Transactions. Vol. xxxvii. Parts iii.-iv. (1893-95) : Vol. xxxviii. Parts i.-ii. (1894-95). *From the Society.*

L'Académie Royale des Sciences, &c., de Danemark, Copenhague—Bulletin, 1896. No. 4. *From the Academy.*

Australasian Journal of Pharmacy. Vol. xi. No. 130 (Oct., 1896). *From the Editor.*

Department of Agriculture, Brisbane — Bulletin. No. 11. Second Series (1896). *From the Secretary for Agriculture*

CLASSIFICATION.

In dealing with the insects in this remarkable family, we are met with the difficulty that, while standing alone, in several respects they combine the characteristics of two distinct orders; and though classified by most of our leading entomologists among the Neuroptera or Pseudo-Neuroptera, there are almost as valid reasons for placing them in the Orthoptera, while in their social habits they conform to the ants and bees among the Hymenoptera. It is well known that the termites come from a very ancient stock, a great number of species having been found in the fossil state in Europe and America. Brauer* considers that they are highly modified forms of a type which departed little from the ancestral simple Orthoptera.

In working out the development of a species from Jamaica (*Eutermes rippertii*) Dr. Knower, in a preliminary abstract, says† · "I think that the Termite and those Orthoptera having a superficial embryo beginning in a disc which must elongate considerably to attain the definite number of segments, have most nearly adhered to the typical method of development for arthopods, and probably best represent the development of the ancestral insects."

---

* F. Brauer, "Systematisch-zoologische Studien." Sitzungsberichte d. Kaiserlichen Akad. d. Wissenschaften, Wien. Band xci. 1885.

† H. Mc E. Knower, "The Development of the Termite." Johns Hopkins University Circulars. Vol. xv. No. 126, 1896, p 87

Dr. Packard, who has given the termites a considerable amount of attention,* in his Entomology for Beginners has erected the Order *Platyptera* (insects with wings flat upon the back) in which he places them with the *Psocidæ* and *Perlidæ*; but they seem to have little affinity in other respects with the stone-flies and the book-louse.

If the wings and the tip of the abdomen be removed from one of the larger termites it might be very easily mistaken for an earwig; and one of our greatest authorities† on the Neuroptera actually described a supposed "wingless termite" from Japan under the name of *Hodotermes japonicus*, but in the following volume appeared a note from the author, stating that upon comparison with a Japanese *Forficula* he had found that the supposed termite proved to be a damaged earwig   Dr. Hagen also remarks that in his opinion "the three families *Termitina*, *Blattina*, and *Forficulina* are co-ordinated, and very nearly allied" (p. 139).

If the wings of the larger termites are compared with those of several of our cockroaches, it will be found that there is a marked resemblance in the form of the parallel nervures with the recurrent forks without any true cross veins running to the extremities of the wings in the cockroaches, while in the termites they generally turn downward, but this is not always the case, for in the wings of a very large termite from Northern Australia (for which I propose the name *Mastotermes darwiniensis*) and some species of *Calotermes*, the parallel veins are stout and thick, forking again and again till they run out at the tips, while in *Mastotermes* the fore wings have several more stout nervures than the hind pair.

Termites do not closely resemble any of the lace-winged insects in their perfect state; their metamorphosis is incomplete, as they pass from the egg to the active little larvæ with perfect propor-

---

* Notes on the external anatomy will be found in Third Report U S. Entom. Commission, 1883, pp. 326-329.

† Dr. Hagen, Proc. Bost Soc. of Nat. History. xi. p 399, 1868.

tions, increasing in size with each successive moult, but always little termites from birth, even the soldiers in some species showing the elongated form of the head long before they reach maturity.

I consider they have a greater affinity to the Orthoptera than the Neuroptera, and, without going into the anatomy of the family, which I leave to an abler pen, would suggest that they form a natural link between the two orders, coming after the *Forficularidæ* and *Blattidæ*.

I have followed Dr. Hagen in the terms used for the venation of the wings and general structure. I try also to describe each species with its habits and life history when obtainable, so that our coming entomologists will be able to recognise the species without much difficulty. In a few instances I have described winged forms only, in the hope of afterwards getting the other forms to complete their life-histories. I have a great number of winged specimens evidently belonging to different species that I retain till I have completed the series for the various localities from which they were taken.

Family TERMITIDÆ.

Perfect insects slender, with a rounded head, and large compound eyes more or less projecting on the sides of the head; ocelli two or in some groups wanting; antennæ long and slender, consisting of from 9-31 or more moniliform joints; jaws stout and short, with a number of pointed or angular teeth covered from above with a large rounded labrum.

The head is attached to the thorax by two very large sclerites placed on either side of the under portion of the head. Thorax moderately large, with the prothorax very distinct and characteristic in the different genera, sometimes heart-shaped, lobed on either side, or saddle-shaped; meso- and metathorax each bearing a pair of flat wings of uniform size resting over each other, and extending beyond the tip of the abdomen. Their venation is simple, consisting of four main parallel nervures, termed the costal, subcostal, median and submedian, which send out a number of short concave or sloping transverse veinlets very variable in number and disposition. The remarkable transverse suture near the base of the wings causes them to drop off at the slightest obstruction, leaving behind attached to the thorax a small slender flap (which I have termed the scapular shield) In the legs the coxæ are large, with a transverse trochanter at the base, to which the thighs are attached and not to the coxæ; the femora are generally stout and short; the tibiæ slender and cylindrical, with two or more stout spines at the tip, the tarsi consist of four joints, the first three round, with the terminal one slender, armed with sharp curved claws, at the base of which there is sometimes a plantula.

The abdomen consists of ten segments, forming an elongated rounded body with a pair of cerci at the base of the 9th segment, and in many species there are sometimes two other slender jointed appendages known as the anal appendices

The integument consists of chitinous plates, generally very thin and delicate, but in some of the larger species of considerable strength.

Termites live in social communities, either constructing distinct nests, earthy mounds covering a woody nucleus, known as a Termitarium, or else simple tunnels or galleries under logs, stones, or in the timbers of houses. Each community consists, broadly speaking, of three castes or classes. Firstly, the winged males and females, which are found in great numbers only at certain seasons of the year, but always in the nests in a larval or imperfect form. Secondly, the workers, aborted males and females, wingless, pale yellow, or white, with a large oval body and no very distinctive characters in most species; these do all the work of the nest, building the walls, gnawing out the wood, and looking after the eggs and young larvæ. Thirdly, the soldiers, also aborted males and females, which have the jaws produced into long scissor-like projections, closing over or meeting at the tips like a pair of shears, very constant in form in the different species, and of use in classification.

In the genus *Eutermes* the soldiers (nasuti) have a nose-like process of the head which is hollow, and connected with a large retort-shaped vesicle occupying the upper portion of the head. From it is discharged a thick honey-like fluid, forming a globule at the tip of the snout; and this is used as a means of defence.[*]

This protective fluid is also made use of among some of the two-jawed soldiers, and when this is the case the opening is above the base of the clypeus, and the ejected fluid is thick and milky.

The abdomen of the soldiers is more slender than that of the workers. Their duties are to protect the workers from any enemies when the walls of the galleries are broken into, and also to direct them at their work.

These are the first three primary forms found in the nest, but there is a great number of secondary ones. First in importance among these is the queen, produced from a winged female fertilised by a male (both of whose wings have either dropped or been pulled off, and who after their flight with the other winged forms) from the parent nest, have been taken care of by some workers who have probably in the first instance found them under a log After fertilization the body swells out into an immense, white, elongate, cylindrical sac the original chitinous plates of the segments forming black bars across the intersegmental membrane of the abdomen, now consisting of a mass of egg tubes rendering the queen incapable of active locomotion.

Next come the complementary queens, another form of the female termite which seems to have reached a secondary stage, with an enlarged corrugated abdomen, and though not ordinarily egg-producing they are capable of becoming so if required, and appear to be "kept in stock," so to speak, to replace the true queen should she be killed or become incapable, so that egg-production may not be checked. Unlike the queen they are produced direct from the female larvæ, and have no sign of

---

[*] H. Mc E. Knower, "Origin of the Nasutus (Soldier) of *Eutermes*." Johns Hopkins University Circulars. Vol. xiii. No. iii. p. 58, 1894.

rudimentary wings as she has. I have as many as ten supplementary queens taken from a single mound. Muller was the
first to notice the forms when working out the life-histories of the
termites of Santa Catherina in Brazil*; in one nest he found 31
complementary queens. Besides these there are larvæ in all stages
of growth, from minute little creatures just emerged from the
eggs to pupæ with the wing-cases extending half way down the
back; as well as young workers and soldiers, the latter showing
the alteration in the form of the head before the last moult.

Lately near Newcastle when turning over some logs I found a
nest of *Eutermes fumigatus*, Brauer, in which the queen was
exposed in the centre of the irregular galleries damaged by the
removal of the log; and among the Eutermes I found six or seven
reddish-brown perfect insects (excepting that they were minus
their wings) of some undetermined species of *Calotermes*; these
did not seem to be quite at home, but had evidently crawled in
under the log for shelter, and thus found their way into the nest.

The family *Termitidæ* has been divided into seven genera, and
four subgenera, several comprising both fossil and existing species,
others only modern forms, and three fossil species only.

Though a good deal of work has been done by entomologists
upon this family it has always been upon different genera   The
late Dr Hagen's Monograph upon the *Termitidæ* is our only
guide to the general classification of the family, and this was
published nearly 40 years ago. His proposed Monograph upon
their anatomy was never published, beyond a short paper on
Eutermes rippertii.† His classification is chiefly founded upon
the structure of the wings, the ocelli, the number of joints of the
antennæ, the shape of the prothorax, and the tibial spines.

Following this very natural classification, I have considered
his four subgenera as genera, and further grouped them into

---

* Fritz Müller, "Beitrage zur Kenntniss der Termiten." Jen. Z. Nat. vii.
pp. 337, 463.

† Psyche v. pp. 203-8, 1889.

subfamilies based upon the neuration of the wings, also taking into account the habits, and the form of the soldiers, which seem to be very similar in most of the genera I have observed. In the case of the genus *Hodotermes* and the two subgenera *Stolotermes* and *Porotermes* I have been somewhat puzzled. In Hagen's definition of *Hodotermes* he says "ocellis nullis," but in his figure of *H viator* (Tab. iii. fig. 8) he shows lateral ocelli, and in the Cambridge Natural History, published last year (Vol. v. p. 556), a figure of *Hodotermes mossambicus* is given "after Hagen," in which the lateral ocelli are most distinctly drawn. The only species of this group that I have in my collection is a doubtful species of *Stolotermes ruficeps*, Brauer, which has no ocelli, and among all my Australian specimens I have not yet found any that can be placed in this group, but an allied group for which I propose the name of *Glyptotermitinæ* takes their place in the Australian fauna. I have placed the genus *Rhinotermes* after the *Calotermitinæ* from a careful study of their habits and the robust form of the wing. I was acquainted with a very curious white ant with two very different-looking kinds of soldiers, but of which I had never seen winged forms among the New South Wales specimens; but in a collection from Queensland I found a number of winged specimens that on comparison with a co-type of Brauer's *Rhinotermes inter-medius* (for which I am indebted to the Director of the Austrian Museum) turned out to be this species. I also have another species of the genus with identical habits which has been sent to me from Kalgoorlie, West Australia, by my father, with a full account of its habits.

In the term used for the venation of the wings I have followed Hagen. But when using the words "scapular shield," I mean that portion of the wing between the body and the cross suture (the "basal scale" of Scudder); its form and structure appear to be very consistent in the different genera.

Family TERMITIDÆ.

i. Subfamily CALOTERMITINÆ.

Wings robust; scapular shield broad, with five or more branches from the base Costal and subcostal nervures connected by sloping cross nervures forming a network of smaller·ones at the tip. Fore wings differing from the hind pair in the venation in , many species.

1. Genus MASTOTERMES, g.n.

Head large, flattened on the summit; eyes large; ocelli small; antennæ 30-jointed; prothorax large, with the sides turned up; scapular shield with more than five branches.

2. Genus CALOTERMES, Hagen. (Recent and fossil.)

Head round; eyes large, projecting; ocelli small, antennæ 16-20-jointed; prothorax large and broad.

3. Genus TERMOPSIS. (Recent and fossil)

Head large, broadest behind; eyes small, oval; ocelli wanting; antennæ long, 23-37-jointed; prothorax small, not as wide as the head.

4. Genus PAROTERMES. (Fossil.)

Head rather large; eyes small; ocelli wanting; antennæ 20-jointed, prothorax subquadrate, not broader than the head.

5. Genus HODOTERMES. (Recent and fossil)

Head large, circular; eyes small, not projecting, facets coarse; ocelli wanting ; antennæ 25-27-jointed ; prothorax small, broader than long.

6. Genus POROTERMES. (Recent.)

Head small; eyes small, facets fine; ocelli wanting. Venation of the wings very fine.

7. Genus STOLOTERMES. (Recent.)

Head large, circular; eyes small, facets coarse; ocelli present; antennæ 12-14-jointed; prothorax heart-shaped.

8. Genus MIXOTERMES. (Fossil.)

Founded by Sterzel upon a fossil wing from Lugau. Allied to *Calotermes* (Berichte der Naturwissenschaftlen Gesellschaft zu Chemnitz. 1878-80).

35

ii. Subfamily RHINOTERMITINÆ.

Scapular shield broad, slightly convex at the cross suture, with four branches. Costal and subcostal nervures very stout, running to the tip of the wing, and joined at the extremity with short irregular thick nervures; median and submedian nervures slender, with a great number of fine oblique nervures, and all the wing thickly covered with fine furrows.

1. Genus RHINOTERMES.   (Recent.)

Head broad; eyes small, projecting and coarsely faceted; ocelli small; antennæ 20-jointed; prothorax not as wide as the head, rounded in front.

iii. Subfamily GLYPTOTERMITINÆ.

Scapular shield slender, angular, with the cross suture transverse, with four or more branches. Costal and subcostal nervures running very close to each other, the latter often merging into the former in the centre, median nervure running through the upper half of the wing, and the submedian about the middle, the latter and the oblique nervures often formed of fine spots or scars

1. Genus GLYPTOTERMES.   (Recent).

Head broad, rather flat and quadrate; eyes moderately large, slightly projecting, coarsely faceted; ocelli rather large, close to the eyes; prothorax long, broadest and concave in front, rounded on the sides

2. Genus HETEROTERMES.   (Recent.)

Head very large, longer than broad; eyes small, not projecting; ocelli wanting; antennæ 16-jointed; prothorax quadrate, with base and apex arcuate.

iv. Subfamily TERMITINÆ.

Scapular shield angular, slightly rounded above but transverse below, showing four branches. Costal and subcostal nervures running parallel, but widely separated from each other; median and submedian slender, the former divided into one or more forks at the extremity.

1. Genus TERMES. (Recent and fossil)

Head large, rounded; eyes large, and prominent, finely faceted, ocelli present; antennæ 13-20-jointed; prothorax heart-shaped; flattened, smaller than the head.

2. Genus EUTERMES. (Recent and fossil.)

The form of head and thorax very similar to that of *Termes*; wings always dark coloured, with the base of the nervures in the scapular shield not as robust as in the latter. Soldiers always nasuti.

3. Genus ANOPLOTERMES (Recent.)

A genus formed by Müller on the internal anatomy of a Eutermes from Brazil *(A. pacificus)* He also places *Eutermes ater*, Hagen, and *E. cingulatus*, Burm., with the new species.

CALOTERMITINÆ.

MASTOTERMES, g n.

Head large, nearly as broad as long, flattened upon the summit; eyes large, projecting, ocelli prominent; antennæ 30 jointed, clypeus large, labrum rounded at the apex. Prothorax shaped like that of *Calotermes*, except that it is turned up on the outer edges, with the scapular shield as long as the meso- and metathorax. Fore wings differing from the hind pair in venation in having fewer parallel nervures between the costal and subcostal, the upper portion of the wings crossed with stout nervures, with the whole of the wing finely reticulated with smaller veinlets. Tibiæ with four spines at the apex; claws large with a small plantula.

This genus is founded upon a species from Port Darwin, W Australia, and is allied to *Calotermes*.

MASTOTERMES DARWINIENSIS, n.sp.

(Pl. xxxv. figs. 3-3a.)

Head castaneous, thorax dark ferruginous; legs, under side and abdomen dark brown, antennæ yellow, wings, scapular shield and nervures ferruginous; the rest yellowish-brown. Length to tip of the wings 16, body 8 lines.

Head large, nearly as broad as long, rounded and broadest behind, rounded on the summit, flattened and rugose in front, truncate across in line with the eyes. Eyes large, circular, projecting, very finely faceted; ocelli large, oval, close to inner margin of the eyes. Antennæ long and slender, 30-jointed, springing from a depression in front of the eyes ; 1st joint large, cylindrical, broad at apex; 2nd nearly as thick but shorter; the others moniliform to near the tip where they become more stalked, the last being the smallest. Clypeus arcuate and broad behind the sides forming little angular flanges, with the middle quadrate and lobed in the centre; labrum broader than long, almost quadrate, with the sides rounded and flattened, shell-shaped; palpi long, with the base of each joint white; jaws broad and rounded, with two small angular teeth at the tip, and a flattened untoothed edge to the base slightly hollowed out in the middle. Prothorax as wide as the head, wider than long, concave in front, rotundate, with the sides and apical margin forming a half circle, depressed in the centre, with the edges (particularly on the sides) turned up. Legs short, thighs stout, with the tibiæ covered with fine hairs, and four stout spines at apex; tarsi short, having the terminal joint slender, with four small sharp spines and a small plantula.

Wings large, thrice as long as broad; scapular shield large, cross suture convex, with eight stout parallel nervures branching out of it: venation of the fore and hind wings different. Fore wings with costal nervure slender, running round to tip of wing, receiving four stout parallel nervures merging from the scapular shield, turning up at regular intervals before the middle of the wing into the costal; subcostal branching out into four stout transverse parallel nervures turning up into the costal beyond the tips of the former ones, forming a regular pattern; median nervure closely parallel with the subcostal, bifurcated in a line with the third fork in the subcostal, the two branches running out into a network of finer nervures at the margins; submedian nervure fine, irregular, running through the middle of the wing, with six short stout oblique nervures at the base, and seven or more slender nervelets running out towards the edge and forming a network all over the wing. Hind wings with only two parallel nervures between the costal and subcostal, one bifurcation less on the subcostal; median forked in the middle of the wing, upper branch bifid at tip, lower one turning downward and again branching; upper one bifid, lower one simple; submedian as in the fore wings, but irregular in the neuration of the oblique nervelets Abdomen short, broad, and rounded at the tip, with short cerci; anal appendices small, slender, close together, near the tip of the abdomen.

Hab —Port Darwin, N.T. (Mr. N. Holtze); Northern Territory (Mr. J. G. O. Tepper).

Among a number of pinned specimens of termites sent to me by Mr. Tepper was a single specimen of this species, which was very noticeable from the network of veins along the costal margin, as well as its large size. During the summer of the following season, Mr. Holtze sent me seven specimens in spirits, taken "flying round the lamp at night" in the Botanic Gardens, Palmerston.

There are two specimens in the Macleay Museum, one of which is labelled Cleveland Bay (Townsville), N.Q., collected, Mr. Masters thinks, by Mr. Spalding; and another from King's Sound, N.W. Australia, taken by myself, flying round the lamp, at a station about 100 miles inland from Derby.

Genus CALOTERMES, Hagen, 1853.

Hagen, Bericht d. K. Akad. Berlin, 1853, p. 480 , Linnæa, xii. p. 33.

Head rather small, triangular or rounded; eyes large and projecting from the sides of the head ; ocelli small , clypeus small, flattened, labrum small, quadrangular, antennæ as long as the head, 16-20-jointed, antennal cleft small ; jaws short, stout and blunt. Prothorax large, as wide or nearly as wide as the head, broader than long, truncate or arcuate in front, with the sides and apical edges forming a semicircle. Legs stout, the tip of the tibiæ with three or four spines; tarsi with plantula. Wings large, narrow, twice or thrice as long as the body; subcostal narrow, widening out towards the tip and connected to costal by five or six veins, irregular in number, forming a network between the two; median nervure slender, running through the middle of the wing, with irregular cross veinlets, the whole of the outer portion of wing showing an irregular network : scapular shield as long as mesothorax in the fore pair, and about half the length of metathorax in the hind pair. Abdomen small, a little wider than the thorax; cerci stout, short, and jointed.

Soldiers short and stout. Head large, cylindrical, flattened in front and rugged or truncated before the jaws, which are stout and strong, about one-third the length of head, almost straight, flattened towards tips, close at the base, with short stout teeth, irregular on opposite jaws; labrum small, short, and transverse or quadrangular.

These termites do not construct regularly formed nests, but live in small communities in logs, timber, beams of houses or under stones; many nests contain under a hundred individuals, chiefly workers or immature nymphs, and sometimes only half a dozen soldiers, though in others these are more numerous. I have never found a queen among any community of the genus.

Calotermes has a pretty wide distribution, ten species being described from America and the West Indies, one from Europe and the north of Africa, four from Australia and Tasmania, one each from Madiera and the Isle of France, and three from Europe known only as fossils.

There are probably many species of this genus to be found, but from their retiring habits they are seldom met with unless closely looked for.

CALOTERMES CONVEXUS, Walk.

Termes convexus, Walker, Brit. Mus. Cat. Neuroptera, p. 527;
Hagen, Mon. Linnæa, xii. p. 45.

Ferruginous; lower surface, abdomen, antennæ, and feet fulvous, smooth and shining; wings subfuscous. Length of body $1\frac{1}{2}$ lines.

Head elliptical, much longer than broad, scarcely smaller than the thorax. Antennæ shorter than the head, probably 13-jointed. Ocelli close to the eyes. Jaws small, two-toothed, with dark points. Prothorax with an indistinct suture in the centre, much broader than long, concave anteriorly, sides convex, flattened behind; body scarcely longer than the thorax. Legs stout, with the 4th joint of tarsi as long as the first three combined. Wings pale brown, costal and subcostal nervures ferruginous, with about 12 oblique branches; the other nervures very pale and indistinct, with rows of finer ones between them, from the lower side about 12 oblique branches, the wings generally feeble and wrinkled.

Soldier greyish, hairy, shining. Length 3 lines. Head oval, reddish-yellow, flat on the summit, ferruginous in front, longer and broader than the thorax; jaws blackish, robust, almost straight, bent in at the tips and armed with two broad teeth. Antennæ shorter than the head, the extremity of each segment light coloured, shorter towards the tip. Prothorax twice as broad as long, anterior angles concave, sides and posterior angles convex, body club-shaped, broader and longer than the thorax, 3 lines in length.

Worker grey. Head small, with a pitch-coloured spot between the antennæ, the latter almost as long as the head; body almost club-shaped, very much broader and longer than the thorax. Length 3 lines.

Hab—Tasmania, and Swan River, W.A.

This description is taken from Hagen's Monograph. He says: "In comparison with the type, the somewhat larger *Termes obscurus* from Swan River (long corp. 2½, exp. alar. 7 lines), is not otherwise different from *T. convexus*. Between the claws is seen a plantula. This species closely resembles *Calotermes improbus*, and whether it should remain separate is a matter for further consideration, though it is much smaller The workers and soldiers described by Walker (Brit. Mus. Cat. p. 52) as belonging to *Termes australis*, are very probably those of *C. improbus*."

CALOTERMES INSULARIS, White.

Calotermes insularis, White, Voy. Erebus & Terror, Zool. Pl. ii.
(Pl. xxxv. fig. 4.)

General colour bright ferruginous, wings hyaline, nervures light brownish-yellow. Length to tip of wings 11, to the tip of body 5 lines.

Head longer than broad, rounded behind, widest behind the eyes, sloping on sides to apical margin, truncate in front, convex on the summit, sharply sloping down on the forehead. Eyes moderately large, round, coarsely faceted, projecting slightly on the sides; ocelli large, round, contiguous to front of the inner margin of the eyes. Antennæ broken (probably about 20-jointed), springing from a cleft in front of the eyes ; joints all parti-coloured, the apical edges barred with pale yellow ; 1st-3rd cylindrical, basal ones largest, 4th orbiculate, the remaining ones turbinate, lightly fringed with hairs. Clypeus wide at base, but very narrow, sloping on the sides to rounded tips at the centre; labrum broad, rounded in front. Prothorax very large, broader than long, deeply concave in front, rotundate and rounded behind, showing faint median suture , meso- and metathorax much narrower. Legs short, thighs broad and rounded; tibiæ short, with three stout spines at apex, terminal joint of the tarsi about as long as the first three combined; plantula and claws large.

Wings very long, four times as long as broad, rather sharp at the tip ; scapular shield large, rounded, with the cross suture rounded to subcostal, but transverse below, showing five parallel nervures; subcostal nervures slender, running round the tip, with a stout parallel-oblique nervure branching out of it in the scapular shield, connected with it beyond the scapular shield by two short oblique nervures and joining it again about one-quarter from the base; subcostal nervure running parallel with five stout parallel-oblique nervures turning up into the costal and several smaller ones forming a network between them at the tips, median nervure running close to the subcostal with no upward cross nervure till

near the tip, where several short ones form an irregular network,
but having a number of short spine-like nervures along the lower
margin; submedian nervure running through the middle of the
wing, turning downwards before reaching the tip, with six stout
oblique unbranched nervures at the basal portion, and nine fine
oblique nervures beyond ; the whole wing finely covered with
indistinct veinlets giving it a frosted appearance Abdomen very
short and thick, smooth and shining, with the cerci of usual size;
anal appendices undistinguishable.

Hab.—Melbourne, Victoria (Mr. Kershaw).

Only one dry pinned specimen, from the National Museum,
Melbourne, but very distinct from any of my other species, and
remarkable for the very long wings.

CALOTERMES IRREGULARIS, n.sp.

(Pl. xxxv. figs. 1, 1*a*, 1*b*).

Head ferruginous; thorax and abdomen ochreous, antennæ, legs
and under surface lighter coloured; wings pale ochreous, with the
nervures fuscous.   Length 8 lines to tip of wings, body 4½ lines.

Head rounded behind, longer than broad, sloping in from the
eyes to the clypeus, lightly clothed with a few scattered hairs.
Eyes very large, projecting; ocelli large, rounded oval, contiguous
to the centre of inner margin of eyes.   Antennæ 19-jointed,
hirsute; 1st joint large, cylindrical, springing from a shallow
antennal cleft below the eyes; 2nd cylindrical, smaller, and half
the length; 3rd more rounded at the tip; 4th shortest, 5-12 moniliform, slightly increasing in size toward the extremity; 13-18
longer, turbinate, with the last elongate-oval.   Clypeus small,
rounded in front, sloping on sides, broadest behind, labrum large,
shell-shaped, rounded in front; jaws large, stout, with the apical
tooth large, curved inwards, a short conical one below, with two
stout angular ones towards the base.   Prothorax as broad as the
head, slightly concave in front, rounded on the sides, truncate
behind, showing a slight median suture; mesothorax narrow, with

rounded base, a slight median suture; metathorax smaller, rounded behind. Legs short, rather hairy; thighs short and stout; tibiæ moderately long, with three short stout reddish spines at the apex; tarsi with the terminal joint not quite twice as long as the three preceding ones combined, tarsal claws long and slender; plantula oval. Wings more than twice as long as wide, broad and rounded at tips; scapular shield long; costal, subcostal and median nervures running parallel to each other at equal distances apart to the tip of the forewing; subcostal with five oblique veins running upwards into costal; median furcate at the tip; sub. median nervure slender, with about 13 oblique nervures, the last four furcate; median with a number of short irregular veinlets along the lower edge, and a faint irregular network of nervelets over the whole wing. Hind wing: costal and subcostal nervures running into each other in the middle of wing ; median furcate a short distance from shoulder, the upper branch dividing into five oblique veins, turning upwards into the costal; the lower branch running parallel, straight out to extremity of wing the rest of wing as in fore pair.   Abdomen large, smooth, shining, rounded at tip; cerci short, stout and hairy.

Soldier.—Head rufous, jaws black; legs, antennæ, and prothorax pale ochreous; the rest dirty white.   Length 6, from tip of jaws to base of head 2¾ lines.   Head longer than broad, rounded at base, straight on sides, emarginate in front and sloping down behind the clypeus, with a median furrow: clypeus hidden, labrum short, rather broad and almost truncate in front; antennæ as in winged insect, only more slender; jaws very stout and thick, ochreous at base, the rest black, rather straight on sides, meeting at the curved-in tips, with several small teeth along the left jaw and two large angular ones on the right side; thorax and abdomen elongate-oval, nearly as broad as the head, tapering to the rounded tip.

Worker.—Head pale yellow, jaws black, the rest dull white. Length 5 lines.   Head nearly spherical; jaws stout, with two sharp-pointed teeth at apex, and angular ones below; clypeus large, rounded in front, with a dark spot on either side; labrum rather long, narrow and truncate in front and straight on the sides : anal appendices large, at right angles to each other; cerci as in others; body long and cylindrical.

Hab.—Mackay, Queensland. (Mr. G. Turner).

CALOTERMES IMPROBUS, Hagen.

Hagen, Mon. Linnæa, xii. p. 44.

Chestnut brown, head somewhat darker; antennæ, legs, and underside bright yellow; head and thorax smooth, not hairy. Length 6½ mm.

Head oblong, quadrangular, almost half as long again as broad, rounded posteriorly. Eyes small, projecting slightly, well in front of head; ocelli large, away from the eyes, a small central mark or false ocellus almost in a line with the hind margin of the eyes. Antennæ short and stout, longer than the head, 20-jointed, coalesced, round; first joint larger than the following ones, 4th and last smallest. Labrum short, oblique below the jaws; labial palpi thicker and shorter than in the other species. Prothorax large, broader than the head, rounded and flat, sides turned down in front, concave, rounded posteriorly, the angles rather truncate behind. Scapular shield of forewings large, round and truncate, longer than the mesothorax. Wings wanting. Legs short, with three spines at apex of tibiæ; the only existing claw is short, sharp, and curved; if a plantula is present it is not noticeable in this specimen   Body egg-shaped, broad; abdominal appendices very small, two small cerci.

The above description is taken from Hagen's Monograph. He described this species from one imperfect specimen, without wings, and with only one imperfect leg.

Hab.—Tasmania.   It does not agree with any of my species from Australia.   But in the case of a species known only from a single imperfect individual it would be hard to identify it without a good series of specimens collected in the same locality.

CALOTERMES LONGICEPS, n.sp.

(Pl. xxxv. fig. 7.)

(Immature). Head pale yellow, jaws black, rest of insect dull white. Length 6 lines.

Head spherical, a little longer than broad. Eyes indistinct; ocelli (?). Antennæ 20-jointed; 1st stout, cylindrical; joints 2-7 very short, orbiculate, the rest moniliform, towards the tip becoming broader at apex; the last smaller, elongate-oval. Clypeus truncate behind, rounded in front, narrow: labrum large, convex in front jaws short and stout, with three teeth above and two angular ones at base. Prothorax as broad as head, slightly concave in front, broadly rounded on sides, and nearly truncated at apex, with a median suture extending through the rest of the thorax; wing covers extend down to the third segment of the abdomen, slender and pointed. Legs rather short; thighs small, slender, tibiæ short and thick, with three stout ferruginous spines at apex; tarsi short, terminal joint large, with plantula and stout claws Abdomen long, cylindrical, rounded at the tip, with very small anal appendices, and the cerci small and hairy.

Soldier.—Head bright ferruginous, jaws black, the rest light yellowish brown. Length 6½ lines, the head and jaws as long as thorax and abdomen. Head very large, longer than broad, convex above but sloping to base of the jaws, slightly rounded on the sides, and emarginate in front of the antennæ: antennæ very slender, 20-jointed; 1st stout, cylindrical; 2nd smaller; the rest moniliform to the tip: clypeus narrow, truncate behind, rounded in the front, labrum spade-shaped, straight on the sides, rounded in front, and projecting to the base of first large tooth; jaws very stout and large, straight on the sides, curving inwards slightly and meeting at the tips, with three small irregular teeth on the left hand jaw and one large angular one on the opposite side; a slight median and cross suture on head. Prothorax not as wide as the head, short, concave in front, truncate behind and rounded on the sides : legs short, thighs thick : abdomen short, and very broad in proportion, flattened, anal appendices showing at tip of abdomen, cerci small.

Hab —Sydney, N S.W. (W. W. Froggatt).

This species lives in dead logs, in small communities of fifty or a hundred, and in several that I have cut out of firewood they have consisted of immature winged ones, with only one soldier, and one or two workers. I have never been able to breed the perfect insects, though a number of them lived for some months in a tin.

CALOTERMES ROBUSTUS, n sp.

(Pl. xxxv. fig. 8.)

Head and prothorax dark ochreous, the upper surface of the rest of the thorax and abdomen lighter coloured; antennæ, under surface and basal portion of legs light ochreous, with the tibiæ and tarsi slightly ferruginous; wings semi-opaque, with the nervures ferruginous. Length to tip of wings 9; to tip of body 5½ lines.

Head orbiculate, about as long as broad, convex, and rounded on summit. Eyes large, coarsely faceted, projecting; ocelli large, oval, contiguous, and in line with the front of the eyes. Antennæ 19 jointed, long and slender towards the tips, springing from a circular antennal cleft in front of the eyes; 1st and 2nd joints large, cylindrical, 3rd-8th short, moniliform; 9th-12th turbinate, 13th-18th more stalked and elongate; terminal one much smaller, slender, elongate, oval. Clypeus rounded in front, very prominent, divided in the centre by a suture forming two convex lobes; labrum large, rounded in front. Thorax with a fine dark median line running down to apex of metathorax; prothorax much broader than long, as broad as the head, truncate at both sides, slightly depressed in the middle of each, and rotundate on the sides, smooth and shining. Legs rather long, thighs comparatively slender, tibiæ short and rather bent, with four stout spines at the apex; tarsi long, claws stout, plantula small.

Wings large, more than thrice as long as broad, rather pointed towards the tips; fore and hind wings differing in the neuration : scapular shield short, rounded, with the cross suture curving round showing the base of the six branching nervures; costal more robust than usual, receiving two stout parallel nervures running out of the scapular shield and sloping up into it; subcostal sending out four other cross nervures sloping into the costal beyond them, and a number of more transverse ones forming numerous short cells towards the tip of the wings; median nervure running close to subcostal and connected with it at irregular intervals by a number of transverse nervures most numerous towards the apex; submedian running through the middle of the wing, with six oblique short thick opaque nervures at base, and five slender nervures branching out, turning downwards and again dividing before reaching the margins; the whole wing thickly reticulated with finer veinlets : hind wing with only one parallel sloping nervure between the costal and subcostal, but connected to the costal with two very short oblique nervures as well as at the tip; subcostal nervure running parallel and sending out three oblique nervures running into the costal, and ending in a regular network at the tip; there is no true median nervure, but a branch emerging from the subcostal, in a line with the base of the 7th oblique nervure of the submedian, takes its place and is connected with short transverse nervures to the tip; the rest of the wing as the forewing. Abdomen elongate, oval, rounded at the tip, with the anal appendices stout, but hidden when viewed from above; cerci stout, conical.

Hab.—Sans Souci, Sydney (Mr. J. L. Bruce).

I have only one spirit specimen, but in perfect condition, taken by Mr. Bruce in the house flying to the lamp. It is somewhat like *Calotermes insularis*, White, in size and colour, but differs in having the head convex and not flattened in front, the smaller prothorax, neuration of wings and other important points.

(Pl. xxxvi. figs. 1-1a.)

General colour dark reddish-brown, with the wings fuscous and the nervures chocolate-brown. Length to tip of wings 5, length to tip of body 3 lines.

Head longer than broad, rounded from the base to the front of the eyes, flattened on the summit and arcuate on the forehead. Eyes large, oval, not projecting very much, finely faceted; ocelli large, reniform, contiguous to the inner margin of the eyes. Antennæ springing from a cleft in front of the eyes, (?) 14-jointed, 1st joint large, cylindrical; 2nd and 3rd of equal length; 4th smallest; the rest broadly pyriform, more truncate on the apical edge towards the tip. Clypeus small, labrum large, quadrate, with the sides rounded in front; jaws stout, with two teeth at the tip, the others indistinct; palpi short and stout. Prothorax broad, truncate in front, slightly concave behind the head, sloping on the sides, slightly concave behind. Wings slender, more than thrice as long as broad; scapular shield large, with five branches, and one parallel vein running into the costal behind the second transverse from the subcostal; subcostal nervure sending out seven transverse nervures running into the costal, and irregularly forked at the tip, median nervure running parallel to subcostal, but merging into it before reaching the tip either in the last fork or the seventh transverse nervure of the subcostal, with three or four oblique irregular slender nervures turning downwards, sub-median nervure with five thick oblique nervures at the base, and six slender ones all forked at the tips; the whole wing finely reticulated between the nervures. Legs short; thighs very thick, tibiæ short and stout, with the apical spines very large, terminal claws of the tarsi large; plantula small. Abdomen short, cylindrical, rounded at the tip, with stout conical cerci.

Soldier —The head ochreous, more ferruginous towards the jaws; antennæ bright yellow, with the apex of the joints pale, the rest dull white. Length 3 lines. Head long, cylindrical, rounded behind, nearly twice as long as broad, sloping down on the fore-head, rugose behind the clypeus; antennæ 13-jointed, springing from a cleft on the sides of the head; 3rd joint shortest, the rest broadly pyriform, the last elongate-oval, clypeus small, truncate upon the sides, labrum large, rounded on the sides and tip; palpi slender, short; jaws broad and stout, curved and slender at the tips, with two angular teeth about the centre, rugose to a large angular tooth at the base; jaws crossing over each other to the centre, left jaw with only one tooth in the centre. Prothorax rounded on the sides, concave in front, abdomen elongate-oval; anal appendices long and hairy, cerci short and stout.

Worker with the head only pale yellow; length 2 lines. Head spherical; antennæ shorter and thicker than those of the soldier; thorax not quite as broad as the head; abdomen long, cylindrical, pointed at the apex.

Hab.—Drury, New Zealand (Captain Thomas Broun).

Spirit specimens of this species were sent to me by Captain Broun under the impression that it was *Calotermes australis,* White It is, however, a very different form, differing both in size, colour, and other details. I am also indebted to Captain Broun for the following information :—"This species originally inhabited the 'Puriri' *(Vitex littoralis)* in our northern forests, where I have frequently cut out the nests containing only a small family. This species has been found in buildings as far south as Tauranga, and is widely distributed throughout the Auckland district even where the 'Puriri' does not grow. This is accounted for by the practice of using blocks of this wood for foundations sometimes infested with the termites; when they have eaten through the blocks they attack the kauri flooring boards, and in some cases eat their way through the wall studs to the roof. In the softer 'Wauri' timber the communities become much more numerous and destructive."

CALOTERMES ADAMSONI, n.sp.

(Pl. xxxv. figs 2, 2*a*, 2*b*.)

Head ferruginous, thorax ochreous, with darker markings at the base of wings; upper surface of abdominal segments darker ochreous; antennæ, legs, and all the under surface lighter coloured; wings pale fuscous with the nervures reddish-brown. Length $7\frac{1}{2}$ to tip of wings, 3 lines to tip of body.

Head broad, rounded behind, flat on summit, longer than broad, blackish and rugose along the front margin, with a small rounded pit in centre behind the clypeus. Eyes very small, round and standing out; ocelli wanting. Antennæ 16-jointed, antennal cleft deep; 1st joint large, broadest at apex; 2nd smaller; 3rd smallest; 4th 5th short; 6th-15th turbinate; 16th elongate-oval, smaller than the others. Clypeus small, pale yellow, truncate behind, rounded in front; labrum large, pale yellow, contracted at base, broad and rounded in front; jaws stout, with two sharp-pointed teeth at tip, and two large flat ones at base. Prothorax short, nearly as broad as the head, almost truncate in front, with a depression in the centre, rounded on sides, slightly arcuate behind, flattened on summit, with the edges slightly turned up; meso- and metathorax large, with a dark median suture, round at apical margin. Legs moderately long; thighs thick, short; tibiæ long, slender, with three stout spines at base, first three joints of tarsi short, 4th twice as long as the three others combined; claws large; plantula wanting. Wings large, slender, rounded at tips, thrice as long as broad; scapular shield small, round at base; tinted with ochreous yellow which extends slightly into the base of the wing: costal and subcostal nervures running parallel to each other and turning round the tip, a stout parallel nervure running out of the scapular shield and turning into the costal about the first quarter; four stout oblique nervures running upwards into the costal, with a network of more irregular shorter ones round the tips, forming irregular cells; median slender, running out towards the tip and branching out into three slender nervures turning downwards; submedian stout at base, slender beyond and turning downwards a little beyond the middle of the wings, with nine oblique nervures, the first six short and thickened, the whole wing covered with an irregular dainty network of nervelets; hind wing with the oblique nervures fewer than in the former, the median nervure running out to tip of wing, dividing into a single

36

fork, the submedian extending nearly the whole length of wing, with eleven oblique nervures, the 6th large and broadly furcate. Abdomen broad oval; cerci large, long and hairy.

Soldier.—Head bright reddish-brown, with front part and jaws black; antennæ and palpi dark reddish-brown at base of segment, giving them a variegated appearance; the rest of the body pale ochreous, with the legs rather darker. Head longer than broad, broadest at base contracting slightly behind the base of antennæ, flattened on the summit, a faint median suture with a transverse one turning down on either side into a raised knob above the antennal cleft: clypeus large, with a black protuberance on either margin; labrum contracted at base, rounded on sides, and turned downwards in front: antennæ more slender, and moniliform from the third joint to tip; palpi very long, extending nearly to the tip of jaws; jaws short and stout, slightly curved in at the tips, with three sharp incised teeth on the upper portion and one large one below; right jaw with one curved fang at tip, and a broad angular tooth below; prothorax more sharply rounded at tips, not as wide as the head, with median suture extending through it to base of metathorax; abdomen large, elongate-oval, narrowest at tip; cerci large; anal appendices large, close together, standing out perpendicularly.

Worker.—Head pale ochreous-yellow, with a dark ferruginous spot in front on either side of clypeus, the rest dirty white to pale yellow; length 4½ lines; head large, orbiculate, as long as broad; abdomen large, cylindrical, rounded at tip.

Hab.—Uralla, N.S.W. (Mr. G. McD. Adamson).

This termite differs from the other members of the genus in having no ocelli, but the wings are so typical that I hesitate to remove the species from the genus *Calotermes.*

This species is rather common in the Uralla District, Mr. Adamson having sent me several families taken in December in dead stumps and logs, with the winged termites in their galleries Of two different lots from different nests, one has the soldiers darker coloured and somewhat larger, but otherwise they agree.

Genus TERMOPSIS, Heer.

Heer, Insektenfauna von Oeningen, 1848.

Head large, rather oval, broadest behind and suborbiculate; eyes small, oval, not very prominent; ocelli wanting; antennæ long, 23 27-jointed. Prothorax small, not wider than the head, semicircular, flat. Legs long, robust, furnished with tibial spines and plantula Wings as in *Calotermes*. Abdomen egg-shaped, anal appendages long, 6-jointed.

This genus contains three species described by Heer and Hagen from fossil specimens in Prussian amber; and two existing species, one from Manitoba and California, and the other from the west coast of South America

Nothing particular is known about the habits of the existing species, but the genus is evidently closely allied to *Calotermes*.

Genus PAROTERMES, Scudder.

Proc. Amer. Acad. of Arts and Science, 1883.

This genus was formed by Scudder for the reception of three fossil species found in the American Tertiaries of Colorado, U.S. He says, "These species are most nearly allied to *Termopsis* and *Calotermes*, but differ from each of them in points wherein they differ from each other, and have some peculiarities of their own. They differ from *Calotermes* in their shorter wings (relative to the length of the body), which lack any fine reticulation, and in their want of ocelli. From *Termopsis* they differ in the slenderer but yet shorter wings without reticulation; their uniform scapular (subcostal ?) vein running parallel to the costa throughout, and provided with fewer and straight branches. From both they differ in the presence of distinct inferior branches to the scapular vein, but especially in the slight development of the intermedian vein and the median vein, the excessive area of the externomedian vein, and the course of the latter, which is approximated much more than usual to the scapular vein and emits branches having an unusually longitudinal course."

Genus MIXOTERMES, Sterzel.

This genus is founded upon the fossil wing of a termite from Lugau. From the description given of the wing it is probably allied to *Calotermes.*

Genus HODOTERMES, Hagen.

Bericht d. K. Akad. Berlin, 1853.

Head large, circular, with the median suture behind branching across towards the eyes; eyes oval, small, facets coarse, not projecting on the sides of the head; ocelli wanting : clypeus short, convex; labrum small, shell-shaped; antennæ a little longer than the head, 21-27-jointed; jaws short, powerful, toothed. Prothorax small, as large as the head, broader than long, saddle-shaped. Wings small, four times as long as broad, twice the length of the body. Tibiæ with five spines. Venation of the wings similar to that of *Calotermes*, broad from the base. Abdomen somewhat broader than the thorax, flattened on the dorsal surface; anal appendages cone-shaped.

In their habits the species resemble *Calotermes.* Seven species have been described from Africa; four fossil species from Europe and one from America. As yet I have found no Australian species of this genus.

The soldiers are remarkable for having true faceted eyes. Dr. Sharp has figured the soldiers of a remarkable species, *H. havilandi,* from Africa, which move about in the sun without any protection.

Genus POROTERMES, Hagen.

Mon. Linn. Ent. xii. 1858.

Head smaller than that of *Hodotermes;* eyes small, facets fine; no ocelli; venation of the wings similar but much finer.

This is one of Hagen's subgenera, and was formed for three species from Chili, S. America.

Genus STOLOTERMES, Hagen.

Mon. Linn. Ent. xii. 1858, p. 105.

Allied to *Hodotermes*, but having only about half the number of joints in the antennæ. Ocelli present. Prothorax heart-shaped; first tarsal joint as long as those following Venation of the wings as in *Hodotermes*, but the straight median nervure somewhat like that of *Eutermes*. Habits resembling *Calotermes*.

STOLOTERMES BRUNEICORNIS, Hagen.

Mon. Linn. Ent. xii. 1858, p. 105, Tab. ii. f 5.

Dark brown, mouth parts, basal joints of antennæ, under surface of head and legs lighter coloured; wings fuscous, with the nervures a little darker; head and thorax smooth and shining: the whole insect rather long and thickly covered with hairs. Length to tip of wings 6½, to tip of body 3 lines

Head small, circular, sloping in front, with a distinct median suture, summit rugose. Eyes round, large, ocelli in front of the inner margin of the eye; a large indistinct central false ocellus-like spot. Antennæ 16-jointed; first two cylindrical, of equal length; the last oval, the rest cone-shaped. Clypeus small, short, labrum circular, mussel-shaped Prothorax much smaller than head, broader than long, flat, rounded behind, contracted slightly in front. Wings long, four times as long as broad, scapular shield truncate, with five branches : costal and subcostal nervures connected by 7-9 very sharp transverse parallel nervures, some-times forked; first two basal ones not springing from subcostal; median nervure running through the centre of the wing, with from 7-9 oblique nervures; submedian nervure very short, turned down, with four short thick nervures. Legs robust, thighs broad; tibiæ long, with two spines at the apex, tarsi one-third the length of the tibiæ, the last joint a little longer than the first three combined; plantula present. Abdomen broader than thorax, oval, cerci large, cone-shaped; anal appendices in the male long, slender.

Hab.—Tasmania.

The above description is compiled from Hagen, who states that he has seen three dried specimens in the Berlin Museum.

STOLOTERMES RUFICEPS, Brauer.

Reise Novara, Zool. Th., Neuroptera, p. 46.

(Pl. xxxvi. figs. 2-2a.)

General colour dark reddish-brown, the under surface much lighter, base of the joints of antennæ fuscous.   Length to the tip of wings 5½, to the tip of body 3½ lines.

Head spherical; convex on the summit, rounded from the base to behind the eyes.   Eyes large, projecting, coarsely faceted; ocelli wanting.   Antennæ long, thickest towards the tips, 15-jointed, springing from cleft in front of the eyes; 1st and 2nd joints stout, cylindrical; 3rd very short; 4th-6th truncate at the extremities, narrowest at the base, 7th to tip broad oval; the last rounded at apex.   Clypeus small, rounded in front, labrum large, broad, rounded at tip; palpi rather short; jaws large, stout, with three small rather blunt teeth near the tip and one similar some distance lower down, the base rounded.   Prothorax narrow, not as broad as the head, broader than long, almost truncate in front, rounded on the sides, sloping to the hind margin, which is slightly arcuate in the centre, flattened on the summit, with a median suture running to base of abdomen, forming a dark line through the centre of the meso- and metathorax.   Legs moderately long; thighs very stout; tibiæ long, slender, and cylindrical, with the tibial spines stout ; tarsi long, claws large.   Abdomen long, slender, cylindrical, broadest at the base, rounded at the apex ; cerci short

Soldier.--Head bright yellow, ferruginous towards the apex; jaws black, upper surface of the thorax brownish yellow, the rest dull white.   Length 3½ lines.   Head longer than broad, rounded behind, flattened on the summit, straight on the sides, contracted from the antennal cleft to the base of the jaws, sloping down, and rugose upon the forehead : with very distinct faceted oval eyes upon the sides of the head behind the antennæ: antennæ springing from a cleft in front of the head, 15-jointed, the basal joints as in the winged insect, with the apical joint stouter and not so stalked; clypeus small; labrum broadest at base, rounded on the sides to a rounded tip; jaws stout at the base, curved in at the tips, and crossing each other in the middle, with two broad angular teeth in the centre. Prothorax not as broad as the head, arcuate and broadest in front, rounded and sloping sharply on the sides to the apical margin; legs short; thighs very thick; tibiæ slender, with the two inner spines at base very close together; abdomen rather large, oval; cerci small.

Hab.—Drury, New Zealand (Captain T. Broun).

I have no workers in my collection, all other examples sent with the soldiers being pupæ with short wing-cases.

Spirit specimens of this species were sent to me by the Government Entomologist of New Zealand, but without any notes upon their habits.

The soldiers are remarkable for their distinctly faceted eyes, though some species of the Hodotermes group are also known to have soldiers provided with eyes. In an African termite *(Hodotermes havilandi)* which is figured in the Cambridge Natural History, and described as going about in the bright sunlight, similar eyes are very distinct.

RHINOTERMITINÆ.

Genus RHINOTERMES, Hagen.

Head as broad as long; forehead flattened, with a parallel cleft through the centre of the rhinarium, which projects slightly in front, forming with the lobed clypeus a snout-like process. Eyes small, coarsely faceted; ocelli present, with a circular false ocellar spot in the base of the cleft: antennæ 20-jointed. Prothorax not as wide as the head, rounded in front. Legs stout, with two spines at the apex of the tibiæ; plantula wanting. Wings short and broad, rounded at the tips; scapular shield short and broad, swelling out and slightly convex at the cross suture; costal and subcostal nervures stout, well separated at the base, slightly connected towards the tips with irregular cross veins; median nervure fine, but irregularly branched at the tip; submedian with a great number of fine bifurcated oblique nervures, the whole wing reticulated with furrows and ramulose veinlets

Hagen placed in this subgenus of *Termes* three species from Cuba, Surinam, and Brazil. A fourth species was described by Brauer from Australia. The members of the Australian species live in communities like *Calotermes*. On account of the irregular veins between the costal and subcostal nervures I have placed them in a separate subfamily.

RHINOTERMES RETICULATUS, n.sp.

(Pl. xxxvi. figs. 3, 3*a*, 3*b*, 3*c*.)

Upper surface pale ferruginous, ventral surface pale yellow; wings light reddish-brown, semitransparent, nervures tawny Length to tip of wings 5½, to tip of body 3 lines.

Head slightly broader than long, broadest behind, sloping in on the sides in front of the eyes, and truncate across, rather flattened on the summit. Eyes small, not projecting, coarsely faceted; ocelli very small, in front of the eyes, elongate-oval Antennæ 20-jointed, springing out of a deep antennal cleft, 1st joint large, cylindrical, 2nd about half the length, 4th the smallest; 5th-20th moniliform, increasing slightly in length, and more stalked to the tip; the terminal one round; all of them rather hairy. Clypeus large, truncate behind, divided by a deep cleft which proceeds from the front of the forehead where it commences in a small rounded spot in a line with the ocelli; labrum spade-shaped, rounded at the tip, longer than broad, jaws thick and stout, sharply curved in at the tip, with four sharp angular teeth, and a rounded edge at the base Prothorax not as wide as the head, rounded in front, rotundate and depressed on apical margin. Legs robust, thighs short and broad; tibiæ long, slender, hairy, with two long spines at the tip; tarsi very spiny. Wings thrice as long as broad; scapular shield short, rounded above, with the suture slightly convex; costal and subcostal nervures thick, running parallel to each other and curving round at the tip, without true cross veins, but with a number at the extreme tip forming irregular cells ; median nervure slender, irregular, crossing the middle of the wing, turning downward and branching into three oblique forks, the first again bifurcated, the second simple and the last again forked ; submedian running parallel with median to middle of wing, turning downwards, with eight oblique branching veinlets not always regular. Abdomen short, broad, rounded at the tip; cerci short and stout.

Soldier.—Head pale yellow, darkest towards jaws which are ferruginous; the rest dull white. Length 3 lines. Head large, short and broad, flattened on the summit, rounded on the sides, and sloping up in front from the deep antennal cleft to the base of jaws; forehead truncate, with a sharp canal cut out in the centre, forming a short gap with a circular spot or opening at the base clypeus concave behind, rounded on the sides and narrowest in front, labrum very long, reaching to the tip of the closed jaws, broad at base, contracted towards the middle and swelling out into a rounded spatulate lobed tip, jaws short, stout, sharply turned over each other at the apex, with two sharp teeth below on the left fang and a single one on the right. Thorax smaller than head, with the prothorax more saddle-shaped than that of the winged ones; legs rather slender; abdomen short and broad, the slender anal appendices showing beyond the tip; cerci hairy.

Soldier (minor).—In this species a second form of soldier is always present in about equal numbers with the larger ones. In general structure they are similar, but with all the parts more slender and elongated; length 2 lines Apical portion of head bright yellow, base much lighter; head broad at the base, sloping to base of the jaws, of a somewhat elongated pear-shape ; jaws much elongated, slender, turning over at the tips; palpi nearly as long as jaws ; antennæ 16-jointed ; labrum very slender, but similar to that of the large soldier.

Worker dull white, lightly tinted with yellow behind the jaws, 2 lines in length. Head very large and broad, sloping round at the jaws, with a curious bilobed pattern above the clypeus; antennæ very slender, 18- to 20-jointed; clypeus large, convex, rounded behind, with a deep median suture, a ferruginous spot on either side at the base of the antennæ; labrum small. Prothorax much smaller than the head, with a fine median suture running from the base through the meso- and metathorax ; abdomen oval, swollen in the middle, broadly rounded at the tip.

Hab.. Kalgoorlie, W.A (Mr. G. W. Froggatt; from nest); Palm Creek, Central Australia (Prof. Spencer, Horn Expedition).

Specimens of these termites were taken by my father in a dead sheoak (Casuarina) stump towards the end of March; and at this time the winged ones were more plentiful than the workers and soldiers. In their habits and general appearance they resemble *Calotermes*, and take the place of the eastern species, *R. intermedius,* both are plentiful in their districts.

RHINOTERMES INTERMEDIUS, Brauer.

Reise Novara, Zool. Th , Neuroptera, p 49.

Upper surface pale ochreous, lighter coloured at the base of head and thoracic segments under side, legs, and antennæ pale yellow; wings pale ferruginous, semitransparent, nervures darker Length to tip of wings 7, to tip of body 4 lines.

Head similar to that of *R. reticulatus,* but with the eyes much larger and more prominent; ocelli larger. Antennæ very hairy, 20-jointed. Clypeus broader and not quite so convex. Prothorax broader and more deeply concave in front behind the head. Legs longer and tibiæ more slender. Wings thrice as long as broad, larger, and lighter coloured, but with the venation identical.

Soldiers and *workers* as in the former species

This species is not very common about Sydney. I found one small colony in the stem of a dead honeysuckle tree *(Banksia serrata)* near Sydney, but at Wallsend, near Newcastle, small colonies under dead logs are common, gnawing irregular passages along the grain of the wood, and retreating into the log when disturbed. They are at once recognised by the large broad heads of the soldiers and the presence of two different forms of soldier.

The soldiers, like those of *Calotermes*, are very timid, never showing fight, but hurrying away to shelter when disturbed, the little soldiers being much the braver. I had never been able to find the winged forms in our nests, but my friend Mr. Gilbert Turner, of Mackay, was more fortunate, sending me down several winged ones with workers and soldiers

Early last year Mr. N. Holtze sent me a small bottle full of winged ones that had been taken flying round the lamps at Palmerston, Pt. Darwin. This species was described by Brauer, the locality given being Sydney, N.S.W., but in a specimen sent from the Vienna Museum, where his types are, the label attached says, " Thorey, Cape York, 1868."

Hab.—Sydney and Newcastle, N.S.W. (W. W. Froggatt); Mackay, Queensland (Mr. G. Turner); Port Darwin, N.T. (Mr. N. Holtze, Botanic Gardens).

GLYPTOTERMITINÆ.

Genus GLYPTOTERMES, g.n.

Head broad; eyes moderately large, coarsely faceted; ocelli close to the eyes; antennæ short, 13- to 15-jointed, springing from a circular cleft in front of the eyes. Prothorax convex in front, rounded on the sides and convex behind, with a slight median suture. Legs stout and rather short, with short thick spines at apex of tarsi; plantula small Wings slender, thrice as long as broad; scapular shield small and angular showing the base of four nervures: costal, subcostal and median nervures running close to each other through the upper half of wing, subcostal generally merging into the costal in the centre, but always separated at the extremities; submedian running through the centre of the wing; it and the oblique nervures often composed of fine dots.

Small dark-coloured termites, with clouded opaque wings, living in small communities in the trunks and bark of trees; soldiers very few; these and the workers slender and cylindrical.

GLYPTOTERMES TUBERCULATUS, n.sp

(Pl. xxxv. figs. 9, 9a.)

General colour pale ochreous; legs and antennæ paler, wings vitreous, with the nervures fuscous at base and light ferruginous towards the tips. Length to tip of wings 6, to tip of body 2½ lines.

Head broader than long, broad behind, almost quadrate, truncate in front, convex on the summit. Eyes standing out on the sides of the head, large and circular, coarsely faceted, ocelli round, in line with the apical margin of eyes. Antennæ short, rather hairy, springing out of a deep antennal cleft in front of the eyes, 15-jointed; 1st stout, cylindrical; 2nd and 3rd shorter, cylindrical, broadest at apex; 4th-14th short, broad, cup-shaped, rather broader towards the extremities, with the last joint oval. Clypeus rounded behind, produced into flanges on the side, narrower, truncate and quadrate in front; labrum broad, rounded in front, shell-shaped; jaws rather stout, with three sharp teeth at the tip; palpal joints very short and oval. Prothorax quadrate, slightly turned up on the edge, slightly concave in front, straight on the sides, truncated behind, with a depression in the centre and a slight suture, rough and flattened on summit. Legs short, thighs thick, rather cylindrical, tibiæ with three or four stout spines at apex; tarsi with the last joint very large; plantula present. Wings more than thrice as long as broad, rounded at tip, whitish when dry, very thin and easily torn; scapular shield narrow and angular, the cross suture truncate, the base of the first three nervures meeting on the upper edge, the lower submedian indistinct; costal, subcostal, and median nervures (the first strengthened at the extreme base by a short transverse parallel nervure running from the subcostal into the costal) running parallel and close together to the tip of wing; submedian nervure weak and irregular, running through the middle of the wing, without any stout oblique nervures at the base, but with some 12 or 14 irregular nervures turning downwards : costal and subcostal of hindwings as in forewings, but with median emerging from subcostal at some distance from the scapular shield, and running parallel with it to the tips; the whole of the wings thickly covered with scars or pustules. Abdomen elongate-oval, slender; cerci short and stout, well under the abdomen; anal appendices wanting.

Soldier.—Head bright reddish-brown, jaws black, labrum luteous, prothorax ochreous, the rest dull yellow. Length 3 lines. Head a little longer than broad, cylindrical, sides straight, sloping in from behind the base of the antennæ to the centre where the forehead is deeply cleft, forming a rounded hollow with a stout knobbed protuberance on either side, and truncate below, and overhanging clypeus, which is small and indistinct; labrum large, flattened, spatulate, finely fringed with hairs; antennæ springing out of a circular pit in line with the base of jaws, 15-jointed; jaws short, ferruginous and very stout at the base, meeting at the tips, with two stout angular teeth below the tip on the left side, the jaw on the right side smooth to apex of labrum, where there is one large tooth. A stout cylindrical finger-like projection stands out on either side of the apical margin of head in front of the antennal cleft. Prothorax saddle-shaped, slightly arcuate in front, rounded on sides, and sloping back to apical edge which is slightly concave in the centre; a fine median suture running through the head and whole of the thorax; thorax and abdomen forming a cylindrical body, narrowing towards the tip, rather hairy; legs short and stout.

Worker about the same length and shape as the soldier, with the exception of the head, which is almost spherical; labrum quadrate; anal appendices very fine, slender, projecting beyond the tip of the abdomen; general colour dull white.

Hab —Uralla, N.S.W. (Mr. G. McD. Adamson).

Described from specimens received from the collector in spirits, and obtained by him in a log.

GLYPTOTERMES IRIDIPENNIS, n.sp.

(Pl. xxxvi. figs. 5, 5a.)

Castaneous to piceous, antennæ and legs dark ochreous, the
wings deeply clouded with pale reddish-brown, nervures reddish-
brown.   Length to tip of wings 5½, to tip of body 2¾ lines

Head longer than broad, widest behind, convex on the summit,
and sloping down on forehead.   Eyes small, round, rather coarsely
faceted, on the sides of the head projecting very slightly; ocelli
round, not contiguous but in line with centre of eye.   Antennæ
short, stout, and rather hairy, springing from a circular antennal
cleft in front of eyes, 15-jointed; 1st stout, cylindrical; 2nd and
3rd smaller; the rest thickened, stout, pyriform; terminal joint
oval.   Clypeus large, quadrate; labrum convex on summit, broader
than long, rounded in front.   Prothorax rather broader than
head, deeply concave in front, rotundate with the sides flanged
and the apex rounded.   Legs short, thighs broad and stout;
tibiæ stout, cylindrical, broadest at the tips, with three short
stout spines beautifully serrate on the edges; tarsi rather long,
the terminal joint as long again as the first three combined, claws
slender, plantula small.   Wings slender, four times as long as
broad, rather pointed at the tip; scapular shield long, narrow,
with five nervures, cross suture transverse; costal nervure slender,
with a stout, parallel, oblique nervure branching out of scapular
shield and running up into costal; subcostal and median branching out together, the latter slightly angular, but running close
together to the tip of wing; subcostal merging into costal about
the middle of wing, but emerging again before the tip is reached;
submedian branched on the scapular shield, with seven thick
unbranched oblique nervures at the base, running through the
centre of the wing, but slender and formed of fine spots like lace
work, with five main similarly formed oblique nervures; the whole
membrane of the wing thickly covered with fine tubercles, forming a network of irregular pattern.   Abdomen short and stout,
rounded behind; cerci small.

Hab.—Frankston, Victoria (Mr. W. Kershaw, National Museum).

This species is described from a single pinned specimen in good preservation; and is very distinct from any other species known to me.

GLYPTOTERMES BREVICORNIS, n.sp.

(Pl. XXXVI. figs. 6, 6*a*).

Upper surface pale ochreous; wings semitransparent, nervures brown tinged with yellow; under surface, legs, and antennæ stramineous. Length to tip of wings 5, to tip of body 2½ lines.

Head a little longer than broad, rotundate, broadest between the eyes, rounded on the summit, with a slight median suture at the base. Eyes small, circular, not very prominent; ocelli oval, contiguous · and in a line with the apical margin of the eyes. Antennæ 13-jointed, 1st joint large, cylindrical; 2nd shorter, cylindrical; 3rd-4th orbiculate; 5th-12th turbinate; the terminal one oval. Clypeus widest behind, narrow, truncate in front, sloping back on the sides; labrum broad, rounded on the sides, and rather truncated in front; jaws broad, with three short blunt teeth at apex, the edge roughened towards base. Prothorax not as broad as head, concave in front, rotundate on the sides and behind, with a slight depression at the apex, a dark median line running from the base through the meso- and metathorax. Legs short and thick, thighs large; tibiæ slender, armed with five stout spines at the apex; terminal joint of tarsi large; claws large; plantula small. Wings slender, twice as long as broad; scapular shield slender, rounded at the cross suture, clouded with fuscous extending into the base of the wing, costal, subcostal, and median nervures running parallel, close together, the last extending a little further round the tip of wing, submedian opaque at base, running through middle of wing, with three stout oblique nervures at the base, the apical one indistinct, about eleven in number, forming slender dotted nervelets turning downwards; the whole of the wings covered with minute spots or scars. Abdomen broad, elongate, rounded at the tip; cerc short and stout.

Soldier.—Head pale ferruginous at base, becoming much darker towards the antennæ; jaws castaneous at base to black at tips; upper surface of thorax and legs pale ochreous, the rest dirty white. Length to tip of body 3½ lines. Head twice as long as broad, rounded behind, straight upon the sides, broadest at base of jaws, flat on the summit and sloping down sharply in front, irregularly roughened; with a median suture dividing in front and running out on either side at base of antennæ; antennæ 13-jointed, short, not reaching beyond tip of jaws; clypeus small, flattened, slightly rounded in front; labrum almost quadrate, lying between the base of jaws, thin and shell-like; jaws very short, broad at the base, irregularly toothed, straight on the sides, curved at tip and just crossing each other, with three small angular teeth below on the left jaw and two larger ones on the right. Body long and cylindrical.

Worker.—Head and prothorax pale yellow, the rest white. Length to tip of body 3 lines. Head spherical, showing pale median and transverse sutures, and a dark mark along apical margin on either side in front of base of antennæ. Body long, cylindrical and rather hairy.

Hab.—Mackay, Queensland (Mr. Gilbert Turner).

GLYPTOTERMES EUCALYPTI, n.sp.

(Pl. xxxv. figs. 5-5*a*.)

The entire insect dark castaneous, antennæ ochreous, under surface somewhat lighter coloured; wings semi-opaque brown, nervures darker, covered with fine dots or scars iridescent in sunlight. Length to tip of wings 3½, to tip of body 2¼ lines.

Head slightly longer than broad, broadly rounded at base, arcuate in front of eyes, rounded on summit, sloping down to clypeus. Eyes moderately large, projecting slightly; ocelli oval, contiguous to front margin of eyes. Antennæ 14-jointed, springing from an antennal cleft between the eyes; 1st joint stout, cylindrical; 2nd shorter; 3rd rather pear-shaped; 4th-13th larger, orbiculate, becoming more turbinate towards the tip; terminal one rounded Clypeus broad and short, truncate behind, overlapping the broad bilobed labrum; jaws small, straight on the sides, with the tip curved in, a sharp tooth below, widely separated from the third. Prothorax nearly as broad as head, broader than long, concave in front, rotundate on the sides and slightly hollow behind, a slender median suture at base to the apex of metathorax. Legs short and thick; thighs broad, rounded; tibiæ with three stout spines at apex. Wings slender, four times as long as broad; scapular shield small and slender, fuscous, the colour extending into the base of wings, the cross suture straight: base of subcostal in forewings robust, with a short nervure running out of scapular shield and turning up into costal just beyond the suture, costal and subcostal only separated from each other at the extremities; submedian stout at base, running through the middle of wing, with five or six opaque oblique nervures emerging from basal portion and six or seven finer and longer ones towards apex, all these more or less irregular from the many little dots covering the wings. Abdomen long, slender, rounded at tip; anal appendices very long and slender, close to the tip of abdomen; cerci short and stout.

Soldiers.—Head pale reddish-yellow, the rest white. Length 3½ lines. Head longer than broad, rounded behind and straight on the sides, emarginate in front at the base of jaws, truncate on forehead and rugose above clypeus; median and transverse sutures distinct, the latter running out on either side to base of antennæ; clypeus hidden; labrum broad, rounded in front and on sides, depressed in the centre and fringed with fine hairs; jaws very broad at base, short, rounded, turning over each other at the tips, with three sharp angular teeth. Abdomen long, slender, and cylindrical, tapering at the tip; cerci short and stout.

Worker of a general dull white colour; head faintly tinged with yellow; abdomen in life reddish-brown from the food eaten

37

showing through the semitransparent skin : head spherical, show-ing two lobes on forehead, rounded towards the base of antennæ, with a dark spot on either side of clypeus; prothorax smaller than head, the rest of thorax and abdominal segments rounded, slender, and cylindrical to the tip.

Hab —Sydney, Botany and Hornsby (W. W. Froggatt).

About Sydney this species is only found by cutting off the loose bark upon the trunks of *Eucalyptus robusta*. The insects feed upon the inner bark, and sometimes on the living sap wood, evidently as a general rule gnawing a passage through from behind, as there are always several tunnels leading inwards in the trunks, which are nearly always rotten and decayed in the centre. They live in small communities of from fifty to a few hundred individuals, the majority being workers or larvæ, with sometimes only one or two soldiers in the colony. Except in the head, the soldiers closely resemble the workers, and try to hide as soon as they are exposed. They form very slender tubular tunnels in all directions in the bark, each individual burrowing on his own account, no room being left to allow of their passing each other. The winged ones are very small in comparison with the workers and soldiers. Some well developed pupæ were obtained in a rather numerous colony in a dead tree (the only time I ever found them away from the living trees), and these matured to the perfect insects in December.

HETEROTERMITINÆ.

HETEROTERMES, g.n.

Head large, longer than broad, nearly quadrate; eyes very small, not projecting; ocelli wanting : clypeus large; labrum broad, antennæ 16-jointed. Prothorax not as broad as head, truncated on the sides; legs stout, with four or five stout spines at the apex of tibiæ. Wings nearly thrice as long as broad ; scapular shield small and angular; costal, subcostal and median nervures running very close to each other, but the costal and sub-costal distinctly separated from each other; submedian and oblique nervures slender.

HETEROTERMES PLATYCEPHALUS, n.sp.

(Pl. xxxv. fig. 10; Pl. xxxvi. fig. 4.)

General colour castaneous, legs brown, labrum ochreous; antennæ barred with white at the apex of each segment; wings pale fuscous with the nervures brown. Length to tip of wings 6, to tip of body 2½ lines.

Head very large, longer than broad, almost quadrate, rounded behind and straight on the sides to well in front of the eyes, flattened upon the summit, slightly arcuate behind the clypeus. Eyes small, circular, well down on the sides of the head, not projecting; the ocelli wanting. Clypeus large, prominent, and rounded on the sides and apex, very slightly concave in front, with a median suture through the centre dividing it into two lobes; labrum broad, rounded in front. Antennæ 16-jointed, long, with large thickened segments, springing from in front of eyes; 1st joint long, cylindrical; 2nd and 3rd very small; 4th-15th increasing slightly in size towards the tip; terminal joint oval. Thorax covered with long scattered grey hairs; prothorax not as broad as head, truncated on the sides, rounded and arcuate in the centre of both base and apex. Legs short, robust; tibiæ broad at tip, with four slender spines; tarsi slender. Wings nearly thrice as long as broad, rounded at the tip; scapular shield slender, hairy, angular, showing the base of four nervures; costal and subcostal nervures running very close together to tip, median nervure very fine, running close to subcostal, divided and turning down at the tip: submedian fine, with seven thickened oblique nervures; the first two very small; the 3rd, 4th, 6th, and 7th furcate, with four or five slender oblique apical nervelets. Abdomen short, elongate and oval at the tip.

Hab.—Kangaroo Island, S.A. (Mr. J. G. O. Tepper).

I have one mounted specimen from the Adelaide Museum. It is a very curious form differing from all other species in the long quadrate head and thick antennæ. There are also four specimens of this termite in the Macleay Museum, labelled South Australia.

EXPLANATION OF PLATES.

PLATE XXXV.

Fig. 1. —Forewing of *Calotermes irregularis*, n.sp.
Fig. 1a. —Hindwing of   ,,   ,,
Fig. 1b. —Head of soldier of *Calotermes irregularis*, n.sp.
Fig. 2. —Forewing of *Calotermes adamsoni*, n.sp.
Fig. 2a. —Hindwing of   ,,   ,,
Fig. 2b —Head of soldier of *Calotermes adamsoni*, n.sp.
Fig. 3. —Forewing of *Mastotermes darwiniensis*, n.sp.
Fig. 3a. —Head of   ,,   ,,
Fig. 4. —Forewing of *Calotermes insularis*, White.
Fig. 5. —Forewing of *Glyptotermes eucalypti*, n.sp.
Fig. 5a. —Head of soldier,   ,,   ,,
Fig. 7. —Head of soldier, *Calotermes longiceps*, n.sp.
Fig. 8. —Forewing of *Calotermes robustus*, n.sp
Fig. 9. —Forewing of *Glyptotermes tuberculatus*, n.sp.
Fig. 9a. —Head of soldier,   ,,   ,,
Fig. 10. —Head of *Heterotermes platycephalus*, n.sp.

PLATE XXXVI.

Fig. 1. —Forewing of *Calotermes Brouni*, n.sp.
Fig. 1a. —Head of soldier,   ,,   ,,
Fig. 2. —Forewing of *Stolotermes ruficeps*, Brauer.
Fig. 2a. —Head of soldier,   ,,   ,,
Fig. 3. —Forewing of *Rhinotermes reticulatus*, n.sp.
Fig. 3a. —Jaw of   ,,   ,,
Fig. 3b. —Head of soldier (major), *Rhinotermes reticulatus*, n.sp.
Fig. 3c. —Head of soldier (minor),   ,,   ,,
Fig. 4. —Forewing of *Heterotermes platycephalus*, n.sp.
Fig. 5. —Head of *Glyptotermes iridipennis*, n.sp.
Fig. 5a. —Wing of   ,,   ,,
Fig. 6. —Wing of *Glyptotermes brevicornis*, n sp.
Fig. 6a. —Head of soldier,   ,,   ,,

THE OCCURRENCE OF RADIOLARIA IN PALÆOZOIC ROCKS IN N.S. WALES.

By Professor T. W. Edgeworth David, B.A., F.G.S.

(Plates XXXVII.-XXXVIII.)

CONTENTS.

1. Bibliography.
2. Localities and Geological horizons of radiolarian rocks in N.S.W.
3. Macroscopic and microscopic description of the radiolarian rocks.
4. Summary
5. Deductions.

1. BIBLIOGRAPHY.

The first reference known to me as to the occurrence of radio-larian rocks in Australia is in a paper by Dr. G. J. Hinde, F.R.S [*]

This rock was obtained by Capt. Moore, of H.M.S. "Penguin," about 1891, from Fanny Bay, Port Darwin. "The rock in question is of a dull white or yellowish white tint, in places stained reddish with ferruginous material; it has an earthy aspect like that of our Lower White Chalk, but it is somewhat harder than chalk, though it can be scratched with the thumb-nail. There are no signs of stratification, and it appears as a fine-grained homogeneous material." Under the microscope the groundmass is seen to be made up of minute granules and mineral fragments, isotropic for the most part, being probably amorphous silica. The minute grains, however, and angular particles polarize : some appear to be quartz, others rutile. The organic structure

---

[*] Q.J.G.S. Vol. xliv. No. 194 May 1st, 1893 Dr. G. J. Hinde Note on a Radiolarian Rock from Fanny Bay, Port Darwin, Australia

of the granules is only very faintly marked. The orders of Prunoidea, Discoidea and Cyrtoidea are all represented The geological horizon to which they belong is very probably that of the Desert Sandstone Formation (Upper Cretaceous).

What is probably an equivalent of this rock has been described by the Rev. J. E Tenison Woods* as follows :—

"What we find whenever a good section is exposed is this—a layer of loose white, or red, decomposed rock or rubble, some 3 or 4 feet thick, lies on the upturned edges of the slates. Above this a layer some 2 feet thick of loamy earth, which has been surface soil. Above this from 14 to 120 feet of magnesite or carbonate of magnesia, more or less impure, with silicates of alumina and iron, and mere traces of lime. Not often is it pure white, for the stains of brown, red and purple, from iron oxide, permeate the whole."

The above statement by the Rev. J. E. Tenison-Woods, as far as can be ascertained, refers to a rock identical with that which has now been proved to be, not a magnesite, but a radiolarian rock.

Reference may here be made to a note by Dr. Hinde† in which he describes a cherty rock from South Australia, which although derived from sponge spicules rather than radiolaria, yet contains globules of opal silica which might easily be mistaken for radiolaria.

The rock described in the note referred to above appears to be of Tertiary age. The specimens were collected by Mr. H. Y. L. Brown at Yorke's Peninsula, near Adelaide. Dr. Hinde states (op. cit. p. 115), "The principal feature is the occurrence of detached sponge-spicules which in places are heterogeneously crowded together in the rock. . . . The matrix in which the

---

* Report on Geology and Mineralogy of the Northern Territory, South Australia, p. 5. By authority. Adelaide, 1886.

† "Note on Specimens of Cherty Siliceous Rock from South Australia." Geol. Mag. New Series Dec. iii. Vol. viii. 1891. pp. 115-116.

spicules and quartz grains are imbedded appears to be mainly of amorphous or opal silica, nearly entirely neutral to polarized light between crossed Nicols, and it is principally in the form of very minute globules or discs usually aggregated together so as to exhibit a microscopic botryoidal appearance, the globules or discs varying from 01 to ·03 mm. in diameter. The globular form of opal silica is similar to that which occurs in many of the sponge-beds of the Upper Greensand in this country, and there can hardly be any doubt that in this Australian Chert it is due, as in the Chert of this country, to the solution and redeposition of the organic silica of the sponge-spicules."

As far as I am aware, the above are the only references to the occurrence of radiolarian rocks in Australia; and in both cases it would appear that the rocks mentioned are of late Mesozoic age.

Before proceeding to describe the horizons where radiolaria have recently been observed by me in Palæozoic rocks in N.S W., it might be of interest, in view of the grand scale on which the radiolarian rocks are now known to be developed in this colony, and in view also of the fact that some of the literature relating to radiolaria is rather inaccessible to Australian geologists, to briefly summarize the more important works relating to Palæozoic and Mesozoic radiolaria in Extra-Australian areas.

Radiolaria have been described by Dr. D. Rüst* from Mesozoic rocks, the Gault of Zilli, and the Neocomian of Gardenazza. The radiolaria in the best state of preservation were those found in the Cretaceous Coprolite Beds of Zilli, in Saxony. These radiolaria have been admirably figured and described by this observer.

Dunikowski has described perfect forms from the Lower Lias of the Austrian Alps; while Hantken believes that certain siliceous limestones with *Aptycus*, of Upper Jurassic age, in Central Europe are almost entirely formed of radiolaria.

---

* Palæontographica. Vol. xxxi. 1885, and *ibidem* Vol. xxxiv. pp. 181-213. Pls. xxii-xxix., 1888, and Vol. xxxviii., 1892.

Gunibell cites them from the St. Cassian beds; and Waters has detected their remains in the Infra-Lias.

Radiolaria have been described by Dr. Geo. J. Hinde and Mr. F. L. Ransome* from Angel Island from Mesozoic (?) rocks.

Radiolaria have been described from Jurassic or older rocks in the coast ranges of California by Fairbanks.†

Radiolaria have been described from Palæozoic rocks by the following :—Shrubsole has recorded them from the Carboniferous rocks of Great Britain.

Dr. G. J. Hinde‡ has described radiolaria from the Llandilo-Caradoc rock at Corstorphane, in the S of Scotland.

The same author has described radiolaria from Ordovician cherts at Mullion Island, Cornwall, England.§

Perhaps the most important contribution to our knowledge of the Palæozoic radiolaria is that of Dr. Rust,‖ and, as much of it has an important bearing on the radiolarian rocks of Australia, I take the liberty of making abstracts from it.

In the *phosphorite* from the Petschora in the S. Urals occur well preserved radiolaria in the form of deep black flinty shells, in a bright brown translucent base. Flinty material and iron are present in the phosphatic limestone. In cases the radiolaria are represented by casts only. In the whetstone and adinole radiolaria are badly preserved.

Radiolaria are beautifully preserved as dark black shells in a cryptocrystalline quartz groundmass in the Lydian-stone of Teufelsecke at Lautenthal.

---

* The Geology of Angel Island University of California. Bulletin of the Department of Geology. Vol. i No 7, pp. 193-240. Pls. 12-14.

† "Stratigraphy of the Californian Coast Ranges "—Journal of Geology, Chicago. Vol. iii., 1895, p. 415.

‡ Geol. Mag. New Series. Dec. iii. Vol. vii., 1890, p. 144, and Ann. & Mag. Nat. Hist. Ser 6, Vol. vi. (1890), p. 40.

§ Q.J.G.S. Vol. xlix., 1893, pp. 215-220. Pl. iv.

‖ Palæontographia. Vol. xxxviii , 1891-92. Beiträge zur Kentniss der fossilen Radiolarien aus Gesteinen der Trias und der Palæozoischen Schichten. Von. Dr. Rust in Hanover.

The red jasper from Sicily contains numberless radiolarian shells, coloured red, in a translucent siliceous groundmass.

Fairly well preserved radiolaria have been found in red jasper of Lower Devonian age.

At Cabrières, in Languedoc, a very hard black siliceous schist of Ordovician age contains radiolaria, mostly in a bad state of preservation. In the phosphorite of Cabrières, however, dark, porous to dense, concretions contain numerous radiolaria.

The following is an analysis of the phosphorite :—

Water..	1·08
Lime phosphate...	73·65
Silicate alumina.	25·27
	100·

The radiolarian shells were black, yellow, or colourless. No sponge spicules were present. In pieces of rock (siliceous shale) from Saxony, poor in radiolaria, fragments of graptolites are numerous.

Black radiolarian fragments have been observed in fairly hard clay shale of Cambrian age. Others occur in flinty pebbles, but not sufficiently well preserved to admit of the species being determined. Fragments of graptolites and graptogonophores were associated.

The fact must be emphasized that it is chiefly in concretions containing phosphoric acid that the radiolaria are best preserved.

It often happens in all flinty rocks, not only Palæozoic but also Mesozoic, that the quartz filling the original hollows of the radiolarian shells shows a radial habit, and has the form of perfect spherulites exhibiting dark fixed interference crosses in polarized light when the objective is rotated.

In most cases the latticed shell has disappeared. Occasionally, however, the pore openings of the shell are preserved, or one sees a dark circle bounding a clear space, with small regularly placed dark indentations on the inner side.

Very often perfect crystals are developed inside and around these little quartz spheres. Generally these are opaque octahedra of magnetite and clear or dark yellow rhombohedra of calcite. These crystals are seldom observable in the Silurian forms, and are not visible in the Devonian. Very little other organic remains are associated with the radiolaria   Only sponge spicules, belonging to the Hexactinellidæ, are found associated with the radiolaria, sometimes in great numbers.

Isolated examples only of foraminifera are met with in the siliceous limestone of the Muschelkalk.   In the Silurian siliceous shales of Langenstriegis, Rehan and Steben fragments of grapto-lites and gonophores are not infrequent.

Plant remains.—Prickly macrospores occur in the radiolarian rocks of the Jura as well as in the Carboniferous siliceous schists of the Hartz Mts.   These were found in great abundance in a Lower Silurian limestone from Koneprus in Bohemia, in which hitherto radiolaria have not been detected.

Another important contribution to the knowledge of Palæozoic radiolaria is that by Hinde and Fox[*], from which the following abstracts may be made.

Radiolaria occur at Codden Hill.   The Codden Hill beds have a baked appearance, are whitish, buff, or dark grey in colour, and have frequently a chertoid texture, consisting of thick shales and fine-grained grits.

In places in the radiolarian chert wavellite is developed along the joint planes.   Sponge spicules are associated with the radiolarian rock.   The radiolarian series of the Culm is probably at least 200 ft. in thickness, if the intercalated fine shales be included.

Individual beds usually are from 2-4 inches thick, rarely as much as 1 foot.

The beds are intersected by numerous fine and even joint planes, which have the effect of dividing the rock up into com-paratively small rectangular or rhombohedral fragments with smooth flat surfaces.

---

[*] Q J.G.S. Nov. 1895, Vol. 1 G. J. Hinde and Howard Fox. "On a well marked Horizon of Radiolarian Rocks in the Lower Culm Measures of Devon, Cornwall, and West Somerset."

The radiolarian beds are composed of dark to black chert with a hackly fracture.   Other portions are dull grey to white, or the rock is made up of alternate light and dark bands, so as to be striped

In places the rock is platy, siliceous, or mottled white and black   The soft grey to white beds are very rich in radiolaria. They disintegrate in some cases in water into a fine cream-coloured mud

The soft beds are of much less frequent occurrence than the hard cherts.

The individual radiolarian beds are minutely laminated.

Microscopic character.—Carbonate of lime is conspicuous by its absence.   The radiolarian rock generally shows a siliceous ground mass, in some cases clear and transparent, in others dark and turbid from the presence of fine particles of carbonaceous or ferrous minerals, and minute crystal needles of rutile and zircon. The siliceous groundmass shows between crossed Nicols the faint speckled appearance of cryptocrystalline silica, like flint from chalk   When radiolaria are abundant chalcedonic tints prevail. The radiolaria in the rock have been filled with clear nearly transparent silica free from the rutile crystals and from the dark substances disseminated in the groundmass, and either micro-crystalline or cryptocrystalline.   Within the radiolarian casts the silica is often fibrous radial, and so shows a black cross in polarized light.

The more distinctly crystalline character of the radiolarian casts facilitates their recognition in the rocks with a clear ground-mass where in ordinary light they are scarcely visible, but between crossed Nicols they appear as so many circles of speckled or bright light on a nearly dark ground.

Minute casts of rhombohedral crystals are frequently present, probably of calcite or dolomite, sometimes inside the radiolarian casts.   A similar occurrence has already been referred to in the Hartz Mountains.   Microscopic cubes of iron pyrites are present in some of the rocks.

In some of the harder and more cherty beds very minute bodies like those in the Pre-Cambrian phthanitic quartzite of Brittany are noticeable, ·006 to ·013 mm There is no evidence to show that these are organic.

Under favourable conditions of light the latticed structure of the radiolarian shells can be distinctly seen in the coarse material resulting from the disintegration of the soft shales in water

A few minute dentated plates, perhaps radulæ of gasteropods, of dark brownish tinge are associated with the radiolaria. Detrital fragments, except mica flakes, are either wholly wanting or extremely minute, ·03 to ·065 mm in diameter.

Rarely limestone is associated with the radiolarian rock, and in the limestone are casts of radiolaria in calcite and also of sponge spicules. Entomostraca, crinoids, and *Endothyra* contribute to form limestones near this radiolarian horizon.

In the majority of the Culm siliceous rocks the radiolaria are now in the condition of solid casts of the original forms; their skeletal walls have entirely disappeared, and the individual casts are only bounded by the siliceous matrix of the rock, and are without definite even outlines. In such instances only the size and general form *with the radial spines* can be distinguished.

In some cases the tests have been naturally stained a brown or amber tint, and in such cases the latticed character of the shell is quite visible.

Mr. Fox in a later paper* thus summarizes the evidence :— "These radiolarian rocks of Cornwall may be compared with similar rocks of S. Scotland and with those described by Rust from the Hartz, as well as those from the coast ranges of California, of Jurassic age or older. . . . It is evident from these examples that in the process of the formation of chert the finer structures and the more delicate forms of the microscopic organisms disappear nearly entirely, so that it is but rarely that traces of them are now to be seen in the older cherts."

---

* "The Radiolarian Cherts of Cornwall." Trans. Roy. Geol. Soc. Corn. read Nov 8, 1895.

2. Localities and Geological Horizons of Radiolarian
Rocks in New South Wales.

With the exception of the opal rocks which contain numerous
spherical casts, possibly of radiolaria, all radiolarian rocks at
present known in N S. Wales are of Palæozoic age. Radiolarian
rocks have so far been discovered by me in N.S. Wales at four
different localities—(1) Bingera, (2) Barraba, (3) Tamworth, (4)
Jenolan Caves. (See Map, Plate xl., fig. 3.)

Devonian. (?)—(1) Bingera and (2) Barraba. In my Address* to
this Society in 1894, I stated "in the New England District of
N S. Wales possibly the red jasperoid shales of the Nundle and
Bingera Districts with the associated serpentines may represent
altered abysmal deposits, as has been suggested by Captain
Hutton for similar rocks in the Maitai Series of New Zealand,
unless the red claystone represents rock locally metamorphosed
where in contact with the serpentines."

Since reading the above Address, as opportunity offered, I have
from time to time studied the red jaspers of Barraba and Bingera,
by means of microscope sections. These revealed the presence of
numerous spherical bodies composed of translucent chalcedony,
distributed through an opaque groundmass of red jasperoid material.
It appeared probable that these were internal casts of radiolaria,
but the evidence was inconclusive. Last January, through the
kindness of Mr. J. J. H. Teall, F.R.S., I was allowed to examine
his carefully prepared microscopic sections of the Lower Silurian
radiolarian cherts from Mullion Island, off Cornwall, and from
the Culm of Devonshire, as well as sections of red radiolarian
jasper from the Antarctic regions. It was at once obvious that
the last mentioned rock in particular closely resembled the
Bingera and Barraba red jaspers. On my return to Sydney, last
March, with the help of the third year University students, I
resumed my examination of the New England red jaspers. Dr.

* P.L.S.N.S.W. Ser. 2, Vol. viii. p. 594.

G. J. Hinde had placed at my disposal, on leaving England, a valuable collection of British Palæozoic radiolarian rocks, which proved of the utmost use for purposes of comparison. A large number of sections of the red jasper proved conclusively that radiolarian rocks were developed on a large scale both at Barraba and Bingera. It is the opinion of Mr. E. F. Pittman, the Government Geologist, that the red colour of the jaspers was the original colour of the beds at the time of their deposition and that it is not due simply to contact metamorphism. A collection of specimens kindly made for me by Mr. Pittman confirms this theory. The question as to whether these red jaspers are altered "red clays" of deep sea origin will be discussed later. The geological horizon of the red jasper may be provisionally placed somewhere in the Devonian System, perhaps in the Middle Devonian, homotaxial with the Burdekin formation of Queensland.

Lepidodendron Australe occurs in some quantity in rocks which seem to be somewhat newer than the radiolarian beds; but it appears to be represented sparingly, almost, if not quite, as low down as the horizon of the radiolarian rock. This, however, is not yet an established fact.

(3) Tamworth.—Traced southwards, the radiolarian beds have recently been found by me to attain a remarkable development in the neighbourhood of Tamworth. They there consist of siliceous, dark bluish-grey, calcareous rocks, fine-grained blackish-grey claystones and cherts, and coralline siliceous limestone. The coralline limestone beds, of which there appear to be at least two, are from 100 to 1000 ft. in thickness, and are composed chiefly of the following fossils :—*Stromatopora, Cyathophyllum, Diphyphyllum Porteri, Cystiphyllum, Favosites gothlandica,* and *F. grandipora* or *Pachypora* (the latter very abundant and characteristic), *Alveolites* (also very abundant), and *Heliolites.*

Mr. Donald A. Porter, of Tamworth, conducted me to the spots where these limestones can be studied to best advantage, and he concurs with me in my provisional deductions with regard to the Tamworth rocks.

The limestones have been considerably altered by contact with
the New England granite.   The claystones and cherty rocks both
above and below the limestones have also been much altered by
innumerable granite sills for a zone over five miles in width,
measured at right angles to the junction line between the
sedimentary rocks and the granite.   A lamination, coincident
with the planes of bedding, has been superinduced in the claystones.   The sills vary from a fraction of an inch up to several
feet in thickness, and at first sight had every appearance of being
regularly interstratified with the sediments.   A careful examination, however, at once revealed their intrusive character, as they
trespass slightly across the planes of bedding and have slightly
altered by indurating and developing chiastolitic minerals, the
sedimentary rocks both above and below them.   The claystones
and cherts dip chiefly westwards at angles of from 45 to 60°.   At
Tamworth Common the dip is W. 20° S. at 52°.   Radiolaria are
abundantly distributed through these claystones and cherts in
the form of chalcedonic casts   Associated with the claystones is
the siliceous calcareous rock previously referred to.   A good
section shewing it *in situ* is exposed at the quarries on the Tamworth Temporary Common.   The chief bed is about 18 inches in
thickness.   It weathers superficially into a soft brown friable
rock of the colour of Fuller's earth, much resembling bath-brick.
Fresh fractures, of unweathered portions, shew the rock to be
bluish-grey and compact.   If a surface of the unweathered portion
be smoothed and polished and then etched with dilute hydrochloric or acetic acid, interstitial carbonate of lime is dissolved
out, and well preserved siliceous shells of radiolaria become visible.
These will be described in detail later.   A second bed of siliceous
radiolarian limestone occurs at a point about a mile easterly from
the preceding.   It is a few inches only in thickness.   For the
general appearance of this rock see Plate xxxvii.   The radiolarian
rocks are probably at least 2000 feet thick at Tamworth.   The
distance from Bingera on the north to Tamworth on the south is
85 miles.   Barraba, intermediate between these two places, is 34
miles south of Bingera and 51 miles north of Tamworth.   The

radiolarian rock is almost certainly continuous from Bingera to Tamworth.

(4) Jenolan Caves —This locality is about 200 miles south by west from Tamworth.   The rocks developed in this neighbour-hood are the Cave Limestone, thin grey argillites and dark grey and reddish-purple shales and black cherts with numerous dykes and sills of quartz-felsite, and basic dykes rendered porphyritic by augite.   The Cave Limestone is a somewhat massive rock from 380 to 420 feet in thickness.   Stratification is well marked at its upper surface.   It dips W. 10° S. at 60° as shown by me this year in my Address to the Royal Society of N S. Wales, Plate II.

The following fossils have been recorded as occurring in it by Mr. R. Etheridge, junr.*.— *Pentamerus Knightii*, J. Sowerby; *Palæoniso Brazieri*, Eth. fil.; *Loxonema antiqua*, De Kon., and a large *Favosites*.

Mr. Etheridge considers that the occurrence of the large varieties of *Pentamerus Knightii* in this Cave Limestone renders it not improbable that it approximates in age to the Aymestry Limestone of England.   At the same time he comments on the fact that *Pentamerus Knightii* has not yet been discovered in the Yass beds of N S. Wales, the horizon of which is almost certainly Upper Silurian, and *Mucophyllum crateroides*, a very characteristic and abundant coral in the Yass beds has not yet been observed in the Jenolan Cave Limestone.   *Stromatopora*, on the other hand, is very abundant, as it is in the Tamworth Limestone.   On the whole, I am of opinion that the Jenolan Cave Limestones and their associated radiolarian beds are somewhat newer than the Yass beds, so that if the Yass beds are Upper Silurian, the Jenolan Cave Limestones may be of Lower or Middle Devonian Age.   Immediately overlying the limestone are fine-grained dark clay shales and argillites and black cherts.   Mr. Voss Wiburd, the guide to the caves, informs me that these must be at least

---

* Records Geol. Surv. N S Wales. Vol. III. Part ii. 1892, p. 57, and Annual Report Dep. Mines, N.S. Wales, 1893, p. 128. By authority. Sydney, 1894.

1000 feet in thickness. They are capped by basalt. Near their junction with the limestone they are seen to be very much intersected by eruptive dykes, porphyritic by augite. It may be inferred from the circumstance that nearly all the dykes to the east of the limestone are felsitic, while no felsite dykes occur to the west of the limestone, that the basic character of the former group of dykes is due to the eruptive rock having assimilated much lime in its passage through the limestone bed, for as the dip of the limestone is westerly at an angle of 60°, and the dykes are nearly vertical, they could not have reached the surface without first passing through the limestone bed The dark shales are not distinctly cherty except where they are in close proximity to the dykes. The cherty character of the beds in this case is due therefore, I think, to contact metamorphism rather than to silica derived from radiolarian shells. Both the black cherts and the softer and less siliceous dark grey shales abound in casts of radiolaria. The casts are in the best state of preservation in the cherty bands Below the Jenolan Cave Limestone are several hundred feet of dark indurated shales, greenish-grey argillites, reddish-purple shale and coarse volcanic agglomerates with large lumps of *Favosites*, *Heliolites*, &c. The argillites and grey shales contain numerous casts of radiolaria, but in a very bad state of preservation.

3. MACROSCOPIC AND MICROSCOPIC DESCRIPTION OF THE RADIOLARIAN ROCKS.

The radiolarian rocks from Bingera and Barraba are hard red jaspers, the base of which is very opaque even in thin section. In places the red jaspers pass into a nearly white quartzite. Such portions of the rock as approach quartzite and chalcedony in character show scarcely any trace of radiolaria, probably owing to the shells having been completely dissolved during the metamorphism of the rock. The opaque red jaspers, however, especially those which have not undergone much metamorphism, contain very abundant casts of radiolaria, so abundant as to make it

38

evident that the radiolaria must in this case have contributed
very largely to form the rock.

Under the microscope numerous spherical or oval bodies, from
·05 mm. to ·215 mm. in diameter, are seen to be distributed through
the base.   The outlines of the larger casts are jagged, the projecting points representing casts in chalcedony of the openings in the
original latticed shell.   Most of the smaller casts are probably
those of the medullary shell.   The larger casts very frequently
occur in pairs.   Only in one instance was the original outer shell
of a radiolarian organism noticed.   It was separated by an inner
ring of red jasper from the cast of the medullary shell.   The
form appeared to be allied to *Carposphæra.*   Some of the largest
of the casts, about ·215 mm. in diameter, are probably referable
to *Cenosphæra.*   Many of the radiolarian casts have participated
in the numerous minute faults to which the rock has been sub-
jected.   The Tamworth radiolarian rocks, as already mentioned,
are partly thin siliceous limestones, partly argillites and black
cherts, partly massive coralline limestones.

The black cherts do not appear to owe their silica entirely to
the radiolaria, but to have derived it largely from the thousands
of granitic sills with which they are so regularly intersected as to
give the appearance of interstratification.

The casts of radiolaria in these cherty argillites are much
better preserved than those in the red jaspers, and also than
those in the black cherts of Jenolan.

Many of them show distinct traces of the latticed structure of
the shell.   The radiolaria, however, are in a far better state of
preservation in the thin siliceous limestones, which weather into
a kind of "rottenstone."   On the weathered surface of this rock
the radiolaria can be very easily distinguished with a pocket lens
Thin sections of the rock do not show much of the structure of
the shells under the microscope on account of the difference in the
respective refractive indices of quartz and calcite being insufficient
to show up plainly the structure of the radiolarian shells.   The
best results were obtained by thinning slices of the rock to the
thickness of the full diameter of the larger radiolarian shells, and

then etching the slice with dilute hydrochloric acid. Much of the structure can be developed in this way as shown on Plate xxxvii., from a microphotograph kindly taken for me by Mr. W. F. Smeeth, M.A., B E., Assoc. R.S.M.

As I have forwarded some of this material to Dr. Hinde, who has kindly undertaken to describe the radiolaria specially, I will not attempt to do more than mention that some of the commonest forms in the Tamworth rock are figured on Plate xxxviii.

It is obvious that the legion of the Spumellaria is much better represented than that of the Nassellaria. Fig 7, Plate xxxviii. appears to represent a *Xiphosphœra*, but the spines appear to be perforated by openings, giving the shell somewhat the appearance of *Pipettetella* (Challenger Reports, Radiolaria, Vol. xviii Pl. 39, Fig. 6). Fig 2 shows the inner and outer shells fairly well preserved, and is probably a *Haliomma*. Fig 5 perhaps represents a *Theodiscus*; and Fig. 9 perhaps a *Staurolonche* or an *Astromma*.

As regards the state of preservation of the shells the original siliceous skeleton is for the most part represented, but is sometimes replaced by iron pyrites. Often internal casts alone, in chalcedony, are all that remain to tell of the former presence of the radiolaria. Spicules of hexactinellid sponges are visible in places, in this rock The radiolaria are so abundant as to give this rock, when etched, the appearance of a Barbadoes earth. It was probably in its original condition a radiolarian ooze.

At the Jenolan Caves, as already stated, the radiolarian casts are best preserved in the black cherts, where they are very numerous. Numerous traces of radiolaria can also be detected in the soft argillites and hardened clay shales

The radiolarian casts are in a better state of preservation in the black cherts than in the red jaspers of Barraba and Bingera. Latticed structure is, however, scarcely anywhere to be seen. Such slight traces of it as do occur are preserved in the form of opaque black fragments of network entangled in a sub-translucent cryptocrystalline base, as seen in thin sections under the microscope.

Casts of the inner and outer shells are well preserved in the form of a nucleus of translucent chalcedony separated by a zone of the grey base from an outer ring of clear chalcedony.

Radial spines are indistinctly visible in many of the specimens, and can be seen best under crossed Nicols. Most of the casts are spherical, and vary in diameter from ·05 mm. to ·2 mm

Internal casts of the medullary shell are more frequent than casts of the outer shell.

Sponge spicules were not observed.

4. SUMMARY.

The radiolarian rocks, as yet discovered in New South Wales, range for at least 285 miles, from the Jenolan Caves on the south to Bingera on the north. Their total thickness has not yet been ascertained, but at Tamworth it appears to amount to at least 2,000 feet, and at Jenolan to not less than 1,000 feet The radiolarian rocks consist of red jaspers, black cherts, thin siliceous limestones, and thin bedded argillites. The radiolaria hitherto discovered are in the best state of preservation when enclosed in the siliceous limestone. For the most part, however, they are represented merely by chalcedonic casts, the casts of the medullary shell being more frequently preserved than those of the outer shell. In the thin siliceous limestones of Tamworth the radiolarian shells frequently have the original substance of the skeleton fairly well preserved in the form of sub-translucent to translucent silica. Rarely the original siliceous skeleton is found to be replaced by iron pyrites. In the Jenolan Cave Cherts the radiolarian skeletons show obscure traces of latticing in the form of fragments of opaque black nets.

At Tamworth and Jenolan the radiolarian rocks have beds of coralline limestone interstratified with them, probably over 1,C00 feet thick at the former, and over 400 feet thick at the latter locality.

At the Jenolan Caves a volcanic agglomerate containing blocks of coral is associated with the radiolarian shales.

The associated fossils prove the radiolarian rocks, at Tamworth at all events, to be homotaxial with the Burdekin Formation of Queensland. Mr. R. L. Jack, the Government Geologist of Queensland, and Mr R. Etheridge, Junr., consider the age of the Burdekin beds to be Middle Devonian.

5. DEDUCTIONS.

(i) In New South Wales there is a great development of rocks, chiefly argillites, cherts and jaspers, formerly considered to be unfossiliferous, but now proved to be formed largely of the shells of marine organisms, the radiolaria.

(ii) The geological horizon of these rocks is probably Middle or Lower Devonian, perhaps Siluro-Devonian.

(iii) The cherty character of some of the rocks containing the radiolarian casts is due rather to the introduction of silica secondarily from eruptive dykes and sills than to the silica contained in the radiolarian shells.

(iv.) The preservation of the radiolarian casts in the black cherts is chiefly due to the silicification and induration superinduced by contact metamorphism.

(v.) This contact metamorphism took place some time between the close of the Carboniferous Period and the commencement of the Permo-Carboniferous Period, and was the result of the intrusion of sills and dykes of granite

(vi.) (a) The presence of thick beds of coralline limestone interstratified with the radiolarian rocks, and (b) the vast thickness of the radiolarian beds (several thousand feet being formed within a single epoch of one period of geological time) render it improbable that the rocks were formed in very deep seas. This agrees with Professor Sollas' recent observations on the 'Soapstone' of Fiji, considered by Brady to be of deep sea origin, but now proved to have been deposited in shallow water. At the same time the absence of conglomerates (with the exception of the volcanic agglomerate at Jenolan) from the radiolarian beds and the abundance of interstratified limestone indicates deposition in tranquil water at some distance from the shore.

(vii.) The red jaspers of Barraba and Bingera may possibly be of deep sea origin, and represent consolidated " red clays," but this is not as yet proved.

My thanks are specially due to Dr. G. J. Hinde for the very valuable collection of radiolarian rocks which he has given me for comparison. I am also much indebted to Mr. J. J. H. Teall and to Mr. Howard Fox, as well as to Mr. Voss Wiburd, of Jenolan Caves, and to Mr. Donald A Porter, of Tamworth.

I would also beg to acknowledge the kind assistance given me throughout the year in the preparation of thin slides of the radiolarian rocks by the following students :— Alice Cooley, Isabella E. Langley, Marion C. Horton and Bertha V. Symonds. I have

NOTE ON THE OCCURRENCE OF CASTS OF RADIO-LARIA IN PRE-CAMBRIAN (?) ROCKS, SOUTH AUSTRALIA.

By Professor T. W. Edgeworth David, B A., F.G S.,
and Walter Howchin, F G.S.

Postscript to

Note on the Occurrence of Casts of Radiolaria in Pre-Cambrian (?) Rocks, South Australia. By Professor David, B.A., F.G.S., and Walter Howchin, F.G.S. (p. 571).

Since this paper was written the authors, when examining a supposed Pre-Cambrian Area at Normanville, about 35 miles southerly from Adelaide, discovered a great number of *Archæocyathinæ* in a thick bed of limestone previously supposed to be unfossiliferous. This limestone dips at from 60° to over 80°, and appears to be conformable to strata which must resemble those in which the radiolarian casts have been observed at Crystal Brook and Brighton in South Australia. This discovery renders it highly probable that most of the rocks in the Mt. Lofty Range, in some of which the radiolarian casts have been found, will prove to be Lower Cambrian or referable to passage beds at the base of the Cambrian rather than Pre-Cambrian.

(To face p. 570.)

2. BIBLIOGRAPHY.

Previous to our discovery of radiolaria in Pre-Cambrian (?) rocks in South Australia, we are not aware that any undoubted radiolaria have been observed elsewhere in rocks having so high a geological antiquity, unless an exception is made in the case of those recorded and figured by M. L. Cayeux,* from the Pre-Cambrian graphitic phthanites of Brittany.

M. L. Cayeux refers the radiolaria to no less than nineteen genera, in which both *Spumellaria* and *Nassellaria* are well represented. He states that the predominant genus is *Cenosphæra* The 45 figures given in his plate, drawn by an artist who had never figured radiolaria, but who simply drew what he saw, are certainly extremely suggestive of the radiolarian types to which he refers them, Pl. xi, fig. 1a, in particular, having a decided organic appearance.

Dr. G. J. Hinde† has reviewed this paper by M. Cayeux.

He comments specially on the exceedingly small size of the radiolaria, ·001 to ·022 mm. in diameter.

He says (*op. cit.* p 418), "The difference is very striking under the microscope, and it may be expressed by the fact that the average diameter of the 44 figured forms of which the dimensions are given is ·0115 mm., whilst the average diameter of 44 of the Palæozoic Radiolaria figured by Dr. Rust (taking the 44 species first described) is ·2 mm.; thus it would require the combined diameters of 17 of the Pre-Cambrian bodies to reach the average diameter of one of the Palæozoic Radiolaria."

Dr. Rust, on the other hand, is inclined to refer the forms figured to detached chambers of foraminifera, related to some genus allied to *Globigerina.* It is clear from these criticisms

---

* Les preuves de l'existence d'organismes dans le terrain pré-cambrien. Première note sur les Radiolaires pré-cambriens, in Bull Soc. Géol. Fr. 3ᵉ Série, t. xxii., pp. 197-228, pl. xi. (1894). See also C.R. Ac. Sc., 3ᵉ Série, t. xxii., p. lxxix.

† Geol. Mag. New Series.—Dec. iv. Vol. i. No. 9. September, 1894, pp. 417-419.

that some of the leading authorities on the radiolaria are not convinced as to the structure of the forms figured by M. L Cayeux being correctly referred to the above group, and his further descriptions of the Brittany rocks are anxiously awaited. Reference may be made here to what have been described as other micro-organisms associated with the Pre-Cambrian radiolaria, or occurring alone.

M. L. Cayeux has described and figured what he believes to be foraminifera from Pre-Cambrian rocks at Saint Lô, at Lamballe (Côtes-du-Nord).*

He has also recorded the occurrence of remains of sponge spicules in the Pre-Cambrian rocks of Brittany.†

These were found by M. Ch Barrois, who also discovered the radiolaria in the Pre-Cambrian rocks of Brittany, from Ville-au-Roi, near Lamballe. These remains are in the form of monaxial spicules, some being probably referable to the *Monactinellidæ*. Others M. L. Cayeux refers respectively to the *Tetractinellidæ*, *Lithistidæ*, and *Hexactinellidæ*. The spicules are from ·05 mm to ·35 mm. in length, mostly ·1 mm. to ·15 mm. The spicules are replaced by pyrites : the particles of pyrites are held together in a siliceous setting. The canal is not preserved.

The occurrence of spicules of fossil sponges in Archæan rocks has been recorded by Mr. G. F. Matthew.‡

These are referred to *Cyathospongia* (?) *Eozoica*, and to *Halichondrites graphitiferus*. They are stated to occur in Upper Laurentian rocks.

The authenticity of these remains has been called in question by Mr. Herman Rauff.§

* C. R. Ac. Sc. Janvier-Juin 1894, pp. 1433-1435.
† Société Géologique du Nord. Annales xxiii. 1895, pp. 52-64. pls. i.-ii. L. Cayeux.—De l'existence de nombreux débris de Spongiaires dans les phthanites du Pré-Cambrien de Bretagne. C.R. Ac. Sc. T cxx. pp. 279-282.
‡ On the Occurrence of Sponges in Laurentian rocks at St. John, N.B. Bull. Nat. Hist. Soc. New Brunswick, No 9, pp. 42-45.
§ H. Rauff. *Ueber angebliche Spongien aus dem Archaicum*, Neues Jahr. für Min., Geol. und Pal. II. Bd. 1893, pp. 57-67, and *Palæospongiologie*, Palæontographica, 1893, Bd. 40, p. 233.

If *Eozoon Canadense* and allied forms be left out of consideration, the above comprise, as far as we are aware, references to all the
more important papers relating to the microzoa of the Pre-Cambrian Rocks.

3. DESCRIPTION OF THE RADIOLARIA.

Obviously the two most important points to be proved in this
note are (*a*) that the supposed organisms are referable to radiolaria,
and (*b*) that the rocks which contain them are of Pre-Cambrian
Age.

If direct proof of the first is wanting, the question as to the
age of the rocks does not so much matter.   We shall, therefore,
proceed first to quote evidence which, in our opinion, is strongly
in favour of the structures about to be described being referred
to the radiolaria, and afterwards we will deal with the question
of the geological horizon of the rocks which contain the radiolaria

Traces of the organisms referred by us provisionally to the
radiolaria occur at two localities, (*a*) Brighton, about 10 miles
S.S.W. from Adelaide; and (*b*) Crystal Brook, about 140 miles
N. of the same city.   At (*a*) Brighton the forms provisionally
referred to the radiolaria occur scattered in great numbers
throughout a greenish siliceous limestone   This limestone in
places exhibits well marked oolitic structure.

Thin sections of these rocks prepared by the students at the
Geological Laboratory, at the University of Sydney, show that
these supposed casts of radiolaria are partly chalcedonic and
opaque, partly replaced by lime and translucent   The latter
types are invested in places with a black network, chiefly composed of iron pyrites, the intimate structure of which is hard to
determine.   Casts of what we consider to be the medullary shells
are most frequent, and are best preserved.   A careful examination,
however, of the material surrounding these spherical translucent
bodies frequently reveals the presence of an outer nebulous ring,
sometimes showing a denticulated margin in cross section.   (See
Pl. xxxix. figs. 5-6.)   That these bodies are radiolarian casts and

not spherulites nor oolitic granules, is rendered probable by the following facts :—

(1). In the Pre-Cambrian oolitic limestone of Hallett's Cove the nuclei of the grains are shaped irregularly, whereas the small translucent bodies inside the nebulous rings in the Brighton lime-stone are perfectly round or oval, and in some cases spinous.

(2). Distinct black netted material envelopes the spherical or oval bodies

(3). The translucent material enclosed inside the rings does not show a dark cross, seen in polarised light, though, even if it did, this would not of course be an insuperable objection to its radiolarian origin.   It proves, however, conclusively that they are not spherulites.

(4). They are probably not oolitic grains, not only on account of many of them possessing an external black network, but also because they are of exactly the same shape, size, and structure as similar bodies in the Pre-Cambrian cherts of Crystal Brook, and oolitic structure, as far as we know, has not been observed in cherts.

(5). Many of the casts very closely resemble those of Mullion Island, Cornwall, and those of the Jenolan Caves and of Bingera in New South Wales.

A considerable variety of forms appear to be present, most of which seem to belong to the Legion *Spumellaria*.

Figs. 5-6 of Pl. xxxix. exhibit forms resembling *Carposphæra*, or possibly *Cenosphæra* with the internal cavity partly filled with chalcedony.

Fig. 7 of Pl. xxxix is suggestive of the genus *Cenellipsis*. It is possible, however, that the netted forms like those in the figures last referred to, are of inorganic origin, the pyrites filling in the interspaces between small crystalline aggregates partly of silica, partly of calcite

The spherical chalcedonic bodies, surrounded by the outer chalcedonic rings, appear to us, however, to be very probably casts of the medullary and cortical shells of radiolaria   The diameters of these bodies vary from ·1 mm. up to ·22 mm

(b) *Crystal Brook.*—In the black chert of Crystal Brook, the radiolarian casts are chiefly in the form of small spherical or oval nuclei of chalcedony, with a more or less distinct partially translucent outer ring of chalcedony.   Much black opaque matter is present in this rock, as well as small spherical developments of iron pyrites, very suggestive of being inner casts of radiolaria.

The Crystal Brook forms, as to the radiolarian character of which we think there can be very little question, are shown on Figs. 1-3 of Pl. xxxix   Their diameter varies from ·1 mm. to ·2 mm.   Figs. 1-3 are very suggestive of forms allied to *Carpo-sphæra.*

4. Geological Horizon of the Radiolarian Rock.

As already stated, the two chief localities in South Australia where the supposed radiolarian casts have been met with are (a) Brighton and (b) Crystal Brook.   These localities merit separate descriptions.

(a) Brighton.—The rocks from Brighton which have yielded the casts above referred to were taken from the quarries of the South Australian Portland Cement Company, situated at Brighton, about 10 miles S S.W. from Adelaide, on a spur of the Mt. Lofty Ranges, which at this point describe a curve to the seashore, marking the southern boundary of the Adelaide plains

The limestones worked by this company form outcrops rising from beneath the Pliocene clays of the plain, and can be traced for miles over the low hills to the south in a line almost parallel to the coast.   The workings extend at intervals for a distance of about 200 yards across the outcrop, and about a quarter of a mile along the line of strike.   The succession of beds can be easily traced, and is as follows, in descending order :—

1. *Buff-coloured Limestone.*—The uppermost bed exposed in the workings.   It is very persistent and maintains its characteristics for a long distance.   Distinguished by its colour, contains a considerable proportion of magnesium carbonate, is very tough and hard.   This bed is not quarried for cement, and marks the horizon

above which no limestones, serviceable for cement or lime, are met with.

2. *Pink-coloured Limestone.*—This bed is sharply defined from the preceding by a bedding plane. It is about 15 feet in thickness, of a pale pinkish colour, and carries about 86 per cent. of carbonate of lime—the purest limestone in the group. The weathered faces of the vertical joints exhibit lines of false bedding.

3. *Blue siliceous Limestone.*—This immediately underlies the pink-coloured limestone, and in the upper portions of the bed is frequently mottled by various sized pinkish patches. It contains forty per cent. or more of silica. The pink-coloured patches contain a lower proportion of silica and correspondingly higher proportion of carbonate of lime, than the distinctly blue limestone.

4. *Very siliceous dark-coloured Limestone* of variable composition, but carrying more silica than No. 3. This bed, as well as the one immediately above it, is strongly laminated. Whenever this feature is present it is said to be an indication of a high proportion of silica in the stone. This limestone is the lowest horizon worked for cement, but the stone used by the company is chiefly won from beds Nos. 2 and 3. Immediately above this bed is a calcareo-siliceous shale of very close texture.

The beds have a strike about N. 12° E. The dip varies from about 50° to 80° in a direction about W. 12° N. These Brighton rocks may be considered the foothills of the Mt. Lofty Range, towards and under which they appear to dip. Whatever, therefore, be the age of the Mt. Lofty Range, the Brighton rocks will prove to be of at least as high a geological antiquity.

The Mt. Lofty and associated ranges form the backbone of the southern portions of South Australia, from Lake Eyre to Kangaroo Island. In the neighbourhood of Adelaide, the western flanks of the ranges show alternations of clay-shales (often micaceous or chloritic), quartzites, and siliceous limestones, with an average dip of about 45°, and are considerably folded. At Hallett's Cove, about five miles south from Brighton, several sharp anticlinal folds occur near the coast and in the gorge of

Field River.   A few miles further south the rocks forming the
sea cliffs are contorted and overthrust from E. to W. in a very
striking manner.   If the coastline be followed to Normanville,
48 miles south from Adelaide, the crystalline and highly metamorphic beds of the eastern flanks of the ranges are met with.
The marked lithological distinction between the western and
eastern sides of the Mt Lofty Ranges is an interesting feature.
The greater part of the ranges, including the western flanks and
highest portions of the watershed, show a series of sedimentary
rocks metamorphosed to only a slight degree, with a general
easterly dip at a steep angle of from 40° to 80°.   The eastern
flanks are composed of highly crystalline metamorphic rocks,
felsites, hornblendic and micaceous schists, gneiss and granites,
which give distinctive features to this side of the ranges for over
200 miles in length.   Intrusive granites are extensively associated
with this zone of extreme metamorphism.

Professor R Tate * regards the Mt. Lofty Ranges throughout
their entire width as forming one great conformable system, the
aggregate thickness of which he estimates cannot be less than ten
miles.   Further, as the dip of these beds is in the main a south-
easterly one, it follows upon the above assumption that the highly
crystalline rocks of the eastern side of the watershed are actually
superimposed on the less metamorphosed shales, limestones, and
quartzites of the western portions   If this reading of the stratigraphical features be the correct one, the Brighton limestones
must rank amongst the oldest rocks exposed in the Mt. Lofty
series, as shown on Fig 1, Plate XL.

The geological age of these old rocks is a subject of great
interest.   Selwyn, and other early observers, regarded them as
Silurian, although the entire absence of fossils from the series
left the question an open one.   The discovery by Mr. Otto
Tepper and Professor R. Tate in 1879† of a fossiliferous horizon
near Ardrossan, Yorke's Peninsula (subsequently determined by

---

* Presidential Address Aust Assoc. Ad. Sc. Vol. V. (1893), p. 47, et seq.
† Trans. Philosop. (Royal) Society S. Aust. Vol. ii. 1879, p. 71.

Mr. R. Etheridge, Junr., to be of Cambrian age),[*] resting unconformably on an older series of mica slates and talcose schists, supplied new data bearing on the possible age of the Mt. Lofty formation. The basal or Pre-Cambrian beds at Ardrossan, exhibit a close lithological resemblance to many portions of the Mt. Lofty series, and may provisionally be considered to be homotaxial with the latter. Unfortunately, in no other place in South Australia, that we know of, are the Cambrian and Pre Cambrian rocks seen in juxtaposition, but they have been observed in the Flinders Ranges in close proximity to the Pre-Cambrian rocks, and it has been noticed that the two groups exhibit strongly marked lithological differences as well as probable unconformity (Pl. XL fig. 2).

Prof. R Tate has for many years advocated the Pre-Cambrian (or Archæan) age of the Mt. Lofty formation.[†] The chief considerations for this view are based on—

(a) The evidence afforded by the unconformity between the Lower Cambrian and the Pre-Cambrian rocks near Ardrossan, and the general resemblance of the inferior rocks of that section to the Mt. Lofty beds (Pl. XL fig. 1), (and so to the Brighton rocks).

(b) In the Flinders Range two formations have been noted (although not seen in contact) in which the less altered beds with lower angle of dip have been determined by their included fossils (Archæocyathinæ, Olenellus, Salterella, &c.) to be Cambrian; and it has been inferred that the more highly metamorphic rocks with higher angle of dip are unconformable and consequently Pre-Cambrian. The Mt. Lofty beds are continuous with those of the Flinders Range.

(c) The absence of fossils (macroscopic) throughout the whole of the Mt. Lofty series, even in places where limestones and shales occur so little metamorphosed that we have no reason to think that organic remains, if originally present, have been obliterated by molecular rearrangement.

---

* Roy. Soc. S. Aust. 1890, p. 10, and R. Tate ibidem 1892, pp 183-189.
† Roy. Soc. S. Aust. Vol. xiii. 1890, p. 20: Aust. Assoc 'Ad Sc. Op. cit. ante.

Mr. H. Y. L Brown, Government Geologist of South Australia, holds, however, a somewhat different view from the above. Mr. Brown considers that the low degree of metamorphism present in the rocks of the western flanks of the Mt. Lofty range indicates an age not earlier than the Cambrian, and that the Flinders and Mt. Lofty beds really form one series. In his official Geological Map of South Australia, published in 1886, Mr. Brown recognises three older formations in the ranges, as follows :—

(1). PALÆOZOIC (LOWER SILURIAN).— Comprising the less altered shales, sandstones, and limestones of the western portions.

(2). PALÆOZOIC, or AZOIC.—The micaceous, talcose, and hornblendic schists, quartzites and crystalline limestones —a middle series towards the eastern side of the ranges

(3) ARCHÆAN.—Metamorphic granite, gneiss, syenite, hornblendic and mica schists, crystalline limestones, quartzites, &c , with igneous intrusions, rising beneath group No. 2 on the eastern flanks.

It will be observed from this table that the succession is interpreted by Mr Brown in an opposite way from that in which it is explained by Prof. Tate, for whilst the latter considers the highly metamorphic group the highest in the series, Mr. Brown places this group at the base.

On the whole it appears to us that Professor Tate's interpretation is probably the correct one, and if so the Brighton rocks must be low down in the Pre-Cambrian group.

(b). Crystal Brook.—The rocks containing the casts of radiolaria, at this locality, are thin laminated limestones, sandy calcareous layers alternating with thin bands richer in lime. Quartzite and banded argillites overlie the laminated limestones. Lenticular beds of black chert or chalcedony occur on at least 15 horizons in the limestone series. They appear to be of later origin than the enclosing rocks, like the flints in the Chalk Formation of Europe. The portion of the limestone series measured by us is at least 1000 feet in thickness. The series is highly folded, and vertical dips are not uncommon.   We think it probable on this account, as well as on account of its lithological character, that this series is also Pre-Cambrian, perhaps on about the same horizon as the siliceous limestones exposed in the vineyards at Burnside, near Adelaide.   Moreover, no macroscopic fossils have been observed by us in these limestones, in spite of their having suffered extremely little through metamorphism, whereas the local Lower Cambrian limestones are abundantly fossili- ferous, and only slightly inclined, without distinct folding.   At the same time, the fact must be mentioned that the Crystal Brook radiolarian locality lies directly in the trend of the Cambrian rocks from Yorke's Peninsula N. by E towards the Blinman Mine to the N.N.E of Port Augusta.   On the whole, however, we think that the evidence is in favour of the radiolarian rock at Crystal Brook being Pre-Cambrian.

5. Summary and Provisional Deductions, &c.

(i.) At Brighton and Crystal Brook in South Australia (their respective positions are shown on Pl. XL. fig. 3), rocks are developed which contain what appear to be casts of radiolaria At the latter locality there can be little doubt, in our opinion, as to the identity of the casts with those of radiolaria.

(ii.) That the age of these rocks is Pre-Cambrian is rendered highly probable by the following considerations :—

(a). The local Lower Cambrian rocks are gently inclined at angles of from 8° to 15°, and they are not folded, whereas the radiolarian rocks dip at 45° to 80°, are considerably folded, and seem to underlie unconformably the Lower Cambrian formation.

(b) The Lower Cambrian rocks of South Australia are pure and massive pteropod limestones, whereas no such beds of pure thick limestones are to be noticed in the radiolarian group.

(c) The Lower Cambrian limestones of South Australia contain a rich and abundant macroscopic marine fauna, whereas no macroscopic fossils have ever been found amongst the Brighton and Crystal Brook radiolarian rocks, although the rocks at both

39

these localities are very well adapted for preserving macroscopic fossils, had they ever existed in them.

(iii.) The evidence on the whole is decidedly in favour of the existence of radiolaria in Pre-Cambrian rocks in South Australia.

(iv.) Such radiolaria appear to differ very little in size from the forms described from Palæozoic, Mesozoic, Tertiary and Post-Tertiary rocks, as their diameters appear to range from about ·1 to ·22 mm.

(v.) Forms allied to *Carposphæra* and *Cenosphæra*, and possibly to *Cenellipsis*, appear to have been represented in Pre-Cambrian time.

We desire to express our thanks to Mr. Stanley Fraser, the manager of the South Australian Portland Cement Company, at Brighton, who has kindly given all the help in his power to facilitate our researches at Brighton. We have also to thank Mr. W. Lewis, of Brighton, for kind guidance and assistance. To Mr. J. W. Jones, the Conservator of Water, we are much indebted for the excellent arrangements which he made for our geological examinations of Crystal Brook and Ardrossan. We also desire to thank for much useful aid given us in the field the following: Mr. Hicks, Mr. C. C. Buttfield and Mr. E. S. A. Willis. Mr. W. S. Dun, the Librarian and Assistant Palæontologist to the Geological Survey of N.S. Wales, we also desire to thank for having obligingly supplied us with most of the references quoted in the bibliography.

———

EXPLANATION OF PLATES.

Casts of Radiolaria from Pre-Cambrian (?) Rocks, Brighton and Crystal Brook, South Australia.

(All the figures × 200.)

PLATE XXXIX.

Figs. 1 and 3.—Internal cast of form perhaps allied to *Carposphæra*, from black chert, Crystal Brook.

Fig. 2.—Internal cast from Crystal Brook, genus not determinable.

Fig. 4.—Internal cast in siliceous limestone, perhaps referable to the Radiolaria; Brighton, near Adelaide.

Figs. 5 and 6.—Internal casts in siliceous limestone, perhaps related to *Carposphæra*; from Brighton, near Adelaide.

Fig. 7.—Form doubtfully referable to the Radiolaria. from siliceous limestone, Brighton, South Australia; possibly allied to *Cenellipsis*.

Fig. 8.—Internal cast in siliceous limestone, perhaps referable to the Radiolaria; Brighton, South Australia.

PLATE XL

Fig. 1.—Sketch Section from near Ardrossan, Yorke's Peninsula, to Murray Bridge, South Australia.

Fig. 2.—Section showing probable junction between the Lower Cambrian and the Pre-Cambrian Rocks near Ardrossan, Yorke's Peninsula, S.A.

Fig. 3.—Map showing positions of chief localities where fossil Radiolaria have been found in S.E. Australia.

Mrs. Kenyon contributed a Note in support of a contention that *Cyprœa caput-anguis*, Philippi, was entitled to independent specific rank, and should not be merged in *C. caput-serpentis*, Linn.

Mr. Brazier exhibited, for Mrs. Kenyon, a series of specimens of *Cyprœa* mentioned in her Note, namely, an adult specimen of *Cyprœa caput-anguis*, Philippi, from Maldon Island, and of its fine variety *C. Sophia*, Braz., as well as of a large variety, a small solid specimen of *Cyprœa tigris*, Linn., and a large but young specimen of the same species showing the spots in four rows of transverse bands. Also a young specimen of *C. tigris* received from Mrs. Waterhouse. Two specimens of a supposed new species of *Pectunculus*, from an unknown locality, were also exhibited.

Mr. Froggatt showed a large series of spirit specimens of the Termites treated of in his paper, together with slides of mounted wings, &c.

Professor David exhibited, in illustration of his paper, photographs, rock specimens, and, under the microscope, rock sections showing Radiolaria.

Mr. Ogilby exhibited specimens of two small Clupeids, and stated that from an examination of a number of specimens he was convinced of the necessity for forming a third genus of "Rough-backed Herrings." The three genera, will be described in full in an early number of the Proceedings. Mr. Ogilby proposes to segregate all the Rough-backed Herrings, recent and fossil, under the common name *Hyperlophinœ*, and points out that the name *Diplomystus* (Cope, 1877) is hardly tenable, Bleeker having used *Diplomystes* for a South American Nematognath in 1863. Bleeker's name—which was arbitrarily changed by Gunther to *Diplomystax*—is still in use and gives the title to the family *Diplomystidœ* of Eigenmann & Eigenmann.

On behalf of Miss Georgina King, Mr. Fletcher communicated several letters written during the last fortnight of September, accompanied by sketches, from Baron von Mueller, on the subject of *Boronia floribunda* referred to in a Note read at the last Meeting. The letters were expressive of the pleasure with which the Baron had seen for the first time specimens of the Boronia in question. These were obtained by Miss King from the Hawkesbury during last month, and forwarded to Melbourne. The species was described by Sprengel in 1827, from specimens obtained by Sieber in 1823, somewhere in the neighbourhood of Sydney or on the Blue Mountains. By Mr Bentham it was considered to be a dimorphic form of *B. pinnata*, but by Prof Urban of Berlin it has been restored to independent specific rank. As compared with *B pinnata* its chief distinguishing characters are that four of the eight stamens are shorter and have smaller anthers, the style is short, and the stigma large and globular. The wish was also expressed by the Baron that as the characters of the fruit are yet unrecorded, an effort might be made during the present season to obtain them for comparison with those of *B. pinnata*

Mr. Fletcher exhibited a series of water-colour drawings of Australian animals, of great intrinsic merit as well as of historical interest. They were the artistic work of Dr. J. Stuart, an army surgeon, who from time to time for some years (circa 1834-37 or even later) undertook the duties of Medical Officer at the Quarantine Station, Port Jackson. They are referred to in one of his papers (Ann. Mag. Nat. Hist. viii. 1842, p. 242) by the late Mr. W. S. Macleay, into whose possession they subsequently passed. Eventually they came to Sir William Macleay, who handed them over to the Society.

WEDNESDAY, NOVEMBER 25TH, 1896.

The Ordinary Monthly Meeting of the Society was held at the Linnean Hall, Ithaca Road, Elizabeth Bay, on Wednesday evening, November 25th, 1896.

The President, Mr. Henry Deane, M.A., F.L.S, in the Chair.

DONATIONS.

Pharmaceutical Journal of Australasia. Vol. ix. No. 10 (Oct., 1896). *From the Editor.*

Société d'Horticulture du Doubs, Besançon—Bulletin. Série Illustrée, No. 9 (Sept., 1896). *From the Society.*

Perak Government Gazette. Vol. ix. Nos. 22-24 (Sept.-Oct., 1896). *From the Government Secretary.*

Société Impériale des Naturalistes de Moscou—Bulletin. Année 1896. No. 1. *From the Society.*

U. S. Department of Agriculture—Division of Ornithology and Mammalogy—North American Fauna. Nos. 10 and 12 (1895 and 1896) : Division of Entomology—Technical Series. No. 3 (1896). *From the Secretary of Agriculture.*

Société Scientifique du Chili—Actes T. ii (1892). 5ème Liv.: T. vi. (1896), 1re Liv. *From the Society.*

Report on the Work of the Horn Scientific Expedition to Central Australia. Part i. Introduction, Narrative, &c. : Part iv. Anthropology. *From W. A. Horn, Esq., per Professor Baldwin Spencer, M.A.*

Proceedings of the American Association for the Advancement of Science. Vols. xxxviii and xl.-xliii. (1889 and 1891-94): Bulletin of the Essex Institute. Vol. i. (1869), Nos. 1-2, 4-6, and 12. ii. 3-5 and 7-9; iii. 3 and 8; iv. 9; v. 1-5 and 11-12; vi.; vii. 1-3, 5, and 9-12; viii.-ix.; x. 7-12; xi. 1-6 and 10-12; xii.; xiii. 19; xiv. 1-6; By-laws, 1876: Proceedings of the American Philosophical Society. Vol xi. (1870), No. 85; Vols. xii-xiv. (1871-75): Science. Vol. iii. No. 49 (Jan., 1884); Vol iv. No. 99 (Dec., 1884); Vol. v. No. 100 (Jan., 1885); Vol. vii. from No. 157 (Feb., 1886); Vols. viii-xxii (complete except title pages and indexes to Vols. xiii. xiv. and xviii); and Vol. xxiii. Nos. 570-581 (Jan.-March, 1894): Annual Reports of Geological Survey of (a) Indiana, ii.-viii. (in six vols.) [1870-78]; (b) Wisconsin, 1877; (c) New Jersey, 1887: Biennial Report of the State Mineralogist of Nevada for 1873-74: Tenth Annual Report of the California State Mining Bureau for 1890: Report of the Geological Survey of the Oil Islands of Japan (1877): General Report on the Geology of Yesso (1877); Report of the Geological Survey of Kentucky. Vol. v. 2nd Ser. Parts viii. and x.: Featherstonhaugh's Report of Geol. Reconnaissance made in 1835 to the Coteau de Prairie: Bulletin of U.S. Nat. Mus. No. 6 (1876). *From the Connecticut Academy of Arts and Sciences.*

Asiatic Society of Bengal—Journal. Vol. lxiv. (1895). Title page and Index to Part i: Vol. lxv. (1896). Part i. Nos. 1-2: Part ii. No. 2. Proceedings, 1896. Nos. ii.-v. (Feb.-May). *From the Society.*

Bombay Natural History Society—Journal. Vol. x. No. 3 (Sept., 1896). *From the Society.*

Johns Hopkins University—Hospital Bulletin. Vol. vii. Nos. 66-67 (Sept., 1896). *From the University.*

American Naturalist. Vol. xxx. No. 358 (Oct., 1896). *From the Editors.*

Victorian Naturalist. Vol. xiii. No. 7 (Oct., 1896). *From the Field Naturalists' Club of Victoria.*

L'Académie Impériale des Sciences de St. Pétersbourg–
Annuaire du Musée Zoologique, 1896. No. 3 *From the
Academy.*

Museo Nacional de Montevideo—Anales v. (1896). *From the
Museum.*

Zoologischer Anzeiger. xix. Band. Nos. 513-514 (Sept.-Oct
1896) *From the Editor.*

Konink Natuurk. Vereeniging in Nederl.-Indie—Tijdschrift·
Dl. ii. Afl. 6 (1851): Dl. iv. Afl 5 and 6 (1853): Dl. vi. Afl. 5
and 6 (1854): Dl. vii. Afl. 1-2 and 5-6 (1854): Dl viii. Afl. 1-4
(1855) Dl. ix. (1855): Dl. xvi. (1858-59): Dl. xvii. Afl 5 and
6 (1858) Dl. xx. Afl. 1-3 (1859): Dl. xxx. Afl 1 and 2 (1867)
Dl. xxxii. Afl. 4-6 (1873): Alphabetisch Register op Dl. i-xxx
(1871), xxxi.-l. (1891)· Naamregister op Dl. i.-xxx. *From the
Society.*

British Museum (Nat. Hist)—Catalogue of Birds. Vol. xxiv
(1896)· Catalogue of Snakes. Vol. iii. (1896): Catalogue of
Madreporarian Corals. Vol. ii. (1896): Catalogue of Jurassic
Bryozoa (1896) *From the Trustees*

Royal Society, London—Proceedings. Vol. lix. No. 358 (Sept·
1896). Vol. lx. No. 359 (Sept., 1896). *From the Society*

L'Acad. Royale Suédoise des Sciences—Bihang. Vol. xxi
(1895-96). Sections 1-4. *From the Academy.*

Revista de Sciencias Naturaes e Sociaes. Vol. iv. No. 16, and
Title page and Index (1896). *From the Editor.*

Pamphlet entitled "Note on the Discovery of Organic Remains
in the Cairns Range, Western Queensland." By R. L. Jack,
F G S., F.R.G.S. *From the Author.*

Société Géologique de Belgique—Annales. Tome xxiii. 2ᵉ
Livraison (1895-96). *From the Society.*

Societas pro Fauna et Flora Fennica—Acta. Vols vi. and
vii. (1889-90). *From the Society.*

Pamphlet entitled "Notes on Rare Lepidoptera in Wellington " By W. P. Cohen. *From the Author.*

Senckenbergische Naturforschende Gesellschaft, Frankfurt a.M.—Bericht, 1896. *From the Society.*

Journal of Conchology. Vol viii. No. 8 (Oct , 1896). *From the Conchological Society of Great Britain and Ireland.*

Entomological Society of London—Transactions, 1896. Part iii. (Sept). *From the Society.*

California Academy of Sciences—Memoirs. Vol. ii. No. 5 (4to Feb., 1896) : Proceedings. Second Series. Vol. v. Part 2 (Jan., 1896). *From the Academy.*

American Museum of Natural History, New York—Bulletin. Vol. viii. Sig. 6 (pp. 81-96. May, 1896) Twenty-seventh Annual Report (1895). *From the Museum.*

Wagner Free Institute of Science of Philadelphia—Transactions. Vol. iv. (Jan., 1896). *From the Institute.*

Cincinnati Society of Natural History—Journal. Vol. xviii. Nos 3 and 4 (Oct., 1895-Jan , 1896.) *From the Society.*

Field Columbian Museum, Chicago—Botanical Series. Vol. i. No 2 (Jan., 1896) : Report Series, Vol. i. No. 1. (Annual Report for 1894-95). *From the Director.*

Academy of Natural Sciences of Philadelphia—Proceedings, 1896. Part i. (Jan.-March). *From the Academy.*

Boston Society of Natural History—Proceedings. Vol. xxvii pp 1-6 (March, 1896) *From the Society.*

Naturwissenschaftlicher Verein für Schleswig-Holstein— Schriften. x. Band, 2 Heft (1895). *From the Society*

Gesellschaft für Erdkunde zu Berlin—Verhandlungen. Band xxiii. (1896), Nos. 1-3 : Zeitschrift. Band xxx. (1895), No. 6 : Band xxxi. (1896), No. 1. *From the Society.*

Zoologische Station zu Neapel—Mittheilungen. xii. Band. 3 Heft (1896). *From the Station.*

Cambridge Philosophical Society—Proceedings.   Vol. ix. Part 3 (1896) : Transactions.   Vol. xvi.   Part i. (Oct., 1896).   *From the Society.*

Geological Survey of Queensland—Bulletin.   No. 4 (1896). *From the Government Geologist.*

Bureau of Agriculture, Perth, W.A.—Journal.   Vol iii.   Nos. 25-26 (Oct.-Nov., 1896).   *From the Bureau.*

Three Conchological Pamphlets.   By E. A. Smith, F.Z.S. (1896).   *From the Author.*

Australasian Journal of Pharmacy.   Vol. xi.   No. 131 (Nov., 1896).   *From the Editor.*

Department of Lands and Survey, New Zealand—Report for the year 1895-96.   *From H. Farquhar, Esq.*

ON THE COMPARATIVE ANATOMY OF THE ORGAN OF JACOBSON IN MARSUPIALS.

By R. ' Broom, M.D., B.Sc.

(Plates XLI.-XLVIII.)

Although the researches of Gratiolet, Balogh, Klein, and others had made us familiar with the structure and relations of Jacobson's organ in a number of the principal types of higher Mammals, until very recent years no examination appears to have been made of the organ in any of the Marsupials.

In 1891, Symington published a paper "On the Organ of Jacobson in the Kangaroo and Rock Wallaby," in which he points out the main features of the organ and its relations, and gives figures of transverse sections at the opening of the organ and also at its most developed part. He concludes that the Marsupial organ agrees very closely with the Eutherian type, and differs markedly from that found in the Prototherian Ornithorhynchus. It is unfortunate that when his paper was written only the aberrant Platypus type had been carefully studied, for had he compared the Marsupial organ with the simpler Monotreme type as found in Echidna, his conclusion would probably have been different.

In 1893, Rose, apparently ignorant of Symington's work, published a very short paper on the organ in the Wombat and Opossum. He gives two good figures of the organ in the young Wombat, but makes no remarks on the peculiarities of the organ or its relations.

The only other papers, as far as I am aware, in which the Marsupial arrangement is touched on are, Symington's recent paper "On the Homology of the Dumb-bell-shaped Bone in

Ornithorhynchus," and some papers of my own where various references are made to points in the Marsupial anatomy for purposes of comparison.

In the present paper I shall confine myself mainly to the consideration of the general morphology of the organ and its duct, with their cartilaginous and bony relationships, and their vascular and glandular connections in typical members of the chief groups of Marsupials, and to the morphological significance of the various peculiarities met with   In discussing the various forms, I shall adopt tentatively the classification as given in Thomas' " British Museum Catalogue of Marsupials and Monotremes", and as the polyprotodont Marsupials have long been recognized as the more generalised—a view which is confirmed by the study of the region under consideration—it will be convenient to examine these first.

DASYURIDÆ.   (Plate XLI.)

Of this group I have studied, (1) Early mammary fœtal *Phascologale penicillata*, (2) mammary fœtal *Dasyurus viverrinus*, (3) two-thirds grown *D. viverrinus*, and (4) adult *D maculatus*

If a series of transverse sections be made of the anterior part of the snout of Echidna, it will be found that there passes out from each side of the base of the septum a flat cartilage, forming a floor to each nasal cavity.   In the very young animal, as shown by Newton Parker, this cartilage is well developed, but in the adult it only remains as a floor to the inner half of the nasal cavity On reaching the plane of the naso-palatine canal, this nasal floor cartilage is found to divide into an inner and an outer part.   The inner becomes the cartilage of Jacobson's organ, while the outer, much reduced just behind the region of the naso-palatine canal, on passing backwards becomes more developed and passes inwards below Jacobson's organ, uniting with the corresponding cartilage of the other side   Although there is no similar development of the posterior outer part of the nasal floor cartilage in any Marsupial yet examined, the mode of division of the two parts and the structure and relations of the anterior part of Jacobson's cartilage will be found to have an almost perfect counterpart in the corresponding structures of the Daysure and its allies.

Phascologale penicillata, Shaw, (mammary fœtus, head length 9 mm.). The nasal-floor cartilage in front of the naso-palatine canal is present as a well developed, slightly curved plate of cartilage passing outwards from the base of the septum and forming a complete floor to the nasal cavity, uniting laterally with the alinasal. On nearing the naso-palatine canal, its inner end becomes detached from the septum and curves upwards and slightly outwards (Pl. XLI. fig. 10). The naso-palatine canal passes somewhat obliquely backwards, as well as upwards, so that in vertical section it is seen connecting the nasal cavity with the mouth. On its passing upwards the premaxillary is seen to separate from its palatine process as if to make a passage (fig. 10), and a little behind this the nasal-floor cartilage divides into its inner and outer parts. The outer part, which is small, disappears almost immediately behind this plane; but the inner part, or Jacobson's cartilage, is well developed and appears as an upright plate with a large process passing outwards from its upper end and forming a support to the *inferior septal ridge*.* The lower part is supported on its lower and inner side by the developing palatine process of the premaxillary.

In fig. 11 the naso-palatine canal has lost its connection with the mouth, and above is seen to receive the opening of Jacobson's duct on the inner side, and on its outer side to be connected with the nasal cavity. Jacobson's cartilage is here well developed, receiving Jacobson's duct or organ in its concave outer side. If this section be compared with the similar section in the young

---

* This ridge, which extends along on each side of the base of the septum, has been generally referred to as the "glandular ridge." The term, however, is inappropriate, as the ridge is often quite devoid of glandular tissue, and I have therefore proposed the above term instead and in contradistinction to a much more typically glandular ridge frequently present in the upper and middle septal region, which may be called the "superior septal ridge." In the present paper, as only the lower septal region is under consideration, when the term "septal ridge" occurs, the inferior septal ridge will be understood.

Echidna as figured by N. Parker, or in the adult as figured by myself, the striking agreement will be manifest.

In fig. 12 is seen the condition of the organ and its relations in the region of its greatest development. The organ is almost oval in section, there being but a very slight indentation of the outer wall : the inner and lower walls of the organ are about 4-6 times the thickness of the outer. Jacobson's cartilage is a curved plate which supports the organ on its inner and lower sides. The palatine process of the premaxilla, here just commencing to ossify, occupies the lower and inner side of Jacobson's cartilage

Near its posterior part the organ is reduced to a duct with simple columnar epithelium, and the cartilage is present as a narrow thick plate passing more outwards than downwards, and forming a floor to the duct and its neighbouring developing glands.

Dasyurus viverrinus, Shaw, (mammary fœtus, head length 15 mm.). In the somewhat older fœtus of the common Dasyure we have the same type, but with the later stage of development the details are better seen. The nasal-floor cartilage is very similar to that seen in the fœtal Phascologale, but an additional feature is revealed From the point where the ascending inner plate of the nasal-floor cartilage sends out the plate to support the basal ridge a detached process of cartilage passes forward supporting the feeble anterior part of the ridge. This is better seen in the adult, and is interesting from the fact that a similar precurrent process has not been found in any other form, except Didelphys

Figs. 1 and 2 illustrate sections in the anterior part of the nasal-floor cartilage. In fig. 2 the outer part of Jacobson's cartilage is seen detached from the inner on one side. This little detached bar is seen in fig 4 to become connected with the lower part of Jacobson's cartilage, and from its being almost invariably present throughout the Marsupialia connecting the upper with the lower parts on the outer side, it will be referred to in the following descriptions as the "outer bar of Jacobson's cartilage"

In fig. 3 the naso-palatine canal is seen, on the right side opening into the anterior end of Jacobson's organ. The organ has a very short duct lined with squamous epithelium. On the left side, which is further back, the opening of the organ into the naso-palatine canal is closing, while the connection between the canal and the nasal cavity is seen. Immediately beyond this plane Jacobson's organ is closed and the lower part of the inner plate of Jacobson's cartilage becomes connected with the outer bar, forming a floor to the organ; and what was the naso-palatine canal becomes lost in the general nasal cavity.

Fig. 4 represents a section through the body of the organ The cartilage on section assumes the appearance of an irregular L or a U with the outer side shorter than the other—an appearance very common in Marsupial types. It is supported on its lower and inner sides by the scroll-like palatine-process of the pre-maxilla. The organ on section is kidney-shaped, with a much indented hilus, which accommodates the rather large blood vessel.

Dasyurus viverrinus, Shaw, (two-thirds grown). In the grown Dasyure the condition of parts is essentially similar to that in the young Fig. 5 shows a section in the region of the hinder part of the papilla—a portion of the papillary cartilage being seen. The nasal-floor cartilage is moderately flat, and somewhat above its inner end by the side of the septum is seen the small precurrent process of cartilage supporting the septal ridge In fig 6 the pre-maxillary is about to give off its palatine process. The naso-palatine canal is seen cut across below the isthmus, while above it the nasal-floor cartilage is dipping down into the hollow. The outer part of the nasal-floor cartilage behind this becomes lost in *D. viverrinus*, though in *D. maculatus* it is seen for a short time as a very small fragment on the outer side of the naso-palatine canal. The organ opens into the naso-palatine canal almost immediately behind the plane of fig. 7. Fig. 8 is just behind the opening of the organ and immediately in front of the plane where the naso-palatine becomes part of the general nasal cavity. Here the organ is roofed over by the union of the inner plate of Jacobson's cartilage with the outer bar. In fig. 9, a little further back, the upper union with the outer bar is lost and the lower connection complete, giving the cartilage the typical appearance on section.

The organ itself at its best developed part has on section a moderately regular kidney shape, the hilus being directed almost quite upwards and having in it a single large blood vessel. There is extremely little glandular tissue in connection with the anterior and middle part of the organ. The sensory layer is unusually well developed, being about $3\frac{1}{2}$ times as thick as the nasal epithelial layer. The outer wall of the organ has small columnar cells only about half the size of those of the nasal epithelium.

Dasyurus maculatus, Kerr, (adult). The organ in this species differs considerably in a number of ways from that of *D. viverrinus*. In almost all large animals the organ is less developed proportionately, and appears to have less of a sensory function, and to become to a greater extent a glandular duct; and yet with the difference in the character of the organ the cartilaginous relations remain very constant in allied species and genera. The only difference in the cartilaginous developments of the two species of Dasyurus is a very slight one of degree; e.g., in *D. maculatus* the cartilage is rather more developed in front, and rather less posteriorly than in the smaller species. As regards the organs, however, the differences are marked. The sensory layer is present quite characteristically, but much less developed than in *D. viverrinus*, while the whole organ is absolutely smaller in lumen, which means that it is relatively only about half the size. Instead of occupying almost the whole of the cartilaginous hollow as in the smaller species, it fills only about one-third the available space, the rest being almost quite filled up by a great development of mucous gland tissue, except that occupied by the large hilar vessel.

DIDELPHYIDÆ.   (Plate XLII.)

In the American carnivorous genus Didelphys, we have a number of points of close agreement with Dasyurus, and also a few features suggesting a considerable gap between them. This genus I have been able to study through the kindness of Sir. W.

Flower in supplying me with three mammary fœtuses—one small and two moderate-sized—of which I have sectioned the small one and one of the large.

Didelphys murina, L., (mammary fœtus, head length 14 mm.). In the young fœtal Opossum the anterior portion of the nasal-floor cartilage agrees very closely with the condition in the Dasyure, not only is it comparatively flat, but from its ascending inner plate it gives off a precurrent process to support the anterior part of the septal ridge. In the plane of the papilla (fig. 1) the premaxilla is seen giving off its palatine process. The nasal-floor cartilage is here curved, the inner end passing up by the side of the septal base into the septal ridge, while it is slightly depressed into the hollow between the premaxilla and its palatine process. A broad but not very thoroughly chondrified papillary cartilage is seen in the section; and by its edge the naso-palatine canal is seen opening. In fig 2—a little distance behind—the nasal-floor cartilage is found to have become divided as in Dasyurus, the inner part having become a well developed Jacobson's cartilage, while the outer part has on this plane become lost. If this figure be compared with fig. 2 of the Dasyure the close agreement between the forms will be seen in the structure of Jacobson's cartilage. There is, however, a slight difference in the relations borne by the developing palatine processes to the cartilages. In Dasyurus the palatine process is mostly inferior; while in this form it lies within the lower half, the bottom end of the cartilage being unsupported by bone. This though apparently a small matter will be seen to be of considerable interest in connection with the condition in the other forms to be described. In *Didelphys murina* the septal ridge is more marked, the lower corner of the nasal cavity passing well in below it. The naso-palatine canal will be noticed to have an almost vertical direction, the obliquity being very slightly marked. The connections of the canal with Jacobson's organ and with the nasal cavity are as in Dasyurus, except that in *Didelphys murina* the organ becomes constricted into a little roundish duct-like canal before opening into the naso-palatine canal. This little constricted part is not a

40

true Jacobson's duct, as it is lined with columnar epithelium. The organ where best developed, as seen in fig. 3, almost completely fills the large hollow cartilage. On section it is kidney-shaped, but the two poles are approximated so as to give the organ an almost circular appearance, folding the small outer wall closely on itself. The cartilage is supported by the small curved palatine process at its lower and inner side.

Didelphys marsupialis, L., (?)* (large mammary fœtus, head length 37 mm.). Between this form and the fœtal *D. murina* there are a number of little differences, in addition to what can be accounted for by difference of age. The nasal-floor cartilage is nearly flat, and on passing backwards turns up at the base of the septum as in *D. murina*. The inferior septal ridge is here less developed, and the precurrent cartilaginous process, present in *D. murina*, is practically absent. In fig. 4 is shown a section in the plane of the opening of the naso-palatine canal. Though the papilla is well developed there is no trace of a papillary cartilage, which is interesting as this is the only Marsupial I have met with where it is quite absent. In fig. 5 the nasal-floor cartilage is found divided and the premaxilla distinct from its palatine process; and in the space between the divided structures is seen the anterior part of the almost vertical naso-palatine canal. The outer part of the nasal-floor cartilage is still distinct. A few sections behind this plane, as seen in fig. 6, show the outer end . of the upper part of Jacobson's cartilage becoming detached, forming the outer bar. The palatine process will be seen to bear the same relation to the cartilage as in *D. murina* (fig. 2). In fig. 7 the naso-palatine canal is seen opening into the nasal cavity, as well as into Jacobson's organ. This last connection is effected by means of a very short duct of Jacobson. In the next figure the organ is closed from the canal which still connects the nasal cavity with the mouth.

---

* The species of this specimen was unknown, but there is very little doubt that it is the young of the Common Opossum, *Didelphys marsupialis*.

The organ itself in the region of best development (fig. 9) has on section the usual kidney shape. There is some resemblance to the organ in Dasyurus, with which it agrees in having a single vessel along the hilus; in Didelphys, however, the blood vessel is considerably smaller. The sensory region is well developed, the upper and lower ends of which curve towards each other constricting the hilar region slightly. In the hilar region are a few mucous glands which open into the organ at the point of union of the upper end of the sensory wall with the non-sensory. The main nerves lie as usual in the little triangular space above the organ.

PERAMELIDÆ. (Plate XLIII.)

In the Bandicoots I have confined myself to the study of one species, *Perameles nasuta;* of which I have examined—(1) a young mammary fœtus; (2) a half grown specimen; and (3) an adult. To Mr. A. G. Hamilton, of Mt. Kembla, N.S.W., I am indebted for the fœtus and the adult specimen.

Perameles nasuta, E. Geoff., (mammary fœtus, head length 21 mm.). In a section through the developing first upper incisor, and also a little in front of and behind this plane, the nasal-floor cartilage will be found to be well developed and moderately flat. By each side of the base of the septum is a rather large inferior septal ridge, and into the base of it, at least, passes an ascending plate of the nasal-floor cartilage, lying close to the septum. This ascending plate is better developed anteriorly in this genus than in either Dasyurus or Didelphys. On reaching the papillary planes the septum is found to have retreated, and its place to have become occupied between the two ascending plates of the nasal-floor cartilage by the two palatine processes of the premaxillary (fig. 1). This very marked retreating of the base of the septum is greater than in the other Marsupials, and recalls the condition in the Insectivora. In fig. 1 is shown the moderately developed papillary cartilage, by the edge of which the naso-palatine canal is seen entering. Here also the well developed nasal-floor cartilage is seen passing up and curving round into the septal ridge forming its support. In the immediately succeeding planes the inner plates of the nasal-floor cartilages about to become Jacobson's cartilages are seen approaching somewhat and the palatine processes becoming more curved along their inner sides; while the process of cartilage supporting the ridge becomes a detached bar. This bar thus becomes detached further forward than in either Dasyurus or Didelphys. A very short distance behind the plane of the posterior part of the papilla, the naso-palatine canal is found passing inwards below the lower edge of Jacobson's cartilage and even below the lower edge of the palatine process. From this point it passes outwards, upwards, and slightly forwards into the hollow of the lower half of Jacobson's cartilage, where it meets a short but distinct Jacobson's duct. It also passes outwards and backwards, as seen in fig. 2, opening into the nasal cavity. On this plane the short duct of Jacobson is replaced by the lower part of the organ proper, which is almost shut off from the naso-palatine canal. In the relations of the canal to the lower part of the palatine process and of the cartilage of Jacobson there is a marked agreement with Didelphys, though the lower unsupported part of Jacobson's cartilage is much greater here than in that genus, and clearly suggest the development met with in both the Phalangers and the Kangaroos Almost immediately beyond the plane of the closing of the organ the lower end of the inner plate of Jacobson's cartilage curves round and unites with the outer bar, giving on section the usual U-shaped hollow trough.

The organ itself closely resembles that in Didelphys in the folding together of the feeble outer wall. There is, however, a marked difference in the support the cartilage obtains from the palatine process. In Perameles the palatine process is largely developed, and forms a bony support to almost the whole inner and lower sides of the cartilage. About the middle region of the organ, in fact, with the exception of a very small portion at the upper angle, the palatine process not only completely surrounds it, but at its outer edge even replaces the cartilage. On nearing the posterior end of the organ the cartilage becomes completely lost in the whole lower region being replaced by the palatine process, and ultimately all that is left of it is a small plate lying over the upper and inner side of the reduced posterior end of the organ.

Perameles nasuta, E. Geoff., (half grown and adult).   Between the adult and half grown condition the chief differences are due to the fact that in the adult the bony development is greater and the cartilaginous elements more degenerate.   In the following account it is the half grown specimen that is being described unless otherwise stated.

In the region immediately in front of the incisor teeth, the nasal septum is rather broad and at its base has on each side a well developed inferior septal ridge.   The nasal-floor cartilage is relatively feeble on the whole, but its inner part is better developed and turns up close against the septum, then curves outwards to form the support of the septal ridge.   On reaching the plane of the first pair of incisors, the only difference worth noting is that the septum has retreated somewhat, and only the inner part of the nasal-floor cartilage remains.

In the adult, even in the region of the predental portion of the premaxillary, the nasal-floor cartilage is represented by little more than the inner part.

In the plane of 2nd incisor in the half grown specimen the nasal-floor cartilage is represented only by the skeleton of the ridge, while on the same plane the premaxilla is seen sending up a process towards the base of the septum.   In the anterior papillary region, as seen in fig. 5, the cartilage is found present as an inner plate and an outer bar.   Though this is in front of the naso-palatine canal, as there is no outer part of the nasal-floor cartilage, it will be better to call it Jacobson's cartilage, for though there is no organ at this point, from the condition of the cartilages and other structures it is highly probable that the organ once extended forwards considerably in advance of its opening into the naso-palatine canal, as is the case in Ornitho-rhynchus.   As it is, the organ still extends some little way in front of its opening into the naso-palatine canal, and on one side of fig. 6 the anterior extension is seen cut across.

In fig. 6 and fig. 7 the very short naso-palatine canal is seen first opening into Jacobson's organ and then connecting the nasal cavity with the mouth in the usual manner.   In both figures the enormous development of the palatine processes is the most noticeable feature.   On the outer side of the outer bar of Jacobson's cartilage is seen in section a precurrent process from the outer part of the palatine process of the premaxillary.   On the left side of fig. 7 the inner plate of Jacobson's cartilage is seen sending down a process by the side of the canal; on the right side, which is a little further back, the inner plate of Jacobson's cartilage has united with the outer bar.

In the adult in the region just considered the palatine process of the premaxillary is very similar, but the cartilage has degenerated into a few irregular patches.   It is interesting that the downward process of Jacobson's cartilage by the side of the naso-palatine canal is persistent (fig. 9).

In the region of greatest development the organ is very similar to that in the other Polyprotodonts.   In the adult the cartilaginous capsule is scarcely observable, the organ being almost entirely supported by the well developed palatine process.   The sensory wall is fairly well developed, though less so than in either *Dasyurus viverrinus* or Didelphys.   Along the hilus there runs a single moderate-sized vessel, and a rather large vein runs along the inner and under side of the organ.   There are no glands in connection with the anterior part of the organ.

PHALANGERIDÆ.   (Plates XLIV.-XLVI.)

Although the Phalangers are probably not the Diprotodonts most nearly related to the Polyprotodonts, yet as they represent most distinctly the typical differentiation of the structures found in the Diprotodonts, it will be more convenient to consider them first.

Sub-family P H A L A N G E R I N Æ.   (Plates XLIV.-XLVI., figs. 1-6.)

Of this group I have examined, (1) early mammary fœtus, *Pseudochirus peregrinus*, (2) adult *P. peregrinus ;* (3) adult

Petauroides volans; (4) adult *Petaurus breviceps;* (5) very early mammary fœtus, *Trichosurus vulpecula;* (6) early mammary fœtus, Trichosurus ; (7) large mammary fœtus, Trichosurus ; and (8) adult Trichosurus.

In all these genera the same type is followed, and the close agreement between the different genera is remarkable.

Pseudochirus peregrinus, Bodd., (mammary fœtus, head length 8·5 mm). In the anterior papillary plane and a little in front the nasal-floor cartilage is well developed, but not of very great lateral extent. The nasal septum comes well down and anteriorly the nasal-floor cartilage abuts squarely against it; but in the middle region of the papilla the septum has begun to retreat, and the inner end of the nasal-floor cartilage curves up towards it somewhat. There is on each side a well developed septal ridge, and the nasal-floor cartilage sends a feebly developed process towards it. In Pl. xliv. fig. 1, the ridge process is not so well developed as just in front. In this section will be seen a feature which is developed in all the Diprotodonts as distinguished from the Polyprotodonts, in the great lateral development of all the structures. The inferior septal ridges project more, making the base of the septal region much broader; the nasal-floor cartilages are further apart at their inner ends, and the palatine processes which are developed in connection with Jacobson's cartilages are, in their early development instead of closely together as in the Polyprotodonts, widely apart. The naso-palatine canal passes obliquely upwards and backwards, and opens into Jacobson's organ on practically the same plane as that in which it becomes part of the general nasal cavity. In Pl. xliv. figs. 2 and 3, the nasal-floor is found divided Jacobson's cartilage is hollowed slightly on the inner side, and in the hollow lies the palatine process of the premaxilla. In the region of best development Jacobson's cartilage is present as a slightly concave plate, which inclines markedly outwards as well as downwards from the base of the septum. The palatine process is present as a small ossified bar lying along the middle of the inner side. The organ itself is almost oval on section; the inner wall of which is more than half the diameter, while the lumen is slightly crescentic, owing to the outer wall being much better developed at its central than lateral portions.

Pseudochirus peregrinus, Bodd., (adult), *Petauroides volans*, Kerr, (adult), and *Petaurus breviceps*, Waterh., (adult). These three genera agree with each other so markedly that it will only be necessary to describe the condition in one—Petaurus—and call attention to the points in which the others differ from it.

In a plane immediately in front of the papilla, the condition of the nasal-floor cartilage is found to agree very closely with that described in Perameles, each inner end having an ascending plate closely placed against the sides of the base of the septum. The only marked difference is that the lateral part of the cartilage is much curved, this, however, is rendered necessary by the largely developed first incisors. In the plane passing through the middle of the papilla the inner ascending plate of the nasal-floor cartilage is much shorter, but has become broadened out, while the inferior septal ridge, which anteriorly was developed considerably vertically, is here a much more defined ridge, and from the outer angle of the irregular square-shaped inner part of the nasal-floor cartilage a slight process passes into the ridge. The outer part of the nasal-floor cartilage becomes almost entirely lost Pl XLIV. fig. 10 represents a section through the third incisor or the posterior part of the papilla Here the nasal-floor cartilage assumes an appearance which may be regarded as typical of the Phalangers. The inferior septal ridge is removed from any direct connection with the septum, and the process from the inner part of the nasal-floor cartilage (which may even here be regarded as Jacobson's cartilage) supporting it, instead of coming from the inner part of the cartilage, springs from a point considerably farther out, while an independent continuation of the nasal floor cartilage extends on to the base of the septum. In Petaurus Jacobson's cartilage lies very obliquely outwards on the palatine process, but in Petauroides and Pseudochirus the cartilage is much more vertical (*cf.* fig. 4); otherwise, however, the structures are similar. Inferiorly the cartilage plate extends downwards considerably past the lower edge of the palatine process, a condition more apparent in Pseudochirus than in Petaurus. On passing backwards the outer part of the cartilaginous process of the ridge becomes detached as the outer bar of Jacobson's cartilage. In Pl. xLIV. fig. 11 the anterior part of Jacobson's organ is indicated, with the naso-palatine canal connected with the short duct of the organ. In Pl. xLIV. fig. 12 the organ communicates freely with the nasal cavity at the plane where the naso-palatine canal becomes part of the cavity.

From Pl. xLIV. figs. 5 and 6 it will be seen that in Pseudochirus the opening of the organ is more directly into the upper part of the canal, while in Petauroides (fig. 8) the condition agrees more nearly with that in Petaurus. The difference, however, is only a very slight one of degree.

After the closing of the organ the lower part of Jacobson's cartilage unites with the outer bar in the usual manner. In Pseudochirus the ridge is considerably lower than in the other Phalangers, so that when the lower part of Jacobson's cartilage is complete, instead of an irregular U-shaped appearance we have a very regular L, as in Pl. xLIV. fig. 7. In Petauroides (fig. 9) the cartilage has the more usual appearance.

The organ in all these genera is well developed, and has on section a rather elongated kidney shape. In the small Petaurus the sensory wall is larger proportionally than in the other two genera. The hilus is very broad and only but slightly depressed, leaving a larger lumen to the organ In all three genera there is a distinct venous plexus usually composed of one, two, or three vessels anteriorly, which branch into six or more posteriorly. There are but few glands in connection with the organ, except at the posterior part.

Trichosurus vulpecula, Kerr, (mammary fœtus, head length 7·5 mm.). In this very small mammary fœtus, which may be taken as the size at birth, the cartilages are all fairly well developed, and the ossification of the premaxillary bones quite distinctly marked. In the plane of the developing incisors the nasal-floor cartilage is very well developed, as seen in Pl. xLV. fig. 1.

At its inner end it sends up a process by the side of the septum, which latter at this early stage descends down between the inner ends of the premaxillaries. At the outer ends the nasal-floor cartilage unites with the alinasal. On reaching the plane of the papilla the nasal-floor cartilage divides into its inner and outer parts; before dividing, however, the downward process of the inner part makes itself manifest. On the left side of Pl xLv. fig. 2, representing the plane a little behind the division of the nasal-floor cartilage, Jacobson's cartilage is seen as a curved plate with, near the middle of the inner concave side, the developing palatine process, present as a minute spicula of bone. The downward process, it will be seen, is more marked than in the young *Pseudochirus*. The naso-palatine canal has the usual relations, opening first into Jacobson's organ and then becoming merged in the nasal cavity. The organ is present as a small oval tube with the inner wall considerably thicker than the outer.

Trichosurus vulpecula, Kerr, (mammary fœtus, head length 10·5 mm.). In this more developed mammary fœtus the relations of parts are better seen. In Pl. xLv. fig. 4 is shown the complex structure of the inner part of the nasal-floor cartilage just before division. From this figure it will be seen that the descending process is a structure superadded to the simple nasal-floor cartilage as seen in the Dasyure. The same can probably also be said of the internal ascending process. In Pl. xLv. fig. 6 Jacobson's cartilage is an almost vertical plate with the rod-like palatine process along the middle of the inner side. The organ is here very large.

Trichosurus vulpecula, Kerr, (mammary fœtus, head length 20 mm.). In the series of sections from this specimen we have the steps intermediate between the condition in the early fœtus and the adult. The nasal-floor cartilage before division as seen in Pl. xLv. fig. 7 may be compared with Pl. xLiv. fig. 4, illustrating the similar part in Pseudochirus. The only marked difference is due to the unusually well developed posterior outer part of the nasal-floor cartilage. In the Ringtail the outer nasal-floor cartilage is only a rudiment, but here it is larger than the inner part.

The ridge process, on the other hand, so large in the Ringtail and Flying Phalangers is only slightly developed in Trichosurus. The descending process is very distinct; and the palatine process more developed vertically than in the younger fœtuses. In Pl. xlv. fig. 8 the naso-palatine canal passes up almost vertically and opens into Jacobson's organ. At this stage there is no chondrification of the outer bar. In the following figure the organ is closed, and the naso-palatine canal is merged in the nasal cavity. Even in this plane the outer part of the nasal-floor cartilage is still well developed. Jacobson's cartilage is an almost vertical plate, and the organ lies against it much flattened from side to side.

Trichosurus vulpecula, Kerr, (adult). In the adult common Phalanger there is considerable agreement with the condition in the adult Petaurus. All the main peculiarities are due to two facts—(1) a much less degree of development of the inferior septal ridge in Trichosurus; and (2) a greater development of the outer nasal-floor cartilage.

In Pl. xlvi. fig. 1 through the posterior papillary region, the inner part of the nasal-floor cartilage is very similar to that in Petaurus, except that the ridge process is more feeble here; the outer part of the nasal-floor cartilage though small is, however, better developed than in Petaurus. The papillary cartilage is well seen in this plane and is interesting from its having a distinct median ridge. In Pl. xlvi. figs. 2, 3 and 4, is seen the mode of division of the nasal-floor cartilage, which is more complicated than in any of the other common Marsupials. In the most anterior part of the gap between the premaxilla and its palatine process there is a most distinct, rather large, descending process filling up the whole gap. On the naso-palatine canal passing up, and on the premaxillary being farther removed from the palatine process, the descending cartilaginous process remains only as a narrow internal plate lying close against the palatine process (Pl. xlvi. fig. 2). In this plane the ridge process though small is distinct, and is connected with both the inner plate of Jacobson's cartilage and the outer part of the nasal-floor cartilage. In Pl. xlvi. fig. 3, a very little behind the previous plane, an anterior prolongation of

Jacobson's organ makes its appearance between the outer end of the ridge process and the inner plate of Jacobson's cartilage, dividing the one from the other; but though the outer part of the ridge process—clearly the outer bar of Jacobson's cartilage—becomes detached from the inner plate, it still retains its connection with the outer part of the nasal-floor cartilage. In Pl. XLVI. fig. 4, however,—a little further back still—the outer bar is free from the nasal-floor cartilage which is now lost. On this plane the appearance quite agrees with that in the Ringtail—the organ connecting with the naso-palatine canal in quite a similar way. In Pl. XLVI. fig. 5 the organ is closed, and the naso-palatine canal is merged in the nasal cavity. In the following figure the usual appearances are presented. The inner plate of Jacobson's cartilage has united below with the outer bar, and an irregular U-shaped hollow is formed for the reception of the organ.

The organ is large and has an irregular crescentic shape; with a well developed sensory wall. The hilus is large and contains two or three large veins and one or two small; while all along the outer side of the organ is an enormous amount of glandular tissue, in which it differs from that of the other Phalangers.

Subfamily PHASCOLARCTINÆ. (Plate XLVI. figs. 7-9.)

Phascolarctus cinereus, Goldf., (two-thirds grown). In Phascolarctus we have a very highly modified type which differs in many ways from that of the Phalangers just described.

The naso-palatine canal is very long and oblique. In Pl. XLVI. fig. 7 we have represented a section through the plane a little in front of the point where the premaxillary gives off its palatine process. In this and the following sections the most striking peculiarity is the depth of the secondary palate. The nasal-floor cartilage is well developed, but with the narrowing of the nasal cavity only a very small portion is really a floor. At its inner end it is very simple and abuts against the base of the septum. Below the septum will be seen the vomer, a most exceptional occurrence, this being the only Marsupial known in which the vomer is directly in contact with the body of the premaxillary.

In the lower part of the section the naso-palatine canal is seen cut across.

On reaching the plane where the premaxillary gives off its palatine process the nasal-floor cartilage is found to bend down into the gap formed, as seen on the left side of Pl. xlvi. fig. 8. There is no more than a slight indication of a downward process apart from the general dipping down and thickening of the nasal-floor cartilage. The palatine process is by the side of the lower third of the downward bent cartilage; while the naso-palatine canal is seen almost in contact with the lower part of the cartilage. On the right side of the same figure is seen the condition a little farther back. The large solid downward extension has given way before the ascending naso-palatine canal, and there is formed a well marked inner plate, extending from the side of the base of the septum, down past the vomer and along the upper half of the palatine process From the upper end of this plate there passes an outward and downward process which becomes continuous with the outer part of the nasal-floor cartilage. In Pl. xlvi. fig 9 we see the inner part of the nasal-floor cartilage or Jacobson's cartilage separated from the outer. It has a well developed inner concave plate, with above a downward and outward sloping roof. In the hollow is the anterior part of Jacobson's organ connected with the naso-palatine canal near the point where it merges into the nasal cavity.

Beyond this plane there is found passing up from the lower edge of the inner plate a process meeting the lower edge of the roof and forming a complete cartilaginous tube for the organ.

The organ itself, however, is very feebly developed relatively, though it possesses the usual sensory wall There are very few glands in the tube; but it is extremely interesting to find a plexus of five or six large veins on the outer side of the organ The whole length of the organ is somewhat less than 10 mm.

MACROPODIDÆ. (Plate xlvii.)

Of the Kangaroo group, Symington, as already stated, has examined the small mammary fœtus of *Macropus giganteus* and of *Petrogale penicillata*, and found that the condition in both forms is "practically identical." Of this group I have examined (1) a series of sections prepared by Prof. Wilson, of a very small mammary fœtus of *Macropus sp.?*; (2) a large mammary fœtus of *M. ualabatus;* and (3) a small mammary fœtus of *Æpyprymnus rufescens.*

Sub-family MACROPODINÆ.   Plate (XLVII. figs. 1-9.)

Macropus sp.? (mammary fœtus, total length 29 mm.).   In this very young fœtus the condition of parts agrees very closely with that in Trichosurus. The nasal-floor cartilage is well developed in the anterior part (Pl. XLVII. fig. 1), but before reaching the upper opening of the naso-palatine canal the outer part is lost. There is a distinct though small downward process. The naso-palatine canal passes up almost vertically, and the organ of Jacobson opens into it on the same plane as that in which it unites with the nasal cavity (fig 2). The palatine process is represented as in Trichosurus by an ossifying rod near the middle of the inner plate of Jacobson's cartilage. Posteriorly the condition agrees with that in the early fœtal Trichosurus.

Macropus ualabatus, Less. & Garn., (large mammary fœtus, head 50 mm.).   This specimen may be taken as the type of the Kangaroo.

In front of the naso-palatine canal (Pl. XLVII. fig. 3) the nasal-floor cartilage is rather feebly developed and very simple in structure. There is no distinct septal ridge, and in consequence the inner end of the floor cartilage remains more simple than in the Phalangers. Inferiorly a broad papillary cartilage is seen. In fig. 4, where the naso-palatine canal begins to be seen, the nasal-floor cartilage becomes very much thickened and dips down in the hollow formed where the palatine process is about to divide off from the premaxilla. The condition resembles in general appearance that of Phascolarctus more than that of the Phalangers. On reaching the plane where the palatine process becomes quite distinct from the premaxilla the following condition is seen on section (Pl. XLVII. fig. 6). The large dipping down portion of the nasal-floor cartilage is hollowed out to accommodate an anterior projection of Jacobson's organ, but we are thereby enabled to understand the different parts. If this section be compared with Pl xlv. fig. 3, the Trichosure condition, there is no trouble in making out the homology of the different parts. The inner plate corresponds to that in Trichosurus, except that it does not curve downwards at its lower end, but retains its connection with the outer part of the nasal-floor cartilage. On the outer side of the opening in the cartilage above the organ is seen a distinct knob attached to the outer nasal-floor cartilage; this is unquestionably the outer bar of Jacobson's cartilage, agreeing closely with the condition in Trichosurus, while the upper opening in the cartilage is due to the customary detachment of the outer bar from the inner plate of Jacobson's cartilage. In Pl. xlvi. fig 4 we have the more usual condition revealed; almost the only difference, in fact, from the similar section in Trichosurus (Pl. xlv. fig. 4) is due to the absence or reduction of the inferior septal ridge in Macropus. The naso-palatine canal opens into the organ and the nasal cavity in the usual way.

At its hinder end, as seen in Pl. xlvii. fig. 9, the organ is situated well up the side of the septum, a condition recalling the appearance in the human fœtus.

The organ itself is on the whole rather feebly developed, and has the appearance of a degenerate Phalanger type. There are few glands anteriorly, and in the hilus are only a few small blood vessels.

Sub family POTOROINÆ. (Plate xlvii. figs. 10-12.)

Æpypiymnus rufescens, Gray, (mammary fœtus, head length 15·5 mm.). In the Rat-Kangaroo, though we have a fairly close agreement with the condition in Macropus, we have some remarkable differences. Pl. xlvii. fig. 10 represents a section in the plane of the 2nd upper incisors. The nasal floor cartilage is well developed, and at its inner part is found turning round to support the inferior septal ridge more after the manner of the Polyprotodonts than of the Phalangers. In the plane through the point where the palatine process is first seen distinct from the pre-maxilla, the inner part of the nasal-floor cartilage curves markedly upwards and sends out a well marked though feeble plate into the inferior septal ridge.   At the lower angle of the nasal-floor cartilage there is sent down a short process into the gap between the premaxilla and its palatine process.

Immediately following this plane we have the remarkable condition shown in Pl. XLVII. fig. 11. The outer part of the nasal-floor cartilage is detached from Jacobson's cartilage, which is present as an inner plate and an outer bar.   In the hollow is found the anterior portion of Jacobson's organ opening directly into the anterior part of the nasal floor, and in no way directly connected with the naso-palatine canal.   It is only some sections posterior to this, after the organ is quite closed, that the naso-palatine canal unites with the nasal cavity.   In other respects the ordinary arrangement is followed.

The relation of the palatine process to the cartilage is more like that found in Petaurus than in Macropus.

In the early fœtal specimen the vascular and glandular relation of the organ cannot be made out very satisfactorily, but there is apparently nothing remarkable about the organ itself.

PHASCOLOMYIDÆ.   (Plate XLVIII).

Of the Wombat I have only had an opportuning of examining the condition in a half grown specimen, but Rose has fortunately published two very good sections of an early mammary fœtus, which I have taken the liberty of reproducing.

Phascolomys wombat, Per. & Less., (very early fœtus, body length 19 mm.) [after Rose].   In this early fœtus the condition most strikingly resembles that in the Dasyure. Indeed, if Pl. XLVIII. fig. 1 be compared with Pl. XLI. fig. 3, illustrating the fœtal Dasyure, there is not a single feature of importance in which any difference can be detected.   The organ opens similarly, the cartilage of Jacobson is similar, the palatine processes exactly agree, and further bear the same relations to the cartilages. Pl. XLVIII. fig. 2, which apparently is a section through the posterior part of the organ, shows some of the Diprotodont characters, *e.g.*, the cartilages being considerably apart, and the organ having a large gland duct entering it from above.

Phascolomys mitchelli, Owen, (half grown specimen). In this specimen, which may be taken as the adult type, we have a great similarity in many ways to the condition in Phascolarctus Here there is, however, but a very feeble development of the outer nasal-floor cartilage, and in this resembling Macropus.

In Pl. XLVIII. fig. 3 we have a section through the posterior part of the very large papilla—a portion of the papillary cartilage being still seen. At this plane the septum dips considerably below the level of the nasal floor, and has by the side of the deep portion a descending plate from the nasal-floor cartilage, or possibly rather an enormously thickened inner end of the cartilage In fig. 4 this large inner part of the nasal-floor cartilage becomes still more developed and extends down into the hollow formed between the premaxillary and its palatine process, about to become detached in section. Below the bony isthmus is seen the very long and oblique naso-palatine canal. In fig. 5 the palatine process is detached from the premaxilla, and in the gap between is a distinct descending plate which almost meets the naso-palatine canal and rests on the palatine process. The cartilage is excavated in the middle for the anterior part of the organ, but its roof is entire and united with the feeble outer portion of the naso-palatine canal. Fig 6 shows the anterior part of the organ situated in the hollow of Jacobson's cartilage and opening into the naso-palatine canal exactly as in Macropus Here the outer part of the roof cartilage has become detached from the outer nasal-floor cartilage. A little behind this plane the lower part of Jacobson s cartilage passes up and forms a complete tube for the organ as in Phascolarctus The palatine process is situated very much as in Macropus, but more inferiorly

The organ is fairly developed, and more than half fills the cartilaginous tube. At its upper inner angle it receives a number of gland ducts, the glands lying at the inner side of the upper

41

end of the tube. Two large nerves lie at the upper end of the tube, and on the inner side are two or three moderately large veins. There is, however, no hilar plexus as in Phascolarctus.

COMPARATIVE OBSERVATIONS.

From the examination of Jacobson's organ in the various types of Marsupials, it will. be noticed that although there are many variations of details, the same general plan is followed in all; though the habits of the different animals vary greatly and with the habits are very distinct differences of tooth structure; though some of the animals are nocturnal and others lovers of the light, some gregarious and others solitary; all possess moderately developed organs of Jacobson, and in all have we the one main type of structure followed. Studies in Eutherian forms lead to the same conclusions, viz., that the type of organ does not vary with the habits, but remains constant throughout large groups of apparently not very nearly related animals. For example, we have one type in such dissimilar forms as the Ox, Sheep, Horse, Dog, Cat, and Hedgehog, but quite a different type in the Rodents. From this constancy of type followed by the organ it is manifest that it must be a very valuable factor in the classification of groups—apparently of more importance than even the dentition.

Before considering the morphological importance of the different varieties in the Marsupialia, a few general observations may be well. In Mammals generally it would seem that the organ is best developed in small forms, and that in animals which have increased much in size from what may be considered the ancestral type, the organ is not found to have increased proportionally, and though still retaining the typical sensory character it is in a measure degenerate. Then, again, in all forms apparently there are mucous glands in connection with the organ and which discharge into it. In small forms, e.g., Mus, Petaurus, Miniopterus &c., the glands are few and mostly situated at the posterior end of the organ; while in relatively larger forms as Lepus, Trichosurus &c., the glands are numerous and open into the organ along nearly its whole extent. This peculiarity is well seen in the two species of Dasyurus; in the small *D. viverrinus* the glands are few, while in the large *D. maculatus* they are very numerous. I am not aware that sex has anything to do with the peculiarities of this remarkable organ, concerning the function of which we know so little.

In the three Polyprotodont genera the nasal-floor cartilage and its inner division or Jacobson's cartilage are very simple in structure and, as already pointed out, bear considerable resemblance to the simple Monotreme type of Echidna. In Echidna, however, the organ is much better developed, as is also the cartilage. By comparing the series of sections of the anterior region of Jacobson's organ in Echidna, given in my paper on the organ of Jacobson in the Monotremes, with the similar series from Dasyurus (Pl. XLI.) there will be found no difficulty in tracing the homology of the parts. In fig. 5 of the Echidna sections Jacobson's cartilage is found on section to be C-shaped, with the upper outer end much thickened. By comparing this with Pl. XLI. figs. 2, 8 and 11 from Dasyurus and Phascologale, it will be seen that it is this thickened outer rim of the cartilage in Echidna that becomes the outer bar of Jacobson's cartilage in Dasyurus. In Echidna, on passing backwards, the lower part of the C joins the upper outer thickened bar (fig. 6), and a complete capsule is formed; and on tracing the outer thickened bar still further back it is found to be continuous with the turbinal plate, and represents probably the rudiment of a turbinal which once extended right to the front of the organ, as is still seen in Ornithorhynchus. In Dasyurus and other Polyprotodonts the main differences are due apparently to the feebler cartilaginous development. The outer bar is present at first in connection with the upper part of Jacobson's cartilage as in Echidna, and almost immediately behind the opening of the organ the lower border of Jacobson's cartilage sweeps round and becomes attached to it, but there is the difference in Marsupials that as a rule before the lower connection is established the upper has given way, so that there is usually for a short distance a detached bar, which on section is apparently neither attached to upper or lower borders.   In Echidna, at the posterior part of the organ, the upper connection gives way and we have the irregular U-shaped appearance as in Marsupials We may thus conclude that we have in the simple Marsupials a somewhat degenerate Monotreme type, the outer bar being the rudimentary remains of a primitive turbinal.

In Didelphys and Perameles we have a short almost vertical naso-palatine canal , while in Dasyurus it is rather long and oblique In Perameles there is a small yet distinct downward process of Jacobson's cartilage in the notch between the pre-maxillary and its palatine process, a process which is more or less developed in all the Diprotodonts, and apparently the forerunner of the long anterior process which supports Jacobson's duct in the higher mammals of the Cat or Sheep type. In Didelphys there is only a slight indication of this process, and in Dasyurus it is absent From this we may consider that Dasyurus is the more primitive. As regards the portion of Jacobson's cartilage supported by the palatine process all three genera differ. In Dasyurus the support is on the lower edge and lower inner third, in Didelphys on the lower inner half ; while in Perameles the whole inner side of the cartilage is supported by the palatine process In neither of the latter two genera, however, is the lower edge of the cartilage completely supported by bone as in Dasyurus In all three genera there is but a single hilar vessel, and as a rule the supply of mucous gland is scanty Perameles is peculiar in having a small anterior prolongation of the organ in advance of the opening, as well as in the extreme shallowness of the secondary palate.

In the Phalangers we enter on a well differentiated type The most remarkable points of difference from the previous forms are to be found in the complex nature of Jacobson's cartilage in the anterior region There is a well developed inferior septal ridge into which is sent a cartilaginous process from the ascending inner part of the nasal-floor cartilage, and which is unquestionably homologous with the similar process in the Polyprotodonts. In addition, however, there is an ascending process, only rudimentary

in the carnivorous Marsupials and but feebly indicated in Perameles, and there is also a very marked descending process by the
side of the naso-palatine canal in the notch. The ascending and
descending processes are well seen in their adult condition in
Pl. XLIV. fig. 4, representing the condition in the adult Pseudochirus, while their mode of development is well seen in Plate XLV.
representing the different stages of the young Trichosurus. By comparing Pl. XLIV. fig 4. with, say Pl. XLIII. fig 1,—the condition in
the young Perameles, and fixing the two unquestionably homologous parts—the processes passing into the inferior septal ridges
—the two additional processes will be readily seen. In the
primitive condition of the palatine processes there is also a marked
difference from that of any of the Polyprotodonts. In those
latter it is always apparently developed as a small curved splint,
supporting a considerable-area of the cartilage. In the Phalangers it is developed as a rod along the middle of the inner side of
Jacobson's cartilage. This would lead one to assume that the
middle region of Jacobson's cartilage in the Phalanger is probably
homologous with the lower third of the cartilage in Dasyurus,
which is the region where the palatine process first developed.
If this be so the downward process in the Phalangers would
become the more manifestly an additional development.

In its posterior parts Jacobson's cartilage follows much the
same lines as in the Polyprotodonts. The outer part of the ridge
process very early becomes separated into the outer bar of
Jacobson's cartilage, which, after being isolated for a short distance, becomes attached to the under part of Jacobson's cartilage,
and the condition differs little from that of the Polyprotodonts
The organ itself is very similar to that in Dasyurus or Didelphys;
there is, however, one very constant difference in that while in
the Polyprotodonts there is only a single blood vessel running
along the hilus, in the Phalangers there is a distinct plexus At
the extreme anterior end there is usually one or two large veins,
and these on passing backwards divide into four or five large subequal branches which run parallel along the hilus This is a
character met with in the Monotremes, but it is probably not of

any very deep significance, as in the Mouse there is but a single hilar vessel, while in the allied Guinea-pig there is a regular plexus   Still it is interesting to note that the plexus is constant among the Phalangers, so far as known.   The arrangement of mucous glands is very variable anteriorly; in Petaurus, Pseudo-chirus and Petauroides they are absent or scanty, while in Trichosurus they are abundant.   As already observed, this is a point of little importance.

In Phascolarctus, not having examined the early conditions of the parts, it would be rash to say much on the relationships of the organ.   Apparently the adult organ and cartilage differ very considerably from those in the Phalangers.   Its most interesting points are – (1) the large proportional development of the nasal-floor cartilage; (2) the low position relative to the cartilage of Jacobson occupied by the palatine process ; (3) the anterior development of the vomer; (4) the persistence of the cartilaginous roof; (5) the complete tube formed by Jacobson's cartilage; and (6) the presence of a plexus on the outer side of the organ. Whether as a parallel development or as indicating an affinity it is difficult to say, but there is a very decided resemblance in many ways to the condition in the Wombat.

In the Macropods, though there are features of resemblance to the Phalangers, both the ascending and descending processes of the inner parts of the nasal-floor cartilage are less marked.   In Macropus the descending process is due more to a bending down of the nasal-floor cartilage than to a distinct downgrowth, though in Æpyprymnus the downgrowth though short is quite distinct, at least in the fœtus.   In Macropus the relations of the naso-palatine canal to the opening of the organ and the nasal cavity follow the usual type.   In Æpyprymnus, however, there is, with practically no difference in other details, the remarkable and, so far as my studies go, unique condition of the organ opening out to the anterior nasal floor, and not into the naso-palatine canal This is practically the condition which we find in an extreme degree in the Rodentia.   If the section (Pl. XLVII. fig. 11) illustrating the condition in the Rat-Kangaroo be compared with the similar section in Didelphys (Pl. XLII. fig. 7) it will be seen that the peculiarity is only due to a slight difference in the relative position of the naso-palatine canal In the low position occupied by the palatine process and the simple condition of the nasal-floor cartilage the Rat-Kangaroo comes considerably nearer the Poly-protodonts than does Macropus.

The Wombat in its early condition shows a very marked agreement with Dasyurus, and also considerable agreement with Æpyprymnus, though the organ opens in the usual way. In the adult the cartilaginous development is on the type of the Macro-pods, though the perfect cartilaginous tube formed by Jacobson's cartilage gives it more of the appearance of Phascolarctus.

CONCLUSION.

From the study of this limited region in the snout of the Marsupials we get a number of interesting suggestions in the way of apparent affinities. In the first place there can be little doubt in placing Perameles with Dasyurus and Didelphys and away from the Phalangers, and though it is more differentiated than either it seems to retain certain primitive characters lost in the others. The Phalangers are all closely allied, though it would seem that Trichosurus is a little further differentiated than Pseudochirus and Petaurus. Phascolarctus is a much modified and aberrant form, and it seems probable that a study of the fœtus will reveal that it is not so near the Phalangers as has been supposed. The Kangaroo group though allied to the Phalangers is, as regards the region under consideration, nearer the Polypro-todonts; and the Rat-Kangaroo, though slightly aberrant, helps to bridge over the gap. The Wombat is a very near ally of the primitive or ancestral Macropods apparently, though it has become much modified along an independent line.

I must acknowledge my indebtedness to Sir William Flower for the specimens of Didelphys examined; to Mr. A. G. Hamilton, of Mt. Kembla, N.S.W., for the young and adult Perameles; and to Prof. Wilson for the permission to examine his sections of the fœtal Macropus. In addition I am indebted to Sir William Turner, Prof. Wilson, and Dr. Elliot Smith for assistance with literature.

PRINCIPAL BIBLIOGRAPHY ON THE LOWER MAMMALIAN ORGAN AND RELATED STRUCTURES.

1. BALOGH, C. ... "Das Jacobson'sche Organ des Schafes." Sitz. Akad. Wien. 1862

2. BROOM, R. ... "On the homology of the palatine process of the Mammalian premaxillary." Proc. Linn. Soc N.S.W. 1895.

3. ————— ... "On the Organ of Jacobson in the Monotremata" Journ. Anat. and Phys. 1895.

4. ————— ... "On the Organ of Jacobson in an Australian Bat (*Miniopterus*)." Proc. Linn. Soc. N.S.W. 1895.

5. ————— ... "Observations on the relations of the Organ of Jacobson in the Horse." Proc. Linn. Soc. N.S.W. 1896.

6. FLEISCHER, E. "Beitr. zu der Entwickl. des Jacobson'sche Organs, &c." Sitzungsber. Phys.-Med. Soc. Erlangen. 1878.

7. GRATIOLET ... "Recherches sur l'Organe de Jacobson." Paris. 1845.

8. HARVEY, R. ... "Note on the Organ of Jacobson." Q J.M.S 1882.

9. HERZFELD, P. "Ueber das Jacobson'sche Organ des Menschen und der Saugethiere." Zool. Jahrb. Bd. 3. 1888.

10. HOWES, G. B. "On the probable existence of a Jacobson's organ among the Crocodilia, &c." P.Z.S. 1891.

11. JACOBSON   ... "Rapport de M. Cuvier sur un Mémoire de M Jacobson." Ann. du Mus. d'Hist. Nat. 1811.

12. KLEIN, E.   ... "Contrib. to the Minute Anat. of the Nasal Mucous Membrane" Q.J.M.S. 1881.

13. ————   ... "A further Contrib. to the Minute Anatomy of the Organ of Jacobson in the Guinea-Pig." Q.J.M S. 1881.

14. ————   ... "The Organ of Jacobson in the Rabbit" Q.J.M.S. 1881.

15. ————   ... "The Organ of Jacobson in the Dog." Q.J.M.S. 1882.

16. LEGAL, E.   ... "Die Nasenhohlen und der Thranennasengang der Amnioten Wirbelthiere." Morph. Jahrb. Bd. 8. 1883.

17. PARKER, W. N. "On some Points in the Structure of the young *Echidna aculeata.*" P.Z.S. 1894.

18. RANGE, P.   ... "Le canal incisif et l'organe de Jacobson." Arch. Internat. de Laryngolog. 1894.

19. RÖSE, C.   ... "Ueber das Jacobson'sche Organ von Wombat und Opossum." Anat. Anz 1893.

20. SCHWINK, F. ... "Ueber den Zwischenkiefer und seine Nachbarorg. bei Saugethiere." München. 1888

21. SMITH, G. ELLIOT "Jacobson's Organ and the Olfactory Bulb in Ornithorhynchus." Anat Anz. xi. Band, Nr. 6, 1895, p. 161.

22. SYMINGTON, J. "On the Nose, the Organ of Jacobson, &c., in Ornithorhynchus" P Z S. 1891.

23. ————   ... "On the Organ of Jacobson in the Kangaroo and Rock Wallaby." Journ of Anat. and Phys. 1891.

24. SYMINGTON, J.   "On the homology of the dumb-bell-shaped bone in Ornithorhynchus." Journ. of Anat. and Phys. 1896.

25. ZUCKERKANDL, E. " Das peripherische Geruchsorg. der Säugethiere." Stuttgart. 1887.

REFERENCES TO PLATES.

a.J.o., anterior prolongation of Jacobson's organ; a n., alinasal; gl., gland ; J.c., Jacobson's cartilage ; J.o., Jacobson's organ ; l d., lachrymal duct ; Mx., maxilla; n., nerve ; n.f.c., nasal-floor cartilage ; n.g.d., nasal gland duct; n.p.c., naso-palatine canal; n.s., nasal septum; o.b.J.c., outer bar of Jacobson's cartilage; o n.f.s., outer nasal-floor cartilage; p.c., papillary cartilage; Pmx., premaxilla; p.Pmx., palatine process of premaxilla; r.p., ridge process of Jacobson's cartilage; r., vein; Vo., vomer

PLATE XLI.

Dasyurus and Phascologale.

Figs. 1 - 4.—Transverse vertical section of Jacobson's organ and relations in D. viverrinus (mam. fœt., head length 15 mm.), × 27

Figs. 5 - 9.—The same in D. viverrinus (two-thirds grown), × 12.

Figs. 10.12.—The same in Phascologale penicillata (mam. fœt., head length 9 mm.), × 36.

PLATE XLII.

Didelphys.

Figs. 1 - 3.—Transverse section of region of Jacobson's organ in Didelphys murina (mam. fœt, head length 14 mm.), × 34.

Figs. 4 - 8.—The same in D. marsupialis (mam. fœt., head length 37 mm.). × 14.

Fig.   9.—Transverse section of Jacobson's organ in D. marsupialis (mam. fœt.), × 33.

PLATE XLIII.

Perameles.

Figs. 1 - 3.—Transverse vertical section of region of Jacobson's organ in Perameles nasuta (mam. fœt, head length 21 mm.), × 30.

Figs. 4 - 7.—The same in P. nasuta (two-thirds grown), × 17.

Fig.1 8.—Transverse section of Jacobson's organ in *P. nasuta* (two-thirds grown), × 27.

Figs. 9-11.—Transverse section of region of Jacobson's organ in *P nasuta* (adult), × 14.

PLATE XLIV.

Pseudochirus, Petauroides, and Petaurus.

Figs. 1 - 3.—Transverse section of region of Jacobson's organ in *Pseudochirus peregrinus* (mam. fœt., head length 8·5 mm.), × 40.

Figs. 4 - 7.—The same in *P peregrinus* (adult), × 11.

Figs. 8 - 9.—The same in *Petauroides volans* (adult), × 10.

Figs. 10-12.—The same in *Petaurus breviceps* (adult), × 16.

PLATE XLV.

Trichosurus.

Figs. 1 - 3.—Transverse section of region of Jacobson's organ in *Trichosurus vulpecula* (mam. fœt., head length 7·5 mm.), × 36.

Figs. 4 - 6.—The same in *T. vulpecula* (mam. fœt., head length 10 5 mm.), × 42.

Figs. 7 - 9.—The same in *T. vulpecula* (mam. fœt., head length 20 mm.), × 18.

PLATE XLVI.

Trichosurus and Phascolarctus.

Figs. 1 - 6.—Transverse section of region of Jacobson's organ in *Trichosurus vulpecula* (adult), × 10.

Figs. 7 - 9.—The same in *Phascolarctus cinereus* (half grown), × 7.

PLATE XLVII.

Macropus and Æpyprymnus.

Figs. 1 - 3.—Transverse section of region of Jacobson's organ in *Macropus* sp? (early fœtus, body length 29 mm.)

Figs. 4 - 9.—The same in *M. ualabatus* (mam. fœt., head length 50 mm), × 10.

Figs. 10-12.—The same in *Æpyprymnus rufescens* (mam. fœt., head length 15·5 mm.), × 25.

PLATE XLVIII

Phascolomys.

Figs. 1 - 2.—Transverse section of region of Jacobson's organ in *Phascolomys wombat* (fœtus, body length 19 mm.), after Rose, × 37.

Figs. 3 - 7.—The same in *P. mitchelli* (half grown), × 6.

Figs. 8.—The same in *P. mitchelli* (half grown), × 18

ON A NEW SPECIES OF *MACADAMIA*, TOGETHER WITH NOTES ON TWO PLANTS NEW TO THE COLONY.

By J. H. MAIDEN, F.L S., and E. BETCHE

MACADAMIA INTEGRIFOLIA, sp.nov.

Small bushy tree, glabrous except the inflorescence and young shoots. *Leaves* petiolate, irregularly whorled in threes, oblong-lanceolate, entire, obtuse, about 5 to 7 inches long, strongly reticulate. *Flowers* in axillary simple racemes often as long as the leaves, generally in pairs irregularly clustered on the rhachis. *Pedicels* about 2 lines long, minutely pubescent. *Corolla* 2 to 3 lines long, nearly glabrous *Hypogynous glands* united in a ring. *Ovulary* hairy, style glabrous or nearly so, with a clavate stigmatic end. *Fruit* globular, with a coriaceous exocarp and a hard endocarp, about $\frac{3}{4}$ inch diameter.

Hab.—Camden Haven, New South Wales. Collected about 30 years ago either by Mr. Charles Moore or Mr. Carron, a former Botanical Collector of the Sydney Botanic Gardens.

Closely allied to the Nut-tree, *Macadamia ternifolia*, F.v.M. (of New South Wales and Queensland), from which it is readily distinguished by the petiolate entire leaves, rather smaller fruits and less hairy flowers and inflorescence

It may be pointed out that the sucker leaves have occasionally leaves with toothed margins, and shorter petioles, somewhat resembling the leaves of *M. ternifolia*, which shows the ancestral relationship of both species of *Macadamia*, but as the full grown leaves are constant in the characters indicated, and for other reasons, we have no hesitation in keeping the two species separate.

The following notes in regard to *Macadamia* and *Helicia* may be convenient for reference.

Bentham (Flora Australiensis, v. 406) recognises 3 species of *Macadamia*, viz., *M. Younqiana*, F.v.M , *M. ternifolia*, F.v.M., and *M. verticillata*, F.v.M.

Bentham and Hooker (Genera Plantarum, iii. 178) reduce these to two, pointing out that *M. verticillata* has been erroneously described as a *Macadamia* from a cultivated plant in the Botanic Gardens, Sydney, which has been proved to be a South African plant *Brabejum stellatifolium*, Linn. The species has since been lost to the Garden.

F. v. Mueller (Census of Australian Plants) recognises but one species of *Macadamia*, viz., *M. ternifolia*,—*M. Youngiana* being transferred to *Helicia*.

Baillon unites *Macadamia*, as well as several species hitherto described under *Helicia*, with the American genus *Andripetalum*, Schott (Baill. Vol. ii. p. 414). The characters of *Andripetalum* are ovules 2, descending, suborthotropous

A Engler (Die natürlichen Pflanzen-familien) recognises *Macadamia* 1 species in Australia; *Helicia* 25 species in Asia, Malayan Archipelago, and Australia; *Andripetalum* is not mentioned. We are, however, of opinion that Engler probably followed Baron von Mueller with regard to Australian plants of these genera.

———

Note on a Plant, hitherto only recorded from New Guinea, found in New South Wales.

Cheirostylis grandiflora, Blume, "Collection des Orchidées les plus remarquables de l'Archipel Indien et du Japon," Plate 13.

" In moist forests between rocks on the coast of New Guinea."

A plant of this species was collected by Dr. W. Finselbach on rocky hills "in a shady locality in the dense scrub," on the Richmond River, near Lismore. It will be seen that in New South Wales it grows under conditions practically identical with those under which it occurs in New Guinea It is a very pronounced saprophyte, growing on dead leaves In fact some of the Richmond River specimens were living on a layer of leaves only ¼ inch thick, and under this layer was the bare rock. The upper side of the creeping rhizome is nearly always exposed to the light, or at all events to the air, and when it is found between stones the rhizome is always fixed to dead leaves

The discovery of this New Guinea plant in New South Wales adds a genus to the flora of Australia. It is perhaps identical with *Gastrodia ovata*, F. M. Bailey (Botany Bulletin, No. xiv. p. 13, Dept. Agriculture, Queensland, 1896), and possibly identical with the *Anœctochilus* ("species unascertained ') recorded as having been found in Queensland. See Mueller's Census (2nd edition, p. 188).

The genus *Anœctochilus* resembles *Cheirostylis* closely in habit, and the two genera may be easily confounded from imperfect material.

A shortened translation of Blume's original description of *Cheirostylis grandiflora* is given herewith, as a matter of convenience.

CHEIROSTYLIS GRANDIFLORA, Blume.

Herb with a creeping fleshy rhizome, constricted between the nodes. Scape ascending, terete, minutely glandular-hairy in the upper part and with two distant sheathing bracts between the flowers and leaves. Leaves generally 4, ¾ to above 1 inch long, and ½ to ¾ inch broad, 3- to 5-nerved and faintly reticulate, brownish-green and somewhat purplish above, pale-purpurascent underneath. Flowers generally 3 on the scape, rarely solitary, shortly pedicellate and with a bract on the base of the pedicel. Sepals connate to above the middle, with a gibbous base, pale rose-coloured and minutely glandular-hairy outside. Petals adnate to the limb of the dorsal sepal. Labellum white, with a canaliculate gibbous base, adnate to the column, the erect concave base with inflexed margins and 4 filiform appendices inside on each side, the exserted limb dilated, 2-lobed, with cuneate lobes laciniate at the end. Column short, thick, with 2 erect appendages in front, about as long as the 2-cleft rostellum. Anthers short, acuminate, caudicle elongated.

Note on Grevillea alpina, *Lindl., new for New South Wales.*

This species has hitherto only been recorded from Victoria (B. Fl. v. 441). It was collected by Major (afterwards Sir

Thomas) Mitchell in his celebrated exploration of what is now the sister colony, and was described by Lindley. Our New South Wales specimens came from Albury, and were communicated by Mr. T. C. Burnell in August last. The flowers of our N.S.W. specimens are orange-red, merging into yellow in the upper half (" brownish-red," Mitchell), and nearly glabrous outside, as figured in *Bot. Mag.* t. 5007, and not villous outside as described by Bentham. Nevertheless Lindley's type specimens already referred to have villous flowers, and are somewhat different in general appearance from the Albury specimens. It might be a matter for further investigation to ascertain to what extent the species is variable before proceeding to name a variety. The species itself is readily recognised by the remarkably long hypogynous gland which projects almost horizontally into the gibbosity of the corolla (perianth).

DESCRIPTIONS OF SOME NEW ARANEIDÆ OF NEW SOUTH WALES. No. 7.

By W. J. RAINBOW.
(ENTOMOLOGIST TO THE AUSTRALIAN MUSEUM).

(Plate XLIX., figs 1, 2, 3, 3a.)

The present paper contains descriptions of three species new to science, and which, taken collectively, must form a valuable addition to our knowledge of the Araneidan fauna of this continent Of these, *Epeira coronata* is exceedingly interesting on account of its extraordinary structure; the second—*Pachygnatha superba*,--one of a small collection taken by Mr Ogilby during an excursion to Cooma, is a remarkably beautiful spider, the silvery granules that decorate the superior surface of the abdomen appearing like jewels against the back-ground of dark brown. The most important of the present series, however, is a new species of "flying" spider, for which I propose the name *Attus splendens*. In 1874 the Rev O. P. Cambridge, F.Z.S., described and figured in " Annals and Magazine of Natural History,"[*] an Attid for which he proposed the name *A volans*. From that singular spider the one now described, although possessing a remarkable affinity, is nevertheless sufficiently distinct to warrant the creation of a new species Each is beautifully coloured, but the scheme of ornamentation is widely different.   In *A. volans* the caput is ornamented with three longitudinal bars of soft greyish-green and two of scarlet, whereas *A splendens* has a curved transverse bar of scarlet but no longitudinal bands; then again the scheme of ornamentation on the abdomen of each is also different   But the chief reasons for describing this species, and which must have the weightiest considerations in such cases, are to

---

[*] Vol xiv 4th Series, pp. 178 180, Plate xvii. figs 4-4d.

be found in the fact that not only are the corpulatory organs somewhat more complicated than in *A. volans,* but the legs of *A. rplendens* are more numerously spined. When immersed in spirit the bright colours entirely disappear, but upon being withdrawn from the tube, and exposed to the atmosphere, the spider soon redisplays its gorgeous livery

Family EPEIRIDÆ.

Genus E P E I R A, Walck.

EPEIRA CORONATA, sp. nov.

(Plate XLIX., fig. 1.)

♀ Cephalothorax 4 mm. long, 3 mm. broad; abdomen 12 mm. in circumference.

Cephalothorax dark brown, convex, longer than broad. *Caput* moderately hairy, prominently elevated, summit surmounted with two lateral coniform tubercles, seated about four times their individual diameter from lateral eyes; normal grooves and indentations distinct. *Clypeus* moderately convex, dark brown, with faint lateral grooves radiating from the centre. *Marginal band* narrow.

Eyes black, the four comprising the central group forming a square or nearly so, and elevated upon a high and prominent tubercle, lateral pairs minute, placed obliquely on tubercles, and not contiguous.

Legs reddish-brown, hairy, moderately long, robust; relative lengths 1, 2, 4, 3; the first and second pairs are considerably the longest, and co-equal, and the third pair the shortest.

Palpi moderately long, robust, reddish-brown, and hairy.

Falces concolorous, robust, hairy; a row of three teeth on the margins of the furrow of each falx, fangs strong, reddish-brown at their base, wine-red at the points.

Maxillæ club-shaped, pale yellow, inclining inwards, a few short hairs at extremities.

42

Labium broad, short, rounded off at apex, reddish-brown at base, pale yellowish at tip.

Sternum shield-shaped, brown, moderately clothed with long coarse hoary hairs.

Abdomen somewhat spherical, projecting over base of cephalo-thorax, moderately clothed with short hairy pubescence, and surrounded with a corona of large and prominent tubercles; inferior surface shiny black at anterior extremity, sides and posterior extremity yellowish, with hoary pubescence.

Epigyne a transverse curved slit, the curvature directed forwards.

Hab.—New England; collected by Mr. A. M. Lea.

Family PACHYGNATHIDÆ.

Genus PACHYGNATHA, Sund.

PACHYGNATHA SUPERBA, sp. nov.

(Plate XLIX., fig. 2.)

♀. Cephalothorax 2 mm. long, 1½ mm. broad; abdomen 4 mm. long, 2 mm. broad.

Cephalothorax dark mahogany-brown. *Caput* slightly elevated, arched, normal grooves distinct; a few long hairs surrounding ocular area. *Clypeus* broad, arched. *Marginal band* broad.

Eyes of an opaline tint, arranged in two rows, slightly curved the curvature directed forwards; the two centre eyes of the front row are rather close together, and are separated from each other by a space equal to once their individual diameter, and those of the second row by a space equal to one diameter and a half, lateral eyes close to each other also.

Legs moderately long, pale yellow, clothed with long yellow hairs, and armed with a few long, fine spines.

Palpi similar in colour and armature to legs.

Falces dark mahogany-brown, divergent.

Maxillæ concolorous, arched, inclining inwards.

Labium concolorous also, rather broader than long, arched.

Sternum cordate, concolorous, smooth, slightly arched and furnished with a few short yellowish hairs.

Abdomen ovate, boldly projecting over base of cephalothorax. Colours: running down the centre from anterior, and terminating close to posterior, extremity is a broad pale yellowish patch, slightly broadest in front, and moderately and finely punctated; the patch is broadest at its anterior extremity and bordered in front and laterally with a sinuous line of bright silvery granules; laterally the colour is dark mahogany-brown; inferior surface brown, but a shade lighter in tint.

Epigyne a simple transverse slip.

Hab.—Cooma*; collected by Mr. J. D. Ogilby.

The position of the genus *Pachygnatha* in the system of the classification of the *Araneidæ* is not yet finally determined. Certain authors, as Westring, Ohlert, Simon, Lebert, and others associate it with the family *Theridiidæ*, but Thorell points out that the spiders of the genus *Pachygnatha* deviate from the typical *Theridiidæ;* Bertkau considers the genus as representing an independent group, to which he also refers the genus *Tetragnatha;* Menge, that it forms an independent family, of which it is the sole representative; finally, Staveley associates the genus *Pachygnatha* with the family *Linyphiidæ.* In commenting upon this question Wagner remarks that the study of these spiders, which is very incomplete, has led him to the conclusion that the grouping of Menge is the nearest approach to the truth, but in adopting Menge's classification, he does not consider the question settled, and accepts provisionally the position allotted by that author to this genus.† After giving the subject considerable thought and study, I have also come to the conclusion that Menge's elucidation of the position is the most correct, and consider it not unlikely that it will ultimately be accepted.

---

* This species appears to have a very wide range. Since the above was written I have received a specimen from Gisborne, Victoria, Mr. George Lyall, Junr., having collected it at that locality.

† Mém. de l'Acad. Imp. des Sci. de St. Pétersbourg, vii⁰ Série, Tome xlii., No. 11. L' Industrie des Araneina: Recherches de Woldemar Wagner, 1894, p. 150.

Family SALTICIDÆ

Genus ATTUS, Sim.

ATTUS SPLENDENS, sp. nov.

(Plate XLIX. figs. 3, 3a.)

♂. Cephalothorax 2½ mm. long, 2 mm. broad; abdomen 2½ mm. long, 2 mm. broad.

Cephalothorax steel-blue, broad, glossy. *Caput* steel-blue banded across the front with a broad curved bar of bright scarlet granules and scale-like hairs, the curvature directed forwards; in front, and surrounding the anterior row of eyes, there is a brush of short tawny hairs. *Clupeus* broad, high, rather flat, narrowest at its posterior extremity; at the junction of the cephalic and thoracic segments there is a broad but somewhat shallow depression, surrounded by a series of four white tufts or hairy brushes, the outer margins of which are surrounded with tawny hairs; sides steel-blue moderately clothed with tawny hairs. *Marginal band* fringed with hoary pubescence.

Eyes arranged in three rows, and nearly forming a square, those of the front row of a bright emerald green; of these the two median eyes are sensibly the largest; the two comprising the second row are much the smallest of the group and are also of a bright emerald green; the third row are somewhat smaller than the lateral eyes of the anterior series, and are of an opaline tint.

Legs moderately long and strong, yellow-brown, clothed with hoary hairs, and armed with short stout spines; relative lengths 3, 4, 2, 1.

Palpi concolorous, short; radial joints rather longer than cubital, thickly clothed with long white hairs on the upper surface, and very sparingly clothed with exceedingly short white hairs on the under side; copulatory organ a large, oblong corneous lobe hollowed on the under side and rather complicated.

Fulces dark brown, conical, divergent at apex, seated well back behind the frontal margin.

Maxillæ, *labium*, and *sternum* concolorous.

Abdomen oblong, narrowest in front, slightly overhanging base of cephalothorax, truncated at posterior extremity; upper side furnished (as in *A. volans*, Camb.) with an epidermis, which is continued laterally on either side to an extent considerably exceeding the width of the abdomen, and of an elliptical form; the outer portion of this epidermis on either side is capable of being depressed and folded round beneath the abdomen, or elevated and expanded to its full width after the manner of wings. The whole of the epidermis is densely covered with short and scale-like hairs, which give the different tints and hues to the abdomen; in the front and at the sides the colour is bright green; upon the upper surface there is a large oval ring of scarlet, the inner margins of which are bordered with bright green granules; in the centre there is a large patch of reddish-grey, surrounding a smaller and somewhat oval patch of scarlet; immediately below posterior margin of the scarlet oval ring there is a short, broad transverse patch covered with green granules, and fringed sparingly at ultimate extremity with scarlet scale-like hairs; lateral flaps furnished with bright green granules and scale-like hairs, becoming less brilliant towards their ultimate extremities; under side of a greenish grey colour, thickly clothed with short scale-like hairs.

Hab.—Sydney.

_____

EXPLANATION OF PLATE.

Fig. 1 —*Epeira coronata*, ♀.
Fig. 2 —*Pachygnatha superba*, ♀.
Fig. 3 —*Attus splendens* ♂.
Fig. 3a— ,,   ,,   showing epidermis folded under.

CONTRIBUTIONS TO A KNOWLEDGE OF THE ARACH-NIDAN FAUNA OF AUSTRALIA. No. 1.

By W. J. RAINBOW.

(ENTOMOLOGIST TO THE AUSTRALIAN MUSEUM).

(Plate XLIX., figs. 4, 4a, 4b.)

In this series of papers it is my intention from time to time to work out such material as could not well be considered or studied under the same headings as those papers already published by me. In the first instance, the title restricted me solely to the *Araneidæ*, and in the second place confined my attentions, so far as descriptions were concerned, to New South Wales. The title of the new series will, therefore, give me a much wider field, both from a zoological and geographical point of view, and enable me to record species and discuss questions appertaining to other groups, such as scorpions, pseudo-scorpions, &c., but also of *Araneidæ* from other Australian colonies than New South Wales. The present paper contains a description of a new species of *Buthus*; this was one of a collection of *Arachnida* obtained by Mr. Ogilby during a visit to Cooma about twelve months ago, and which that gentleman handed over to me shortly after his return to Sydney.

Order SCORPIONIDÆ.

Family ANDROCTONIDÆ.

Sub-family ANDROCTONINI.

Genus BUTHUS, Leach.

BUTHUS FLAVICRURIS, sp. nov.

(Plate XLIX., figs. 4, 4a, 4b.)

Colour: yellowish-brown above and laterally, pale yellowish underneath; palpi yellow-brown; tail yellow-brown above, laterally, and beneath; aculeus glossy, yellow-brown at base, deepening to dark brown at ultimate extremity; eyes dark brown; legs yellow.

Cephalothorax strongly arched, glossy, rather longer than broad, narrowest in front; anterior margin strongly indented; a deep longitudinal groove runs down the centre from anterior to posterior extremity, and separates the median eyes; these latter are seated on dark brown tubercles; the surface is smooth above, and has but few punctures; the sides are rather thickly furnished with minute granules; near the posterior extremity there are deep lateral compressions and grooves, and the minute darkish granules produce rather a dull tint; a few very fine yellowish hairs fringe the anterior extremity. *Marginal band* narrow and free from hairs.

Tergites keeled in the median line, minutely granulated, and fringed with a few short yellowish hairs; the final tergite is also keeled both above and laterally, the lateral keels seated low down.

Sternites glossy, with deep median and lateral depressions and minute punctures, the final sternite keeled laterally.

Tail long, glossy, almost parallel-sided, the segments deeply grooved, and strongly keeled and granulated laterally; sides and inferior surface strongly keeled and granulated; the segments vary in length, each succeeding one being longer than its predecessor, and the final one much the longest of any; each segment sparingly fringed laterally and underneath with rather long and fine yellowish hairs. *Vesicle* flat and glossy above, strongly arched, keeled and grooved laterally, the keels granulated; inferior surface sparingly furnished with yellow hairs, strongly keeled and grooved, the keels granulated. *Aculeus* moderately long and strong, gently incurved; *vesicle* and *aculeus* taken together are considerably shorter than the fifth caudal segment.

Legs yellow, sparingly clothed with long yellow hairs; femora and trochanters firmly keeled and granulated underneath; tibiæ, metatarsi, and tarsi armed with short strong spines.

Palpi long, powerful, fringed with short yellowish hairs; superior surface of humerus, brachium, and manus keeled and

granulated; of these the first two joints are much more strongly
granulated than the latter; lower surface keeled and granulated
laterally; the back of humerus and brachium moderately so;
humerus keeled laterally on inner side, sparingly granulated in
the median line; brachium granulated laterally, deeply grooved
down the middle; manus thick, moderately long, powerful, keeled
and granulated underneath; hand-back keeled, broad, moderately
granulated; fringes short, powerful, incurved, the keels and
granules giving them a somewhat darker appearance than the
hand; movable finger somewhat the longest.

Pectines long, somewhat tapering, and furnished with 16 teeth.

Measurements (in millimeters):—Total length, 52; length of
cephalothorax 6, width in front 3, behind 5; length of tail, 27½—
first segment 3, second 3½, third 4, fourth 5, fifth 6, vesicle and
aculeus 5; length of humerus 5; of brachium 5; hand, 6; hand-
back, 6; movable finger, 4½; width of humerus, 2; of brachium
2½, of hand (at base) 4, at apex 3; of hand-back, 3.

Hab.—Cooma.

EXPLANATION OF PLATE.

Fig. 4. --*Buthus flavicruris.*

Fig. 4a.—   ,,   ,,   tail, profile.

Fig. 4b.—   ,,   ,,   first and second caudal segments, ventral
surface.

REVISION OF THE GENUS PAROPSIS.

BY REV. T. BLACKBURN, B.A., CORRESPONDING MEMBER.

PART I.

Paropsis is probably the most numerously represented in Australia of the Coleopterous genera, and there is certainly no genus in greater need of revision or presenting greater difficulties to the task of revision. In attempting the task I cannot hope to execute it in a final manner owing to the large number of species that have been described in such fashion that it is impossible to identify them without seeing the types, and of the types there is little doubt many have perished, while the rest are so scattered over public and private collections as to preclude the examination of them by any individual reviser.

The species of this genus are extremely difficult to identify for another reason, viz, their great variability in respect of colour and markings. There is no species of which I have seen a long series in which I do not find more or less variability, and therefore it is necessary for the describer, if his work is to be of value, to base his specific distinctions almost entirely on structural characters, on form, and on sculpture.

In dealing with the enormous mass of species constituting the genus *Paropsis* the first step must necessarily be to divide the species into primary groups, and for this division I have come to the conclusion that in the main the best character to rely upon is that which Dr. Chapuis proposed for the purpose (Ann. Soc. Ent. Belg. xx.), viz., the sculpture of the elytra, for the adoption of any other character (that I have experimented with) disregards too radically the obvious affinities of species or fails by merely separating a few groups of very small extent and leaving the

great majority of the species to form one vast group. I think,
however, that there is one character founded on form that may
be profitably employed in constituting primary groups, viz., the
shape of the prothorax, as there is a large number of species
obviously allied *inter se*, the sides of whose prothorax are mucronate in front (in many instances bisinuate) and very few indeed
possessing this character which there can be any hesitation in
regarding as naturally allied to them. In following Dr. Chapuis'
system of groups I have, however, found it desirable to modify
it by somewhat increasing. the number of primary groups, and
also transposing the position of some of his groups, as I feel con-
fident that the natural place of his fourth group is immediately
after his first group. I propose, therefore, the following division
of the genus into primary groups :—

A. Sides of the prothorax mucronate in front (in many species
    bisinuate) .... Group ı

AA. Sides of the prothorax evenly arched.
    B. Puncturation of the elytra without any linear arrange-
        ment..... .... Group ii.
    BB. Puncturation of the elytra more or less linear in
        arrangement.
        C. About 20 more or less regular rows of punctures
            on each elytron.
            D. Elytra verrucose................... ... . Group iii.
            DD. Elytra devoid of verrucæ................. .... Group iv.
        CC. The linear arrangement is very partial and merely
            the result of several longitudinal unpunctured
            spaces. .... Group v
        CCC. About 10 defined rows of punctures on each
            elytron...   .... Group vi.

In the above scheme Groups i. and ii. together include almost
exactly the species of Dr. Chapuis' Group i. ; Groups iii. and iv.
together equal (again almost exactly) Dr. Chapuis' Group iv. ;
Groups v. and vi. equal Dr. Chapuis' Groups ii. and iii. respectively.

The present memoir begins with Group iii. I have already read a paper to the Royal Soc. of S.A. (Tr. Roy. Soc. S.A. 1894) on Dr. Chapuis' Group i. (my i. and ii.), but during the interval since its publication so many new species of that group have come into my hands that it will be desirable to deal with it afresh, and as the new material throws fresh light on and modifies a considerable part of the work there seems to be almost a necessity for rewriting my paper on it. This, however, I purpose postponing until I have finished my work on the other groups, and, therefore, I begin with the first group that has as yet received no systematic treatment.

The section of *Paropsis* to be now dealt with,—that containing the species with about 20 rows of punctures, and also with *verrucæ*, on each elytron,—is for more than one reason, the most difficult in the genus to treat satisfactorily. It is one of the two sections containing a very large number of species, the species appertaining to it are mostly obscure, closely allied and very variable, and many of those already named are described in a manner that completely defies identification.

Dr. Chapuis (*loc. cit.*) enumerates 42 species as forming this group, but there are doubtless others among the 43 species enumerated by him as unable to be referred to a definite place in *Paropsis*. Since the publication of Dr. Chapuis' memoir only 5 species have been added. Dr. Chapuis' descriptions are far from satisfactory, because they are mere diagnoses without any notes of comparison between one species and another, and because they deal with colour and marking to an extent that is misleading in dealing with variable insects. I have, however, been fortunate enough (through the courtesy of M. Sevrin, of Brussels) to secure a considerable collection of types and named specimens from Dr. Chapuis' collection, without which I could not have ventured on the present work, but even with this assistance there is an unsatisfactory number of names that I have been compelled to disregard totally as incapable of identification with any particular species; many of the descriptions annexed to them might refer to almost any species of the group.

spι o ⸚, np.

. *papulenta*, Chp., (*papulosa*, Stäl, n
          on the same insect as *ru*
tion is insufficient to furnish grou.
ss.

. *atomaria*, Oliv., is possibly a m
opsis but cannot be identified by th
ot certain that it was taken in Aus
Islands of the South Seas."

. *aspera*, Chp., attributed by its a
·emely anomalous species of which I
puis collection.   I have, however, re
account of the front angles of its pr
wing to the variability and close
opses of this group I have found
·itic distinctions almost entirely on st
iminary remarks on the nature of t
ienclature I have employed in recorc
.   After long and careful study I h
t important and constant character
ies.   This is a character particu

upon a method of characterising the form that will render it
practically available. The difference of form between one species
and another is best observed by looking at the specimen from the
side, and when a number of species of this group are examined
they are found to present two very different types of outline, the
one in which the arch of the upper outline has its summit near
the front of the elytra and thence curves away continuously
downwards to the apex, the other in which the summit is considerably further back. To express this distinction clearly I have
called this summit of the curve the point at which the insect is at its
"greatest height;" and as it is easier for the eye to determine the
middle of a straight line than of a curve I have called the middle
of the *lower* outline (as viewed from the side, whence it appears
as a straight line) "the middle of the elytral margin." Thus I
have formed two main divisions of the *Paropses* of this group on
the position of the "greatest height' in relation to the "middle
of the elytral margin;" it being in the one case opposite a point
considerably in *front* of the "middle of the elytral margin," in
the other case opposite a point just about (or a little behind) the
middle. It must be noticed that this character is slightly affected
by sex, the "greatest height" being usually a little further back
in the female than in the male, but this does not invalidate the
divisions founded upon it, as I find that even in the females of
the one group the "greatest height" is markedly nearer the base
of the elytra than in the males of the other group, and there are
very few species sufficiently intermediate to cause any difficulty.
With a little practice and comparison of specimens I think this
character will be found quite easy to appreciate. This difference of
form then I take as the character on which primary divisions of
this group of *Paropsis* should be based, after first eliminating from
the crowd of species a few possessing altogether exceptional
characters on the strength of which I treat them as forming a
separate division. These exceptional characters need no explanation and will be easily recognised by the student; the aggregate
that they bring together is entirely artificial, but the convenience
of forming it is obvious.

For secondary and tertiary divisions I have found the most valuable characters in the margins of the prothorax, and the structure of the humeral regions of the elytra In many species the transverse convexity of the prothorax is even (independently of the evenness or otherwise of the *sur,face*), *i e.*, the convexity of the disc continues unchanged to the extreme lateral margin; in the rest of the species the convexity becomes less strong on a more or less wide marginal space. In these latter I call the prothorax "explanate at the sides." On the elytra the area between the humeral callus and the lateral margin presents two aspects,—in some species being flattened (or even concave) so that (looked at from a certain point of view) there appears to be a space (roughly triangular, the humeral angle of the elytron being the apex of the triangle) on a more or less different plane from that of the general surface; in the other species this portion of the elytra continues quite uninterruptedly the general plane of the surface. I characterise the former of these aggregates as "depressed under the humeral callus."

Another character calling for remark is the relation of the marginal portion (which is the external surface of the epipleuræ) of the elytra to the disc. In most species the distinction between these is indicated by a lightly impressed ill-defined longitudinal concavity (generally most noticeable for a short space near the apex). I have called this concavity the "submarginal sulcus."

And yet another character requires comment, viz., the structure of the epipleuræ of the elytra. These consist of an inner more or less horizontal piece (generally a mere fine line in its apical half) and an external more or less vertical piece. The height of the external piece varies greatly in different species, but is very constant in the individuals of a species. Its height, however, is so difficult to express profitably in words that I have had to fall back upon characterising its indication on the upper surface According as it is more or less high, the lateral margin is further from or nearer to the humeral callus, so that in species with the external piece of the epipleuræ greatly elevated the inner edge of the humeral callus is as far from the external margin of the elytra as from the suture, while in others it is much nearer to the external margin.

It will be observed that in the following descriptions I have in some instances mentioned only characters in respect of which a species differs from some other to which it is closely allied and added the statement "cetera ut . . . " (an instance of this occurs in the description of *P. extranea*). I have adopted this course to avoid needless repetition, but it will be well to state explicitly here that in every such case I have carefully compared the insect on which the abbreviated description is founded with the detailed description preceding it (in the case of *P. extranea*, *e.g.*, with the description of *P. sternalis*), and ascertained that the whole of the detailed description applies to it except in respect of the characters noted in the abbreviated description.

I divide this group of *Paropsis* (distinguished by having the sides of the prothorax neither mucronate in front nor bisinuate, and each elytron with about 20 rows of punctures and also some verrucæ) then into subgroups as follows :—

A. Species with strongly marked characters (as detailed in
    the tabulation of species)...................................... Subgroup i.
AA. Species not referable to Section A.
    B. The greatest height of the insect (viewed from the
        side) not or scarcely in front of the middle of the
        elytral margin.
        C. Elytra depressed under the humeral callus......... Subgroup ii.
        CC. Elytra not depressed under the humeral callus. Subgroup iii.
    BB. The greatest height of the insect (viewed from
        the side) considerably in front of the middle of
        the elytral margin...................................... Subgroup iv.

This first part of my "Revision of the genus *Paropsis*" deals with the first three of the subgroups into which I divide the group. I begin with a tabulated statement of the distinctive characters of the species in Subgroup i., and then proceed to furnish descriptions of the new species enumerated in the tabulation. Afterwards I treat Subgroups ii. and iii. similarly. The names printed in italics are the names of those species which I have etermined by studying the descriptions without having

seen an authentic type.   It is possible that there may be incorrect
identifications among these; but I think not since they are all
species described as presenting well marked characters.

I have to thank many friends for their courtesy in lending me
their collections for study and comparison, especially Mr Masters,
to whom I fear I have given much trouble by my enquiries
regarding types in the Macleay Museum, and who has done me
the great favour of sending me specimens carefully compared with
those types, whereby the reliability of my memoir has been vastly
increased, making him really a co-worker with me in the production.   I have had the privilege also of examining the following
collections, viz., S.A. Museum, Agricultural Department of New
South Wales and Agricultural Bureau of W. Australia, together
with the collection of Mr. A. M. Lea; also numerous specimens
forwarded by Mr. A. Simson, Mr. C. French, Mr. W. W. Froggatt, and the late Messrs Olliff and Skuse.

TABULATION OF THE SPECIES FORMING SUBGROUP I.

A. Prosternum not sulcate down the middle. .... insolens, Blackb.
AA. Prosternum sulcate down the middle; but very
    wide, and scarcely narrowed in front.
  B. Colour testaceous or red, elytra moderately
    punctured.
    C. Prothorax at its widest much behind the
      middle.
      D. Sides of elytra nearly vertical, a slight
        subhumeral depression...... .. .. extranea, Blackb.
      DD. Sides of elytra slope obliquely outward,
        no subhumeral depression ..   .. sternalis, Blackb.
    CC. Prothorax at its widest scarcely behind
      the middle. .... funerea, Blackb.
  BB. General colour black ; elytra coarsely punc-
    tured. .... squiresensis, Blackb.
AAA Prosternum normal, but other characters
    exceptional, as follows : –
  B. The humeral calli elevated into large ear-like
    processes. .... papuligera, Stäl.
  BB. A well- defined antemedian discal exca-
    vation on the elytra.

C. Form oblong, very little convex............... scabra, Chp.
CC. Form broadly ovate, strongly convex.. . . rugosa, Chp.
BBB The exceptional characters lie in the
    elytral epipleuræ
  C. Epipleuræ subhorizontal .... armata, Blackb.
  CC. Inner (horizontal) part of epipleuræ
    nearly reaches the apex as a distinct ledge.
    D. Basal ventral segment coarsely punctured.
      E. Sides of prothorax strongly explanate.
        F. Underside testaceous..................... Chapuisi, Blackb.
        FF. Underside black.
          G. Interstices of elytral punctures
            but little rugulose . ... latipes, Blackb.
          GG Interstices of elytral punctures
            strongly rugulose, almost con-
            cealing the punctures .... raucipennis, Blackb.
      EE. Sides of prothorax only slightly ex-
        planate .... Karattæ, Blackb.
    DD. Basal ventral segment feebly punctu-
    late.
      E. Elytra with a postbasal discal im-
      pression.
        F The marginal part of elytra mode-
        rately wide and more or less vertical.
          G. Size very large (Long. 6 lines)
            suture and some vittæ black . *graphica*, Chp
          GG. Size much smaller (Long. 5 l.)
            suture concolorous with gene-
            ral surface. .. .. rustica, Blackb.
        FF. The marginal part of the elytra
        very wide and very strongly out-
        sloped .... læviventris, Blackb.
      EE. Elytra without any postbasal im-
      pression on disc.......................... sublimbata, Chp.

P. INSOLENS, sp.nov.

♀. Elongato-ovalis vel sat late subparallela, modice convexa, altitudine majori (a latere visa) contra elytrorum marginem medium posita; subnitida; rufa, hic illic picescens; capite fortius minus crebre punctulato; prothorace quam longiori ut 2⅓ ad 1 latiori, ab apice longe ultra medium dilatato, pone
43

apicem haud impresso, grosse vermiculato-ruguloso et sparsim
punctulato, lateribus sat arcuatis haud deplanatis, angulis
posticis rotundatis; scutello lævi fortiter convexo; elytris sub
callum humeralem vix depressis, pone basin haud impressis,
antice suturam versus subseriatim vermiculato-rugulosis
(latera versus crebre confuse verrucosis), partibus elevatis
quam depressæ magis rufis, parte marginali a disco haud
distincta (margine summo nihilominus præter modum lato),
calli humeralis margine interno a sutura quam ab elytrorum
margine laterali multo magis distanti, segmento ventrali
basali subtiliter sparsissime punctulato; elytrorum epipleuris
subhorizontalibus ; prosterno medio haud longitudinaliter
concavo   Long 6, lat 4⅓ lines.

Quite incapable of confusion with any other *Paropsis* known
to me.

W. Australia , sent to me by Mr. French.

P. STERNALIS, sp.nov.

♀. Ovalis, modice convexa, altitudine majori (a latere visa) ante
elytrorum marginem medium posita; minus nitida; flavo-
castanea, in prothorace maculis 4 (transversim positis) et
in elytris verrucis numerosis nigris, capite crebre fortius,
prothorace sat crebre fortiter (ad latera grosse), punctulatis.
hoc quam longiori plus quam duplo (ut 2⅘ ad 1) latiori, ab
apice longe ultra medium dilatato, pone apicem transversim
vix impresso, lateribus leviter arcuatis haud deplanatis,
angulis posticis rotundatis, scutello nitido fere lævi: elytris
sub callum humeralem vix depressis, paullo pone basin
transversim vix impressis, crebre subseriatim fortiter (quam
prothorax paullo magis, ad latera quam in disco vix magis,
fortiter) punctulatis, interstitiis vix rugulosis, parte marginali a disco vix distincta (sulculo submarginali sub-
obsoleto, apicem summum haud attingenti, apicem versus
leviter impresso), calli humeralis margine interno a sutura
quam ab elytrorum margine laterali multo magis distanti.
segmento ventrali basali sparsim subfortiter punctulato,
prosterni parte concava mediana lata.   Long. 4, lat. 3 lines.

Easily distinguishable by its uniform flavo-castaneous colour interrupted only by the black spots on the prothorax and verrucæ on the elytra together with its very broad prosternal longitudinal furrow, which is quite as wide as in *P. geographica*, Baly. The humeral callus is extremely feeble.

N. Territory of S. Australia.

P. EXTRANEA, sp.nov.

♀. Altitudine majori ad medium (vel fere pone medium) elytrorum posita; obscure brunneo-rufa, ut *P. sternalis* nigronotata; prothorace in disco minus crebre punctulato, antice fortiter angustato, lateribus fortiter rotundatis; elytrorum callo humerali sat prominenti, puncturarum interstitiis apicem versus sat rugulosis; cetera ut *P. sternalis*.

Very like *P. sternalis* but at once distinguishable from it (apart from colour) by its greatest height being not at all in front of the middle, by its prothorax being much less closely punctulate on the disc with its sides much more strongly rounded and its front part much more narrowed, and by its much better developed humeral calli.

N. S Wales ; I do not know the exact habitat.

P. SQUIRESENSIS, Blackb.

♂. Leviter ovata ; minus lata; modice convexa, altitudine majori (a latere visa) contra elytrorum marginem medium (vel etiam magis retro) posita; sat nitida; nigra vel nigropicea, capite antennis pedibus (elytrorumque verrucis nonnullorum exemplorum) plus minusve rufescentibus, capite crebre subtilius punctulato; prothorace quam longiori ut $2\frac{1}{4}$ ad 1 latiori, ab apice ultra medium dilatato, pone apicem transversim impresso, inæqualiter (in disco puncturis majoribus cum aliis minoribus intermixtis, ad latera confertim grosse) punctulato, lateribus minus arcuatis nullomodo deplanatis, angulis posticis obtusis; scutello lævi vel vix punctulato; elytris sub callum humeralem leviter depressis, pone basin transversim impressis, crebre fortiter sat seriatim

punctulatis, verrucis numerosis sat magnis plus minusve
rufescentibus confuse instructis, interstitiis antice minus
(feminæ quam maris magis distincte) postice magis rugulosis,
parte marginali angustissima modice distincta, calli humeralis margine interno a sutura quam ab elytrorum margine
laterali multo magis distanti, segmento ventrali basali
(maris sat fortiter feminæ subtilius) punctulato; prosterni
parte concava mediana lata.   Long. 3-3¾, lat. 2⅓-2⅖ lines.
Femina quam mas paullo magis convexa.

Easily distinguishable (among the species with the median
space of the prosternum exceptionally wide) by the nearly black
colour of the general surface, the elytral verrucæ being slightly
reddish but not conspicuously different in colour from the derm
The elytral margin viewed from the side is very sinuous (as in
strigosa and a few other species).   I have thought it well to re-
describe this species as the acquisition of more specimens shows
some variation from the type, especially in colour.

N. W. Australia ; sent to me by Mr. Masters.   [Also previously taken by the Elder Exploring Expedition.]

P. ARMATA, sp.nov.

♀. Sat late subovata, minus convexa, altitudine majori (a
latere visa) haud ante elytrorum marginem medium posita,
minus nitida; supra rufo-aurantiaca, prothorace (lateribus
exceptis) scutello et elytrorum tuberculis maculisque
picescentibus; subtus picescens, antennarum basi tarsisque
rufis, capite sat fortiter ruguloso; prothorace quam longiori
ut 2⅓ ad 1 latiori, ab apice vix ultra medium dilatato, pone
apicem transversim vix impresso, grosse vermiculato-
ruguloso et sparsim punctulato, lateribus modice arcuatis
haud deplanatis, angulis posticis obtusis; scutello sublævi in
medio convexo, elytris sub callum humeralem depressis, pone
basin transversim vix manifeste impressis, subseriatim sat
fortiter punctulatis et tuberculorum conicorum seriebus 9
armatis, parte marginali angusta a disco (per sulculum sub-
obsoletum continuum) divisa, calli humeralis margine in-

terno a sutura quam ab elytrorum margine laterali multo
magis distanti; segmento ventrali basali sparsius minus
subtiliter punctulato; epipleuris subhorizontalibus Long. 5,
lat. 3⅘ lines.

Somewhat resembles *P. insolens*, its most striking character
consists in the structure of the epipleuræ; in most *Puropses* these
(as noted above) consist of an inner horizontal ledge and an external almost vertical piece, but in the present species (and even
more markedly in *P. insolens*) the two pieces are narrow and
scarcely distinct *inter se* and form an almost evenly continuous
surface outturned so as to be obliquely subhorizontal.

N. S. Wales.

P. CHAPUISI, sp.nov.

♂. Late ovalis, modice convexa, altitudine majori (a latere visa)
sat longe ante elytrorum marginem medium posita : minus
nitida, castanea, antennis ultra medium prosterno elytrorumque verrucis infuscatis; capite crebre subtiliter punctulato;
prothorace quam longiori plus quam duplo (ut 2½ ad 1) latiori,
ab apice longe ultra medium dilatato, crebre sat subtiliter
subæqualiter (sed ad latera subgrosse) punctulato, pone
apicem transversim distincte impresso, lateribus sat late
deplanatis sat fortiter arcuatis, angulis posticis nullis ;
scutello leviter sparsissime punctulato; elytris sub callum
humeralem triangulariter distincte depressis, paullo pone
basin leviter distincte transversim late impressis, crebre sat
fortiter sat æqualiter (latera versus vix magis crasse) punctulatis, verrucis parvis nonnullis apicem versus instructis, parte
marginali lata a disco (sulculo manifeste impresso sed paullo
ante medium interrupto hinc ad apicem continuo) divisa,
calli humeralis margine interno a sutura quam ab elytrorum
margine laterali haud magis distanti ; epipleurarum parte
interna (horizontali) fere ad apicem (ut dorsum distinctum)
continua; segmento ventrali basali fortiter subgrosse punctulato, apicali emarginato, incisuræ facie postica subverticali.
Long. 5, lat. 4¼ lines.

Very distinct among its near allies by its entirely (the infuscate prosternum excepted) pale castaneous under surface in combination with a coarsely punctured basal ventral segment and widely explanate sides of prothorax. I have seen only a single specimen, which is from Dr. Chapuis' collection, and is ticketed "*papulosa*." *P papulosa*, Er., however, is a much smaller and very differently sculptured insect, while *P. papulosa*, Stal, is also much smaller and very differently sculptured (especially in having the whole of the elytra thickly studded with verrucæ). I think Dr. Chapuis was certainly mistaken in calling this species *papulosa*

Australia.

P. RAUCIPENNIS, sp.nov.

♀. Late ovalis, valde convexa, altitudine majori (a latere visa) vix ante elytrorum marginem medium posita; minus nitida, castanea, prothoracis maculis nonnullis elytrorum sutura (verrucisque nonnullis) et corpore subtus (coxis abdominisque apice exceptis) nigris, antennis (basi excepta) infuscatis, capite crebre minus subtiliter punctulato ; prothorace quam longiori multo plus quam duplo (fere ut $2\frac{3}{4}$ ad 1) latiori, ab apice paullo ultra medium dilatato, crebre minus subtiliter (in disco paullo minus crebre, ad latera sat grosse) punctulato, cetera ut præcedentis (*P. Chapuisi*) ; scutello medio opaco confertim punctulato; elytris crebre granulato-rugulosis (sicut puncturæ vix manifestæ sunt), pone basin vix distincte impressis, cetera ut præcedentis ; epipleuris et segmenti basalis ventralis sculptura ut præcedentis. Long. 5, lat $4\frac{1}{2}$ lines.

Differs from the preceding (apart from colour) chiefly by its evidently more transverse prothorax and the very different sculpture of its elytra, which are covered with rugulosity (chiefly transverse wrinkles and confused granules) in such fashion that the puncturation is very little noticeable except in the marginal portion. Unfortunately I have seen only a female of this species and a male of the preceding, but I have little doubt that the female of the preceding is a markedly less convex insect than this with the summit of the upper outline of the elytra (viewed from the side) evidently nearer to the base. In both this species and the preceding the continuance of the shallow sulciform impression (which marks the distinction between the discal and marginal regions of the elytra) to the actual apex causes the appearance, when the insect is viewed from the side, of the suture being produced hindward in a short mucro.

S Australia.

P. KARATTÆ, sp.nov.

♀. Late ovalis, modice convexa, altitudine majori (a latere visa) sat longe ante elytrorum marginem medium posita, minus nitida; castanea (prothoracis maculis nonnullis, elytrorum sutura disci margine externo et verrucis numerosis regulariter seriatim positis, corporeque subtus maculatim, nigris), antennis apicem versus infuscatis; capite prothoraceque (colore excepto) fere ut *P. Chapuisi*, sed hujus lateribus vix manifeste deplanatis, scutello puncturis nonnullis impresso; elytris sub callum humeralem triangulariter distincte depressis, pone basin vix manifeste impressis, crebre subreticulatim rugulosis sed minus distincte punctulatis, sulculo subhumerali minus determinato et ante apicem ipsum toto deficienti, calli humeralis margine interno a sutura quam ab elytrorum margine laterali manifeste magis distanti epipleuris et segmenti basalis ventralis sculptura ut *P. Chapuisi*. Long. 5, lat. 4⅕ lines

Distinguished among its near allies (apart from probably variable characters) by the sides of its prothorax markedly less explanate, the feebleness of the distinction between the elytral disc and margins (the submarginal sulcus failing entirely before the apex so that viewed from the side there is no appearance of a sutural projection), and the humeral callus with its inner margin considerably nearer to the lateral margin than to the suture. The sculpture of the elytra resembles that of *P. raucipennis* in consisting of rugulosity mostly concealing the puncturation but it is feebler and less granulose than in that species so that the puncturation is not quite so much obscured.

Kangaroo Island.

P. RUSTICA, sp.nov.

♀. Ovalis, minus lata; modice convexa, altitudine majori (a latere visa) longe ante elytrorum marginem medium posita; minus nitida; rufo-brunnea (elytrorum verrucis numerosis sat æqualiter, nec regulariter seriatim, dispositis, vittulisque nonnullis indeterminatis et sternis epipleurisque, nigris; antennis apicem versus infuscatis); capite prothoraceque fere ut *P. læviventris* sed hoc magis transverso (ut 2¾ ad 1) lateribus vix deplanatis minus fortiter arcuatis; elytrorum depressione humerali, sulculo submarginali (hoc in medio minus abrupte interrupto), impressione subbasali et epipleuris ut *P. Chapuisi*; elytris sat fortiter subseriatim sat crebre punctulatis, interstitiis in disco vix (verrucis neglectis) rugulosis, parte marginali sat grosse rugulosa; segmento ventrali basali subtiliter punctulato.   Long. 5, lat. 4 lines (vix).

Near *P. sublimbata*, Chp., but at once distinguishable by the very much coarser puncturation of the elytra as well as by their greatest height (viewed from the side) being markedly nearer to the front and by the elytral verrucæ being manifestly larger, more conspicuous, more numerous, and less regularly seriate. The elytral apex (viewed from the side) projects as in *P. Chapuisi*.

N. S. Wales; taken by Mr. Lea at Forest Reefs

P. LÆVIVENTRIS, sp.nov.

♂. Sat late ovalis, minus convexa, altitudine majori (a latere visa) paullo ante elytrorum marginem medium posita; sat nitida; castanea (elytrorum macula elongata communi anteriori verrucis nonnullis exemplorumque nonnullorum vitta indeterminata submarginali posteriori nonnullorum exemplorum scutello et sternis, nigris; antennis apicem versus infuscatis); capite dupliciter (subtiliter et minus subtiliter) sat crebre punctulato; prothorace quam longiori plus quam duplo (ut 2½ ad 1) latiori, ab apice sat longe ultra medium dilatato, inæqualiter (in disco medio subtilius minus crebre in lateribus sat grosse, alibi magis crebre) punctulato, pone apicem transversim distincte impresso, lateribus leviter deplanatis sat fortiter arcuatis, angulis posticis nullis, scutello sublævi, elytris sub callum humeralem triangulariter leviter depressis, paullo pone basin leviter distincte transversim impressis, sat crebre sat distincte subseriatim (latera versus vix magis fortiter) punctulatis, interstitiis sat fortiter rugulosis, verrucis sparsis minus conspicuis series duas (in interstitiis circiter 5° 9°que positis) formantibus, parte marginali callo humerali et epipleuris ut *P. Chapuisi,* segmento ventrali basali minus perspicue punctulato.

♀. Manifeste magis convexa (exempli typici sternis piceis potius quam nigris). Long. $3\frac{1}{5}$-$4\frac{1}{2}$, lat 3-$3\frac{2}{5}$ lines.

Smaller and more nitid than any of its immediate allies. Easily distinguishable by the characters specified in the tabulation and by the large blackish blotch resembling a more or less wide dilatation of the anterior one-third portion of the suture. Viewed from the side the apex of the elytra appears to project as in *P. Chapuisi.*

S. Australia ; near Adelaide.

TABULATION OF THE SPECIES FORMING SUBGROUP II.

A. Inner edge of humeral callus distinctly nearer
 to lateral margin of elytra than to suture.
 B. Sides of prothorax more or less explanate.
  C. Elytra not having well-defined continuous
   costæ.
   *D. Puncturation of elytra not particularly
    fine.
    E Upper surface of elytra in general, or
     at least the verrucæ, black or nearly so.
     F. Explanate margins of prothorax wide
      (each about ⅓ of width of discal part).
      G. Postbasal impression of elytral
       disc feeble.

---

* In *P. exsul* the elytral puncturation is not very much finer than in the species under this letter.

H. Prothorax at its widest notably
behind the middle.
  I. Elytral puncturation (or at
least its seriation) much obscured, especially behind, by
close rugulosity of the inter-
stices................................ *explanata.* Chp.
  II. Elytral puncturation well
defined, and seriate to apex.
  J. Legs testaceous.
    K. Form very wide; elytra
strongly rounded at sides regularis, Blackb.
    KK. Form much less wide,
elytra less rounded at
sides .... comma, Blackb.
  JJ. Legs dark.. .... sylvicola, Blackb.
HH. Prothorax at its widest at the
middle............................ *melanospila*, Chp.
GG. Postbasal impression of elytral
disc very strong.................... baldiensis, Blackb.
FF. Explanate margins of prothorax
much narrower.
  G. Median verrucæ of prothorax
scarcely defined.
  H. Prothorax dark in the middle,
the sides pallid in strong con-
trast.. .. .. piceola, Chp.
  HH. Prothorax not coloured as in
piceola.
    I. Elytral verrucæ large, all iso-
lated, nowhere confused with
interstitial rugulosity.
    J. Puncturation of prothorax
not asperate.
      K. Puncturation of prothorax
sparse, coarse and irregu-
lar .... caliginosa, Chp.
      KK. Puncturation of protho-
rax much finer, closer
and more even.. .... pustulosa, Blackb.
    JJ. Puncturation of prothorax
very close and asperate *serpiginosa*, Er.

II. Elytral verrucæ much less distinct, confused (especially
in front) with interstitial
rugulosity.
   J. Puncturation of prothorax
    close and asperate, form
    strongly convex.....   . . mixta, Blackb
   JJ. Puncturation of prothorax
    not close and asperate;
    form much less convex.
     K. Postbasal impression of
      elytra almost wanting . sordida, Blackb.
     KK. Postbasal impression of
      elytra well defined   foveata, Blackb.
  GG. Median verrucæ of prothorax
   tuberculiform   .... *verrucicollis*, Chp.
EE. Upper surface (including verrucæ,
which are very large) red or brown
  F. Prothorax not much narrowed in
   front, widest at the middle   montuosa, Blackb.
  FF. Prothorax much narrowed in front,
   widest considerably behind middle rosea, Blackb.
DD. Puncturation of elytra decidedly fine.
  E. Prothorax not much narrowed in front,
   widest at middle   .... exsul, Blackb.
  EE. Prothorax much narrowed in front,
   widest considerably behind middle.
   F . Size moderate (Long. 3¾ l) .... simulans, Blackb.
   F F. Size very small (Long. 2½ l) . abjecta, Blackb.
CC. Elytra with well defined continuous costæ *ferrugata*, Chp.
BB. Sides of prothorax not at all explanate.
C. Elytra not having a well defined transverse
  wheal-like ridge.
  D. Form nearly circular, elytra wider than
   long. .... mediocris, Blackb.
  DD. Form less wide, elytra not wider than
   long.
   E. Prothorax with somewhat evenly
    rounded sides, only moderately nar-
    rower in front than at base.
    F. Puncturation of elytra not particu-
     larly fine and close

G. Disc of prothorax closely and
    evenly punctulate.
  H. Prothorax at its widest markedly
    behind the middle .. .. **ruficollis, Blackb.**
  HH. Prothorax at its widest at the
    middle . . .. **propria, Blackb.**
  GG. Disc of prothorax (especially in
    in the middle) considerably less
    closely punctulate .... **whittonensis, Blackb.**
FF. Puncturation of elytra exception-
  ally fine and close.
  G. Submarginal part of elytra very
    distinct near apex .. . . **cribrata, Blackb.**
  GG. Submarginal part of elytra not
    distinct .... **declivis, Blackb.**
EE. Prothorax widening from apex almost
  to base, base much wider than front
  margin.
  F. Puncturation of elytra not particu-
    larly fine.
    G. Elytral verrucæ large, scarcely
      elevated, isolated, very nitid and
      black .. .. **Tatei, Blackb.**
    GG. Elytral verrucæ not as in *Tatei.*
    H. Surface of elytra (disregarding
      the verrucæ) only moderately
      rugulose.
      I. The elytral verrucæ incon-
        spicuous, darker than derm
        and tending to be trans-
        versely elongated.
        J. The humeral calli in their
          normal position.
          K. Upper outline of elytra
            (viewed from the side) a
            strong regular curve ..... **punctata, Marsh.**
          KK. Upper outline of elytra
            (viewed from the side)
            somewhat flattened ... **alticola, Blackb.**
        JJ. The humeral calli excep-
          tionally near lateral mar-
          gins of the elytra ..... . **Victoriæ, Blackb.**

II. The elytral verrucæ very con-
　　spicuous and pallid　　...... solitaria, Blackb.
HH. Surface of elytra (disregarding
　　the verrucæ) closely granu-
　　lose-rugulose even at the base lima, Blackb.
　　FF. Puncturation of the elytra excep-
　　　tionally fine .... invalida, Blackb.
CC. Elytra having a well-defined transverse
　　wheal-like ridge............ .... transversalis, Blackb.
AA. Inner edge of humeral callus equidistant
　between suture and lateral margin of elytra oxarata, Chp.

P. COMMA, sp.nov.

Sat late subovata, modice convexa, altitudine majori (a latere
visa) contra marginem medium (vel paullo magis antice)
posita; sat nitida; ferruginea, capite postice prothoracis
maculis 2 (his figuram comma simulantibus) et elytrorum
verrucis nigris, lateribus dilutioribus, corpore subtus nigro
(rufo-variegato) antennis basi excepta piceis; capite subtilius
subrugulose punctulato; prothorace quam longiori ut $2\frac{2}{5}$ ad
1 latiori, ab apice sat longe ultra medium dilatato, pone
apicem transversim minus perspicue impresso, sat fortiter vix
confertim (ad latera grosse rugulose) punctulato, lateribus
fortiter arcuatis late leviter deplanatis, angulis posticis nullis;
scutello sublævi; elytris sub callum humeralem leviter depressis, pone basin transversim leviter impressis, fortiter sat
crebre subseriatim (ad latera paullo magis, postice paullo
minus, grosse) punctulatis, verrucis (his a basi ad apicem
continuis) elongatis cum aliis rotundatis instructis, interstitiis minus rugulosis, parte marginali lata a disco (per
sulculum ante medium vix interruptum) divisa, calli humeralis
margine interno a sutura quam ab elytrorum margine
laterali vix magis distanti; segmento ventrali basali (hoc
rufo) sparsim subtilius punctulato; antennarum articulo 3^{u}
quam 4^{ns} sat longiori.　Long. $4\frac{1}{5}$-$4\frac{1}{2}$, lat $3\frac{1}{3}$-$3\frac{1}{2}$ lines.

Femina quam mas paullo magis convexa.

This species is superficially very much like *P. serpiginosa*, Er.,
from which it differs i ter alia by its larger size, evidently greater

convexity, more widely (though not more strongly) explanate
sides of prothorax, different prothoracic markings, and especially
by the extra-discal part of the elytra much wider and evidently
sloping outward (in *serpiginosa* it is nearly vertical) with the
humeral callus considerably more distant from the lateral margin
of the elytra, as well as by the considerably longer third antennal
joint (in *serpiginosa* this joint is scarcely longer than the fourth)
If an example be looked at with the head directed towards the
observer the mark on the observer's right resembles a comma
(that on the left being of course reversed).   The tails of the two
marks are confluent in some examples   In *serpiginosa* the prothorax is usually without markings, but in some examples there
are four more or less conspicuous blackish spots placed in a
transverse row.   This species is also very near *P. regularis*,
Blackb., differing by its smaller size, evidently narrower form,
less closely punctulate prothorax with different markings, &c.

Tasmania; sent by Mr. Simson from Launceston.

P. SYLVICOLA, sp.nov.

♀. Late ovalis; minus convexa, altitudine majori (a latere
visa) contra vel paullo pone elytrorum medium posita; minus
nitida; picea,¹ capite prothorace (hoc plus minusve piceo-
adumbrato) elytrorum maculis nonnullis (his præsertim ad
latera positis) antennisque (his apicem versus infuscatis)
rufo-aurantiacis; capite crebre subtilius subrugulose punctulato; prothorace quam longiori ut 2⅔ ad 1 latiori, ab apice
paullo ultra medium dilatato, pone apicem transversim
distincte impresso, crebre rugulose subfortiter (ad latera
valde rugulose) punctulato, lateribus modice arcuatis haud
deplanatis angulis posticis obtusis; scutello plus minusve
punctulato; elytris sub callum humeralem distincte depressis, pone basin transversim vix impressis, crebre fortiter
subseriatim (postice magis subtiliter, ad latera magis rugulose) punctulatis, verrucis lævibus sat numerosis sat seriatim
(hic illic in costis minus distinctis positis) antice quam
postice minus perspicue instructis, interstitiis subrugulosis

(ad latera, vix ad apicem, magis rugulosis), parte marginali minus lata sed (parte submediana excepta) a disco per sulculum sat distinctum divisa, calli humeralis margine interno a sutura quam ab elytrorum margine laterali multo magis distanti; segmento ventrali basali minus sparsim minus subtiliter punctulato; antennarum articulo 3° quam 4um sat longiori. Long. 4$\frac{1}{4}$-4$\frac{3}{4}$, lat. 3-3$\frac{1}{2}$lines.

In general appearance much like *P. sordida*, but with the third joint of the antennæ considerably longer, the elytral puncturation stronger, the verrucæ more conspicuous (especially behind), the submarginal sulculus of the elytra strongly interrupted in front of the middle, &c. Also resembles *P. punctata*, Marsh., but differs by sides of prothorax distinctly flattened, coarser puncturation of elytra, narrower form, &c.

N. S. Wales ; taken by Mr. Lea near Forest Reefs.

P. BALDIENSIS, sp nov.

♂. Sat late ovata, modice convexa, altitudine majori (a latere visa) contra elytrorum marginem medium (vel etiam magis retro) posita; nitida; subtus picea hic illic rufescens, capite prothoraceque rufis, (nonnullorum exemplorum plus minusve infuscatis) elytris piceo rufoque incerte variegatis pedibus antennisque rufis, his apicem versus infuscatis ; capite crebre subtilius punctulato; prothorace quam longiori ut 2$\frac{1}{2}$ ad 1 latiori, ab apice ad medium dilatato, pone apicem transversim minus distincte impresso, minus æquali, subtilius minus crebre (ad latera grosse rugulose) punctulato, lateribus sat æqualiter arcuatis late fortiter deplanatis, angulis posticis rotundatis; scutello fere lævi; elytris sub callum humeralem distincte depressis, pone basin transversim late fortiter impressis, sat grosse sat crebre subseriatim (ad latera paullo magis, postice multo minus fortiter) punctulatis, verrucis sat numerosis nitidis nigris sat inæqualibus in dimidia parte posteriori instructis, interstitiis (præsertim postice) rugulosis, parte marginali lata et sat late extrorsum directa a disco (per sulculum continuum) bene divisa, calli humeralis margine

interno a sutura quam ab elytrorum margine laterali parum
magis distanti; segmento ventrali basali sparsius minus
subtiliter punctulato.   Long 3½, lat. 2½ lines

The widely explanate and evenly rounded sides of the prothorax
are the conspicuous character of this species, which is also notable
for the strong postbasal impressions of the elytra.   Two examples
from Mt. Kosciusko in N S. Wales are smaller with the prothorax a trifle more closely punctulate, but I do not think them
distinct specifically.   The intermediate verrucæ of the prothorax
are fairly well defined.

Victoria ; M. Baldi

P. PUSTULOSA, sp.nov.

♀   Ovalis, minus convexa, altitudine majori (a latere visa) vix
ante elytrorum marginem medium posita; nitida; subtus
nigra, ferrugineo-variegata; capite prothoraceque rufis, hoc
transversim nigro 4-maculato, scutello obscuro; elytris rufis
seriatim verrucis magnis rotundatis (sed parum elevatis)
nigris ornatis; antennis pedibusque obscuris, illis basin
versus rufis, capite subtiliter sat crebre punctulato; prothorace quam longiori plus quam duplo latiori (fere ut 2½ ad
1), ab apice paullo ultra medium dilatato, pone apicem transversim vix impresso, sparsius subtilius (ad latera sat grosse)
punctulato, lateribus sat arcuatis sat anguste deplanatis,
angulis posticis valde obtusis; scutello fere lævi; elytris
fortiter subseriatim sat crebre punctulatis (latera versus perspicue magis grosse), interstitiis (etiam ad apicem) parum
rugulosis, sub callum humeralem distincte depressis, pone
basin transversim late leviter impressis, parte marginali a
disco vix distincta, calli humeralis margine interno a sutura
quam ab elytrorum margine laterali multo magis distanti;
segmento ventrali basali sparsissime subtilissime punctulato.
Long 4, lat. 2⅔ lines

A very nitid species, notable for the very large nitid flattish
verrucæ distributed somewhat sparsely in a subseriate fashion

over the whole of its elytra, the largest of them scarcely smaller than the black spots on the prothorax.

Victoria.

P. MIXTA, sp.nov.

♀ Sat late ovata, sat convexa, altitudine majori (a latere visa) contra elytrorum marginem medium posita, subnitida, nigra, capite prothoraceque rufis plus minusve nigro notatis, elytris nigro rufoque variegatis, antennarum basi rufa, capite crebre subaspere punctulato; prothorace quam longiori fere triplo latiori, ab apice fere ad basin dilatato, pone apicem transversim parum distincte impresso, confertim sat aspere minus subtiliter (ad latera magis grosse) punctulato, lateribus modice arcuatis anguste deplanatis, angulis posticis rotundatis, scutello punctulato, elytris sub callum humeralem fortiter depressis, pone basin transversim vix manifeste impressis, sat crebre sat fortiter subseriatim (ad latera magis, postice minus, fortiter) punctulatis, verrucis nigris numerosis sat distinctis subseriatim instructis, interstitiis rugulosis, parte marginali minus (apicem versus paullo magis) distincte a disco divisa, calli humeralis margine interno a sutura quam ab elytrorum margine laterali multo magis distanti; segmento ventrali basali sparsius sat subtiliter punctulato. Long. $3\frac{4}{5}$, lat. $2\frac{1}{5}$ lines.

Notable among its immediate allies by its very strongly transverse prothorax with close asperate even puncturation, the extremely strong depression of the elytra outside the humeral callus and the absence of any distinction between the discal and marginal parts of the elytra (except for a short distance near the apex).

Victoria; Alpine region.

P. SORDIDA, sp nov.

Sat late ovata, minus convexa, altitudine majori (a latere visa) ad vel paullo pone elytrorum marginem medium posita; sat nitida; picea, hic illic (praesertim in capite et ad elytrorum prothoracisque latera) rufescens, antennarum basi rufa; capite

44

aspere sat crebre punctulato; prothorace quam longiori fere
ut $2\frac{1}{2}$ ad 1 latiori, ab apice paullo ultra medium dilatato, pone
apicem transversim distincte impresso, crebrius aspere subfortiter (ad latera magis grosse) punctulato, lateribus sat
arcuatis vix deplanatis, angulis posticis obtusis; scutello fere
lævi; elytris sub callum humeralem distincte depressis pone
basin transversim vix impressis crebre sat fortiter subseriatim
(ad latera parum fortius, apicem versus magis crebre) punctulatis, verrucis nonnullis parvis minus distinctis confuse
instructis, interstitiis distincte (præsertim apicem versus)
rugulosis sed rugulis in disco puncturas haud obscurantibus,
parte marginali sat angusta sed a disco (per sulculum continuum) bene divisa, calli humeralis margine interno a sutura
quam ab elytrorum margine laterali multo magis distanti;
segmento ventrali basali sparsim subtiliter punctulato.

Mas quam femina paullo magis depressus, hujus antennis paullo
minus elongatis. Long. 4-$4\frac{1}{2}$, lat. 3-$3\frac{3}{10}$ lines.

The narrow lateral portion of the elytra divided from the discal
by a continuous furrow in combination with the prothorax at its
widest not much behind the middle, and the inconspicuous small
verrucæ (concolorous with the derm) of the elytra forms the leading characteristic of this species among its near allies. In the
female the greatest height of the elytra is a little further back
than in the male.

S Australia; Mt. Lofty, &c.

P. FOVEATA, sp.nov.

♀. Sat late ovalis (fere ovata), minus convexa, altitudine majori
(a latere visa) paullo pone elytrorum marginem medium
posita; sat nitida; ut *P. sordida* colorata; capite prothoraceque crebre subtilius leviter (in hoc ad latera puncturis sat
grossis intermixtis) punctulatis; hoc quam longiori ut fere
$2\frac{1}{4}$ ad 1 latiori, ab apice paullo ultra medium dilatato, pone
apicem transversim parum impresso, lateribus sat arcuatis
anguste deplanatis, angulis posticis sat rotundatis; scutello
plus minusve punctulato; elytris sub callum humeralem dis-

tincte depressis, pone basin transversim sat fortiter impressis, sat crebre fortius subseriatim (ad latera magis grosse) punctulatis, verrucis nonnullis minus distinctis confuse instructis, interstitiis rugulosis (in partis impressæ subbasalis fundo opacis nec rugulosis), parte marginali minus lata a disco per sulculum sat distinctum (hoc ante medium et ad apicem summum interrupto) divisa, calli humeralis margine interno a sutura quam ab elytrorum margine laterali multo magis distanti; segmento ventrali basali sparsim fortius punctulato. Long. 4, lat. 2⅛ lines.

Resembles *P. sordida* but is readily separated from it *inter alia* by the strongly marked subbasal impression on the elytral disc (which has somewhat the appearance of a subrotundate large shallow fovea suggestive of, though very different from, the deep fovea of *P. fossa* and *scabra*), and by the submarginal sulculus being interrupted in front of its middle and not reaching the extreme apex.

N.S. Wales; taken by Mr. Lea near Forest Reefs; also from Inverell.

P. MONTUOSA, sp.nov.

P. baldiensi affinis; quam hæc magis lata et multo magis convexa ; elytris rufo-brunneis vix piceo-variegatis, pedibus obscuris; prothoracis disco magis crebre punctulato; elytris antice manifeste costatis, verrucis multo majoribus (cum superficie concoloribus) instructis, parte marginali minus fortiter extrorsum directa; abdomine magis crebre magis fortiter punctulato; cetera ut *P. baldiensis*. Long. 3⅗, lat. 3 lines (vix).

Femina quam mas etiam multo magis convexa.

Rather closely allied to *P. baldiensis* structurally, though to a casual glance more suggestive of *P. rosea* and *P. impressa*, Chp. Its wider and very much more strongly convex form together with the very much larger and more elevated verrucæ of its elytra render it impossible to be confused with *baldiensis*. The greatest height of *P. baldiensis* is considerably less (of *P. montuosa* decidedly more) than half the length of the elytra. From *P.*

rosea the present species differs *inter alia* by the greatest width of its prothorax being at the middle, as well as by its very different colour, while from *P. impressa*, Chp., it differs very widely in form.   The elytral verrucæ have more or less tendency to run together into transverse ridges, especially on the lateral declivity.

Victoria; Alpine region.

*P. ROSEA, sp.nov.

♀. Ovata, modice lata, altitudine majori (a latere visa) contra elytrorum marginem medium (vel etiam magis retro) posita, minus nitida; læte rosea, antennis apicem versus et corpore subtus plus minusve infuscatis ; capite crebre subfortiter punctulato; prothorace quam longiori ut 2½ ad 1 latiori, ab apice sat longe ultra medium dilatato, pone apicem transversim vix perspicue impresso, minus æquali, sat fortiter sat crebre ,ad latera crebre grosse) punctulato, lateribus postice sat fortiter arcuatis late minus fortiter deplanatis, angulis posticis nullis; scutello fere lævi, vel subtiliter coriaceo , elytris sub callum humeralem sat fortiter depressis, pone basin transversim fortiter impressis, sat grosse sat crebre subseriatim (postice minus grosse) punctulatis, verrucis sat magnis inæqualibus (his hic illic transversim subconjunctis) sat numerosis confuse instructis, interstitus (præsertim transversim) inæqualiter rugulosis, parte marginali modice lata a disco (per sulculum paullo ante medium anguste interruptum) bene divisa, calli humeralis margine interno a sutura quam ab elytrorum margine laterali manifeste magis distanti; segmento ventrali basali sparsius subfortiter punctulato.   Long. 3¾, lat. 2½ lines.

Notable for its (probably constant) uniform bright rosy-red colouring on the upper surface.   The coarse uneven verrucæ of

---

* A male example received from Mr. Masters since this description was written scarcely differs from the female except in respect of sexual characters common to all species of *Paropsis*.   A female sent by Mr Masters is somewhat larger than the type.

the elytra (tending to run together here and there into transverse ridges on the laterally declivous portions) are suggestive of *P. impressa*, Chp , from which, however, the present species differs *inter alia* by its much less convexity, its elytra at their highest much further from their base, and the much less strongly elevated verrucæ and ridges of the elytra. The intermediate verrucæ of the prothorax are fairly well-defined.

Victoria , Black Spur : also from the Blue Mountains (Mr. Masters)

P. EXSUL, sp.nov.

♂ Late ovata, sat convexa, altitudine majori (a latere visa) contra elytrorum marginem medium posita; sat nitida; picea, rufo-variegata (præsertim in capite fere toto, in prothoracis lateribus, in elytrorum marginibus et maculis indistinctis nonnullis, in antennarum basi, et in abdominis lateribus), capite crebre aspere punctulato; prothorace quam longiori ut fere 2⅜ ad 1 latiori, ab apice ad medium dilatato, pone apicem transversim vix perspicue impresso, crebre minus subtiliter (ad latera sat grosse) punctulato, lateribus sat arcuatis distincte sat anguste deplanatis, angulis posticis obtusis ; scutello subtiliter punctulato, elytris sub callum humeralem distincte depressis, pone basin subrotundatim impressis, crebre sat subtiliter subseriatim (ad latera paullo minus, postice paullo magis, subtiliter) punctulatis, verrucis nonnullis vix perspicuis subseriatim instructis, inter-stitiis leviter (apicem versus magis perspicue) rugulosis, parte marginali modice lata a disco (per sulculum ante medium late interruptum pone medium sat profundum) bene divisa, calli humeralis margine interno a sutura quam ab elytrorum margine laterali paullo magis distanti ; segmento ventrali basali sparsius subfortiter punctulato. Long. 3⅕, lat. 3 lines. Easily distinguishable among its near allies by the fine puncturation of its elytra (the verrucæ of which need looking for) in combination with the subquadrate prothorax (which is at its widest at the middle).

N S. Wales; Richmond R. district, I believe.

P. SIMULANS, sp.nov.

♀. Subovata; sat lata; minus convexa, altitudine majori (a
latere visa) contra elytrorum marginem medium posita,
modice nitida; castanea, antennis apicem versus et sternis
picescentibus; capite subtilius sat crebre vix aspere punctulato, prothorace quam longiori ut 2¾ ad 1 latiori, ab apice
sat longe ultra medium dilatato, pone apicem transversim
impresso, sat crebre subtilius haud rugulose (sed ad latera
grosse rugulose) punctulato, lateribus sat arcuatis sat anguste deplanatis, angulis posticis fere nullis; scutello lævi,
elytris sub callum humeralem depressis, pone basin transversim leviter impressis, subtiliter (puncturis etiam magis
subtilibus intermixtis, ad latera paullo minus postice paullo
magis subtiliter) subseriatim punctulatis, verrucis sat numerosis (his minus elevatis) sparsim seriatim (basin versus
obsoletis) instructis, interstitiis haud (apicem versus vix
manifeste) rugulosis, parte marginali angusta a disco (per
sulculum continuum) manifeste divisa, calli humeralis margine interno a sutura quam ab elytrorum margine laterali
paullo magis distanti; segmento ventrali basali crebrius
subfortiter punctulato.   Long. 3¾, lat. 2⅕ lines.

This species bears a remarkable superficial resemblance to *P.
castanea*, Marsh., which however belongs to the last subgroup on
account of its different form.   Besides the difference of form
from *castanea* it is distinguished *inter alia* by the much closer and
more even puncturation, and much less widely explanate sides of
its prothorax and by the well-marked depression below its
humeral calli.

N. S. Wales ; near Sydney.

P. ABJECTA, sp nov.

♀. Subovata; sat lata; modice convexa, altitudine majori (a
latere visa) contra elytrorum marginem medium posita,
modice nitida; obscure rufa, corpore subtus elytrisque
piceo-adumbratis, antennis (exempli typici) carentibus;

capite crebre rugulose punctulato, prothorace quam longiori
ut $2\frac{1}{2}$ ad 1 latiori, ab apice sat longe ultra medium dilatato,
pone apicem transversim impresso, sat crebre subrugulose
subtilius (ad latera paullo magis grosse) punctulato, lateribus
sat arcuatis sat anguste deplanatis, angulis posticis fere
nullis; scutello subtiliter ruguloso, elytris sub callum humeralem leviter depressis, pone basin rotundatim impressis,
subtilius sat crebre subseriatim (ad latera vix magis, postice
vix minus, fortiter) punctulatis, verrucis sat numerosis
minus distinctis subseriatim instructis, interstitiis sat rugulosis, parte marginali a disco vix distincta, calli humeralis
margine interno a sutura quam ab elytrorum margine laterali
sat multo magis distanti; segmento ventrali basali sparsius
subtilius punctulato. Long. $2\frac{1}{2}$, lat. $1\frac{4}{5}$ lines.

This is an inconspicuous species bearing much superficial resemblance to *P. foveata* and *sordida* from both of which it differs
by its much smaller size and the considerably finer puncturation
of its elytra. It also superficially resembles *P. mediocris, whit-
tonensis* and *opacior* but differs from them *inter alia* by the very
distinctly though narrowly explanate sides of its prothorax.

N. S. Wales.

P. MEDIOCRIS, sp nov.

♂. Latissime ovata, modice convexa, altitudine majori (a latere
visa) contra elytrorum marginem medium posita; sat nitida;
ut *P. exsul* colorata; capite crebre aspere punctulato; prothorace quam longiori fere triplo latiori, ab apice fere ad
basin dilatato, pone apicem transversim impresso, sat crebre
subfortiter (ad latera grosse) punctulato, lateribus leviter
arcuatis haud deplanatis, angulis posticis nullis; scutello
medio leviter punctulato; elytris sub callum humeralem
manifeste depressis, pone basin transversim late distincte
impressis, fortiter crebre subseriatim (ad latera paullo magis,
postice paullo minus, fortiter', punctulatis, verrucis nonnullis
modice distinctis nigris (his in lateribus transversim plus
minusve confluentibus) instructis, interstitiis sat rugulosis
(postice subgranuliformibus), parte marginali a disco (per

sulculum paullo ante medium angustius interruptum) sat bene divisa, calli humeralis margine interno a sutura quam ab elytrorum margine laterali paullo magis distanti, segmento ventrali basali sparsim subtilius punctulato.   Long. 3, lat. 2½ lines.

Notable among its immediate allies for its extremely wide form and very strongly transverse prothorax.   The humeral callus is more distant from the lateral margin than in most of its immediate allies.

N S. Wales; Richmond R district, I believe.

P. RUFICOLLIS, sp.nov.

Ovata, modice lata, modice convexa, altitudine majori (a latere visa) contra elytrorum marginem medium (vel paullo anterius) posita; sat nitida; picea, capite prothorace antennarum basi scutello elytris (horum verrucis parte suturali antica et margine summo, piceis) et corporis subtus pedumque partibus nonnullis rufis; capite crebre subtilius vix aspere punctulato, prothorace quam longiori ut 2¾ ad 1 latiori, ab apice sat longe ultra medium dilatato, pone apicem transversim distincte impresso, minus fortiter sat crebre (ad latera grosse) punctulato, lateribus sat arcuatis haud deplanatis, angulis posticis fere nullis; scutello coriaceo vel fere lævi; elytris sub callum humeralem distincte depressis, pone basin transversim impressis, crebre minus fortiter subseriatim (ad latera magis grosse) punctulatis, verrucis sat numerosis sat seriatim instructis, interstitiis minus rugulosis, parte marginali a disco (per sulculum ante medium late interruptum) sat distincte divisa, calli humeralis margine interno a sutura quam ab elytrorum margine laterali sat multo magis distanti, segmento ventrali basali subfortiter minus sparsim punctulato.

Femina quam mas paullo magis convexa, ejus antennis paullo brevioribus.   Long 3⅕, lat. 2⅔ lines.

The (apparently constant) uniform red colouring of the head and prothorax in contrast to the much darker elytra distinguishes this species among its immediate allies.   In the male the greatest

height of the elytra is a trifle nearer the front than in the female.
N.S. Wales; taken by Mr. Lea.

P. PROPRIA, sp. nov.

♂. Sat late ovata, sat convexa, altitudine majori (a latere visa)
contra elytrorum marginem medium posita, sat nitida; obscure
rufo-castanea (ad latera fere sanguinea), corpore subtus
antennisque plus minusve infuscatis, capite crebre sat fortiter
punctulato; prothorace quam longiori ut $2\frac{1}{2}$ ad 1 latiori, ab
apice vix ultra medium dilatato, pone apicem transversim
impresso, crebre sat fortiter (ut caput, sed ad latera grosse
rugulose) punctulato, lateribus sat arcuatis haud deplanatis,
angulis posticis distinctis obtusis; scutello crebre subtiliter
punctulato, elytris sub callum humeralem distincte depressis,
pone basin transversim leviter impressis, crebre fortiter subseriatim (ad latera paullo magis, postice minus, fortiter)
punctulatis, verrucis nonnullis minus perspicuis (his cum
superficie concoloribus) subseriatim instructis, interstitiis
minus rugulosis, parte marginali sat lata a disco (per sulculum in medio sat late interruptum) sat distincte divisa, calli
humeralis margine interno a sutura quam ab elytrorum
margine distincte magis distanti; segmento ventrali basali sat
crebre sat fortiter punctulato.

♀. Quam mas magis convexa. Long. $3\frac{1}{5}$-$3\frac{4}{5}$, lat. 3 lines.

Decidedly near *P. ruficollis*, but very distinct from it (apart
from colour) *inter alia* by its prothorax at its widest *at* the middle.
South Australia, widely distributed; also Kangaroo Island.

P. WHITTONENSIS, sp. nov.

♂. Ovalis, minus convexa; altitudine majori (a latere visa) ad
vel paullo pone elytrorum marginem medium posita; sat
nitida, supra obscure rufa, capite antice piceo, prothorace
nigro- vel piceo-notato, elytris plus minusve piceo-adumbratis
et verrucis nigris variegatis, subtus picea plus minusve
rufescens, pedibus concoloribus, antennis pallide rufis apicem
versus infuscatis; capite crebre minus subtiliter vix rugulose

punctulato; prothorace quam longiori ut $2\frac{1}{3}$ ad 1 latiori, ab apice paullo ultra medium dilatato, pone apicem transversim distincte impresso, in disco minus fortiter minus crebre haud rugulose (ad latera grosse rugulose) punctulato, lateribus sat arcuatis haud deplanatis, angulis posticis obtusis; scutello punctulato: elytris sub callum humeralem leviter depressis, pone basin vix impressis, sat crebre fortius subseriatim (ad latera magis grosse) punctulatis, verrucis sat distinctis seriatim instructis, interstitiis latera apicemque versus sat rugulosis (rugulis nonnullis transversis plus minusve elongatis et continuis latera versus intermixtis), parte marginali ut *P. foveatæ*, calli humeralis margine interno ut *P. foveatæ* posito; segmento ventrali basali sparsim subtilius punctulato. Long. $3\frac{1}{5}$, lat. $2\frac{1}{5}$ lines.

Very much like *P. foveata* superficially, but differing from it *inter alia* by its considerably smaller size, the sides of its prothorax not at all explanate, the much more numerous and better defined verrucæ of its elytra and the extreme faintness (almost absence) of the subbasal impression of the elytra. The transverse rugulosities of the elytra have a slight tendency to simulate the continuous wheal-like ridge that forms a conspicuous character in some species of *Paropsis* (e.g., *transversalis.*)

N S Wales; taken by Mr. Lea near Whitton.

P. CRIBRATA, sp.nov.

P. propriæ simillima, differt corpore minus nitido, elytris ad latera quam in disco vix magis rufis, horum verrucis nigris magis numerosis magis perspicue seriatis; prothorace aspere multo magis fortiter punctulato, elytris pone basin haud distincte impressis, his multo magis subtiliter (ad latera quam in disco haud magis fortiter) punctulatis, cetera ut *P. propria*. Long. $4\frac{1}{2}$-$4\frac{4}{5}$, lat. $3\frac{1}{2}$-$3\frac{3}{5}$ lines.

Near *P. propria*, but very readily separable from it by the characters cited above; in *P. propria* the puncturation in the subbasal impression of the elytra is conspicuously coarser than on the general surface of the disc, while in the present species (there being

no distinct subbasal impression) there is no discal space notable
for the coarseness of its puncturation. The whole puncturation of the elytra is manifestly finer. I have two examples of
Paropsis differing from *P. cribrata* in their smaller size and less
numerous elytral verrucæ which, moreover, are concolorous with
the derm. I have little doubt that they represent a distinct
very close species, but I refrain from naming them without
observing more specimens.

S Australia; Yorke's Peninsula.

P. DECLIVIS, sp.nov.

♀. Sat late ovata; sat fortiter convexa, altitudine majori (a
latere visa) contra elytrorum marginem medium posita;
minus nitida; obscure rufa, antennis (basi excepta) corpore
subtus pedibus capitis parte antica scutello et elytrorum
verrucis piceis; *P. propriæ* affinis; differt prothorace ab apice
manifeste ultra medium dilatato, pone apicem (hoc magis
angustato) haud impresso, angulis posticis magis rotundatis;
scutello sublævi; elytris paullo magis crebre magis subtiliter
punctulatis, pone basin haud impressis, verrucis vix elevatis
nigris ut superficies punctulatis, parte marginali a disco haud
distincta; cetera ut *P. propria.* Long. 4, lat. 3⅓ lines.

Resembles *P. propria* and *P. cribrata* but differs from both
inter alia by the marginal portion of its elytra (especially behind)
continuous with the discal portion so that there is no longitudinal
concavity but the lateral and apical declivous parts descend quite
evenly without being outturned at the margin. The puncturation
of the elytra continuous over the verrucæ is also a notable
character and very rare in *Paropsis.*

N. S. Wales ; near Sydney.

P. TATEI, sp.nov.

♂. Ovalis, minus convexa, altitudine majori (a latere visa) ad
elytrorum marginem medium posita; nitida; fere ut *P. pustulosa* colorata, sed antennis rufis apicem versus vix infuscatis
et elytrorum verrucis multo minoribus elongatis; capite

subtilius crebrius subrugulose punctulato; prothorace quam longiori ut $2\frac{2}{5}$ ad 1 latiori, antice fortiter angustato, ab apice longe pone medium dilatato, pone apicem transversim haud impresso, crebrius sat fortiter sat rugulose (ad latera valde rugulose) punctulato, lateribus modice arcuatis nullo modo deplanatis, angulis posticis valde obtusis; scutello lævi fortiter convexo; elytris subgrosse seriatim minus crebre punctulatis (ad latera etiam grossius, apicem versus multo magis crebre), interstitiis in disco haud (ad latera et versus apicem sat perspicue) rugulosis, sub callum humeralem distincte depressis, pone basin transversim late vix impressis, parte marginali a disco haud distincta, calli humeralis margine interno a sutura quam ab elytrorum margine laterali multo magis distanti, segmento ventrali basali sublævi. Long. $4\frac{1}{5}$, lat. 3 lines.

Rather closely resembling *P. pustulosa* superficially but readily distinguishable from it *inter alia* by its prothorax being not at all explanate laterally, much narrower in front and much more strongly and less smoothly punctulate; also by its elytra being evidently more coarsely and less closely punctulate, more regularly seriate, with much smaller and differently shaped verrucæ, and having their marginal part not distinct from the discal (in *pustulosa* there is an evident though very narrow lateral outturned portion especially noticeable near the apex). It should be noted that in this species the third antennal joint is slightly longer than the fourth, but too slightly to justify placing it among the species with the third joint "markedly" longer.

Victoria; presented to me by Professor Tate.

P. ALTICOLA, sp.nov.

Late ovalis (fere subcircularis), minus convexa, altitudine majori (a latere visa) contra elytrorum marginem medium posita; sat nitida; subtus nigro-picea; supra (antennis pedibusque inclusis) rufescens, elytris plus minusve obscure piceo-adumbratis, capite sat crebre vix aspere punctulato; prothorace quam longiori ut $2\frac{2}{3}$ ad 1 latiori, ab apice sat longe ultra medium dilatato, pone apicem transversim distincte impresso, crebre aspere minus fortiter (ad latera grosse nec vel vix confluenter) punctulato, lateribus fortiter arcuatis nullo modo deplanatis, angulis posticis nullis; scutello sat opaco, dupliciter (sparsim fortius et confertim subtiliter) punctulato; elytris sat distincte sub callum humeralem depressis (et pone basin transversim impressis), crebre fortius subseriatim (ad latera multo magis grosse, postice magis crebre magis subtiliter) punctulatis, verrucis nonnullis parvis minus distinctis confuse instructis, interstitiis antice modice (postice crebre sat aspere) rugulosis, parte marginali sat angusta a disco (per sulculum antemedium anguste interruptum) bene divisa, calli humeralis margine interno a sutura quam ab elytrorum margine laterali vix multo magis distanti; segmento ventrali basali sat sparsim subfortiter punctulato.

Mas quam femina nonnihil magis depressus, hujus antennis paullo minus elongatis. Long $3\frac{1}{5}$-$4\frac{1}{5}$, lat. $2\frac{3}{5}$-$3\frac{1}{5}$ lines.

Resembles *P. sordida* superficially but differs from it by a multitude of characters, conspicuous among which are its distinctly more convex form, more transverse differently shaped prothorax, scutellum so closely punctulate as to be subopaque, and humeral callus distinctly more distant from the lateral margin of the elytra.

Also near *punctata*, Marsh., but of considerably more depressed form.

S. Australia; on the hills near Adelaide, &c ; also Kangaroo Island.

P. VICTORIÆ, sp nov.

♀. *P. alticolæ* simillima; subtus nigra, pedibus obscuris, prothorace paullo magis crebre punctulato; scutello fere lævi; elytris ad latera quam in disco vix magis fortiter punctulatis; calli humeralis margine interno a sutura quam ab elytrorum margine laterali multo magis distanti; cetera ut *P. alticola*. Long. $3\frac{1}{5}$, lat. $2\frac{1}{5}$ lines.

Another species very close to *P. alticola* but differing from it in the evidently closer puncturation of its prothorax; the extremely

fine and sparse puncturation of its scutellum; a slight difference
(mentioned above) in the elytral puncturation; and especially in
the external (vertical) part of the elytral epipleuræ being less
elevated, so that the humeral callus is nearer to the lateral margin
of the elytra (being placed as in *P. sordida*). This latter character
inter alia forms a good distinction from *P. punctata*, Marsh. I
have not seen a male of this species. In the type the scutellum
is very nitid, convex and scarcely punctulate; in a second example
(possibly representing a distinct species) the scutellum is subopaque, being very finely coriaceous, but both examples are devoid
of the comparatively coarse punctures with which the scutellum
is impressed in *P. alticola* and *punctata*, Marsh. In the "second
example" the elytral verrucæ are a trifle more conspicuous and
less tending to run together transversely.

Victoria.

P. SOLITARIA, sp.nov.

♀. Elongato-ovalis, modice convexa, altitudine majori (a latere
visa) paullo pone elytrorum marginem medium posita; sat
nitida; subtus nigra; capite prothoraceque brunneo-rufis
nigro-adumbratis; elytris piceis, verrucis numerosis seriatim
positis sordide testaceis et vittis concoloribus circiter 10
ornatis; pedibus antennisque nigris, his basin versus sordide
testaceis; capite subtilius sat crebre punctulato; prothorace
quam longiori plus quam duplo (ut $2\frac{3}{5}$ ad 1) latiori, ab apice
longe ultra medium dilatato, pone apicem transversim vix
impresso, in disco sat subtiliter minus crebre (ad latera crebre
crasse) punctulato, lateribus sat arcuatis vix deplanatis,
angulis posticis rotundatis; scutello lævi, elytris sat crebre
subfortiter subseriatim (ad latera quam in disco vix magis
fortiter) punctulatis, interstitiis in disco leviter (postice magis
fortiter) rugulosis, sub callum humeralem distincte depressis,
parte marginali a disco vix distincta (ut *P. sternalis*), calli
humeralis margine interno a sutura quam ab elytrorum margine laterali multo magis distanti; segmento ventrali basali
sparsim subtiliter punctulato; antennarum articulo 3° quam

4^{ms} vix longiori; epipleurarum parte externa (verticali) minime elevata. Long. 5, lat. 3½ lines.

The most striking character in this species is the external (vertical) part of its elytral epipleuræ being very narrow [scarcely so wide as is the internal (horizontal) part where the latter is at its widest]. The colouring of the elytra in the unique type is also very remarkable, the derm being of a pitchy colour traversed by a number of dull testaceous vittæ on which are placed rather closely numerous concolorous verrucæ.

Victoria; Black Spur.

P. LIMA, sp.nov.

♀. *P. alticolæ* affinis sed magis convexa; pedibus antennisque (harum basi excepta) obscuris; elytris crebrę granuloso-rugulosis. Long. 4, lat. $2\frac{9}{10}$ lines.

Another near ally of *P. alticola* but incapable of confusion with it on account of its much more convex form (at any rate in the female) and the strong close granule-like rugulosity of its elytral interstices which is so prominent as greatly to obscure the puncturation except in the subbasal impression. In the type this subbasal impression is almost circular, but I hesitate to attach much value to this character since the corresponding impression in *P. alticola* shows some approach (though less marked) to a similar form, the impression being subinterrupted in the middle so that its inner part (regarded separately) is scarcely transverse. From *P. punctato*, Marsh., it differs by its still more convex form, more nitid surface, and much more rugulose elytral interstices.

Victoria; sent to me by Mr. Billinghurst.

P. INVALIDA, sp nov.

♀. Ovalis, parum convexa; altitudine majori (a latere visa) paullo pone elytrorum marginem medium posita; sat nitida; ut *P. sordida* colorata; capite minus crebre minus subtiliter punctulato, interstitiis valde distincte subtiliter punctulatis, prothorace fere ut *P. sordidæ* sed in disco sparsius sat leviter haud aspere (ad latera sat grosse sat crebre) punctulato,

lateribus haud deplanatis, angulis posticis magis rotundatis,
scutello punctulato, elytris sub callum humeralem leviter
depressis, pone basin vix impressis, seriatim sat subtiliter
(latera versus magis fortiter) punctulatis, verrucis parvis
modice distinctis seriatim instructis, interstitiis sat planis
(apicem versus magis rugulosis), parte marginali a disco vix
(apicem versus subdistincto) distincto, calli humeralis
margine interno a sutura quam ab elytrorum margine laterali
haud multo magis distanti; segmento ventrali basali sparsim
subtilius punctulato.   Long. 3¾, lat. 2⅔ lines.

Also resembling *P. foveata* superficially, but at once distinguishable from it and its other near allies *inter alia* by the very much
finer puncturation of its elytra, and by the inner edge of the
humeral callus being very little nearer to the lateral margin than
to the suture.   Also resembles *P. seriata*, Germ., but differs from
it *inter alia* by the presence of a depression below the humeral
callus

N S Wales; taken by Mr. Froggatt on the Blue Mountains

P. TRANSVERSALIS, sp nov.

Ovata; sat convexa, altitudine majori (a latere visa) contra
elytrorum marginem medium (vel paullo magis antice) posita,
nitida, subtus rufa vel rufo-picea; capite prothoraceque rufis
hoc plus minusve piceo-adumbrato, elytris piceis rufo-variegatis et nigro verrucatis, antennis pedibusque rufis (nonnullorum exemplorum magis obscuris); capite crebre subtilius
punctulato, prothorace quam longiori ut 2½ ad 1 latiori, ab
apice ad vel paullo ultra medium dilatato, pone apicem transversim manifeste impresso, sat crebre subirregulariter (ad
latera grosse rugulose) punctulato, lateribus sat fortiter
arcuatis nullo modo deplanatis, angulis posticis rotundatis;
scutello fere lævi; elytris sat fortiter sub callum humeralem
depressis (et pone basin late transversim fortiter impressis,
fortiter sat crebre subseriatim (ad latera magis, postice minus,
fortiter) punctulatis, verrucis nitidis sat magnis instructis

(his in parte impressa postbasali carentibus, et pone hanc partem ut ruga transversa fere a sutura ad marginem lateralem continua confluentibus), interstitiis vix rugulosis, parte marginali minus lata a disco (per sulculum ante medium late interruptum) divisa; calli humeralis margine interno a sutura quam ab elytrorum margine laterali sat multo magis distanti; segmento ventrali basali sparsim minus subtiliter punctulato. Femina quam mas magis convexa. Long 3-3½, lat. 2⅕-2⅗ lines. At once distinguishable from all its allies by the tendency of the elytral verrucæ to coalesce into coarse nitid ridges, the most conspicuous of which is placed at about the middle of the elytra and runs from near the suture almost to the lateral margin.

S. Australia; widely distributed.

TABULATION OF THE SPECIES FORMING SUBGROUP III.

*A. Elytra with a distinct postbasal impression on disc.
  B. Elytral margin (viewed from the side) straight or but little sinuous.
    C. Elytral puncturation (and especially its seriation) much obscured by irregular transverse rugulosity.
      D. Elytra not marked with a common dark blotch behind the scutellum.
        E. Elytral verrucæ of hind declivity all closely placed in rows... ... . granaria, Chp.
        EE. Elytral verrucæ of hind declivity sparse and confused.
          F. Inner edge of humeral calli evidently nearer to lateral margin than to suture. . . . . rugulosior, Blackb.
          FF. Inner edge of humeral calli equidistant between lateral margin and suture...................................... morosa, Blackb
      DD. Elytra with a conspicuous common dark blotch behind scutellum.................... stigma, Blackb.
    CC. Elytral interstices not, or but very feebly, rugulose, not obscuring the punctures.

* The impression is less marked in *granaria*, Chp., than in its allies

D. Prothorax strongly rugulose, even more so
    than in *P. serpiginosa.* . . .. Sloanei, Blackb
DD. Prothorax not, or but little, rugulose.
    E. Depressed species, upper outline (viewed
        from side) more or less straight,
        humeral callus exceptionally near
        lateral margin.
        F. Elytral margin (viewed from side) dis-
            tinctly though not strongly sinuous;
            form wide .... grossa, Blackb.
        FF. Elytral margin (viewed from side)
            straight; form notably less wide.... seriata, Germ.
    *EE Species of more convex form; upper
        outline (viewed from side) a contin-
        uous curve
        F. Prothorax closely punctulate.
            G. Prothorax with black markings
                H. Underside testaceous (here and
                    there infuscate) .. .. interioris, Blackb
                HH. Underside black .. .. tincticollis, Blackb.
            GG. Prothorax without markings (size
                small, scarcely 3 lines)............. malevola, Blackb
        FF. Prothorax sparsely punctulate ..... Leai, Blackb
BB. Elytral margin (viewed from the side) strongly
    sinuous
    C. Elytra furnished with strongly defined inter-
        rupted costæ . ... *costipennis*, Chp
    CC. Elytra without costæ .. . . . strigosa, Chp.
†AA. No postbasal impression on disc of elytra.
    B. Elytral verrucæ concolorous with or darker than
        general surface.
        C. Puncturation of prothorax more or less close
            and at most moderately strong.
            D. Seriate arrangement of elytral punctures
                and verrucæ well defined
                E. Head marked with black, elytral verrucæ
                    concolorous with general surface. .... maculiceps, Blackb.
                EE. Head unicolorous, elytral verrucæ
                  quite black . ... pustulifera, Blackb.

---

 * *P interioris* is somewhat intermediate between this and the more depressed form.
     † In *P. inornata* there are some traces of an impression.

DD. Seriate arrangement of elytral verrucæ
and especially the punctures scarcely
evident.

  E. Elytra exceptionally finely punctulate.

    F. Form exceptionally wide, elytra by
measurement wider than long   .... alta, Blackb.

    FF. Form notably less wide, elytra longer
than wide. .... inornata, Blackb.

  EE Elytra much more coarsely punctulate inæqualis, Blackb.

CC. Puncturation of prothorax very coarse.

  D Inner edge of humeral calli much nearer
to lateral margin of elytra than to suture alpina, Blackb.

  DD. Inner edge of humeral calli equidistant
between lateral margin of elytra and
suture... asperula, Chp.

CCC. Puncturation of prothorax very sparse and
fine. .... borealis, Blackb.

BB. Elytral verrucæ conspicuously paler in colour
than the general surface

C. Form oval and depressed........................... notabilis, Blackb.

CC. Form subcircular and strongly convex.. ... vomica, Blackb.

P RUGULOSIOR, sp.nov.

♂. Latissime subovalis, subcircularis; modice convexa, altitudine majori (a latere visa) contra elytrorum marginem
medium (vel paullo magis antice) posita; sat nitida; ferruginea, corpore subtus pedibus elytrisque plus minusve fusco-
adumbratis, horum verrucis piceis, capite crebre subaspere
punctulato; prothorace quam longiori ut $2\frac{2}{3}$ ad 1 latiori; ab
apice longe ultra medium dilatato, pone apicem transversim
leviter impresso, crebrius subfortiter subrugulose (ad latera
grosse rugulose) punctulato, lateribus modice arcuatis haud
deplanatis, angulis posticis nullis; scutello nitido vix punctulato, elytris sub callum humeralem haud depressis, pone
basin transversim impressis, crebre minus fortiter subseriatim
(ad latera multo magis grosse, postice magis subtiliter)
punctulatis, verrucis modice magnis sat numerosis confuse
instructis, interstitiis (parte subbasali impressa excepta) confertim granuloso-ruguloso (præsertim apicem versus), parte

marginali a disco vix distincta, calli humeralis margine
interno a sutura quam ab elytrorum margine laterali paullo
magis distanti; segmento ventrali basali punctulato.   Long
$2\frac{4}{5}$, lat. $2\frac{2}{5}$ lines.

An inconspicuous species chiefly notable for its wide form,
almost entire absence of distinction between the discal and marginal parts of the elytra and fine close but not strongly elevated
granulosity of the interstices of the elytral puncturation,—such
that the rugulosity of the elytra (especially behind) is more conspicuous than the puncturation.

S Australia; Adelaide district.

P. MOROSA, sp.nov.

P. rugulosiori affinis; valde convexa; colore magis obscura,
nonnullorum exemplorum prothorace nigro-maculato; prothorace quam longiori ut $2\frac{1}{2}$ ad 1 latiori, in disco magis
subtiliter magis æqualiter nullo modo rugulose punctulato;
elytris subtiliter punctulatis, magis crebre et subtiliter
rugulosis, ad latera quam in disco vix magis grosse sculptu-
ratis, calli humeralis margine interno a sutura quam ab
elytrorum margine laterali haud magis distanti; cetera ut
P. rugulosior.   Long. $3\frac{1}{2}$, lat. 3 lines.

Femina quam mas etiam magis convexa.

This is a somewhat isolated species owing to its great convexity
(the "greatest height," viewed from the side being distinctly
greater than half the length of the elytral margin, at any rate in
the female).   Most of the species of similar convexity have the
"greatest height" much nearer to the front.   Its fine puncturation is also a notable character, and the great elevation of the
vertical part of its epipleuræ, owing to which the inner edge of
the humeral callus is unusually distant from the lateral margin.

Kangaroo Island.

P. STIGMA, sp.nov.

Ovata; sat fortiter convexa; altitudine majori (a latere visa)
contra elytrorum marginem medium (vel paullo magis antice)

posita; sat nitida; ferruginea, prothoracis maculis nonnullis elytrorum maculis nonnullis (præsertim macula sat magna communi antemediana) et corporis subtus partibus nonnullis piceis; capite crebre subtilius punctulato; prothorace quam longiori ut 2⅔ ad 1 latiori, ab apice longe ultra medium dilatato, pone apicem transversim impresso, sat crebre minus fortiter (ad latera grosse rugulose) punctulato, lateribus sat fortiter arcuatis nullo modo deplanatis, angulis posticis nullis, scutello fere lævi; elytris sub callum humeralem haud depressis, pone basin transversim leviter impressis, sat crebre sat fortiter vix seriatim (ad latera multo magis grosse) punctulatis, verrucis minus numerosis minus ordinatim instructis, interstitiis sat fortiter (præsertim transversim) rugulosis, parte marginali sat lata a disco minus (prope apicem magis perspicue) distincto; segmento ventrali basali subfortiter punctulato. Long 2⅓, lat. 2⅓ lines.

Feminæ quam maris altitudine majori paullo magis postice posita.

The dark markings on the prothorax of the type consist of several small ill-defined blotches which in some examples coalesce into a large and better defined blotch on each side   In the type the common blotch on the elytra is accompanied by several small spots in the basal region, but in some examples it is the only dark mark except the verrucæ ; I have not seen any example of the species in which the common elytral blotch is altogether wanting. In some examples the verrucæ are scarcely darker than the derm.

Victoria; N S.W.; S. Australia.

P. SLOANEI, sp.nov.

♀. Ovata minus lata, minus convexa, altitudine majori pone elytrorum marginem medium posita ; sat nitida , testacea, corpore subtus piceo-vario, prothorace elytrisque tortuose nigro-notatis, horum verrucis nigris; capite fortius subrugulose punctulato; prothorace quam longiori ut 2½ ad 1 latiori, ab apice ultra medium dilatato, pone apicem transversim impresso, fortiter (ad latera grosse) rugulose punctulato,

lateribus sat arcuatis nullo modo deplanatis, angulis posticis
nullis; scutello punctulato; elytris sub callum humeralem
haud depressis, pone basin parum perspicue impressis, sat
crebre subgrosse subseriatim (postice minus grosse) punctulatis, verrucis numerosis sat æqualiter seriatim instructis,
interstitiis vix (postice magis perspicue) rugulosis, parte
marginali a disco vix distincta, calli humeralis margine
interno a sutura quam ab elytrorum margine laterali multo
magis distanti; segmento ventrali basali sparsim subtiliter
punctulato.   Long 4, lat. 2⅘ lines.

A conspicuous species, notable for the sharply defined contrast
between the testaceous derm and the intricate sinuous black
markings and verrucæ of its upper surface, also for the strong but
somewhat fine rugulosity of the disc of its prothorax, the coarse
puncturation of its elytra, &c.

N.S. Wales; sent to me by Mr. Sloane.

P. GROSSA, sp.nov.

♀. Ovata, sat depressa, modice nitida; ferruginea, corpore
subtus pedibus prothorace elytrisque plus minusve piceo-
adumbratis; capite subtilius sat crebre punctulato; prothorace
quam longiori ut 2½ ad 1 latiori, ab apice ultra medium
dilatato, pone apicem transversim impresso, dupliciter (sc.
subtiliter et magis fortiter), ad latera grosse rugulose, minus
crebre punctulato, lateribus modice arcuatis nullo modo
deplanatis, angulis posticis rotundatis; scutello punctulato;
elytris sub callum humeralem haud depressis, pone basin
leviter impressis, sat crebre sat grosse subseriatim (postice
paullo minus grosse) punctulatis, verrucis piceis irregularibus
(his hic illic ut costæ conjunctis) instructis, interstitiis leviter
rugulosis, parte marginali a disco vix distincta, calli humeralis
margine interno a sutura quam ab elytrorum margine laterali
multo magis distanti; segmento ventrali basali sparsim subtiliter punctulato.   Long. 3⅗, lat. 2⅘ lines.

This species bears much general resemblance to *P. alticola* and
its allies, but may be at once separated from them by the entire

absence of any depression below the humeral callus, as well as by its more depressed form, differently sculptured prothorax, &c.

N.S. Wales; Tweed River district.

P. INTERIORIS, sp.nov.

♀. Subovata; modice convexa, altitudine majori (a latere visa) contra elytrorum marginem medium posita; rufo-ferruginea, prothoracis maculis nonnullis et elytrorum maculis nonnullis verrucisque nigro-piceis; capite crebre minus fortiter punctulato; prothorace quam longiori ut 2$\frac{4}{7}$ ad 1 latiori, ab apice fere ad basin dilatato, pone apicem transversim impresso, sat crebre subaspere (ad latera grosse rugulose) punctulato, lateribus minus arcuatis nullo modo deplanatis, angulis posticis rotundatis; scutello fere ut prothorax punctulato sed minus crebre; elytris sub callum humeralem haud depressis, pone basin transversim impressis, crebre fortiter subseriatim (ad latera magis, postice minus, fortiter) punctulatis, verrucis sat numerosis (per totam superficiem, parte postbasali impressa excepta, distributis) seriatim instructis, interstitiis antice vix (postice manifeste) rugulosis, parte marginali a disco vix distincta, margine ipso angusto manifeste extrorsum inclinato, calli humeralis margine interno a sutura quam ab elytrorum margine laterali multo magis distanti ; segmento ventrali basali sparsim subtilius punctulato. Long. 4$\frac{1}{4}$, lat. 3$\frac{1}{2}$ lines.

A species without any very strongly marked structural characters, a little less markedly convex, moreover, than the other species with which I have associated it. The presence of about four ill-defined blackish marks on the prothorax and the regular seriation of the elytral verrucæ together with the blackish stains on the elytra, especially about the middle of the suture, are superficial characters (probably not very variable) by which the species may be somewhat easily recognised among its near allies. It is not unlike P. funerea, Blackb., which, however, is very easily recognised by the great width of its prosternal ridge.

Central Australia.

P. TINCTICOLLIS, sp.nov.

♂. Late subovata, modice convexa, altitudine majori (a latere
visa) contra elytrorum marginem medium posita; sat nitida,
testacea, corpore subtus prothoracis maculis 4 transversim
positis sat parvis elytrorum verrucis sat magnis parum elevatis
sat numerosis nigris, antennis apicem versus paullo infuscatis:
capite crebre subtilius punctulato; prothorace quam longiori
ut fere 3 ad 1 latiori, ab apice ultra medium dilatato, antice
minus angustato pone apicem transversim vix impresso, sat
crebre minus fortiter (ad latera grosse rugulose) punctulato,
lateribus sat fortiter arcuatis nullo modo deplanatis, angulis
posticis rotundatis; scutello vix punctulato; elytris sub
callum humeralem haud depressis, pone basin transversim
impressis, sat crebre fortiter subseriatim (ad latera multo
magis grosse) punctulatis, verrucis sat numerosis seriatim
instructis, interstitiis (nisi ad latera) vix rugulosis, parte
marginali a disco minus distincta, calli humeralis margine
interno a sutura quam ab elytrorum margine laterali paullo
magis distanti; segmento ventrali basali sparsim minus
fortiter punctulato. Long. 3½, lat. 2⅝ lines.

Resembles *P. granaria*, Chp., in colour and markings of upper
surface, but differs by its black underside (the legs nevertheless
testaceous), considerably wider prothorax much less narrowed in
front, discal interstices of elytra scarcely at all rugulose even
close to the apex, &c.

W. Australia; taken by E. Meyrick, Esq.

P. MALEVOLA, sp.nov.

♀. Subovata; minus lata; sat convexa; *P. stigmati* affinis;
elytris macula communi suturali haud ornatis; prothorace
quam longiori ut 2⅖ ad 1 latiori, magis crebre magis rugulose
punctulato; scutello rugulose ut prothorax punctulato; elytris
propter interstitia minus (præsertim transversim) rugulosa
magis perspicue seriatim punctulatis; cetera ut *P. stigmatis*.
Long. 3, lat. 2³⁄₁₀ lines (vix).

A species quite capable of being confused with several others, especially *P. rugulosior* and *P. stigma*. From both these it may be at once distinguished by the evidently more conspicuous and regularly seriate puncturation of its elytra, from the former also by its much narrower form and strongly rugulose scutellum, and from the latter also by its rugulose scutellum and the entire absence of any blackish patch on the sutural region.

S. Australia, near Adelaide.

P. LEAI, sp.nov.

♂. Ovata; modice lata; sat convexa, altitudine majori (a latere visa) contra vel fere ante elytrorum marginem medium posita, sat nitida; subtus piceo- rufoque-variegata; supra testaceo-brunnea, prothoracis maculis 4 parvis (his transversim in disco dispositis) et elytrorum verrucis obscuris, antennis rufis apicem versus piceis, pedibus piceis plus minusve rufo-variegatis; capite crebrius minus subtiliter punctulato, prothorace quam longiori ut 2⅔ ad 1 latiori, ab apice sat longe ultra medium dilatato, pone apicem transversim leviter impresso, subtilius sat sparsim (ad latera grossius nec confluenter) punctulato, lateribus sat arcuatis nullo modo deplanatis, angulis posticis rotundatis; scutello sparsissime punctulato; elytris sub callum humeralem haud depressis, pone basin transversim impressis, fortius minus crebre subseriatim (ad latera vix magis, postice vix minus, fortiter) punctulatis, verrucis parvis sat numerosis sat regulariter seriatim instructis, interstitiis haud rugulosis, parte marginali sat lata a disco vix perspicue (apicem versus magis distincte) divisa, calli humeralis margine interno a sutura quam ab elytrorum margine laterali sat multo magis distanti; segmento ventrali basali sparsim subtiliter punctulato. Long. 3½, lat. 2½ lines.

This species is rather closely allied to *P. interioris*, which it greatly resembles in markings and colour except in the underside being much darker and the patches of dark colour on the elytra being absent. It is, however, very much smaller, with the prothorax very much less closely and more finely punctured.

N. S. Wales; sent to me by Mr. Lea.

P. STRIGOSA, Chp.

I have an example named as this species from Dr. Chapuis' collection, and there is also before me an example belonging to Mr. Lea which I cannot distinguish from it. Chapuis' locality is "Parao River," Mr. Lea's "Swan River." It is, of course, possible that the species is found in these two very distant localities, but I think it more probable either that Dr. Chapuis' locality is wrong, or my example is not really conspecific with the type but represents a closely allied species.

P. MACULICEPS, sp.nov.

♀. Subovata, modice lata; sat convexa, altitudine majori (a latere visa) contra elytrorum marginem medium posita, minus nitida; obscure ferruginea; capite antice, antennis apicem versus, nonnullorum exemplorum vittis elytrorum et (in his) verrucis, pedibus plus minusve, et nonnullorum exemplorum sternis, piceis, capite sat crebre subrugulose punctulato; prothorace quam longiori ut 2⅔ ad 1 latiori, ab apice longe ultra medium dilatato, pone apicem transversim impresso, sat crebre subfortiter sat rugulose (ad latera grosse) punctulato, lateribus sat fortiter arcuatis nullo modo deplanatis, angulis posticis rotundatis; scutello leviter punctulato, elytris sub callum humeralem haud depressis, pone basin transversim haud impressis, subfortiter subseriatim (ad latera paullo magis, postice paullo minus, fortiter) punctulatis, verrucis parvis sat crebre seriatim instructis, interstitiis modice rugulosis, parte marginali a disco vix manifeste divisa, calli humeralis margine interno a sutura quam ab elytrorum margine laterali paullo magis distanti; segmento ventrali basali sparsius subfortiter punctulato. Long. 4, lat. 3 lines.

Among its allies structurally (having no subbasal elytral impression) this species is superficially distinct by its subseriate elytral puncturation together with the almost regular rows of small rather closely placed verrucæ, which are concolorous with the derm. There is, however, a tendency to the elytra being marked with dark vittæ (which in some examples are very well-defined), and on these vittæ the verrucæ are concolorous with them and not with the general surface.

S. Australia, Yorke's Peninsula.

P. PUSTULIFERA, sp.nov.

P. alticolæ affinis; differt colore toto (prothoracis maculis non-nullis, et elytrorum verrucis, nigris exceptis) testaceo-castaneo; prothorace in disco magis fortiter minus crebre (ad latera grosse confluenter) punctulato; scutello nitido sparsim fortiter punctulato; elytris in disco magis fortiter punctulatis, verrucis valde perspicuis (haud transversim elongatis) in seriebus integris circiter 9 sat crebre sat regulariter dispositis; cetera ut *P. alticola.*

Femina quam mas paullo magis convexa. Long. 4, lat. $2\frac{1}{5}$ lines.

Although superficially very different from *P. alticola*, this species is structurally very close to it. The notably coarser punctaration of its upper surface, however, forms a reliable distinction, and the colour and markings are so different that it is unlikely any varieties approximate much to *alticola*. With the exception of some black marks on the prothorax (a longitudinal blotch on either side of the middle and a few small spots nearer the margins, in the type) and numerous small round black verrucæ (about 15 in a series) placed in about 9 series very evenly over the whole elytra, the entire insect is of a uniform pale chestnut colour. There is, in the type, also a common dark blotch on and around the suture a little in front of its middle, apparently caused by the intervals between two or three verrucæ being stained with dark colouring similar to that of the verrucæ.

N. W Australia; sent to me by Mr. Froggatt.

P. ALTA, sp.nov.

♀. Ovata, latissima; valde convexa, altitudine majori (a latere visa) pone elytrorum marginem medium posita; modice nitida; castaneo-brunnea, antennis apicem versus pedibus in parte et corpore subtus piceis ; capite sat crebre ruguloso-punctulato; prothorace quam longiori ut 2¼ ad 1 latiori, ab apice sat longe ultra medium dilatato, pone apicem transversim leviter impresso, crebre subfortiter aspere sat æqualiter (parte laterali sat grosse rugulosa excepta) punctulato, lateribus sat fortiter arcuatis nullo modo deplanatis, angulis posticis nullis; scutello ut prothorax punctulato; elytris sub callum humeralem haud depressis, pone basin transversim haud impressis, confertim dupliciter (subtilius et magis subtiliter) sat aspere vix subseriatim (latera versus paullo magis, postice vix minus, fortiter) punctulatis, verrucis parvis nonnullis parum perspicuis instructis, interstitiis minus rugulosis, parte marginali a disco vix distincta, calli humeralis margine interno a sutura quam ab elytrorum margine laterali paullo magis distanti; segmento ventrali basali subfortiter vix crebre punctulato; antennarum articulo 3º quam 4ᵘ distincte longiori.   Long. 3½, lat. 3 lines.

A somewhat isolated species on account of its resembling by its great convexity the species of the next subgroup, but differing from them by the greatest height of the elytra being very far back.   On careful examination it is seen that the 3rd joint of the antennæ is distinctly longer than the 4th, but the difference in length is not marked enough to associate the species with P. *regularis* and its allies, and its natural place is certainly near P. *inornata*, Blackb.

S. Australia; Adelaide; also Murray Bridge.

P. INORNATA, sp.nov.

♂. P. *altæ* affinis ; minus lata, multo minus convexa; picea, antennis basin versus rufis; prothorace ab apice paullo minus longe ultra medium dilatato; scutello lævi; elytris pone basin

transversim vix penitus æquali; antennarum articulo 3°
quam 4ᵘˢ haud longiori; cetera ut *P. alta.* Long. 4, lat. 3½
lines.

♀. Quam mas subconvexiori.

Except in respect of a few well-marked characters this species
is so close to *P. alta* that it seems unnecessary to repeat the whole
of the description of the latter which (modified by the characters
noted above) applies exactly to this insect. The much less convexity and the antennal difference at once separate *P. inornata*,
as also the absence of puncturation on the scutellum, but this
latter character I do not so absolutely rely upon, as I find that
there is a slight tendency to variation in the puncturation of the
scutellum of many species of *Paropsis.* I do not think, however,
that any specimen of *P. inornata* would have anything like the
strong scutellar puncturation of *P. alta,* which is quite continuous
with the puncturation of the prothorax. Indeed, I have before
me some examples of *Paropsis* from Yorke's Peninsula and from
Eucla which I believe to be *P. inornata,* in which the scutellum
bears some fine punctures. It is possible that they represent a
distinct very close species, but the point could not be certainly
decided without the examination of more examples from the same
locality as the type of *P. inornata,* from which locality I have
seen only one female, and that one is in bad condition.

W. Australia; Eyre's Sand Patch.

P. INÆQUALIS, sp.nov.

♂. Late ovata; minus convexa, altitudine majori (a latere visa)
contra elytrorum marginem medium posita; modice nitida;
nigra, antennarum basi et pedibus maculatim (tarsis totis)
rufis; capite prothoraceque æqualiter (sed hoc ad latera grosse
rugulose) crebre subfortiter fere rugulose punctulatis; hoc
quam longiori ut 2⅔ ad 1 latiori, ab apice ultra medium
dilatato, pone apicem transversim leviter impresso, lateribus
fortiter arcuatis nullo modo deplanatis, angulis posticis nullis;
scutello (exempli typici carente); elytris sub callum humeralem haud depressis, pone basin transversim haud impressis,

sat grosse vix crebre vix subseriatim (ad latera magis, postico
minus, grosse) punctulatis, verrucis sat magnis sat numerosis
minus elevatis sat seriatim instructis, interstitiis parum
rugulosis, parte marginali a disco vix (in parte subapicali
paullo magis distincte) divisa; segmento ventrali basali subfortiter subcrebre punctulato. Long. 3½, lat. 2⅘ lines

A fairly distinct species notable for its black colour and the
coarse puncturation of its elytra, the verrucæ of which are somewhat large and numerous but not strongly elevated. The front
margin of the prosternum is exceptionally wide.

S Australia; Adelaide district.

P. ALPINA, sp.nov.

♀ Ovata, sat fortiter convexa, altitudine majori (a latere visa)
ad elytrorum marginem medium posita; sat nitida; sordide
flavo-brunnea, elytris (parte basali mediana et parte laterali
antica exceptis) nigro-adumbratis et confuse nigro-maculatis,
antennis apicem versus vix infuscatis; capite inæquali fortius
sat rugulose punctulato; prothorace quam longiori plus quam
duplo (ut 2½ ad 1) latiori, ab apice longe ultra medium
dilatato, pone apicem haud transversim impresso, subgrosse
rugulose (ad latera etiam magis grosse) punctulato, lateribus
sat arcuatis haud deplanatis, angulis posticis nullis; scutello
lævi; elytris dupliciter (grosse et minus grosse) sat crebre
subseriatim punctulatis, antice haud (postice vix distincte)
verrucosis, interstitiis antice vix (ad latera vermiculatim
grosse, postice crebre sat granulatim) rugulosis, sub callum
humeralem leviter depressis, parte marginali a disco vix
distincta, calli humeralis margine interno a sutura quam ab
elytrorum margine laterali paullo magis distanti; segmento
ventrali basali sparsius minus subtiliter punctulato; antennarum articulo 3° quam 4ᵗ sat longiori. Long. 4, lat. 2⅘ lines.

At its widest somewhat behind the middle of the elytra; notable
by the 3rd joint of the antennæ markedly longer than the 4th,
also (so far as the unique type is concerned) by the peculiar
colouring of its elytra, which are of a yellow-brown colour with

an ill-defined festoon-like patch of blackish colour a little behind
the base (its extremities on the humeral calli), behind which the
whole surface (except the front half of the marginal portion) is
thickly set with blackish irrorations very various in size. Genuine
verrucæ are almost non-existent except near the apex, and even
there they are so much mixed with confused rugulosity as to need
being looked for.

Victoria; on the higher Alps.

P. BOREALIS, sp nov.

Subovata; sat fortiter convexa, altitudine majori (a latere visa)
contra elytrorum marginem medium posita; nitida, rufa,
prothoracis marginibus scutello elytrorum macula communi
antemediana et utrinque macula prope humerum posita corporeque subtus (hoc maculatim) indeterminate piceis, capite
sparsim subfortiter punctulato; prothorace quam longiori ut
2¼ ad 1 latiori, ab apice vix ultra medium dilatato, pone
apicem transversim haud impresso, sparsim inæqualiter sub-
acervatim (ad latera sat grosse sat crebre nec confluenter)
punctulato, lateribus minus fortiter arcuatis haud deplanatis,
angulis posticis rotundatis; scutello punctulato, elytris sub
callum humeralem haud depressis, pone basin nullo modo
impressis, minus fortiter sat crebre sat æqualiter (antice
suturam versus magis subtiliter) subseriatim punctulatis,
verrucis nonnullis parvis subseriatim dispositis instructis,
interstitiis vix rugulosis, parte marginali a disco haud distincta, calli humeralis margine interno a sutura quam ab
elytrorum margine laterali paullo magis distanti; segmento
ventrali basali sparsim obsolete punctulato. Long 4, lat.
2⅔ lines.

As the type has lost its tarsi, I am not sure of its sex, but have
little doubt of its being a female. The entire absence of any
trace of a subbasal elytral impression and the evenness of the
elytral puncturation are well-marked characters. The inconspicuous verrucæ are concolorous with the derm and run in fairly

regular rows.   The markings resemble those of *P. asperula,* Chp., to which this species is certainly allied, though differing in many characters (*inter alia,* the much finer prothoracic and elytral punctures, with non-rugulose interstices, and the absence of distinction between the discal and marginal parts of the elytra).

N. Territory of S. Australia; taken by the late Dr. Bovill.

P. NOTABILIS, sp.nov.

♂. Ovalis; minus convexa, altitudine majori (a latere visa) contra elytrorum marginem medium posita; nitida; testaceo-brunnea, maculis in capite prothoraceque nonnullis elytris (verrucis exceptis) antennis apicem versus et corpore subtus (hoc maculatim) obscurioribus; capite sparsius subtilius punctulato; prothorace quam longiori fere ut 2⅓ ad 1 latiori, latitudine majori fere ad basin posita, antice minus fortiter angustato, pone apicem haud impresso, fortiter acervatim haud crebre (ad latera sat grosse nec crebre) punctulato, lateribus leviter arcuatis haud deplanatis, angulis posticis obtusis; scutello lævi; elytris sub callum humeralem haud depressis, pone basin nullo modo impressis, fortiter sparsius (ad latera parum magis fortiter) punctulatis, verrucis numerosis magnis parum elevatis instructis, interstitiis haud rugulosis, parte marginali angusta a disco (per sulculum sat distinctum) pone medium divisa, calli humeralis margine interno a sutura quam ab elytrorum margine laterali multo magis distanti; segmento ventrali basali vix manifeste punctulato.   Long. 6, lat. 4⅕ lines.

A remarkable species, with considerable superficial resemblance to *P. solitaria,* Blackb., but differing from it *inter alia* by its much larger size and elytra not depressed below the humeral callus.   Its large, scarcely elevated, numerous elytral verrucæ of pallid colour furnish a notable character.   Its prothorax is suggestive of species of the *variolosa* group, but is neither mucronate at the front angles nor laterally sinuate.

N. S. Wales; in the collection of Mr. G. Masters.

P. VOMICA, sp.nov.

♂ Latissime ovata; fortiter convexa, altitudine majori (a latere visa) anterius quam contra elytrorum marginem medium posita; sat nitida; rufo-brunnea, elytrorum verrucis testaceis vel flavescentibus, corpore subtus in majori parte picescenti; capite sat crebre aspere punctulato; prothorace quam longiori ut 2⅔ ad 1 latiori, ab apice paullo ultra medium dilatato, pone apicem transversim vix impresso, sat crebre dupliciter (subtiliter et sat fortiter, ad latera grosse) punctulato, lateribus sat arcuatis late distincte deplanatis, angulis posticis rotundatis; scutello fere lævi; elytris sub callum humeralem haud depressis, pone basin haud impressis, subtilius vix seriatim (ad latera vix magis grosse) punctulatis, verrucis magnis (minus fortiter elevatis) numerosis seriatim instructis, interstitiis paullo rugulosis, parte marginali a disco (nisi apicem versus) minus distincta, calli humeralis margine interno a sutura quam ab elytrorum margine laterali paullo magis distanti; segmento ventrali basali sublævi; antennarum articulo 3° quam 4us sat longiori.

♀. Quam mas paullo minus lata, segmento ventrali apicali magis perspicue punctulato. Long. 4-4⅕, lat. 3½ lines.

An extremely distinct species, on account of the large moderately elevated verrucæ of the elytra conspicuously more pallid than the general surface and very evenly distributed except on a small roundish common antemedian space. Its strongly convex form suggests alliance with the species of the next subgroup, but the greatest height of its elytra is very little in front of the middle. It seems to be somewhat uncertain in position in the genus, the slightness of the tendency to seriate arrangement in the punctures of its elytra being suggestive of species with the front angles of the prothorax mucronate.

N. W. Australia; sent to me by Mr. Masters.

46

THE SILURIAN TRILOBITES OF NEW SOUTH WALES, WITH REFERENCES TO THOSE OF OTHER PARTS OF AUSTRALIA.

By R. Etheridge, Junr., Curator of the Australian Museum
—and John Mitchell, Public School, Narellan.

Part IV.

The ODONTOPLEURIDÆ.

(Plates L.-LV.)

The next family we propose to take up is that of the Odonto-pleuridæ, adopting this name in preference to Acidaspidæ, because we have every reason to believe it to have precedence. Burmeister used the term in 1843, but we have not been able to ascertain at how early a date Barrande employed that of Acidaspidæ, with which Zittel credits him. It could, however, hardly have been before the date in question. The genera, or sections of the old genus, *Acidaspis*, whichever the idiosyncrasy of the reader may choose to regard them, are the following :—

Ceratocephala, Warder, 1838.
Odontopleura, Emmrich, 1839.
Acidaspis, Murchison, 1839.
Dicranurus, Conrad, 1841.
Selenopeltis, Corda, 1847.
Ancyropyge, Clarke, 1891.

Of these we have been able to recognise in Australia only two, viz. :—

Odontopleura, Emmrich.
Ceratocephala, Warder.

but possibly a third (*Selenopeltis*, Corda) may be represented by our *Ceratocephala longispina*.

None were described by Prof. L. G. de Koninck in his work on the " Palæozoic Fossils of N.S. Wales."

The study of this group has proved an arduous one from the complex nature of the cephalic shield or cephalon, and we may have erred by introducing too much detail; this is, however, an error on the right side.

"Of all the extravagant forms of this curious family of Trilobites," says Salter,[*] "none seem so extravagant in its ornament as the genus *Acidaspis;* the head, thorax, and tail being literally crowded with spines wherever an available angle occurs."

Genus O D O N T O P L E U R A, Emmrich, 1839.

Odontopleura, Emmrich, De Trilobitis, 1839, p. 35.
       ,,   Burmeister, Organization of Trilobites (Ray Soc.),
               1846, p. 61.
               Clarke, 10th Ann. Report State Geol. N. York for
               1890 (1891), p. 67.

Obs.—This genus is distinguished from other Acidaspids by having the occipital ring either with or without a tubercle in the centre, but totally devoid of a spine or spines. The type, according to Mr. J. M. Clarke, is *O. ovata*, Emmrich, a form having some characters in common with our first species, but in others departing widely from it.

The specific history of the Acidaspidæ in Australia is a brief one As recorded by Mr. F. Ratte,[†] Mr. Chas. Jenkins, L.S., appears to have been the first to recognise the presence of the genus in our rocks. He figured the greater portion of a Trilobite that he referred to *Acidaspis Brightii*, Murchison,[‡] from Yass, but during our researches we cannot say that we have met with any Trilobite that would strictly agree with that species; indeed we have not seen a true *Acidaspis*, as now restricted, from Australia. Mr. Jenkins was followed by the late Mr. Felix Ratte, who contributed two papers to the Proceedings of this Society

---

* Brit. Org. Remains, Dec. vii., Pt 6, p. 2.
† Proc. Linn. Soc. N.S. Wales, 1887, ii. (2), p. 99 (footnote).
‡ Proc. Linn. Soc. N.S. Wales, 1879, iii., Pl. 17, f. 5.

dealing with Acidaspids from Bowning. In the first he described species ascribed by him to the following well-known Trilobites*.—

> *A. Verneuili*, Barr., or *A. vesiculosa*, Barr.
> *Acidaspis* near *A. Prevosti*, Barr.
> *Acidaspis* near *A. mira*, Barr.

In the second paper† the following :—

> *Acidaspis* near *A. Dormitzeri*, Corda.
> *Acidaspis* near *A. Leonhardi*, Barr.

At a later period one of us‡ described a new species, also from Bowning, as *A longispinis*. The whole of these will be passed in review in the present paper.

We now recognise the following four species :—

> *Odontopleura bowningensis*, nobis.
> „　　　*Rattei*, nobis.
> „　　　*parvissima*, nobis.
> *Jenkinsi*, nobis.

ODONTOPLEURA BOWNINGENSIS, *sp.nov.*

(Pl. L., figs. 1-3; Pl. LII., fig. 5.)

*Sp. Char.—Body—*Ovoid. *Cephalic shield or cephalon—*Subelliptical, about three times as wide as long measured between the base of the genal spines, very tumid, rising abruptly from the posterior margin, which is unusually straight, ornamented with fine and moderately coarse granules. Glabella quadrate, central lobe small, oblong, very intensely arched transversely, moderately so fore and aft, and almost sloping into the front margin, granulated very distinctly, front lateral expansions distinct; lateral lobes small, granulated and tumid, median pair about half the size of the posterior pair and semiglobular, posterior pair suboval, each pair very distinctly separated from each other by the basal

---

* Proc. Linn. Soc. N.S. Wales, 1886, i. (2), Pt. 1, pp. 1066-69.

† *Loc. cit.* 1887, ii. (2), pp. 96-102.

‡ Mitchell, *loc. cit.*, 1838, iii. (2), p. 398; 1887, ii. (2), t. 16, f. 7-10, 12.

pair of glabella grooves; anterior pair absent; glabella grooves wide and distinct and joining the axial and false furrows; axial furrow faint anteriorly but distinct posteriorly; false furrows very distinct and wide. Fixed cheeks of moderate size; genal lobe subtriangular, very tumid, granulated; ocular bands or ridges very narrow and partly overhung by the genal lobes, and themselves intensely overhanging the free cheeks and bearing a distinct row of granules ; genal or palpebral furrows distinct ; eyelobes small, triangular areas very small. Free cheeks of tolerable proportionate size, intensely tumid, borders intensely thickened, particularly towards the genal angles, each bearing twelve short, acicular, deflected spines exclusive of the genal spines, marginal furrow very distinct. Genal spines short, stout, falcate, and forming obtuse angles with the cephalon. Facial sutures anteriorly appear to be soldered, but their course is indicated along and under the ocular ridges, and they incline towards each other at an angle of 35°, cutting the front margin in a line with the axial furrows; posteriorly they run obliquely to the median point of the lateral extensions of the fixed cheeks, thence parallel with those extensions to the genal angles. Occipital furrow wide and shallow centrally, but deep at the sides, continuing across the sides distinctly and joining the marginal furrows of the free cheeks. Neck or occipital ring strongly arched vertically, only moderately so backwards, sides nodular, no central tubercle. Eyes prominent, as high as the highest part of the central glabella lobe, small, very wide apart, the distance between them being equal to twice the length of the cephalon.

Thorax.—Consists of ten segments, width equal to the combined length of itself and pygidium, granulated. Axis prominent, rather wider than the pleuræ, posterior width half of the anterior width, rings nodular at the sides. Pleuræ flat between the axial grooves and the fulcra, thence short and sharply deflected, sutures distinct, median ridges prominent, tuberculated, one very prominent tubercle on each ridge nearer the fulcra than the axial grooves, forming a longitudinal row along each lateral lobe; ends of pleuræ or median ridges thickened and bispinate, posterior ьpines long and acicular, anterior ones short and lanceolate and serrated, all much deflected. Axial furrows faint.

Pygidium.—Small, about four times as wide as long, strongly tubercular; the anterior margin straight between the fulcra, thence gently turned backwards. Axis prominent, consisting of one highly arched anterior ring and a terminal piece which is ridged and circumfurrowed, and centrally depressed. The lateral lobes are divided into two pairs of pleuræ by one pair of pleural ridges, extending from the axis ring; they are flat, tuberculate and punctate, border much thicker and internally bounded by a distinct furrow. Tail spines fourteen, acicular, four intermediate and four on each side of the axial pleural spines, the latter diminishing rapidly in length from the axial pair outwards, so that the first and second pairs are very short.

Obs.—The striking features of this species are:—(1) The great proportionate width, particularly of the cephalon; (2) the deflected spines and short, jutting, obtuse hornlike genal spines, (3) the very small eyes; (4) the absence of an occipital tubercle; (5) the great width between the eyes and their nearness to the posterior margin of the cephalon; and (6) the excessive tumidity of the cephalon as a whole.

Whilst resembling *O. ovata*, Burmeister,[*] the generic type, in the great proportionate breadth of the body to its length, our form departs very markedly in possessing ten instead of eight thoracic segments, in the very small pygidium, the increased number of spines around the margin of the latter, and in the shorter and stouter genal spines. Similar characters separate it from *O. elliptica*, Burmeister.[†] From an allied American species, *O. crossota* (Locke), Meek,[‡] our species is separated by the size, shape, and segmentation of the pygidium. In this species also the facial suture extends even further laterally than in *O. bowningensis*, and there is no occipital tubercle.

---

[*] Organization of Trilobites, 1846, p. 62, t. 2, f. 11.
[†] *Loc. cit.* p. 63, t. 1, f. 4.
[‡] Ohio Geol. Report, 1873, I. t. 14, f. 10, 10a.

From the American Devonian species *O. callicera*, Hall,[*] our species is equally distinct. It lacks the long genal spines and large eyes of the former and possesses a greater number of cheek spines.

It is with the Bohemian species that the Bowning Trilobite seems to correspond best, although it is a broader form than the majority of the former, if not indeed of all those allied to it.

In *O. Leonhardi*, Barr., the pleuræ are single-spined, in our form double, and the pygidium spines are increased in number and are constant. In the former the genal spines are long and acicular, in the latter short and stout, and the courses of the facial sutures are different in the two species.

From *O. minuta*, Barr., *O. bowningensis* is at one distinguished by the uniformity of the spines extending from the pygidium of the former, and again by the nature of the pleural and genal spines. It may be said also that the same characters separate our form from *O. Dormitzeri*, Barr., and *O. Roemeri*, Barr. In the latter the backward extension of the genal spine is enormous.

The description is taken from decorticated specimens.

Loc. and Horizon—Bowning Creek, near Bowning, Co. Harden, Lower Trilobite Bed—Bowning Series (= *Hume Beds*, Jenkins, and *Yass Beds*, David)—? Wenlock.   *Coll.*—Mitchell.

ODONTOPLEURA RATTEI, *sp.nov.*

(Pl. L., fig. 7; Pl. LI., figs. 8-9; Pl. LII., figs. 1-4; Pl. LIII., figs. 1-3.)

Acidaspis near *A. Leonhardi*, Ratte (non Barr.), Proc. Linn. Soc. N.S. Wales, 1887, ii. Pt. 2, p. 99, Pl. 2, figs. 2-4.

Sp. Char.—*Body*—oval. *Cephalic shield or cephalon*—Subsemicircular, a little wider than twice the length, and straight in front. Glabella quadrate, width between eye lobes equals length, including the neck ring, distinctly and evenly granulate, front margin dentate; central portion suboblong, intensely arched transversely, moderately so from front to back, highest medially

---

[*] Pal. N. York, 1888, vii. t. 16b, f. 1-13.

and bending rapidly to and merging into the front margin, slightly
expanded in front; the first pair of lateral lobes in a rudimentary form (tubercles merely); lateral portions distinctly
bilobed, median pair suboval, very tumid, about half the size of
the posterior pair, and very distinctly separated by the glabella
furrows which join the axial and false axial furrows; false axial
furrows very distinct, particularly at their junctions with the
lateral furrows, passing into the neck furrow; axial furrows distinct and intensely so as they join the neck furrow, faint across
the posterior margins. Fixed cheeks suboblong, tumid; genal
lobes ridged, ocular ridges or bands prominent, each bearing a
row of granules; genal furrows distinct; triangular areas large
and flat. Free cheeks very tumid, granulated, borders thickened,
marginal furrows distinct and terminating at the front angles of
the glabella, the borders bear fourteen acicular spines exclusive of
the genal spines, which are also acicular, strong, slightly falcate
and long, and bear the last two or three cheek spines; facial
sutures anteriorly straight and nearly parallel with the axial
centre, posteriorly parallel with the lateral extensions of the fixed
cheeks. Neck furrow shallow generally, but deep at its junctions
with the axial furrows, its lateral extensions interrupted by the
tumid ends of the neck ring, thence moderately distinct across the
posterior borders of fixed cheeks. Neck or occipital ring strong,
intensely arched backwards, ends nodular, granulated, and central
tubercle present. Eyes prominent, of medium size, conoid and
faceted.

Thorax.—Consists of nine segments, suboblong or subfusiform,
width equal to the combined length of itself and pygidium,
exclusive of the spines; axis prominent, having two rows of distinct dorsal granules, rings arched backwards, ends nodular,
width equal to width of side lobes between the fulcrum and axial
furrows, which are faint; side lobes horizontal between axial
furrows and fulcra, thence moderately deflected, median ridges of
pleuræ strong and prominent, and each at the fulcrum bears a
very distinct tubercle, forming a persistent row on each lobe, on
the deflected ends the ridges widen, and are produced into long

acicular spines, except in the case of the first pair of pleuræ on each lobe, which are very rudimentary; the spines of the third pair equal the length of the thorax and tail together, and are flected backwards at about 45°, each succeeding pair increasing in backward flection till those from the last pair are rectangular to the thorax.

Pygidium.—Widely triangular, rather flat, strongly granulated; front margin straight between the fulcra, thence backwards at an angle of 45° nearly.   Axis short, consisting of one very prominent ring and terminal piece, the latter clearly separated from the former by a furrow, and bearing a small but distinct and persistent granule on each side, and is also nearly circumfurrowed.   From the ends of the axis ring extend a pair of pleural ridges obliquely and distinctly across the lateral lobes, and are produced into the axial or pleural spines. Side lobes divided into two lobes, one pair of pleural furrows present, border bearing twelve to fourteen acicular spines, two intermediate and four to five exterior to the axial pair; the first two on each side adjacent to the anterior face are rudimentary and seldom visible when the tail is attached to the thorax; the pleural pair have a length equal to half the length of the thorax; intermediate pair appear to be about two-thirds as long as the axial pair; all bear a row of granules.

Obs.—This species is one of those figured by the late Mr. Felix Ratte,* and placed by him near *O. Leonhardi*, Barr., although he was careful to point out that it did not strictly accord with that Trilobite.

From the preceding form, *O. bowningensis*, nobis, it may be at once distinguished by possessing a segment less in the thorax, by the presence of frontal spines or serrations to the glabella proper, and so far as we are able to discern, by the thoracic pleuræ being unispinate only; furthermore, it is a more slender species.   The genal spines are very different, as are also the pygidium and other parts.

---

* Proc. Linn. Soc. N.S. Wales, 1887, ii (2), Pt. 2, p. 99, Pl. ii. figs. 2-4.

As regards *O. Leonhardi*, with which Mr. Ratte compared this fossil provisionally, the two are unquestionably near one another. Mr. Ratte appeared to think that a greater breadth existed in the fixed cheeks of the Australian fossil; but we would rather rely on other characters of possible specific value. For instance, the spines of our form are much longer and stouter than those of *O Leonhardi*, the anterior ones of ours, too, are always flected backwards at a greater angle; the genal spines at their bases press on the two anterior pairs of the thoracic pleuræ, and these two pairs of pleuræ have very rudimentary spines, which is a feature of itself that clearly separates it from *O. Leonhardi* and its congeners. The frontal margin of the glabella of *O. Rattei* is spined or serrated, but the margin of *O. Leonhardi* is smooth The pleural spines are more graduated in length from before backwards, producing a remarkable frill-like appearance in *O. Rattei*, whilst the characters of the pygidium are very distinct. In *O. Leonhardi*, between the axial or pleural spines are four peripherals, and exterior to the former two peripheral spines on either side. In *O. Rattei*, on the other hand, there are in a typical specimen two peripheral spines occupying the first position and four to five the second; but in another typical specimen (immature) there are two peripherals in the first and three to four in the second position. We have never seen three spines on the pygidium of *O. Rattei* between the axial or pleural spines, and it is wider and the spines larger, longer and more unequal in length than is the case with those of *O. Leonhardi*.

The normal number of spines that can be seen on a pygidium of *O. Rattei* when attached to the thorax is ten, and the actual number twelve, the one on each angle being mostly very rudimentary, and in some specimens bifurcate. In cases such as the latter, a tail may bear fourteen spines, but not more than ten would probably be visible if the fossil were complete.

Odontopleura pigra, Barr., sp.,* is so far related to the present species that although the pleural spines of the first two thoracic

---

* Novak, Dames & Keyser's Pal. Abh. 1890, v., Heft 3, t. 2, f. 11 & 13.

segments are present, they are so much reduced in size as to indicate a transition towards *O. Rattei.*

Named in honour of the late Mr. Felix Ratte, Mineralogist to the Australian Museum, Sydney.

Loc. and Horizon.—Bowning Village, Co. Harden, Middle and Upper Trilobite Beds—Bowning Series (= *Hume Beds,* Jenkins, and *Yass Beds,* David)—? Wenlock. *Coll.*—Mitchell

ODONTOPLEURA PARVISSIMA, *sp. nov*

(Pl. L., figs. 4-6; Pl. LII., fig. 8.)

Acidaspis near *A. Dormitzeri,* Ratte (non Corda), Proc. Linn. Soc. N.S. Wales, 1887, ii. (2), Pt. 2, p. 96, t. 2, f. 1, 1 bis.

Sp. Char.—Body.—Suboblong-oval. *Cephalic shield or cephalon.* —Subquadrate, twice as wide as long, tumid and strongly tubercled throughout. Glabella quadrate, half as long as the thorax, or including the neck ring its length equals the width between the eyes; central lobe narrow, intensely arched transversely, moderately so fore and aft, extending to the front or limb, which is straight and appears under a strong lens to be delicately dentate ; the lateral lobes mere tubercles ; lateral and false furrows distinct, axial furrows indistinct.　Fixed cheeks very small and tumid; genal lobes very small (practically narrow bands each bearing a row of tubercles) ; genal or palpebral furrows moderately distinct; ocular ridges distinct anteriorly and tubercled. Eyes very small and prominent.　Free cheeks proportionately large, tumid, outer borders thickened, narrow, and each bearing ten short acicular horizontal spines, and on the upper surface a row of prominent tubercles; genal angles produced into long, slender and subfalcate spines.　Facial sutures distinct, anteriorly gently curving towards the axis and passing out at the front angles of the central lobe; posteriorly are parallel with the edges of the lateral extensions of the fixed cheeks, and pass out at the genal angles. Neck furrow distinct, narrow, lateral extensions faint.　Neck ring intensely arched, lateral nodules small, but distinct, tubercled, but no prominent central tubercle.

Thorax.—Possesses nine segments, nearly square, greatest width equal to its length. Axis prominent, wider than the pleural lobes; rings faintly nodular at ends, dorsally each bearing two prominent tubercles. Axial furrows distinct. Lateral lobes narrow; pleural ridges and sutures very distinct, each pleural ridge bearing two very prominent tubercles, one at the fulcrum and the other near the axial furrow; at least seven pairs of pleuræ bear acicular spines, those on the third pair (none visible on the first and second pairs) are short, and at right angles with the axis, each succeeding pair have an increasing backward flexion till the last pair are parallel with the axis, they also increase in length posteriorly; the fifth, sixth and seventh pairs are subfalcate, the eighth and ninth pairs in some specimens show indications of having stood upright.

Pygidium.—Very small, widely triangular, distinctly tubercled Axis very prominent, consists of one ring and small terminal piece; both bear a pair of small tubercles. Lateral lobes divided into two pleuræ by the pleural ridges extending from the ends of the axis ring; these ridges are bituberculate; the border bears eight acicular spines of nearly uniform length, four intermediate and one on each side of the principal pair. Axial furrows distinct.

Obs.—This species was briefly described by Mr. F. Ratte,[*] and determined by him to be near *O. Dormitzeri*, but he pointed out that it did not exactly agree with that or any other species known to him. He noticed the small proportionate length of the tail to the whole body, and the rounded contour of the free cheeks, and with these observations we agree. In his description he apparently fell into an error in assigning an ample genal lobe or triangular area, which we find to be very small, also in locating the eyes much more forward than they are in allied species. In this latter feature it agrees with the *O. Leonhardi* type in having the eyes near the posterior border of the cephalon. It is, however, separated from *O. Leonhardi* and its congeners by possessing no triangular areas on the glabella, in its more rounded and expanded

---

[*] Proc. Linn. Soc. N.S. Wales, 1887, ii. (2), Pt. 2, p. 96, t. 2, f. 1, 1 bis.

free cheeks, and in the structure of the pygidium, in which characters it also differs from *O. Rattei*, nobis. The tuberculation is singular among the known Australian species. It resembles *O. Rattei* in the proportionate length to width of the cephalon, and in the pleuræ being unispinate.

In form it approaches *O. minuta*, Barr, but as the late Mr. Ratte pointed out, it bears only two rows of tubercles on the pleural lobes, while on those of *O. minuta* there are three rows; and the largest of our specimens is not more than half the size of that fossil. The genal and pleural spines are much larger in ours than in the Bohemian species.

Mr. Ratte seems also to have erred in fixing the number of cheek spines at fourteen. We find them to be ten; and they occupy two-thirds of the border, the anterior third being spineless.

From *O. Dormitzeri* our species differs in having a much more quadrate cephalon, a highly granulose pygidium, and an absence of the axial pleural spines. It is much nearer to *O. minuta*, Barr., and this is in all probability its nearest ally. The distinguishing features of *O. parvissima* are—(1) The semicircular curve of the borders of the free cheeks; (2) the fine acicular cheek spines; (3) the subfalcate pleural spines; (4) the tubercled pleural ridges; (5) the uniform tail spines, and absence of strong pleural ridges (pads) on the pygidium; (6) the small central and lateral glabella lobes; (7) the remarkably strong tuberculation of the whole test; (8) its minuteness; and (9) the equality in the length of the thorax and width of the head-shield.

Our Pl. L., fig. 4, is drawn from the same specimen as Mr. Ratte's t. 2, f. 1, bis.

Loc. and Horizon.—Bowning Creek, Co. Harden, Lower Trilobite Bed—Bowning Series (= *Hume Beds*, Jenkins, and *Yass Beds*, David)—? Wenlock. *Coll.*—Mitchell.

ODONTOPLEURA JENKINSI, *sp.nov.*

(Pl. LII., figs. 6-7; Pl. LIII., figs. 4-7.)

Acidaspis Brightii, Jenkins (*non* Murch.), Proc. Linn. Soc. N.S. Wales, 1879, iii., p. 221, t. 17, f. 5.

Acidaspis Prevosti, Ratte (*non* Barr.), *loc. cit.* 1886, I. (2), Pt. 4, p. 1069, t. 15, f. 12 (excl. f. 11).

Sp. Char.—This species is so near *O. Rattei,* nobis, that it will be sufficient for us to state the points of difference between the two fossils on which we rely for justification in separating them. In *O. Jenkinsi* the limb or margin in front of the glabella is smooth instead of being dentated as in *O. Rattei;* each pleura of the thorax bears two prominent tubercles, and some of the anterior pairs four, the axis also appears more prominent. The pygidium carries the same number of spines as that of *O. Rattei,* but four of them are constantly intermediate of the principal or axial pair. The side lobes are more distinctly ridged and furrowed, the ridges are surmounted by very distinct rows of tubercles. The pleural ridges from the axial ring are less prominent than they are in *O. Rattei,* but the tuberculation is more conspicuous throughout.

Obs.—We hesitated very much about according this form specific separation from *O. Rattei,* and we do so only after examining a great number of specimens and finding that the characters already pointed out were constant, and because it comes from a higher horizon and is not found associated with *O. Rattei* in the lower horizon, where that fossil is very numerous.

We believe that this is the Acidaspid described and figured by Mr. Jenkins as *Acidaspis Brightii,* and as pointed out by Mr. Ratte, this appears to have been the first notice of the discovery of a member of this family in Australia. With regard to its identification with *O. (Acidaspis) Brightii,* Murch., we are quite in accord with the doubt expressed by Mr. Ratte, and also admitted by Mr. Jenkins himself. The points which separate *O. Rattei* from the European and other members of the genus known to us, given under its description, will also apply to this, except that it approaches still nearer to *O. Leonhardi* than does *O. Rattei* in having a smooth frontal glabella margin.

Loc. and Horizon.—Bowning Railway-Station Yard, Bowning Village, Co. Harden, Upper Trilobite Bed—Bowning Series (= *Hume Beds,* Jenkins, and *Yass Beds,* David)—? Wenlock. *Coll.*—Mitchell.

Genus C E R A T O C E P H A L A, Warder, 1838.

Ceratocephala, Warder, Am. Journ. Sci., 1838, xxxiv., p. 377.

Trapelocera, Corda, Prod. Bohm. Trilobiten, 1847, p. 158.

Trapelocera, Angelin, Pal. Scandinavia, 1878 (Lindstrom's edit.)
p. 34.

Ceratocephala, Clarke, 10th Ann. Rept. State Geol. N. York for
1890 (1891), p. 67.

Obs—Mr. J. M. Clarke has already indicated the lines on which this name should be used, and it is here adopted by us in conformity with his researches, except that we employ it as one of the genera of the Odontopleuridæ rather than as the typical generic name of the whole group, superseding *Acidaspis*, for reasons already given.

In Australia *Ceratocephala* is represented by four species, so far as we have been able to ascertain, viz. :—

> *Ceratocephala Jackii*, nobis.
> „   *Vogdesi*, „
> „   *impedita*, nobis.
> „   *longispina*, Mitchell.

The last may possibly appertain to the genus *Selenopeltis*, Corda.

CERATOCEPHALA VOGDESI, *sp.nov.*

(Pl. L., figs. 8 and 9; Pl. LI., figs. 1-7; Pl. LIII., fig. 9.)

Acidaspis Verneuili, Ratte (*non* Barr.), or *A. vesiculosa*, Ratte (*non* Beyr.), Proc. Linn. Soc. N.S. Wales, 1886, i. (2), Pt. 4, p. 1066, t. 15, f. 5-10.

Acidaspis Prevosti, Ratte (*non* Barr.), *Loc. cit.*, p. 1068, t. 15, f. 11 (excl. f. 12).

Sp. Char.—Suboblong or oblong-ovoid.   *Cephalic shield or cephalon.*—Suboblong, of complex structure, moderately tumid, rugose and tuberculate throughout, twice as wide as long, front margin rather straight and centrally slightly projecting; tubercles of various sizes, and some very conspicuous.   Glabella:—Central lobe large, suboblong, front lateral expansions very distinct, only

moderately tumid and arched, sloping very gradually into the
front marginal and neck furrows, frontal expansions triangular,
very moderately tumid, and their apices reaching the anterior
points of exit of the facial sutures; first pair of lateral lobes
absent, median pair subconical or subtriangular, of moderate size,
very moderately tumid, basal pair large, subquadrate, with
rounded outer margins; first pair of glabella furrows deep and
wide, second pair shallow towards the axial furrows and deep
towards the false furrows, both pairs uniting the axial furrows with
the false furrows; false furrows wide and deep; axial furrows very
wide, distinct, shallow along the median portions.   Fixed cheeks
large, genal lobes large, ridged, tumid, subtriangular, united to the
lateral lobes of the neck ring by the genal ridge, and falling
abruptly into the lateral extensions of the neck furrow, bearing
some very large tubercles.   Genal or palpebral furrows moderately
distinct and highly tubercled   Eye or palpebral lobes large, very
prominent and triangular.   Ocular ridge very prominent and
overhanging the facial sutures.   Eyes small proportionately,
subpedunculate, fixed obliquely outwards and very slightly
forward, remarkably near the front margin, very wide apart, the
distance between them being equal to the diagonal from the base
of a genal spine, and the point at which the facial suture cuts the
front margin on the opposite side of the glabella, or one and a
quarter times the length of the cephalon.   Neck furrow wide and
shallow behind the central glabella lobe, narrow and deep between
the basal glabella lobes and the lateral lobes of the neck ring; its
lateral extensions (as are the axial furrows also) are interrupted
by the genal lobe ridges, and from the genal lobe ridges they extend
widely and deeply to the bases of the genal spines, thence bend
anteriorly, passing (deeply under the eyes) to the front marginal
furrows   Neck ring very wide and very moderately arched vertically, but greatly so posteriorly.   Occipital spines strong, long,
and originating in the median transverse line of the neck ring,
extending upward and outward for the first part at an angle of
60°, then arching backward and inward and the ends sharply
deflected.   Facial sutures soldered, but indicated anteriorly along

and under the ocular ridges, passing out in a line with the outer edges of the median glabella lobes, and cutting the margins at an angle of about 25°.   Free cheeks subtriangular or subcrescentic, much expanded at the front lateral angles, from thence to the genal angles rather straight and inclining inwards, highly tuberculate and rugose; genal spine ridges strong, very prominent, and vanishing under the eyes; borders distinct, strap-like, smooth and entire: marginal furrows faint; genal angles almost in a line with eyes, axially, bearing strong, suberect, long arching spines, which will apparently reach to the fifth or sixth thoracic segment.

Thorax.—Unknown in a complete state, probably consisting of ten segments, and as wide as long; very conspicuously tuberculated and granulated, and flat.   Axis very distinct, very moderately arched vertically, ends of segments very distinctly separated from the central portion by furrows, strongly inclined forwards, and with a very joint-like character, only moderately tumid; central portion of segments without backward arch, each segment bearing two prominent tubercles, one on either side, about midway between the nodes and the central line; articulating surfaces very large, furrows distinct.   Lateral lobes horizontal; ventral ridges of the pleuræ on the inner halves as wide as the pleuræ, thence contracting to the bases of the pleural spines and leaving low grooved triangular areas on each side, of which the anterior ones are the largest, they are furrowed along the central line from the bases of the spines for about half of their length; the interpleural furrows very deep and wide; sutures distinct, straight and rectangular to the axis   Pleuræ bispinate, principal or upper spines very long, barbed, and on the anterior pleuræ subhorizontal, and subrectangular to the axis, subarcuate with reflected ends, posterior ones having sharply backward and upward directions; posterior pair at least rising perpendicularly from the pleuræ with their extremities converging towards each other, and originating some distance short of the extremities of the pleuræ; the secondary or inferior spines originate almost immediately under the principal spines, are stout, cylindrical, flected sharply downwards and forwards at about 30° and barbed with acicular

47

spines; each pleura bears a number of large tubercles, usually
four, along the front margin of the ridge and two or three on the
posterior margin, two of them very persistent, one on the anterior
angle adjoining the axial furrows, and one on the posterior margin
a short distance from the axial furrow; these large tubercles (as
is the whole surface of the pleuræ) are covered with smaller
tubercles; the tubercles from which the spines arise in the
posterior pair of pleuræ are very large.   Axial furrow very
distinct.

Pygidium.—Proportionately very small and granulate, at least
four times wider than long, arciform.   Axis consists of one rather
intensely arched ring; axial furrows distinct and deeply curving
inwards behind the axial ring; side lobes slender, border indistinct, lateral angles acicular and having a slight forward curve,
spines are seven in number, very strong, cylindrical, long, sub-
uniform, and strongly barbed and granulated, central one projecting from the axial ring.

Obs.—On the nodes of the axis the granules sometimes become
confluent and form ridges parallel to the longer axis, and the
posterior pleural spines when decorticated are fluted longitudinally.

The late Mr. Ratte was right in regarding this species as closely
allied to *C. Verneuili* and *C. vesiculosa,* Barr., but after careful comparison of ours with the figures of those species given by Barrande
we find it possesses so many features peculiar to itself that, in
our opinion, give it indisputable claim to rank as an independent
species.

From *C. Verneuili* it differs (1) by the absence of the spines
along the anterior border of the cephalon and free cheeks; (2) in the
relative position of the genal spines and their much greater extent
and curvature; (3) by the barbed character of both of the pleural
spines, the much greater size of these spines, and the vertical
nature of the last pair of principal pleural spines; (4) by the
contour of the cephalon, which in *C. Verneuili* has sharp
re-entering angles from the free cheeks, while in *C. Vogdesi* the
front margin is rather straight, projecting centrally, with greater
backward curvature at the front angles of the free cheeks.

In *C. Verneuili*, however, the pleuræ are flattened from above quite similar to our figures of *C. Vogdesi*.

The same features separate it from *C. vesiculosa*.

Mr. Ratte referred to the disputed point of the existence of an articulation between the pleuræ and the axial segments, said to exist in some trilobites by Emmrich, and disputed by Burmeister, the latter being upheld by Barrande. Mr. Ratte basing his opinions upon certain features one of our figured specimens exhibits, was inclined to support Emmrich's view. He says :— "One cannot help being struck in examining the specimen in question at the great resemblance to an articulation of the junction of the axis with the pleuræ. It seems as if the test (or its different joints) had been covered by a thin epiderm as admitted by Burmeister,[*] and that this epiderm is wrinkled at the articulation as shown in fig. 5, and especially in the enlarged sketch fig. 8."[†]

Whilst admitting the very joint-like appearance, somewhat exaggerated in Ratte's figure, we do not see any direct evidence of the jointing; but, on the contrary, there is one strong feature we have observed which disposes of the question in favour of the negative, and, that is, in all the many thoracic segments which have come under our notice, we have never seen a specimen divided at this point.

This joint-like appearance at the ends of the thoracic axial segments is also seen in the type of *Selenopeltis* (*S. Buchii*, Barr., sp.)

Ratte figured the principal tubercles of the pleuræ surrounded by a complete circlet of granules in every respect resembling the primary tubercle and its miliary ring on the interambulacral plates of an ordinary Echinid, such as the genus *Cidaris*. His figures correctly represent the specimen used by him, but on no other specimen can we find this feature nearly so distinct.

---

[*] Barrande, loc. cit p. 231.
[†] Proc. Linn. Soc. N.S. Wales, I (2), p. 1068, t. 15.

This is the largest Odontopleurid yet discovered, and seems to agree in size with *C. Verneuili*, its European analogue. When mature it appears to have been from four to five inches long.

We have had the advantage of studying the specimens provisionally referred by Mr. Ratte* to *Acidaspis Prevosti*, Barr. One of these (his fig. 11) we believe to be the present species, although Mr. Ratte represented spines along the frontal borders of the cephalon which do not exist in the specimen, whilst he neglected to figure the genal and occipital spines that are preserved. This specimen also shows the subpedunculate protruding character of the eyes

Named in honour of our valued correspondent, Capt. Anthony W. Vogdes, U S. Artillery, San Francisco, author of the highly useful " Bibliography of the Palæozoic Crustacea."

Loc. and Horizon.--Bowning Creek, Bowning, and Limestone Creek, near Bowning, Co. Harden, Lower Trilobite Bed—Bowning Series (= *Hume Beds*, Jenkins, and *Yass Beds*, David)– ? Wenlock. *Coll.*—Mitchell; Australian Museum, Sydney; Geological Survey of N.S Wales, Sydney.

CERATOCEPHALA JACKII, *sp.nov.*

(Pl. LIII., fig. 8, Pl. L. fig. 6.)

Sp. Char—*Cephalic shield or cephalon*—Greatest width a little more than twice the length (16-7), subelliptical, very moderately tumid, and distinctly granulated. Glabella large, length from centre of the neck furrow equals the greatest width between the axial furrows, central lobe moderately tumid, and arched very gradually, sloping to the front margin and into the neck furrow, front angles slightly expanded; median and basal pairs of glabella lobes distinct, moderately tumid; glabella furrows wide and shallow; false furrows and also the axial furrows wide and shallow, the latter being much less distinct than the former. Fixed cheeks moderately large; genal lobes large, cleaver-shaped, granulated,

---

* Proc Linn. Soc. N.S. Wales, 1886, I. (2), Pt. 4, t. 15, f. 11 and 12.

moderately tumid; ocular ridges filamentous, and distinctly tubercled; genal or palpebral furrows distinct, particularly anteriorly; palpebral lobe very small. Eyes very small, distance between them is to length of the cephalon as 10·7, or a little greater than the distance between a genal spine and the alternate neck spine. Free cheeks of moderate size, moderately tumid, laterally expanded beyond the genal angles, suboval; borders wide, tumid, each bearing a row of four distinct tubercles on the median line, and at least sixteen stout, horizontal spines, all having a forward direction and apparently increasing in length from front to back to the twelfth, from which each succeeding one is a little shorter; marginal furrow wide and distinct between the facial sutures and genal angles, where they terminate. Genal spines straight, acicular, subslender, and forming an angle of 100' with the posterior border of the cephalon, or of 120° with the straight line joining their bases, apparently of moderate length. Facial sutures anteriorly nearly straight, inclining inwards at an angle of 45° and passing out in front of the axial furrows, dividing the greatest width of the cephalon into three equal parts nearly, posteriorly arciform, passing out at the genal angles. Neck furrow wide and shallow, centrally deeper between the false and axial furrows, lateral extensions interrupted by the genal lobe ridges, distinct between the genal lobe and the genal spine ridges. Neck ring indistinctly separated from the neck furrow, very moderately arched, curved sharply backward, side lobes small. Occipital spines subslender, projecting backward, and but slightly raised and curved.

Thorax.—Unknown in a complete state. Pleuræ horizontal, flat, fulcra very indistinct, ends not deflected nor thickened, bispinate; posterior spines strong, and projecting from the posterior angles of the pleuræ; anterior ones swimmeret-like or dagger-shaped, intensely barbed, directed forward and originating in the front angles of the pleuræ, so that the two spines on each of the posterior pleuræ at least have their points widely divergent from each other.

Pygidium.—Unknown.

Obs.—The glabella of this species is very similar to that of *C. longispina*, Mitchell (*Acidaspis longispinis*, Mitchell), but here the specific resemblance of the two species ceases. The cephalon of *C. Jackii* has a greater proportionate width, and its spined free cheeks, shorter and slender occipital, genal and pleural spines, and the very different anterior pleural spines clearly separate it from the former.

From *C. Vogdesi* it is so different that comparison is needless. For the same reason we need not enter into any explanation to differentiate it from *C. Verneuili* and *C. vesiculosa*, Barr. From *C. Dufrenoyi* it is distinguished by the much less quadrate outline of the cephalon in that species, nor does this species possess the expanded anterior lateral portions (free cheeks) of *C. Jackii*. The same feature also distinguishes it from *C. mira*, Barr., and in addition also the highly pedunculated eye of the last named is a strongly differentiating character. On the other hand like *C. Jackii*, Barrande's species possesses the peculiar swimmeret-like spines on the thoracic pleuræ. Lastly, in *C. Prevosti* these spines are replaced by short simple ones, whilst the proportions of the cephalon entirely disagree with those of *C. Jackii*.

Named in honour of Mr. R. L. Jack, Government Geologist of Queensland, who collected the specimens.

Loc. and Horizon.—Bathurst Road, near Bowning, Co. Harden, Middle Trilobite Bed—Bowning Series (= *Hume Beds*, Jenkins, and *Yass Beds*, David)—? Wenlock. *Coll.*—Geological Survey of Queensland, Brisbane; and Mitchell.

CERATOCEPHALA IMPEDITA, *sp.nov.*

(Pl. LIII., figs. 11-13.)

Sp Char.—*Body and cephalon* in a complete form unknown. Glabella highly tumid, tuberculated throughout: central lobe very intensely arched transversely and longitudinally, long, narrow, much higher than the cheeks or lateral glabella lobes, much compressed laterally just behind the frontal expansions, which are very distinct, narrow and each surmounted by two distinct tubercles. Median and basal pairs of lateral glabella lobes only

present, long, narrow, very tumid, and granulated, subequal in
length and not fully separated from each other by the basal
glabella furrows on the outer sides; false furrows very deep and
wide; median glabella furrows very deep, basal pair shallow, wide
an l not quite passing into the axial furrows; axial furrows distinct and narrow, and passing rather clearly over the genal lobe
ridges.   Fixed cheeks of moderate size; genal lobes very tumid,
falling abruptly into the furrows of the lateral extensions and
sloping more gradually anteriorly. ocular ridges indistinct, very
filamentous, palpebral furrows distinct anteriorly, triangular areas
small, lateral extensions short   Neck furrow wide, trough-like,
very deep between the false and axial furrows, faint over the
genal lobe ridges, thence narrow but distinct.   Neck ring robust,
thick, very distinctly arched; side lobes or nodules very small,
and ridged.   Occipital spines acicular and only moderately robust,
arcuate.

Obs.—Thorax, pygidium, and free cheeks are unknown.   It
approaches nearer to *C Jackii* than any other known Australian
species, and from this it is readily separable by the much greater
tumidity of the cephalon and its distinctive granulation, the
longer central glabella lobe and its greater convexity, the longer,
narrower, and more tumid lateral glabella lobes, the shorter
lateral extensions of the fixed cheeks; by the more ridge-like prominent frontal glabella expansions and its prominent tubercle;
and lastly by the very small lateral lobes of the occipital ring.
The proportionate width between the eyes and length of the
glabella is also different in the two species.   From *C. longispina*
it is separated by the same characters

Loc. and Horizon—Bowning Village, Co. Harden, Middle
Trilobite Bed—Bowning Series (= *Hume Beds*, Jenkins, and *Yass
Beds*, David)—? Wenlock.   *Coll.*—Mitchell.

CERATOCEPHALA LONGISPINA, *Mitchell, sp*

(Pl. LIII., fig. 10, Pl. LIV., figs. 1-5.)

Acidaspis near *A mira*, Ratte (*non* Barr.), Proc. Linn. Soc. N.S.
Wales, i. (2), Pt. iv. p. 1069, t. 15, f. 13, 14

Acidaspis longispinis, Mitchell, Proc. Linn. Soc. N.S. Wales,
1888, iii. (2), Pt. 2, p 398, t. 16, figs. 7-12.

Sp. Char.—Body oval, suboblong. *Cephalic shield or cephalon*
only moderately tumid, and distinctly but sparsely granulated.
Glabella with the central lobe suboblong, very moderately arched,
and sloping gradually into the neck furrow and to the front margin,
front angles moderately expanded and bearing distinct tubercles,
three pairs of side lobes present, first very small, depressed,
second and basal pair large, subcircular, moderately tumid and
nearly of equal size, false furrows distinct and very wide; glabella
furrows—first pair faint, second pair deep and distinct, uniting
the axial and false furrows, basal pair very wide and shallow,
also uniting the axial and false furrows; axial furrows very faint
anteriorly and moderately distinct posteriorly; genal lobes small,
distinctly and regularly granulated, prominent posteriorly, inconspicuous anteriorly; palpebral furrows distinct anteriorly, ocular
ridges prominent, filamentous, and distinctly granulated; lateral
extensions of the fixed cheeks robust, having very prominently
thickened borders. Facial sutures anteriorly straight, and making
angles of 120° degrees with the front margin, posteriorly straight,
passing out at the genal angles, and making angles of 35° with
the posterior borders of the cephalon. Eyes prominent, conoid,
numerously and minutely faceted. Free cheeks of moderate size,
borders very wide, moderately tumid, lobe-like, bearing several
rows of granules, one of which is rather distinct, margins very
minutely spinate, spines only visible under a lens, genal spines
very long, strong, arcuate, and diverging from the thorax,
triangular area small. Neck furrow very wide and shallow
centrally, but deep behind the basal glabella lobes, very faint
over the genal lobe ridges, thence distinct to the genal angles,
branches distinct. Neck ring very moderately arched, wide,
surmounted centrally by a very prominent granule, nodules or
side lobes distinct, separated from central portion by the axial
extensions of the neck furrow; occipital spines very robust and
long, arching, the ends divergent and apparently reaching to the
extremity of the pygidium.

Thorax.—Apparently consists of nine segments, length equal to the width, sparsely granulated; axis prominent and as wide as the side lobes, nodules inconspicuous; axial furrows moderately distinct; lateral lobes horizontal, pleural ridges moderately conspicuous, the anterior pleural margins raised into ridges, and giving to the pleuræ the appearance of being centrally furrowed instead of being ridged; pleural spines on the first, second and third pairs of pleuræ moderately reflected and much smaller than those situated more posteriorly; the latter are very long, hastate, robust and intensely flected backwards, centrally fluted when compressed; secondary or anterior spines small, paddle-shaped, subfalcate, having entire margins, and the appearance of articulation to the pleuræ.

Pygidium.—Triangular, two and a half to three times as wide as long, granulated distinctly; axis very prominent, one-half to two-thirds of the length of the pygidium, unsegmented, bearing one prominent ring, axial furrows faint; side lobes flat, undivided, one pair of pleural ridges present, extending from the ends of the axial ring, borders inconspicuous; pleural spines strong, acicular, converging, and about as long as half the pygidial width, but except for these the border is practically entire, although under a *lens* very minute spination or serration is visible along the whole margin.

Obs.—Mr. Ratte figured (*loc. cit.*) two imperfect glabellæ of this species and referred them to *A. mira*, Barr., but as will be readily seen by a comparison of the descriptions and figures of the two fossils they are widely dissimilar. This species was afterwards characterised, fully described and figured by one of us.

C. longispina is so clearly distinct from all the other Australian species of the genus that it is unnecessary to point out the divergencies. Its chief characteristics are : the practically spineless cheek borders; presence of three pairs of lateral glabella lobes; the very large occipital spines which are borne by a cowl-like appendage originating at the back of the central glabella lobe, instead of originating in the occipital ring; the massive principal pleural spines and non-serrated secondary spines; the simple bi-spinate pygidium; prominent and clearly faceted eyes.

Many cephalons occur from which the cowl and spines have separated, and left the occipital ring quite smooth and to all appearance spineless.

C. longispina attains a length of two and two-thirds inches. The pygidium bears a very close resemblance to that of *Selenopeltis Buchii*, Barr., sp., in its spineless margin other than the axial spines.

The cephalon represented in Pl. LIV., fig. 2, possesses occipital spines that exhibit a decided tendency to curl underwards, as do those of *Selenopeltis Buchii*, Barr. More complete examples of our form may determine the necessity of transferring it to Corda's genus.

Loc. and Horizon. — Bowning Village, Co. Harden, Middle and Upper Trilobite Beds — Bowning Series (= *Hume Beds*, Jenkins, and *Yass Beds*, David). *Coll.*—Mitchell.

EXPLANATION OF PLATES.

Plate L.

ODONTOPLEURA BOWNINGENSIS. *E. and M.*

Fig 1.—A nearly complete specimen, but with the genal spines wanting, and the various portions of the cephalon undisturbed (×2½). *Coll. Mitchell.*

Fig. 2 —A cephalon with one genal spine preserved (×3½). *Coll Mitchell.*

Fig. 3.—A cephalon somewhat distorted *Coll. Mitchell.*

ODONTOPLEURA PARVISSIMA, *E. and M.*

Fig. 4.—A nearly complete example (×3). *Coll. Mitchell.*

Fig 5 —Portion of a thorax, and the pygidium (×4). *Coll. Mitchell.*

Fig. 6.—Portion of a cephalon (×2½). *Coll. Mitchell.*

ODONTOPLEURA RATTEI, *E. and M.*

Fig 7.—An almost complete example, with the central lobe of the glabella removed, exhibiting the labrum in position (×3). *Coll. Mitchell.*

CERATOCEPHALA VOGDESI, *E. and M.*

Fig. 8.—Portion of a cephalon, with the right genal spine preserved and the right occipital spine indicated (× 2). *Coll. Mitchell.*

Fig. 9.—Bispinate distal end of a thoracic pleura, the spines barbed; slightly enlarged. *Coll. Mitchell.*

Plate LI.

CERATOCEPHALA VOGDESI, *E. and M.*

Fig 1.—Portion of a thorax showing the peculiar distal termination of the axial segments, tubercles of the pleuræ, and large and strong spines of the latter; somewhat reduced. *Coll. Australian Museum, Sydney.*

Fig. 2.—Portion of another thorax exhibiting the bispinate character of the distal ends of the pleuræ; somewhat reduced. *Coll. Mitchell.*

Fig. 3.—Cephalon showing the nature of the genal and occipital spines and position of the eyes; slightly reduced. *Coll. Mitchell.*

Fig. 4.—The last thoracic segment with its perpendicular spines; slightly reduced. *Coll. Mitchell.*

Fig. 5.—Pygidium with its barbed spines; slightly reduced. *Coll. Mitchell.*

Fig. 6.—A principal tubercle from one of the pleuræ of fig. 1; highly magnified.

Fig 7.—A principal tubercle from a similar position on fig. 2; highly magnified.

ODONTOPLEURA RATTEI, *E. and M.*

Fig. 8.—Glabella without the side lobes, showing granulation and occipital tubercle; slightly enlarged. *Coll. Mitchell.*

Fig. 9.—Free cheeks; somewhat enlarged. *Coll. Mitchell.*

Plate LII.

ODONTOPLEURA RATTEI, *E. and M.*

Fig. 1.—A nearly complete specimen (× 2½). *Coll Mitchell.*

Fig. 2.—A cephalon without the free cheeks, &c. (× 2½). *Coll. Mitchell.*

Fig. 3.—The pygidium with strongly developed spines (× 3). *Coll. Mitchell.*

Fig. 4.—Portion of the two posterior thoracic segments, and the pygidium (× 2½). *Coll. Mitchell.*

ODONTOPLEURA BOWNINGENSIS, *E. and M.*

Fig. 5.—The four posterior thoracic segments and the pygidium (× 2½). *Coll. Mitchell.*

ODONTOPLEURA JENKINSI, *E. and M.*

Fig 6.—Portion of the thorax and pygidium showing very distinctly granulation on the pleural ridges; slightly enlarged. *Coll. Mitchell*

Fig. 7.—Pygidium only, showing pleural ridges and their granulation; slightly enlarged. *Coll. Mitchell.*

ODONTOPLEURA PARVISSIMA, *E. and M.*

Fig. 8.—Free cheek with genal and marginal spines, and eye (× 3). *Coll Mitchell.*

Plate LIII

ODONTOPLEURA RATTEI, *E. and M.*

Fig. 1.—Cast from an impression of an almost perfect individual; spines of pygidium incorrectly shown; somewhat enlarged. *Coll Mitchell*

Fig. 2.—Thorax and pygidium, the two anterior segments of the former devoid of pleural spines, somewhat enlarged. *Coll. Geol Survey Queensland, Brisbane.*

Fig. 3.—Free cheek, slightly enlarged. *Coll Mitchell.*

ODONTOPLEURA JENKINSI, *E. and M.*

Fig. 4.—An almost complete example, with a single series of tubercles on either side; somewhat enlarged. *Coll. Mitchell.*

Fig. 5.—Three thoracic segments with from two to three rows in a similar position; somewhat enlarged. *Coll. Mitchell.*

Fig 6.—Glabella with its lateral lobes and extensions of the neck ring and one free cheek; slightly enlarged. *Coll. Mitchell.*

Fig. 7.—A second glabella; slightly enlarged. *Coll. Mitchell.*

CERATOCEPHALA JACKII, *E. and M.*

Fig. 8.—Impression of the cephalon; slightly enlarged. *Coll. Geol. Survey Queensland, Brisbane.*

CERATOCEPHALA VOGDESI, *E. and M.*

Fig. 9.—Pygidium with its large dentate spines; somewhat enlarged. *Coll. Mitchell.*

CERATOCEPHALA LONGISPINA, *Mitchell, sp.*

Fig. 10.—Pygidium with its axial spines; somewhat enlarged. *Coll. Mitchell.*

CERATOCEPHALA IMPEDITA, *E. and M.*

Fig. 11.—Portion of the cephalon, showing glabella, side lobes and bases of occipital spines; slightly enlarged. *Coll. Mitchell*

Fig 12.—Another and less perfect specimen; slightly enlarged. *Coll Mitchell*

Fig. 13 —A third example; slightly enlarged.   *Coll. Mitchell.*

Plate LIV.

CERATOCEPHALA LONGISPINA, *Mitchell, sp.*

Fig. 1.—Portion of the cephalon and thorax, slightly reduced   *Coll Mitchell*

Fig 2.—Cephalon less the free cheeks, with the occipital spines *in situ*,
the left one showing a tendency to curl under as in the genus
Selenopeltis; somewhat enlarged.   *Coll. Mitchell.*

Fig 3.—A similar specimen, somewhat enlarged   *Coll Mitchell.*

Fig. 4.—Free cheek, with the eye *in situ*, somewhat enlarged.   *Coll.*
Mitchell

Fig 5.—Distal end of one of the posterior thoracic pleuræ with its enormously elongated spine, somewhat enlarged   *Coll. Mitchell*

Fig. 6.—Crushed cephalon and thorax, with the position of the occipital
spines indicated, somewhat enlarged.   *Coll. Geol Survey*
Queensland, Brisbane.

Plate LV.

Structural diagrams of the cephalon of *Odontopleura* and of *Ceratocephala.*

Fig. 1. - *Odontopleura.*

Fig 2.—*Ceratocephala.*

Reference Letters.

aa. Central lobe of the glabella.   *bb.* Anterior lateral lobes of the glabella
(seldom present)   *cc.* Median lateral lobes of the glabella.   *dd.* Basal or
third pair of lateral glabella lobes.   *ee.* Lateral lobes of the neck ring.   *ff.*
Genal or cheek lobes (in *Odontopleura* mostly very rudimentary).   *gg.*
Genal spines.   *hh.* Neck furrow.   *iii.* False furrows   *jj* Front lateral
expansions of the central lobe of the glabella.   *kk.* Lateral cheek furrows.
mm. Cheek borders with spines.   *no.* Ocular ridge.   *on* Posterior extension
of ocular ridge present in some *Ceratocephala.*   *oo* Eyes.   *p.* Palpebral
lobes, very small in *Odontopleura.*   *q.* Genal spine ridges.   *r.* Neck ring.
ss. Occipital spines.   *tt.* Palpebral furrow.   *xxxxx.* Axial furrows.   *yy.*
Lateral extensions of the neck furrow.   *ww.* Genal lobe ridges, joining the
genal lobes to the lateral lobes of the neck rings and interrupting the
lateral extensions of the neck furrow.   *fb.fs.* Front border of the glabella,
sometimes bearing fine spines.   *fs.o fs.* Facial sutures, sometimes not
defined or soldered as in the case of *C. Voydesi*, nobis.   *ix.ix.ix ix.* Lateral
glabella furrows, seldom more than two pairs present   *trt.* Triangular
areas, very small or absent from *Ceratocephala*   *hc.* Branches of the neck
furrow or continuations of the false furrows.   *tu.* Central tubercle of the
neck ring.   *fs.x.* Thickened borders or ridges of lateral extensions of the
fixed cheeks.

TWO ADDITIONS TO THE FUNGI OF NEW SOUTH WALES.

By D. McAlpine.

(Communicated by J. H. Maiden, F.L.S.)

1. Puccinia hieracii, Mart.

Hawkweed Puccinia.

On both surfaces of leaves of *Hypochaeris radicata*, L. October. Wagga Wagga, N.S.W. (Maiden). Not hitherto recorded for New South Wales.

2. Capnodium callitris, McAlp., n.sp.

Murray Pine Capnodium.

(Plate LVI.)

Black, widely effused, not readily separating and then in small particles, giving a sooty appearance to the dark green branches. *Hyphæ* dark brown, creeping, interwoven, branched, septate, moniliform or joints cuboid, up to 14 μ broad; branches rigid, short, usually simple, tapering to about 4 μ. Slender, colourless and pale green filaments also present; often in moniliform chains. *Gonidia* on both brown and colourless filaments: on brown, usually uniseptate and oblong, dark yellow to dark brown, very variable in size, 13-28 × 7-13 μ; on colourless, elliptical, uniseptate, about 11 × 5$\frac{1}{2}$ μ. *Gemmæ* or detached bud-like bodies frequent. *Spermogonia* elongated-fusiform or somewhat hemispherical, very dark brown, greenish at apex, variable in size and shape, 84-130 × 50-66 μ. *Spermatia* hyaline, rod-like, imbedded in gelatinous material, 4-4$\frac{1}{2}$ × 1 μ. *Pycnidia* roughly bottle- or flask-shaped, with bulging part often one-sided, dark brown, with colourless fringe at mouth. *Pycnospores* at first colourless, then greenish, and finally yellowish-brown, end cells often colourless, ellipsoid, 5-septate and septa stout, 22-24 × 9-11 μ. *Perithecia* simple, dark coloured but dark green when crushed, and walls irregularly netted, with more or less globular or oval head, often supported by stout body, papillate at apex when ripe and extruding plug of dirty yellow material, 170-280 × 90-156 μ or even larger. *Asci* fusoid-clavate, sessile, apex rounded, 8-spored (79 × 26 μ). Sporidia at first colourless, then pale green, finally dark brown, oblong, constricted at the middle, 3-septate, and usually longitudinally divided, often in each division, 17-19 × 8-9½ μ.

The various reproductive bodies are intermixed. Pale green glomeruli (*Heterobotrys*) are also present.

On *Callitris robusta*, R.Br. October. Wagga Wagga, N.S. Wales. (Maiden).

Besides the gonidia, detached portions of the hypha probably serve as such, and there are many-celled swollen bodies, between the ordinary cells, which likely have the same function. The spermogonia vary considerably in shape, but the rod-like spermatia are very characteristic. The pycnidia are easily recognised by their long and usually straight neck, composed of elongated twisted filaments and reaching a length of 190 μ, apart from the body. The fringed mouth is in contrast to that of the perithecium which is papillate and splits irregularly. The pycnospores are at first unicellular and colourless, borne at the end of colourless, jointed filaments. They soon develop two or three septa and become greenish, then finally turn brown on maturity, with 5 septa constantly. It is interesting to observe that the same changes of colour are seen in the sporidia. There is a species of Capnodium (*C. australe*, Mont.) found in Australia on Conifers, but it differs from this one in several important respects. The perithecia are dichotomous, but here they are simple; the sporidia are 4-5-septate and not constricted, but here they are 3-septate and constricted.

EXPLANATION OF FIGURES.

Capnodium callitris.

Fig. 1.— Hyphæ branched and unbianched (× 540).

Fig. 2.—Colourless moniliform hypha bearing gonidium (× 1000).

Fig. 3. —Uniseptate gonidia borne by coloured hyphæ (× 1000).

Fig. 4.—Detached brown body germinating and giving rise to colourless
tube (× 1000).

Fig. 5.—Spermogonium with spermatia (× 540).

Fig. 6.—Spermatia (× 1000).

Fig. 7.—Pycnidium with colourless fringe at mouth-opening (× 143).

Fig. 8.— Pycnospores (× 1000).

Fig. 9.—Pycnospores germinating usually laterally, sometimes at end
(× 1000).

Fig 10.—Perithecium (× 270).

Fig. 11.—Ascus with 8 sporidia (× 1000).

ON SOME AUSTRALIAN ELEOTRINÆ.

By J. Douglas Ogilby.

Up to the present time all Australian writers on ichthyology have been content to follow the author of the British Museum Catalogue of Fishes (*1859-1870*) in collecting all the various forms of the Eleotrine Gobies in a single large, heterogeneous, and unwieldy genus; under the common name *Eleotris* this is made to include a number of fishes, which, although having a general resemblance to one another in their habits and mode of life, have developed such widely diverse structural peculiarities that the impossibility of maintaining the intimate connection inaugurated in that work, and subsequently adhered to in other important papers by the same author, becomes immediately apparent to anyone to whom the opportunity of studying the fishes themselves is given.

In the paper here submitted, I have, therefore, endeavoured to separate into natural groups certain of our common south-eastern cismontane species, in the hope that the proposed genera will form a nucleus round which to gather a part at least of our Australasian forms and so facilitate the identification of the remainder.

In undertaking even this partial revision of our *Eleotrinæ*, I am, however, placed at a great disadvantage through my inability to consult Dr. Bleeker's paper on the divisions of the *Gobiidæ*, no copy of which is obtainable in Sydney, nor indeed, so far as I am aware, does one exist in any of the Australian Colonies. It is quite possible, therefore, that one or other of the four genera here proposed may be identical with one of Bleeker's, but the advantage to my fellow-workers in Australia of having a clear

48

and concise definition of certain forms, which are probably distributed over the length and breadth of our faunic region, must be held to outweigh in importance the risk of unnecessarily increasing the synonymy.

The only paper dealing with the divisions of the genus *Eleotris* —as accepted by Australian authors—to which I have access is the "Review of the *Gobiidæ* of North America," by Professor Jordan and Eigenmann,* and I am unable to identify any of the five species described below with the genera there characterised

Though somewhat irrelevant to the subject matter of this paper, as set forth in its heading, a short account of the fish life to be met with in the waterholes near Sydney—everywhere favourite haunts of the fishes of this subfamily—will be both interesting and instructive, as a proof of the vast capabilities which even a small and to all appearance most unpromising puddle may possess towards elucidating some of the problems of our fresh-water fauna ; and the fact of the discovery of so brilliantly coloured yet undescribed a species as *Carassiops longi*, within so short a distance of the metropolis, speaks for itself as to the possibilities dependent on a systematic examination of the waterholes and overflow ponds in the more remote parts of the Colonies, while it is a tangible demonstration of the culpable ignorance which prevails among us in regard to the many curious and interesting forms of animal life which inhabit our streams and ponds.

I shall make, therefore, no further apology for interpolating here the following account of a collecting trip made by me last April in company with Mr J. D. Grant, Inspector of Fisheries, to the Liverpool district, and which produced results quite unexpected by me.

This visit was paid, by invitation, on the 24th of last April to the Hon. Wm. Long's estate of Chipping Norton, and was undertaken principally with the object of obtaining examples of a

---

* Proc. U. S. Nat. Mus ix 1886, p. 477; for a copy of this excellent paper I am indebted to the courtesy of the authors.

Gray Mullet, which was said to be found in the George's River above the weir at Liverpool and in the adjacent waterholes, and which, my informant assured me, differed greatly from any of those inhabiting the estuary, in which it was very rarely obtained, and then only after severe floods, by which a few of these fishes and of the fresh-water Herrings (*Potamalosa novæ-hollandiæ*) are occasionally swept down over the weir from the upper reaches of the river.

The pools which we netted are merely drinking-places for stock, either of artificial construction or natural depressions of the ground, and are fed by the overflow from the river during flood-time supplemented by the rainfall, or in one instance at least by filtration through the sandy ridge intervening between the water-hole and the river, the water always maintaining the same level in the two.

At the time of my visit all the pools were very low in consequence of the long continued drought, only the one to which reference has just been made being anywhere of a greater depth than six feet, and in it, owing to the inequalities of the bottom and the presence of snags, assisted by the clearness of the water—the result of filtration—we were almost quite unsuccessful, our entire capture consisting of a single example of the Smelt (*Retropinna*) and a young Australian River-Perch (*Percalates colonorum*).

The latter of these species is known to occur abundantly along the entire coastal region of south-eastern Australia and northern Tasmania, but the range of *Retropinna* is by no means so well understood, as it has been very generally confounded with *Galaxias*, but, in such opportunities as I have enjoyed for observing our fresh-water fishes in their native haunts, I have not so far succeeded in detecting the two genera as associating in the same waters. In Macleay's Catalogue, No. 840, Vol. ii p 164, (*Proc. Linn. Soc. N.S. Wales*, vi. 1881, p. 228) the only Australian locality given is " Rope's Creek," and we may, therefore, take it for granted that this was the only place known to the author from which the genus had been recorded outside of New

Zealand.   No less than fifteen years previously, however, Dr.
Steindachner, in a paper entitled "Zur Fischfauna von Port
Jackson in Australien" (*Sitzb. Ak. Wien, liii. i. 1866, p 469*), had
recorded the species; again, no further mention of the species is
made in Macleay's Supplement (*1884*), though during the previous
year Johnston's "Catalogue of the Fishes of Tasmania" (*Proc
Roy. Soc. Tas 1882*) had been published, at p. 62 of which the
author states that it is "found in the various estuaries of Tasmania at certain periods of the year."   Personally I have caught
these fishes in the stream which flows from the dam of the Parramatta water supply; in the Nepean River at Menangle; in the
Prospect Reservoir, where they swarm in almost incredible
numbers, and, as above mentioned, in the watershed of the
George's River, it may, therefore, be inferred that *Retropinna*
is an inhabitant of most of our coastal waters, though its exact
northward and southward extension has yet to be determined
On the latter I am enabled, however, to throw some light, as a
small example is present among some fishes forwarded to me by
Mr James A. Kershaw, and the notice accompanying the specimen runs thus—"Pyramid Hill (about 150 miles from Melbourne
and north of Bendigo)"; this extension of range, though in itself
an interesting addition to our meagre knowledge of the species,
is much less important than the fact—of which I have reliable
information—that the section of country in which Pyramid Hill
stands drains into the Murray River, and that, therefore, in one
district at least *Retropinna* has succeeded in crossing the Dividing
Range.

It was in the deep pool that we expected to catch the Mullets
for which we were especially in search, and though, for the
reasons given above, we were unsuccessful on this occasion, there
can be no doubt that the species is *Mugil breviceps*, Steindachner,
a very handsome Mullet, remarkable for its small head and
entirely confined to fresh water, which I subsequently found to be
common in the upper waters of the Nepean River at Menangle,[*]

---

[*] A full account of this species will be given in a paper on the Australian
Mugilidæ now in course of preparation.

at both places I was further assured that there was a second species of Mullet found in the fresh water.

The three other pools which we fished were of much smaller dimensions—the largest about twenty-five yards by ten, the smallest not a third of that size—and nowhere exceeded four feet in depth ; they were, however, crowded with fishes of several kinds, indeed it is difficult to imagine whence food could have been supplied in sufficient quantity to keep so many individuals in the healthy condition in which we found them , the only aquatic animals which I found associated with them were a small shrimp (*Palæmon, sp.*) and a large and handsome water-beetle (*Homæodytes scutellaris*), and though these were brought ashore among the weeds in considerable abundance, their numbers, unless materially supplemented from outside, were quite insuffi- cient to bring about the results which we witnessed.

In point of numbers the ubiquitous Carp (*Carassius auratus*) of course greatly exceeded all the other species together ; they were of all sizes and of all tints, from a dull olive-green or brown to gold, among the latter being some of the largest and most brilliantly coloured individuals that I have ever seen. These pests swarm in most of the fresh waters of the metropolitan and neighbouring districts, usurping the place and consuming the food of better fishes; introduced from abroad like the rabbit and the sparrow, they have similarly thriven and multiplied, and, but for the nature of the element in which they live and their distaste for or inability to live in purely salt water, would doubtless have similarly spread with equally disastrous results to the native fauna; yet in the face of this and of the fact that they are useless as food, the "Fisheries Act" now before the country proposes to protect the "Carp" and makes it penal to offer them for sale if under five ounces in weight or by analogy to destroy them.* In

---

* The true Carp (*Cyprinus carpio*), a species of considerable value as a food fish—and which with the Small-headed Mullet (*Mugil breviceps*), the Tench (*Tinca vulgaris*), and the Gourami (*Osphronemus olfax*) might with advantage be introduced into all Government tanks, especially in the western districts—has never been acclimatised in any part of the colonies.

place of this, it should be made punishable to introduce this pest into any waters of the colony at present free from it, and stringent regulations should be at once issued to all caretakers of Government tanks prohibiting its introduction therein.

Both species of fresh-water Eel (*Anguilla australis* and *reinhardtii*) were taken, the latter being, as is invariably the case in this district, much the larger. The Long-finned or Reinhardt's Eel is the common eel of the New South Wales rivers and estuaries, so that nine out of every ten exposed for sale in the Sydney markets belong to this species, which attains to a weight of at least fifteen pounds, whilst with us a specimen of *australis* exceeding two pounds is a rarity, though, according to Mr. Johnston (*Proc. Roy. Soc. Tas. 1882, p. 61*) that species reaches the enormous weight of thirty pounds in some parts of Tasmania. Both Macleay and Tenison Woods have confounded *reinhardtii* with *australis*, from which it may at once be distinguished by the anterior position of the origin of the dorsal fin, which commences far in advance of that of the anal instead of nearly opposite to it as in *australis*. Roughly speaking, *australis* is the southern form, being the common fresh-water Eel of Tasmania, Victoria, and South Australia, while *reinhardtii* occupies a similar position on the east coast from Sydney northwards to Cape York.

To return to the *Eleotrinæ* :—

The name "Gudgeon" is very generally accepted throughout Australia for these little fishes, having been doubtless given to them by the earlier colonists on account of a certain similarity in their mode of life as well as a fancied resemblance in their appearance to the European Gudgeon (*Gobio fluviatilis*).

Sexual and seasonal differences:—Among bony fishes distinctive characters by which the sexes may be recognised externally are not common, but, as far as the species considered in this paper are concerned, the Australian Gudgeons are an exception to this general rule, the shape and size of the genital papilla being an accurate guide to the sex; in all the Gudgeons proper (*Eleotrii* and *Butii*) of Bleeker, this organ is narrow and triangular in the male, short, broad, and posteriorly emarginate in the female, while in the Carp-Gudgeons (? *Carassiopsi*) it is oblong in both sexes, with the hinder border emarginate, but that of the male is so much the longer that its lobes embrace the origin of the anal fin. In some species, also, there is a marked prolongation of some of the fin-rays in the male fish.

During the spawning season the cheeks in both sexes, but more especially in the males, become to a greater or less extent tumid, while the genital papilla of the female develops one or more series of small supplementary papillæ, forming a fringe.

These facts should be carefully borne in mind by anyone describing or identifying a species from a single individual.

B r e e d i n g —I have been unable to find any account of the breeding habits of the Eleotrids, or the means employed, if any, to ensure the safety of the eggs and newly hatched young and to guard against hybridisation, but the fact that in a single small pool many pairs of these fishes, belonging to three different species, were simultaneously engaged in spawning, and that no hybrid has ever been recognised, clearly suggests that nests of some sort are formed for the reception of the eggs * Where the nests are situated and whether the ova when deposited are watched over by the parents must be left for future investigation to decide, but there was no appearance of any such construction among the weeds drawn ashore by the net.

Appended is a synopsis of the genera proposed in this paper :—

i. Abdominal vertebræ more numerous than the caudal; sexes dissimilar in colour, similar in the shape of the genital papilla.

A. Head deeper than wide, mouth small; outer series of mandibular teeth slightly enlarged; gill-openings narrow; six branchiostegals, genital papilla large; head partially scaly ... ...

CARASSIOPS, p. 732

---

* This is known to be the case with some at least of the allied marine Gobies.

ii. Abdominal vertebræ less numerous than the caudal; sexes similar in colour, dissimilar in the shape of the genital papilla.

A. Head as wide as deep; mouth small; outer series of teeth slightly enlarged; gill-openings narrow; five branchiostegals; genital papilla large; head partially scaly.

    a. First dorsal with 7 rays; fourth ventral ray produced and filiform; pectoral with not more than 16 rays: scales large; cheeks and interorbital space scaly ...

KRLFFTIUS, p. 736

    a'. First dorsal with 6 rays; fourth ventral ray not produced; pectoral with not less than 18 rays; scales moderate; cheeks mostly, interorbital region entirely naked ... ...

MULGOA, p. 740

A'. Head wider than deep; mouth large; gill-openings wide; six branchiostegals; genital papilla small; head almost entirely naked ... ...

OPHIORRHINUS, p. 745

CARASSIOPS, gen. nov.

Eleotris, sp. auctt.

Body oblong and compressed, the back rounded; head rather small, compressed, much deeper than wide, rounded above; mouth small and oblique, the lips thin; premaxillaries protractile, maxillaries narrow, with the distal end exposed and curved downwards; lower jaw but little the longer; jaws with a band of villiform teeth, the outer mandibular series slightly enlarged in front; lower pharyngeals forming together a subtriangular patch, the outer and symphyseal series strong and hooked; nostrils widely separated, the anterior tubular; eyes lateral; none of the bones of the head armed; gill-openings narrow, extending forwards to below the angle of the preopercle, the isthmus narrower than the interorbital regions; six branchiostegals; pseudobranchiæ present, small; gill-rakers short, stout, and simple. Dorsal fins separate, with vi, i 9-10 rays, the spinous ones flexible; anal fin originating behind the second dorsal, with i 10-11 rays ; ventral fins well developed, not in contact basally, inserted behind the base of the pectorals, with i 5 rays, the fourth soft ray produced and filiform, pectoral fins moderate and pointed, with 13 or 14 rays, the middle ones the longest; caudal fin rounded, the peduncle strong. Genital papilla large, scales large and somewhat deciduous, those of the tail a little larger than those of the trunk, head partially scaly; scales of the head and anterior part of the body cycloid, the remainder ciliated   Vertebræ 25 (14 + 11).

E t y m o l o g y.—*Carassius*, a Carp; ὤψ, resemblance.

T y p e.—*Eleotris compressus*, Krefft.

D i s t r i b u t i o n.—Coastal regions of Eastern Australia.

CARASSIOPS LONGI, sp.nov.

Long's Carp-Gudgeon.

D. vi, i 9.   A. i 10.   P. 13-14.   Sc. 27-29/8.   Vert. 14/11.

Body moderate, the tail not conspicuously compressed.   Length of head $3\frac{7}{10}$ to $3\frac{9}{10}$, depth of body $3\frac{4}{5}$ to 4 in the total length; depth of head $1\frac{1}{3}$ to $1\frac{4}{9}$, width of head $1\frac{7}{10}$ to 2, of the slightly convex interorbital region $3\frac{3}{4}$ to $4\frac{1}{3}$,* diameter of eye $3\frac{2}{5}$ to $4\frac{1}{4}$ in the length of the head ; snout much broader than long, very obtusely rounded in front, not depressed, as long as to as much as one-fourth of a diameter longer than the eye.   Maxillary not reaching to the vertical from the anterior margin of the eye, its length $3\frac{1}{4}$ to $3\frac{4}{5}$ in that of the head.   Ten gill-rakers on the lower branch of the anterior arch, all of them simple and tooth-like. The space between the origin of the first dorsal fin and the extremity of the snout is as long as or a little less than its distance from the base of the last soft ray; the fourth spine is the longest, $1\frac{1}{3}$ to $1\frac{3}{5}$ in the length of the head and reaching when laid back beyond the origin of the second dorsal fin in the ♂, $1\frac{4}{5}$ to 2 in the head and not reaching as far as the second dorsal in the ♀; in the ♂ the seventh soft ray is the longest, as long as the head, in

---

* $4\frac{1}{5}$ in one specimen.

the ♀ the second and third are the longest, $1\frac{1}{3}$ to $1\frac{1}{2}$ in the head: the anal fin originates a little behind the second dorsal and is in all respects similar to it: fourth ventral ray considerably longer than the third or fifth, longer than the head and extending well beyond the vent in the ♂, shorter than the head and reaching to or not quite to the vent in the ♀: pectorals rounded, the middle rays the longest, as long as or a little shorter than the first ventral ray, reaching to or beyond the vertical from the origin of the second dorsal in the ♂, to beneath the dorsal interspace in the ♀: caudal fin large and rounded, as long as or a little longer than the head; caudal peduncle shorter and deeper in the male than in the female, as long as or a little shorter than the head, its depth $1\frac{3}{5}$ to $1\frac{3}{5}$ in the ♂, $1\frac{9}{10}$ to $2\frac{1}{10}$ in the ♀ in its length. Genital papilla large and oblong, notched at the extremity, which is simple and passes along either side of the origin of the anal in the ♂, double, papillose, and does not extend as far as the anal fin in the ♀. All the scales imbricate, those of the head (except the opercle), throat, and anterior part of the body smaller than the others.

♂. Greenish-yellow, with the edges of the scales olive, the head, nape, and belly orange; a purple spot on the opercle and another in the axil of the pectoral present or absent; dorsal and anal fins orange, with a wide purple marginal band, the soft dorsal posteriorly with white spots, the extremities of the anal rays white; caudal fin yellowish-gray with irregularly anastomosing series of microscopic spots; pectorals and ventrals gray

♀. Yellowish-green, the upper scales with or without a basal violet spot, which, when present, gradually disappears on the sides; below grayish-white; upper surface of head golden, the opercles gray, both more or less clouded with violet; sometimes with a golden band on the sides of the abdomen; dorsal fins bright yellow, with a wide marginal violet band, the anterior with some scattered dots, the posterior with clouded spots formed by irregular groups of similar dots; anal fin gray washed with yellow posteriorly and widely margined with pale violet; sometimes a dusky axillary spot.

This handsome species can be at once distinguished from *compressus*, of which it is the southern representative, by its more elongate body, that of *compressus*, the type of which I have compared with my specimens, having a depth of $3\frac{1}{8}$ in the length, while the depth of the head is almost equal to its length ; the same measurements are maintained in two examples from the Tweed River in the Macleay collection

In 1867 Dr. Franz Steindachner described a species of *Carassiops* from Cape York, for which he proposed the name of *Eleotris brevirostris*,* and this northern form appears to approach more closely to the Sydney species than to Krefft's; in fact at a later page (325) of the same volume Steindachner himself confuses the northern and southern fishes by recording two examples of *brevirostris* from Port Jackson.

In the Annals and Magazine of Natural History (4) xv. 1875, p 147, Mr O'Shaughnessy states that the *brevirostris* of Steindachner is identical with the *compressus* of Krefft, but for the reasons given above, as well as on account of the larger scales of the former, I cannot agree with him.

Instead of uniting the different forms in a single species of extraordinary variability, I prefer, at least for the present, to recognise four distinct but closely related species of Carp-Gudgeons, namely —(1) *longi*, from the metropolitan district of New South Wales, (2) *compressus*, from the Clarence, Richmond, and Tweed River districts; (3) *brevirostris*, from the Mary River – Australian Museum† and ? Challenger—and Port Denison—Krefft—to Cape York,—Steindachner—and (4) *elevatus*, Macleay, from Port Darwin, North-western Australia.

I obtained nine examples of this handsome species from one of the waterholes on the estate of the Hon. Wm. Long on the 24th of April last, and have much pleasure in dedicating it to that gentleman in remembrance of the pleasant afternoon spent at Chipping Norton.

---

* Sitzb. Ak Wien, lvi. i. 1867, p 314
† Two small bleached specimens in very bad condition.

The difference in colour between the sexes is so marked that it was only when examining my specimens on the following day that I recognised the relationship; this is possibly more apparent during the spawning season than at other times.

The dark purplish ground colour which is so conspicuous a feature, in the males at least, of both *compressus* and *brevirostris* is entirely absent in *longi*, its place being taken by orange, and so brilliant is this colour that it was only with difficulty that I could persuade many persons that they were not Gold-fishes. Curiously enough, a small specimen, which had evidently suffered from an accident in its youth, had partially reproduced the variety of the Golden Carp known as the "Telescope fish," the eyes being produced in front of the head.

The specimens measured from 82 to 100 millimeters and were all full of spawn.

The types are in my possession.

KREFFTIUS, gen nov.

Eleotris, sp. auctt.

Body oblong, compressed posteriorly, the back broad and flat in front of the dorsal fins, rounded behind; head rather large, about as wide as deep, the snout moderate and but little depressed; mouth small and oblique, the lips fleshy, premaxillaries slightly protractile; maxillaries narrow, with the distal end exposed and bent forwards; lower jaw a little the longer. Jaws with a band of small hooked teeth, the outer series enlarged and fixed, lower pharyngeals forming together a subtriangular patch, armed with small acute fixed teeth, the anterior and symphyseal series more or less enlarged, nostrils widely separated, the anterior valvular, eyes lateral, none of the bones of the head armed, gill-openings extending forwards to below the angle of the preopercle, the isthmus a little wider than the interorbital region; five branchiostegals; pseudobranchiæ present, small; gill-rakers short, stout, and serrulate. Dorsal fins separate, with vii, i 8 rays, the spinous ones flexible; anal fin commencing behind the origin of the second dorsal, with i 8 rays; the last soft ray of the second dorsal and anal fins divided to the base : ventral fins not in contact basally, inserted a little behind the root of the pectorals, with i 5 rays, the fourth produced and filiform ; pectoral fins rounded, with 15 or 16 rays, the middle ones the longest; caudal fin rounded, the peduncle strong. Genital papilla large, triangular in the male, oblong in the female. Scales large and adherent, those of the tail not much larger than those of the trunk; head partially scaly, the snout naked, scales of the head and anterior portion of the body cycloid, the remainder ciliated. Vertabræ 28 (13 + 15).

E t y m o l o g y.—Dedicated to the late Mr. Gerard Krefft, to whom belongs the honour of having first pointed out the differences between certain of the Eleotrids of New South Wales.

T y p e.—*Eleotris australis*, Krefft.

D i s t r i b u t i o n.—Coastal region of New South Wales.

KREFFTIUS AUSTRALIS.

Eleotris australis, Krefft, Proc. Zool. Soc. London, 1864, p. 183; Castelnau, Proc. Linn. Soc. N S Wales, iii. 1878, p. 384 (1879); Macleay, Proc. Linn. Soc. N S Wales, v 1880, p. 617 (1881). Ogilby, Catal. Fish. N.S. Wales, p. 36, 1886.

Striped Gudgeon.

D vii, i 8   A. i 8.   P. 15-16.   Sc 31-33/8-9.

Body stout and moderately deep, the tail compressed. Length of head $3\frac{2}{7}$ to $3\frac{3}{4}$, depth of body $3\frac{3}{4}$ to $4\frac{1}{2}$ in the total length, head as deep as or a little deeper than wide, its width $1\frac{1}{4}$ to $1\frac{2}{3}$, that of the slightly convex interorbital region 4 to $4\frac{3}{4}$, diameter of eye $4\frac{1}{7}$ to $4\frac{2}{3}$ in the length of the head, snout much broader than long, very obtusely rounded in front, not or but little depressed, from one-tenth to one-third of a diameter longer than the eye. Maxillary extending to or not quite to the vertical from the anterior margin of the eye, its length 3 to $3\frac{2}{7}$ in that of the head. Eight or nine gill-rakers on the lower branch of the anterior arch, the front ones reduced to spiny knobs. The space between the origin of the first dorsal and the extremity of the snout is a little more than its distance from the base of the last soft ray; outer border of the first dorsal rounded, its height $1\frac{9}{10}$ to $2\frac{1}{10}$ in the length of the head, the last ray reaching when laid back in the $\male$ to, in the $\female$ not quite to the origin of the second dorsal; the rays of the soft dorsal increase in length in the $\male$ to the last, which is $1\frac{1}{10}$ to $1\frac{2}{5}$ in the head, in the $\female$ to the third or fourth—rarely the fifth—which are $1\frac{2}{3}$ to $1\frac{4}{7}$ in the head: the anal fin originates below the second ray of the soft dorsal, and the penultimate ray in the $\male$, the third or fourth in the $\female$ are the longest, as long as those of the soft dorsal: fourth ventral ray considerably longer than the third or fifth and terminating in a filament; in the $\male$ it reaches well beyond the vent and is as long as the head, in the $\female$ to or not quite to the vent and about one-fifth less than the head: pectoral fin rounded, the middle rays the longest, reaching to or not quite to the vertical from the dorsal interspace, its length $1\frac{1}{4}$ to $1\frac{1}{3}$ in that of the head : caudal fin rounded, $1\frac{1}{10}$ to $1\frac{1}{5}$ in the length of the head, its peduncle as long as or as much as one fifth shorter than the head, the depth $1\frac{1}{2}$ to 2 in the length. Genital papilla large; lanceolate, simple, longer than the eye, and nearly twice as long as broad in the $\male$; oblong, truncated, much shorter than the eye, and not much longer than broad in the $\female$. Scales large, not larger on the tail than on the sides of the body; those of the head, nape, thorax, and abdomen smaller, and with very delicate concentric striæ; the remainder with coarser longitudinal striæ; scales of the interorbital region and cheeks smaller than those of the occiput and opercle.

Upper surface rich brown or purple, passing into green or greenish-gold on the sides, gray below, all the scales with a lighter border; each of the lateral scales has a large purple basal spot or short streak forming together longitudinal bands, those which originate behind the pectorals being the most conspicuous and persistent ; between these bands are narrower stripes of bright gold; cheeks and opercles strongly tinged with yellow, the latter often clouded by more or less concurrent groups of microscopic violet dots; a purplish band from the lower angle of the orbit to the base of the pectoral and sometimes a second parallel band to the axil; dorsal rays yellow, the spinous portion with two series of spots, the posterior of which are chestnut; the soft portion with four or five series of subequal chestnut spots or with a basal series of large and numerous small scattered spots, caudal fin violet, the rays with alternate transverse bars of white or yellow and chestnut spots; anal fin orange in the ♂, golden in the ♀ with a broad lilac or gray marginal band, ventral fins violet, with the outer borders white or golden; pectoral fins yellow bordered with gray and with a basal purple band which is succeeded by a conspicuous broad stripe of orange or gold, behind which a more or less distinct dusky band may be present; a large purple spot in the axil of the pectoral and another at the root of the caudal present or absent.

The description of the colouration given above is drawn up from a series of specimens taken during the breeding season, and represents, therefore, the nuptial dress of this fine species.

Irrespective of any difference in colour—which indeed is a mere matter of shade—an analysis of the above description shows that the male fish may at all times be distinguished from the female by the two following characters :—

(1) The shape and size of the genital papilla, and

(2) The greater comparative length of the fin rays, especially those of the posterior portion of the soft dorsal and the anal, and the fourth soft ray of the ventrals.

In addition to these, the caudal peduncle appears to be distinctly shorter and deeper in the adult male than in a female of the same size.

In the metropolitan district these Gudgeons deposit their spawn during the latter half of April and the beginning of May, and as soon as this important function has been completed they retire to their winter quarters and do not again make their appearance until the ensuing spring ; during the intervening months they remain quiescent and cannot be taken either by hook or net, but I am unable to say precisely whether they merely conceal themselves under stones and snags or in holes in the bank or completely

bury themselves beneath the mud; I am, however, inclined to
believe that the latter is the true solution of their disappearance;
that their abstinence, whether enforced or voluntary, has no ill
effects on them is proved by the perfect condition in which they
are when they reappear with the first warm weather.

Krefft's Striped Gudgeon is abundant in all the fresh waters in
the neighbourhood of Sydney, and extends its range northwards
at least as far as the Clarence River, from whence specimens were
obtained by its original describer; it appears to prefer muddy
waterholes and sluggish creeks to clearer and swifter waters, and
is, therefore, more distinctly a denizen of the lower lands in the
vicinity of the coast than is the next species.

My examples were taken from waterholes near Liverpool, in
which I found them abundant, as also they are in the George's
River above the weir. I have also examined specimens from the
neighbourhood of Port Stephens, from Rope's Creek, from Cook's
River, and from Nowra, as well as Krefft's types from Bronte
and the Botany Swamps.

The largest of these examples measured 135 millimeters, and
the description is drawn up from an examination of thirty-five
specimens ranging from that size down to 63 millimeters.

MULGOA, gen.nov.

Eleotris, sp. auctt.

Body elongate-oblong, strongly compressed posteriorly, the
back broad and almost flat in front of the dorsal fins, rounded
behind; head moderate, about as deep as wide, the snout slightly
depressed, mouth small and oblique, the lips fleshy; premaxillaries
slightly protractile; maxillaries narrow, but little arched, with
the distal extremity exposed, lower jaw the longer. Jaws with
a band of small curved teeth, the outer series slightly enlarged
and fixed, lower pharyngeals forming together a subtriangular
patch, armed with small, acute, fixed teeth, the anterior and
symphyseal series enlarged, nostrils widely separated, the anterior
valvular; eyes supero-lateral; none of the bones of the head
armed; gill-openings extending forwards to below the angle of the

preopercle, the isthmus twice as wide as the interorbital region; five branchiostegals; pseudobranchiæ present, small; gill-rakers short, stout, and serrulate.   Dorsal fins separate, with vi, i 8-9 rays, the spinous ones flexible; anal fin commencing well behind the origin of the second dorsal, with i 8-9 rays; the last soft ray of the second dorsal and anal fins divided to the base; ventral fins not in contact basally, inserted below the root of the pectorals, with i 5 rays, the fourth the longest, but not produced into crinoid filaments, pectoral fins rounded, with 18 or 19 rays, the middle ones the longest; caudal fin rounded, the peduncle strong. Genital papilla large, triangular in the male, oblong in the female. Scales moderate and adherent, those of the occiput about as large as those of the tail and a little larger than those of the trunk; head partially scaly, the interorbital region, snout, and anterior portion of the cheeks naked, scales of the head, nape, and throat cycloid, all the rest ciliated and finely carinated ; head with numerous series of small pores.   Vertebræ 28 (12 + 16).

Etymology.—Named after the district in which the typical species was first obtained and where it is abundant.

Type.—*Eleotris coxii*, Krefft.

Distribution.—Coastal region of New South Wales.

MULGOA COXII.

Eleotris coxii, Krefft, Proc. Zool. Soc. London, 1864, p. 183; Macleay, Proc. Linn. Soc. N.S. Wales, v. 1880, p. 618 (1881); Ogilby, Catal. Fish. N.S. Wales, p. 36, 1886.

Eleotris richardsonii, Steindachner, Sitzb. Ak. Wien, liii. i. 1866, p. 455, c. fig.; Ogilby, l.c.

Eleotris mastersii, Macleay, l.c. p. 622; Ogilby, l.c

Cox's Gudgeon.

D. vi, i 8-9.   A. i 8-9.   P. 18-19.   Sc. 37-40/11.

Body stout and moderately elongated, the tail compressed. Length of head $3\frac{2}{3}$ to $3\frac{9}{10}$, depth of body $4\frac{3}{5}$ to $5\frac{1}{2}$ in the total length; head as wide as or a little wider than deep, its width $1\frac{1}{2}$

49

to $1\frac{3}{4}$, that of the flat interorbital region $7\frac{1}{4}$ to $8\frac{1}{4}$, diameter of the
eye 4 to $4\frac{3}{4}$ in the length of the head; snout much broader than
long, rounded in front and slightly depressed, from one-tenth to
two-fifths of a diameter longer than the eye.   Maxillary not
reaching to the vertical from the anterior margin of the eye, its
length $3\frac{1}{3}$ to $3\frac{1}{2}$ in that of the head   Eight or nine gill-rakers on
the lower branch of the anterior arch, the last ones reduced to
serrulate knobs   The space between the origin of the first dorsal
fin and the extremity of the snout is as long as or a little longer than
its distance from the base of the last soft ray; outer border of the
first dorsal fin rounded, the third or fourth ray the longest, $1\frac{1}{8}$ to
$2\frac{1}{10}$ in the length of the head, and the last ray when laid back
reaches in the $\male$ to, in the $\female$ not quite to the origin of the second
dorsal; in the $\male$ the fourth and fifth, in the $\female$ the second and
third rays of the second dorsal are the longest, $1\frac{3}{5}$ to $1\frac{3}{4}$ in the
head : the anal fin originates below the third ray of the second
dorsal; the sixth and seventh rays are the longest, as long as the
soft dorsal rays   fourth ventral ray not reaching to the vent in
either sex, its length $1\frac{1}{4}$ to $1\frac{1}{2}$ in the head : middle pectoral rays
extending to the vertical from the origin of the second dorsal or
not quite so far, their length in the $\male$ subequal to, in the $\female$ about
one-fifth shorter than that of the head : caudal rounded, $1\frac{1}{10}$ to
$1\frac{1}{3}$ in the length of the head; the peduncle stout, as long as or a
little shorter than the head, its depth 2 to $2\frac{1}{3}$ in its length
Genital papilla triangular, as long or nearly as long as the eye,
and much longer than wide in the $\male$; quadrangular, two-thirds
or less than two-thirds of the eye, and as long as or but little
longer than wide in the $\female$, in which the posterior border is
concave.   Scales of the opercle unequal in size, deeply embedded,
and more or less non-imbricate, posterior portion of the cheeks
with rather small, deeply embedded, non-imbricate scales, a series
of small closely set pores from the snout round the upper margin
of the eye, extending backwards between the occiput and opercles
to the shoulder.

Purple to olive-green above, ultramarine-blue to silvery below,
the sides sometimes tinged with yellow ; the lighter-coloured

specimens are everywhere powdered with minute dusky dots; back with or without a series of dark blotches; a similar series of more or less irregularly arranged, often concurrent blotches almost always present along the middle of the sides and ending in a large dark blotch at the root of the caudal fin; side of head generally with two oblique dark bars, the upper from the postero-superior angle of the eye to the axil of the pectoral, forming a conspicuous spot on the upper half of the base; the lower from the snout along the inferior margin of the eye to the edge of the opercle, the interspace sometimes as dark as the bars, chin purple; a dusky blotch on the gill-rakers, dorsal fins, the first with a broad orange to pale yellow or hyaline dark-edged median band, the second with two or three similar but narrow bands near the base, the outer half clouded with purple or violet, caudal yellowish-brown, closely ornamented with a network of more or less regular dark spots; anal stone-gray or vinous, tipped with violet, often with the anterior ray brown and a median posterior golden patch; ventrals violet or gray, sometimes washed with gold towards the tip; pectorals olive-green, with or without a dusky shade on the upper rays and with a more or less brilliant golden basal band. Irides golden brown.

As a rule the more brilliant colours—the purple, blue, and orange—may be taken as the prerogative of the male fish, but this is not always the case, one or two females in my possession being quite as brightly marked as their partners.

All my specimens were obtained during the spring, and I cannot therefore say whether any difference in colouration takes place during the breeding season.

This species has been exceptionally unfortunate in its describers; Krefft—who obtained his examples from Dr. James C. Cox—described them as having seven rays in the anterior dorsal fin; his type specimen, which came from the Mulgoa Creek, a tributary of the Nepean River, into which it falls not far from Penrith, and two others from Rope's Creek in the same district, still bearing labels in Krefft's own handwriting, are fortunately in existence and possess six rays only in every instance; he also describes the head as being scaly, which is misleading, as the greater part of the cheeks, the interorbital region, and the snout are naked

Two years subsequently Dr. Franz Steindachner, in his description of *Eleotris richardsonii*, gives the number of rays in the first dorsal as seven in the letterpress, while in the excellent figure (unnamed and unnumbered) six are correctly shown; there is no other material difference between Steindachner's description and mine except in the comparative measurements of the interorbital region, the width of which according to him is greater than the diameter of the eye, while a reference to the above diagnosis will show that I make it much less at all ages; this, however, may possibly be explained by a difference in the system of measurement employed, the width in my descriptions always being that of the bony space only.

Finally Sir William Macleay, in diagnosing *Eleotris mastersii*, again falls into the same error, giving seven as the number of spinous dorsal rays; of the five examples labelled as above, now in the University Museum and undoubtedly the very ones from which Macleay took his description, not a single one has more than six rays. Rope's Creek, whence the types of *E. mastersii* were brought, is one of the original localities from which *E. coxii* came.

From the shape of the genital papilla, as given by the three authors referred to above, it is evident that their descriptions were taken in every case from female examples.

Cox's Gudgeon is very generally distributed throughout the entire network of streams and ponds connected with the Upper Hawkesbury, and wherever found appears to be abundant. Hitherto I have failed to find it east of the range which divides the Nepean and Wollondilly from the Parramatta and George's Rivers, and am, therefore, sceptical as to its occurrence in the Bronte Lagunes as asserted by Krefft; as far as my experience goes, this species is confined to the upper waters of the Hawkesbury, where it replaces *Krefftius australis*, which is the prevailing species along the littoral zone, the range of the two forms overlapping in a kind of neutral zone which lies somewhere about the altitude of Penrith, where both species occur abundantly.

Besides the specimens enumerated above, I have to thank Mr. W. J. McCooey for three examples obtained in the neighbourhood of Camden; and more especially am I indebted to Mr. M. P. Gorman, of Burragorang, for three magnificent series forwarded during the months of October and November from the Wollondilly and "a small creek in the mountains away from the river altogether." These series are fully illustrative of the growth of the fish between the lengths of 33 and 138 millimeters, and the opportunity of examining them in a fresh condition has enabled me to thoroughly satisfy myself as to the identity of *richardsonii* with Krefft's species.

Fifty-three specimens have been examined in the preparation of this article, the largest measuring just 180 millimeters.

O P H I O R R H I N U S, gen.nov.

Eleotris, sp. auctt

Body rather elongate, compressed posteriorly, the back broad and flat in front of the dorsal fin, rounded behind; head very large and strongly depressed, much wider than deep, the snout short and very obtuse; mouth large and but little oblique, the lips thin, premaxillaries but little protractile; maxillaries narrow, with the distal end exposed and linear; lower jaw much the longer; jaws with a broad band of cardiform teeth, all of which are fixed; lower pharyngeals forming together a subtriangular patch, armed with small, stout, hooked teeth, a few at the apex and along the symphysis somewhat enlarged; nostrils moderately separated, the anterior valvular; eyes sublateral; none of the bones of the head armed; gill-openings extending forwards to below or before the angle of the mouth, the isthmus about half as wide as the interorbital space; six branchiostegals; pseudobranchiæ present, small; gill-rakers short and rather slender, mostly serrulate. Dorsal fins separate, with vii, i 9-10 rays, the spinous ones flexible; anal fin originating behind the second dorsal, with i 9-10 rays; the last soft rays of the second dorsal and anal fins divided to the base; ventral fins small, not in contact basally, inserted beneath or somewhat in front of the base of the pectorals, with i 5 rays, the fourth soft ray the longest, but not produced or filiform; pectoral fins large and pointed, with 18 or 19 rays, the middle ones the longest, caudal fin rounded, the peduncle rather slender. Genital papilla small. Scales moderate and adherent, those of the tail much larger than those of the trunk; entire head, except a portion of the occiput, naked; scales deeply embedded, cycloid and smooth in front, imbricate and feebly ciliated behind, muciferous system of head well developed. Vertebræ 30 (13 + 17)

Etymology.—ὄφις, a snake; ῥίν, snout.

Type —*Eleotris grandiceps*, Krefft.

Distribution.—Coastal region of south-eastern Australia

The following analysis will suffice to distinguish the two species here described :—

Width of head $1\frac{2}{5}$-$1\frac{2}{3}$, of interorbital region $4\frac{2}{3}$-$5\frac{2}{5}$, length of fourth ventral ray $1\frac{3}{5}$-$1\frac{3}{4}$, of caudal peduncle $1\frac{2}{5}$-$1\frac{3}{5}$ in the length of the head, inner series of teeth enlarged, 11-12 gill-rakers; scales 42 or less along the middle of the body

grandiceps, p. 746

Width of head $1\frac{3}{5}$-2, of interorbital region $5\frac{1}{2}$-$6\frac{1}{2}$, length of fourth ventral ray 2-$2\frac{2}{5}$, of caudal peduncle $1\frac{1}{5}$-$1\frac{1}{4}$ in the length of the head ; all the teeth subequal ; 7-9 gill-rakers, scales 43 or more along the middle of the body   ... ...

nudiceps, p. 748

OPHIORRHINUS GRANDICEPS.

Eleotris grandiceps, Krefft, Proc. Zool. Soc. London, 1864, p. 183; Macleay, Proc. Linn. Soc. N.S. Wales, v. 1880, p. 618 (1881); Ogilby, Catal. Fish. N.S. Wales, p. 36, 1886.

Flat-headed Gudgeon.

D. vii. i 9-10.   A. i 9.   P. 19.   Sc. 38-42/12.

Body moderately elongate, tapering from the shoulder, the tail strongly compressed. Length of head $2\frac{9}{10}$ to $3\frac{1}{5}$, depth of body 5 to 6 in the total length; depth of head $2\frac{1}{10}$ to $2\frac{1}{3}$ ($\male$), $2\frac{1}{4}$ to $2\frac{1}{2}$ ($\female$), width of head $1\frac{2}{5}$ to $1\frac{1}{2}$ ($\male$), $1\frac{1}{3}$ to $1\frac{2}{3}$ ($\female$), of inter-orbital region $4\frac{2}{3}$ to $4\frac{1}{7}$ ($\male$), 5 to $5\frac{2}{3}$ ($\female$), diameter of eye $4\frac{2}{3}$ to $5\frac{1}{5}$ in the length of the head; snout broad, rounded in front, and much depressed, one-half to three-fifths of a diameter longer than the eye. Maxillary extending to the vertical from the posterior margin of the eye ($\male$), the middle of the eye ($\female$), its length $1\frac{7}{10}$ to $1\frac{9}{7}$ ($\male$), 2 to $2\frac{1}{4}$ ($\female$) in that of the head. The teeth of the inner series are the largest, those preceding them growing gradually smaller. Eleven or twelve gill-rakers on the lower branch of the anterior arch. The space between the origin of the first dorsal fin and the extremity of the snout is greater than its distance from the base of the last soft ray; outer margin of the spinous dorsal convex, the second or third ray the longest, $2\frac{1}{5}$ to $2\frac{1}{2}$ in the length of the head, and reaching when laid back in the $\male$ to, in the $\female$ not so far as the origin of the second dorsal, in the $\male$ the seventh and eight soft rays are the longest, $1\frac{1}{2}$ to $1\frac{3}{4}$, in the $\female$ the third and fourth are the longest, $2\frac{1}{10}$ to $2\frac{1}{4}$ in the length of the head: the anal fin commences a little behind the origin of the second dorsal and is in all respects similar to it: fourth ventral ray not greatly produced beyond the third or fifth and not nearly reaching to the vent in either sex, its length $1\frac{1}{3}$ to $1\frac{3}{4}$ in that of the head: middle pectoral rays the longest; they are in the $\male$ much longer than the fourth ventral ray, reaching well beyond the vertical from the origin of the second dorsal, and $1\frac{1}{4}$ to $1\frac{1}{3}$ in the length of the head, in the $\female$ subequal to the fourth ventral ray, reach to or not quite to the vertical from the dorsal interspace, and $1\frac{1}{2}$ to $1\frac{2}{3}$ in the head: caudal rounded, $1\frac{3}{10}$ to $1\frac{1}{2}$ in the length of the head; the peduncle rather slender, not differ-ing appreciably in both sexes, its length $1\frac{2}{2}$ to $1\frac{1}{3}$ in that of the head, its depth $2\frac{1}{4}$ to $2\frac{2}{3}$ in its length. Genital papilla very small and triangular in the $\male$, oblong and notched in the $\female$, in which it is somewhat larger. Scales small and irregular anteriorly; those on the tail with an angular border, occipital scales small, deeply embedded, and non-imbricate, extending forwards almost to the eyes.

Pale reddish-brown above, yellowish below, the head darker, everywhere densely punctulated with blackish dots which are often concurrent, forming two more or less conspicuous series of dark spots, one along the dorsal profile, the other along the middle of the body, the latter terminating in a blotch which is always present at the base of the caudal fin; a pair of oblique brown bands from the eye across the opercles generally present; first dorsal pale yellow with a basal, median, and marginal dusky band; the second similar but with four or five narrower bands, caudal with about eight irregular transverse bars, which often form a network; anal and ventrals gray, with or without microscopic dusky dots, pectorals yellow, with a more or less faint darker basal band.

In the breeding season the upper surfaces, dorsal and caudal fins are deeply tinged with salmon colour.

I found this to be the most abundant species in the waterholes near Liverpool on the occasion of the visit above referred to, when, like the two other species obtained at the same time, they were busily engaged in the duties of reproduction. Subsequently I obtained a number of young specimens, under two inches in length from a waterhole at Camden Park, but failed to catch any adults.

The Flat-headed Gudgeon is an inhabitant of the coastal watershed of New South Wales from the Richmond River—whence Krefft records it—southwards; it is abundant in the metropolitan district, and the limit of its range inland appears to be somewhat similar to that of *Krefftius australis* or extending to an altitude of about one hundred feet above the level of the sea; exactly how much further southward it ranges I am unable to say.

This species never attains to the size of the two preceding, the largest example, of twenty three utilised in the preparation of the above description, barely measuring 100 millimeters.

OPHIORRHINUS NUDICEPS.

Eleotris nudiceps, Castelnau, Proc. Zool. & Acclim. Soc. Vict. i. 1872, p. 126 (1873); Macleay, Proc. Linn. Soc. N.S. Wales,

v. 1880, p. 619 (1881); Lucas, Proc. Roy. Soc. Vict. (2) ii. 1890, p. 29.

? *Philypnodon nudiceps*, Bleeker.

Yarra Gudgeon.

D. vii, i 9-10.   A. i 9.   P. 19.   Sc. 43-47/12-13.

Body moderately elongate, tapering from the shoulder, the ˙ tail strongly compressed. Length of head 3 to $3\frac{1}{4}$, depth of body $5\frac{2}{3}$ to $5\frac{1}{3}$ in the total length; depth of head 2 to $2\frac{1}{4}$, width of head $1\frac{1}{3}$ to 2, of interorbital region $5\frac{1}{2}$ to $6\frac{1}{3}$; diameter of eye $4\frac{1}{2}$ to $4\frac{9}{10}$ in the length of the head; snout broad, rounded in front, and moderately depressed, one-third to one-half of a diameter longer than the eye. Maxillary extending to the vertical from the anterior third to the posterior fourth of the eye, its length $1\frac{9}{10}$ to $2\frac{1}{4}$ in that of the head. All the teeth are subequal in size. Seven to nine gill-rakers on the lower branch of the anterior arch. The space between the origin of the first dorsal and the extremity of the snout is greater than its distance from the base of the last soft ray; outer margin of the spinous dorsal gently rounded, the second, third, or fourth ray the longest, $2\frac{2}{5}$ to $2\frac{3}{4}$ in the length of the head, and reaching when laid back nearly to, to, or a little beyond the origin of the second dorsal; the seventh or eighth soft rays are the longest, $1\frac{9}{10}$ to $2\frac{1}{4}$ in the length of the head: the anal fin commences behind the origin of the second dorsal and is in all respects similar to it: fourth ventral ray but little produced beyond the third and fifth, not nearly extending to the vent in either sex, its length $2\frac{1}{10}$ to $2\frac{2}{5}$ in that of the head: middle pectoral rays the longest, reaching nearly to, to, or a little beyond the vertical from the origin of the second dorsal, and are $1\frac{1}{10}$ to $1\frac{3}{5}$ in the length of the head: caudal rounded, $1\frac{1}{2}$ to $1\frac{3}{5}$ in the length of the head; the peduncle rather slender, its length $1\frac{1}{3}$ to $1\frac{1}{4}$ in that of the head, its depth $2\frac{1}{10}$ to $2\frac{3}{5}$ in its length. Genital papilla triangular in the ♂, oblong and crenulate in the ♀. Scales small and very irregular anteriorly, some of those on the tail with an angular border; occipital scales deeply embedded and non-imbricate, extending forwards beyond the preopercle.

Olive-green or brown above, the sides paler, gray below: head purple above, shading into violet beneath; sides with a series of faint dusky blotches, only that in front of and partly on the base of the caudal fin at all conspicuous, first dorsal fin violet with three longitudinal series of purplish or chestnut spots, the second pale grayish-green with four series of similar but smaller spots, caudal, anal, and ventral fins gray, sometimes with the extremities of the rays violet ; pectorals grayish-green, the upper half the darker.

This is the only Eleotrid which has as yet come under my notice from Victoria, nor so far as I know have any of my Melbourne co-workers been more fortunate, though two other species have been recorded from the Yarra by European scientists, namely, *cyprinoides* by Klunzinger and *melbournensis* by Sauvage.*

This Gudgeon is very abundant in the Yarra, and there cannot be any doubt as to the identity of my species with that of Count Castelnau; there are, however, several points of difference which need explanation, as follows :—

(1) In Castelnau's description the interorbital region is said to be "one-third" of the length of the head, while I find it to be only half that width,† this may be explained in a similar manner to that suggested as the cause of difference between Steindachner's description of *Eleotris richardsonii* and mine of *Mulgoa c. sa* (see p. 744).

(2) The apparently larger size of the eye in my examples is easily capable of explanation by the fact that Castelnau's measurement of the length of the head is taken from the extremity of the projecting mandible, mine from that of the snout.

---

* It is one of the most remarkable problems connected with Australian fish literature how the continental naturalists, receiving small collections from such well worked localities as Port Jackson and Hobson's Bay, invariably succeed in obtaining fishes, which we, despite our local knowledge, and despite that having been once recorded they are more carefully sought for, are unable to find

† Castelnau's words—"eyes considerably apart, the distance from one or other being nearly equal to the third of the length of the head "—are rather ambiguous, but there can be little doubt as to what his meaning is

(3) According to Castelnau, "the head has no scales," but those of the occiput must have been overlooked by him, for though they are small and deeply embedded, they are nevertheless plainly visible.

These, however, are but minor discrepancies as compared with (4) the dentition; referring to this Castelnau writes—"the teeth . . . extend on the vomer and the palatines ; the posterior part of the tongue is also covered with them." This is quite the opposite of what I find; in all my examples there is no sign of teeth on any part of the mouth except those on the jaws. If Castelnau's fish really had the subsidiary teeth attributed to it by its describer— which on a review of all the facts of the case I may be permitted to doubt—it would of course be necessary to place it in another genus, and this has possibly been already done by Dr Bleeker, since his *Philypnodon nudiceps* possesses the same dentition as that assigned to his species by Castelnau *

The differences which separate *grandiceps* from *nudiceps* are undoubtedly slight, but those which are noticed in the preceding analysis (*see p.* 746) appear to be constant ; the close affinity between the two species was recognised by Castelnau, who writes "The principal reason for not uniting my sort with Krefft's is, that he says that the pectorals attain the base of the anal; while in my specimens they do not." I consider this elongation of the pectoral fins to be merely a sexual character.

This little fish is abundant in the Yarra, along the banks of which it is known as the "Big-head" according to Castelnau

---

* The want of Bleeker's paper prevents me from ascertaining whether his genus *Philypnodon* is founded upon Castelnau's description of *nudiceps*; if this be the case, Bleeker's genus, being specially formed on account of a character which it does not possess, must if monotypic be suppressed. And this raises another question to which I am unable to find a satisfactory answer, namely--if a genus be founded on a character which is purely mythical, should the name so proposed stand in preference to another correctly characterised from the same species but at a later date? If the practice of forming new genera from descriptions only were discouraged or disallowed, errors of this nature would soon cease.

(fide Lucas), who states that they are very voracious and feed on "fishes as large as themselves and generally of their own species."

Writing of this fish, Mr. T. S. Hall remarks (*in lit.*):—"It differs from Castelnau's *E. nudiceps* in the proportions of the head and especially in the teeth. Locality, "Yarra River at Melbourne (tidal)." Further on he says, "As a boy I have often caught what I imagine to be the same fish in the Barwon near Geelong in fresh water, and have seen a similar looking fish in the crater lake of Bullenmerrie, which is slightly brackish. I cannot vouch for the identity of the three forms. We used to call them ' bullies ' or ' bull-heads,' and regarded them as poisonous." It is hardly necessary to say that the last supposition was erroneous.

My description is founded on an examination of sixteen specimens, ranging in size from 42 to 110 millimeters, for which I have to thank Mr. J. Kershaw, of the National Museum, and Mr. T. S Hall, of the Melbourne University, the latter of whom sent me no less than fourteen fine examples.

The type of *nudiceps* is not, so far as I know, in existence.

In Macleay's Catalogue twenty-nine species of *Eleotris* are included among Australian fishes, but as, since the publication of the Supplement in 1884, this number has been nearly doubled from various sources, I append a list of all the species which have been recorded as occurring within our limits or on the opposite coast of New Guinea up to the present day. As all or almost all these have been described as *Eleotris*, I have drawn up the list in alphabetical order, making no attempt at this stage to segregate the species in natural groups, and even including such synonyms as *mastersii* and the like, so that the present list may partake of the character of an index to the Australian forms.

1. *adspersa*, Castelnau, Proc. Linn. Soc. N.S. Wales, iii. 1878, p. 142.

2. *aporocephalus*, Macleay, Proc. Linn. Soc. N.S. Wales, ix. 1884, p. 33, = *planiceps* (not Castelnau) Macleay, l.c. viii. 1883, p. 206, ? = *porocephalus*, Cuvier & Valenciennes, xii. p. 237, 1837.

3 *australis*, Krefft, Proc. Zool. Soc. London, 1864, p. 183 ; see
p. 737 et seq.

4. *brevirostris*, Steindachner, Sitzb. Ak. Wien, lvi. i. 1867, p. 314.

5. *butis*, Hamilton-Buchanan, Fish. Ganges, pp. 57, 367, 1822.

6. *castelnaui*, Macleay, Proc. Linn Soc. N S. Wales, v. 1880,
p. 620 (1881), = *obscura* (not Schlegel) Castelnau, Proc.
Zool. & Acclim. Soc. Vict. ii. 1873, p. 134 (1874).

7. *compressus*, Krefft, Proc. Zool. Soc. London, 1864, p. 184 ;
see p. 735.

8. *concolor*, De Vis, Proc. Linn Soc. N S. Wales, ix. 1884,
p. 692.

9. *coxii*, Krefft, Proc. Zool. Soc. London, 1864, p. 183; see p. 741
et seq.

10. *cyanostigma*, Bleeker, Kokos, iv. p. 452.

11. *cyprinoides*, Cuvier & Valenciennes, Hist. Nat. Poiss. xii.
p 248, 1837.

12. *darwiniensis*, Macleay, Proc. Linn. Soc. N.S. Wales, ii. 1877,
p. 360 (1878) as *Agonostoma darwiniense*.

13. *derisi* nom nov , = *cavifrons* (not Blyth) De Vis, Proc. Linn.
Soc. N.S. Wales, ix. 1884, p. 693.

14 *elevata*, Macleay, Proc. Linn. Soc. N.S. Wales, v. 1880, p. 622
(1881) = *compressus* (not Krefft) Macleay, l c ii. 1877,
p. 358 (1878); see p. 735.

15. *elongata*, Alleyne & Macleay, Proc. Linn. Soc. N.S. Wales, i.
1876, p 334 (1877).

16. *fusca*, Bloch & Schneider, Syst. Ichth. p. 453, 1801.

17. *gobioides*, Cuvier & Valenciennes, Hist. Nat. Poiss. xii. p. 247,
1837.

18. *grandiceps*, Krefft, Proc. Zool. Soc. London, 1864, p. 183 ;
see p. 746 et seq.

19. *gymnocephalus*, Steindachner, Sitzb. Ak. Wien, liii. i. 1866,
p. 453 (1867); ? *Gymnobutis gymnocephalus*, Bleeker.

20. *gyrinoides*, Bleeker, Sumatra, ii. p. 272, 1853.

21. *humilis*, De Vis, Proc. Linn. Soc N S Wales, ix. 1884, p. 690.

22. *immaculata*, Macleay, Proc. Linn. Soc N.S. Wales, viii. 1883, p. 263.

23. *larapintæ*, Zietz, Rep. Horn Exped. Centr. Austr. Zool. p. 179, 1896.

24. *laticeps*, De Vis. Proc. Linn. Soc. N S Wales, ix. 1884, p. 692

25. *lineolatus*, Steindachner, Sitzb. Ak. Wien, lv. i. 1867, p 13.

26. *longi*, Ogilby, Proc. Linn. Soc. N S. Wales, xxi. 1896, p. 733 et seq.

27. *longicauda*, De Vis, Proc. Linn. Soc N S. Wales, ix 1884, p. 691.

28. *macrodon*, Bleeker, Bengal en Hind. p. 104, 1853.

29. *macrolepidotus*, Bloch, Ausl. Fisch. v. (pt. ix.) p. 35, 1797, not Gunther, Fisch. Sudsee, Heft vi. p 186, which is *tumifrons*, = *aporos*, Macleay.

30. *mastersii*, Macleay, Proc. Linn. Soc, N.S. Wales, v. 1880, p. 622 (1881), = *coxii;* see p. 744.

31. *melbournensis*, Sauvage, Bull. Soc Philom. (7) iv. 1880, p. 57.

32. *mimus*, De Vis, Proc Linn. Soc. N.S. Wales, ix. 1884, p. 690. ? = *adspersus*.

33. *modesta*, Castelnau, Proc. Zool. & Acclim. Soc. Vict. ii. 1873, p. 85 (1874).

34. *mogurnda*, Richardson, Voy. Erebus & Terror, Ichth. p. 4, 1846.

35. *muralis*, Cuvier & Valenciennes, Hist. Nat. Poiss xii. p. 253, 1837.*

36. *nigrifilis*, nom nov., = *lineata* (not *Dormitator lineatus*, Gill, 1863) Castelnau, Res. Fish. Austr. p. 24, 1875.

---

* There is also an *Eleotris muralis*, Sauvage, Bull Soc. Philom. (7) vi. 1882, p. 172; as I have not had an opportunity of consulting this work I cannot say whether he is referring to the above species or describing a new one by the same name.

37. *nudiceps,* Castelnau, Proc. Zool. & Acclim. Soc. Vict i. 1872, p. 126 (1873); see p. 748 et seq.

38. *oxycephala,* Schlegel, Faun. Japon. Poiss. p. 150, 1850.

39. *pallida,* Castelnau, Res. Fish. Austr. p. 24, 1875.

40. *planiceps,* Castelnau, Proc. Linn. Soc. N.S. Wales, iii. 1878, p. 49.

41. *porocephaloides,* Bleeker, Sumatra, iii. p. 514; ?= *porocephalus.*

42. *porocephalus,* Cuvier & Valenciennes, Hist. Nat. Poiss. xii. p. 237, 1837.

43. *reticulatus,* Klunzinger, Sitzb. Ak. Wien, lxxx. i. 1879, p. 385 (1880).

44. *richardsonii,* Steindachner, Sitzb. Ak. Wien, liii. i. 1866, p. 455, = *coxii,* see p. 744.

45. *robustus,* De Vis, Proc. Linn. Soc. N.S. Wales, ix. 1884, p. 692.

46. *selheimi,* Macleay, Proc. Linn. Soc. N.S. Wales, ix. 1884, p. 33, = *planiceps* (not Castelnau) Macleay, o c. vii. 1882 p. 69.

47 *simplex,* Castelnau, Proc. Linn. Soc. N.S. Wales, iii. 1878, p. 49.

48. *striatus,* Steindachner, Sitzb. Ak. Wien, liii. i. 1866, p. 452.

49. *sulcaticollis,* Castelnau, Proc. Linn. Soc. N.S. Wales, iii. 1878, p. 142.

50. *taeniura,* Macleay, Proc. Linn. Soc. N.S. Wales, v. 1880, p 624 (1881).

51. *tumifrons,* Cuvier & Valenciennes, Hist. Nat. Poiss. xii. p. 241, 1837, = *ophiocephalus,* Macleay.

Three of the species included in the above list have so far been found on the opposite coast of New Guinea, but may confidently be expected to occur on our northern shores; they are *butis,* *gyrinoides,* and *immaculatus.*

Of the remaining forty-seven only six—*australis, coxii, grandiceps, compressus, oxycephalus* and *mastersii*—were known to

Macleay as inhabitants of the rivers and estuaries of New South
Wales up to 1884, when his "Supplement." was published, but
two years later I was able to increase this number by four, adding
mogurnda, gymnocephalus, striatus, and *richardsonii;* two of these,
however,—*mastersii* and *richardsonii*—I have shown in the foregoing paper to be identical with *coxii;* a third—*mogurnda*—rests
its claim upon its inclusion by Steindachner in his "Fishes of
Port Jackson (*Sitzb. Ak. Wien, lvi. i. 1867, p. 328*) and the
authority of a single specimen now in the Australian Museum,
and said to have come from the Clarence River, and though this
is very possibly correct, still in the lack of confirmatory evidence
it is safest to look with suspicion on any record of its occurrence
so far south ; a fourth species—*oryiephalus*—I unhesitatingly
reject; this is one of the fishes said to have been obtained by the
collectors of the Novara during the short stay of that war-ship in
the waters of Port Jackson, but which has never been found since;
it is a Chinese and Japanese species, and the improbability of its
occurrence so far from its native shores is obvious.* With the
addition of the new species above described and of *gobioides,*
included by Steindachner in his Port Jackson fishes,† this leaves

---

* The following species, only recorded in the Fishes of the Novara, I
must excise from the New South Wales catalogue until more conclusive
evidence of their occurrence is available :—1, *Mesoprion marginatus;* 2,
Apogon quadrifasciatus; 3, *Chætodon setifer;* 4, *Lethrinus harak;* 5, *Ampha-
canthus hexagonatus;* 6, *Batrachus trispinosus;* 7, *Gobius frenatus;* 8, *Eleotris
oxycephalus;* 9, *Petroscirtes solorensis ;* 10, *Mugil cephalotus;* 11, *Crepido-
gaster tasmaniensis;* 12, *Ophiocephalus striatus;* 13, *Polyacanthus cupanus;*
14, *Platyglossus trimaculatus ;* 15, *Pseudoscarus octodon;* 16, *Rhombosolea
leporina;* 17, *Solea humilis;* 18, *Exocœtus unicolor;* 19, *Balistes maculatus;*
20, *Tetrodon richei;* and 21, *Tetrodon erythrotænia.* I do not, of course,
assert that none of these fishes are found on the New South Wales coast,
some of them—such as 7, 11, 12, 17, and 21—most probably are, but I
distinctly reject them so long as their claim to admission rests solely on
the unsatisfactory evidence adduced.

† This is a New Zealand species, and its occurrence here requires
confirmation.

the New South Wales list with seven good and two doubtful species, namely :—

1. Carassiops compressus.
2. Carassiops longi.
3. Krefftius australis.
4. Mulgoa coxii.
5. Ophiorrhinus grandiceps.
6. Gymnobutis gymnocephalus.
7. ? striatus.
?8. Mogurnda mogurnda.
?9. Gobiomorphus gobioides.

I have been for some time past making special endeavours to obtain examples of *gymnocephalus* and *striatus*, but have failed so far in doing so, nor is either species represented in the collections of the Australian Museum or the Sydney University.

The genus *Gymnobutis* was probably founded by Bleeker with Steindachner's *gymnocephalus* as the type; I am unable to suggest to which of the recent genera *striatus* should be referred.

50

ON DOMATIA IN CERTAIN AUSTRALIAN AND OTHER PLANTS.

By Alex. G. Hamilton.

(Plate LVII.)

Some years ago, when collecting *Pennantia Cunninghamii*, Miers, my attention was attracted by the presence of prominences on the upper surface of the leaves which I at first took to be the nidus of some leaf-mining insect larva. But further observation showed that they always had an opening on the under surface of the leaf, and invariably occurred in the same position, viz.. in the axils of the veins. A short time after, I happened to look at the leaves of the ornamental New Zealand shrub, *Coprosma lucida*, commonly cultivated in gardens, and I was much interested to notice in this plant also the presence of cavities opening to the exterior by conspicuous pores on the lower surface of the leaf. After this I began to examine the leaves of the plants within reach more systematically.

No books that I was able to consult seemed to throw any light on the subject, and as I am distant from libraries, I was glad to make known my needs to several Sydney friends who were at length successful in giving me a clue.

The first important intelligence came from Mr. E. Betche, who sent me the following quotation from Dr. R. Schumann's article on Rubiaceæ, in Engler's "Naturliche Pflanzenfamilien":—"In den Nervenachseln auf der Rückseite der B. befinden sich zuweilen Haarbüschel, welche eine etwas eingedrückte Stelle der Blattspreite umwachsen; man nannte diese Stellen Blatt-scropheln (Scrobiculæ) gegenwartig werden sie als Domatien bezeichnet. Sie sind fur gewisse Arten von *Cinchona* zur Unterscheidung benutzt worden."

The next difficulty encountered was to find definitions of scrobicula and domatium. Eventually Mr. J. J. Fletcher found in Henslow's "Floral Structures" (p. 115) a reference to Dr. Lundstrom's important paper on the subject* (with a copy of which the author most kindly favoured me subsequently). Also that Howard (Illust. N. Quinologia) speaks of "the scrobicules or glands [in Cinchona], as Pavon calls them."

Mr. J. P. Hill sent me Geddes' "Chapters in Modern Botany," on p. 134 of which Lundström's views are mentioned. Mr. C. T. Musson obtained for me the reference to Mr. Cheeseman's paper "On the New Zealand Species of Coprosma,"† and so disposed of any doubt that New Zealand naturalists had failed to notice the structures in question in plants of this genus.

Dr. Lundström was the first naturalist who systematically investigated these structures. The following extracts from the summary of it in the Journ. R. Microscop. Soc. (1888, p. 87) will sufficiently indicate the conclusions at which he arrived in his valuable paper.

"Domatia.—Dr. A. N. Lundström defines as 'domatia' those formations or transformations on plants adapted to the habitation of guests, whether animal or vegetable, which are of service to the host, in contrast to cecidia, where such habitation is injurious to the plant. He describes these domatia in detail on the lime, alder, hazel, and other trees and shrubs, and gives a very long list of species, belonging to a great variety of natural orders, on which they are found.

"The principal types of shelter are as follows :—(1) Hair-tufts, e.g., in *Tilia europœa*; (2) recurvatures or foldings in various parts, e.g., in *Quercus robur* . . . ; (3) grooves without hairs, as in *Coffea arabica* . . . ; with marginal hairs, e.g., *Psychotria daphnoides* . . . ; with basal hairs, as in *Anacardium occidentale* . . . ; (4) pockets, as in *Elæocarpus oblongus* . . . ; (5) pouches, e.g. *Eugenia australis*. These different types of domatia are connected by transition forms. The habit of producing domatia in a species may become hereditary without the actual presence of the predisposing cause. Certain orders,

---

* Nov. Act. R. Soc. Sc. Upsala, (3) xiii. (1887), pp. 1-72 (4 pls.).

† Trans. N. S. Inst. xix. 1886, p. 221 [1887].

e.g., *Rubiaceæ* (famous also for ant-domatia) show a marked predisposition to acaro-domatia. Many groups seem entirely without them, *e.g.*, Monocotyledons and Gymnosperms, and all herbs. They are most abundant and best developed in tropical (and temperate) zones.

"In the second chapter the author discusses in detail the various interpretations which may be put upon domatia. (1) They may be pathological, like galls; (2) they may be for catching insects; (3) they may have only an indirect connection with their tenants; (4) they may be of use to the plant as the dwellings of commensals. He adopts the last interpretation. He draws an interesting parallel, however, between galls and domatia, and is inclined to suppose that the domatia were first directly caused by the insects, but have gradually become inherent, transmitted characteristics. The author gives a clear table, distinguishing the cecidia or galls due to 'antagonistic symbiosis,' either plant or animal, (phyto- and zoo-cecidia), and domatia due to 'mutual symbiosis,' either plant or animal (phyto- and zoo-domatia). Those due to plants are again subdivided into myco- and phyco-cecidia or -domatia."

Mr. Cheeseman's remarks are very interesting, not only because his paper was published in the same year (1887) as Lundstrom's, but also because he, too, noticed that the domatia of *Coprosma* were often tenanted by Acarids. He says: "In nearly all the species except a few of the smaller-leaved ones, curious little pits exist on the under surface of the leaves, in the axils formed by the union of the primary veins with the midrib. They are never more than $\frac{1}{8}$ of an inch in length, and are usually much less. Inside they are lined with numerous stiff white hairs, which on being treated with caustic potash are seen to be composed of two or three cells. So far as I have observed, the pits do not secrete anything, and I am quite unable to guess at their function. They are often inhabited by a minute yellow Acarid, which makes use of them as a home. Sometimes two or three Acarids may be found in the same pit, and they crawl freely about the young leaves and branches." (Trans. N.Z. Inst. Vol. xix. p. 221.)

Lundström, quite reasonably. expresses surprise that domatia have attracted so little notice. And hardly less remarkable is it that up to the present time, the text books have still nothing, or so little to say about them or their significance. Nevertheless, they were long ago noticed in at least one Australian plant, but having been relegated to the category of "glands"—"that word of many meanings," as De Bary remarks—their nature seemed to be looked upon as settled For example, in Vol li. of Curtis's Botanical Magazine, published in 1824, there is a figure (Pl 2488) of *Cissus* [*Vitis*] *antarctica* [= *V. Baudiniana*, F.v M], in which domatia are distinctly shown, while the text mentions "foliis ovatis laxe serratis glabriusculis subtus glandulosis." The synonymy also shows that at a still earlier period Poiret, because of the presence of these supposed glands, had described the species under the name of *C. glandulosa*, "foliis ovatis glabris laxe dentato-serratis nervis basi glandulosis."

In 1879, at a Meeting of the Linnean Society of London, "Mr. R. Irwin Lynch directed attention to a growing example from Kew Gardens, and some of the dried leaves of *Xanthosoma appendiculatum*, on the under surface of which peculiar pouch-like excrescences emanate from the midrib This pseudo-monstrosity is of remarkably constant occurrence.' * If these excrescences be, as I think they are, domatia, the plant (an Aroid) is remarkable as being the only instance known of the occurrence of domatia in the Monocotyledons. Mr. Lynch, too, is the first, apparently, who saw anything uncommon in the structures.

A few other references to what would now be called domatia may be given.

Trimen says of *Psychotria bisulcata*, "Lateral veins often with very deep pits in their axils, which appear as warts on the upper surface." ("Handbook of the Flora of Ceylon.")

---

* Journ. of Bot. April, 1879, p 125, but not noticed in the Proceedings of the Society.

In classifying the Cinchonas, Howard states that Pavon divided the 40 species into two groups: 23 species without glands (sin glandulas) and 12 with glands (con glandulas). ("Illustrations of the Nueva Quinologia of Pavon," 1862) *C. villosa*, one of the second group is thus described : "Folia . . . glandulis nonnullis rotundatis, subtus concavis, marginibus villosis, supra prominentibus, ad nervorum axillas insertis, supra obscure viridia, subtus dilute." . . . "This is a species moderately hairy all over, especially on the under-side veins." From the position of these so-called glands in the nerve-axils, and their appearance in the figures, I have no doubt but that they are domatia. Among the species spoken of as without glands, *C. viridifolia* is described as "At nerve-axils pilose-tomentose," which is one of the forms of domatia. *C. villosa* and *C. conglomerata* are mentioned as hairy. This is contrary to Dr. Lundstrom's experience: his opinion being that domatia do not occur in hairy-leaved plants.

A doubtful species of *Calisaya* known as "naranjada" is spoken of by Howard as having "scrobicules not only at the axils of the veins, but also at their junction with the smaller veins, as in *Olea scrobiculata*." The accompanying figure shows very distinct domatia, which are visible on both sides of the leaf. (Journal of Botany, 1869, p. 3)

Of *Cinchona Ledyeriana*, Trimen says : "scrobicules not conspicuous, mostly confined to the upper vein-angles." (Journ. of Botany, 1881, p. 323.)

Martius in the " Flora of Brazil " refers to these structures in several descriptions of the leaves.

Hooker says of *Elæocarpus dentatus*, "with hollows where the veins meet the midrib." (Handbk. N.Z. Flora, p 34).

F. v. Mueller remarks of *Cupania foveolata* : "The principal veins with dimples in their axils." (Fr. Phy. ix. p. 95)

Bentham describes *Nephelium foveolatum* as "having frequently a cup-shaped cavity in the axils of the primary veins." (Fl Aust. i. p. 466) : *Cupania xylocarpa*, " with hairy tufts almost always conspicuous in the axils of the raised primary veins."

(*ib.* p. 447); the leaves of *Vitis oblongata* " with two large glands underneath in the axils of the lateral veins": the leaflets of *V. sterculifolia* " with glands or foveoleæ in the axils of some of the primary veins underneath" (*ib.* p. 450). He also mentions "glands" on the leaves of *V. Baudiniana.*

Recently G. de Lagerheim has described some new acarodomatia (2) in *Solanum jasminoides* and *S. pseudoquina,* and he refers to the descriptions in De Candolle's Prodromus as evidence of several other species being domatia-bearing : he also discusses a new form of domatium in some plants of the genus *Cestrum.*

A great part of the observations recorded below were embodied in a paper read at the Meeting of this Society in November, 1895, but as at this time I was not aware of Lundström's paper, I was allowed to withdraw it for the purpose of re-writing with a knowledge of that author's work.

The domatia that have come under my notice consist of hollows in the under surface of the leaf, and always occurring in vein axils. They are usually roofed over either by an extension of the leaf tissues, or by hairs. They are distinguished by peculiarities in the minute structure of the part of the leaf lying over them. Those that are known to me I divide into groups according to their outward structure as follows :—

Group i —Circular lenticular cavities on the under side of the leaf, each with a small opening and a thickened rim. Those found in *Pennantia Cunninghamii* present the highest development of this type which I have seen.

Group ii.—Pouches formed by a widening of the principal and lateral veins at the axils, the space being filled in with tissue so as to form a triangular pouch or pocket. To this group belong the domatia in *Dysoxylum Fraserianum.*

Group iii —Depressions or hollows formed by a thinning of the leaf substance at the axils. Of this type *Viburnum chinense* furnishes the best example.

Group iv.—Bunches of hairs in the axils proceeding from the principal and secondary veins, such as are found in *Rubus Moorei.*

Group v.—Thicker bunches of hairs at the axils in plants which have leaves coated with hairs, as in *Psychotria loniceroides*.

In groups i., ii. and iii. there may or may not be hairs in the domatium or round the edge of the orifice, or the hairs may be entirely absent. A regular gradation may be traced between all these forms, and it is sometimes difficult to say in which category a particular domatium should go. I think that this arrangement is the most natural, for as will be seen it is much the same as the consecutive steps in the development of the domatium of *Pennantia.**

Group. i.

PENNANTIA CUNNINGHAMII, Miers.—In this plant, Domatia probably reach their highest development   They occur most commonly at the first axils of the secondary veins, but are sometimes to be found in the axils of the principals and secondaries, and very often on the ramifying veins at junctions (fig. 1). They· vary from 9 to 50, and I have counted more than 100 on one or two leaves. They are very constant in occurrence. But some time ago I found two plants on opposite sides of a creek, and within a few yards of each other, in one of which every leaf had upwards of 30, while many on the other had none, and the majority only a few. But this was the only plant out of some hundreds which I examined that was in this condition. On recently visiting these plants, I find that all the young leaves on the plant formerly without domatia have them in normal numbers and perfect in development on the mature leaves. It will be seen

---

* Since completing this paper I have observed in *Mackaya bella*, Harv., (Acanthaceæ) rows of white hairs with crimson tips on the main and secondary veins at the axils ; and triangular pouches in the leaves of *Eupatorium riparium*, Regel. (Compositæ). This last is interesting, as it is an herbaceous plant, in which Lundstrom supposed domatia did not occur. I have also received, through the kindness of Mr. R. T. Baker, herbarium specimens of *Weinmannia paniculosa*, F.v.M. (=*Ackama Muelleri*, Benth.), which possess the most remarkable domatia I have yet seen. I hope to describe these hereafter when I am able to examine fresh specimens. They certainly differ from the five types enumerated above.

that the absence of cavities in some species is a by no means uncommon occurrence, and Lundstrom and Lagerheim note the same fact

The upper surface of the leaf is extremely glossy and dark green; the under side is duller and lighter in colour. When dipped in water, the upper wets readily, while the water gathers in patches on the under side, as if it were greasy. The pits appear on the upper surface as very distinct, though small, domed protuberances, circular or elliptical in outline (fig. 3); they are flatter on the under side (fig. 2). They vary in size in mature leaves from 1 to 3·5 mm in diameter (outside measurement), and the depth is usually two-thirds of the diameter. The opening is small and usually circular, and in the largest about ·75 mm.; it is surrounded by a thickened rim in which are vascular bundles proceeding from the veins between which it occurs; the rim is lighter-coloured than the rest of the leaf. The interior is usually lined with 1-celled hairs. Stomates occur plentifully on the lower side of the leaf, but they are absent in the domatia and on the upper side of the leaf. The pits are often inhabited by minute Acari, and their ova and excrement are also found in them. The mites sometimes quit the cavities and wander about on the under surface of the leaf. I have also seen similar Acari in the stomatal cavities of *Banksia*, in the rolled leaves of *Ricinocarpus*, &c., and in any other cracks or cavities suitable for shelter in plants.

The microscopic examination of sections of the domatia cut at right angles to the midrib and vertically, as regards the blade of the leaf, shows the structure described below, which is pretty constant in all the domatia I have cut. Beginning at the upper surface of the leaf, *i.e.*, on the upper leaf-surface there are—

(1) The cuticle, which is thin (fig. 5a).

(2) An epidermis composed of one layer of small oblong cells (fig. 5b).

(3) A single layer of hypodermal cells (5e) much larger than those of the epidermis, and from elliptical to oblong in shape, with thickening at the angles. These cells are very thick-walled, and

in other parts of the leaf have little protoplasmic contents; but over the dome they are richer, and often contain chloroplasts. From their varying appearance in leaves of different ages I believe this layer is derived from the next below.

(4) The palisade-tissue (5d) consisting of two rows of short oblong cells, their long diameter being horizontal instead of vertical. These cells contain many (up to seven) very large chloroplasts.

(5) A layer of spongy parenchyma (5f) containing also very large chloroplasts. In this particular region this tissue can scarcely be termed spongy, as it is composed of oblong cells laid over each other like bricks in a wall; but away from the summit the cells are branching and form the usual network, and the most open part lies all round the perimeter of the cavity. The dense layer over the roof is characteristic of the domatia in all the plants I have examined. In the lamina, at a short distance from the cavity, the intercellular spaces are arranged perpendicularly, and extend from the lower epidermis to the palisade-tissue above, the stomata opening as usual into the spaces. All through the leaf in this region there are cells not to be distinguished in a fresh section, but which stain very deeply with any stain, and more especially with hæmatoxylin, they become quite opaque before the rest of the section is sufficiently stained (5e). These cells are very rich in tannin, and with ferric chloride give a greenish-black reaction. There are also ducts in the vascular bundles of the veins on each side of the cavity filled with the same substance. The tannin-sacs are arranged in two parallel layers, one just under the palisade-tissue, the other at the bottom of the layer of spongy parenchyma, and resting on the inferior epidermis. All round the domatia the two layers coalesce, and then open out again the upper set going into the roof and the lower to the floor, and extending right to the rim of the mouth. The lower layer is less continuous, and thinner than the upper.

(6) The inner epidermis of the cavity (5g) continuous with and similar to the epidermis of the rest of the under side, but thinner.

(7) The inner cuticle (5h). Through this penetrate unicellular hairs (fig. 10) which are epidermal outgrowths, and are thick-walled and destitute of contents. They are rarely septate as shown in the figure, but usually resemble those of *Coprosma lucida* (fig. 11). This cuticle, as above remarked, has no stomata.

The same layers, omitting the palisade-parenchyma, are met with in the floor of the cavity, but in reversed order, and in the rim is a vascular bundle composed of five or six vessels.

On examining leaves of various stages of growth, I find that in leaves 5-9 mm. in length, the domatia appear as slight hollows. In leaves 1-9 cm. long I find the hollow deeper, and a little tuft of hairs in the angle. These are of two kinds: the ordinary pointed hair (fig. 10) and short thick ones composed of four almost globular cells. In a leaf of 4 cm. long a thickening is apparent along the sides of the veins, making a triangular pocket as in Group ii, and the hairs project from this. At 5 cm. long the thickening begins to extend across the mouth from the sides, so that there is a hollow surrounded by a ridge Up to this stage the whole of the under side of the leaf is a purplish-brown in colour, but the ridge is a very bright green. The ridge had grown higher all round in leaves 5-5 cm. long, and a few hairs had grown on the front part of the ridge, their points directed towards the centre of the hollow. In leaves 6 cm long the greater height makes the cavity appear much deeper At 8 cm. the ridge has reached its full height, and there are a few hairs on the outside of the ridge—simple and pointed. The domatia are completely formed when the leaf is 11 cm long, and no further alteration takes place except that in leaves a year old there are fewer hairs in the interior of the domatium. The leaves reach a length, when full grown, of 16 cm. and upwards. In examining a large series of young leaves, I found no Acari present until the domatium was fully formed. This fact has an important bearing on Dr. Lundstrom's theory of the meaning of the structures.

COPROSMA LUCIDA, Forst.—This plant also belongs to Group I. The domatia are very large and highly developed. They occur in the axils of the secondary veins and midrib, in pairs, or alternately. They vary in number from 3-8. They rarely occur in the forks of the secondary veins. The leaf is very dark green, and has a varnished upper surface; it is lighter in colour and duller below. It wets readily on the upper side, but is greasy on the under side. It is very thick, fleshy and soft, and the rim of the cavities does not project beyond the veins as in *Pennantia*. They show above as slight rounded projections and have a round orifice below, surrounded by a slightly thickened rim, the thickening being internal. Internally they are lenticular, 2-3 mm. in external diameter and the opening ·5-1 mm. The interior cavity is proportionately smaller than in *Pennantia*. The rim is lighter in colour than the rest of the under surface. The interior is lined with thick-walled unicellular hairs (fig. 11), and hairs of the same kind occur on the midrib below, sparsely on its upper surface, and very plentifully in the channel of the petiole in young leaves. A section of the cavity perpendicular to the plane of the leaf and across the axis of the cavity shows the following structure, beginning on the roof—the upper surface of the leaf :—

(1) The cuticle.

(2) The epidermis, composed of one layer of small elliptical or oblong thick-walled cells.

(3) A single hypodermal layer of oblong cells with thickened walls, and almost always without protoplasmic contents.

(4) The palisade-parenchyma, made up of four or five rows of oblong cells little longer than wide, and very rich in chromatophores, sometimes as many as 20 lining a single cell. Besides these, there is often a highly refractive globule, yellowish-green in colour, and like an oil drop, which dissolves in ether and is probably a resin or oil. The cells of the highest row are much larger than those of the lower ones, each succeeding layer being of smaller cells. The outer cells are here and there empty, and occasionally a whole row is in this state, and then, except for the vertical position, they resemble the hypodermal layer, and as in *Pennantia* the latter appears to be derived from them. Under this lies :—

(5) A thick layer of spongy parenchyma, arranged in a network, but very closely, and with few intercellular spaces, and these very small. The cells of this layer are small. At the sides of the cavity they are larger and looser in arrangement, so that the perimeter of the cavity is surrounded by this more open network of cells, which gradually passes into the ordinary spongy parenchyma of the rest of the leaf. Here the intercellular spaces are regularly arranged, and extend from the lower epidermis to the palisade-tissue. These cells also have very many chloroplasts, and those nearest to the palisade cells have the oil globules above mentioned. But there are none of the tannin-sacs noted in *Pennantia*, and in the densest part they are never arranged like brickwork as in that species.

(6) A single layer of epidermis, the cells thick-walled, and the cavity circular in outline. From this proceed the unicellular thick-walled hairs springing from much enlarged cells, and sometimes but rarely septate.

(7) The cuticle of the inside continuous with that of the lower side of the leaf. No stomata occur in the cavity, but they are found up to the very margin of the orifice. Vascular bundles occur in the spongy parenchyma all round the cavity.

In the floor of the cavity all these layers except the palisade-tissue and the hypoderma occur in reversed order. The development of the domatia in young leaves takes place much as in *Pennantia*, but the unicellular hairs appear later, only the 4-celled hairs being present at first.

The points of resemblance between *Pennantia* and *Coprosma* are the dense spongy parenchyma over the roof and round the cavity, and the epidermal hairs inside and at the mouth. The differences are the occurrence of tannin-sacs in *Pennantia* and not in *Coprosma*, and the non-occurrence of oil globules in the cells, and of hairs on the outside of the leaves in the former.

COPROSMA FŒTIDISSIMA, Forst.—I have seen dried leaves only of this and the following seven species, and am not able therefore to give particulars of the minute structure. In this species the domatia are in the axils of the second and third pairs of veins with the principal vein, and are from 2-4 in number.   They resemble those of *C. lucida* externally.

C. HIRTELLA, Labill.—These resemble the last, but are small.

C. CUNNINGHAMII, Hook. f.—The domatia are small but otherwise like those of *C. lucida*.

C. SPATHULATA, A. Cunn.—As might be expected from the small size of the leaf, the 2-4 domatia are very minute.

C. BAUERIANA, Hook. f.—Dr. Lundström, speaking from observation of cultivated plants, says that the domatia in this species are hairless.   I find that my notes afford no indication of whether hairs are present in the herbarium specimens I examined.   I have simply noted that they resemble those of *C. lucida*.

C. GRANDIFLORA, Hook. f.—The domatia are long and the openings slits parallel to the midrib.

CANTHIUM LUCIDUM, Hook. et Arn.—The pits are situated in the axils of the second pair of veins and the midrib, and rather high up in the forks.   They are two in all I have seen, but Mr. E. Betche informs me that they are often entirely absent.   The leaf is a very glossy one.   The openings are circular, about 1 mm. in diameter, the rim is raised and light-coloured, and vessels occur in it.   So far as I can see there are no hairs present.

C. OLEIFOLIUM, Hook.—The leaf is evidently fleshy, and in the 'Handbook of the Flora of N.S W.' is said to be "scarcely shining."   The pouches are situated in the axils of the first and second pairs of veins and midrib, and are slightly alternate, they are 4 in number, but as in the preceding species are not constant. The opening is triangular or circular, and the rim is thickened and contains vessels.   No hairs were seen in the interior.

RANDIA MOOREI, F.v.M.—The domatia in the leaf of this plant are minute.   They are in the usual position, and are four or five in number   There is a prominence on the upper side of the leaf, and the thickened rim round the orifice on the lower side forms a conical mound, on the summit of which is the small opening.   There are no hairs either round the mouth or in the interior.   The microscopic structure is rather remarkable.   There are :—

(1) The cuticle, which does not differ from that elsewhere on the leaf.

(2) A thick-walled epidermis, the cells often containing protoplasm

(3) A row of bottle-shaped cells, of very large size, arranged touching each other at their large ends, but with spaces between the necks, which point to the mesophyll (fig. 14a). This occurs over all the leaf.

(4) The palisade-tissue which fills in between the necks of the bottle-like cells and below them. This is moderately dense, and the cells full of chloroplasts.

(5) A layer of close spongy parenchyma, which in all parts of the leaf is penetrated a little above the lower epidermis by

(6) A layer of thick-walled apparently empty cells (fig. 14b), which stain very deeply, and are, I think, 4-armed, as whether sections are made parallel, or at right angles to the midrib, cut ends are seen, circular and thick-walled. Both these and the bottle cells give a bright purple with ferric chloride, and are most likely tannin-sacs as in *Pennantia*. In fresh sections both kinds of cells are transparent and colourless, but in old spirit specimens they are bright brown. This layer divides in the same way as that in *Pennantia*, one part going to the roof and the other to the floor of the domatium. Those above are of normal size, while those below are smaller and more scattered.

(7) The epidermis resembling that of the upper surface.

The roof and floor of the domatia are irregular, almost papillose, and stomata occur in great numbers on the elevations. Vessels are present in all the walls.

R. STIPULARIS, F.v.M.—The leaf is very large, thick, fleshy and shining, and has very thick veins. The cavities are small and closely covered inside with hairs like those of *Coprosma*. These all point towards the orifice, so that looking down into it a close mat of points fills up the opening. This last is small and elliptical.

The epidermis is thick-walled, the palisade-parenchyma is composed of 5 or 6 rows of small oval cells closely packed, the spongy parenchyma is also composed of oval cells, with small and few intercellular spaces.   The hairs have an enlarged cell at the base and are thick-walled and destitute of contents.

R. CHARTACEA, F.v.M.—In herbarium specimens imperfect domatia, and bunches of hairs were seen in the axils of midrib and secondary veins, but fresh leaves showed no sign of them. I cut sections through the axils and found a few minute hairs, but no approach to the characteristic structure described in the foregoing species.   I was struck, however, by the packing of large collenchyma cells on the upper side of the midrib and veins. These stained very deeply, and when tested with ferric chloride gave the same purple reaction as *R. Moorei.*

MORINDA JASMINOIDES, Cunn.—This is a climbing plant.   The cavities are usually high up in the axils of the third pair of veins and midrib.   They are opposite or alternate.   There are from one to four, but are sometimes absent.   The leaves are rather thin, dark green, but not very glossy.   The domatia project very much on the upper side of the leaf, and but slightly on the lower.   They are very large, and look like blisters or galls externally.   They vary from 1-5 mm. long.   The openings are sometimes of the full size of the cavity, but usually they are small and circular.   There is sometimes a ridge parallel with the vein, thus forming a channel leading to the orifice.   The rim is slightly thickened and lighter-coloured than the rest of the leaf. Many vessels occur in it and in the roof.   Ordinarily there are no hairs on the interior, which is quite smooth and has large stomata in all parts.   The minute structure, as seen in transverse sections, differs somewhat from that found in the previous plant.   Beginning as before at the summit of the roof on the upper surface of the leaf, we meet with :—

(1) A thin cuticle.

(2) An epidermis here composed of very large oblong cells with thin walls, the longer diameter being horizontal.   But over the rest of the lamina, the cells are longer vertically, and of great depth in proportion to the mesophyll.   The upper and lower

epidermis taken together are as thick as, or thicker than the layers between. The epidermal cells are very clear and free from contents.

(3) The palisade-parenchyma composed of two rows of very small oblong cells. the inner row smaller and rounder than the outer and very closely packed.

(4) A very dense spongy parenchyma, becoming more open near the domatium. Both this and the palisade layer are very dense all through the leaf and very full of chlorophyll bodies, so that it is difficult even in the thinnest sections to make out the structure. I found hydrate of chloral most useful in clearing the sections.

(5) The epidermis of the domatium, in two layers, the inner composed of larger cells.

(6) The inner cuticle, through which stomata open in all parts of the cavity. The same layers occur in reverse order in the floor, and running from the midrib and vein is an extension of the round strengthening cells which occur outside these.

The above is a description of the domatium in an ordinary healthy state. I have rarely seen Acari in them. But some time ago I came across a plant with very large domatia which were evidently in an unhealthy state, being pale or brown, or even black. On examining them, I found that all the unhealthy domatia contained numbers of Acari and their ova. Sections of these showed the palisade and spongy parenchyma cells greatly swollen and very irregular in shape, and undistinguishable from each other. Brownish patches occurred here and there, and also in places a number of cells had taken a bright crimson colour. In some of the cells of the mesophyll there was a deposit of granular matter on the walls. The epidermal cells were normal as to shape, but even larger than ordinary. Where ova rested on the interior of the domatium, the cells were dark-coloured and very closely placed. At the mouth, hairs of the same kind as in *Pennantia* were placed. In three sections from the same domatium I counted ninety-two ova, besides several young and mature Acari.

51

occur in .. ower par

same leaflet some veins are in pairs
the domatia thus are in pairs or sing
by me there were 14, 15 and 17.
and smooth, shining on the upper su
wets readily on this side, but on the
together and passes down the vein c
enter the domatia, as the orifice is to
by a widening of vein and midrib ru
and almost meeting in the centre (fi
sion leading into the domatium. S
meet, and then the mouth is circula
form in Group i. although it re
i. and ii. The thickened part is ligl
the under side of the leaf. There
surface, but they are plentiful below
found in the interior of the domat
in diameter. Vascular bundles are p
The interior is lined with stiff hairs
points all being directed to the ori
the domatia. The layers in a sectio

(3) The palisade-parenchyma, consisting of long cells, arranged in two layers, and very full of chloroplasts.

(4) The spongy parenchyma, denser here than elsewhere in the leaf, but yet more open than in *Pennantia* or *Coprosma*. It has a layer of tannin-sacs, but not very rich in tannin.

(5) The inner epidermis, thick-walled and with brown contents.

(6) The cuticle, through which project hairs, without stomata. The floor has cuticle, epidermis, spongy parenchyma (denser than that in the roof), epidermis, and outer cuticle. The stomata in the lower epidermis extend to the very edge of the mouth.

The brown contents of the epidermal cells are found all over the leaf, and appear solid and squarish in outline. The hairs of the domatium have also brown contents, often broken up so as to resemble a string of beads.

VITEX LITTORALIS, Forst.—Mr. E. Betche discovered that the herbarium specimens of this plant in the museum of the Sydney Botanical Gardens, collected in New Zealand by Mr. T. Kirk, have well marked domatia, but on examining the growing plant in the gardens none could be seen. Many domatia-bearing plants show this inconstancy, but I have not been able to trace the cause. It must be remembered, however, that young leaves show nothing but the depression in the angle, to the naked eye, or even to the hand lens. In this way I think it happened that a plant of *Hodgkinsonia* to be referred to was recorded as being without these structures. From the above causes I am compelled to speak only of dried material of this species. The opening is circular, the rim very much thickened, and the domatium projects beyond the surface of the leaf both above and below. They are placed in the main axils and are 4-8 in number. I attempted sections after prolonged soaking in glycerine with a little spirit, and succeeded in cutting them fairly thin, but the cells were much distorted, and I could only see that the arrangement of layers resembled that in other plants, and that there were no hairs in the cavity or round the orifice.

PSYCHOTRIA CARRONIS, C. Moore, et F.v.M.—I have seen only herbarium specimens of this plant. The domatia occur in the main axils and are very large, with a wide elliptical opening. I could see no hairs present anywhere.

P. CYMOSA, Ruiz et Pav.—The pouches occur in the principal axils and have a circular opening. The microscopic structure resembles that of *Coprosma lucida*, having the same dense layers of palisade-tissue composed of small cells very rich in chromato-phores. The hairs are different, being septate, with as many as thirteen divisions (fig. 9). They have little or no cell contents. There are no stomata in the cavities.

P. BISULCATA,   .—I have not seen this plant, and I am indebted to Mr. E. Betche for the information that Trimen (3) says of the leaf, "Lateral veins often with deep pits in their axils, which appear as warts on the upper surface."

The above-mentioned plants are all in which I have seen this highly developed form of domatium, but Lundstrom (1) describes a large number of other species which have it, mostly Australian.

Group ii.

DYSOXYLUM FRASERIANUM, Benth —The domatia are in the principal axils of the leaf or leaflet; and, so far as I have seen, this form never occurs in the secondaries. Sometimes they are found on only one side of the midrib, but generally on both. They vary in number from one to twelve. The leaves of a plant growing in a shady situation are very dark green and shining, on the under side lighter. From a sunny spot, they are much lighter in colour and smaller. The leaf wets readily on the upper side, but is greasy below. The pits do not appear much on the upper side of the leaf, but on the lower side they are very prominent, some-times projecting above the leaf surface 3 mm., and then are corky and diseased-looking. The size is on an average 3 × 2 mm. The opening is wide and arched (fig. 6). Vessels occur in the walls. The interior is hairy, the points of the hairs projecting from the mouth. There are no stomata inside the domatium.

The substance of the domatium roof consists of—(1) cuticle, (2) epidermis; (3) close palisade-tissue in two layers of very narrow cells, which are nearer the normal shape and arrangement

than any I have seen in other plants, (4) close spongy parenchyma, (5) epidermis; and (6) cuticle. Here and there in the
spongy parenchyma occur spherical interspaces of large size and
destitute of contents   In the diseased-looking domatia of great
thickness I found that the spongy parenchyma layer was of
greater thickness, the hairs absent, and the roof and floor
epidermal cells filled with a red substance which formed a thick
layer on both roof and floor.   I fancy that this diseased state is
caused by some insect (not a mite', taking up its abode in the
domatia as I repeatedly found remains in sections of some rather
large insect.   The mites were found in a few of the domatia, and
in all the domatia were found dust, pollen grains, and both spores
and mycelium of fungi.   It is rather remarkable that these should
be so plentiful, as from the mouth opening towards the apex of
the leaf, and the leaf itself having a horizontal position, they
could scarcely be washed in by rain, especially as they are on the
under side of the leaf.   I did not find such quantities of foreign
matter in any other domatia, even of those with orifices as large.
But Dr. Lundstrom notes the same kind of thing in many species
examined by him.

CEDRELA AUSTRALIS, F.v.M.—The domatia are like those of the
last plant, but flatter, stomata occur in the inside and there are
none of the spherical intercellular spaces mentioned above.

In very young leaves (10 x 1·5 mm) the under side of the leaf
is covered all over with hairs; as the leaf grows older, the hairs
drop off, except those in the axils where domatia are to form.
The hairs are of two kinds, pointed and thin, and short 4-celled
hairs filled with bright brown matter   These persist for some
time on the general leaf surface, and in the axils   They are
probably colleters.   In a leaf 10 x 3 mm. I found the hair tufts
and a slight widening of the veins in the axils, and in larger-sized
leaves the tissue widens progressively.   But the domatia have
not reached their full development even when the leaf is full
grown as to size.   It is only when the leaf has gained its mature
hardness and consistency that the process of growth in the
domatia is complete.

ELÆOCARPUS GRANDIS, F.v.M.

E. CYANEUS, Ait.

ELÆOCARPUS LONGIFOLIUS, C. Moore

{ The domatia of these three species resemble those already described under *Dysoxylum* and *Cedrela*, but *E. grandis* has very long slender hairs, and *E. cyaneus* has none.

HODGKINSONIA OVATIFLORA, F.v.M.—Herbarium specimens of this plant showed very distinct bunches of hairs in the axils, especially one taken from a cultivated plant in the Sydney Botanical Gardens. But on examining fresh leaves (young) from the same plant, no hairs could be seen with a hand lens. I cut some sections of the axils, however, and found that a very small hairy depression did exist, and examination of numbers of sections revealed a slight extension of tissue from vein to midrib. I have no doubt, therefore, that mature leaves of the plant would show that it should be placed in Group ii. The hairs are few in number, straight and septate.

VITIS BAUDINIANA, F.v M.—I have placed this form in Group ii. because though ordinarily it presents a marked difference in the shape of the triangular pouch, yet it is a modification of that shape, and in addition, all stages may be found from the triangular form almost to the sunken cavity with a circular orifice They occur in the axils of the lateral veins and midrib, but are frequently present in the secondary vein-axils also, two or three on one vein. At the base of the leaf there are on each side of the midrib, first a small, and next a large lateral springing from the insertion of the petiole, and here are found four large domatia In the whole leaf they vary from 8 to 30, or probably many more I have never found them entirely absent in any leaf. The leaf is hairy, more especially on the under side and in the younger stages It is easily wetted on both sides, but the water runs into greasy patches on the upper surface. The domatia are formed by the extension of tissue from the midrib and vein, but in the middle the extension grows out into a point, which arches over the mouth (fig 7). In the centre, too, there is a dome formed by the arching of the tissues. There is also sometimes a closed-in cavity on each side of the domatium. This I have seen in *Morinda jasminoides* also. The domatium is 2 mm. high, and the transverse measurement 2·5 mm. in large specimens. The interior is thickly lined with thin cottony hairs, and there are besides stalked T-shaped hairs (fig. 8). Stomata are found only in the lower epidermis, and do not extend to the cavity. I have often found in the domatia small hemipterous insects, which apparently are in the habit of frequenting the cavities, for when driven out of one they go straight to another.

The microscopic structure is much like that in *Dysoxylum*. The palisade-cells occupy half the thickness of the leaf. There is no thickening or thinning of the leaf blade at the domatium, but it curves upward slightly, showing a slight protuberance on the upper surface. Vessels occur in the domatium walls. It is difficult to make out the domatia in young leaves on account of the thick felty layer of hairs But even in the bud stage I could make out that the tissue extension is present. I have not seen this so early in any other plant.

Group III.

VIBURNUM CHINENSE, Hook.—The depressions are large and occur in the axils of midrib and veins. They are 6-14 in number The leaf is thick in texture, light green, but not glossy. The depression is formed by a thinning of the leaf substance, and has sloping sides and an irregular surface. There is a slight thickening of the leaf all round the hollow (fig. 13), and on this and the elevations are tufts of light brown and curled hairs They are thick-walled, and their contents are arranged in globules like a string of beads. On the thinner veins where there are no domatia a few rows of straight hairs grow. The hollows are about 2 mm. in diameter Stomates occur on the lower surface of the leaf and in the hollows. The minute structure is as follows —(1) Cuticle, (2) epidermis of the upper surface with thick walls, the cells containing a considerable amount of light green chlorophyll, (3) palisade-tissue very full of large chromatophores, passing gradually into (4) a very loose spongy parenchyma also rich in chlorophyll, the cells large in size, and staining deeply; (5) a thick-walled epidermis sometimes having brown contents as in *Tarrietia*, out of which grow the hairs, two, three or more hairs springing from one cell (fig 13); (6) the cuticle with stomata.

SLOANEA WOOLLSII, F.v.M.—The depressions are in the axils of the midrib and laterals, and begin at the lowest pair   They number 15-21, and are minute—1 mm. in diameter.   The leaf is hard in texture and smooth; it wets readily above, but on the under side the water runs into patches.   There is not such a decided thinning of the leaf as in *Viburnum*, but the thickened rim runs all round, and few hairs grew on this.   Stomata are found on the under surface, but, so far as I can see, none extend to the hollow.   The microscopic structure is as in the last-named species, except that there are no deeply staining cells, and the spongy parenchyma becomes very dense over the roof.

GARDENIA sp.—In a commonly cultivated species of this plant I found depressions filled in with long straight hairs springing from the vein and midrib : they are roughened on the surface, septate, and have green or brown contents at the tip.   Stomata occur in the pit.

Group iv.

Examples are seen in *Hydrangea hortensis*, Sieb., *Morinda citrifolia*, Linn., and *Mandevillea* sp.hort.   There is nothing resembling the microscopic structure of the cavities, etc., to be seen in these.   The cells from which the hairs spring in *Mandevillea* are bright crimson.   I have also seen them in *Prunus Lusitanica*, Linn., *P. domestica*, Linn., *Rubus Moorei*, F.v.M., *Solanum* sp hort., and some other plants, but I have not made sections of these.

Group v.

The only plants which I have seen, hairy all over but having a thicker tuft in the axils, are *Psychotria loniceroides*, Sieb., and *Diploglottis Cunninghamii*, Hook. f.

I have described the domatia of the above-named species fully as types of the structures in question. The following list of domatia-bearing plants which I have myself examined is arranged according to Natural Orders. I have followed Baron von Mueller's arrangement in the Second Systematic Census of Australian Plants.

MELIACEÆ.

Dysoxylum Fraserianum, Benth. .... ii.

Synoum glandulosum, A. de Juss.. .... ii.

Cedrela australis, F.v.M....... . ... ii.

STERCULIACEÆ.

Tarrietia actinophylla, C. Moore .... ι

TILIACEÆ.

Elæocarpus cyaneus, Ait. .....:.................... ii.

grandis, F.v.M........................ ii.

obovatus, G. Don. .... ii.*

Sloanea Woollsii, F.v.M....... .... iii.

SAPINDACEÆ.

Diploglottis Cunninghamii, Hook. f. . .. . v.

Nephelium foveolatum, F v.M. .... ii.

Beckleri, Benth........................ ii.

Harpullia Wadsworthii, F.v.M. .... ii.†

ROSACEÆ.

Rubus Moorei, F.v.M............................... iv.

Prunus Lusitanica, Linn. .. .. iv.

domestica, Linn.......... . ... iv.

---

* Probably the species *E foveolatus* was named from the presence of domatia. I have not seen it.

† *Cupania foveolata*, F.v M., is described as having dimples in the axils.

SAXIFRAGEÆ.

Hydrangea hortensis, Sieb..........................   iv.

VINIFERÆ.

Vitis Baudiniana, F.v.M. ....   ii.

ARALIACEÆ.

Panax elegans, C. Moore et F.v.M. ....   ii.

OLACINEÆ.

Pennantia Cunninghamii, Miers ....   ɪ.

RUBIACEÆ.

Gardenia sp.hort.	iii.
Randia chartacea, F.v.M.	i.
Moorei, F.v.M.	i.
stipularis, F.v.M.	i.
densiflora, Benth.	iv.
Hodgkinsonia ovatiflora, F.v.M	iv.
Canthium oleifolium, Hook........................	i.
lucidum, Hook. et Arn.	i.
Morinda citrifolia, Linn........	iv.
jasminoides, Cunn........................	i.
Psychotria cymosa, Ruiz. et Pav.	i.
loniceroides, Sieb.	v.
Carronis, C. Moore et F.v.M.	i.
Coprosma lucida, Forst.	i.
robusta, Raoul	i.
grandiflora, Hook. f...................	i.
Cunninghamii, Hook. f...............	i.
foetidissima, Forst....................	i.
hirtella, Labill.	i.
Baueriana, Hook. f.....................	i.
spathulata, A. Cunn.	i.

CAPRIFOLIACEÆ.

Viburnum chinense, Hook...... .... iii.

APOCYNEÆ.

Mandevillea sp.hort.............. .... iv.

SOLANACEÆ.

Solanum sp.hort............ .. .. iv.

BIGNONIACEÆ.

Tecoma Capensis, Lindl........... .... iv.*

VERBENACEÆ.

Vitex littoralis, Cunn.............. .... ι.

I have counted the species of domatia-bearing plants in each order in Lundström's, Lagerheim's, and this paper, and arranged them in descending order.

Rubiaceæ, 107 ; Tiliaceæ, 40 ; Bignoniaceæ, Oleaceæ and Lauraceæ 16 each; Cupuliferæ, 15; Solaneæ, 13, Apocyneæ, 12; Rhamnaceæ, Aquifoliaceæ and Juglandiaceæ, 6 each; Loganiaceæ and Anacardiaceæ, 4 each; Caprifoliaceæ, Bixaceæ, Meliaceæ, and Rosaceæ, 3 each , Compositæ, Ribesiaceæ, and Hamamelideæ, 2 each ; Asclepidiaceæ, Sapotaceæ, Aceraceæ, Myrtaceæ, Magnoliaceæ, Ulmaceæ, Platanaceæ, Sterculiaceæ, Olacineæ, Araliaceæ, Viniferæ, Saxifrageæ, and Verbenaceæ, 1 each. From the above it will be seen that the orders Rubiaceæ and Tiliaceæ are far before the others in domatia-bearing species

There are, however, included in Dr. Lundström's list some plants which are only doubtfully possessed of these structures, and one or two which certainly are not. To take the latter first

TECOMA AUSTRALIS, R.Br.—Dr. Lundström says (1, p. 37)— This plant "has 1-3 dimples which are (always ?) inhabited, but

---

* Remarkable as having branching hairs in the axils

ur in quite an indefinite manner on
cannot assert with positive certain
atia, but I may commend these pa
xamination by those biologists who
ring them in the open." The struc
a number of plants, *e.g.*, *Cedrela*
nd many indigenous Rutaceæ. Tl
, and when young the edges overl
rse section the appearance of such a
osma. But the whole cavity is fillec
tical gland, flat-topped, shining wit
l. Sometimes in old leaves the gla
tly dried up and fallen out. In *C*
n the veins, usually near the top o
e I found one in the hair-tufted d
t stage of a domatium. But ordin
e veins, and I could not find any r
currence. Acarids are sometimes f

ILEX spp.—Dr. Lundström describes backward curls of the edge of the leaf near the base, forming a cylindrical room, and found here the cast skins of mites. But so far as dried material could show, there was not the peculiar structure found in domatia. I have found in *Eupomatia laurina* similar structures, but could find no mites or traces of them.

SCHINUS spp.—These have a wing on the rhachis provided with a small tooth on each side at the insertion of the leaf, which folds over and forms a cavity. I am inclined to think that none of these structures are true domatia, and would restrict that term to cavities or depressions in the leaf surface showing the peculiar appearances described under the types I have taken. But under Dr. Lundstrom's definition of a domatium, viz., all those structures of plants which act as dwellings or shelters for insects and receive in turn some benefit from the latter, all these might be included.

Dr. Lundstrom classifies domatia into the following five groups —(1) Hair tufts at axils; (2) bending back or folding of leaf or edge of rachis; (3) dimples with or without hairs; (4) small pockets; (5) bags, &c His group 1 corresponds with my group 4, his 3rd with my 1st, and 4th with my 2nd. His 2nd and 5th groups I have not taken to be domatia, and he does not particularly notice my 3rd or 5th groups.

I have arranged the groups of types as shown because it indicates the order of development—beginning with the highest. The domatium usually begins either as a small hair-tuft or a depression. Then an outgrowth from the veins begins extending right across the angle Later a ridge thickens up across the open angle and runs round to the sides, so that when all the parts are grown to full height a circular orifice is formed This is well seen at times in *Vitis Baudiniana*, which usually has the triangular pouch, but at times forms the circular cavity in this way. As the order of types, beginning with the 5th, represents the development of the domatia in a single plant, so also it probably brings before us the order of evolution.

So far as I have looked into the matter, it appears to me that domatia are most common in plants of a southern origin. At any rate, it is certain that all the species having very perfect domatia are so, and of these New Zealand, Lord Howe Island and Australia supply a large proportion.   Out of 41 species named in my original paper, 32 were from these localities.

The most interesting question, however, is, What is the meaning and purpose of these structures?   And it is a difficult one to answer.   The first possibility that occurred to me was that they were pathological in their nature.   But prolonged observation of the plants and study of sections convinced me that they were not so.   I have seldom seen the slightest appearance of disturbance of the tissues which form the walls and roof.   In Packard's Forest Insects (4) p. 554, there is a figure of a section of a *Phytoptus* gall from *Fraxinus viridis* which in outline resembles a domatium. But it is only in general features that this resemblance holds—it is quite different in details.   *Phytoptus*, too, spends its life in the gall and can always be found there.   But when a *Phytoptus* is found in a domatium, Dr. Lundstrom observes and my own experience coincides, that pathological changes are always present. In *Panax elegans* and *Morinda jasminoides* I found domatia with many Acarids—not *Phytoptus*—were diseased and altered, but it was from the mites destroying the epidermis in the one case, and in the other the appearance of the tissues was completely changed from the normal state.   At the same time I found leaves with fully formed domatia, on the same twig and even at the same node, opposite, which had no mites and were quite normal. There was no doubt, therefore, that the diseased state was induced in normal tissue by the insects.

It next occurred to me that they might be glands, and as many of the plants bearing the highest type of domatia have extremely glossy leaves (*e.g.*, *Coprosma lucida*), it seemed possible that they might secrete resin or varnish.   But an examination of all stages of growth showed no secretion of any sort, nor did the structure resemble that of any gland I know of, so that I abandoned this line of inquiry.

It was suggested to me that the domatia-bearing plants may at one time have been like *Banksia* and *Nerium*, i.e., that their stomata were contained in crypts in which they were sheltered from excessive transpiration by long hairs, and that under altered climatic and other conditions the stomata passed out to the general surface, leaving the pits as relics of the former state of affairs I made a careful examination of several species of *Banksia* and of *Nerium*, but found the crypts of a totally different character, and in addition, in both genera, the crypts are evenly scattered all over the surface, while in the species under consideration they occur only in the axils of the veins, or rarely (*e.g.*, *Pennantia*) on the course of the veins and appear to have a definite relation to those organs.

Again, the solution was offered that they might be extra growths caused by the superabundance of sap at the axils. But the fact that they are found mostly in the middle axils on the midrib, and not on the lower ones, where the sap would naturally be more plentiful, bears against this, and their regular organisation and appearance I think sufficiently negatives this theory.

The purpose which seemed to me most feasible, and which I took most pains in working out, was that they might perhaps be organs for absorbing gas, vapour or water, and this seemed all the more likely from the fact that the plants possessing them are all inhabitants of moist climates, New Zealand, Norfolk and Lord Howe Islands being their head quarters. Careful experiment showed that they would not fill when the leaf was wetted, the small opening being stopped by an air bubble, nor could I, even by prolonged submersion, succeed in filling them. To be sure I was not mistaken, I tried an alcoholic stain (as it flowed freely and would leave the epidermis stained as a record) and even mopped the cavities out with alcohol to encourage capillary action, but still the liquid would not run in. Mr. Betche tells me he succeeded in filling the pouches of *Dysoxylum Fraserianum* by immersion for some hours, and he thinks the fact that dust is often found inside is an additional proof that rain does run in and carries with it foreign matter. Their position on the under side of the leaf, too, is to some extent unfavourable for their filling, so that on the whole I had to abandon the hypothesis. I also tried experiments by waxing cut petioles, letting the leaf wilt and then weighing and immersing in water, at the same time treating leaves of the same area, weight and consistence in a similar way. I found that both kinds of leaves gained in the same ratio, from 5 to 20 per cent., so that the domatia-bearing leaves had no advantage. I tested them in the same way for the absorption of vapour in closed moist chambers, in sunlight, diffused light, and darkness. The results were contradictory in both kinds of leaves. Some gained 1 per cent., and others lost as much or more. At that time I was under the impression that stomata did not occur in the pits, but as has been shown, this was a mistaken view. For want of a quantity of material I did not experiment on these plants. But it must be remembered in this connection that the stomata are in no way different nor more abundant relatively in the pits, and as there are thousands on the free surface of the leaf, no great advantage could accrue from the presence of a few in sheltered pits.

Dr. Lundström in considering their use took up the possibility of their being connected with motile phenomena, but found that untenable. He also considered them as being perhaps insect traps, but was compelled to abandon that view also, as the mites go in and out freely. In this my experience coincides with his.

The final conclusion he came to was that we have here an instance of symbiosis between the plants and the mites, and he thinks that the production of incomplete domatia has become hereditary in these plants, the stimulus given by the arrival and presence of the mites causing the final development of the domatia. He was led to this from observing the almost universal presence of mites in the cavities—in which I cannot say that my experience coincides. I find mites sometimes, but just as often not, and in the two instances in which I found large numbers (before referred to) I found the domatia damaged by them. He claims that mites of the type figured by him do not damage the cavities, but that *Phytoptus* mites do. But in both the instances I speak of the mites were remarkably like those figured, and most certainly were not *Phytoptus*. In answer to the question of what benefit these little animals may be to the plant, he says they eat, and as a consequence excrete and give off gases, and he thinks it probable that the excreta and gases are absorbed by the plants, which are thus benefited. He also speculates as to whether certain crevices observed in some fruits may not be domatia to shelter the mites till the young plant grows and gives them the leaf-domatia. Still another service they may do is that they may eat the spores and mycelia of noxious fungi which rest and germinate on the leaf, and in support of this he mentions having seen minute rings which were undoubtedly the chewed mycelia, and also digested spores in the excreta. Some of the strongest evidence he has to offer in favour of there being a relation of mutual helpfulness between the two is as follows.

Speaking of *Psychotria daphnoides* he says: "I have kept a specimen of this species for six years in a dwelling room When it was brought thither the domatia were for the most part inhabited, but afterwards the mites almost entirely disappeared, partly because they were swept off with a brush, and partly banished by smoking. It was curious to observe how the unin-habited domatia on the new sprouts altered by degrees, the hair formation almost entirely disappeared, the opening widened, and the inside of the domatium passed into a shallow cup-shaped depression . . . . On some leaves the domatia have almost entirely disappeared, and the epidermis in the vein-axils has by degrees assumed the same appearance usual to the under side of the leaf. At the same time the domatia which remain inhabited retain their normal form. From these facts, it may, in my opinion, be inferred that when the corresponding organs on a sprout find no opportunity for action, *i.e.*, do not become inhabited, the domatia on the following lateral sprouts become more and more rudimentary till they disappear. Whence it follows that the importance of the domatia depends on the little creatures inhabit-ing them " (1, p. 15)

Speaking of the protoplasm in the cuticle of the domatia walls : "It remains to examine more closely how this protoplasm behaves in cells which lie under the excrement of mites; in some sections

52

it seemed considerably browner and thicker, in others again it was not distinguishable from the plasma of cells which were not covered with masses of excrement. . . . Through examination of consecutive sections of an inhabited domatium, I have proved that the inner wall is quite unhurt, not injured by punctures or bites " (1, p 20).

Again, under *Laurus nobilis* :—" On a specimen about 2 met. high which I have kept six years in a room, and from which the mites have been removed partly by smoke and partly by means of a brush, the domatia have become by degrees indistinct, and indeed have quite disappeared from certain boughs. It has been distinctly proved by this, that where mites are absent, there the domatia have not attained their normal development and size, so that the full development of the domatia is in necessary connection with the presence of mites " (1, p 49).

By means of carefully planned culture experiments, he attempted to prove that the domatia only came after the arrival of the mites, but partially failed, as the resulting plants did produce domatia, although fewer in number, smaller and poorer in hairs than normally. On p. 61, he says it has been plainly proved that the domatia in *Psychotria*, *Tilia*, *Laurus* and others can only reach their full development in the presence of mites, and that these being absent, the domatia do not develop fully.

After prolonged consideration of the subject, I cannot consider Dr. Lundstrom's theory as perfectly explanatory of the use of these structures, although I must acknowledge that I have no better solution to offer. Some of the points which have occurred to me as being against his view follow.

The mites are not always to be found in wild plants; even when the domatia are fully developed, they are often absent. Dozens of domatia may be searched and no mites found. In examining large numbers of leaves of *Pennantia Cunninghamii* I found none present in the earlier stages of the development, which is just the time when their presence is needed. I find them also in the rolled leaves. such as *Ricinocarpus pinifolius,* and in the stomatal crypts of *Banksia,* and they seem to be just as much at home there as in the domatia. I have often seen them in cracks and crevices of the plant, as between bud-scales, or in the chink between a petiole and a stem, as has Dr. Lundstrom himself. But I do not think that it is necessary to consider any of these places as dwellings specially prepared for the mites. Indeed Dr. Lundstrom uses an apt illustration of this very point when he says it would be as reasonable to consider a wood where a hare was started as a dwelling specially formed for the hare. The fact that the two plants in which I found great numbers of mites had in the one case diseased and in the other damaged domatia is very important, especially as they were not the hurtful mites, but of the same kind as those figured as domatia-dwellers. Again Dr. Lundström takes the fact of the leaves containing most domatia being very luxuriant in growth and very healthy as proving the benefit derived from the mites. But is it not possible that the Acarids might be attracted by those very states?

On the whole, therefore, while not denying the possibility of Dr. Lundstrom's view being the right one, I am of opinion (and I set forth my opinion in opposition to that of so good an observer with considerable hesitation) that the whole question needs much further observation and research. The following points need special attention :—

(1). The development of the tissues in all stages of the formation of the organs.

(2). The careful determination of the species of mites found in each species of domatia-bearing plants (a) in a state of nature ; (b) in plants cultivated in different countries.

There also remains much to be done in the discovery of other domatia-bearing plants, and in the habitat in which each is found.

I should have mentioned that I have never been able to find either in specimens or in figures of fossil leaves any appearance of these structures.

But Mr. Henry Deane informs me that from Gippsland he has some fossil leaves of a Coprosma-like plant which apparently show decided prominences in the principal vein-axils. As this is the invariable situation of domatia in that genus it is not improbable

that they may be these organs.   That they are of great antiquity
I have no doubt.

I have to thank three lady friends for translating Dr. Lundström's valuable memoir, and also Messrs. E. Betche, J. J.
Fletcher, and J. P. Hill for very material assistance.

REFERENCES.

(1) LUNDSTROM, DR. A. N.—Pflanzenbiologische Studien ü. Die
Anpassungen der Pflanzen an Thiere.   Nova Acta Reg.
Soc. Sc. Upsal.   Ser. Tert.   Vol. xiii., Fasc. ii. (1887), No. 6.

(2) LAGERHEIM, G. DE—Einige neue Acarocecidien und Acarodo-
matien.   Berichte der Deutschen Botanischen Gesellschaft.
Band x., Heft 10, 1892, p. 615.

(3) TRIMEN—Handbook of the Flora of Ceylon.

(4) PACKARD, DR. A.—Forest Insects.   Fifth Report of the U S.
Ent. Comm.

EXPLANATION OF PLATE.

Pennantia Cunninghami (Figs. 1-5).

Fig. 1.—Part of leaf showing arrangement of domatia on veins.
Fig. 2.—Domatium from under side of leaf (× 4).
Fig. 3.—Hair from interior of domatium (× 120).
Fig. 4.—Outline of section of domatium (× 20).
Fig. 5.—Section of roof of domatium (× 120).   .
        a, cuticle ; *b*, epidermis ; *c*, hypodermal layer ; *d*, palisade
        tissue ; *e*, tannin-sacs ; *f*, spongy parenchyma ; *g*, lower
        epidermis; *h*, cuticle.
Fig. 6.—Hair of domatium, *Coprosma lucida* (× 120).
Fig. 7.—Pouch-shaped domatium of *Dysoxylum Fraserianum* (× 10)
Fig. 8.—Outline of section of domatium, *Dysoxylum.*
Fig. 9.—Outline of section of depression in leaf of *Viburnum Chinense*
      (× 10).
Fig. 10.—Tuft of hair in axil, *Synoum glandulosum* (× 10).
Fig. 11.—Hair of domatium, *Psychotria cymosa* (× 120).
Fig. 12.—Section of leaf. *Randia Moorei*; *a* and *b*, tannin-sacs.
Fig. 13.—Hair, *Vitis Baudiniana* (× 120)

NOTES ON TWO PAPUAN THROWING STICKS.

By J. JENNINGS.

(Communicated by C. Hedley, F.L.S.)

(Plate LVIII.)

Preceding volumes of these Proceedings contain a series of articles by Mr. R. Etheridge, junr., describing and figuring in detail numerous varieties of the womerah or Australian throwing stick.*

Only in recent years has it been announced that a like implement is also employed by the Papuans of Northern New Guinea. Finsch figured and described† a specimen which he collected at Venushuk, New Guinea, and Edge Partington illustrates, apparently by a copy of Finsch's figure, this throwing stick ‡ Ratzel in the Natural History of Man also gives figures.§

By far the fullest account of the Papuan form of the throwing stick, however, we owe to Dr. F. v. Luschan, who in "Das Wurfholz in New Holland und in Oceanien," Bastian Festchrift, Berlin, 1896, pp. 131-155, Pl. IX., X., XI., has dealt exhaustively with the subject. Specimens of the Papuan type which have lately been acquired by my friend Mr. Norman Hardy do not exactly coincide with any portrayed by Dr. v. Luschan. I have therefore obtained permission to lay before the Society the following account and accompanying drawings of two specimens, the

---

* Series ii. Vol. vi. p. 699, fig.; Vol. vii. pp. 170, 399, Pls. III. and XI.; Vol. viii. p. 300, Pl. XIV.; Macleay Memorial Vol. p. 236.

† Ann. K.K. Hofmus. Vol. iii. 1888, Pl. xv. f. 5.

‡ Ethnographical Album, 189, Ser 1, Vol. ii. Pl. 37, f 1.

§ Ratzel, "The History of Mankind," English Ed. I. 1896, p. 181.

former of which is said to have come from Berlin Harbour, German New Guinea; the second is without a history.

The first weapon (fig. 1) is made from a piece of nearly straight bamboo, weight 6½oz., 2ft. 2in in length and barely one inch in diameter, embracing three nodes. At a distance of 2¾ inches from the distal end and half an inch from a joint, a transverse incision has been made through two-thirds of the diameter, thence the cut gradually and obliquely ascends to the upper surface, terminating at a point 11½ inches distant, the whole incision somewhat resembling what is technically known to carpenters as a scarf Two inches in front of the above-described incision a slot 4½in by ½in. wide has been excavated for the reception of a piece of hard wood richly carved in high relief and inclined at an acute angle towards the distal end, which evidently was intended as a rest for the spear when being aimed and thrown. To retain this in its place are two rings of split and interwoven bamboo, two inches apart, these being in their turn held together by means of strands of fibre   The entire carving is eight inches long, two broad, and half an inch thick, and the design that of a conventionalised crocodile, the head, body and tail being suggested by appropriate segments.

The head is portrayed with a considerable degree of accuracy, the nasal prominences and eyes being carefully located, on the body seven imbricating scales indicate the dorsal scutes, three concentric grooves divide the sides into oval ridges; on the dorsal surface of the tail scutes are again suggested by a different method of treatment, while the sides harmonise with the body. In dealing with the ventral surface, the carver has allowed his artistic faculty unrestricted scope, the teeth of the reptile being indicated by curved bars which unite the upper and lower surface, the last bar being carried in a bow from the neck to the tail and offering a grasp for the previously mentioned bamboo rings Distal of the spear socket one inch is ornamented by a pattern of a series of circles and conjoined loops containing lozenges; distal of this again it has been cut down so as to make a neck terminating in a knob.

The second weapon (fig. 2) is similar in construction to that above described, but is somewhat longer, being 32 inches from end to end and weighing 4½oz. Rather more than 2 inches from the distal end a sloping groove, as in the previously described implement, has been cut for a distance of 15¾ inches, not as in the first instance in a plane with a carved rest, but inclining to a considerable degree towards the right, thus indicating the side on which the spear was held The carved wooden projection against which the spear was rested is 7 inches long, inclines at the same angle and in the same direction as the former, and is attached to the bamboo shaft at both ends by means of woven bands of split bamboo, midway between which is a third and lighter band. This highly interesting feature differs very much in character from fig. 1, being much flatter, carved in lower relief, and is more conventional in design.

An elongated human (!) head on the upper end is directed from a proximal in a distal direction by a curved and pierced band connected with the body of the implement, this surrounds two intersecting pierced ovals which are proximally attached to an irregular elongated triangular body of which the upper or dorsal edge is unevenly serrated and pierced, the whole forming an acute angle with the main body of the instrument, the flattened sides are decorated in a design formed by successive curved bands, chevrons and dots carved in low relief. A handle convenient for grasping is afforded by a finely plaited bamboo knob or bulb which is fastened in its place by a strong wooden peg. The distal termination is in its main character like that of fig. 1, but for a distance of 2 inches is carved in a series of bands, chevrons and dots harmonising in design with the flattened sides of the spear rest.

Some ethnologists have traced a connection between the Australian Aborigines and the Dravidians of India. It has been suggested to me by my friend Mr. C Hedley, F.L.S, that the isolated occurrence of a womerah on the north coast of New Guinea may indicate a vestige of the emigrants on the line of march, for it is even possible that while the identity of a race might have slowly disappeared through intermarriage, yet a custom or weapon could have descended unchanged. On the other hand, the throwing stick is not the exclusive heritage of the Australians or their kin; indeed, it may have been independently arrived at by various peoples.

The Papuan implement is broadly distinguished from any of the numerous aspects assumed by the womerah in Australia In the former case, the spear end is received into a socket, in the latter the spear is cupped to receive the peg of the womerah * Again, the former is remarkable for the raised, ornately carved crest against which Dr. Luschan states the spear rests, for which no homologue occurs in the Australian type

The Micronesian form may be described as like the Papuan, but without the raised spear rest; in Micronesia Keate† long ago described it from the Pelews, and Luschan figures it both from that Archipelago and from tho Carolines.‡　A mechanical device for propelling spears from a loop of rope has been recorded from New Caledonia §　The Esquimaux possess a form of the throwing stick which has been described at length by Otis T Mason,‖ mention of the use of this instrument by natives of the Polar regions has also been made by Nordenskjold¶ and Nansen.**　Lieutenant W. H. Hooper mentions them as being nsed by the Esquimaux of Icy Reef, Humphreys Point.††

---

* Nevertheless, Mr. Harry Stockdale has informed me that he has observed an exception to this rule in the case of a Northern Territory (Australia) tribe who used a socketed womerah.

† Keate, "An Account of the Pelew Islands," 1788, p. 314.

‡ *Loc. cit.* pp. 133, 132, fig. 9.

§ Edge Partington, *loc. cit.*　Second Series, Pl. 67, f. 11.　Stevens "Flint Chips," 1870, p. 304.

‖ Smithsonian Institute, Ann. Rep. 1884, Part 2, pp. 279-289, Plates.

¶ "Voyage of the Vega," London, 1881, Vol. 2, p. 105, fig. 5.

** "First Crossing of Greenland," Vol. 2, pp. 263, 340.

†† "The Tents of the Tuski," London, 1853 p. 259.

The Central and South American throwing sticks have been dealt with in a most thorough manner by Dr. Ed. Seler* in a paper entitled "Altmexicanische Wurfbretter," which is finely illustrated both by woodcuts and coloured plates. Dr. Hjalmar Stolpe in the same publication† communicates a valuable article on the subject, and furthermore gives illustrations of the weapons used by the Tecunas, Canibos, Quito, Campevas and Chambiriguas tribes of South America, in all ofwhich the spear is kept in place by a peg.

———

EXPLANATION OF FIGURES.

The right hand division of the plate constitutes fig. 1; the left, fig. 2.

---

* Internationales Archiv für Ethnographie, 1890, Band iii. pp. 137-148, Pl. xi.

† Loc. cit pp. 234-238.

OBSERVATIONS ON THE EUCALYPTS OF NEW SOUTH WALES.

By Henry Deane, M.A., F L S., &c , and J. H. Maiden, F.L.S ,&c.

(The Illustrations by R. T. Baker, F.L.S.)

PART II.

(Plates LIX.-LXI.)

The Eucalypts dealt with in this Part fall naturally into a single group distinguished chiefly by the fibrous and tenacious character of the bark, and to some extent by the length and straightness of the grain of the wood. They go under the vernacular names of Stringybark and Messmate

The Stringybarks proper are *E. capitellata*, *E. macrorrhyncha.* and *E eugenioides* These three species, although in their typical forms so distinct, have connecting links, and in the case of some of these varieties it is often difficult to decide under which species to place them

E. obliqua and *E fastigata* (Cut-tail) are Messmates, the former being sometimes called Stringybark.

There is another species in New South Wales, *E. Baileyana*, which is said to be called Stringybark, but we propose to postpone consideration of this species. as well as of other fibrous-barked species, *E. acmenoides*, White Mahogany (sometimes called Stringybark), *E. microcorys*, Tallowwood, and *E. pilularis*, Blackbutt, to some future occasion.

EUCALYPTUS CAPITELLATA, Sm.

Smith's original description and notes on this species are as follows :—

"*Eucalyptus capitellata*, operculo conico calyceque anguloso subancipiti, capitulis lateralibus pedunculatis solitariis

"Lid conical, and, as well as the calyx, angular, and somewhat two-edged. Heads of flowers lateral, solitary, on flower-stalks.

"The *leaves* are ovate-lanceolate, firm, astringent, but not very aromatic We have seen no other species in which the *flowers* stand in little dense heads, each flower not being pedicellated so as to form an umbel. The *lid* is about as long as the *calyx*. *Flower-stalk* compressed, always solitary and simple.

"The fruit of this species, standing on part of a branch whose leaves are fallen off, is figured in Mr. White's ' Voyage,' p. 226, along with the leaves of the next species." ('Botany of New Holland,' p 42).

The description was made from plants procured in the neighbourhood of Sydney.

Vernacular names — " Red Stringybark " is a name generally applied to this species in this colony in allusion to the darker colour of the wood as compared with that of *E*. *eugenioides*, White-Stringybark.   It also goes under the name of " Broad-leaved Stringybark."   In the Walcha district it appears to be confused with Red Mahogany

Seedling or sucker leaves.—These are well represented in Howitt's ' Eucalypts of Gippsland,' Pl. 14 (Trans. Roy Soc. Vict II.).   Like those of *E*. *macrorrhyncha* and *E*. *eugenioides*, they are placed opposite one another at an early stage, but very soon become alternate.   The young shoots are warty.

Mature leaves —They are very coriaceous, even when grown at a considerable distance from the sea   The leaves are larger and coarser than those of two other Stringybarks *(E macrorrhyncha* and *E. eugenioides)*, and very oblique.

Buds.—The buds and peduncles are generally somewhat thick and angular or flattened, and contrast with the neatness of shape of those of *E*. *eugenioides* and *E. macrorrhyncha*   In some cases, however, the buds are round, symmetrical and plump, and resemble more nearly those of *E eugenioides*.

Flowers.—The filaments of the anthers sometimes dry dark.

Fruits.—In consequence of the fruits being sessile or nearly so and crowded into heads, they assume a polygonal shape at the base as if they had been pressed together when in a plastic condition. With this exception, the fruits have the form of a very much compressed spheroid, the horizontal diameter of which is from 1½ times to twice the depth. The fruit is swollen out below the rim, which is sometimes very well defined and of a red or brown colour. The fruit is sometimes truncate, but more frequently the rim is dome-shaped.

There is great variability in the amount of the exsertion of the valves. In an example from Wallsend in which the inflorescence has the same character as the Sydney form, the fruits are smaller, less compressed, and the valves more exserted.

Timber.—The wood, as already stated, is open, somewhat reddish, and darker than that of *E. eugenioides.* It stands well in the ground and is otherwise durable It is very suitable for building purposes, but is very free.

Range.—Howitt states in his 'Eucalypts of Gippsland' that he has not seen it growing at a less elevation than 500 feet, and that it cannot therefore strictly speaking be called one of the littoral species. In this colony, however, it is found growing quite close to the sea; for instance, on the shores of Sydney harbour, and from the coast inland to the summit of the Dividing Range. The most northerly locality from which we have it is the Round Mountain, Guy Fawkes Range, 4250 feet above the sea, and about 50 miles east of Armidale, on the Grafton Road.

The most westerly locality from which it has been obtained is Mudgee, where it is called "Silvertop" according to Mr. R. T Baker, who collected it.

Variations from type.—The most remarkable known to us is perhaps one from the Port Stephens district, where together with the normal form is one apparently similar in every respect except as to greatly diminished size. Variations exist also as to length of pedicel and amount of compression of the fruits into heads

EUCALYPTUS MACRORRHYNCHA, F.v.M.

This in its typical form is a very easily recognised species. The buds are, when fully developed, large, rhomboidal in longitudinal section, with pointed operculum, and the pedicels are long, so that the flowers and fruits form loose heads.

Vernacular names.—It is usually known as "Stringybark" merely, but by comparison with *E. eugenioides* as "Red Stringybark." According to Howitt, it is known as "Mountain Stringybark" in Gippsland, a name to which in this colony the other Stringybarks have also some claim. *E. macrorrhyncha*, however, appears to be quite absent from the coast districts.

Seedling or sucker leaves.—The remarks made under *E. capitellata* apply equally to this species.

Mature leaves.—These are coriaceous and much resemble those of *E. capitellata.*

Buds.—These are strongly pedicellate, and the edge of the calyx tube forms a prominent ring, while the operculum is acuminate and often lengthened out into a point. In the matter of shape one cannot help likening them to those of *E. rostrata*, which, however, are very small in comparison.

Fruits.—These vary somewhat in shape and size, but owing to the long pedicels, the prominent edge to the rim, and the domed top, they can always be recognised. A particularly large-fruited form has been collected by Mr R. T. Baker in the Rylstone district, where trees with fruits of ordinary size are also found. The remark about the buds as to their resemblance in shape to those of *E. rostrata* applies here also.

Timber.—This seems in every respect to resemble that of *E. capitellata*

Range.—In Gippsland this is essentially a mountain species, and Mr. Howitt has not seen it growing at a lower elevation than 200 feet. In this colony it is found along the Dividing Range and Table Land from New England in the north. We have it from Mt. Wilson, from Yass, and from near Delegate. It grows down the western slopes and on the spurs of the main range and on the isolated ranges some distance into the interior. The most westerly localities actually recorded are Mudgee and Grenfell.

That *E. capitellata* and *E. macrorrhyncha* possess points of resemblance is apparent to the most superficial observer. A comparison of the two may be roughly tabulated as follows:—

E. capitellata —Operculum obtuse. Flowers and fruits sessile or nearly so. Fruit expanded below the rim.

E. macrorrhyncha.—Operculum acuminate or conical. Flowers and fruits strongly pedicellate; calyx border prominent.

But these characters are not absolute, and only belong to the types, considerable variation occurring in some specimens.

Baron von Mueller in the 'Eucalyptographia,' under *E. macrorrhyncha*, says:—

"*E. macrorrhyncha* stands nearest to *E. capitellata*, leaves and fruits of both are the same; but the flowers of the latter are always sessile or nearly so and thus crowded into heads as the species name signifies, besides being usually smaller; the lid of *E capitellata* is hemispheric, without any prominent point, and shorter in proportion to the tube, the latter being also more angular and downward less attenuated."

With all respect to the very high authority of Baron von Mueller, we cannot agree that the fruits of *E. capitellata* and *E macrorrhyncha* are the same; and a study of the figures of the two species in the 'Eucalyptographia' will prove the inaccuracy of the statement; we, however, show that there are intermediate forms.

Under *E. macrorrhyncha* in the 'Flora Australiensis' we find:—

"Var. (?) *brachycorys*. Operculum short and obtuse. Fruit of *E. macrorrhyncha*. Expanded flowers not seen, and therefore affinities uncertain. New England. "Stringybark." (B.Fl. iii 207).

The Eucalypt thus referred to by Bentham is evidently one of those connecting links between *capitellata* and *macrorrhyncha*, of which we possess specimens, but we doubt the expediency of giving names to any of these numerous varieties until our knowledge concerning them is more advanced.

Although the fruit of *E. capitellata* is usually sessile, or nearly so, we have specimens which are distinctly stalked. If these forms be examined in fruit only (without reference to the buds), they may be readily mistaken for *E. macrorrhyncha.*

Usually, however, these connecting links between *capitellata* and *macrorrhyncha* show a leaning towards the type of either one species or the other, so that we may conveniently classify them, but in regard to the following tree we are unable to place it with either one species or the other. It is the tree found on the Gulf Road, Rylstone district, and attributed to *E. obliqua* by R. T. Baker, Proc. Linn. Soc. N.S.W. 1896, p. 446.

The buds resemble those of *E. eugenioides* The fruits are shortly pedicellate, and in that respect approach *E. macrorrhyncha*, but otherwise they are hemispherical and flat-topped like many specimens of *E. eugenioides*, but there is a distinct and sharp edge or rim, with a tendency to doming, like *E. macrorrhyncha.* The valves are only slightly exserted. The buds appear to us dissimilar to those of *E. obliqua*, and the fruits are too broad and hemispherical for that species, the only real resemblance to *E. obliqua* existing in the leaves, which, however, equally resemble *E. capitellata.*

We have specimens collected by Mr. Augustus Rudder in the same district and named by him "Mountain Stringybark." They have fruits with slightly longer pedicels and many of them are more of a domed character, but on the same twig with these somewhat dome-shaped fruits are other fruits precisely similar to those from the Gulf Road. We are quite of opinion that they are from identical trees, and would on no account place them under *E. obliqua.*

Should it be found necessary, on account of persistence of characters over a large area, to separate this tree from *capitellata-macrorrhyncha* (it being desirable, in our opinion, to look upon it as a connecting link between these species, for the present), it would perhaps be advisable to give it specific rank.

EUCALYPTUS EUGENIOIDES, Sieb.

Sieber's definition of *E. eugenioides* (Sprengel's Curæ Posteriores IV. 195), is as follows :—

"E. operculo mucronulato, umbellis lateralibus racemosis, ramulis teretibus, foliis inæqualiter oblongo-lanceolatis," a description which would have rendered it impossible to state what species was meant had not a specimen, named by Sieber, been in existence.

Vernacular names —It is usually known as "White Stringy-bark" in this colony, the colour of its timber being paler than that of either *E. capitellata* or *E. macrorrhyncha*.

Seedling or sucker leaves —These are well represented in the 'Eucalyptographia' and in Howitt's 'Eucalypts of Gippsland.' The young shoots are warty and the leaves, which at first are placed opposite to one another, soon become alternate.

Mature leaves.—These are generally much thinner and more delicate in texture than those of *E. capitellata* and *E. macrorrhyncha*. They are also of a richer green, more shapely, graceful and Eugenia-like, a circumstance which led to the adoption probably of the specific name. Exceptions, however, occur, and specimens in our possession from Wallsend and Mudgee are coriaceous and shiny.

Buds.—The buds are clustered and often very much crowded into heads, by which the inflorescence assumes a very marked character. They always have pointed opercula, the points being sometimes so marked as to approach those of *E. macrorrhyncha*, but they are then fuller on the top and do not show such a prominent edge at the base of the operculum.

Fruits.—The fruits are slightly pedicellate, often crowded into more or less globular heads, but not compressed like those of *E. capitellata*. They are much smaller than those of the allied species, somewhat hemispherical in form, with slightly raised rim. Occasionally the fruit is quite flat-topped. The rim is often red.

The plate in the 'Eucalyptographia' shews a very thin rim, which is most unusual and *not* typical of the species, at least in New South Wales, but we have an example of this form from Port Macquarie.

Timber.—The timber of this species is good for building purposes, being strong and durable and not particularly liable to warp. It is often considered, as at Mudgee, superior to " Red Stringybark " (*E. macrorrhyncha*).

Range.—Coast district and tableland throughout, and extending westerly as far as Mudgee, though apparently not so abundant as *E macrorrhyncha.*

In the 'Flora Australiensis' *E. eugenioides* is reduced to a variety of *E. piperita*, but it has since been shown to be an undoubtedly good species, its affinities being more with *E. capitellata* than with *E. piperita*. From the latter it is easily distinguished in the young state by the strong fibrous character of the bark which extends to the small branches, the other species having a bark of the texture of *E. amygdalina*, and being only half-barked in general like *E. pilularis*. The fruits of *E. piperita* are more contracted at the top with a thin rim, whereas those of *E. eugenioides* have a well-marked rim, sometimes flat but generally raised.

We have leaves and fruits of a very interesting Stringybark from the Glen Innes district (Hartley's Mill). We refer the plant to *E. eugenioides* in the absence of complete material. The fruits are larger than those' of *E. eugenioides* usually are, and have a well-defined prominent rim, grooved on the outer edge, and show a tendency to exsertion of the valves.

E capitellata and *E. eugenioides* are very intimately related. Besides their relation as Stringybarks, we have trees with fruits so shaped that it is not entirely satisfactory to refer them to either species.

Some fruits show a tendency to *E. capitellata* in having fruits larger and more "squatty" or compressed than those of *E. eugenioides* But the valves of the fruits are not exserted, nor are the buds so flat and angular as those of *E. capitellata* usually are The buds are, in fact, those of *E. eugenioides*. The precise shape of the fruits will be seen on reference to the figure (Pl. LX. fig. 1). These intermediate forms are common on the Southern Dividing Range and the Blue Mountains. On both ranges we have typical *eugenioides* and *capitellata*, together with the intermediate forms alluded to.

53

E. eugenioides displays a tendency to form globular masses of closely packed sessile fruits, after the manner of *E. capitellata* (see Pl. LX. fig. 5). These globular masses present such a different appearance to the ordinary form of *E. eugenioides* that they may, at first sight, be reasonably supposed to form a variety, but we have many gradations between them and the ordinary form.

This head-flowered form may, perhaps, be looked upon as exuberance of growth arising from unusual vitality of a plant.

At Hilltop, near Mittagong, there is a variety locally known as " Blueleaf Stringybark." It appears to be confined to a few of the gullies about there. It is so called because the leaves, especially in the sunlight, are observed to have a bluish cast, and this bluish appearance (especially noticeable in the young leaves), is largely retained on drying for the herbarium. The tops of the trees can be readily noticed, amongst the other foliage, from a neighbouring eminence. The fruits are in spherical clusters, and if it were desirable to distinguish this tree as a variety of *eugenioides*, the name *agglomerata* would be very suitable (See Agric. Gazette N.S.W. vii 268, May, 1896.)

E. OBLIQUA, L'Her.

Although this species is so well-known in Victoria and Tasmania, its occurrence in New South Wales has scarcely been observed by botanists. Yet it is a fine well-developed forest tree in the south-eastern district, and the timber is sawn up and finds a ready market.

Vernacular names.—It is usually known as " Stringybark " in Tasmania and South Australia, and to a less extent in Victoria; in the last colony, however, it is usually known as " Messmate," because it is associated with other Stringybarks and fibrous barked Eucalypts. The same name is in use in southern New South Wales, as for instance at Sugar Loaf Mountain, Braidwood, and at Tantawanglo Mountain, near Cathcart. Apparently this is the most widely used name for it in New South Wales, and the term " Stringybark " does not seem to be ever applied to it in this colony.

Because it is usually rough-barked to the ends of the branches, it sometimes goes by the name of "Woolly-topped Messmate" in the Braidwood district (Monga, &c.).

Seedling or sucker leaves.—Broadly ovate, somewhat cordate, tending to become unequal, but not always so, and apparently always attenuate, as pointed out by Howitt. Venation well marked and more transverse than in the foliage of the mature tree.

Leaves of mature trees—It is a coarse-foliaged tree, by which characteristic alone it can usually be distinguished from those species with which it is usually associated, or with which it is likely to be confused. Its strikingly oblique, unsymmetrical leaves have no doubt given origin to its name. Obliquity is a character of nearly all Eucalypt leaves, but in the species under consideration and in *E. capitellata* it is particularly observable. The leaves are sometimes dotted and channelled like *E. stellulata* (see Part i. p. 598).

Fruit—A figure of the usual Victorian form will be found in the 'Eucalyptographia;' we give a representation of the fruit as found in the southern mountain ranges in this colony.

The orifice is sometimes a little contracted, reminding one, in this respect, and in its general shape of the capsule, of some forms of *E. piperita*, but it is larger than the fruit of that species. Drying accentuates the contraction of the orifice in both The two may be at once separated by the venation and shape of the leaves, shape of the buds, &c., but the two species approach one another sometimes very closely in the shape of the fruits.

The fruits in the southern parts of this colony are subcylindrical in shape, while those of the Victorian specimens, figured in the 'Eucalyptographia,' are more hemispherical.

The fruits of *E. gigantea*, Hook. f. ('The Botany of the Antarctic Voyage;' Hooker, 'Flora Tasmaniæ,' t. 28) usually referred to *E. obliqua*, and doubtless correctly, are more pear-shaped, and with valves more sunk, than we have observed in the New South Wales specimens.

Bark —Rough-barked to the ends of the branches; the bark of the trunk and branches is decidedly fibrous, but the fibres are not so clean and tenacious as those of the true Stringybarks, and the bark is not so suitable for roofing.

Timber.—Timber from New South Wales localities is a rather inferior, coarse, open-grained porous wood, liable to shrink and warp. It is not esteemed for public works. Its open nature may be, at least in part, a consequence of rapid growth, for which, according to several authorities, *E. obliqua* has the reputation.

It has been used in the Braidwood and Cooma districts for many years for building purposes. In Victoria and Tasmania it is largely used, and a recent official publication of the latter colony states "It is our most valuable wood." In considering the value of this statement it should, of course, be borne in mind that neither of these colonies possesses a series of excellent timbers such as New South Wales can boast of.

Range —Chiefly a Tasmanian and Victorian tree, it is abundant in many places along the top of the eastern slope of the coast range from Braidwood south. Its northernmost limit is a matter for further investigation, but it extends nearly to the Clyde River. It is found growing in company with *E. goniocalyx* and other species on the Irish Corner Mountain, Reidsdale, Sugar Loaf Mountain, and around Monga, both on the eastern and western fall of those mountains. The trees are fairly abundant, and are to be found growing to a height of from 100 to 150 feet, with a girth of from 6 to 10 feet.

Howitt (Trans Roy. Soc. Vict. ii. Pt. i, 1890, p. 92) makes the statement, as regards Gippsland, that "It appears to be essentially a littoral form, but ascends the mountains, &c."

The first part of this statement does not appear to hold true in New South Wales The tree grows right on the top of the southern range with us and never in the littoral lands, as far as observed. It frequents situations where it can be reached and enveloped in the sea-fogs, in this remote sense alone can the word "littoral" be applied to the trees with us.

On the Tantawanglo Mountain it grows abundantly in company with "Cut-tail" and other Eucalypts at a height of about 3000 feet above the sea   At Reidsdale it occurs at an elevation of from 2000 to 2500 feet.

E. obliqua has never been positively recorded from north of Sydney; in fact, its recognised localities are many miles to the south. Nevertheless, we have a specimen undoubtedly, in our opinion, belonging to this species, obtained by an experienced collector in the ranges in the Upper Williams River district. The precise locality is unfortunately lost, and therefore we do not wish to do more than invite the attention of botanists to the desirability of searching for *E obliqua* in the district named. The collector is Mr Augustus Rudder, formerly forester of the district, whose recollection is perfectly clear in regard to the specimen referred to.

The Eucalypt from Gulf Road, Rylstone district (R. T. Baker, 'Proc. Linn. Soc. N S W.' 1896, p. 446) we have discussed under *E macrorrhyncha (ante,* p. 803).

The following description of *E. obliqua* from Sir J. E. Smith's 'Specimen of the Botany of New Holland,' p. 43 (London, 1793), is interesting, and may be convenient for reference : —

"*Eucalyptus obliqua,* operculo hemisphærico mucronulato, umbellis lateralibus solitariis, pedunculis ramulisque teretibus.

"Lid hemispherical, with a little point   Umbels lateral, solitary, flower-stalks and young branches round.

"Syn. *E obliqua,* Ait Hort. Kew. v. 2, 157 ; L' Herit. Sert. Angl t. 20."

"From the only specimen we have seen of this, which is in Sir Joseph Banks' herbarium, it appears the *branches* are all round to the very top.   *General flowering-stalks* round, the *partial ones* only slightly angular, not compressed.   *Bark* rough from the scaling off of the cuticle, but this may be an unnatural appearance   *Leaves* ovate-lanceolate, aromatic, but without the flavour of peppermint."

E FASTIGATA, n sp

Introductory.—While dealing with the Stringybark group we draw attention to a tree which is very closely related to one of them, and is, to all intents and purposes, a Stringybark.   We

allude to the forest tree known as Cut-tail in the southern part of
the colony. It attains a height of 60-100 feet and more, and a
diameter of at least 4 feet Its affinities to other species will be dealt
with under various headings, but we may point out that it strongly
resembles *E. obliqua* in bark and wood, while the two species
have very dissimilar buds and fruits. The only point of resemblance to *E. amygdalina* lies in the fruits, which are rather like
those of our variety *latifolia* figured in our former paper of this
series.

We do not hesitate to say that "Cut-tail" cannot be included
under any existing species, and therefore propose the name
fastigata for it, in allusion to the shape of the operculum and
leaves.

Vernacular names.—Several names are more or less in use in
different places. The one most in use, where also the tree is
best developed, is "Cut-tail," and inasmuch as this name is not
applied to any other tree, so far as we are aware, we would suggest
that all other English names be dropped as far as possible in
favour of this We have made many enquiries as to the meaning
of the term "Cut-tail," but without success, and can only suggest
that it has reference to the rough bark on the branches which in
comparison with *E. obliqua*, which it so much resembles in general
appearance, it is cut-tailed or curtailed.

Other names that have been mentioned to us for this tree are
"Blackbutt," on the Nimbo Station, Braidwood-Cooma Road, and,
on the Tantawanglo Mountain, "Messmate." "White-topped
Messmate" and "Silvertop" at various places, and "Brown-
barrel" at Queanbeyan."

Seedling or sucker leaves.—Ovate-lanceolate, early becoming
oblique, scattered, in this respect very dissimilar to those of *E.
amygdalina*, the leaves of which remain opposite until the tree
has attained some size. The veining of the under side prominent.
The twigs rusty tuberculate like *E. amygdalina* and some other
species.

Leaves of mature trees.—Lanceolate, and when fully grown
narrow-lanceolate. Often more or less ovate-lanceolate, and

always more or less attenuate. They are rather coriaceous, smooth and rather shining. They possess no odour of peppermint.

Buds.—The chief characteristic is the shortly acuminate operculum, which is much accentuated in dried specimens. In *E. obliqua* the operculum is blunt, and the whole bud club-shaped, very different to those of the species now under review.

The anthers are partly folded in the bud.

Fruits.—The figure (Pl. LXI.) will make the shape clear. They are pear-shaped, have a conical or domed rim, with the valves somewhat exserted. They are always 3-celled as far as seen. Diameter of rim $2\frac{1}{2}$ to nearly 3 lines. Length from end of pedicel to rim $2\frac{1}{2}$ lines.

The fruit differs from that of *E. obliqua* in being more or less conical, while that of *E. obliqua* is subcylindrical. The latter species has no well defined rim and the valves are sunk, whereas in the tree now under consideration there is a prominent rim, while the valves are somewhat exserted. The fruits of *E. obliqua* are also larger than those of our species and have shorter stalks. In the latter species the peduncles are elongated over half an inch in fruit, and are distinctly pedicellate, about $1\frac{1}{2}$ lines.

Bark.—It resembles closely that of *E. obliqua*, the principal difference between the two trees, in this respect, consisting in the fact that the tops and the branches of "Cut-tail" are smooth, while those of *E. obliqua* are the reverse

Timber.— It has all the characteristics of the timber of *E. obliqua*, from which it is scarcely, or not at all, to be distinguished. At Montgomery's mill on the Tantawanglo Mountain, near Cathcart, the two trees are considered of equal value, and the timbers of the two cut up and sold as one and the same.

Range.—The coast range from Tantawanglo Mountain to near Braidwood, so far as observed at present. Specific localities are :—Tantawanglo Mountain, growing with *E obliqua* and *E. goniocalyx;* Nimbo (head of Queanbeyan River), mixed with stellate variety of *E. goniocalyx ;* Braidwood district (Reidsdale, Irish Corner Mountain), with *E. obliqua* and *E. goniocalyx.*

We have not yet determined whether it occurs to the west of the Dividing Range.

EXPLANATION OF PLATES.

PLATE LIX.

E. macrorrhyncha.

Fig. 1.—Fruit from Yass.

Fig. 2.—Fruit from Bendigo, Victoria.

Fig. 3.—Fruit from Albury.

Fig. 4. ⎫
Fig. 5. ⎬ Fruits from Rylstone ; No. 5 is especially large.

Fig. 6.—Umbel and young buds.

Fig. 7. ⎫ Types of the angular buds, with beaked opercula. From
Fig. 7A. ⎭ Rylstone.

E. capitellata.

Figs. 8 and 8A.—Fruits and buds of common Sydney form (Mosman's Bay).

Fig. 9.—Fruits from Kalgoola, Mudgee district.

Fig. 10.—Fruits from Mt. Victoria, showing flattened top or truncate rim and lateral compression.

Fig. 11.—Fruits from Round Mountain, New England.

Fig. 12. ⎫ Fruits intermediate in character between E. capitellata and E
Fig. 13. ⎭ eugenioides, from Stroud and Hill Top (Mittagong) respectively

Fig. 14.—Buds of E. capitellata, showing a less flattened form than usual

Fig. 15.—Fruits depicted in White's 'Voyage,' p. 226, as E. piperita, but described by Smith, Trans. Linn. Soc. iii. 285 (1797), as E. capitellata.

PLATE LX.

E. eugenioides.

Fig. 1.—Fruits from Mt. Victoria.

Fig. 2.—Fruits from Tweed River, showing slightly exserted valves.

Fig. 3.—Fruits from Ulladulla, showing hemispherical shape.

Fig. 4.—Fruits from Bega, showing sessile character.

Fig. 5.—Fruits from Cabramatta, near Sydney, showing disposition into a dense globular head.

Fig. 6.—Fruits from Homebush, near Sydney, showing pilular shape and sunk rim.

Fig. 7.—Fruits from Hogan's Brush, near Gosford, unusually large in size, and with well-defined rim. Intermediate in character between this species and E. capitellata : Cf. Plate LIX.. figs. 12 and 13.

Fig. 8.—Normal buds of E. eugenioides.

E. obliqua.

Fig. 9.—Fruits.

Fig. 10.—Leaf showing oblique outline, and venation.

Fig. 11. ⎫ Fruits and buds of the Eucalypt provisionally placed between *E.*
Fig. 12. ⎰ *capitellata* and *E. macrorrhyncha* (Gulf Road, R. T. Baker;
also Mr. Rudder's specimen).

PLATE LXI.

E. fastigata, sp.nov.

Fig. 1.—Seedling foliage.
Fig. 2.—Twig in bud.
Fig. 3.—Mature leaf, showing venation.
Fig. 4.—Fruit, showing exserted valves.
Fig. 5.—Transverse section of fruit.

DESCRIPTION OF A NEW SPECIES OF PUPINA FROM QUEENSLAND.

By C. E. Beddome.

Pupina bidentata, sp.n.

Jaw consisting of a chitinous, transparent membrane covering the greater part of the lips, minutely reticulated; under high magnifying power the membrane appears to be composed of very numerous rectangular plaits.

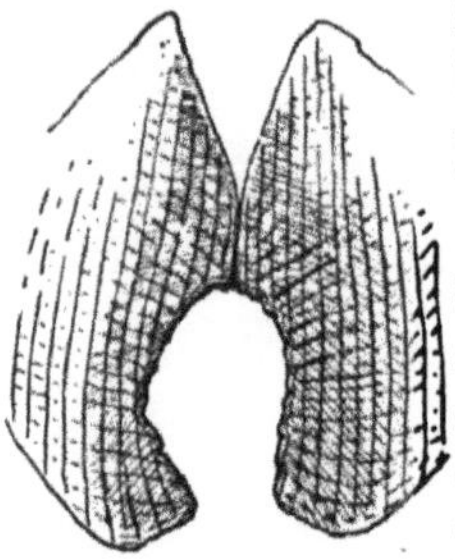

1
Fig. 1.

Radula strap-shaped, with about 75 oblique rows of teeth; formulæ 2-1-1-1-2. Rhachidian tooth with its base constricted in the middle, posterior and concave; there are three rather small cusps, the median one much larger than the laterals, with small, broadly rounded cutting points. Lateral teeth elongated, with three cusps, the median with a large blunt cutting point. First marginal with three, second with two cutting points.

The dentition is that characteristic of the *Cyclophoridæ*, and the peculiarity of the jaw, if that term may be applied, is shared by the arboreal Achatinellas.

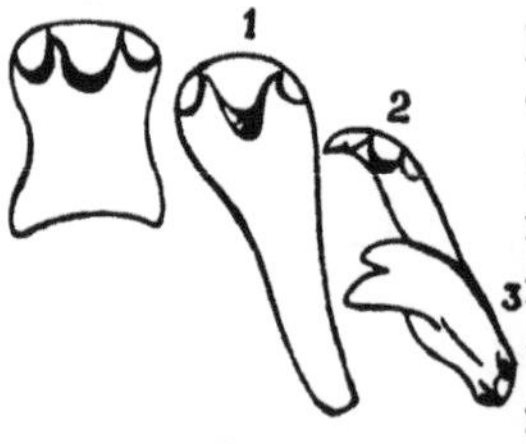

2
Fig. 2.

Shell pupiniform, shining translucent, pale horn colour. Whorls 6½, gently convex. Aperture circular, the greater part of which is encircled by less inflated margin of inner lip; peristome elsewhere markedly thickened and reflexed with whitish sinuous porcelain-like margin, notched or slit narrowly and tortuously anteriorly and posteriorly; the thickened and backward bent extremities of inner and outer lip slightly diverge, forming two ribs on body whorl enclosing a narrow triangular area which is crossed near the thread-like slit of rounded aperture by a tooth-like process. There is a finer tooth on lower extremity of outer lip which further constricts the fine slit at aperture.

Fig. 3.   Fig. 4.

Operculum concentric, concave, shining, straw colour.

Length 10 mil., diam. 4 mil., breadth of aperture 1½ mil.

Hab.—Near Cairns, Queensland. The type specimens are in C. E. Beddome's Collection.

EXPLANATION OF FIGURES.

Pupina bidentata.

Fig. 1 —Jaw (x 50).
Fig. 2.—Part of radula (× 240).
Figs. 3-4.—Front and back views of shell.

(Figs. 1-2 drawn from nature by Mr. H. Suter; Nos. 3-4 by Mr. C. Hedley.)

NOTES AND EXHIBITS.

Mr. Fred. Turner sent for exhibition a specimen of *Lepturus cylindricus*, Trin, one of several plants recently found by him near Hay. This very rare grass in New South Wales has not hitherto been found growing away from the coast; only once before had he seen it, growing in company with *L. incurvatus* on the shores of Port Jackson. Also specimens of two West Australian leguminous plants (*Brachysema undulatum*, Ker, and *Isotropis juncea*, Turcz), forwarded from the Bureau of Agriculture of West Australia, as being plants supposed to be poisonous to stock.

Mr. Edgar R Waite exhibited a lizard, *Nephrurus lævis*, De Vis, received by the Australian Museum some months ago. Suspecting that its characters were common to both *N. lævis* and *N. platyurus*, Blgr., Mr. Waite examined the type of the former species, kindly lent by Mr. De Vis, when it became apparent that the two descriptions applied to the same species—a conclusion in accordance with the views of Messrs. Lucas and Frost, from the examination of a series of specimens from Central Australia. ("Report of the Horn Expedition" ii p. 116). The exhibited specimen was shown to record a locality intermediate between the known habitats, Queensland and South Australia, the example having been obtained at Bathurst, New South Wales.

Some varieties of Australian Mollusca were shown by Mr. Hedley. On behalf of Mr. Whitelegge an example was exhibited of *Pleurobranchœa luniceps*, Cuvier, collected by him at Maroubra Bay. Though this remarkable species, apparently a pelagic form, was described in 1817, so little is known about it that its exact locality has not been before announced. Mr. Pilsbry writing on this form in the present year [Man. Conch. (1) xvi. p. 229], proposes for it the subgeneric name *Euselenops*, in lieu of *Neda* preoccupied in the Coleoptera.

By the courtesy of the Curator of the Australian Museum Mr. Hedley further exhibited examples of *Monodonta Zeus*, Fischer, a series described without locality in the Journ. de Conch. 1874, p. 372. Dr. Fischer's shrewd guess that it was of Australian origin is for the first time confirmed by the receipt of instances collected by Mr. Moore at Dongara, near the mouth of the Irwin River, West Australia. In the same parcel were also *Monodonta carbonaria*, Philippi, and *Haliotis elegans*, Koch, both noteworthy and of interest as extending the geographical range of these shells.

Mr. Ogilby exhibited for Dr. Cox a small Sole received from Mr. J. K. Larner, Public School, Codrington, caught in fresh water about 58 miles above the mouth of the Richmond River; he identified it with *Aserrayodes macleayanus*, Ramsay, which had previously been recorded from fresh water in the Hunter River, as *Solea fluviatilis*, Ramsay.

Mr. Brazier read the following

Note on the Shells found in Kitchen Middens at Bondi Bay.

The following is a list of the species of Mollusca found in Kitchen Middens accumulated by the Aborigines under rock shelters at Bondi Bay (Boondi of the Aborigines). *Triton Spengleri*, Chem , (some specimens broken off at the apex, others with the back of the shell broken, to allow of the extraction of the animal) ; *Purpura succincta*, Martyn ; *P. striata*, Martyn ; *Ninella straminea*, Martyn (the opercula of the same very plentiful); *Lunella undulata*, Martyn; *Monodonta zebra*, Menke; *M. multicarinata*, Chenu; *Scutus anatinus*, Donov.; *Nerita nigra*, Gray (= *N. atrata*, Reeve, *non* Chem) , *Natica plumbea*, Lam.; *Patella tramoserica*, Martyn, and *P. aculeata*, Reeve (both species very plentiful); *P. costata*, Sowb , (= *alticostata*, Angas,—very few specimens); *Haliotis naevosa*, Martyn; *Plaxiphora petholata*, Sowb , (the foot of this Chiton must have been much in request as an article of food, the shell-plates occurring in countless numbers in

the midden near the baths); and *Anomalocardia trapezia*, Desh,
the common mud-cockle (scarce, evidently brought over from the
mud-flats at Rose Bay). Portions of the skull of the schnapper
(*Pagurus unicolor*, Quoy) were also met with.

Mr. Brazier also exhibited (1) a fine specimen of *Cyprae
vitellus*, Linn., of unusual coloration (light brown, the margins
thickened with enamel of a dark fawn colour, the dorsal surface
showing bluish-white lines in splashes, and a few white dots in
place of the ordinary large white spots), dredged alive at Little
Coogee : and (2) a perfect adult specimen of the shell described
at the July Meeting as *Clathurella Waterhousae*, which must now
be referred to the genus *Cantharus*, the lip of the type specimen
having been broken ; it is larger than the type (length 15,
diameter 5½, length of aperture 6 mm.); it was found at Vaucluse,
Port Jackson, in possession of a hermit crab, and is allied to
C. australis, Pease, and *C. unicolor*, Angas, both from Port Jackson.

Mrs. Kenyon sent for exhibition a series of specimens of *Conus
Rutilus*, Menke, and five varieties, *C. Macleayana*, T. Woods,
C. Smithi, Angas, *C. Grayi*, Reeve, *C. maculatus*, Sowb., and *C.
Anemone*, Lam., with young and distorted examples of the same,
and communicated a Note thereon,

Mr. Darley exhibited an apparently ancient boomerang found
in a deposit of hardened mud, 10 feet below the present bed of
the River Darling, 2½ miles below Bourke, when excavating for a
coffer-dam.

Mr. Darley also communicated some interesting particulars as
to the reported occurrence of Teredo and Rock Oyster for the
first time at the mouth of the Gippsland Lakes about four years
ago, whereas previously both were said to be unknown in the
locality.

Mr. W. S. Dun exhibited on behalf of the Mining and Geological Museum, a beautiful series of Trilobites from the Upper
Silurian of Dudley, England, lately presented by Mr. C. Holcroft.

Mr Deane exhibited a number of rock specimens from Victoria
Lakes east of Menindie, Lake Boolaboolka, and Kilfera.

Dr. Norton communicated a Note recording an instance in which an ant-resembling spider was observed to attack fatally one of the community in a nest of the so-called bull-dog ants.

The Rev. J. Milne Curran exhibited a fine series of enlarged photographs and numerous rock-specimens illustrative of the physiography and geology of the Mt. Kosciusko Plateau, especially in relation to the so-called evidences of glaciation. Having been over the same ground as Dr. Lendenfeld and Mr. Helms, Mr. Curran could not but agree with Mr. Helms as to the absence of any evidence of glaciation in the Wilkinson Valley such as Dr. Lendenfeld had reported. But he also felt compelled to differ from Mr. Helms in respect of the other localities in which this observer thought he had detected evidence of glacial action, as indicated on the map accompanying his paper; and he was forced to the conclusion that the evidence adduced is wholly insufficient, and that no striæ, groovings, or polished faces due to ice action, or roches moutonnées, perched blocks, moraine-stuff or erratics are to be met with. Only one example of anything like a polished block was noted, and in this case the polishing and striæ-like markings were clearly due to a "slicken-side." Most of the granite is of a gneissic character, but normal granites are also present, the latter weathering into spheroidal masses, the contours of which in a few cases are suggestive of ice action. It had been stated that the rocks on the plateau are not such as would preserve glacial striæ. With this Mr. Curran did not agree, as he found porphyries, diorites and basalts, the latter belonging to the non-felspathic section of these rocks, specimens of which were exhibited. Apart from local evidence the general contour of the valleys is not in the least suggestive of glaciers. He there-fore concluded that (1) there is no satisfactory evidence of glaciers in the present valleys. (2) There is absolutely no evidence of extensive glaciation on the Kosciusko Plateau. (3) The "glacial epoch of Australia" in Post-Tertiary times as described by Dr. Lendenfeld, has no foundation in fact.

Mr. Hardy exhibited two examples of the Nerrum, or Knarrarm (Loddon River Tribe), or strangulation cord. They were originally obtained by Mr. John R. Peebles from either the Watty-Watty or the Litchoo-Litchoo tribe at Tyntynder, near the River Murray in the year 1857.

WEDNESDAY, MARCH 31st, 1897.

The Twenty-Third Annual General Meeting of the Society was held in the Linnean Hall, Ithaca Road, Elizabeth Bay, on Wednesday evening, March 31st, 1897.

The President, Mr. Henry Deane, M A., M. Inst. C.E., F.L S , in the Chair

The Minutes of the previous Annual General Meeting were read and confirmed.

The President then delivered the Annual Address.

PRESIDENT'S ADDRESS.

I have the honour once more to address you from this Chair.

The year just concluded has been one of fair activity, and the papers read before the Society have been of an important character. There have been nine ordinary meetings, and at these forty-four papers have been read

Some of the papers have had to me a particular interest as bearing on one of the subjects which I took up for special treatment in my Address last year. These are as follows :—Captain Hutton communicated a paper on the probability of a former land connection between Australia and South America. Mr. Ogilby presented some observations on groups of fishes the distribution of which can scarcely be understood except on the supposition of a former Antarctic continent. Professor David has contributed valuable information on the occurrence of diatomaceous earth and Radiolaria, and the Rev. J. M. Curran read some notes, which are, as I understand and hope, preliminary to a paper, on the supposed glaciation of Mt. Kosciusko.

Mr. Maiden and I have been working at Eucalypts and have presented a contribution on the subject. It is one that has

53A

already been dealt with by such competent authorities as the late Baron von Mueller, Mr. Howitt, and the late Rev. Dr. Woolls, but many species have at present been considered only from a Victorian standpoint, and demand investigation as to habits and variation when found in New South Wales.

The difficulty of defining what is a species among Eucalypts or indeed in any large and variable genus is very great. It is very easy to make very serious mistakes by grouping some that ought to be kept separate, or in the case of very variable forms, giving specific rank to mere varieties.

A curious example of errors that may be committed before sound knowledge is acquired I find in a Report to the Lieutenant Governor by Mr. William Swainson, F R S., in 1853. This gentleman divided up what he called the "Eucalyptidæ" into seven genera and 1520 species and varieties, while of the genus *Casuarina* he found 213 species, some of which he was obliged to leave unnamed, having exhausted his vocabulary. The most difficult species of *Eucalyptus* are probably those with smallish fruits, for there is then so little opportunity to seize hold of distinguishing characters, and it is only by taking into account forms of buds, anthers, fruits, leaves, seedling and mature, bark and perhaps the wood itself that anything like certainty can be arrived at.

What an opportunity is here for some of the spare energy of the old country which spends itself on monographs of a small and variable genus of Compositæ! What scope of useful investigation exists in the study of the variation of vegetable forms on a large continent like our own, which has been altogether free from the destructive and thinning out action of Glacial Periods and catastrophes, and where opportunities of almost unlimited variation exist!

During the year four ordinary Members and one Associate have been added to the Roll, and one Member has resigned.

The Society has lost the services of Mr. U. A. Henn and Dr Martin on the Council, but I am glad to be in a position to say that they will continue their Membership although circumstances have induced them to take up their residence outside New South Wales. Messrs. Brazier and Whitelegge have also resigned from the Council.

We have to deplore the loss of our oldest Honorary Member, Baron F. von Mueller, who was elected on the 22nd January, 1876. To this event I shall take the opportunity of referring presently.

The distinguished Algologist, Professor G. B. Toni, of Padua, has been elected an Honorary Member of the Society.

In accordance with the resolution passed at the beginning of last year, a sound investment having been found for the funds left by the late Sir William Macleay, the Council took steps to invite applications in England and the Colony for the position of Macleay Bacteriologist. Five applications were received, but after considering the qualifications of the applicants, the Council has decided not to appoint any of them, but to give a wider publicity to the Society's requirements and advertise afresh later on with a view to obtaining a better selection. In the meantime, the principal will be increased by the year's interest, so that pecuniarily the delay will not be a loss.

BARON F. VON MUELLER.

I must now take the opportunity of saying a few words in tribute of respect to the memory of the late Baron F. von Mueller, whose friendship and good qualities many of us learnt to appreciate.

I do not intend to offer a lengthy account of the Baron's life, as that has already been done by others far more fitted to the task than myself. I may refer to the interesting account given in the "Sydney Mail" of the 17th October last, written some time ago by the late Rev. Dr. Woolls, and to that published in the "Victorian Naturalist," (No. 7, Vol. xiii.), which is due to the able and sympathetic pen of Professor Baldwin Spencer.

Baron F. von Mueller is a fit compeer of such men as Robert Brown, Dr. Hooker, and Mr. Bentham. He was a man of indomitable energy and perseverance, and during his 44 years of official life he achieved such results as few can boast of.

His reputation was a world-wide one, and there are few countries in which he has not at some time or other found correspondents.

Baron F. von Mueller arrived in Adelaide in 1847 and imme diately set himself to prosecute his favourite study of botany. In 1852 he was appointed Government Botanist of Victoria, and was thus enabled to commence his investigations on the flora of a part of Australia which was untouched by Robert Brown. At that time he commenced a series of most arduous journeys to the Australian Alps and elsewhere, often unattended, and what that meant in those days can be imagined only by those older residents of this country now living who have had experience of the inhospitable character of the Australian bush and the dangers connected with it In 1855, one of the most important journeys was made; he then accompanied Mr. Gregory to the north and north-west of Australia, and the expedition undoubtedly until the time of the recent Horn Expedition stood out above all others for its valuable scientihc results, and in general interest and import-ance of discovery it was second only to Leichhardt's.

In the earlier part of his career Baron F. von Mueller was much in the field and had opportunities of studying the forms and habits of living plants which later in life he missed

Included in the vast collections which enabled Bentham to carry out that unique work, the "Flora Australiensis," the only complete continental Flora written, were more than fifty "large cases" of specimens collected or forwarded by the Baron to Kew; and to his assistance was the success of the work largely due. There are now more than double the species of vascular plants described compared with those known to Robert Brown Mr. Bentham in his eulogy on Robert Brown (Proceedings of the Linnean Society of London, May 24th, 1888), gives 4,200 as his total. In the "Flora Australiensis" are described 7,000 species. The second edition of the "Census" published in 1889 includes 8,839 species, distributed among 1,409 genera and 156 orders.

It was to be expected that a large amount of work would remain unfinished. The complete investigation of the flora of a continent is a work not of one generation nor of two, but the foundation has been laid for the carrying on of the study of various important groups, and among the most interesting of the subjects to which the Baron devoted his attention are those of genera and orders possessing in Australia peculiar characters and forming often a special feature of the flora. I refer in particular to his Monograph entitled "Eucalyptographia," consisting of descriptions, with plates, of 100 species of the genus Eucalyptus, and to the series of illustrations of Acacia, consisting of 13 decades or 130 species, Salsolaceæ of 9 decades or 90 species, and of Candolleaceæ 1 decade only. A work on the Myoporineæ containing figures of a large number of the species of *Myoporum* and *Eremophila* was also begun and one volume completed When it is considered that there are probably at least 150 species of *Eucalyptus* and that only 100 are given in the "Eucalyptographia," and that out of more than 300 species of the genus Acacia only 130 are figured, it will be seen that a large amount of work remains to be done with those groups alone.

The Baron's note on *Boronia floribunda*, read at the meeting of this Society on September 30th last, is believed to be his last scientific contribution.

A fitting memorial to the late Baron would be the publication of a supplemental volume to the "Flora Australiensis." As he took so important a part in furnishing material for the seven existing volumes, it would be a graceful tribute to his memory to dedicate the supplement to him. This work should of course be carried out on the lines and according to the same system as that adopted in the "Flora," which, whatever its objections may be, has very much to recommend it, not only on account of its being that made use of in the "Genera Plantarum," but chiefly because a supplement could only thus be of real utility. It would, however, be a convenience if at the end of the volume a reference in tabular form to the system and nomenclature of the Baron's Census were supplied. It is to be hoped that in whatever way the work may be carried out, all jealousies will be laid aside and the greatness of the man to whose memory the tribute is offered alone remembered. This volume might well be made a joint work subscribed for by the four Governments of South Australia, Victoria, New South Wales and Queensland, and jointly edited by the four representative botanists of those colonies.

I should now like to add a few words on the subject of nomenclature, but I do not wish that these remarks should be taken in any way as disparaging to the late Baron's work. Every man has a right to his own views, and certainly none more so than the late leading botanist of Australasia, but there are few who agreed with him on certain points, and some who have followed his methods during his lifetime will probably feel themselves justified in now throwing off the restraint previously imposed upon them

Many of the well known names of the "Flora Australiensis" were dropped by the Baron and do not appear in their expected places in his "Census of Australian Plants." Other names which he considered to have the right of priority have instead been adopted by him, to the great discomfort of most of us. Under one large genus, many generic names with which we are familiar have been grouped For example, such genera as *Crowea*, *Phebalium*, *Asterolasia* and many others are thrown into *Eriostemon*, and *Astroloma*, *Leucopogon*, *Melichrus*, *Acrotriche*, *Monotoca*, *Lissanthe* and a host of others are suppressed and the species placed under *Styphelia*. The annoyance is great enough when in looking a plant up you miss its generic designation, but if, as in the case of many, you lose the specific name as well, it is confusion worse confounded. Priority should not be the only guide when adopting a name, but use must be taken into consideration. Mr. Thistleton Dyer in his Address to Section K of the British Association for the Advancement of Science, 1895, says that to him "botanists who waste their time over priority are like boys who when sent on an errand spend their time in playing by the roadside. By such men even Linnæus is not to be allowed to decide his own names." And in another part of the same address he makes the pertinent remark that "if science is to keep in touch with human affairs, stability in nomenclature is a thing not merely to aim at but to respect. Changes become necessary, but should never be insisted

on without grave and solid reason;" and in a note he calls attention to Darwin's saying, "I cannot yet bring myself to reject any
well known names." No doubt the Baron thought he had grave
and solid reason to change some names, and we should be loth to
charge him with loitering on his errand like the schoolboy, but I
am sure all of us prefer the names we became used to through
the "Flora Australiensis"; let us therefore adhere to them as
much as possible.

Mr. R. D. Fitzgerald's "Australian Orchids" consisted at his
death of one Volume of seven parts, and four other parts towards
a second Volume. One hundred and eighty-three species were
figured and described, with interesting notes on their habits
and modes of fertilisation by Mr. Fitzgerald himself. Seeing
the number of fine drawings still unused, it was proposed to
continue the publication. The assistance of Mr. A. J. Stopps
was secured for the lithographic work, and I was asked to work
up the text. Many friends came forward to help with information, and Part 5 of the second Volume was brought out under the
editorship of Dr James Norton in 1895. About half the plates
required for Part 6 and some notes for the text are ready, but
there is no money to go on with the publication. Only a small
sum is really necessary to complete this part, but the Government
steadfastly refused last year to place any money for the purpose
on the Estimates. It will be a great pity if this part cannot be
finished, and also Part 7, which would make up the second
Volume. I hope a renewed effort may be made some day to
induce the Government to provide the requisite funds for carrying
out this essentially Australian object.

One of the scientific events chronicled for the past year is the
ineffectual attempt to execute a wish of Charles Darwin to pierce
a coral island to its foundation and, by bringing up a core, test
the mystery of its origin. A committee appointed by the Royal
Society of London for the purpose of this investigation had a
man-of-war placed at their disposal by the Admiralty. The
New South Wales Government further assisted them with a loan

of boring machinery. The expedition, of which Professor Sollas was placed in charge, sailed from Sydney in May, to Funafuti, an atoll lying half way between Fiji and the equator which had been selected as the scene of operations.

The impression usually prevailing that a coral reef was a dense and homogeneous mass was soon dispelled when the diamond drill began to work, and it was shown instead to contain caves and fissures filled with quicksand The latter proved too much for the apparatus at command, and after penetrating to a depth of 105 feet in the first instance and 72 feet in the second, further attempts were abandoned.

Although the chief line of inquiry was defeated, important results were achieved by the officers of H.M.S. Penguin, whose soundings perfectly develop the submarine slope and contour lines.

These observations and some made later in the year by officers of the same vessel on the Alexa Bank throw much light on the conditions under which coral formations appear to take their rise, which are apparently not those assumed by Darwin.

The naturalists attached to the Funafuti expedition amassed collections illustrative of the ethnology, zoology and botany of the island, and a memoir based on the gatherings and observations of Mr. Hedley, who was attached to the expedition, is now in course of publication by the Australian Museum. The reports so far reflect immensely to the credit of Mr. Hedley, who seems to have allowed no detail, however apparently unimportant, to elude his observation. The general account of the atoll and of the manners and customs of the inhabitants, written by himself, will be read with interest by all, and the hope will naturally arise that Mr. Hedley may have opportunities accorded to him of taking part in future investigations of the kind.

A reprint of Professor Sollas's Report, published in " Nature," of February 18th, came to hand yesterday.

Another expedition for the investigation of coral reefs—I refer to Professor Agassiz's visit last winter to the Great Barrier Reef had, owing to bad weather, which is a most unusual occurrence at the time of year, to be given up before its main object had been attained.

News has just been received that another scientific excursion to the Pacific has met with some success. After enduring considerable toil, hardship and danger, Dr. Willey has, in the Loyalty Islands, succeeded in obtaining eggs of the Nautilus, but unfortunately these have failed to develop.

A remarkable discovery in morphological botany has recently been made in Japan of another connecting link between flowering and flowerless plants. The discoverers are Professor Ikeno and Dr. Hirase, who have found in *Cycas* and *Ginkgo* the fertilisation of the ovule effected by a partial penetration of pollen tubes, and a subsequent development of antherozoids for the completion of the process.

With regret we learn from "Nature," of February, 18th, that the veteran palæontologist and botanist, Baron Constantine von Ettingshausen, had died at Graz at the age of 71.

HORN EXPEDITION.

In my Address of last year lengthy reference was made to the first instalment of the "Report of the Horn Scientific Expedition to Central Australia"—Part ii. Zoology, then just published. Three additional parts—Part i. Introduction, Narrative, Summary, &c, with Map, by Professor Baldwin Spencer, M.A.; Part iii. Geology and Botany, by Professor Tate, and J. A. Watt, M.A., B.Sc.; and Part iv. Anthropology, by Professor Stirling and Mr. Gillen—have since been issued under the able editorship of Professor Spencer, completing this important work. The Report in its complete form, as a contribution to Australian scientific literature, has fully justified our expectations of its importance, and it demands a further expression of our indebtedness to Mr. Horn, the promoter, and to all who have shared in its production.

A very substantial increase of knowledge in all departments has been gained, but Professor Spencer has so ably summarised the results that it is needless to attempt a re-summary. I will merely refer to his remarks on the relations of the Autochthonian

Flora to the assumed early "Cosmopolitan Flora" on pages 174 and 175, in which he points out that the autochthonian flora of the west was, from very early times, to a large extent shut off by barriers from an immigration of other types, and that it is difficult to see how, if the autochthonian has been derived from the cosmopolitan, representatives of typical Australian genera only are found, and not a trace of such doubtful forms as *Quercus, Betula, Salix,* &c., upon the presence of which in fossil remains the theory of the cosmopolitan flora in Australia really rests.

Professor Spencer's "Narrative" is of special interest. We have many narratives of Australian travel and exploration, but these have been written by the leaders of expeditions, much of whose time and attention was necessarily devoted to administrative details, and absorbed by the anxiety unavoidably connected with these; but we have here a narrative from the pen of an expert biologist, well versed in the subject of the natural history of Australia, with a keen eye and a ready pencil, and pursuing his work undistracted by drawbacks such as those alluded to above. And the work is rendered additionally attractive by an admirable series of topographical and other views reproduced from actual photographs. Nature was unfortunately in a very dry mood, and the opportunity of witnessing the advent of good rains, and the circumstances attendant on a Central Australian flood, did not present themselves. Floods and droughts have, however, to be taken as they come.

The experiences of the expedition have afforded Professor Spencer opportunity for a masterly exposition of some of the probable former relations of Australia, and a comparison of the special features of its botanical and zoological subdivisions, resulting in the conclusion that these are not coincident. The names applied to his Subregions—Torresian, Eyrian and Bassian —are, I think, particularly happy, as avoiding all objections of implied theory and dogmatism.

Professor Spencer, in discussing the question of the route by which Australia received its mammals of characteristic types decides in favour of a former land connection between South-eastern Australia and South America, through what is now Tasmania, and thus adds his support to a theory, the objections to which are continually losing weight.

In my Address last year I pointed to the necessity of this connection in former times in order to account for the affinities of a portion of the floras of Australia, New Zealand and South America, and the occurrence in a fossil state in South America of marsupials allied to our own. The chief objections are—first, that an ocean of considerable depth lies between these countries, the bottom of which, it is therefore supposed, could never have been above the surface. As a matter of fact, even if Wallace's 1000 fathom limit of possible elevation or depression could be acknowledged, it is to be remarked that not enough soundings have been taken in the higher latitudes to prove the non-existence of submerged plateaux. The lowest continuous line of soundings seems to have been made by the officers of the Challenger; it lies near latitude 50°, and there is to the south of that parallel plenty of room for extensive plateaux to show themselves or even quite shallow depths when soundings are taken. The other objection, that the temperature and climate would have been too severe, can scarcely have weight. In the early and middle Tertiary mild temperatures existed in the northern hemi-sphere up to latitude 79° in Spitzbergen, and 81¾° in Grinell Land, and there is no reason why, at the same epochs, if the disposition of the land was suitable, there should not have been temperatures favourable to life in the corresponding latitudes near the south pole. Fossil remains from the Straits of Magellan indicate tropical conditions. During the Pliocene, temperature generally became lowered, and the vegetation of the temperate zone had begun to retreat from the North Pole; but even if the same process took place at the South Pole, there might still be abundant warmth between, say, 55° and 70°, to permit of the existence of a luxuriant vegetation and fauna.

I have been unable through lack of time to carry out my intention of completing the comparison of the tertiary fossil leaves with those of the existing vegetation. Sufficient has, however, been done with the assistance of Mr. R. T. Baker, F.L.S. to confirm me in the opinion I last year expressed, that the fossil flora would find its representatives in the existing coast vegetation. Some of the fossil fruits of the Pliocene Gold Leads also closely resemble those of to-day on the coast, but as structure has been almost entirely lost, there is not that certainty that one would like to find. It seems, however, quite clear that before seeking for analogies in distant countries careful comparisons with the existing flora should be made, and this is not the method that the eminent palæontologist, in whose hands the fossil plant remains from Dalton, Vegetable Creek and Oxley were placed, adopted Taking into consideration the difference between the Eocene and Miocene climate and that of the present period, we might expect to find existing types a few degrees further south in the fossil state, but that is quite a different idea from going to the other side of the earth for analogies.

I can find little or no information about the fossil tertiary floras of Western Australia, South Africa and South America. This is much wanted, as also further information about fossil remains of the tertiary beds of Kerguelen Island.

Some months ago, when on a visit to South Gippsland, Mr. J. H Wright took me to some leaf beds lying on a horizon above the "Lower Basalt." The most interesting finds on that occasion were leaves in all respects resembling those of a species of *Coprosma*, "domatia" and all. This is a curious indication of the antiquity of these peculiar structures. *Eucalyptus* was not noticeable in these beds, and the vegetation appeared to have such a character as would imply humid atmospheric conditions. Mr. Wright showed me some beautiful leaf remains from the siliceous shales below the "Lower Basalt," in which there were a good many leaves of the *Fagus* type, as well as what might be taken for an early form of Eucalypt.

Through the kindness of Mr. R. L Jack, Government Geologist of Queensland, I have received a number of samples from the Oxley beds, referred to in my Address of last year. The impressions are very fragmentary, and thus very difficult to make out. They seem to me as a whole to be rather conspicuous for the scarcity of Eucalypts and Proteads as we know them, a circumstance which, as I have already indicated, we need not be at all surprised at

AFFINITIES OF THE SOUTH AFRICAN FLORA.

The belief in the former connection between Australia and South America is continually obtaining more adherents, but the possibility of a land bridge having ever existed between South Africa and Western Australia is treated with much greater incredulity. The affinities of the existing floras, however, seem to point to it as the only possible explanation. Strong evidence of a connection in the Carboniferous Period has already been adduced by Dr. Blandford and others, on the ground of a common flora, which flourished not only in South Africa and Australia, but in Southern India and South America as well.

Had we not this evidence from Carboniferous times, we must acknowledge that the resemblance between the existing floras of the south-west region of South Africa and that of Australia, and particularly of Western Australia, is too remarkable to be accounted for by saying that they are relics of a once cosmopolitan flora, and that their peculiarities have been produced by the selective action of the floral climates. Those botanists who have closely studied them would not be contented with any other explanation than that of actual land connection, or at least of a former tolerably close proximity of the land areas, after the peculiarities of the flora had become developed. Strips of deep sea now separate the two countries, but it does not follow that there was never any land bridge between them. It is certain that parts of the ocean where now there are depths of 1500 fathoms have been land in the Miocene—for example, that from New Zealand northwards. Could we not allow of a local sub-

sidence of, say, 2500 fathoms since the Cretaceous ? That is all
that is necessary.

Last year I referred to the comparison made by Dr. Hooker in
his Introduction to the "Flora of Tasmania." The following
additional particulars from Dr. Harry Bolus's article in the "Cape
Handbook" will be of interest.

The region over which the Proteaceæ are found, and to which
they are practically confined, is the south-west region. It is a
narrow strip about 400 miles long, extending from and including
Cape Town to Port Elizabeth, when it gradually but rapidly
merges into the tropical African region. The vegetation of this
latter region, like the luxuriant vegetation on our own east coast,
extends southwards from the tropics far into temperate latitudes.
The width of the south-west African region is about 50 or 60
miles on the average, and its northern boundary is a very sharply
defined one   To the north is the Karroo region, a particularly
remarkable one also as will be seen. The flora of the south-west
region is characterised by abundance of Rutaceæ, Bruniaceæ,
Ericaceæ, Proteaceæ, Restiaceæ, Leguminosæ, and some others.
The Karroo region which adjoins it on the north is noted for the
complete absence of the orders named, and for the scarcity of
Leguminosæ.   The other regions of South Africa mentioned by
Mr. Bolus are the Composite and the Kalahari, but these do not
interest us to the same extent.

South Africa is, in Mr. Bolus's paper, assumed to be limited
by the Tropic of Capricorn. It exhibits a most remarkable
variety of plant life, and a comparison with Australia presents
some remarkable analogies :—

Australia contains 152 orders and 1300 genera.
S. Africa   „   142 „   „   1255 „
In Australia there are 520 endemic genera.
S. Africa   „   446   „   „
But it is to be noticed that the area of Australia is five times
that of South Africa, and it extends northwards to 10° south
latitude, instead of being limited by the Tropic of Capricorn.

The south-west region possesses the following orders in the greatest abundance :—

1. Compositæ.	8. Cyperaceæ.
2. Leguminosæ.	9. Restiaceæ.
3. Ericaceæ.	10. Liliaceæ.
4. Proteaceæ.	11. Orchideæ.
5. Irideæ.	12. Rutaceæ
6. Geraniaceæ.	13. Scrophularineæ.

A comparison with the most abundant Australian orders shows that *Irideæ, Geraniaceæ, Restiaceæ, Liliaceæ, Rutaceæ,* and *Scrophularineæ,* although existing, are not so prominent, and would have to take a lower place, and the orders *Myrtaceæ* and *Goodenovieæ* would be substituted. The order *Ericaceæ* is represented by the closely allied order *Epacrideæ.*

With regard to the other orders, it is to be noticed that *Restiaceæ,* although not so abundant, are peculiarly Australian; that the suborder *Boronieæ* of *Rutaceæ* is peculiarly Australian, like the *Diosmeæ* of the same order in South Africa; and that among *Liliaceæ* there is a peculiar genus—*Nauolirion*—which is closely allied to *Herpolirion* of Australia, Tasmania and New Zealand.

The study of geological phenomena and the distribution of life on the earth lead to two important conclusions: *first,* that the earth's surface has been subject to repeated and extensive deformation, implying a considerable amount of flexibility of the earth's crust, whereby the land connections have been varied at different times; and *secondly,* that over portions of the earth's surface extraordinary changes of climate have taken place, so much so that glacial and temperate, subtropical and even tropical conditions appear to have become interchanged.

PERMANENCE OF OCEAN BASINS.

In spite of the undoubted truth of the first of the above propositions, the theory of the permanence of ocean basins and continental areas holds still a very strong position in the minds of many. The chief argument in its favour lies in the supposed absence of deep sea deposits on dry land.

Speaking on this subject, Professor H. Alleyne Nicholson in his Presidential Address to the Royal Physical Society of Edinburgh, 1894, points out that the deepest deposits are necessarily thin, scanty and of limited area. Radiolarian deposits, which are supposed to indicate deep sea, have been discovered of various ages. In Lanarkshire they are accompanied by green and red mudstone, a forcible reminder of modern deep sea deposits.

Professor David's observations tend to shew that radiolarian deposits do not necessarily indicate deep sea. Probably in this case we should have to judge by the circumstances under which the Radiolaria are found, and it is to be remembered that land drifts and vegetable débris may be found mixed with deep sea deposits in the most incongruous manner. The dredging operations between the west coast of Central America and the Galapagos carried out between February and May, 1891, with the U S. Fish Commission steamer Albatross, under charge of Alexander Agassiz,* showed together with characteristic globigerina ooze. a large amount of decayed vegetable matter. Terrigenous material was dredged up from depths of over 2,000 fathoms. and with it logs, branches, twigs, and decayed vegetable matter. Off the West Indies immense quantities of vegetable matter had also been obtained from depths of over 1,500 fathoms. It is evident that if such materials were found fossilised in such surroundings they would be thought to indicate shallow depths.

Professor Poulton in his Address to the Zoological Section of the British Association last year, refers to the results of some of the Challenger dredgings from great depths as follows.— "These most interesting facts prove furthermore that the great ocean basins and continental areas have occupied the same relative positions since the formation of the first stratified rocks, for no oceanic deposits are found anywhere in the latter." This

---

* Bulletin Museum Comparative Zoology, Harvard College, Vol. xxiii. p. 12.

is curiously at variance with what Mr. Marr was saying at or about the same hour of the same day in a neighbouring hall. Mr. Poulton's statement is an expression of the theory of the permanence of ocean basins and continental areas, and it is that neither more nor less. It is desirable therefore to inquire what is meant. As it stands, it is a general statement too vaguely put to be of much use. Does it mean that the whole of the great ocean basins and the whole of the continental areas have always occupied the same relative positions? Clearly not, for we know that nearly or quite all existing land has at some time or other been under the water, and there have been land connections where there is now sea. The proposition must then be reduced to this, that *portions* of the great ocean basins and *portions* of the continental areas have occupied the same relative position. In other words, some portion or other of the great ocean basins has always been under the water and that some portion or other of the existing continental areas has always been above the sea. The statement thus corrected is useless to us; it affords no explanation of the distribution of life on the earth, for it may be true that some areas of existing land and water have always been land and water respectively, and yet we know that continental areas have been differently divided and cut up, and the same is the case with the seas. If it was intended to mean that the continents and oceans had been practically the same through all time as they are now, it is incorrect. For example, we have very good reason from the study of the flora to believe that in Permo-Carboniferous times South Africa, Southern India, Australia and South America formed part of one continent, and that in the early Tertiary Period North and South America were broken up into quite distinct land masses, and that in the same period and earlier Europe and Western Asia were indented and crossed by seas in a way that would make that part of the world quite unrecognisable now.

Mr. J. E. Marr in his opening address to Section C (Geology) of the British Association, 1896, says:—"We have been told that our continents and ocean basins have been to a great extent

53B

permanent as regards position through long geological ages; we now reply by pointing to deep sea sediments of nearly all geological periods, which have been uplifted from the ocean-abysses to form portions of our continents; and as the result of study of the distribution of fossil organisms we can point almost as confidently to the sites of old continents now sunk down into the ocean depths. It seems clear that our knowledge of the causes of earth movements is still in its infancy and that we must be content to await awhile until we have further information at our disposal."

Captain Hutton says :—"We know as a matter of fact that continental areas are liable to subsidence, and that oceanic areas are liable to elevation ; and we cannot as yet place a limit on the possible amount of continental depression or of oceanic elevation."* Further on (p. 411) he says :—

" We certainly do find a large number of geological periods represented in Europe, Asia, America, Australia, and New Zealand, but in all cases there are also long periods unrepresented, especially in the Palæozoic era, when there are many physical breaks in continuity, accompanied by an almost complete change in animal life, and Sir A. Ramsay says that these breaks may each indicate a period of time as great as the vast accumulations of the whole Silurian series. The question is, What was the condition of these areas during the unrepresented periods ? Certainly they might have been land, but also they might, in some cases at least, have been deep ocean."

RIGIDITY OF THE EARTH.

As the facts of the deformation of the earth's surface as well as that of alterations of climate depend largely upon the flexibility of the crust, it will be interesting to consider shortly the conclusions that have been arrived at as to the rigidity of the earth.

---

* "Has the Deep Ocean ever been Land ?" New Zealand Journal of Science, Vol. i. p. 410.

Lord Kelvin and Mr. G. H. Darwin, from a study of the long period oceanic tides, conclude that the earth's mass as a whole is more rigid than steel but not quite so rigid as glass. Such a degree of rigidity would at first sight appear to preclude any alteration of the levels of the land with respect to the ocean; we know, however, that certain tracts of the earth's surface are rising and others falling, so that the question arises what such an amount of rigidity implies.

Mr. R. L. Woodward in a paper entitled "The Mathematical Theories of the Earth," published in the American Journal of Science, Vol. 138, p. 343, says :—" Whatever may have been the antecedent condition of the earth's mass, the conclusion seems unavoidable that at no great depth the pressure is sufficient to break down the structural characteristics of all known substances and hence to produce viscous flow whenever and wherever the stress difference exceeds a certain limit, which cannot be large in comparison with the pressure " Internal fluidity is therefore not a necessary condition to account for movements of the crust.

Roche considered that geological phenomena were best explained by postulating a solid nucleus with a zone of fusion separating the crust from the nucleus.

In a paper entitled "An elementary proof of the earth's rigidity," published in the American Journal of Science, Vol. 139, p 336, the author, Mr. George F. Becker, points out that although the earth is a very rigid body, it does not necessarily follow that it is solid. The assumption of solidity is objected to by geologists as opposed to the possibility of the occurrence of geological phenomena. There is, however, no conflict between geology and physics. He says :—" Time enters into the expression of viscosity, and the fact that the earth behaves as a rigid mass to a force which changes its direction by 360° in 24 hours is not inconsistent with great plasticity under the action of small forces which maintain their direction for ages. For a considerable number of years I have constantly had the theory of the earth's solidity in mind while making field observations on upheaval and

subsidence, with the result that, to my thinking, the phenomena
are capable of much more satisfactory explanation on a solid globe
than on an encrusted fluid one."

CHANGES OF CLIMATE.

The changes of climate, which occurred in the Carboniferous
period, if the phenomena are rightly interpreted, are much more
extraordinary than those of the Pleistocene when the so-called
Glacial period or periods set in, for the latter appear to have
been chiefly due to a general cooling of the poles and a consequent enlargement of the ice caps.   The latter phenomena are
visible both in the northern and southern hemispheres, whereas
the glacial action which appears to be traceable in the Carboniferous period extended over Southern India, South Africa,
Australia and South America only.   At this time Dr. W. T.
Blandford (Part 2, Vol. xxix of the Records of the Geological
Society of India) says that these countries formed a continent,
judging from the peculiar flora which characterises them.   In
each case a boulder bed "undoubtedly glacial in origin" has been
found associated with them.   Dr. Feistmantel states that the
Lepidendron flora was swept away at the ushering in of the new
conditions and gave way to the *Glossopteris* and *Gangamopteris*
flora.   He shows that a shifting of the pole would not account
for the new conditions, as on the opposite side of the earth the
vegetation remained unaffected, and the difficulty of imagining so
large an area of the earth's surface influenced by the advance of
the polar cap is all the greater seeing that since the writing of Dr.
Feistmantel's report South America has been added to the territory.
Dr. Blandford points out how this area must have been cut off from
the rest of the world by sea, so that once the vegetation had
changed it was preserved from immigration of the old flora; but
how did it become changed? could it have been by cold, seeing
that the other side of the earth was unaffected?   The phenomena
of stranded boulders, groovings and scratchings are extraordinarily
like what glaciation produces, but can they only be accounted for
by ice?   Assuming the glacial phenomena to have existed in

certain districts only over this large area, can these local conditions be considered to have been sufficient to produce a complete
change in the flora? Mr. Dubois in "The Climates of the
Geological Past," attributes the alteration to a general raising of
the land, but it still seems rather strange that all the land should
be raised, and although coal was still formed, no suitable positions
should be left for the old flora. He says:—"Just as during the
Carboniferous Age an extensive lowland, cut up by the sea into a
large marshy archipelago, accounts for the formation of coal over
nearly the whole of the northern hemisphere, to such an extent
that comparison can only be made with the extensive deposits of
Jurassic coal, extending from Western Asia to Australia, it seems
that a large mountainous continent ("Gondwána Land" of
Suess), at the south of the equator, has caused extensive accumulations of ice in suitable places. A great uniformity of orographic
conditions over extensive continental parts of the earth's crust
seems to have been characteristic of the Coal period. It is thus
possible, and even probable, that by a gradual upheaval of such
a continent, the changed conditions of existence caused the
development of a new flora, which only much later, in the
beginning of the Mesozoic period, should find in Europe, in the
higher upheaval of the ground, conditions it was better fitted for
than was the older Palæozoic flora which in consequence would
suffer extermination. Traces of glaciation are believed to have
been actually found in the Permian formation of Europe. From
those high centres of acclimatisation the new flora, accommo-
dating itself to a higher temperature, could then have gradually
spread over the lowlands."

Up to quite recently there were, and perhaps even at the
present time, there are geologists who hold that the Glossopteris
Flora belongs to a much later period of the world's history than
the Lepidodendron Flora of the Coal Measures, but representatives
of the two floras have been found associated in the same beds,
which must be accepted as a final and conclusive proof of their
contemporaneous existence. (Rec. Geol. Sur. of India, Vol. xxix.
Part 2, p. 58).

Glacial phenomena are reported from various ages between the Carboniferous and Pleistocene Formations, and the phenomena as exhibited in Australia are well set forth in Professor David's Address to Section C. of the Aust. Assoc. Brisbane, 1895.

The most important and tangible of the recorded phenomena in the northern hemisphere are those of the "Great Ice Age," as it is called, in the title of Dr. James Geikie's book. The temperature of the earth, the north pole at least, was cooling down at the end of the Tertiary Period, and the cold culminated in the Pleistocene. Dr. Geikie says that at least six different periods can be proved during which the cold advanced and retreated, and between which mild conditions prevailed. Other geologists count these to be less in number.

Various explanations have been given for the spread of Arctic conditions from the pole, the most noted being probably that known as Croll's theory. Dr. Croll argues that the orbit of the earth, in consequence of the varying positions and attractions of the planets, increases in eccentricity at long intervals of many hundreds of thousands of years. The Glacial epoch occurred in one of these periods. High eccentricity would, when the axis of the earth was inclined in the line of the major axis of the orbit, cause long mild summers and short winters in one hemisphere and short summers and long cold winters in the other. Under the latter conditions, great cold and accumulation of snow and ice and what is called a glacial period would result. With the precession of the equinoxes, the conditions would alternate in the northern and southern hemispheres till the orbit of the earth lost its extreme eccentricity.

Major-General Drayson considers that the pole describes a circle round a point 6° from the pole of the ecliptic, so that, 13,700 B.C., the angular distance of the two poles would be such as to bring England within the Arctic circle.

"Professor G. H. Darwin has considered the possibility of the pole having worked its way in a devious course 10° to 15° from its present geographical position, but points out that such a movement would require extensive surface deformation and shifting of surface weights not easy to understand." (Great Ice Age, Note p. 791).

Sir Charles Lyell considered that all climatic changes could be explained by gradual changes in the distribution of land and water. There are few that now hold this view. It is to be remarked that in Pleistocene times the distribution of land and water was practically the same as now, and yet it was just in that period that the most remarkable oscillations of temperature conditions occur.

Dr. Geikie in the work already referred to points out that there are oscillations of temperature and rainfall shown by advance and retreat of glaciers, rising and falling of level of lakes and inland seas, and asks whether these may not be due to cosmic causes, and whether such causes may not have to do with the larger and more extensive oscillations producing glaciation or mild temperatures up to near the pole.

As regards the question of the geographical shifting of the pole, I find in "Nature," of September 25, 1884, a letter by Mr. Flinders Petrie referring to an Address by Professor Young, which stated that a change of one second per century had been noted at Pulkowa in the earth's axis. Other corroborations of the same fact exist. He says :—"Such a change might be effected by causes which are beyond our observation ; as, for instance, unbalanced ocean circulation equal to a ring of water only 4 square miles in section moving at a mile an hour across the poles." Mr. Petrie refers to the Gizeh Pyramids ; these structures, the errors of which are but a few seconds of angle, agree in standing as much as 4' or 5' to the west of the present north.

Professor Newcomb some years ago, from observations of the transit of Mercury, concluded that the rotational period of the earth was not a fixed quantity, and it has since been amply shown from the study of the same phenomena that the period is subject to variation, increasing for a number of years and then diminishing again, and so on   I do not know whether any explanation has been offered of this phenomenon, but may it not indicate movements of the viscous interior, more or less independent of that of the crust ?

Some of the peculiarities of the distribution of temperature in the Tertiary seem to be more easily explained on the assumption of a geographical shifting of the pole, and as a slow shifting seems to be going on at the present moment, it may be looked upon as helping to solve the difficulty.

Mr. Marr says in his Address previously referred to that Dr. Neumayr in his work (Ueber Klimatische Zonen während der Jura und Kreidezeit) has, in the opinion of many geologists, established the existence of climatic zones in former times. This may be the best way of testing any supposed extensive shifting of the pole, although it is to be observed that up till the late Tertiary actual polar conditions must have been confined to a very few degrees round the pole, and may be, therefore, difficult to identify.

With regard to the possible geographical shifting of the axis, it has seemed to me that somewhat extensive changes could have taken place in former times when the earth was less rigid and the interior more closely resembling a fluid, in the following manner. We believe that the rotation of the earth is being slowly but surely retarded by the action of the tides. If the interior were fluid or thinly viscous, the retardation of the crust would not immediately affect the interior, as it would take time to communicate the retardation—that is to say, the interior would always rotate at a slightly greater speed than the crust. Now the solid crust would not be smooth underneath : if corrugations form exteriorily, through cooling or other causes, the under side would be roughened too. If the fluid or viscous interior were not absolutely homogeneous, and it is not likely ever to have been so, it would contain masses of solid matter, or of matter of at least firmer consistence than the rest. These floating masses on the under side of the crust would come in contact with the ridges, and would tend to produce—away from the equator—an acceleration at that spot which would cause the rotation of the whole to be modified and the axis shifted.

A general alteration of climate over the surface of the earth might be caused by an alteration in the constitution of the atmosphere. Mr. H. C. Russell at a meeting of the Royal Society of New South Wales in 1892 pointed out, when giving some particulars of probable life conditions on the Planet Mars, that the existence of a thin layer of olefiant gas in the atmosphere of this planet would allow the sun's heat to enter, but would prevent its radiation again into space, so that the existence of the addition of small quantities of such a gas if liberated by extensive volcanic disturbances from coal strata below would be the cause of materially raising the general temperature of the earth's surface. On the other hand, if the earth with the sun passed into regions of space which happened to be crowded with meteoric matter, the power of the sun's rays would be so much diminished that a considerable enlargement of the polar area and an extension of glacial phenomena into temperate regions would result.

In "The Climates of the Geological Past," Mr. Eugene Dubois shows how that in all ages up to the end of the Tertiary Period mild temperatures have been proved to exist up to within 10 or 15 degrees of the North Pole, and in the Eocene we have such in Grinell Land at $81\frac{3}{4}°$ N., 95° W.; Spitzbergen $77\frac{1}{2}°$ to 79° N., about 20° E., while in the Island of New Siberia in latitude $75\frac{1}{4}°$ and 140° east longitude deposits of brown coal are found. In the southern hemisphere it has not been possible to penetrate so far, but in Kerguelen, which now has a rigorous climate, *Cupressoxylon* has been found, while at Punta Arenas, in the Straits of Magellan, $53\frac{1}{2}°$ S., the conditions appear to have been tropical. The author concurs with Heer in disputing the fact of any indication of geographical shifting of the pole, as the vegetation follows close on the pole all round, and if the ancient conditions seem to have been warmer on the Atlantic side, it is only similar to what is the case now. In the early Tertiary especially this intensity of conditions producing warmth might well have been even greater than now, as Europe consisted of islands and peninsulas, with inland seas and large bays, and there is little doubt that the Arctic Ocean was at that time connected with the warm seas of Europe and Southern Asia. The author argues for a gradual cooling down of the sun as producing all the phenomena observable. The sun is now in the condition of a yellow star; all through the Palæozoic, Mesozoic and part of the Tertiary it was a white star, thus the heat conditions were more intense, and although the tropics need not have been hotter, the heat would be better distributed towards the poles. He points to the more ancient types of plants and animals (reptiles) as requiring warmer conditions, while warm blooded mammalia and birds are adapted to the cooler conditions now prevailing.

As a rule every writer looks to his own theory as being all-sufficient, whereas probably there has been a combination of conditions producing the effects, so that not only may we conclude that the reduction of the sun's radiating power may have had much to do with the present less favourable conditions, but that some of the intermediate changes may have been contributed to by various causes—namely, small shiftings in the geographical position of the earth's axis, increase in the eccentricity of the orbit, to some extent by an alteration of the distribution of land and water and the induced air and ocean currents, and also by cosmical causes and intercepting of the sun's heat by diffused inter-stellar matter.

Insular Floras and Oceanic Islands.

This subject is one the consideration of which cannot be separated from that of the permanence of oceanic beds.

Wallace divides islands into three classes:—Recent continental islands, ancient continental islands and oceanic islands, the latter being generally understood to be those surrounded by seas of more than 1000 fathoms, although as an exception it is acknowledged that some islands belong to the continental classes notwithstanding that the ocean barrier is now over 1000 fathoms. I think that Wallace scarcely sufficiently allows for the effect of long periods of time in altering depths. Time may be all that is wanted to permit of a connection in the past of the remotest group of islands with the mainland.

There seems to be an argument in a circle as far as oceanic insular floras are concerned. First of all it is assumed that if the depth is over a certain amount—say, 1000 fathoms—former land connection was not possible; then comes the study of the flora and fauna of those islands which are thus situated, and those are then looked upon as characteristic of such islands—other islands have these characteristics—the conclusion is drawn that they also have never been connected with the land.

I shall not attempt to prove that important oceanic groups like the Sandwich Islands and the Galapagos Islands were once connected with any of the continental areas. I leave that to abler debaters than myself—like Captain Hutton and Dr. von Jhering—but I wish merely to draw attention to some of the difficulties that the holders of the oceanic insular theory have to contend with.

First let me say that there are many islands, formerly held to be oceanic islands, which are now acknowledged to have had a former continental connection—such as New Zealand, the Fiji and the Solomon Islands. Atolls and coral islands, and some islands of volcanic origin are probably acknowledged by every one to be truly oceanic, and about these there is no dispute. The difficulty lies in the determination whether such groups as the Samoan, Tongan, Marquesan and other groups of the Western and Central Pacific, the Sandwich Islands, Galapagos and some detached islands like Pitcairn and Easter Islands come under this category.

It is well known and acknowledged that there are about 200 species of plants the seeds of which stand immersion in salt water for a certain time, and are, therefore, capable of germination if thrown up by the sea on to a favourable spot, and out of these there is a smaller number which do not lose their germinating powers after prolonged immersion. Then, again, there are some seeds with a hard testa surrounded by pulp, which, after being eaten by birds, may be conveyed to islands at short distances, or perhaps for 50 or 100 miles, as the birds may be in the habit of

visiting them.   There are also plants which have extremely light
or small seeds, or, as in the case of most Compositæ, possessing a
pappus, by means of which they are borne by the wind over long
distances.   Again, there are seeds with barbed hooks which may
adhere to the feathers of birds, or others of small size produced by
plants growing on the margin of water or elsewhere which may be
taken up with particles of mud, and be thus conveyed over considerable distances.   But when this list is exhausted there are still many
plants growing on the larger islands the presence of which cannot
be accounted for.

In the Hawaiian or Sandwich Islands, according to the late
Dr. Hillebrand's investigations, there are 999 species of
phanerogams and vascular cryptogams.   After deducting from
this number the usual littoral and drift species, and many
useful and ornamental plants probably introduced by the natives,
and even allowing a margin for endemic evolution of new species,
after introduction of those from elsewhere, it must be acknowledged that a great power of belief is required to satisfy one that
the balance are all introduced.

The situation of the islands is this :—They are 2,040 miles
from the coast of America, 1,860 from the Marquesas, and 2,190
miles from Tahiti.   It can be seen how small a chance there is
for winds, waves and birds to bring together the component parts
of this rich flora from elsewhere.   There are a few plants representative of the Eastern Australian Region, a few with Southern South
American affinities.   Most of the plants are allied to those on the
nearest coast of America.   Dr. Hillebrand divides the flora
according to the zones which they inhabit on the islands.   These
are as follows :—

1. Lowland zone.

2. Lower forest zone—1,000 or 2,000 feet.

3. Middle   „   —up to 5,000 to 6,000 feet, this being the
   most luxuriant.

4. Upper forest zone—up to 8,000 or 9,000.   On Mauna Kea
   shrubby vegetation extends to 11,000 feet.

5. Bog flora of high table land of Kauai, and of the broad top
   of Mt. Ecka or West Maui.   Here are representatives
   from Antarctica (New Zealand, Falkland Islands,
   Southern Andes, &c.).

It is to be noted that there are 40 endemic or peculiar
genera, one of which is the curious Lobeliaceous tree *Sclerotheca.*

It is most difficult to understand how winds, waves and
birds could have combined to bring the seeds of all these plants
together and pop them down just on the right spot where germination could take place.

The Galapagos Islands are another example; but here the
distance from the mainland is much less, and the number of
species smaller, so that the possibility of accidental introduction
is largely increased; but it is curious that the different islands
possess different species, and those chiefly distinct from the
mainland.   This remark applies to the land snails as well as the
plants.*   The affinities of the endemic flora are entirely American.
A few plants such as *Lipochœta laricifolia,* have congeners in
the Sandwich Islands, and not in America, but the arboreous
Lobeliaceæ are absent.   There are only five species noticed
common to all islands, two species in four islands, and six in
three, according to Mr. Botting Hemsley's account in the "Botany"
of the Challenger.   If species have drifted from the mainland, or
been conveyed by birds or otherwise, why should the same species
not have been conveyed to all islands, or those on one island not
have been transferred to the others?

The floras of the larger islands of the south-western Pacific have
a decidedly Malayan character, and there is not the development
of endemic genera which would lead to the certain conclusion that
the islands were relics of a former more extensive land area.

In the "Botany" of the Challenger Expedition, p. 68, there is
an interesting and instructive remark on the Flora of the Eastern

---

* See Mr. Dall's paper in the Proc. Acad. Nat. Sci. Philadelphia, 1896,
p. 395.

Pacific 'Islands, which runs thus: —"The Australasian genus
Metrosideros penetrates as far eastward as Pitcairn, where, as in
the Sandwich Islands, it forms large woods ; and the prominence
of such other Australasian or Asiatic genera in the Sandwich
Islands as *Pittosporum, Alphitonia, Cyathodes, Scævola* and
Cyrtandra is noteworthy.   On the other hand, the peculiar Sandwich Island types seem to have had a former wider extension, as
is indicated by the Lobeliaceous arboreous genus *Sclerotheca* and
a species of *Phyllostegia* in Tahiti."

When treating of Tristan d'Acunha in the South Atlantic, Mr.
Botting Hemsley says (Appendix, p. 313) :—"Whether the
present distribution of *Phylica nitida* was brought about by the
agency of birds is highly problematical.   The distribution of the
genus, like that of many others of the African region, points
rather to a former greater land connection."

The scientific methods of the present age, starting with Darwin
and Wallace, have been chiefly directed towards discrediting the
miraculous and catastrophal, and towards accounting for all
phenomena by means of existing mechanical causes.   The old
method of explaining facts is admittedly unscientific, but are we not
tempted under modern methods to press the argument just a little
too far the other way ; and having found, for instance, that some
plants, and even some animals, can be dispersed by winds, waves,
birds, &c., assume that all have arrived on the scene by the same
group of chances ?   Is it unscientific to assume the existence in the
past of larger land areas in the Pacific and elsewhere than now
exist ?

Captain Hutton says :—" In the distribution of reptiles and of
some birds in Polynesia, we have evidence of the existence of a
former continent.   The brush turkeys or megapodes are birds
that are unable to fly, and yet they are found in Borneo, Celebes,
the Philippine Islands, Australia, New Guinea, New Caledonia,
the Marion Islands, the Samoan Islands and others in the Pacific.
Reptiles are widely spread throughout the islands of Polynesia,
and we can only account for it by supposing a former land communication.   Mr. Wallace, in his 'Island Life,' attempts to

explain the fact by suggesting that reptiles have some unknown and exceptional powers of dispersal. But if so, why is the phenomenon limited to Polynesia? And why should Mr. Wallace himself explain the small number of reptiles in Great Britain and Ireland by the supposition that they are unable to cross the English and Irish Channels?"[*]

The results of the Challenger dredgings seem to show that the principal part of the Pacific was ocean during the Tertiary period, but it is not impossible that chains of volcanic islands or masses of land may have existed during or before that period and that these, being of a shifting character, at first connected with a continent and afterwards cut off, might preserve the relics of a continental fauna and flora. A continent properly so called can scarcely have existed. The difficulties are too great in the way of such a supposition, but only connections similar to that which we are certain existed between New Zealand, New Caledonia, the Fijis, and the main land which was perhaps at its period of greatest development in a state of oscillation need be conceded.

Captain Hutton's theory of a bridge for the migration of marsupials to Patagonia across the Pacific presents too many difficulties, and my remarks above are by no means intended to support the idea, for the absence of relics on the road is a strong argument against it. Neither on the islands nor on the mainland of Asia between Europe and the Malay Peninsula have at present any fossil remains been found of those animals which alike are represented in Tertiary Europe and Patagonia.

The facts seem rather to point to the conclusion that the Australian Marsupials were derived either from an ancient and extended Patagonia or that the ancestors in both countries were developed previously in some Antarctic region now submerged.

Some ight on the subject of the former distribution of land and water is thrown by Dr. H. von Jhering, who has kindly

---

[*] New Zealand Journal of Science, Vol. i. 1883, p. 411.

furnished me with a copy of his Treatise "Das neutropische Floren-
gebiet und seine Geschichte," (Engler's Jahrbuch, 1893). This able
treatise deals with matters of special interest to us, and therefore
deserves notice in this place, but I find that I have not space to
refer to it at the length which it deserves, and I must therefore
now confine myself to stating his main arguments, at the same
time recommending those interested to study the original work.
The author sets himself to upset Wallace's axiom of the permanence of continents and oceans which would, if true, require
that South America was always cut off from connection with
south-eastern Asia as it is at present, and he disputes the validity
of the assertion that the bottoms of oceans over 1000 fathoms in
depth could never have been dry land. He says that greater
depths only indicate longer time for subsidence. The effect of
separation at different epochs would be that we should find the
fauna limited to the groups which had reached their development
before then, and he points to the Pacific Islands, where the
tertiary fauna are absent altogether, as proof of their isolation
in Mesozoic times, while on the other hand lizards, ancient types
of mollusks and insects are found.

The author divides South America into three regions. The
northernmost has affinities with North America, the middle one
with Africa, Madagascar and Bengal. These regions he suggests
after an investigation of the fresh water fauna. They were in
the Cretaceous and early Tertiary separated by ocean. He
concludes that a great continent which he calls "Archhelenis"
extended across the Atlantic to Africa and beyond; this possessed
probably no mammalia, but a rich fresh water fauna and identical
reptiles and amphibia. The lower region he calls "Archiplata,"
it was formerly connected with New Zealand, and partly
with Australia and Tasmania. The early Tertiary mammals
existed in this region, but not in Archhelenis. The Dasyure
group connects with Australia. The *Anoplotheridæ* and *Thero-
domyidæ* have affinities with the Eocene fauna of the old world.
Argentina can only have received her Eocene mammals across
antarctic lands. In the Pliocene North and South America

became united, and an interchange of forms thereafter took place. The land connecting La Plata and Patagonia with South Eastern Asia he calls "Archinotis." He says the bridge between South America and Africa broke up before that between India and Africa, so that when the middle and southern South American regions became united no neotropical African types could migrate to Australia.

The author then discusses the various methods by which plants and animals are understood to be transported across the ocean, and throws doubt upon the whole theory of oceanic islands. Speaking of the island group of Ferdinand Noronha, he says, "It is certain that on the main island birds scatter the seeds of berries, fruits, &c., but when wind and birds do not cause the spread of the plants even from one island to another the distance of a gunshot, how can one believe that this means of distribution is effective across gaps of hundreds or thousands of kilometers?" The author disputes the fact of the Andean migration; he says there is not a species common to the Californian Sierra Nevada and the Andes. With regard to the exchange of plants of higher latitudes north and south of the equator, he is of opinion that formerly these must have been capable of existing in warm regions as well as in cold. Even now *Ranunculus, Polygonum, Stellaria media, Samolus Valerandi, Veronica anagallis, Parietaria debilis,* &c., are not sensitive to climate. He says that formerly plants were not so restricted by climate, so that the following genera are found together in the Upper Pliocene of Niederrad and Hochst am Main : *Juglans, Aesculus, Carya, Liquidamber, Corylus avellana, Betula alba, Picea vulgaris,* and the alpine *Pinus cembra* and *Pinus montana.* The author then discusses the distribution of various genera, *Podocarpus* and other southern *Coniferæ, Cocos, Nipa* and other Palms, *Cupuliferæ,* &c. He is of opinion that the completeness of the Indo-Australian territory must have been longer retained than the connection of Australia and New Zealand, and he says that if the genera *Canis* and *Sus,* the *Muridæ,* &c , could push into New Guinea and Australia, the connection with Asia must have lasted into the Miocene. During the whole

53c

Tertiary period there was a constant change of mammals between North America and Europe, but it was not complete, probably those that could not face a temperate climate could not pass. This might explain the fact of the *Anoplotheridæ* and *Theridomyidæ* being found in the Argentine beds and Europe but not in North America. The author then discusses the fresh water flora and finds the conclusion derived from their consideration to fit in with that deduced from the fresh water fauna.

The South American Mammalia—Recent and Extinct

I cannot conclude my Address without making special reference to the wonderful discoveries of fossil mammals recently made in South America. The importance of these discoveries to us is that in this region not only placental mammals of very peculiar types have been found differing in important respects from allied forms in other parts of the world, but that marsupials of distinctly Australian affinities also occur. Here I should like to refer to a most interesting find in Ecuador of a living animal of a strange type, and the proof that it is marsupial in character. Before this the only living representative in America was the Opossum (that is the true Opossum or Didelphys which belongs to the Polyprotodont group). This new animal called *Cænolestes* resembles the group of Kangaroos and Australian Opossums (properly called Phalangers) in being diprotodont, but differs from them in not being syndactylous.

The work of describing the fossil mammals is being carried out by Lydekker and Osborne; Scott and F. Ameghino have also written on the subject. I do not propose, however, to go into details, which indeed would be premature, seeing that the whole subject has only been partially investigated, and I would rather refer to the works of the above authors. I wish, however, to call attention to what appears to be the latest deliverance on this subject by Florentino Ameghino, which has been translated by Arthur Smith Woodward, and published in the Geological Magazine for January of this year under the title—" Notes on the Geology and Palæontology of Argentina." This is a very important paper, because if the views as to the age of the beds and the affinities of the remains are corroborated, Patagonia must have been a centre of distribution of mammals not only for the Antarctic regions of the time, but also for Europe and, perhaps, North America.

Mr. F. Ameghino shows that beds exist—red sandstones— containing remains of Dinosaurs and undoubtedly of Upper Cretaceous Age. Above those and quite continuous with them comes the Pyrotherium Formation, containing armoured and unarmoured Edentates, peculiar Carnivora, Plagiaulacidæ, Hystricomorphous Rodentia, peculiar Ungulates and primitive forms of Primates. Ameghino includes *Pyrotherium* among the Ungulates, and considers it allied to the Proboscidea, but Woodward asks in a note at the end whether it may not be allied rather to *Diprotodon*. Ameghino says that if these beds are not Cretaceous, then Dinosaurs lived in Patagonia until a more recent epoch than in other portions of the globe.

Above the Pyrotherium Formation comes the Patagonian Formation, which has been erroneously confounded with the marine formations of Parana. The mollusca of the Patagonian Formation have been stated by D'Orbigny, Sowerby, Philippi, Hupé, Remond de Corbineau and Steinman to be partly of Eocene and partly of Upper Cretaceous Age. The objection to this antiquity is the presence of remains of Cetacea, which only appear in Europe during the Miocene, but F. Ameghino thinks the group might well have originated earlier in the southern hemisphere, and says their remains are more primitive in type, as has been recognised by Lydekker.

Next above comes the Santa Cruz Formation, which was at one time supposed to be anterior to the Patagonian, on account of the latter having been confused with the Parana. There are here numerous remains of extinct mammals, gigantic birds and reptiles. There are marsupials of the Diprotodont group, which like the living *Cænolestes* above referred to, and unlike the Kangaroos, are not syndactylous. These are stated to resemble the Plagiaulacidæ. This formation comprises a marvellous collection of animals including *Homunculus*. Philippi considers it to be of Miocene Age.

Above this lies the Boulder or Tchuelche Formation, which, as Darwin has shown, is of marine not glacial origin. This is stated to be of Miocene Age.

Later signs of geological phenomena are the transverse valleys of Patagonia and the Pampean Formation, which latter contains six or seven successive mammalian faunas. Dr. von Jhering says of the mollusca that almost all the species live still on the shores of Brazil.

There are numerous plant remains in the beds included in this formation, and it is to be hoped that investigation of the same may be made without delay.

TERTIARY PLANT REMAINS IN AUSTRALIA.

Mr. T. S Hall and Mr. G. B. Pritchard have done much to unravel the difficulties of determination of the age of the Tertiary beds of Victoria.

Much confusion had previously resulted from a misunderstanding of the position of what is termed the Older Basalt, which was considered Miocene by Professor McCoy, on account of its being supposed to overlie beds of Miocene Age. Messrs. Hall and Pritchard have shown this view to be erroneous, and the date, instead of being Miocene, to be early Tertiary,* as it has been found to be overlapped by acknowledged marine Eocene strata.

Underneath the Lower Basalt lie in various localities of the colony indurated clays cemented with ferruginous or siliceous material and containing beautifully preserved plant remains, and the conclusion seems almost forced upon us that these are Upper Cretaceous in age.

---

* On the Age of certain Plant-bearing beds in Victoria. Aust. Assoc. 1893, Adelaide.

Messrs. Hall and Pritchard in the same paper suggest that the beds at Dalton and Vegetable Creek, which have the same lithological character, and which Baron Ettingshausen considered Eocene, may have to be referred back to the Cretaceous also.

Messrs. Hall and Pritchard have written several valuable papers discussing the age of the Tertiary strata of Victoria, and Mr. T. H. Wright has in the most painstaking manner investigated the geological features of an area of Gippsland, and proved the true sequence of the beds, in some cases entirely reversing previously received ideas. Unfortunately I am unable through lack of time and space to enter into these matters as I should like, and can, therefore, only refer to the papers read by those gentlemen before the Royal Society of Victoria and Australasian Association, and in the case of Mr. Wright's investigation, to the 8th Report of the Geological Survey of Victoria.

EARLIEST DICOTYLEDONS IN THE NORTHERN HEMISPHERE.

In the Report of the United States Geological Survey (Vol. xvi. Part 1), just received, there is a paper by Professor Lester F. Ward entitled "Some Analogies in the Lower Cretaceous of Europe and America."

Up to 1888 the oldest known dicotyledon was one from the Middle Cretaceous of Greenland, which was described by Heer under the name of *Populus primaeva*.

Professor Fontaine in 1888 found in some of the Lower Potomac Series, in what was supposed to be Jurassic, some portions of leaves resembling dicotyledons, but not easily distinguishable from the lower groups, ferns, cycads and other gymnosperms.

In the Report to which reference is now made Professor Ward says :—" On numerous occasions, dating as far back as 1878, I have expressed the opinion that the dicotyledons could not have had their origin later than the Middle Jura, and it will not surprise me if the final verdict of science shall place the Potomac formation, at least the lower member, in which the plants occur, with that geologic system "

Since then the known flora of the Potomac formation has been greatly increased by further discoveries, and an unbroken series from the oldest to the newest beds brought to light—in the latter the dicotyledonous element largely predominates.

Marquis Saporta called attention not long after Prof. Fontaine's discovery to the existence of peculiar forms in the Lower Cretaceous of Portugal, some of which he referred to his group of Proangiosperms while others represented true Dicotyledons. These beds are probably of the age of the Gault, that is Middle Cretaceous. It was found that other collections from older beds also contained dicotyledons, and in 1891 Saporta published a paper on the subject.

Professor Ward, comparing the Jurassic flora of Portugal with the Potomac beds, concludes as follows :—" But the special interest which these comparisons have in this place is the intimate bond which they furnish between the late Jurassic of Portugal (supposed to correspond closely with the Kimmeridge Clays of England, but perhaps running up into the Portland beds and thus closely approaching the Purbeck, which has been treated in this paper as part of the Wealden) and the oldest Cretaceous of America, which some geologists in this country make to extend some distance into the Jurassic, but which is here treated as a Cretaceous deposit."

EARLIEST DICOTYLEDONS IN AUSTRALIA.

The fossils of the Oxley beds are well developed dicotyledons, quite equal in development to those found in the Upper Cretaceous in Europe and North America. The Oxley beds are near the top of the Ipswich Coal Measures, which are supposed to be at latest Jurassic in age. The difficulty of reconciling the fact of the full development of the dicotyledonous type in Australia with the very archaic rudimentary types of the same age in North America which are mentioned by Lester Ward, struck me very forcibly, and as in the western parts of the Colony it had been shown that the Lower Cretaceous beds lie, conformably, or at an angle not distinguishable, upon the beds below them, 1 thought it desirable to inquire of Mr. R. L Jack whether it was not possible that the same condition existed on the coast side of the Dividing Range, and that thus the beds in question might really be of Lower Cretaceous Age.

Mr. Jack's reply is as follows :—" I cannot see my way to putting the Oxley beds on a higher horizon than the rest of the Ipswich formation. Stratigraphically it would not work They form an integral part of the formation which from top to bottom yields the assemblage of plants on which the Triasso-Jurassic age of the whole was founded. They are pretty well up in the series, but what evidence there is is all against their being the uppermost part or anywhere near it. I believe them to be *below* the thick Murphy s Creek Sandstone and the Clifton Coals and Shales which give the same fossil plants as the shales associated with the coal seams of Ipswich proper."

If Mr. Jack's views as to the age of the beds is correct, they point undoubtedly to the conclusion that at an age when European and American dicotyledons exhibited a rudimentary or transition character, the southern hemisphere already possessed types of high development. Before this becomes an accepted fact, it is needless to say that some further corroboration of the conclusions as to the correspondence in age of the so-called Jurassic beds of Australia and those of the northern hemisphere should be sought.

I wish to take this opportunity of expressing my best thanks to Messrs. R. Etheridge, Junr., T. W. E. David, E. F. Pittman, R. L. Jack, T. S. Hall, G. B. Pritchard, J. H. Wright, H. C. Russell, C J. Merfield, C. Hedley, R. T. Baker, H. C. L Anderson, J. J. Fletcher and others for the assistance they have given me in the preparation of this Address and that of last year by placing books and facts at my disposal.

On the motion of Professor Haswell, seconded by Mr. W. S. Dun, a very hearty vote of thanks was accorded to the President for his interesting Address.

The subjoined financial statement for the year ending March 31st, 1897, was presented by the Hon. Treasurer, and adopted.

CAPITAL ACCOUNT.

	£ s. d.		£ s. d.
Endowment: Received from Sir William Macleay in his life-time 14,000 0 0		Loan A 3,000 0 0	
Further sum bequeathed by his will £6,000, less probate duty deducted by his executors 5,700 0 0		Loan B 5,000 0 0	
Bacteriology: Legacy of £12,000 bequeathed to the University of Sydney, less £600 duty; paid by the University into Court and ordered to be paid to the Society 11,400 0 0		Loan C (together with £900 out of income to make up £24,000)... ...23,100 0 0	
£31,100 0 0		£31,100 0 0	

March 30th, 1897.
Audited and found correct.

HUGH DIXSON
E. G. W. PALMER } Auditors.

JAMES NORTON, Hon. Treasurer,

INCOME ACCOUNT.

	£	s.	d.		£	s.	d.
Balance from 1895	51	9	7	Printing Proceedings and Sundries ...	311	4	2
Entrance Fees from four Ordinary, and one Associate Members ...	9	9	0	Plates and Drawings for Proceedings ...	141	0	6
				Stationery	3	13	0
Subscriptions from 78 Members ...	94	16	6	Postage (Hon. Treasurer's) ...	1	6	0
Interest received	2,670	15	4	Rent, Rates and Insurance ...	83	1	2
Sales of Publications, including 100 copies of Proceedings sold to the Government	109	5	1	Law Charges	2	2	0
				Bank Charges for Collecting Cheques, &c.	0	13	10
				Petty Cash (Advertisements, Gas, Distribution of Publications, Postage, and Sundries)	30	0	0
				Salaries and Wages	478	0	0
				Dulau & Co., to cover advertisements for Bacteriologist	10	0	0
				Interest re-invested included in Loan C for £24,000	900	0	0
				Balance in Commercial Bank ...	974	14	10
	£2,935	15	6		£2,935	15	6

March 30th, 1897.
Audited and found correct.

HUGH DIXON
E. G. W. PALMER } Auditors.

JAMES NORTON, Hon. Treasurer.

BACTERIOLOGY BEQUEST.

INCOME ACCOUNT.

	£ s. d.		£ s. d.
Interest on £11,400 received from the Supreme Court...	1,446 0 1	Dulau & Co., to cover advertisements for Bacteriologist	10 0 0
Interest at 4 per cent. on £11,400 (part of £24,000: Loan C); seven months' interest, 30th March to 30th October	260 11 5	Balance covered by £900 (part of £24,000: Loan C), and £817 2s. 11d., part of £974 14s.10d. in Commercial Bank	1,717 2 11
Interest on £900, part of same £24,000 (which £900 is a re-investment of part of the £1,446 0s. 1d. received from the Court for interest)... ...	20 11 5		
	£1,727 2 11		£1,727 2 11

March 30th, 1897.
Examined and found correct.

HUGH DIXSON
E. G. W. PALMER } Auditors.

JAMES NORTON, Hon. Treasurer.

The following gentlemen were elected

OFFICE BEARERS AND COUNCIL FOR 1897.

PRESIDENT :

PROFESSOR J. T. WILSON, M.B., CH.M.

VICE-PRESIDENTS :

J. C. COX, M.D., F.L.S.

HENRY DEANE, M.A., M. INST. C.E., F.L.S.

PROFESSOR T. W. E. DAVID, B.A., F.G.S.

HONORARY TREASURER :

HON JAMES NORTON, LL.D., M.L.C.

COUNCIL :

RICHARD T. BAKER, F.L S.

CECIL W. DARLEY, M. INST. C.E.

THOMAS DIXSON, M.B., CH M.

JAMES R. GARLAND, M A

PROFESSOR W. A. HASWELL, M.A , D.SC.

CHARLES HEDLEY, F.L.S.

A. H. S. LUCAS, M.A., B.SC.

J. H. MAIDEN, F.L.S., &c.

PERCEVAL R. PEDLEY.

THOMAS STEEL, F.C.S

PROSPER N. TREBECK, J.P

FRED. TURNER, F.L.S.

AUDITORS :

HUGH DIXSON, J.P.

EDWARD G. W. PALMER.

INDEX.

(1896.)

Names in Italics are Synonyms.

PROCEEDINGS

OF THE

LINNEAN SOCIETY

OF

NEW SOUTH WALES.

SUPPLEMENT TO PROCEEDINGS, 1896.

CATALOGUE OF THE DESCRIBED COLEOPTERA OF AUSTRALIA. SUPPLEMENT, PART II.

Dytiscidæ, Gyrinidæ, Hydrophyllidæ, Staphylinidæ, Pselaphidæ, Paussidæ, Silphidæ, Scaphidiidæ, Histeridæ, Phalacridæ, Nitidulidæ, Trogositidæ, Colydiidæ, Cucujidæ, Cryptophagidæ, Lathridiidæ, Mycetophagidæ, Dermestidæ, Byrrhidæ, Parnidæ, Heteroceridæ.

By George Masters.

Family DYTISCIDÆ.

Sub-Family DYTISCIDES.

CANTHYDRUS, Sharp.

7581. Bovillæ, Blackb., P.L.S.N.S.W. (2) iv. 1889, p. 446.
S. Aust.; N. Territory.

Part ii. of the Catalogue is contained in Vol. x., Part 4, pp. 583-672 (published April 3, 1886).

The left hand number continues the pagination of the Catalogue; the right hand that of the Supplement.

HYPHYDRUS, Illiger.

7582. LYRATUS, Swartz, Schonh. Syn. Ins. ii. p. 29, t. 4, f. 1, = Sp. 1030, *Hydroporus fossulipennis*, Macl.; Sharp, Trans. R. Dubl. Soc. (2) ii. 1882, p. 997.

Australia; widely distributed.

Sp. 968. H. AUSTRALIS, Clark = *H. Blanchardi*, Clark; Sharp, l.c p. 1000 = Sp. (probably) 1028, *Hydroporus bifasciatus*, Macl.; Sharp, l.c. p. 789.

Australia; widely distributed.

HYDROCANTHUS, Say.

7583. WATERHOUSEI, Blackb., Trans. Roy. Soc. S. Aust. x. 1886-7 p. 65.

. S. Australia.

STERNOPRISCUS, Sharp.

Sp. 999. S. MULTIMACULATUS, Clark = *Hydroporus sinuaticollis*, Clark; Sharp, l.c. (2) ii. 1882, p. 999.

S. and W. Australia.

MACROPORUS, Sharp.

Sp. 1022. M. GARDNERI, Clark = *Hydroporus brunnipennis*, Macl.; Sharp, l.c (2) ii. 1882, p. 996.

Australia and Tasmania.

Sp. 1024 M HOWITTI, Clark = *Hydroporus foveiceps*, Macl.. Sharp, l.c. p. 997.

Australia and Tasmania.

NECTEROSOMA, Macleay.

7584. UNDECIMLINEATUS, Babing., (*Hydroporus*) Trans. Ent. Soc. Lond. iii. 1841, p. 13; Sharp, l c. (2) ii. 1882, p. 414.

N.S. Wales; Clyde River.

Sp. 1038. N. PENICILLATUS, Clark = *N. vittipenne*, Macl., Sharp, l.c. p. 823.

Australia; widely distributed.

PLATYNECTES, Sharp.

Sp. 1047. P. DECEMPUNCTATUS, Fab. = *P. lugubris*, Blanch.,
Mastersi, Macl., and *spilopterus*, Germ.; Sharp, l.c. (2)
ii. 1882, p. 988.

Australia; widely distributed.

LANCETES, Sharp.

7585. OCULARIS, Lea, P.L.S.N.S.W. (2) x. 1895, p. 224.

W. Aust.; Donnybrook.

COPELATUS, Erichson.

Sp. 1064. C. AUSTRALIS, Clark = *Celina australis*, Clark;
Sharp, l.c. (2) ii. 1882, p. 564.

Australia; various localities.

RHANTATICUS, Sharp.

7586. SIGNATIPENNIS (HYDATICUS), Lap., Etud. Ent. p. 95, Aubé,
Spec. p. 158; Sharp, l.c. (2) ii. 1882, p. 691.

Australia.

HYDATICUS, Leach.

7587. CONSIMILIS, Régimb., Notes Leyd. Mus. ix. 1887, p. 224.

Queensland.

7588. GORYI, Aubé, Spec. p. 174 = Sp. 1114, *H. ruficollis*, Fab.;
Sharp, l.c. (2) ii. 1882, p. 656, No. 1020.

Australia; N.S. Wales and Queensland.

7589. PARALLELUS, Clark, Trans. Ent. Soc. Lond. 1864, p 219;
Sharp, l.c. p. 653.

N.S. Wales.

Sp. 1060. H. PULCHER (COLYMBETES), Clark; Sharp, l.c.
p. 665.

Australia; widely distributed.

CYBISTER, Curtis.

7590. GRANULATUS, Blackb., P.L.S.N.S.W. (2) iii. 1888, pp. 812 and 1393.

S. Aust.; N. Territory.

7591. TRIPUNCTATUS (DYTISCUS), Oliv., Ent. iii. 1795, p. 14, t. 3. f. 4 = Sp. 1087, *C. gayndahensis*, Macl ; Sharp, l c. (2) ii. 1882, p. 1140, Olliff, Memoirs, Aust. Mus. ii. 1889, p. 80. Australia; widely distributed; Lord Howe Island.

ERETES, Castelnau.

Sp. 1108. E. AUSTRALIS, Erich. = *Eunectes punctipennis*, Macl.; Sharp, l.c. (2) ii. 1882, p. 994.

Australia; widely distributed.

Family GYRINIDÆ.

DINEUTES, W. S. Macleay.

7592 INFLATUS, Blackb., Trans. Roy. Soc. S.A. xix. 1895, p. 28, Victoria.

GYRINUS, Geoffroy.

7593. STRIOLATUS, Guér, Voy. Coquille, 1830, ii. Col. p. 62; Boisd., Voy. Astrol. ii. p. 66.

Australia.

MACROGYRUS, Régimbart.

Sp. 1119. M. OBLONGUS (ENHYDRUS), Boisd.; Régimb., Ann. Soc. Ent. Fr. 1882 (6) ii. p. 449.

Sp. 1120. M. REICHEI (ENHYDRUS), Aubé; Rég., l.c. p. 453.

Sp. 1130. M. OBLIQUATUS (GYRINUS), Aubé; Rég., l.c. p. 443.

Sp. 1132. M. VENATOR (GYRINUS), Boisd; Rég., l.c. p. 443.

Sp. 1151. M. RIVULARIS (ENHYDRUS), Clark; = (probably) *M. longipes*, Régimb.; Régimb, l.c. p. 450.

Family HYDROPHYLLIDÆ.

HYDROPHILUS, Geoffroy.

7594. BREVISPINA, Fairm., Journ. Mus. Godeff. 1879, p. 80.
Queensland, Moreton Bay.

STETHOXUS, Solier.

7595 PEDIPALPUS, Bedel., Rev. d'Ent. x. 1892, p. 312.
Australia.

STERNOLOPHUS, Solier.

7596. TENEBRICOSUS, Blackb., P.L.S.N.S.W. (2) iii. 1888, p. 813.
N. Aust.; Palmerston.

HYDROBIUS, Leach.

7597. MACER, Blackb., l.c. (2) iii. 1888, p. 819.
Victoria.
Sp. 1142. H. ASSIMILIS, Hope; Blackb., l.c. p. 818.

HYDROBIOMORPHA, Blackburn.

7598. BOVILLI, Blackb., l.c. (2) iii. 1888, p. 816.
N. Aust.; Palmerston.
7599. TEPPERI, Blackb., l.c. p. 817.
N. Aust.; Palmerston.
7600. HELENÆ, Blackb., l.c. (2) iv. 1889, p. 741.
S. Aust.; N. Territory.

PARACYMUS, Thomson.

7601. LINDI, Blackb., l.c. (2) iii. 1888, p. 821.
S. Aust.; Port Lincoln.
7602. METALLESCENS, Fvl., Rev. d'Ent. ii. p 352.
Australia.

7603. NIGERRIMUS, Blackb., Trans. Roy. Soc. S.A. xiv. 1891,
p. 66.

Mountains of Victoria.

7604. NITIDIUSCULUS, Blackb., P.L.S.N.S.W. (2) iii. 1888, p. 820.

S. Aust. and Victoria.

7605. SUBLINEATUS, Blackb., l c. p. 821.

S. Aust.; Roseworthy.

PHILHYDRUS, Solier.

7606. BURRUNDIENSIS, Blackb., P.L.S.N.S.W. (2) iv. 1889, p. 447.

S. Aust.; N. Territory.

7607. EYRENSIS, Blackb., Trans. Roy. Soc. S.A. xix. 1895, p. 29.

S. Aust.; Eyre's Peninsula.

7608. LÆVIGATUS, Blackb., P.L.S.N.S.W. (2) iii. 1888, p. 822;
Trans. Roy. Soc. S. Aust. xv. 1892, p. 207.

S. Aust. and Victoria.

LACCOBIUS, Erichson.

7609. AUSTRALIS, Blackb., Trans. Roy. Soc. S.A. xiv. 1891, p. 67.

Victoria ; Ovens River.

7610. MONTANUS, Blackb., l.c. p. 67.

Mountains of Victoria.

HYDROBATICUS, Macleay.

7611. AUSTRALIS, Blackb., P.L.S.N.S.W. (2) iii. 1888, p. 823.

S. Aust. and Victoria.

7612. CLYPEATUS, Blackb., l.c. (2) v. 1890, p. 305.

S. Aust.; N. Territory.

Sp. 1148. H. TRISTIS, Macl.; Blackb., l.c. (2) vii. 1892, p. 99.

Sp. 1149. H. LURIDUS, Macl.; Blackb., l.c. p 99.

BEROSUS, Leach.

7613. APPROXIMANS, Fairm., Journ. Mus. Godeff. xiv. 1879, p. 82.

Queensland ; Peak Downs.

7614. AURICEPS, Blackb., P.L.S.N.S.W. (2) iv. 1889, p. 447.

S. Aust.; N. Territory.

7615. DECIPIENS, Blackb., l.c. (2) iii. 1888, p. 827.

S. Aust.; N. Territory.

7616. DISCOLOR, Blackb., l.c. p. 829.

S. Aust.; Port Lincoln.

7617. DUPLO-PUNCTATUS, Blackb., l.c. p. 828.

S. Aust.; Adelaide, Port Lincoln, &c.

7618. EXTERNIPENNIS, Fairm., Journ. Mus. Godeff. xiv. 1879, p. 81.

Queensland ; Rockhampton.

7619. FLINDERSI, Blackb., l.c. p. 831.

S. Aust.; Port Lincoln.

7620. GRAVIS, Blackb., l.c. p. 826.

S. Australia.

7621. MAJUSCULUS, Blackb., l.c. p. 824 ; Trans. Roy. Soc. S.A. xv. 1892, p. 207.

S. Australia.

7622. MUNITIPENNIS, Blackb., l.c. xix. 1895, p. 30.

S. Aust.; near Lake Callabonna

7623. OVIPENNIS, Fairm., Journ. Mus. Godeff. xiv. 1879, p. 83.

Queensland ; Port Mackay.

7624. PALLIDULUS, Fairm., Journ. Mus. Godeff. xiv. 1879, p. 81.

Queensland ; Peak Downs.

7625. SIMULANS, Blackb., P.L.S.N.S.W. (2) iii. 1888, p. 832.

S. Aust.; Adelaide.

7626. STICTICUS, Fairm., Journ. Mus. Godeff. xiv. 1879, p. 82.
Queensland ; Peak Downs.

7627. STIGMATICOLLIS, Fairm., l.c. p. 82.
Queensland ; Peak Downs.

NOTOBEROSUS, Blackburn.

7628. ZIETZI, Blackb., Trans. Roy. Soc. S.A. xix. 1895, p. 30.
S. Aust.; near Lake Callabonna.

SPERCHEUS, Kugelann.

7629. MULSANTI, Perr., Ann. Soc. Linn. Lyon, 1864, p. 91.
Australia.

7630. FRISCUS, Sharp, Ent. Mo. Mag. xi. p. 250 ; Fvl., Rev
d'Ent. ii. 1883, p. 351.
Australia.

VOLVULUS, Brullé.

7631. PUNCTATUS, Blackb., P.L.S.N.S.W. (2) iii. 1888, p. 839.
S. Aust.; N. Territory.

7632. SCAPHIDIFORMIS, Fairm., Journ. Mus. Godeff. xiv. 1879, p.
83.
Queensland ; Rockhampton.

HYDROCHUS, Leach.

7633. ADELAIDÆ, Blackb., P.L.S.N.S.W. (2) iii. 1888, p. 832.
S. Aust.; Adelaide.

7634. OBSCURO-ÆNEUS, Fairm., Journ. Mus. Godeff. xiv. 1879, p.
80.
Queensland ; Port Mackay.

7635. REGULARIS, Blackb., l.c. (2) iii. 1888, p. 833.
Victoria.

7636. VICTORIÆ, Blackb., l.c. p. 835.
Victoria ; Ararat.

OCHTHEBIUS, Leach.

7637. AUSTRALIS, Blackb., P.L.S.N.S.W. (2) iii. 1888, p. 835.
S. Aust.; Adelaide.

HYDRÆNA, Kugelann.

7638. ACUTIPENNIS, Fairm., Journ. Mus. Godeff. xiv. 1879, p. 81.
Queensland ; Brisbane.

7639. TORRENSI, Blackb., l.c. (2) iii. 1888, p. 837.
S. Aust.; Adelaide.

CYCLONOTUM, Erichson.

7640. ABDOMINALE, Fabr. Syst. El. i. p. 94 ; Muls., Ann. Soc.
Agr. Lyon, 1844, p. 179 ; Blackb., P.L.S.N.S W. (2)
ix. 1894, p. 91.
Queensland ; Brisbane.

7641. AUSTRALIS, Blackb., l.c. iii. 1888, p. 839.
S. Australia.

Sp. 1157. C. PYGMÆUM, Macl.; Blackb., Trans. Roy. Soc.
S.A. xviii. 1894, p. 203.

CERCYON, Leach.

7642. FLAVIPES, Fabr., Ent. Syst. i. p. 81 ; Blackb , Trans. Roy.
Soc. S.A. xiv. 1891, p. 68.
Mountains of Victoria.

7643. FOSSUM, Blackb., P.L.S.N.S.W. (2) iii. 1888, p. 839.
S. Australia.

Family STAPHYLINIDÆ.

Sub-Family ALEOCHARIDES.

FALAGRIA, Mannerheim.

Sp. 1160. F. FAUVELI, Solsky = *Myrmecocephalus cingulatus,* Macl.; Oll., P.L.S.N.S.W. (2) i. 1886, p. 410. ·
Queensland ; Gayndah.

7644. BICINGULATUS (MYRMECOCEPHALUS), Macl.; Oll., P.L.S.
N.S.W. (2) i. 1886, p. 411.

Queensland ; Gayndah.

7645. PALLIPES, Oll., l.c. p. 411.

Tasmania ; Gould's Country.

BOLITOCHARA, Mannerheim.

Sp. 1161. B. DISCICOLLIS, Fvl.; Oll., l.c. (2) i. 1886, p. 413.
S. and W. Australia.

SILUSA, Erichson.

Sp. 1162. S. MELANOGASTRA, Fvl.; Oll., l.c. (2) i. 1886, p. 450.
Victoria and Tasmania.

Sp. 1163. S. PALLENS, Fvl.; Oll., l.c. p. 450.
N.S. Wales ; Sydney.

ALEOCHARA, Gravenhorst.

7646. ACTÆ, Oll., P.L.S.N.S.W. (2) i. 1886, p. 458.
Sydney.

7647. BALIOLA, Oll., l.c. p. 462.
Tasmania.

7648. INSIGNIS, Blackb., Trans. Roy. Soc. S.A. x. 1887, p. 47.
S. Aust.; Port Lincoln.

7649. INSUAVIS, Oll., l.c. (2) i. 1886, p. 460.
N.S. Wales ; Monaro.

7650. LÆTA, Blackb., Trans. Roy. Soc. S.A. x. 1887, p 46.
S. Aust.; Port Lincoln.

7651. OCCIDENTALIS, Blackb., l.c. p. 46.
W. Australia.

7652. PELAGI, Blackb., l.c. p. 45.
S. Aust.; Port Lincoln.

Sp. 1164. A. BRACHIALIS, Jekel ; Oll., P.L.S.N.S.W. (2)
i. 1886, p. 459.

Queensland ; Wide Bay. N.S. Wales ; Sydney.

Sp. 1165. A. HÆMORRHOIDALIS, Guér ; Oll., l.c. p. 459.

Australia ; widely distributed.

Sp. 1166. A. MASTERSI, Macl.; Oll., l.c. p. 460.

Queensland ; Gayndah

Sp. 1167. A. PUNCTUM, Fvl.; Oll, l c. p. 456.

Tasmania. N.S. Wales ; Illawarra.

Sp. 1168. A. SEMIRUBRA, Fvl.; Oll., l.c. p. 465.

Queensland ; Gayndah.

Sp. 1169. A. SPECULIFERA, Erichs ; Oll., l.c. p. 455.

N.S. Wales ; Hunter River. Tasmania ; Port
Frederick.

Sp. 1191. A. (OXYPODA) ANALIS, Macl.; Oll., l c. p. 461.

Queensland ; Gayndah.

Sp. 1192. A (OXYPODA) BISULCATA, Redtenb.; Oll., l.c. p.
457 ; Fvl., Ann. Mus. Genov. x. 1877, p. 289.

N.S. Wales ; Sydney. S. Aust.; Port Lincoln.

7653. CROCEIPENNIS, Mots., Bull. Mosc. xxxi. 1858, p. 238 ;
Fvl., Ann. Mus Genov. x. 1877, p. 292.

sanguinipennis, Kraatz, Wiegm. Arch 1859, p. 17.

maculipennis, Kraatz, l c. p. 17 ; Oll., P.L.S.N.S.W. (2) i.
1886, p. 463.

Queensland ; Gayndah.

7654. MARGINATA, Fvl., l c. x. 1877, p. 291 ; Oll, l.c. p. 463.

N. Australia ; Cape York.

7655. PUBERULA, Klug, Col. Madag. p. 139 , Kraatz, l.c. xxv.
1859, p. 16.

decorata, Aubé, Ann. Soc. Ent. Fr. (2) xix. 1850, p. 131.

Armitagei, Woll., Ins. Mad. 1854, p 599.

dubia, Fvl., Ann. Soc. Ent. Fr. (4) iii. 1863, p. 429; Oll., P.L.S.N.S.W. (2) i. 1886, p. 464.

Australia ; widely distributed.

7656. VICINA, Oll., l.c. p. 464.

W. Australia ; K. G. Sound.

CORREA, Fauvel.

Sp. 1170. C. OXYTELINA, Fvl.; Oll., l c. (2) i. 1886, p. 466
S. Aust.; Adelaide.

POLYLOBUS, Solier.

7657. ACCEPTUS, Oll., P.L S.N.S.W. (2) i. 1886, p. 441.

N.S. Wales ; Watson's Bay, Sydney, &c.

7658. FUNGICOLA, Oll., l.c. p. 442.

N.S. Wales ; Sydney.

7659. LONGULUS, Oll., l c. p. 440.

N.S. Wales ; Shelley's Flats.

7660. NOTUS, Oll., l.c. p. 440.

N S. Wales ; Sydney.

7661. OBESUS, Oll., l c. p. 442.

N.S. Wales ; Sydney.

7662. SODALIS, Oll , l.c. p. 438.

N.S. Wales ; Sydney.

7663. TASMANICUS, Oll , l.c. p. 444.

Tasmania.

7664. USITATUS, Oll., l.c. p. 443.

N.S. Wales ; Sydney.

Sp 1171. *P. apicalis*, Fvl.; Oll., l.c. p. 443.

Victoria.

Sp. 1172. *P. aterrimus*, Fvl.; Oll , P.L.S.N.S.W. (2) i. 445.
W. Aust.; K. G. Sound.

Sp. 1173. P. CINCTUS, Fvl.; Oll., l c. p. 436.
Victoria.

Sp. 1174. P. FLAVICOLLIS, Macl.; Oll., l.c. p. 438.
Queensland ; Gayndah.

Sp. 1175. P. INSECATUS, Fvl.; Oll., l c. p. 439.
N.S. Wales ; Queensland.

Sp. 1176. P. PALLIDIPENNIS, Macl.; Oll , l.c. p. 437.
Queensland ; Gayndah   N.S. Wales.

Sp. 1177. P. PARVICORNIS, Fvl.; Oll , l.c. p. 444.
Victoria.

MYRMEDONIA, Erichson.

Sp. 1179. M. CLAVIGERA, Fvl.; Oll., l c. (2) i. 1886, p. 448.
N.S. Wales ; Sydney, &c.

Sp. 1180. M. INSIGNICORNIS, Fvl., Oll , l.c. p. 448.
Australia.

BARRONICA, Blackburn.

7665. SCORPIO, Blackb , Trans. Roy. Soc. S.A. xix. 1895, p. 203.
N. Queensland ; Barron River District.

PELIOPTERA, Kraatz.

7666. ASTUTA, Oll., P.L.S N.S.W. (2) i. 1886, p. 424.
Tasmania.

Sp. 1181. P. SPECULARIS, Fvl.; Oll., l.c. p. 424.
N.S. Wales; Sydney.

CALODERA, Mannerheim.

7667 AGLAOPHANES, Oll., l.c. (2) i. 1886, p. 430.
S. Aust.; Port Lincoln.

7668. ATYPHA, Oll., P.L.S.N.S.W. (2) i. 1886, p 433.
  Tasmania.

7669. CARISSIMA, Oll., l.c. p. 426.
  Tasmania.

7670. ERITIMA, Oll., l.c. p. 429.
  N.S. Wales; Wagga Wagga.

7671. PACHIA, Oll., l.c. p. 432.
  Tasmania.

7672. PYRRHA, Oll., l.c. p. 429.
  N.S. Wales ; Upper Hunter.

7673. SIMSONI, Oll., l.c. p. 432.
  Tasmania.

  Sp 1182. C. ABDOMINALIS, Fvl.; Oll. l.c. p 427.
  Australia.

  Sp. 1183 C. AUSTRALIS, Fvl.; Oll., l.c p. 427.
  S. Aust.; Adelaide. Victoria.

  Sp. 1184. C. CORACINA, Macl.; Oll., l.c. p 431.
  Queensland; Gayndah.

  Sp. 1185. C. CRIBELLA, Fvl.; Oll., l.c. p. 431.
  N.S. Wales. S. Australia. W. Australia.

  Sp. 1186. C. INÆQUALIS, Fvl.; Oll., l c. p. 425
  Victoria.

  Sp. 1187. C. MACILENTA, Fvl.; Oll., l c. p. 428.
  Victoria.

  Sp. 1188. C. RUFICOLLIS, Fvl.; Oll., l.c. p. 428.
  N.S. Wales; Sydney.

    MYRMECOPORA, Saulcy.

  Sp. 1189. M. SENILIS, Fvl.; Oll., l.c. (2) i. 1886, p. 434.
  Victoria.

APPHIANA, Olliff

7674. VERIS, Oll., P.L.S.N.S.W. (2) i. 1886, p. 422, t 7. f. 1.

N.S. Wales; Wagga Wagga, Sydney.

GNYPETA, Thomson.

Sp 1190 G. FULGIDA, Fvl.; Oll., l.c. (2) i. 1886, p. 421

Victoria.

OXYPODA, Mannerheim.

Sp. 1193. O. VARIEGATA, Fvl.; Oll., l.c. (2) i. 1886, p. 435.

N S. Wales; Sydney.

Sp. 1194. O. *vincta*, Fvl., Oll., l c. p. 435.

N.S. Wales.

HOMALOTA, Mannerheim.

7675. ATYPHELLA, Oll., l.c. (2) i. 1886, p. 416.

N.S. Wales; Tasmania.

7676. CHARIESSA, Oll., l.c. p. 418.

Tasmania.

7677. CORIARIA, Kraatz, Ins. Deutsch. ii. p. 282; Sharp, Trans.
Ent. Soc. Lond. 1869, p. 204; Fvl., Ann. Mus. Genov.
x. 1877, p. 283.

australis, Jekel, Col. Jek. i. 1873, p. 47; Oll., l.c. p. 415.

N.S. Wales. S. Australia.

7678 INDEFESSA, Oll., l.c. p. 420.

Tasmania.

7679. MOLESTA, Oll., l.c. p. 415.

N.S. Wales; Sydney.

7680. PAVENS, Erichs., Käf. Mark. i. p. 689; Sharp, Trans. Ent.
Soc. Lond. 1869, p. 98; Fvl., Ann Mus. Genov. xiii.
1878, p. 578; Oll., l.c. p. 469.

Victoria.

7681. PSILA, Oll., P.L.S.N.S.W. (2) i. 1886, p. 416.
Tasmania.

7682. SORDIDA, Marsham, Ent. Brit. 1802, p. 514; FvL, Ann.
Mus. Genov. xiii. 1878, p. 576; Oll., l.c. p. 419
S. Aust.; Adelaide.

Sp. 1178. AUSTRALIS (MYRMEDONIA), Macl.; Oll., l.c. p. 417.
Queensland, Gayndah.

Sp. 1196. H. GENTILIS, Fvl.; Oll., l.c. p. 418.
N.S Wales; Sydney.   Melbourne.

Sp. 1197. H. PICEICOLLIS, Fvl.; Oll., l.c. p. 414.
N.S. Wales; Sydney.

Sp. 1198. H. POLITULA, Fvl.; Oll., l.c. p. 417.
S. Aust ; Adelaide.

Sp. 1199. H. ROBUSTICORNIS, Fvl.; Oll., l.c. p. 420.
N.S. Wales; Sydney.

PLACUSA, Erichson.

Sp. 1200. P. TENUICORNIS, Fvl.; Oll., l.c. (2) i. 1886, p 452.
Australia.

Sp. 1201. P. TRIDENS, Fvl.; Oll., l.c. p. 451.
N.S. Wales; Sydney.

PHLŒOPORA, Erichson.

Sp. 1202. P. GRATIOSA, Fvl.; Oll., l.c. (2) i. 1886, p. 447
W. Australia.

Sp. 1203. P. LÆVIUSCULA, Fvl ; Oll., l.c. p. 446.
N.S. Wales.   Victoria.

DABRA, Olliff

7683. CUNEIFORMIS, Oll , l.c. (2) i. 1886, p. 454.
W. Aust.; K. G. Sound.

7684. MYRMECOPHILA, Oll., P.L.S.N.S.W. (2) i. p. 453, t. 7, f. 2.
    W. Aust., Fremantle, K. G. Sound.

OLIGOTA, Mannerheim.

Sp. 1204. O ASPERIVENTRIS, Fvl.; Oll., l.c. (2) i. 1886, p. 467.
    Victoria.

GYROPHÆNA, Mannerheim.

Sp. 1205. G. CRIBROSA, Fvl.; Oll., l.c. (2) i. 1886, p. 468.
    N.S. Wales; Sydney.

BRACHIDA, Mulsant et Rey.

Sp. 1206. B. ANNULATA, Fvl.; Oll., l.c. (2) i. 1886, p. 471.
    N.S. Wales; Sydney.

Sp. 1207. B. ATRICEPS, Fvl ; Oll., l.c. p. 470.
    Victoria.

Sp. 1208. B. BASIVENTRIS, Fvl.; Oll., l c. p. 470.
    N.S. Wales; Sydney.

Sp. 1209. B. SUTURALIS, Fvl.; Oll., l.c. p. 469.
    N.S. Wales. S. Australia.

MYLÆNA, Erichson.

7685. INTERMEDIA, Erichs., Kaf Mark. i. 1857, p. 383; Matthews,
    Cist. Ent. iii. 1883, p 37 bis; Oll., l.c. (2) i 1886, p. 472.
    Victoria.

DINOPSIS, Matthews.

Sp. 1210. D. AUSTRALIS, Fvl.; Oll., P.L.S.N.S.W. (2) i.
    1886, p. 472.
    Victoria.

Sub-Family TACHYPORIDES.

LEUCOCRASPEDUM, Kraatz.

Sp. 1211. L. SIDNEENSE, Fvl.; Oll., l.c. (2) i. 1886, p. 903.
    N.S. Wales; Sydney.

CILEA, Jacquelin-Duval.

7686. LAMPRA, Oll , P.L.S.N.S.W. (2) i. 1886, p. 900.

Queensland; Ipswich. N.S. Wales; Tarcutta.

Sp. 1213. C. DISCIPENNIS, Fvl ; Oll., l.c. p. 901.

N.S. Wales; Sydney.

TACHINUS, Gravenhorst.

7687. MARGINELLUS, Fabr., Spec. Ins. i. p. 337; Erichs., Gen.
Staph 1840, p. 263; Kraatz, Nat. Ins. p 412; Oll., l.c.
(2) i. 1886, p. 902.

N.S. Wales; Sydney.

7688. NOVITIUS, Blackb., Trans. Roy. Soc. S.A. xiv. 1891, p. 68.
Mountains of Victoria.

TACHYPORUS, Gravenhorst.

7689. VIGILANS, Oll , l.c. (2) i. 1886, p. 899.
Tasmania.

Sp. 1214. T. RUBRICOLLIS, Macl.; Oll., l.c. p 900.
Queensland; Gayndah.

Sp. 1216. T. TRISTIS, Macl.; Oll., l.c. p. 899.
Queensland; Gayndah..

MYCETOPORUS, Mannerheim.

7690. FLORALIS, Blackb., Trans. Roy. Soc. S.A. x. 1887, p. 3
S. Aust.; Port Lincoln.

CONOSOMA, Kraatz.

7691. ACTIVUM, Oll., P.L.S.N.S.W. (2) i. 1886, p. 891.
Tasmania.

7692. AMBIGUUM, Oll., l.c. p. 894.
S. Aust.; Adelaide.

7693. ENIXUM, Oll., l.c. p. 896.
N.S. Wales. Tasmania.

7694. EXIMIUM, Oll., P.L.S N.S W. (2) i. 1886, p. 896.

     Victoria; S. Australia.

7695 INSTABILIS (CONURUS), Blackb., Trans. Roy. Soc. S.A. x.
     1887, p. 3.

     S. Aust.; Port Lincoln.

7696. PHOXUM, Oll., l.c. (2) i. 1886, p. 894.

     S. Aust.; Adelaide.

     Sp. 1212. C. ATRICEPS (CONURUS), Macl.; Oll., l.c. p. 895.

     Queensland; Gayndah.

     Sp. 1219. C. ELONGATULUM (CONURUS), Macl.; Oll., l.c.
     p. 893.

     Queensland; Gayndah.

     Sp. 1221. C. AUSTRALE (CONURUS), Erichs., Gen. Staph.
     1840, p. 221; Fvl., Ann. Mus. Civ. Genov. x. 1877,
     p. 479; Oll., l.c. p. 890.

     Tasmania. Victoria.

     Sp. 1222. C. DISCUS, Fvl.; Oll., l.c. p. 897.

     Victoria.

     Sp. 1223. C. FUMATUM, Erichs., l c. p. 228; Fauvel, l c.
     p. 280; Oll., l.c. p. 893.

     Tasmania.

     Sp. 1224. C. IMPENNE, Fvl.; Oll., l.c. p. 892.

     W. Aust.; K.G. Sound.

     Sp. 1225. C. PERSONATUM, Fvl., Oll., l.c. p. 897.

     N.S. Wales.

     Sp. 1215. C. RUFIPALPE (CONURUS), Macl., = Sp. 1226. C.
     stigmalis, Fvl.; Oll., l.c. p. 891.

     Australia; widely distributed.

     Sp. 1227. C. TRIANGULUM, Fvl.; Oll., l.c. p. 892.

     Victoria. S. and W. Australia.

TACHYNODERUS, Motschulsky.

Sp. 1218. T. AUSTRALIS, Fvl.; Oll., P.L.S.N.S.W. (2) L
1886, p. 889.

Queensland; Cairns, Rockhampton, Wide Bay.

Sp. 1220. T. HÆMORRHOUS, Fvl.; Oll., l.c. p. 888.

North Aust. N.S. Wales. Tasmania.

BOLITOBIUS, Stephens.

7697. FAUVELI, Oll., l.c. (2) i. 1886, p. 905.

N.S. Wales; Sydney.

7698. SHARPI, Oll., l c. p. 906.

N.S. Wales; Sydney.

Sub-Family STAPHYLINIDES.

ACYLOPHORUS, Nordmann.

7699. INDIGNUS, Blackb., Trans. Roy. Soc. S.A. x. 1887, p. 4
S. Aust.; Adelaide.

QUEDIUS, Stephens.

7700. ANDERSONI, Blackb., l.c. x. 1886-7, p. 6.

S. Aust.; Port Lincoln District.

7701. DIEMENENSIS, Blackb, P.L.S.N.S.W. (2) ix. 1894, p. 91.

Tasmania.

7702. FEROX, Blackb., Trans. Roy. Soc. S.A. x. 1886-7, p. 66.

S. Aust.; near Adelaide.

7703. FULGIDUS, Fab., Mant. Inst. i. p. 220; Fvl., Ann Mus.
Civ. Genov., x. 1877, p. 268.

Australia.

7704. HYBRIDUS (PHILONTHUS), Grav., Mon. p. 71 ; Erichs., Gen.
Staph. p. 432; Fvl., Ann. Mus. Civ. Genov. x. 1877,
p. 270.

Tasmania.

7705. INCONSPICUUS, Blackb , Trans Roy. Soc. S.A. x. 1886-7, p. 5.

S. Aust.; Wallaroo.

7706. KOEBELEI, Blackb., l.c. xix. 1895, p. 203.

N. Queensland.

7707. MESOMELINUS, Marsh.; Fvl., Ann. Mus. Civ. Genov. xiii. 1878, p. 552.

Australia.

7708. RUFICOLLIS (PHILONTHUS', Grav., Mon. p. 71; Erichs., Gen. Staph. p. 431 ; Kraatz, Berl. Zeit. 1859, p. 14, nota.

N.S. Wales. Victoria. S. Australia.

7709. TAURUS (HETEROTHOPS), Blackb., l.c. x. 1886-7, p. 4 ; l.c. xiv. 1891, p. 69.

S. Aust.; Port Lincoln.

7710. TEPPERI, Blackb., l.c. x. 1886-7, p. 6.

S. Aust.; Mount Lofty.

Sp. 1242. Q. CUPRINUS, Fvl. (var. [?] *baldiensis*) ; Blackb , l.c. xiv. 1891, p. 69.

Mountains of Victoria.

MYSOLIUS, Fauvel.

7711. CHALCOPTERUS, Oll., P.L.S.N.S W. (2) ii. 1887, p. 497.

N. Queensland ; Mulgrave River.

ACTINUS, Fauvel.

7712. MACLEAYI, Oll., l.c. (2) ii. 1887, p. 495.

N. Queensland ; Cairns.

OXYPORUS, Fabricius.

7713. RUFUS, Linn., Faun. Suec. nr. 844 ; Blackb., Trans. Roy. Soc. S.A. x. 1886-7, p. 6.

Australia.

COLONIA, Olliff.

7714. REGALIS, Oll., P.L.S.N S.W. (2) ii. 1887, p. 494.

N.S. Wales; Richmond River.

CREOPHILUS, Mannerheim.

Sp. 1262. C. ERYTHROCEPHALUS, Fabr.; Oll., l.c. (2) ii. 1887, p. 492.

Norfolk and Lord Howe Islands.

Sp. 1263. C. LANIO, Erichs.; Oll., l.c. p. 492.

PHILONTHUS, Curtis.

7715. ÆNEUS, Rossi, Faun. Etr. i. p. 249.

Australia.

7716. DISCOIDEUS, Grav., Micr. p. 38.

Australia.

7717. HEPATICUS, Erichs., Gen. p. 451.

Australia.

7718. LONGICORNIS, Steph., Ill. Brit. v. p. 237.

Australia.

7719. NIGRITULUS, Grav., Micr. p. 41.

Australia.

7720. ORNATUS, Blackb., Trans. Roy. Soc. S.A. x. 1886-7, p. 47.

S. Australia.

7721. SORDIDUS, Grav., Micr. p. 176.

Australia.

7722. VENTRALIS, Grav., l.c. p. 174; Blackb., l.c. p. 48.

Australia.

Sp. 1268. P. PACIFICUS, Erichs.; Fvl., l.c. x. 1877, p. 254; Oll., l.c. (2) ii. 1887, p. 504.

Sp. 1255. P. SUBCINGULATUS (QUEDIUS), Macl.; Fvl., l.c. 1877, p. 270; Blackb., l c. xviii. 1894, p. 203.

CAFIUS, Stephen?.

7723. AMBLYTERUS, Oll., P.L.S.N S.W. (2) ii. 1887, p. 502.
Tasmania.

7724. AUSTRALIS (OCYPUS), Redt., Reise Novara, Zool. ii. 1867,
p. 28; Fvl., l.c. x. 1877, p. 251; Oll., l.c. p. 500.
N.S. Wales; Sydney.

7725. DENSIVENTRIS, Fvl., l.c. p. 258; Oll., l.c. p. 507.
Queensland; Port Mackay.

7726 LÆTABILIS, Oll., l.c. p. 501.
S. Aust. Tasmania.

7727. LAEUS, Oll., l.c. p. 503.
N.S. Wales. S. Aust. Tasmania.

7728. SERICEUS (REMUS), Holme, Trans. Ent. Soc. Lond. ii. 1837,
p. 64; *Philonthus sericeus*, Erichs., Gen. Staph. 1840,
p. 509; Fvl., l.c. xiii. 1878, p. 542; Oll., l.c. p. 507.
S. and W. Australia.

7729. OCCIDENTALIS, Blackb., Trans. Roy. Soc. S.A. x. 1877,
p. 48; Oll., l.c. p. 508.
W. Australia.

Sp. 1276. C. VELUTINUS, Fvl.; Oll., l.c. p. 506.
N.S. Wales. Victoria. W. Aust

HESPERUS, Fauvel.

7730. PACIFICUS, Oll., P.L.S.N.S.W. (2) ii. 1887, p. 509.
Lord Howe Island.

7731. PULLEINEI, Blackb., l.c. x. 1887, p. 7; Oll., l.c. p. 512.
S Aust.; Burnside.

Sp. 1278. H. HÆMORRHOIDALIS, Macl. = Sp. 1179, *H. mira-
bilis*, Fvl.; Oll., l.c. p. 508.
N.S. Wales. Queensland.

Sp. 1180. H. SEMIRUFUS, Fvl.; Oll., P.L.S.N.S.W. (2) ii.
1887, p. 591.

Queensland; Cairns, Port Denison, &c.

Sp. 1277. H. AUSTRALIS, Macl.; Oll., l.c. p. 510.

Queensland. N.S. Wales.

XANTHOLINUS, Serville.

7732. ALBERTISI, Fvl., Ann. Mus. Civ. Genov. x. 1877, p. 246;
l.c. xii. 1878, p. 245, t. i. f. 26; Oll., l.c. (2) ii. 1887,
p. 489.

Northern Queensland.

7733. CYANOPTERUS, Erichs., Gen Staph. 1840, p. 311; Oll., l.c
p. 488.

Tasmania. Victoria.

7734. HOLOMELAS, Perr., Ann. Soc. Linn. Lyon, xi. 1864, p. 84:
Fvl., Ann. Soc. Ent. Fr. 1874, p. 436; Ann. Mus. Civ
Genov. x. 1877, p. 244: Oll., l.c. p. 488.

Australia; widely distributed.

7735. LORQUINI, Fvl., Ann. Mus. Civ. Genov. x. 1877, p 241.
l.c. xii. 1878, p. 245, t. 1, f. 25; Oll., l.c. p. 481.

N.S. Wales. Queensland.

7736. OLLIFFI, Lea, P.L.S.N.S.W. (2) ix. 1895, p. 589.

N.S. Wales; Tamworth.

7737. ORTHODOXUS, Oll., l.c. (2) ii. 1887, p. 484.

N.S. Wales; Sydney, Port Hacking.

Sp. 1285. X. CHALCOPTERUS, Erichs. = Sp. 1289, X. cyaneipennis, Macl.; Oll., l.c. p. 486.
Australia; widely distributed.

Sp. 1286. X. CHLOROPTERUS, Erichs.; Oll., l.c. p. 483.
Tasmania and Australia; widely distributed.

Sp. 1287. X. CŒLESTIS, Fvl.; Oll., l.c. p. 487.
Victoria.

Sp. 1288. X. CRIBRATUS, Fvl ; Oll., P.L.S.N.S.W. (2) ii. 1887, p. 490.

Victoria.

Sp. 1291. X. ERYTHROPTERUS, Erichs., Gen. Staph. 1840, p. 320; Fvl., Ann. Mus. Civ. Genov. x. 1877, p. 240; Oll., l c. p. 480.

Australia; widely distributed.

Sp. 1292. X. HÆMORRHOUS, Fvl.; Oll., l c. p. 480.

Queensland; Rockhampton.

Sp. 1293 X. PHŒNICOPTERUS, Erichs., Gen Staph. 1840, p. 314; Oll., l.c. p. 483.

Australia; widely distributed.

Sp. 1294. X. RUFITARSIS, Fvl.; Oll., l c. p 481.

N.S. Wales. Queensland.

Sp. 1295. X. SIDERALIS, Fvl.; Oll., l.c. p. 486.

W. Australia.

Sp. 1296. X. SOCIUS, Fvl. = *Leptacinus picticornis*, Blackb , Trans. Roy. Soc. S.A. x. 1887, p. 7; l.c. p. 190: Oll., P.L.S.N.S.W. (2) ii. 1887, p. 476; l.c. p. 490.

Australia; widely distributed.

LEPTACINUS, Erichson.

7738. FILUM, Blackb., Trans Roy. Soc. S A. x. 1887, p. 7 ; Oll , P.L.S.N.S W. (2) ii. 1887, p. 477.

S. Aust.; Port Lincoln.

7739. LINEARIS, Grav., Micr. p. 43 ; Blackb , l.c. p 7 ; Oll., l c. p. 476.

S. Aust.; Port Lincoln.

7740 PARUMPUNCTATUS, Gyll., Ins. Suec. iv 1808, p. 481 , Erichs , Gen. Staph. 1840, p. 335 ; Fvl., Ann. Mus. Civ. Genov. xiii. 1878, p. 537 ; Oll., l.c. p. 474.

Victoria.

Sp. 1297. L. NOVÆ-HOLLANDIÆ, Fvl.; Oll., P.L.S.N.S.W. (2)
ii. 1887, p. 475.

Queensland ; Rockhampton.   Victoria.   W. Aust.

Sp. 1299. L. LURIDIPENNIS, Macl.; Oll., l.c. p. 474.

Queensland ; Gayndah.

METOPONCUS, Kraatz.

7741. CAIRNSENSIS, Blackb., Trans. Roy. Soc. S.A. xix. 1895, p
204.

N. Queensland

7742. ENERVUS, Oll., l.c. (2) ii. 1877, p. 478.

Tasmania.

7743. FUGITIVUS, Oll., Mem. Aust. Mus. ii. 1889, p. 81.

Lord Howe Island.

Sp. 1298. M. CYANEIPENNIS, Macl.; Oll., P.L.S.N.S.W. (2)
ii. 1887, p. 477.

N.S. Wales.   Queensland.   Lord Howe Island.

DIOCHUS, Erichson.

Sp. 1301. D. DIVISUS, Fvl.; Oll., l.c. (2) ii. 1887, p 473
N.S. Wales.

Sp. 1302. D. OCTAVII, Fvl.; Oll., l.c. p.   472.

Queensland ; Wide Bay.   Victoria.

Sub-Family PÆDERIDES.

LATHROBIUM, Gravenhorst.

7744. ADELAIDÆ, Blackb., Trans. Roy. Soc. S.A. x. 1887, p. 8.

S. Aust.; Adelaide.

7745. EXIGUUM, Blackb., l.c. p. 66.

S. Australia.

7746. VICTORIENSE, Blackb., l.c. p. 71.

Mountains of Victoria.

Sp. 1304. L. AUSTRALICUM, Solsky = Sp. 1374. *Notobium australicum*, Solsky; Fvl., Ann. Mus. Civ Genov. x. 1877, p. 227.

HYPEROMA, Fauvel.

7747. ABNORME, Blackb., Trans. R Soc. S.A. xv. 1892, p 22.
Victoria; Alpine District.

S. 1317. H. LACERTINUM, Fvl.; Blackb., l c. xiv. 1891, p 71; l.c. xv. 1892, p. 21.
Victorian Alps.

SCYMBALIUM, Erichson.

7748. AGRESTE, Blackb., l.c. x. 1887, p. 8.
S. Aust.; Port Lincoln, &c.

7749. LÆTUM, Blackb , l c. p. 9.
S. Aust.; Henley Beach and Woodside.

DICAX, Fauvel.

Sp. 1335. D. LONGICEPS, Fvl. = Sp. 1310. *Lathrobium longiceps*, Fvl.

CRYPTOBIUM, Mannerheim.

7750. ADELAIDÆ, Blackb., Trans. Roy. Soc S A x. 1887, p. 69.
S. Aust.; Adelaide.

7751. DELICATULUM, Blackb., l.c. p. 69.
S. Aust.; Port Lincoln.

7752. ELEGANS, Blackb., l.c. p. 70
S. Aust.; Port Lincoln.

7753. VARICORNE, Blackb., l.c. p. 68.
S. Aust ; Port Lincoln.

STILICUS, Latreille.

Sp. 1341. S. OVICOLLIS, Macl., = *Scopæus ruficollis*, Fvl , Blackb., l.c. xviii. 1894, p. 203.

SCOPÆUS, Erichson.

7754. DUBIUS, Blackb , Trans. R. Soc. S.A. xiv. 1891, p. 37.
Victorian Alps.

7755. FEMORALIS, Blackb., l.c. xv. 1892, p. 22.
N.S. Wales; Blue Mountains.

7756. LATEBRICOLA, Blackb., l.c. x. 1887, p. 71.
S. Australia.

7757. OBSCURIPENNIS, Blackb., l.c. xiv. 1891, p. 73.
Victoria; Wandiligong.

LITHOCHARIS, Lacordaire.

7758. CINCTA, Fvl., Ann. Mus. Civ. Genov. x. 1877, p. 222.
Australia.

7759. DEBILICORNIS, Woll., Cat. Col. Mader. 1857, p. 194; Fvl,
l.c. 1878, p. 215.
Australia.

7760. LINDI, Blackb., l c. x. 1886-7, p. 48.
S. Aust.; Port Lincoln.

7661. OBSOLETA, Nordm., Symbol. p. 146; Fvl., l.c. x. 1877, p. 221.
Australia.

7762. VARICORNIS, Blackb., l.c. xiv. 1891, p. 72.
Victorian Alps.

DOMENE, Fauvel.

7763. TORRENSENSIS, Blackb., Trans. Roy. Soc. S.A. xiv. 1891,
p. 75.
S. Aust.; Torrens River.

PÆDERUS, Fabricius.

7764. ADELAIDÆ, Blackb., l.c. x. 1887, p. 10.
S. Aust.; Torrens River.

7765. MEYRICKI, Blackb., Trans. R. Soc. S.A. xiv. 1891, p. 72.
W. Australia.

7766. SIMSONI, Blackb., P.L.S.N.S.W. (2) ix. 1894, p 91.
Tasmania.

Sp. 1355. P. CRUENTICOLLIS, Germ.; Blackb., Trans. Roy.
Soc. S.A. xiv. 1891, p. 72 = Sp. 1354, *P. cingulatus*,
Macl.; Fvl., Ann. Mus. Civ. Genov. x. 1877, p. 223.

SUNIUS, Stephens.

7767. ÆQUALIS, Blackb., Trans. Roy. Soc. S.A. x. 1887, p. 9.
S. Aust.; Port Lincoln.

PALAMINUS, Erichson.

768. NOVÆ-GUINEÆ, Fvl.; Blackb., l c. xix. 1895, p 204.
N. Queensland; Barron River.

7769. VITIENSIS, Fvl.; Blackb., l.c p. 204.
N. Queensland.

Sp. 1358. P. AUSTRALIÆ, Fvl.; Blackb., l.c. xiv. 1891,
p. 75.
Queensland.

ŒDICHIRUS, Erichson.

7770. ANDERSONI, Blackb., l.c. x. 1887, p. 10.
S. Aust.; Port Lincoln.

PINOPHILUS, Gravenhorst.

7771. LATEBRICOLA, Blackb., l.c. x. 1887, p. 10.
S. Aust.; Henley Beach.

P. AUSTRALIS, Har., = Sp. 1370. *P. opacus*, Redt.; Fvl.,
Ann. Mus. Civ. Genov. x. 1877, p. 213. (nom. præocc.)

Sub-Family STENIDES.

7772. AUSTRALICUS, Blackb., P.L.S.N.S.W. (2) v. 1891, p. 780.
Mountains of Victoria.

Sub-Family OXYTELIDES.

BLEDIUS, Stephens.

7773. ADELAIDÆ, Blackb., Trans. Roy. Soc. S.A. x. 1877, p 49
    S. Aust.; Adelaide.

7774. CAROLI, Blackb., l.c. p. 13.
    S. Aust ; Port River.

7775. INFANS, Blackb, l c. xiv. 1891, p. 76.
    Victoria; Ovens River.

7776. INJUCUNDUS, Blackb., l c. x 1887, p. 14.
    S. Aust ; Port Lincoln.

7777. INSIGNICORNIS, Blackb., l.c. xiv. 1891, p. 75.
    Victoria ; Ovens River.

7778. MINAX, Blackb, l.c. x. 1887, p. 14.
    S. Aust.; Port Lincoln.

7779. OVENSENSIS, Blackb., l c xiv. 1891, p. 76.
    Victoria ; Ovens River.

TROGOPHLŒUS, Mannerheim.

· 7780. BILINEATUS, Steph , Ill. Brit. v. p. 324, t. 27, f. 4 ; Frl.,
    Ann. Mus. Civ. Genov. xiii. 1878, p. 489.
    Australia.

7781. EXIGUUS, Erichs., Käf. Mark. i. p. 604 ; Fvl., l.c. 1877, p.
    195.
    Australia.

7782. PALLUDICOLA, Blackb., Trans. Roy. Soc. S.A. x. 1887, p.
    49.
    S. Aust.; Port Lincoln.

7783. SIMPLEX, Motsch., Bull. Mosc. 1857, iv. p. 505 ; Fvl, l c.
    xiii. 1878, p. 490.
    Australia.

OXYTELUS, Gravenhorst.

7784. SCULPTUS, Grav., Mon. p. 191, Fvl., Ann. Mus. Civ.
Genov. x. 1877, p. 200.
Australia.

Sub-Family OMALIDES.

AMPHICHROUM, Kraatz.

7785. ADELAIDÆ, Blackb., Trans. Roy. Soc. S.A. xv. 1892, p. 23.
S. Aust.; near Adelaide.

OMALIUM, Gravenhorst.

7786. ADELAIDÆ, Blackb., l.c. x. 1887, p. 191.
S. Aust.; Torrens River.

Sub-Family PIESTIDES.

ELEUSIS, Castelnau.

7787. PARVA, Blackb., l.c. xv. 1892, p. 24.
N.S. Wales; Blue Mountains.

LEPTOCHIRUS, Germar.

7788. SAMOENSIS, Blanch., Voy. Pôle Sud, p. 54, t. 4, f. 11; Fvl.,
l.c. xiii. 1878, p. 480.
N. Queensland.

Family PSELAPHIDÆ.

Sub-Family PSELAPHIDES.

CTENISTES, Reichenbach

7789. ADELAIDÆ, Blackb, Trans. R. Soc. S. A. xii. 1889, p. 136.
S. Aust.; Adelaide.

7790. ANDERSONI, Blackb., l.c. xiv. 1891, p. 77.
S. Australia.

7791. TENEBRICOSUS, Blackb., Trans. R. Soc. S. A. xii. 1889,
p. 137.

S. Aust.; Port Lincoln.

Sp. 1438. C. KREUSLERI, King; Blackb., l.c. p. 137.

TYROMORPHUS, Raffray.

7792. COMES, Schauf., Tijdschr. Ent. xxix. 1886, p. 284.
Australia.

7793. CONSTRICTINASUS, Schauf., l.c. p. 285.
Australia.

EUDRANES, Sharp.

7794. CARINATUS, Sharp, Ent. Mo. Mag. xxviii. 1892, p. 242.
N. W. Australia.

DIDIMOPRORA, Raffray.

Sp. 1455. TYRUS VICTORIÆ, King; Raff., Rev. d'Ent. ix.
1890, p. 148.

TYRAPHUS, Sharp.

7795. PROPORTIONALIS, Schauf., Tijdschr. Ent. xxix. 1886, p. 261.
Australia.

7796. SOBRINUS, Schauf., l.c. p. 262.
Australia.

7797. UMBILICARIS, Schauf., l c. p. 261.
Australia.

RYTUS, King.

7798. GEMMIFER, Schauf., Tijdschr. Ent. xxix. 1886, p. 287.
Australia.

7799. ORIENTALIS, Schauf., l.c. p. 287.
Australia.

7800. PROCURATOR, Schauf., l.c. p. 286.
Australia.

GONATOCERUS.

7801. TERTIUS, Schauf., Tijdschr. Ent. xxix 1886, p. 279.
Australia.

PSELAPHUS, Aubé.

7802. BIPUNCTATUS, Schauf., Tijdschr. Ent. xxix. 1886, p. 250.
Australia.

7803. FRONTALIS, Schauf, l c. p. 251.
Australia.

7804. INSIGNIS, Schauf., l.c. p. 249.
Australia.

7805. LONGEPILOSUS, Schauf., l.c. p. 248.
Australia.

7806. SQUAMICEPS, Schauf., l.c. p. 252.
Australia.

7807. TRIPUNCTATUS, Schauf., l.c. p. 252.
Australia.

TOSIMUS, Schaufuss.

7808. GLOBULICORNIS, Schauf., Tijdschr. Ent. xxix. 1886, p 295.
Australia.

7809. LONGIPES, Schauf, l.c. p. 294.
Australia.

7810. MODESTUS, Schauf., l.c. p. 295.
Australia.

Spp. 1477 + 1478 to be placed in this genus.

TYCHUS, Leach.

7811. POLITUS, Schauf., Tijdschr. Ent. xxix. 1886, p. 260.
Australia.

F

7812. TASMANIÆ, Schauf , Tijdschr. Ent. xxix. 1886, p. 260.
  Tasmania.

CURCULIONELLUS.

7813. ANOPUNCTATUS, Schauf., Tijdschr. Ent. xxix. 1886, p. 254.
  Australia.

7814. BICOLOR, Schauf., l.c. p. 253.
  Australia.

7815. SEMIPOLITUS, Schauf., l.c p. 255.
  Australia.

DURBOS.

7816. AFFINIS, Schauf., Tijdschr. Ent. xxix. 1886, p. 291.
  Australia.

7817. CRIBRATIPENNIS, Schauf., l.c. p 292.
  Australia.

7818. INTERMEDIUS, Schauf., l.c. p. 292.
  Australia.

7819. INTERRUPTUS, Schauf., l.c. p. 291.
  Australia.

MESOPLATUS, Raffray.

  Sp. 1487. BATRISUS BARBATUS, King; Raff., Rev. d'Ent. ix.
  1890, p. 103.

BRYAXIS, Leach.

7820. HARTI, Blackb., Trans. R. Soc. S. A. xiv. 1891, p. 78.
  S. Aust.; near Adelaide.

7821. INUSITATA, Blackb., l.c. p. 79.
  S. Aust.; near Port Lincoln.

7822. LINDENSIS, Blackb., l.c. p. 77.
  S. Aust.; near Port Lincoln.

7823. OVENSENSIS, Blackb., Trans. R. Soc. S. A. xiv. 1891, p. 80.
Victoria; Ovens River.

7824. PALUDIS, Blackb., l.c. p. 81.
S. Aust.; near Adelaide.
Sp. 1825. B. HYALINA, Schauf.; Blackb., l.c. p. 79.

EUPINES, King.

7825. MILITARIS, Blackb., Trans. R. Soc. S. A. xiv. 1891, p. 85.
S. Aust.; near Port Lincoln.

7826. NAUTA, Blackb., l.c. p. 83.
S. Aust.; near Port Lincoln.

7827. NAUTOIDES, Blackb., l.c. p. 84.
S. Aust.; near Port Lincoln.

7828. RELICTA, Blackb., l.c. p. 292.
Victoria; Mordialloc.

7829. SORORCULA, Blackb., l.c. p. 82.
Australian Alps.

7830. SPINIVENTRIS, Blackb., l.c. p. 84.
S. Aust.; near Port Lincoln.

CYATHIGER, King.

7831. REITTERI, Schauf, Tijdschr. Ent. xxix. 1886, p. 242.
Australia.

ABASCANTUS, Schaufuss.

7832. SANNIO, Schauf., Tijdschr. Ent. xxix. 1886, p. 258.
Austra'ia.

ARTICERUS, Dalman.

7833. ASPER, Blackb., Trans. R. Soc. S. A. xii. 1889, p. 138.
S. Aust.; Adelaide.

7834. FOVEICOLLIS, Raffr., Rev. d'Ent vi p. 18.
    Australia.

ED.ÆRANES, Reitter.

Wien. Ent. Zeit. iv. p. 228, for *Narcodes* (nom. præoc.).

Family PAUSSIDÆ.

PAUSSUS, Linné.

7835. AUSTRALIS, Blackb , Trans. R. Soc. S A. xiv. 1891, p. 68.
    Queensland ; Mt. Bartle Frere.

ARTHROPTERUS, W. S. Macleay.

7836. FOVEIPENNIS, Blackb., Trans. R. Soc. S. A. xv. 1892, p 24.
    S. Aust.; N. Territory near Palmerston.

7837. KINGI, Macl., Trans. Ent Soc. N. S. W. ii. 1871, p. 154
    Queensland; Gayndah.

7838. OCCIDENTALIS, Blackb , Trans. R. Soc. S. A. xv. 1892, p. 25.
    W. Aust.; Yılgarn.

    Sp. 1591. A DENUDATUS, Westw , = Sp. 1584. *A. angusti-
    cornis*, Macl., Gestro, Ann. Mus. Civ. Genov. 1884, p. 5.

    Sp. 1603. A. MELBOURNI, West., = Sp. 1582. *A. angulatus*,
    Macl.; Gestro, l.c. p. 4.

Family SILPHIDÆ.

Sub-Family SILPHIDES

ANISOTOMA, Illiger.

7839. TASMANIÆ, Olliff, Proc. Linn. Soc. N. S. W. (2) iii. 1888,
    p. 1513.
    Tasmania.

COLON, Herbst.

7840 MELBOURNENSE, Blackb , Trans. R. Soc. S. A. xv. 1892, p. 25.
Victoria; near Melbourne.

CHOLEVA, Latreille.

7841. ADELAIDÆ, Blackb., Trans. R. Soc. S. A. xiv. 1891, p. 87.
S. Australia.

7842. ANTIPODUM, Blackb., l.c. p. 87; l.c. xviii. 1894, p. 139.
Victorian Alps.   Tasmania.

7843. MINUSCULA, Blackb., l c. p. 88.
S. Australia.

7844. VICTORIENSIS, Blackb., l c. p. 88.
Victorian Alps.

Sp. 1648. C. AUSTRALIS, Erichs.; Blackb., l c. p. 67

CHOLEVOMORPHA, Blackburn.

7845. PICTA, Blackb., Trans. R. Soc. S. A. xiv. 1891, p. 90
Mountains of Victoria.

Family SCAPHIDIDÆ.

SCAPHIDIUM, Olivier.

7846. ALPICOLA, Blackb., Trans. R. Soc. S. A. xiv. 1891, p. 90.
Victorian Alps.

SCAPHISOMA, Leach.

7847. NOVICUM, Blackb., Trans. R. Soc. S. A. xiv. 1891, p. 91.
Victorian Alps.

Family HISTERIDÆ.

Sub-Family HOLOLEPTIDES.

HOLOLEPTA, Paykull.

Sp. 1667. H. SIDNENSIS, Mars., = Sp. 1666. *H. Mastersi,*
Macl.; Lewis, Ann. Nat. Hist. (6) xi. 1893, p. 418.

PLATYSOMA, Leach.

7848. BIIMPRESSUM, Schmidt, Ent. Nachr. xviii. 1892, p. 21.
Queensland.

7849. CONDITUM, Mars., Ann. Mus. Civ. Genov. 1879, p. 268.
Australia.

7850. CONSTRICTUM, Lewis, Ann. Nat. Hist. (6) vii. 1891, p. 385.
N. W. Australia.

7851. MOLUCCANUM, Mars., Ann. Mus. Civ. Genov. 1879, p. 268.
Australia.

7852. PAUGAMI, Mars., l.c. p. 266.
Australia.

7853. ROBUSTUM, Schmidt, Ent. Nachr. xviii. 1892, p. 22.
Australia.

7854. SEMILINEATUM, Schmidt, l.c. p. 22.
Australia.

Sub-Family HISTERIDES.

CARCINOPS, Marseul.

7855. PUMILIO, Erichs., Jahrb. 1834, p. 119; Mars., Mon. 1855,
p. 91, t. 8, nr. 22, f. 4; Ann. Mus. Civ. Genov. 1879,
p. 272.
Australia.

PAROMALUS, Erichson.

7856. LUDOVICI, Blackb., Trans. R. Soc. S. A. xv. 1892, p. 26.
N. S. Wales; Blue Mountains.

EPIERUS, Erichson

7857. BISERIATUS (Stictostix), Schmidt, Ent. Nachr. xvi. 1890,
p. 39.
Australia.

CHLAMYDOPSIS, Marseul?

7858. INÆQUALIS, Blackb., Trans R. Soc. S. A. xiv. 1891, p 94

S. Aust.; near Woodville.

7859. STERNALIS, Blackb., l.c. p. 93.

S. Aust.; near Woodville.

Sp. 1695. C. STRIATELLA, Westw. = Sp. 1914. *Byzenia jormicicola*, King; Lewis, Ann. Nat. Hist. (6) xiv. 1894, p. 113.

TERETRIOSOMA, Marseul?

7860. SOMERSETI, Mars., Ann. Mus. Civ. Genov. 1879, p. 281.

N Queensland; Somerset.

TERETRIUS, Erichson.

7861. AUSTRALIS, Lewis, Ann. Nat. Hist. (6) xi. 1893, p. 428.

Queensland.

7862. BASALIS, Lewis, l.c. (6) iii. 1889, p. 286.

S. Australia. ?

7863 WALKERI, Lewis, l.c. ix. 1892, p. 353.

Tasmania.

SAPRINODES, Lewis.

7864. FALCIFER, Lewis, Ann. Nat. Hist. (6) viii. 1891, p. 396.

Queensland; Rockhampton.

SAPRINUS, Erichson.

7865. SPECIOSUS, Erichs., Jahrb. 1834, p. 179; Mars., Mon. 1855, t. 16, f. 23; Ann. Mus. Civ. Genov. 1879, p. 280.

Australia.

ACRITUS, Leconte.

7866. TASMANIÆ, Lewis, Ann. Nat. Hist. (6) ix. 1892, p. 357.

Tasmania.

Family PHALACRIDÆ.

LITOCHRUS, Erichson.

7867. ALPICOLA, Blackb., Trans. R. Soc. S.A. xiv. 1891, p. 98.
Victorian Alps.

7868. ALTERNANS, Blackb., l.c. p. 95.
Victorian Alps.

7869. COLORATUS, Blackb., l.c. xix. 1895, p. 207.
N. Queensland, near Cairns.

7870. CONSORS, Blackb., l.c. xvii. 1893, p. 295.
N. Queensland; near Cairns.

7871. FRIGIDUS, Blackb., l.c. xiv. 1891, p. 97.
Victorian Alps.

7872. KOEBELEI, Blackb., l.c. xix. 1895, p. 208.
N. S. Wales; Blue Mountains.

7873. LÆTICULUS, Blackb., l.c. xiv. 1891, p. 95.
Victorian Alps.

7874. LATERALIS, Blackb., l.c. p. 97.
S. Aust.; near Port Lincoln.

7875. MACULATUS, Blackb., l.c. p. 96.
S. Australia.

7876. MAJOR, Blackb., l.c. xix. 1895, p. 208.
S. Australia.

7877. NOTEROIDES, Blackb., l.c. p. 208.
N. Queensland; near Cairns.

7878. PALMERSTONI, Blackb., l.c. xiv. 1891, p. 95.
N. Territory of S. Aust.

7879. PULCHELLUS, Blackb., l.c. xix. 1895, p. 207.
N. Queensland; near Cairns.

7880. SUTURELLUS, Blackb., Trans. R. Soc. S. A. xiv. 1891, p. 96.
      W. Australia.

7881. SYDNEYENSIS, Blackb, l.c. xv. 1892, p. 26.
      N.S. Wales; near Sydney.

7882. TINCTUS, Blackb, l c xix. 1895, p. 208.
      N. Queensland; near Cairns.

7883. UNIFORMIS, Blackb., l.c. xiv. 1891, p. 98.
      S. Aust.; near Adelaide.

PARASEMUS, Guillebeau.

7884. COMES, Blackb., Trans. R. Soc. S.A. xix. 1895, p. 212.
      N. Queensland; near Cairns.

7885. DISCOIDEUS, Blackb., l c. p 211.
      N. Queensland; near Cairns.

7886. DOCTUS, Blackb., l.c. p. 212.
      N. S. Wales; Blue Mountains.

7887. GROUVELLI, Guill., Ann. Soc. Ent. Fr. 1894, p. 300.
      Australia.

7888. INTERNATUS, Blackb., Trans. Roy. Soc. S. A. xix. 1895,
      p. 213.
      S. Aust.; Petersburg.

7889. MODESTUS, Blackb., l.c. p. 212.
      N. Queensland; near Cairns.

7890. OBSOLETUS, Blackb., l.c. p. 213.
      N. Queensland; near Cairns.

7891. TORRIDUS, Blackb., l.c. p. 211.
      N. Queensland; near Cairns.

PHALACRINUS, Blackburn.

7892. AUSTRALIS, Blackb., Trans. R. Soc. S.A. xiv. 1891, p. 99.
      S. Aust ; Port Lincoln, &c.

7893. COMIS, Blackb., l.c. xix. 1895, p. 215.
      Victoria.

7894. NOTABILIS, Blackb., l.c. p. 215.
      N. Queensland; near Cairns.

7895. OBTUSUS, Blackb., l.c. xiv. 1891, p. 100.
      S. Aust.; Port Lincoln.

7896. ROTUNDUS, Blackb., l.c. p. 100.
      S. Aust.; near Port Lincoln.

PHALACRUS, Paykull.

7897. BURRUNDIENSIS, Blackb., Trans. R. Soc. S. A. xiv. 1891,
      p. 101.
      N. Territory of S. Aust.

7898. CORRUSCANS, Payk., Faun. Suec iii. 1798, p. 438; Blackb.
      l.c. p. 100.
      S. Australia.   Victoria.   .

MICROMERUS, Guillebeau.

7899. AMABILIS, Guill., Ann. Soc. Ent. Fr. 1894, p. 396.
      Australia.

OLIBRUS, Erichson.

7900. VICTORIENSIS, Blackb., Trans. R. Soc. S.A. xiv. 1891, p. 101.
      Victorian Alps, and N.S. Wales.

Family NITIDULIDÆ.

MELIGETHES, Stephens.

7901. NIESLII, Reit., Verh. Ver. Brünn, 1872, p. 161.
      Australia.

CARPOPHILUS, Stephens.

7902. MUTILATUS, Erichs., Germ., Zeitschr. iv. p. 258, Murray,
      Mon. 1864, p. 378.
      Australia.

7903. DIMIDIATUS, Fabr., Ent. Syst. i. p 261; Murray, Mon. 1864,
p. 379.

Australia.

CIRCOPES, Motschulsky.

Sp. 1762. C. (POCADIUS) PILISTRIATUS, Macl.; Reit Verh.
Ver. Brunn, xii. 1873, p. 80.

MIMEMODES, Fairmaire?

7904. LATICEPS (PROSTOMIS), Macl., Trans. Ent. Soc. N S W ii.
1871, p. 167; Fairm., Ann. Soc Ent. Fr. (6) i. 1881,
p. 257; Reit., Wien. Ent. Z. iii. p. 272.

Queensland; Gayndah, Wide Bay, &c.

NOTOBRACHYPTERUS, Blackburn.

7905. AUSTRALIS, Blackb., Trans. R. Soc. S.A. xv. 1892, p 27.

W. Australia.

7906. BIFOVEATUS, Blackb., l.c. p. 28.

S. Aust.; near Adelaide.

7907. CREBER, Blackb., l.c p. 27.

S. Aust , Port Lincoln District.

7908. LILLIPUTANUS, Blackb., l.c. p. 29.

S. Australia.

7909. NITIDIUSCULUS, Blackb , l.c. p. 28.

W. Australia.

BRACHYPEPLUS, Erichson.

Sp. 1722. Murrayi, Macl., = B. Haagi, Reit.; Blackb , Trans.
R. Soc. S.A. xviii. 1894, p. 203.

IDÆTHINA, Reitter.

7910. CINCTA, Blackb., Trans R. Soc. S. A xiv. 1891, p. 107

S. Aust.; near Victor Harbour

MACRURA, Reitter.

7911. BAILEYI, Blackb., Trans. R. Soc. S. A. xiv. 1891, p. 108;
xix. 1895, p. 31.

Queensland; Mt. Bellenden-Ker.

7912. DECEPTOR, Blackb., l.c. p. 108.

N. Territory of S. Australia.

Sp. 1735. CARPOPHILUS LURIDIPENNIS, Macl.,=*C. excellens*
Reit.; Blackb., l.c. xviii. 1894, p. 203.

ERICMODES, Reitter.

7913 AUSTRALIS, Grouv., Trans. R. Soc. S.A. xvii. 1893, p. 14
S. Australia.

NITIDULA, Fabricius.

7914. QUADRIPUSTULATA, Fab., Ent. Syst. i. p. 255; B
Trans. R. Soc. S.A. xiv. 1891, p. 105.

S. Aust.; Adelaide (probably introduced).

ÆTHINODES, Blackburn.

7915. MARMORATUM, Blackb., Trans. R. Soc. S.A. xiv 18
p. 109.

Tropical Australia.

LASIODACTYLUS, Perty.

7916. CALVUS, Oll., P L S.N.S.W. (2) ii. 1887, p. 1003.

Norfolk Island.

Sp. 1742. L. MARGINATUS, Reit., var. (?) OBSCURUS, Black
Trans. R. Soc. S.A. xiv. 1891, p. 106.

Queensland.

CRYPTARCHA, Shukhard.

7917 DEPRESSA, Grouv., Trans. R. Soc. S.A. xvii. 1893, p. 142.
S. Australia.

7918. NITIDA, Reit , MT. Münch. Ent. Ver. i. 1877, p. 129.
    S. Aust.; Adelaide.

SORONIA, Erichson.

7919. SIMULANS, Blackb ,'Trans. R. Soc. S.A. xiv. 1891, p. 105.
    Victorian Alps.

THALYCRODES, Blackburn.

Sp. 1753. T. AUSTRALE (? Germ.), Blackb , l.c xiv., 1891,
    p. 110
    S. Australia.

7920. CYLINDRICUM, Blackb., l.c. p. 112.
    Victorian Alps.

7921. PULCHRUM, Blackb., l.c. p. 111.
    S. Aust : near Port Lincoln.

HAPTONCURA, Reitter

7922. LINDENSIS, Blackb., Trans. R. Soc. S.A. xiv. 1891, p. 103.
    S. Aust.; near Port Lincoln.

7923. MEYRICKI, Blackb., l.c. p. 104
    W. Australia.

7924. UNIFORMIS, Blackb , l.c. p. 104.
    Victorian Alps.

7925. VICTORIENSIS, Blackb., l.c. p. 103.
    Victorian Alps. Tasmania.

OMOSITA, Erichson.

7926. COLON, Linn., Faun. Suec. p. 151; Erichs ,'Nat. Ins. iii.
    p. 167.
    N.S. Wales (introduced).

Family TROGOSITIDÆ.

TROGOSITA, Olivier.

7927. PUNCTULATA, Reitt., (Tenebrioides) Verh. Ver. Brünn, xiii. 1875, p. 74.

S. Australia.

LEPERINA, Erichson.

7928. CONSPICUA, Oll., P.L.S.N.S.W. x. 1886, p. 704.

Lizard Island, N.E. Australia.

7929. FRATERNA, Oll., l.c. p. 707.

W. Aust.; Salt River.

7930. SEPOSITA, Oll., l.c. p. 702.

King George's Sound.

Sp. 1780. L. TURBATA, Pasc., = Sp. 1775. *L. fasciculata* Redtenb.; Oll., l.c. p. 705.

Sp. 1774. L. DECORATA, Erichs., = 1776. *L. Ganglbauensis*, Macl.; Oll., l.c. p. 702.

LATOLÆVA, Reitter.

7931. CASSIDIOIDES, Reitt., Verh. Ver. Brünn, xiv. 1876, p. 50.

N. Queensland; Cape York, &c.

NEASPIS, Pascoe.

7932. PUSILLA, Blackb., Trans. R. Soc. S.A. xiv. 1891, p. 112.

S. Aust.; near Adelaide.

Sp. 1752. N. (SORONIA) VARIEGATA, Macl., = Sp. 1782. *N. subtrifasciata*, Reitt.; Oll., P.L.S.N.S.W. x. 1885, p. 709.

OSTOMA, Laicharting.

7933. PUDICUM, Oll., Mem. Aust. Mus. ii. 1889, p. 82.

Lord Howe Island.

PELTOSCHEMA, Reitter.

7934 FILICORNIS, Reitt., Verh. Ver. Brünn, xviii. 1880, p. 5.
Australia.

LOPHOCATERES, Olliff.

7935. IVANI, Allib., (OSTOMA), Rev. Zool. 1847, p 12, Oll., P.L.S.
N.S.W. x. 1885. p. 715.
Sydney.

ANCYRONA, Reitter.

7936. ÆGRA, Oll., P.L.S.N.S.W. x. 1885, p. 711.
Sydney.

7937. AMICA, Oll., l.c. p. 713.
W. Aust.; Albany. S. Aust.; Port Lincoln.

7938. LATEBROSA, Oll., l.c. p. 712.
Queensland; Wide Bay.

7939. LATICEPS, Oll , l.c. p. 710.
N.S. Wales and Queensland.

7940. VESCA, Oll., l.c. p. 713.
N.S. Wales: S. Aust.: W. Aust.

PELTONYXA, Reitter.

7941. AUSTRALIS, Blackb., Trans. R. Soc. S.A. xiv. 1891, p. 113
S. Aust.; Adelaide District.

7942. PUBESCENS, Blackb., l.c. p. 113.
Victoria; Alpine District.

PHYCOSECIS, Pascoe.

Removed from *Tenebrionidæ* to *Trogositidæ;* Champ.
Trans. Ent Soc. Lond. 1894, p. 364.

Family COLYDIIDÆ.

SPARACTUS, Erichson.

7943. COSTATUS, Blackb., Trans. R. Soc. S.A. xiv. 1891, p. 117.
S. Australia.

7944. ELONGATUS, Blackb., l.c. p. 116.
S. Aust.; Port Lincoln District.

7945. PROXIMUS, Blackb., l.c. p. 116.
S. Aust.; near Adelaide.

7946. PUSTULOSUS, Blackb., l.c. p. 116.
S. Australia.

DITOMA, Herbst.

7947. HILARIS, Blackb., Trans. R Soc. S.A. x. 1887, p. 194.
S. Aust ; Port Lincoln, &c.

7948. LINEATOCOLLIS, Blackb , l.c. p. 195.
S. Aust.; Port Lincoln, &c.

7949. NIVICOLA, Blackb., l.c. xiv. 1891, p. 114.
Victorian Alps.

7950. OBSCURA, Blackb., l c. x. 1887, p. 193
S. Aust ; Roseworthy.

7951. PARVA, Blackb., l.c. p. 193.
S. Aust.; Woodville.

7952. PERFORATA, Blackb , l.c. p. 193.
S. Aust.; Adelaide.

7953. PULCHRA, Blackb., l.c. p. 191; l.c. xiv. p. 114.
S. Aust.; Mount Lofty.

7954. TORRIDA, Blackb., l.c. xiv. 1891, p. 114.
N. Queensland.

MERYX, Latreille.

7955. ÆQUALIS, Blackb., Trans. R. Soc S.A. xiv. 1891, p. 115.
S. Aust.; near Port Lincoln.

SARROTRIUM. Illiger.

7956. AUSTRALE, Blackb., Trans. R. Soc. S.A. xiv. 1891, p. 115.
Victorian Alps.

PHORMESA, Pascoe.

7957. EPITHECA, Oll., Mem. Aust. Mus. ii. 1889, p. 83
Lord Howe Island.

GEMPYLODES, Pascoe.

7958. TMETUS, Oll., Mem. Aust. Mus. ii. 1889. p. 83.
Lord Howe Island.

PYCNOMERUS, Erichson.

7959. LONGULUS, Sharp, Trans. R. Dubl. Soc. (2) iii. 1886,
p. 389, t. 12, f. 21; Oll., Mem. Aust. Mus. ii. 1889,
p. 84.
Queensland; Pine Mountain, near Ipswich.

7960. MŒSTUS, Oll., l.c. p. 83.
Lord Howe Island.

MINTHEA, Pascoe.

7961. SIMILATA (?), Pasc., Journ. of Ent. ii. 1863, p. 141, t 8, f.
10; Blackb., Trans. R. Soc. S.A. xiii 1890, p 121.
Adelaide (probably introduced).

TRISTARIA, Reitter.

7962. FULVIPES, Reitt., Stett. Ent. Zeit xxxix. 1878, p. 322
Australia.

7962 *bis*. GROUVELLEI, Reitt., l.c. p. 321.
Queensland; Rockhampton.

7963. LABRALIS, Blackb., Trans. R. Soc. S.A. xv. 1892, p. 30.
Victoria; near Cheltenham.

G

TODIMA, Grouvelle.

7964. FUSCA, Grouv., Trans. R. Soc. S.A. xvii. 1893, p. 143.
W. Aust.; K. G. Sound.

7965. RUFULA, Grouv., l.c. p. 144.
. W. Australia.

SYMPANOTUS.

7966. AUSTRALIS, Grouv., Trans. R. Soc. S.A. xvii. 1893, p. 144.
Mountains of Victoria.

BOTHRIDERES, Erichson.

7967. COSTATUS, Blackb., Trans. R. Soc. S.A. x. 1887, p. 197.
S. Aust., Port Lincoln.

7968. TIBIALIS, Blackb., l.c. p. 196.
S. Aust.; Victoria.

7969. VARIABILIS, Blackb., l.c. p. 196.
S. Aust., Victoria.

7970 VICTORIENSIS, Blackb., l.c. xiv. 1891, p. 117.
Victorian Alps.

Sp. 1815. B. MERUS, Pasc.; Blackb., l.c. p. 118.

Sp. 1806. B. (DERETAPHRUS) PUTEUS, Newm.; Pasc., Journ.
of Ent. i. p. 240.

NEOTRICHUS, Sharp.

7971. LUCIFUGUS, Oll., Mem. Aust. Mus. ii. 1889, p. 82.
Lord Howe Island.

DASTARCUS, Walker.

7972. POROSUS, Walk., Ann. Nat. Hist. (3) ii. 1858, p. 209; Pasc.,
Journ. of Ent. i. p. 108; ii. p. 138.
Australia.

Family CUCUJIDÆ.

Sub-Family CUCUJIDES.

LEMOPHLÆUS, Castelnau.

7973. AUSTRALASIÆ, Blackb., Trans. R. Soc. S.A. xv. 1892, p. 30.

Victoria; Dandenong Ranges.

7974. DIFFICILIS, Blackb., P.L.S.N.S.W. (2) iii. 1888, p. 840.

S. Aust.; Port Lincoln.

7975. LINDI, Blackb., l.c. p. 841.

S, Australia.

7976. PUSILLUS (CUCUJUS), Schön., Syn. Ins. i. 3, p. 55.

L. (CUCUJUS) *testaceus*, Steph , (*nec* Fab.) Ill. Brit. Ins. iv. p. 224, t. 21, f. 9; Blackb., Trans. R. Soc. S.A. xiii. 1890, p. 121.

Australia (probably introduced).

Sub-Family HEMIPEPLIDES.

INOPEPLUS, Smith.

7977. OLLIFFI, v. d. Poll, Notes Leyden Mus. 1887, p. 140.

N. Queensland.

Sub-Family TELEPHANIDES.

CRYPTAMORPHA, Wollaston.

7978. DELICATULA, Blackb., Trans. R. Soc. S.A. x. 1887, p. 200.

S. Aust.; Port Lincoln.

7979. LINDI, Blackb., l.c. p. 198.

S. Aust.; Port Lincoln.

7980. MACLEAYI, Blackb., l.c. xv. 1892, p 31.

N.S. Wales; Blue Mountains.

7981. BRÉVICORNIS (DENDROPHAGUS), White, Voy. Ereb. Terr.
Ent. p. 18; Waterh., Ent. Mo. Mag. xiii. p. 124.
Australia.

7982. OLLIFFI, Blackb., Trans. R. Soc. S.A. x. 1887, p. 199.
S. Aust.; Port Lincoln, &c.

7983. VICTORIÆ, Blackb., l.c. p. 199.
Western Victoria.

Sub-Family SILVANIDES.

SILVANUS, Latreille.

7984. ADVENA, Waltl., Faunus, i. 1832, p. 169; Blackb., Trans.
R. Soc. S.A. x. 1887, p. 200.
Australia (introduced).

7985. ARMATULUS, Blackb., l.c. xiv. 1891, p. 118.
Victorian Alps.

7986. MONTICOLA, Blackb., l.c p. 118.
Victorian Alps.

7987. UNIDENTATUS, Oliv., Ent. ii. 18, p. 12, t. 1, f. 4; Blackb.,
l.c. 1887, p. 200.
S. Aust. and Victoria.

MYRABOLIA, Reitter.

7988. LINDENSIS, Blackb., Trans. R. Soc. S.A. xv. 1892, p. 31.
S. Aust.; near Port Lincoln.

7989. PARVA, Blackb., l.c. p. 32.
· N.S. Wales; near Sydney.

Sp. 1876. M. HAROLDIANA, Reitt.; Blackb , l.c. p. 32.

Family CRYPTOPHAGIDÆ.

TELMATOPHILUS, Heer.

7990. BREVIFORMIS, Blackb , Trans. R. Soc. S.A. xix. 1895, p. 218.
N. Queensland.

7991. CAIRNSENSIS, Blackb., Trans. R. Soc. S.A. xix. 1895, p. 217.
    N. Queensland; near Cairns.

7992. KOEBELEI, Blackb., l.c. p. 217.
    Queensland.

7993. SHARPI, Blackb., l.c. p. 216.
    N. Queensland; near Cairns.

7994. SINGULARIS, Blackb., l.c. p. 218.
    N. Queensland.

7995. STYGIUS, Blackb., l.c. p 218.
    N. Queensland.

ATOMARIA, Stephens.

7996. AUSTRALIS, Blackb., Trans. R. Soc. S.A. xiv. 1891, p. 119
    S. Australia.

7997. EUCALYPTI, Blackb., l.c. xv. 1892, p. 33.
    N.S. Wales; Blue Mountains.

7998. LINDENSIS, Blackb., l.c. xiv. 1891, p. 119.
    S. Aust.; near Port Lincoln.

CRYPTOPHAGUS, Herbst.

7999. AFFINIS, Sturm., Ins. xvi. p. 79, t. 314, f. c. C.; var. ?
    AUSTRALIS, Blackb., Trans. R. Soc. S.A. x. 1887, p. 201.
    S. Australia (probably introduced).

8000. GIBBIPENNIS, Blackb., l.c. xv. 1892, p. 32.
    Victoria and Tasmania.

8001. LINDENSIS, Blackb., l.c. xiv. 1891, p. 119.
    S. Aust ; Port Lincoln District.

Family LATHRIDIIDÆ.

LATHRIDIUS, Herbst.

8002. APICALIS, Blackb., Trans. R. Soc. S.A. x. 1887, p. 204.
    S. Aust.; Port Lincoln.

8003. AUSTRALICUS, Belon., C.R. Ent. Soc. Belg. 1889, p. xix.
Australia.

8004. BIFASCIATA, Reitt., MT. Münch. Ent. Ver. 1887, p. 138, n. 22.
Australia.

8005. COSTATIPENNIS, Blackb., Trans. R. Soc. S.A. x. 1887, p. 202.
Western Victoria; Tasmania.

8006. MINOR, Blackb, l.c. p. 204.
S. Aust.; Port Lincoln, &c.

8007. NIGROMACULATUS, Blackb., l.c. p. 203.
S. Aust.; Woodville.

8008. NODIFER, Westw., Introd. Class. Ins. i. p. 155, t. 13, f. 23;
Blackb., l.c. p. 201.
Tasmania.

8009. PUNCTIPENNIS, Blackb., l.c. p. 204.
S. Aust.; Port Lincoln.

8010. SATELLES, Blackb., l.c. p. 202.
S. Aust.; Port Lincoln.

8011. SEMICOSTATUS, Blackb., l.c. p. 203.
S. Aust.; Port Lincoln.

MONOTOMA, Herbst.

8012. RUFA, Redtenb., Faun. Austr. i. p. 203; Blackb., l.c. x.
1887, p. 201.
S. Aust.; Port Lincoln.

CORTICARIA, Marsham.

8013. ADELAIDÆ, Blackb., Trans. R. Soc. S.A. xiv. 1891, p. 120.
S. Australia and Tasmania.

8014. ALUTACEA, Blackb., l.c. p. 121.
S. Australia.

8015. ANDERSONI, Blackb , Trans. R. Soc. S.A. xiv. 1891, p. 121.
     S. Aust.; Port Lincoln District.

8016. AUSTRALIS, Blackb., l.c. p. 120.
     S. Aust., Victoria, Tasmania.

8017. CONFERTA, Reitt., Verh. Ver. Brünn, xviii. p. 5.
     Victoria.

8018. DILATIPENNIS, Reitt., Deutsche Ent. Zeit. 1878, i. p. 96
     (= foveola, Beck.).
     Australia.

8019. LINDENSIS, Blackb., Trans. R. Soc. S.A. xiv. 1891, p. 120.
     S. Aust ; Port Lincoln District.

8020. SUBTILISSIMA, Reitt., Mitth. Münch. Ent. Ver. 1887, p. 139.
     n. 23.
     Australia.

Family MYCETOPHAGIDÆ.

TRIPHYLLUS, Latreille.

8021. INTRICATUS, Blackb., Trans. R. Soc. S.A. xiv. 1891, p. 122.
     Australian Alps.

8022. MINOR, Lea, P.L.S.N.S.W. (2) x. 1895, p. 226.
     N.S. Wales; various localities. Queensland; Brisbane.

8023 MULTIGUTTATUS, Lea, l.c. p. 225.
     N.S. Wales; Richmond River.

MYCETÆA, Stephens.

8024. PILOSELLA, Blackb., Trans. R. Soc. S.A. xiv. 1891, p. 122.
     S. Aust ; near Port Lincoln.

DIPLOCŒLUS, Guérin.

8025. ANGUSTULUS, Blackb , Trans. R. Soc. S.A. xiv. 1891, p. 122.
     S. Australia.

8026. EXIGUUS, Blackb., Trans. R. Soc. S.A. xiv. 1891, p. 123.
S. Aust.; near Port Lincoln.

8027. LATUS, Lea, P.L.S.N.S.W. (2) x. 1895, p. 228.
W. Aust.; Donnybrook.

8028. LEAI, Blackb., Trans. R. Soc. S.A. xviii. 1894, p. 204.
N.S. Wales and Queensland.

8029. PUNCTATUS, Lea, P.L.S.N.S.W. (2) x. 1895, p. 227.
N.S. Wales; Richmond River.

TYPHÆA, Stephens.

8030. FUMATA, Linn., Syst. Nat. I. 2. p. 564; Blackb., Trans. R
Soc. S.A. x. 1887, p. 205.
S. Aust.; Port Lincoln.

Family DERMESTIDÆ

CRYPTORHOPALUM, Guérin.

8031. AUSTRALICUM, Blackb., Trans. R. Soc. S.A. xiv. 1891, p. 130.
S. Aust.; near Port Lincoln.

8032. INTERIORIS, Blackb., l.c. p. 131.
S. Aust.; Basin of Lake Eyre.

8033. QUORNENSE, Blackb., P.L.S.N.S.W. (2) ix. 1894, p. 93.
S. Aust.; near Quorn.

8034. WOODVILLENSE, Blackb., Trans. R. Soc. S.A. xiv. 1891, p. 130.
S. Aust; Woodville.

TROGODERMA, Latreille.

8035. ADELAIDÆ, Blackb., Trans. R. Soc. S.A. xiv. 1891, p. 125.
S. Aust.; Adelaide District.

8036. ALPICOLA, Blackb, l.c. p. 124.
Victorian Alps.

8037. ANTIPODUM, Blackb., Trans. R. Soc. S.A. xiv. 1891, p. 128.
S. Aust.; near Adelaide.

8038. BALDIENSE, Blackb., l.c. p. 127; l.c. xv. 1892, p. 208.
Victorian Alps.

8039. DIFFICILE, Blackb., l.c. p. 126.
S. Aust.; near Port Lincoln.

8040. EYRENSE, Blackb., l.c. p. 124.
S. Aust.; Basin of Lake Eyre.

8041. FROGGATTI, Blackb., l.c. xv. 1892, p 34.
N.S. Wales; near Yass.

8042. LINDENSE, Blackb., l.c. xiv. 1891, p. 125.
S. Aust., near Port Lincoln.

8043. MACLEAYI, Blackb., l.c. p. 126.
S. Australia and Victoria.

8044. MEYRICKI, Blackb., l.c. p. 128.
W. Australia.

8045. OCCIDENTALE, Blackb., l.c. p. 127.
W. Australia.

8046. REITTERI, Blackb., l.c. xv. 1892, p. 207.
N.S. Wales; near Sydney.

8047. SINGULARE, Blackb., l.c. xiv. 1891, 128; l.c. xv. 1892, 34
S Aust.; near Port Lincoln.

8048. VARIPES, Blackb., l.c. xv. 1892, p. 208.
S. Aust.; near Adelaide.

8049. YORKENSE, Blackb., l.c. xiv. 1891, p. 127.
S. Aust ; Yorke's Peninsula.

ADELAIDIA, Blackburn.

8050. RIGUA, Blackb., Trans. R. Soc. S.A. xiv. 1891, p 130.
S. Australia.

H

ANTHRENUS, Geoffroy.

8051. FLINDERSI, Blackb., Trans. R. Soc. S.A. xiv. 1891, p. 132.
S. Aust.; near Port Lincoln.

8052. OCELLIFER, Blackb., l.c. p. 132.
S. Australia.

8053. VARIUS, Fab., Syst. Ent. p. 60; Erichs, Nat. Ins. III. p.
455; Blackb, l.c. p. 132.
Australia (probably introduced).

8054. SOCIUS, Lea, P.L.S N.S.W. (2) x. 1895, p. 228.
N.S. Wales; Sydney.

Family BYRRHIDÆ.

BYRRHUS, Linné.

8055. RAUCUS, Blackb., Trans. R. Soc. S.A. xiv. 1891, p. 133.
Victorian Alps.

8056. TORRENSENSIS, Blackb., l.c. xii. 1889, p. 138.
S. Aust.; Torrens River.

ASPIDIPHORUS, Latreille,

8057. HUMERALIS, Blackb., P.L.S.N.S.W. (2) ix. 1894, p. 92.
Tasmania.

Family PARNIDÆ.

ELMIS, Latreille.

8058. TASMANICUS, Blackb., P.L S.N.S.W. (2) ix. 1894, p. 94.
Tasmania.

8059. V-FASCIATUS, Lea, l.c. x. 1895, p. 590.
N.S. Wales; Tamworth.

Family HETEROCERIDÆ.

HETEROCERUS, Fabricius.

8060. FLINDERSI, Blackb., Trans. R. Soc. S.A. x. 1887, p. 205.
S. Aust.; Port Lincoln, &c.

8061. INDISTINCTUS, Blackb., l.c.'xiv. 1891, p. 134.
Victoria; Ovens River.

8062. MULTIMACULATUS, Blackb., l.c. x. 1887, p. 205.
S. Aust.; Torrens River.

8063. VICTORIÆ, Blackb., l.c. xiv. 1891, p. 133.
Victorian Alps.

Family LUCANIDÆ.

Sub-Family LUCANIDES.

PHALACROGNATHUS, Macleay.

8064. WESTWOODI, Shipp., Trans. Ent. Soc. Lond. 1893 p. 428.
N. Queensland; Cape York.

CLADOGNATHUS, Burmeister.

8065. LIMBATUS, C. O. Waterh., Ann. Nat. Hist. 1887, p. 381.
N. Queensland; Cape York, &c.

CERATOGNATHUS, Westwood.

8066. FROGGATTI, Blackb., P.L.S.N.S.W. (2) ix. 1894, p. 94.
N.S. Wales; Botany.

8067. GILESI, Blackb., Trans. R. Soc. S.A. xix. 1895, p. 215.
Victoria.

FIGULUS, W. S. Macleay.

8068. TRILOBUS, Westw., Ent. Mag. v. 1838, p. 263.
N.S. Wales.

Family SCARABÆIDÆ.

Sub-Family COPRIDES.

CEPHALODESMIUS, Westwood.

8069. CORNUTUS, Macl., P.L.S.N.S.W. (2) ii. 1887, p. 220.
N. Queensland; Mossman River.

EPILISSUS, Reiche.

8070. GLOBULUS, Macl., P.L.S.N.S.W. (2) ii. 1887, p. 222.
N. Queensland; Cairns.

GESSERODON, Hope.

8071. GESTROI, Lansb., Ann. Mus. Civ. Genov. (2) ii. 1885, p. 375.
N. Queensland; Cape York.

8072. VARIOLOSUS, Macl., P.L.S N.S.W. (2) iii. 1888, p. 897.
N.W. Aust.; King's Sound.

TEMNOPLECTRON, Westwood.

8073. DIVERSICOLLE, Blackb., Trans. R. Soc. S.A. xviii. 1894,
p. 204.
N. Queensland.

8074. LUCIDUM, Macl., P.L.S.N.S.W. (2) iii. 1888, p. 898.
N.W. Aust.; King's Sound.

8075. OCCIDENTALE, Macl., l c. p. 898.
N.W. Aust.; King's Sound.

8076. POLITULUM, Macl., l.c. (2) ii. 1887, p. 221.
N. Queensland; Cairns.

8077. PYGMÆUM, Macl., l.c. iii. 1888, p. 898.
N.W. Aust.; King's Sound.

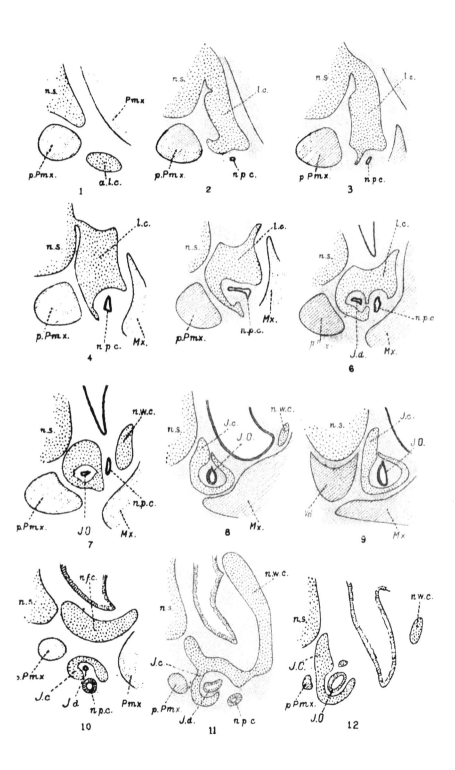

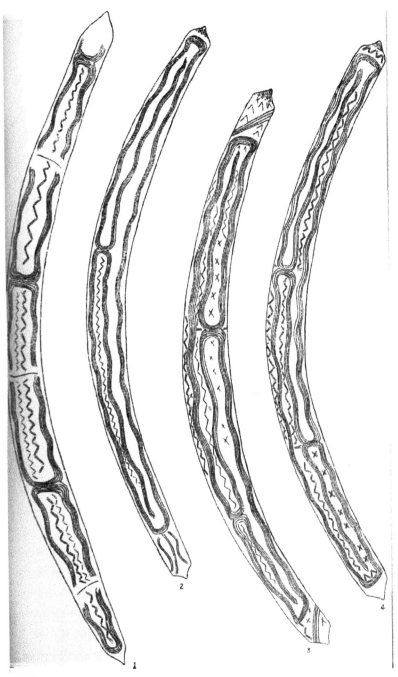

C R del

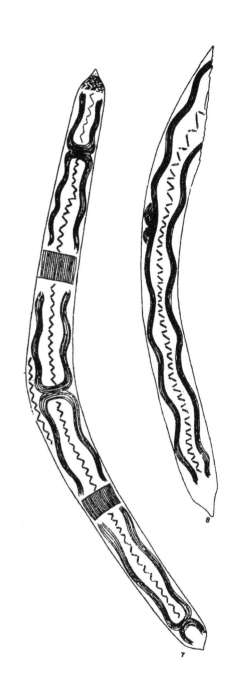

7

8

5

Pl IV

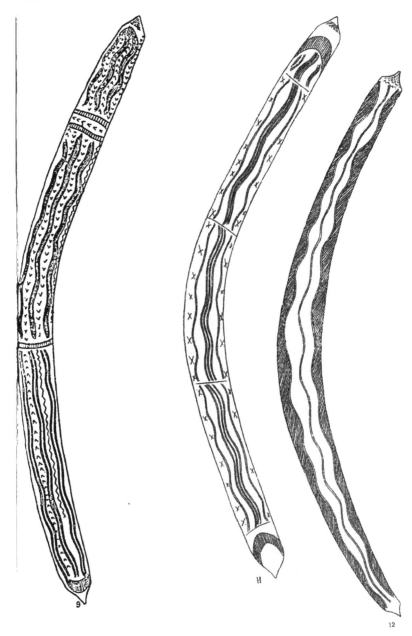

9

11

12

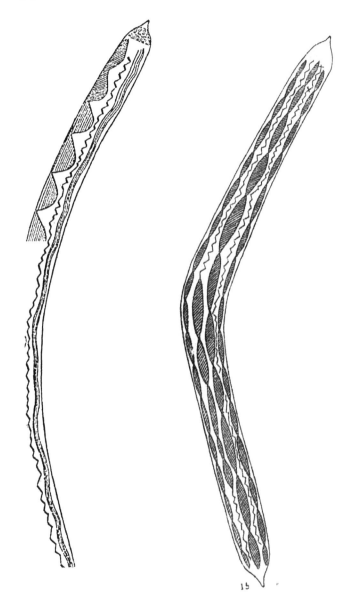

15

13

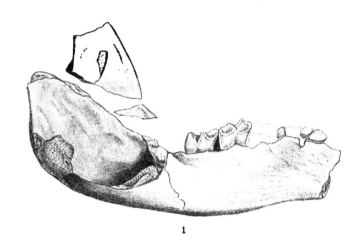

1

2

3

m^2 . m^1 dp^4

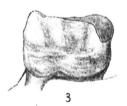

5

6

7

ꞂB. del ad nat

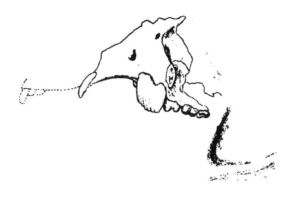

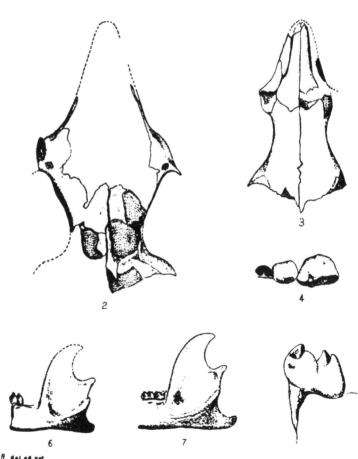

2

3

4

6

7

RB del ad nat

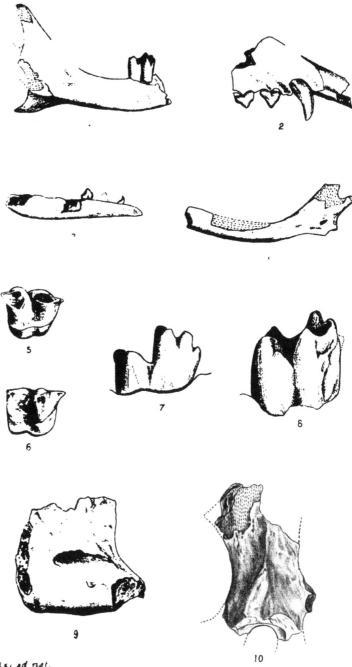

2

5

6

7

8

9

10

del. ad nat.

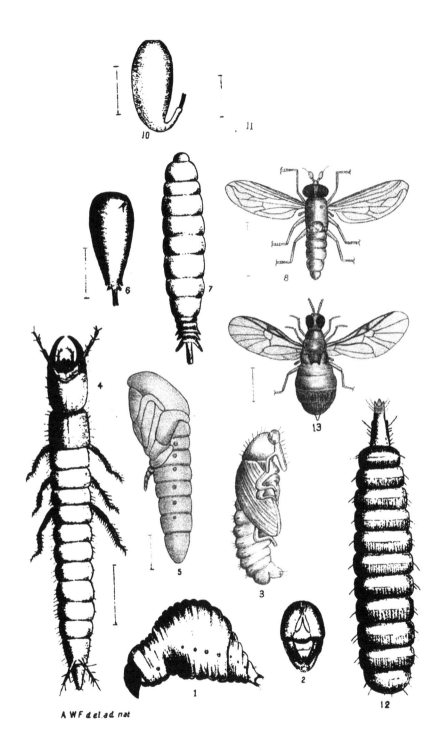

A W F del.ad nat

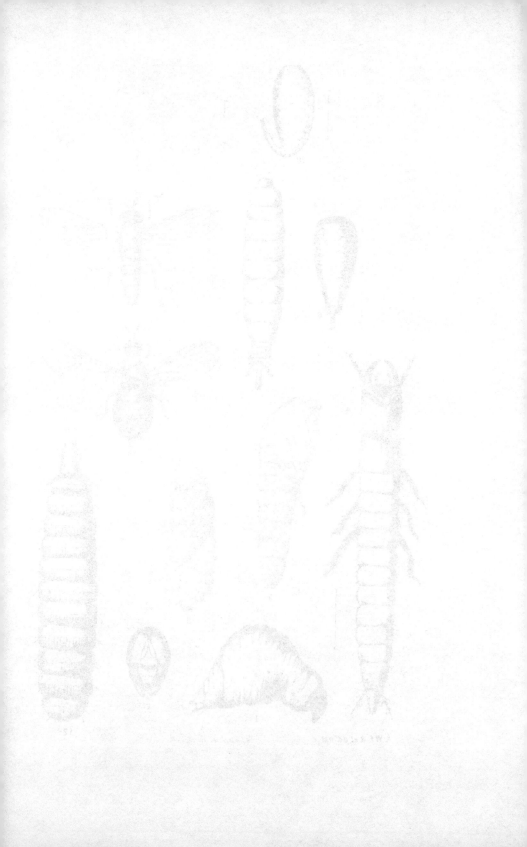

Pl. X.

1

2

2b

2a

a   5

b

c

d

e

3

6

4

32 x 14 μ

8

9

25 x 24 μ   12   27 x 24 μ

10

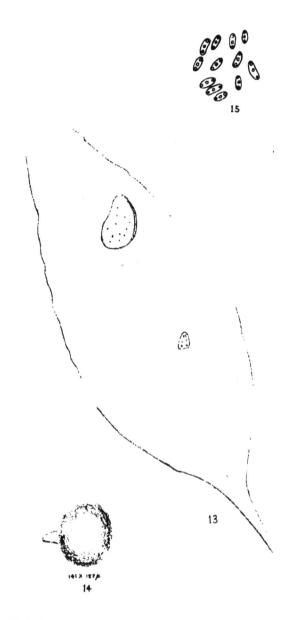

141 X 127μ

15

13

14

D Mᶜ.A. del.

PHOMA STENOSPORA

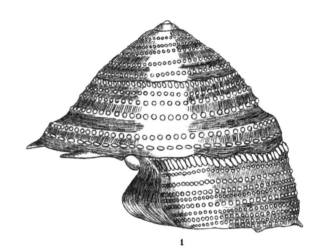

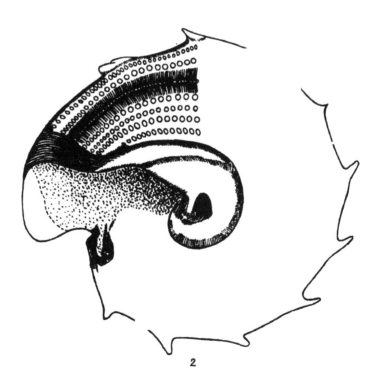

1

2

Pl. XIII

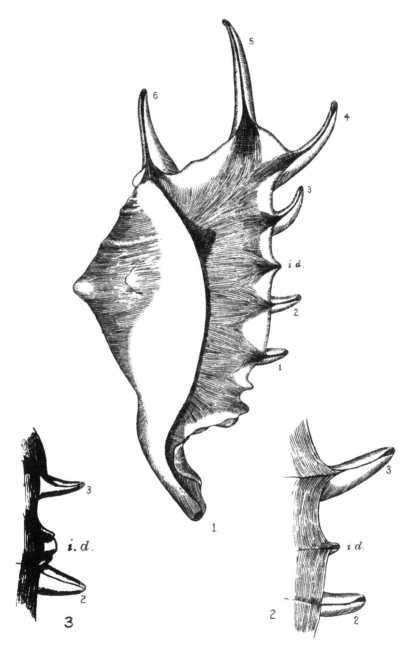

A.W. del.

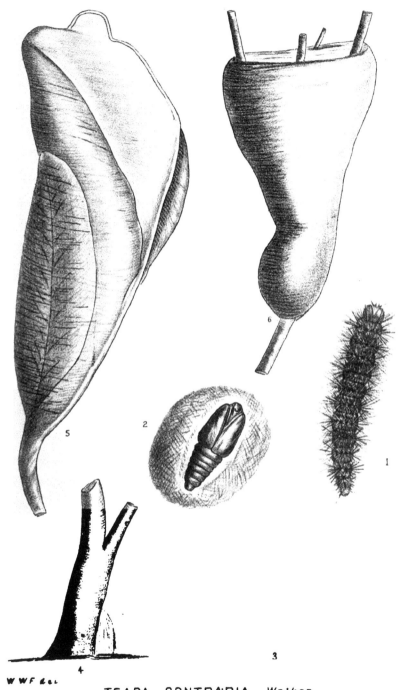

WWF del.

TEARA CONTRÁRIA, *Walker*

the Warru
:ary of Ua
n South Wa

eet

(few inches) Rotten n
decomposed

6 ft p
and ch

clay shales

SECTION   in Wantialable Creek
Near Tooraweena,   Warrumbungle Mountains,
showing   intercalation   of Diatomaceous   Earth
in   the   Trachyte   Series

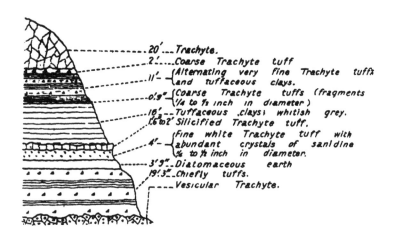

20′ ... Trachyte.
2′ ... Coarse Trachyte tuff
11′ — {Alternating very fine Trachyte tuffs and tuffaceous clays.
0′.9″ {Coarse Trachyte tuffs (fragments ¼ to ½ inch in diameter)
10′ ... Tuffaceous clays; whitish grey.
1.6 to 2′ Silicified Trachyte tuff.
4′ — {Fine white Trachyte tuff with abundant crystals of sanidine ¼ to ½ inch in diameter.
3′ 9″ Diatomaceous earth
19′ 3″ Chiefly tuffs.
... Vesicular Trachyte.

SECTION   in Wantialable Creek,
Near Tooraweena,   Warrumbungle Mountains,
showing diatomaceous earth in association
with Cinnamomum Leichhardtii

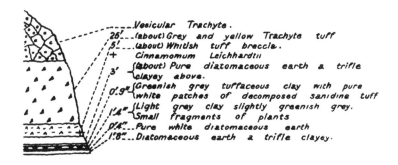

... Vesicular Trachyte.
26′ ... (about) Grey and yellow Trachyte tuff
5′ ... (about) Whitish tuff breccia.
1′ + Cinnamomum Leichhardtii
3′ {(about) Pure diatomaceous earth a trifle clayey above.
0′.9″ {Greenish grey tuffaceous clay with pure white patches of decomposed sanidine tuff
1′4″ {Light grey clay slightly greenish grey. Small fragments of plants
0′4″ ... Pure white diatomaceous earth
1′8″ ... Diatomaceous earth a trifle clayey.

Vertical   Scale   0   10   20   30   40   50   60   feet

Pl XVII

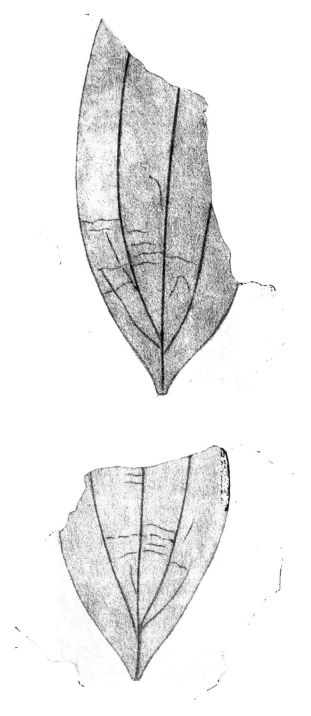

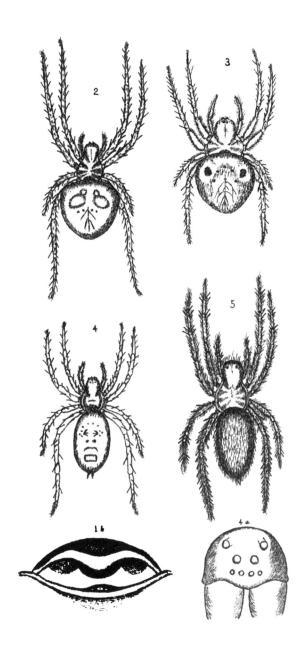

2

3

4

5

2 a

1 b

4 a

p del

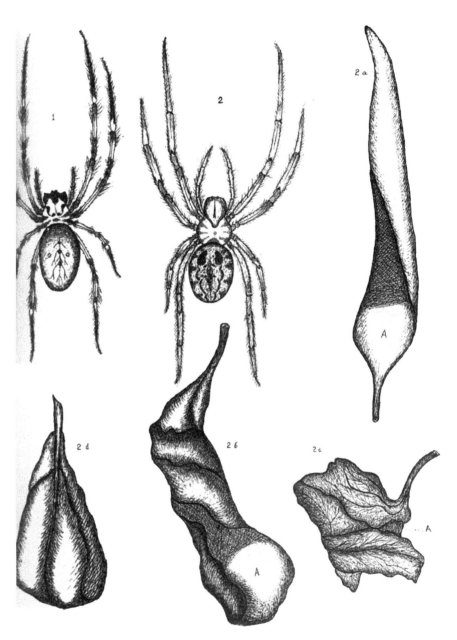

WJR del

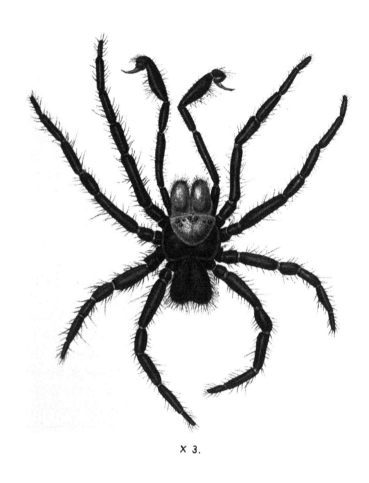

X 3.

ACTINOPUS FORMOSUS, *Rainbow.*

Pl XXI.

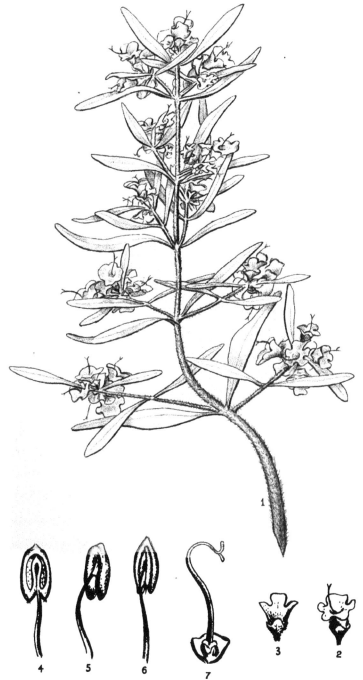

Pl XVII.

R.T.Baker del ad nat.

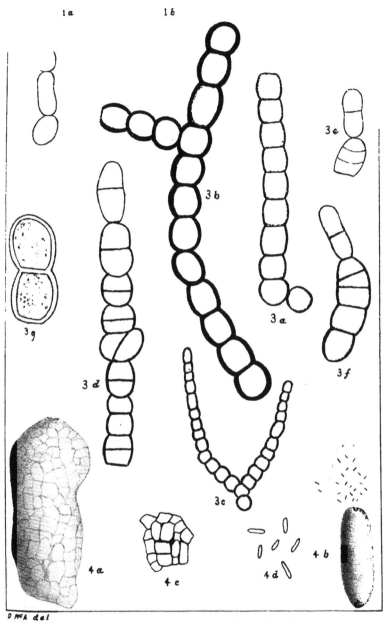

1 a   1 b

3 e

3 b

3 a

3 f

3 d

3 g

3 c

4 a

4 c

4 d

4 b

D M^c A del

CAPNODIUM CITRICOLUM, M^c ALP

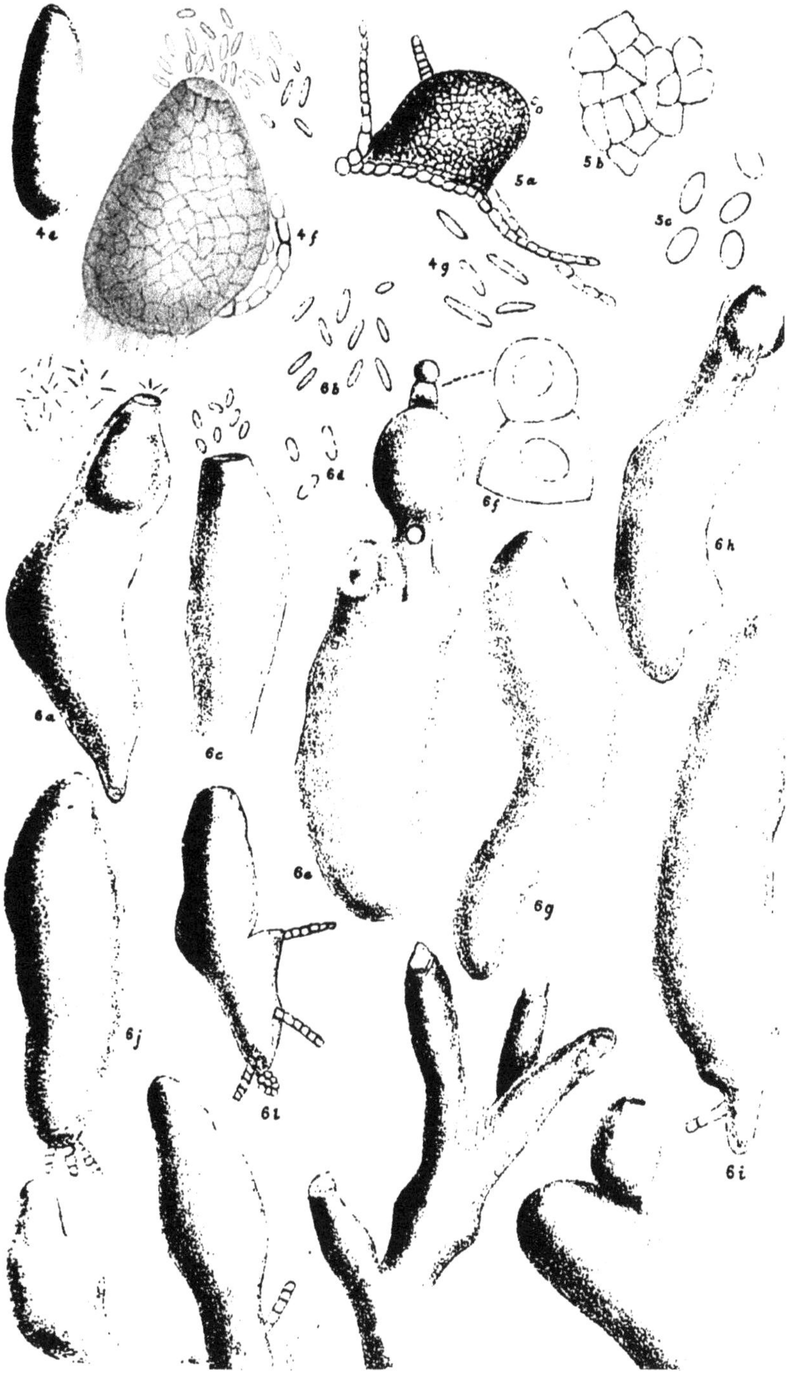
4e
4f
4g
5a
5b
5c
6a
6b
6c
6d
6e
6f
6g
6h
6i
6j
6l

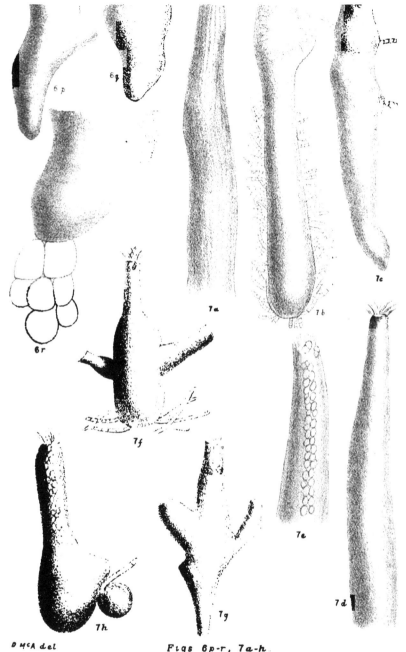

6 p

6 q

7 a

7 b

7 c

6 r

7 f

7 e

7 d

7 h

7 g

D McA del

Figs 6p-r, 7a-h

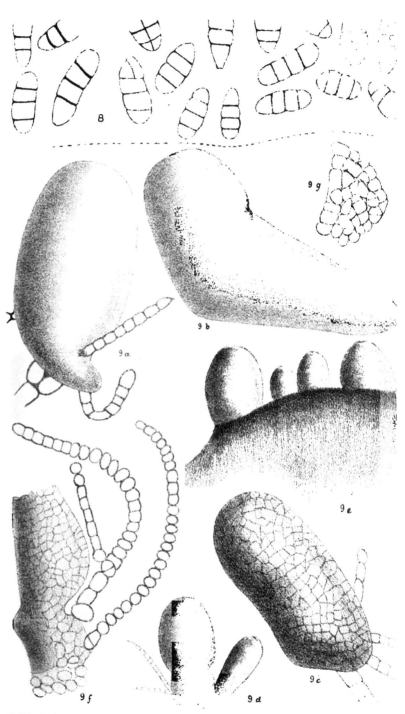

8

9 g

9 b

9 a

9 e

9 f

9 c

9 d

O.McA del.

Fi

Pl xxvii.

Pl. X

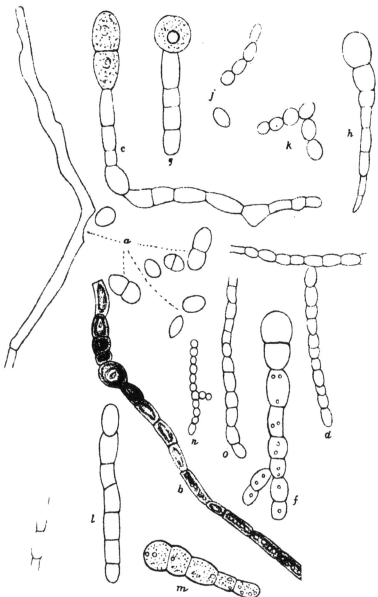

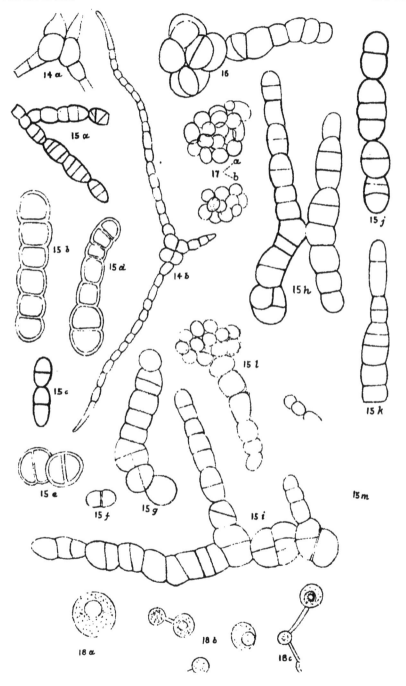

Pl. XXX.

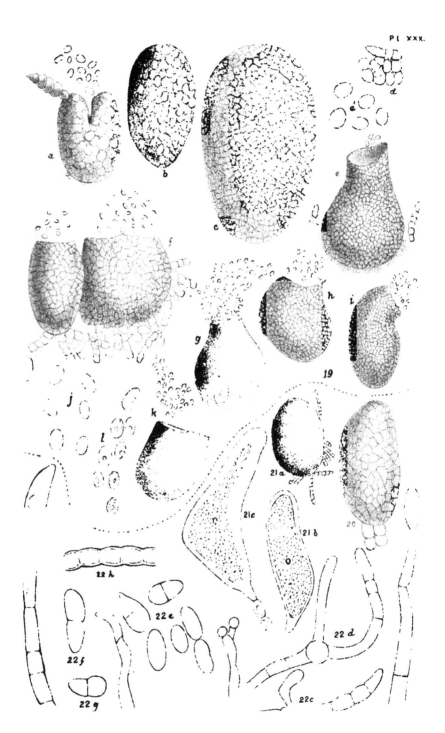

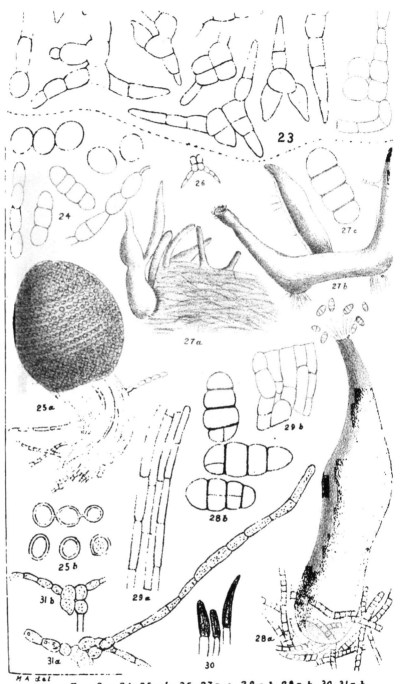

23

24 26 27c

27b

27a

25a

29b

28b

29a

25b

31b

31a 29a 30 28a

MA del

Figs 2.; 24. 25 a-b; 26; 27a-c; 28 a-b; 29 a-b, 30; 31 a-b.

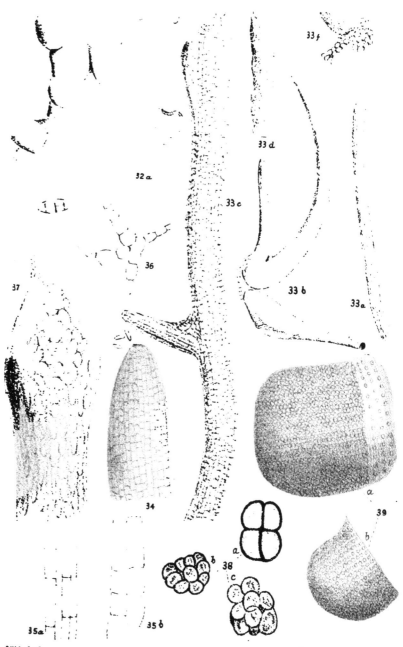

33 f

33 d

32 a

33 c

36

37

33 b

33 a

34

a

39

b

a

38

b

c

35 a

35 b

BM A.del.

32 a-b 33 a-g, 34, 35 a-b, 36, 37, 38 a-c, 39 a-b

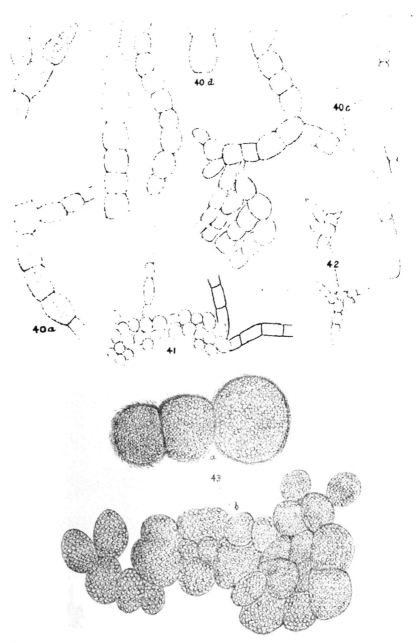

40 d

40 c

42

40 a

41

a

43

b

McA. del.

Figs. 40 a-d, 41, 42 a-b, 43 a-b.

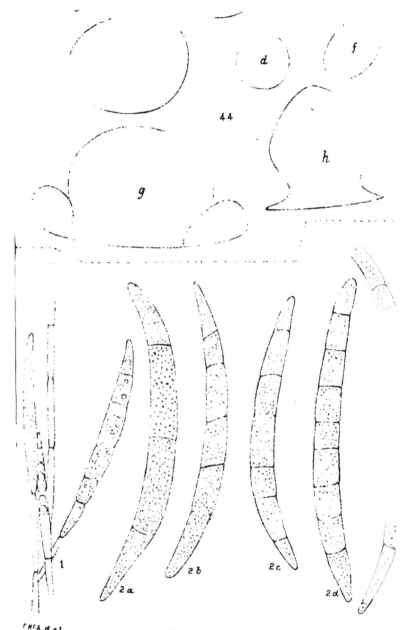

d

f

44

h

g

1

2a

2b

2c

2d

Figs. 44 a-h. Figs 1, 2 a-e.

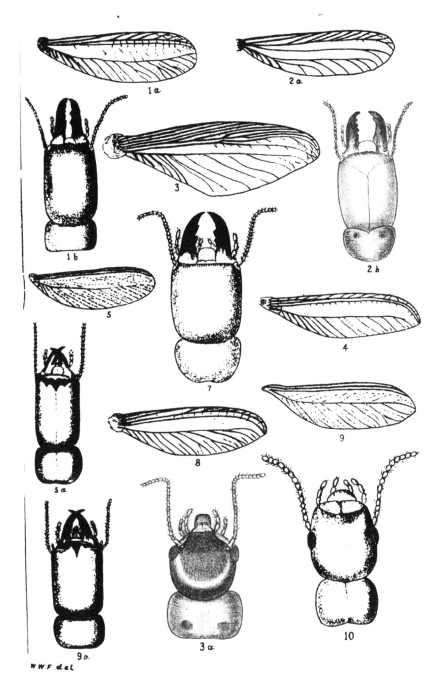

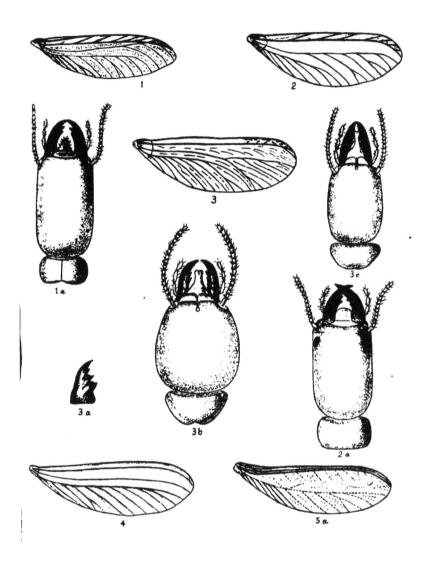

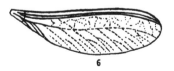

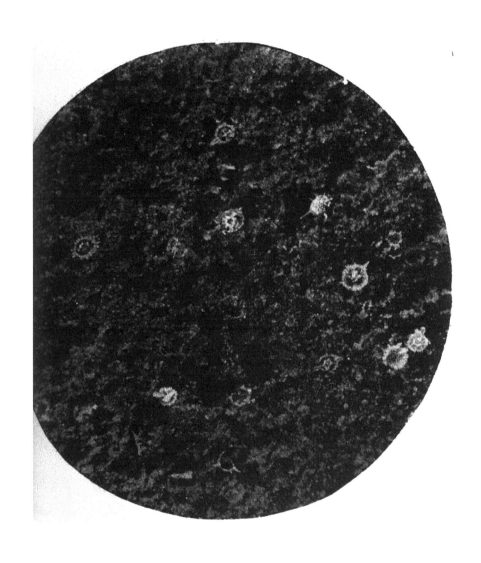

Radiolarian   Rock

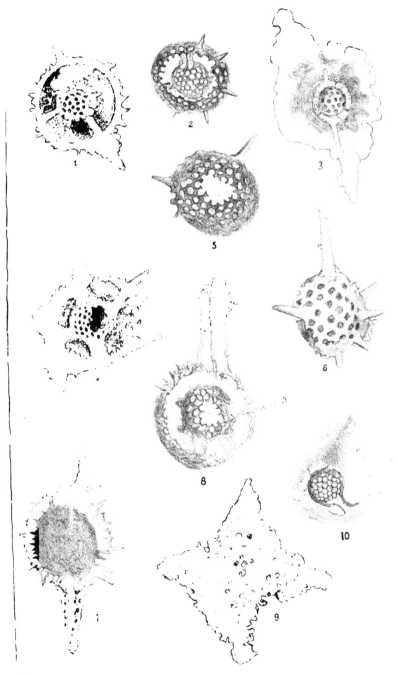

T.W.E.D. del.

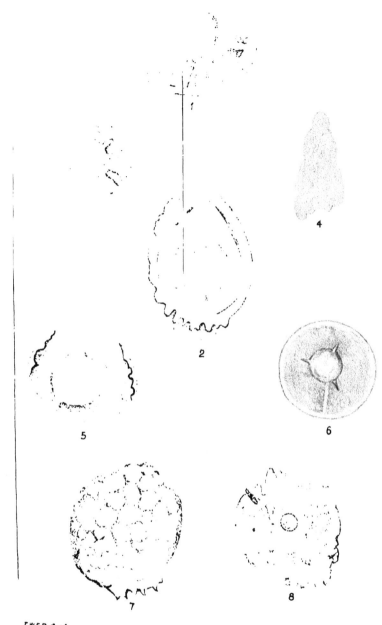

1

2

4

5

6

7

8

TWED del

CASTS OF RADIOLARIA

Pl XL.

to MURRAY BRIDGE, SOUTH-AUSTRALIA

n? Rocks, which latter contain casts of Radiolaria at

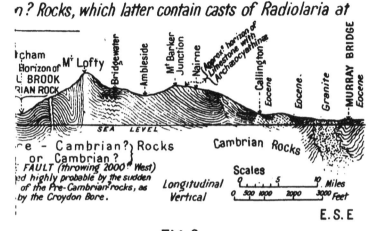

tcham
Horizon of
L. BROOK
RIAN ROCK

Mt Lofty
Bridgewater
Ambleside
Mt Barker Junction
Nairne
Approx^e horizon of Limestone with Archæocyathinæ
Callington
Eocene.
Eocene.
Granite
MURRAY BRIDGE
Eocene

SEA LEVEL

re - Cambrian?) Rocks
or Cambrian?
FAULT (throwing 2000' West)
ed highly probable by the sudden
of the Pre-Cambrian? rocks, as
by the Croydon Bore.

Cambrian Rocks

Scales

Longitudinal
Vertical

0 _____ 5 _____ 10 Miles

0 500 1000 2000 3000 Feet

E.S.E

Fig 2

Showing probable junction between the
Lower Cambrian and the Pre-Cambrian Rocks
near ARDROSSAN, Yorkes Peninsula. S. A.

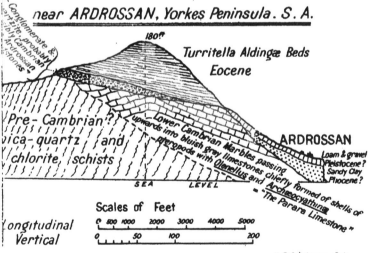

Conglomerate &
quartzite probably
possibly Cambrian
Ardrossan
stones?

180^{ft}

Turritella Aldingæ Beds
Eocene

Pre- Cambrian?
ica-quartz and
chlorite, schists

Lower Cambrian Marbles passing
upwards into bluish grey limestones
pteropods with Olenellus and

ARDROSSAN

Loam & gravel
Pleistocene?
Sandy Clay
Pliocene?

chiefly formed of shells of
Archæocyathinæ
= "The Parara Limestone"

SEA LEVEL

Scales of Feet

Longitudinal
Vertical

0 500 1000 2000 3000 4000 5000

0 50 100 200

H. E. C. Robinson. Delt.

9 780331 506242